Subcellular Biochemistry

Volume 100

The book series SUBCELLULAR BIOCHEMISTRY is a renowned and well recognized forum for disseminating advances of emerging topics in Cell Biology and related subjects. All volumes are edited by established scientists and the individual chapters are written by experts on the relevant topic. The individual chapters of each volume are fully citable and indexed in Medline/Pubmed to ensure maximum visibility of the work.

Tapas Kumar Kundu • Chandrima Das
Editors

Metabolism and Epigenetic Regulation: Implications in Cancer

Springer

Editors
Tapas Kumar Kundu
Transcription and Disease Laboratory,
Molecular Biology and Genetics Unit,
Jawaharlal Nehru Centre for Advanced
Scientific Research
Bangalore, India

Chandrima Das
Biophysics and Structural Genomics Division
Saha Institute of Nuclear Physics
Kolkata, India

ISSN 0306-0225 ISSN 2542-8810 (electronic)
Subcellular Biochemistry
ISBN 978-3-031-07633-6 ISBN 978-3-031-07634-3 (eBook)
https://doi.org/10.1007/978-3-031-07634-3

This Springer imprint is published by the registered company Springer Nature Switzerland AG
The registered company address is: Gewerbestrasse 11, 6330 Cham, Switzerland

This volume of Subcellular Biochemistry is dedicated to the loving memory of Shri Ashim Kumar Das, Retired IAS, Commissioner of Health and Family Welfare, Government of West Bengal, India. His constant inspiration, support, and lesson of service from his own life were the driving force to complete this work, even during the most challenging period of Humanity.

Foreword

Chromatin consumes a significant amount of cellular energy for its dynamic function. There is a cyclical impact of the metabolic milieu on the transcription programs of a cell and vice versa. With an increase in the landscape of epigenetic modifications found within a cell, it is becoming quite evident that most of the chemical reactions that modify chromatin require metabolites as coenzymes or cofactors. Thus, the metabolic status of the cell plays a key determining role in the epigenetic spectrum including histone post-translational modification (PTM), in DNA methylation, and also in regulating the function of noncoding RNA (lncRNA or miRNA). These epigenetic modifications in turn impact the chromatin states and gene expression programs.

The metabolic regulation of a normal cell is distinctly different from that of a cancer cell. In order to cater to extensive energy demands, cancer cells turn on several nonconventional programs. Interestingly, from the initial state of uncontrolled cellular proliferation to turning on the EMT program leading to the establishment of the distant metastatic niche, a myriad of distinct metabolic conditions are apparent in cancer. In addition to the dysfunction of metabolic enzymes, there is the presence of oncometabolites which are specifically seen in cancer cells. Thus, a comprehensive understanding of the interconnection of the metabolite pools and the cross talk between metabolism and chromatin function through epigenetic mechanisms is immensely important in the context of cancer.

This book entitled *Metabolism and Epigenetic Regulation: Implications in Cancer* is a confluence of these concepts where a group of established researchers have attempted to cover all the contemporary aspects in a unified version. The book has three parts. In the first part, the key metabolic pathways that are implicated in cancer including carbohydrate, lipid, and amino acid metabolism and the role of certain metalloenzymes are covered. In addition, the role of tumor microenvironment and cell death pathways like autophagy in cancer are discussed from the perspective of metabolic regulation. In the second part, the fundamental epigenetic programs that regulate metabolic pathways in cancer are summarized, where the role of lncRNA, DNA/RNA methylation, epigenetic reprogramming by the chromatin modifying enzymes, and readers/effectors are highlighted. Here, special emphasis is placed on nutritional epigenetics and stress-response pathways like hypoxia. In the final part, the role of epigenetic regulation from the perspective of cancer signaling pathways and potential therapeutic strategies are highlighted. In the latter context, the contemporary problems of therapeutic resistance and how small molecule modulators of epigenetic enzymes can be used to combat cancer are discussed. Further, the structural basis of targeted imaging and immunotherapy in cancer is also covered in this section.

Overall, the book provides some compelling evidence that epigenetic regulation of metabolic programs in the context of cancer is an emerging paradigm that is generating attention for a better understanding of the complexity of the disease. The book is a stimulating overview of the field aimed for a broad readership from students to established investigators.

Stowers Institute for Medical Research,
Kansas City, MO, USA

Jerry L. Workman Ph.D.

Preface

Epigenetics is a fundamental phenomenon of life which connects the genes with environment, nature, and nurture. Thus, health and disease of life always are under the remote control epigenetic machineries. Cancer as a disease is diverse in nature and shows variable response from originating tissue to individuals. Metabolic programs of an individual are a key determinant to disease susceptibility and its response, in epigenetic context. In recent years, a huge body of knowledge has been generated in the area of epigenetic signaling that regulates metabolic programs in the context of cancer. There is tremendous potential for exploring the metabolic homeostasis of an individual through epigenetic perspective and to understand how it goes awry during a multifactorial disease like cancer. The major goal of *Metabolism and Epigenetic Regulation: Implications in Cancer* is to provide a holistic connectivity between epigenetic regulation of metabolic pathways during cancer.

As a *Subcellular Biochemistry* volume, this book is intended to provide an understanding of the complex metabolic programs in cancer cells through an epigenetic perspective. The intracellular signaling pathways play an important role in regulating cancer cell metabolism which is crucial for maintenance of their energy homeostasis as well as regulating the growth and proliferation. The metabolic hallmark of cancer cells is aerobic glycolysis as proposed by Otto Warburg in 1926. Although early-on hypothesis shows a dependency of the cancer cells on the glycolytic ATP production due to an impaired mitochondrial oxidative phosphorylation mechanism, later studies revealed that cancer cells can indeed utilize mitochondrial respiration for ATP synthesis. This increased glucose utilization by the cancer cells is a potential aspect that could be exploited during therapeutic intervention. However, apart from carbohydrate metabolism, the lipid, amino acid, and nucleic acid metabolism also play a crucial role in meeting up the energy demand of cancer cells. It is also important to understand that these metabolic pathways are interconnected into complex networks and the overall energy metabolism landscape has immense contributions toward cancer cell proliferation and metastasis. Hence, a fundamental understanding of the metabolic programs of cancer cells and their modulation through epigenetic perspective has immense importance in devising therapeutic strategies.

This book is an attempt to understand the connectivity between epigenetic regulation of metabolic pathways during cancer and is contributed by a cohort of authors who have fundamental contribution in their respective domains. We look forward for understanding cancer through a nonconventional approach of epigenetic reprogramming of the metabolic pathways and believe it would have a great contribution in the present-day context. The book has been divided into three main parts: Part I: *Regulation of Key Metabolic Pathways in Cancer*, Part II: *Epigenetic Regulation of Cellular Metabolic Pathways*, and Part III: *Epigenetic Regulation in Cancer*.

In the first part of the book, we intended to highlight the key metabolic pathways that are implicated in cancer. To this end, we aimed to give comprehensive information on the carbohydrate, protein, and lipid metabolic pathways that go awry during cancer. In this section, we have also focused on the altered tumor microenvironment that is influenced by metabolic milieu. Additionally, we have emphasized upon a fundamental connection between cellular metabolic environment and hallmarks of cancer including cell death mediated by autophagy.

In the next part of the book, we have elucidated the epigenetic regulations that are implicated in steering the metabolic programs in the cell. In this context apart from carbohydrate, amino acid, and lipid metabolism, we are also highlighting on the nucleic acid metabolism and the known epigenetic cross talk. We have covered aspects of epigenetic regulations of long noncoding RNA as well as DNA/RNA methylation signaling mediating metabolic homeostasis in cancer. We are introducing the concept of nutritional epigenetics with an emphasis on amino acid metabolism. We have also discussed aspects of reprogramming of host metabolic pathways by crucial epigenetic readers/effectors, which have direct implication in tumorigenicity.

In the final part, we are focusing to understand the fundamental epigenetic regulations which implicate in cancer and have covered several nonconventional aspects in this regard. We have covered epigenetic regulations with a special emphasis on oral cancer which is one of the most fatal cancers and how small molecule modulators of epigenetic enzymes can be employed as a therapeutic tool in this context. We further cover the role of these modulators to target the metabolic programs of the cell. We have also included the epigenetic regulations in cancer signaling pathway as well as generation of resistance phenotypes in cancer. We have also covered structure-guided imaging techniques with epigenetic drugs which have potential therapeutic implication. Finally, we have addressed the cross talk between epigenetics and immunometabolism as a strategy to combat cancer.

Thus, the proposed book encompasses certain fundamental aspects of epigenetic regulations of metabolic programs that are implicated in cancer. We strongly believe that this book with its proposed content would be a fundamental contribution in today's hunt for understanding cancer better. Its timeliness would have a direct bearing on the present-day research and would be a seminal contribution in the field.

Bangalore, India
Kolkata, India
September 3, 2022

Tapas Kumar Kundu
Chandrima Das

Contents

Part I
Regulation of Key Metabolic Pathways in Cancer

Chapter 1
Reprogramming Carbohydrate Metabolism in Cancer and Its Role in Regulating the Tumor Microenvironment

Swagata Adhikari, Deblina Guha, Chitra Mohan, Shravanti Mukherjee, Jessica K. Tyler, and Chandrima Das

Abstract Altered metabolism has become an emerging feature of cancer cells impacting their proliferation and metastatic potential in myriad ways. Proliferating heterogeneous tumor cells are surrounded by other resident or infiltrating cells, along with extracellular matrix proteins, and other secretory factors constituting the tumor microenvironment. The diverse cell types of the tumor microenvironment exhibit different molecular signatures that are regulated at their genetic and epigenetic levels. The cancer cells elicit intricate crosstalks with these supporting cells, exchanging essential metabolites which support their anabolic processes and can promote their survival, proliferation, EMT, angiogenesis, metastasis and even therapeutic resistance. In this context, carbohydrate metabolism ensures constant energy supply being a central axis from which other metabolic and biosynthetic pathways including amino acid and lipid metabolism and pentose phosphate pathway are diverged. In contrast to normal cells, increased glycolytic flux is a distinguishing feature of the highly proliferative cancer cells, which supports them to adapt to a hypoxic environment and also protects them from oxidative stress. Such rewired metabolic properties are often a result of epigenetic alterations in the cancer cells, which are mediated by several factors including, DNA, histone and non-histone protein modifications and non-coding RNAs. Conversely, epigenetic landscapes of the cancer cells are also dictated by their diverse metabolomes. Altogether, this

Swagata Adhikari and Deblina Guha contributed equally with all other contributors.

S. Adhikari · C. Das (✉)
Biophysics and Structural Genomics Division, Saha Institute of Nuclear Physics, Kolkata, India

Homi Bhaba National Institute, Mumbai, India
e-mail: chandrima.das@saha.ac.in

D. Guha · S. Mukherjee
Biophysics and Structural Genomics Division, Saha Institute of Nuclear Physics, Kolkata, India

C. Mohan · J. K. Tyler
Department of Pathology and Laboratory Medicine, Weill Cornell Medicine, New York, NY, USA

T. K. Kundu, C. Das (eds.), *Metabolism and Epigenetic Regulation: Implications in Cancer*, Subcellular Biochemistry 100,
https://doi.org/10.1007/978-3-031-07634-3_1

metabolic and epigenetic interplay has immense potential for the development of efficient anti-cancer therapeutic strategies. In this book chapter we emphasize upon the significance of reprogrammed carbohydrate metabolism in regulating the tumor microenvironment and cancer progression, with an aim to explore the different metabolic and epigenetic targets for better cancer treatment.

Keywords Carbohydrate metabolism · Glycolytic flux · OXPHOS · Hypoxia · Tumor microenvironment · Metastasis · Metabolic reprogramming · Acetyl-CoA · SAM · DNMTs · HATs · HDACs · HMT · HDM · Epigenetics · Oncometabolites

List of Abbreviations

2-OGDO	2-oxoglutarate-dependent dioxygenases
AIF	Apoptosis inducing factor
AMPK	AMP-activated protein kinase
CAFs	Cancer-associated fibroblasts cells
CAIX	carbonic anhydrase IX
CTC	Circulating tumor cells
CTLA-4	Cytotoxic T lymphocyte associated antigen
DNMT	DNA methyltransferase
DZNep	3-Deazaneplanocin-A
EGF	Epidermal growth factor
EMT	Epithelial to Mesenchymal Transition
ENO-1	Enolase-1
ERAD	ER associated degradation pathway
FBP1	Fructose 1,6-bisphosphatase-1
FH	Fumarate hydratase
G6P	Glucose-6-phosphate
GAPDH	Glyceraldehyde 3-phosphate dehydrogenase
GLUT	Glucose transporter
HATs	Histone acetyltransferases
HDACs	Histone deacetylases
HIF1α	Hypoxia inducible factor 1α
HK2	Hexokinase
HMT	Histone methyltransferase
IDH	Isocitrate dehydrogenase
IDH3α	Isocitrate dehydrogenase 3α
IGF-1	Insulin-like growth factor
KDMs	Histone lysine demethylases
LDHA	Lactate dehydrogenase A
LKB1	Liver kinase B1
MCT	Monocarboxylate transporter
MDSCs	Myeloid derived suppressor cells
NAD	Nicotinamide adenine dinucleotide

NDUFA4L2	NADH dehydrogenase ubiquinone 1 alpha subcomplex, 4-like 2
OXPHOS	Oxidative phosphorylation
PD-1	Programmed death 1
PDK1	Pyruvate dehydrogenase kinase-1
PDPK	Phosphoinositide dependent protein kinase-1
PFK	Phosphofructokinase
PFKFB-3/4	6-phosphofructo-2-kinase/fructose-2,6-bisphosphatases-3 and -4
PFKL	Liver-type PKF
PGAM1	Phosphoglyceratemutase-1
PGK1	Phosphoglycerate kinase-1
PHD2	Prolyl hydroxylase domain-containing protein-2
PKM2	Pyruvate kinase M2
PPP	Pentose phosphate pathway
SAH	S-adenosyl-homocysteine
SAM	S-adenosyl methionine
SCO2	Synthesis of Cytochrome C oxidase 2
SDH	Succinate dehydrogenase
TAM	Tumor-associated macrophages
TETs	Ten-eleven translocation hydroxylases
TIGAR	TP53-inducible Glycolysis and Apoptosis Regulator
TME	Tumor microenvironment
α-KG	α-ketoglutarate
α-KGDDs	α-ketoglutarate-dependent dioxygenases

Introduction

It is now well-accepted that cell division anomalies are not the sole cause of tumorigenesis and cancer progression. Rather, the underlying altered energy landscape in these cells is the key driver. Normal cells produce ATP to meet their energy need via glycolysis followed by the citric acid cycle and the oxidative phosphorylation (OXPHOS) which are the fundamental carbohydrate metabolic pathways. However, in 1956, Otto Warburg observed an unusual behavior whereby even in the presence of oxygen, cancer cells were found to solely depend on "aerobic glycolysis" as the major energy producing metabolic pathway, known as the Warburg effect (Warburg 1956). Furthermore, these cells utilize glycolysis-derived lactate for ATP generation under oxygen deficiency. Although, the OXPHOS mediated ATP production is highly efficient, even then the cancer cells prefer to undergo aerobic glycolysis (DeBerardinis et al. 2008) to fulfil their energy requirements for achieving increased proliferation and tumor growth which leads to metastatic progression (Liberti and Locasale 2016).

The concept of metabolism as a cancer hallmark was highlighted by Weinberg, where in contrast to the normal cells, the cancer cells were able to support their metabolic requirement through alteration of coordinated gene transcription (Hanahan and Weinberg 2011). In this context, involvement of transcription factors in regulating the metabolic switch of cancer cells has gained significant attention. Notably, among the transcription factors, the hypoxia inducible factors (HIF-1α and HIF-2α) are known to play critical role in regulating carbohydrate metabolism. They are known to positively regulate the glycolytic gene expression and thereby promote cancer (Hanahan and Weinberg 2011). Later six important features of metabolic regulation of cancer cells were proposed by Pavlova and Thompson to shed light on how cancer cells maintain their viability and biomass in a nutrient-poor environment (Pavlova and Thompson 2016). These include, (i) Deregulated glucose and amino acid uptake, (ii) An opportunistic mode of nutrient acquisition, (iii) Utilizing glycolysis and TCA cycle intermediates, (iv) Increased Nitrogen demand, (v) Metabolite driven gene expression changes, and (vi) Metabolic interactions between the tumor cell and its microenvironment. It is noteworthy that intra/extracellular metabolites and tumor microenvironment play a crucial role in these processes.

Metabolic regulation within the tumor cells is influenced by the heterogeneous network of cells within the tumor microenvironment (TME) (Lyssiotis and Kimmelman 2017), where the malignant cells functionally interact with other cell types and experience an intricate signaling crosstalk. Tumor cells and the multitude of cells constituting the TME benefit each other by exchange of metabolites which is required for their proliferation and tumor growth. Altered pH, concentration of glucose/lactate and accumulation of different metabolic intermediates induce various stress responses leading to cancer progression. Interestingly, cancer cell metabolism is intricately connected to its other properties including proliferation, angiogenesis, metastasis and immune escape. Notably, the TME within a solid tumor is often hypoxic which escalates the angiogenic and metastatic potential of cancer cells (Lyssiotis and Kimmelman 2017).

In vitro and in vivo studies suggest the key role of genetic and epigenetic changes in contributing towards tumor metabolic heterogeneity (Kim and DeBerardinis 2019). Importantly, different epigenetic events, including DNA methylation, histone and non-histone protein modification, and non-coding RNA mediated changes on the chromatin landscape are involved in deregulation of metabolic gene expression in cancer (Wong et al. 2017). Moreover, epigenetic dysregulation of gene expression in both tumor and other cells in TME (fibroblasts, immune cells, endothelial cells) contributes to the altered phenotype of tumor microenvironment which is empowered by their metabolic milieu. The metabolic intermediates can also serve as cofactors and substrates for the epigenetic regulators (Wong et al. 2017).

In this chapter, we highlight the role of reprogramming of carbohydrate metabolism in cancer progression and explore the epigenetic - metabolic interplay that have a great therapeutic potential.

Molecular Cues Influence Metabolic Phenotype in Different Cancers

Role of Hypoxia in Tumor Microenvironment

Under the metabolically compromised conditions generated by the tumor microenvironment, cancer cells alter their metabolic needs to survive and proliferate. To thrive under such compromised conditions, cancer cells need a continuous source of fuel as nutrients. The TME is a complex niche, comprising a wide variety of components including blood vessels, cancer associated fibroblasts (CAF), inflammatory cells, immune cells, stromal cells and extracellular matrices that significantly modulate the metabolic programs of cancer cells (Lyssiotis and Kimmelman 2017). In addition, factors such as nutrient availability, oxidative stress and hypoxia significantly impact the metabolism of cancer cells.

Hypoxic condition is very common in TME which plays an essential role in causing malignancy. Generally, the oxygen diffusion limit is 100–200μm, which indicates that the cells need to stay within this radius for adequate oxygenation (Yu et al. 2014; Carmeliet and Jain 2000). In a solid tumor, hypoxia is one of the stress factors that has significant role to play in cancer progression. Due to the oncogene-driven uncontrolled proliferation, cancer cells exhaust its nutrient and oxygen supply which in turn elicits their hypoxic nature. Hypoxia is known to be largely regulated by hypoxia-inducible factor-1 (HIF-1α), which has been implicated to induce several metabolic (glucose, lipid, amino acid) pathways of cancer cells to meet the nutrient and energy demand (Huang et al. 2014a). Hypoxia regulates the entry of glucose into TCA cycle and favors glycolysis over OXPHOS, although less efficient but more rapid way of ATP generation. This makes it essential to understand the role of hypoxia in glucose metabolism of cancer cells, where energy production is sensitive to altered oxygen tension. Hypoxia changes the redox status of mitochondria which eventually slows down the electron transport chain (ETC) as there is a reduction in NADH oxidation (Eales et al. 2016). As a consequence, Warburg effect is more pronounced, and it results in excess lactate production. Therefore, hypoxia directly promotes extracellular acidification which triggers malignancy in cancer cells (Eales et al. 2016). Moreover, hypoxia-induced metabolic alterations in cancer are important for designing the novel therapeutic targets.

The co-ordination among the TME components with the aforementioned physiological conditions fine-tunes the metabolic reprogramming of cancer cells within a heterogeneous tumor cell population. In this context, the role of the hypoxic nature of the TME and its different cell types are discussed below.

Metabolic Adaptation Due to Hypoxia in the Tumor microenvironment

Cancer cells adapt themselves metabolically to hypoxia by increasing cellular energy production. The most prominent metabolic alterations associated with hypoxia are enhanced (i) glucose utilization and (ii) lactate production. Such changes are catalyzed by HIF-1α, which is activated during hypoxia (Gordan and Simon 2007).

Role of HIF1α in Regulating Glucose Metabolism
HIF-1α remains inactive under normoxic condition which is mediated by prolyl hydroxylation followed by ubiquitin-mediated proteasomal degradation. However, low oxygen levels in cells reduces hydroxylation which stabilizes and promotes association of HIF-1α protein with the HIF-1β subunit thereby increasing HIF-1α mediated transcription. In addition, generation of reactive oxygen species (ROS) from oxidative phosphorylation also encourages HIF-1α stabilization (Brunelle et al. 2005). Once stabilized, HIF-1α upregulates its target genes including those related to aerobic glycolysis and glucose metabolism such as glucose transporter enzymes GLUT1 and GLUT3 as well as glycolytic enzymes (Semenza et al. 1994). Increased expression of both mRNA and protein levels of GLUT1 and GLUT3 have been reported in papillary carcinoma and in poorly differentiated endometrial and breast cancer (Moldogazieva et al. 2020). Expression of HIF1-α has been found to be highly correlated with that of GLUT1 in ovarian and colon cancers (Moldogazieva et al. 2020). Also, expression of 6-phosphofructo-2-kinase/fructose-2,6-bisphosphatases-3 and -4 (PFKFB-3/4) is positively correlated to that of GLUT1 and vascular endothelial growth factor (VEGF) in pancreatic and gastric cancers. Higher GLUT3 expression has been reported in cancers, over-expressing glycolytic enzymes hexokinase-2 (HK2) and pyruvate kinase M2 (PKM2) (Moldogazieva et al. 2020). HIF-1α over-expression in hepatocellular carcinoma (HCC) cells results in enhanced expression of a variety of enzymes which include GLUT1, GLUT3, HK1, HK2, carbonic anhydrase IX (CAIX), Glyceraldehyde 3 Phosphate Dehydrogenase (GAPDH), liver-type PKF (PFKL), phosphoglycerate kinase 1 (PGK1) and PFKFB3 (Guo et al. 2020). Increased glycolytic enzymatic activity escalates the glycolytic flux ultimately leading to HCC growth and progression (Guo et al. 2020). Furthermore, metastasis in primary HCC tissues is reported to be associated with increased levels of PKM2, HK1 and GLUT1 enzymes.

One of the rate determining enzymes of the glycolysis pathway is phosphofructokinase (PFK) which converts fructose-6-phosphate to fructose 1,6-bisphosphate. There is a correlation between HIF-1α and activity of PFK-1 and PFK-2 enzymes as HIF-1α along with Ras/Src oncogenic signaling pathway, regulate expression of the PFK isozymes and hence, promote tumor cell survival and proliferation (Moldogazieva et al. 2020). In addition, isoform PFKFB-3/4 is also transcriptionally activated by HIF-1α, which catalyzes fructose-6-phosphate to fructose-2,6-bisphosphate formation during glycolysis (Moldogazieva et al. 2020). The PFKFB group of enzymes are known to be associated with tumor growth, metastasis and invasion as evident from the fact that silencing of PFKFB2 gene significantly represses tumor growth in ovarian and breast cancer (Yang et al. 2019). An important outcome of hypoxia induced metabolic reprogramming in cancer cells is glycosylation of the PFK-L and PFK-M enzymes which inhibit PFK-1 activity, thereby redirecting glucose flux towards biosynthesis pathways such as the pentose phosphate pathway (PPP) (Yi et al. 2012). Increased expression of PPP enzymes along with HIF-1α stabilization is a predictor of poor cancer prognosis (Moldogazieva et al. 2020). All these evidence emphasize the role of HIF1α in modulating glucose metabolism in cancer.

Role of HIF1α in Regulating Lactate Metabolism

HIF-1α transcriptionally activates pyruvate dehydrogenase kinase 1 (PDK1), which inactivates the mitochondrial enzyme pyruvate dehydrogenase thereby suppressing the conversion of pyruvate to Acetyl-CoA (Kim et al. 2006). As a result, HIF-1α effectively shunts pyruvate as a substrate from the mitochondrial oxidation pathway towards lactate and NAD+ production. Further, the lactate transporter enzyme monocarboxylate transporter 4 (MCT4) and lactate dehydrogenase A (LDH-A) enzyme is positively regulated by HIF-1α in multiple cancers. In breast cancer cells, endogenous as well as exogenous lactate is responsible for activation of HIF-1α, as well as oncogenes including, Ras and Myc, metalloproteinases, and tumor suppressors BRCA1 and BRCA2, thereby promoting tumor cell migration and invasion (Moldogazieva et al. 2020). Conversion of lactate to pyruvate is catalyzed by LDHA which is also induced by HIF-1/2α. Cui et al. reported that knockdown of endogenous HIF-1α and HIF-2α in pancreatic cells downregulates LDHA level which subsequently decreases lactate production and glucose consumption (Cui et al. 2017). Furthermore, LDHA promotes growth and migration of pancreatic cells.

Notably, HIF-1β deficient hepatoma cells have reduced expression of many glycolytic enzymes, however, the glycolytic flux remains unaltered, although there is. This is due to high AMP/ATP ratio that allosterically activates PFK1, which sustains glycolysis in HIF-1β deficient condition (Masson and Ratcliffe 2014). Here, HIF-1β deficient hepatoma cells display unaltered glucose uptake or lactate synthesis despite the presence of low expression of the pathway enzymes. Also, it is reported that there is no alteration in glucose uptake in HIF-1β silenced colon adenocarcinoma cells, although hypoxia-induced lactate production is significantly lowered (Masson and Ratcliffe 2014). Taken together, these studies reveal that HIF-1 α acts as a potent metabolic inducer of glycolysis in a wide variety of cancer types.

Role of AMP-Activated Protein Kinase (AMPK) During Hypoxia

Besides HIF-1α, AMPK is also activated during hypoxic stress (Laderoute et al. 2006). It plays an important role in maintaining cellular energy homeostasis by sensing the AMP: ATP ratio (Laderoute et al. 2006). Laderoute et al. demonstrated how HIF-1α and AMPK activation in low-oxygen or ischemic-tissue microenvironments are important for the growth of solid tumor. Increased AMPK activity is known to be linked with cell proliferation as well as cell cycle transition in both normal and cancer cells and is also highly relevant for cellular glucose uptake (Carling 2017). Recently, it was suggested that for cancer cell survival there is enhanced HIF-1α stability and decreased AMPK activity leading to chemoresistance of breast cancer cells. Tamoxifen resistant LCC2 and LCC9 breast cancer cells show hypoxia independent HIF-1α activation and increased glycolysis, where HIF-1α is regulated by Akt/mTOR and phosphor AMPK activity in the chemoresistant cells. Inhibition of HK2 along with increased phospho-AMPK activity reduces cell survival and prevents lactate accumulation in the resistant cell population (Moldogazieva et al. 2020). Such example highlights the role of AMPK in cancer linked to other metabolic reprogramming factors such as HIF-1α.

Tumor Microenvironment Associated Cells

Residing within a heterogeneous population of cells, the cancer cells not only favor their own sustenance but also modulate the behavior of other cell types located in their proximity. These cell types reside in the TME of a developing tumor include cancer-associated fibroblasts cells (CAFs), endothelial cells, tumor-associated macrophages (TAM) and immune cells (Hanahan and Weinberg 2000).There is a metabolic interplay between these cells and cancer cells which serves as a steering factor for promoting the proliferation of the cancer cells (Fig. 1.1), as discussed in the sections below.

Cancer-Associated Fibroblasts (CAFs)

One of the important members of the TME are the CAFs, which are differentiated and activated form of normal fibroblast cells (Kalluri 2003). CAFs secrete a wide variety of growth factors, such as EGF, HGF, cytokines such as CXCL12 and IL-6, ECM components such as collagen, fibronectin and ECM-remodeling enzymes such as MMPs, LOX and TGMs. These factors modulate the behavior of the surrounding cells in the TME including cancer cells, indicating that CAFs have the potential to alter the biochemical properties of the TME during tumor progression (Santi et al. 2018). Usually, cancer cells obtain their energy from glycolysis instead of using OXPHOS (Heiden et al. 2009). However, in a solid tumor environment, CAFs perform glycolysis while the cancer cells upregulate the OXPHOS pathway (Pavlides et al. 2009). This "reverse Warburg's hypothesis" contradicts the conventional one. Elevated levels of the glycolytic enzymes HK2 and PFKL in solid tumors ensures high glycolysis level in CAFs (Zhang et al. 2015). There are multiple mechanisms that support CAF driven glucose metabolism which are discussed further.

CAFs associated with prostate cancer cells exhibit higher expression of the GLUT1 transporter which increase glucose uptake (Fiaschi et al. 2012). On the contrary, lactate produced by CAFs is exported out to the cancer cells through high levels of MCT4. This is accompanied by release of protons, thereby making the TME acidic, which in turn favors EMT by inducing the expression of matrix metallo-proteinase-9 (MMP-9) (Peppicelli et al. 2014; Baltazar et al. 2014). Increased export of lactate along with elevated MCT4 expression have also been observed in vitro co-culture system, where CAF cells grow in association with breast and bladder cancer cells (Whitaker-Menezes et al. 2011). However, prostate cancer cells are also reprogrammed towards reduced GLUT1 expression and enhanced lactate uptake via MCT1 overexpression. Moreover, lactate importing function via MCT-1 has also been observed in osteosarcoma (Baltazar et al. 2014; Bonuccelli et al. 2014). This lactate utilization by cancer cells as an energy source or a

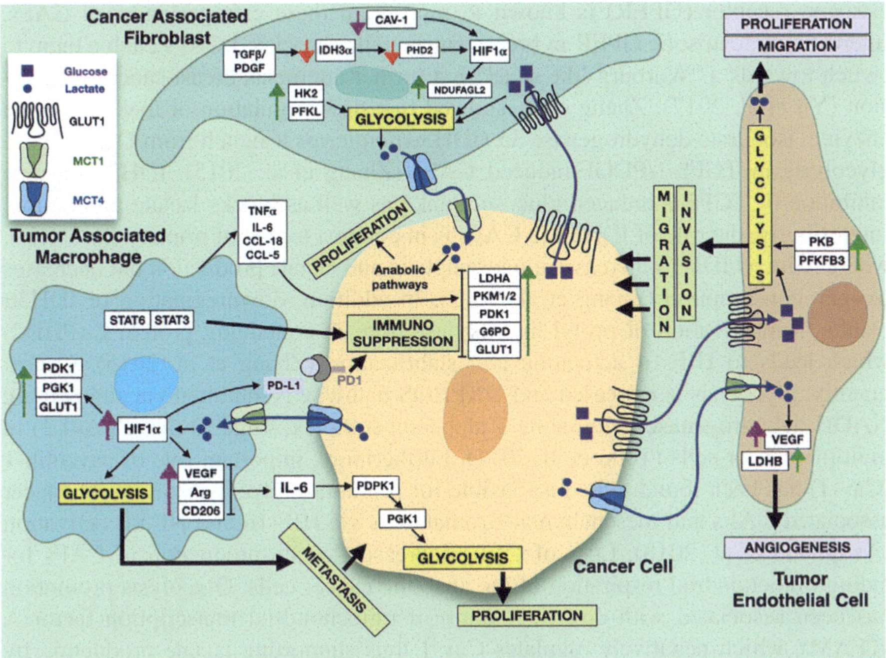

Fig. 1.1 Metabolic crosstalks between the cancer cells and the heterogenous cell types, cancer associated fibroblasts (CAFs), tumor associated macrophages (TAMs) and tumor endothelial cells (TECs) within the tumor microenvironment. This figure depicts an interdependence between the heterogeneous cell population of tumor microenvironment and cancer cell through metabolites exchange, in turn supporting the survival and proliferation of each other. The cancer cells export glucose through GLUT1 receptor which is imported into CAFs and serves as a substrate for glycolysis. Lactate produced here is exported out through MCT4 and imported into cancer cells via MCT1 which is utilized in anabolic pathways. Glycolysis in CAF is promoted by the upregulation of several glycolytic enzymes as well as the downregulation of other enzymes and proteins. The release of cytokines, chemokines and other factors from TAMs into the cancer cells induces immunosuppression and increases enzymatic activity related to glycolysis. VEGF produced in hypoxic environment of TME recruits TAM which favors HIF-1α expression, followed by enhanced glycolytic flux. Lactate generated by the cancer cells also stabilizes HIF-1α expression in TAMs and leads to increased activation of glycolytic enzymes, ultimately leading to tumorigenesis. Tumor ECs are mostly dependent on glycolysis, promoted by the enhanced expression of PFKFB3 enzymes which supports cancer cell invasion and migration. Lactate accumulation favors proliferation and migration of ECs. The cancer cells enable increased lactate uptake by ECs which in turn augments VEGF level and promotes angiogenesis. Green arrows indicate upregulation of metabolic enzymes; red arrows indicate downregulation of metabolic enzymes; Levels of other proteins or factors are indicated by purple arrows

metabolite further confirms reprogramming by the TME. It is known that the TME acidification promotes tumor progression and is linked to metabolic reprogramming that favors Warburg's mechanism (Fiaschi et al. 2013).

Several other studies have highlighted the involvement of CAFs in support of Warburg's theory. As observed in breast cancer associated CAFs, G-protein coupled estrogen receptor (GPER) is known to reprogram these estrogen-induced CAFs. Interestingly, cytosolic GPER in breast cancer cells associated CAFs enable them to switch towards a "Warburg-like state" that further augments breast cancer progression (Yu et al. 2017). Zhang et al. showed that downregulation of the TCA cycle enzyme, isocitrate dehydrogenase 3α (IDH3α) indicates a switch from OXPHOS to glycolysis in TGFβ-1/PDGF-induced CAFs (Zhang et al. 2015). IDH3α ensures inhibition of TGFβ-1 induced glucose uptake as well as blocks lactate production. Interestingly, the role of IDH3α in CAFs is in contrast to that of primary fibroblasts, where reduced IDH3α increases glucose uptake and lactate production but decreases oxygen consumption (Zhang et al. 2015). In addition, downregulation of IDH3α results in inactivation of prolyl hydroxylase domain-containing protein 2 (PHD2) which leads to HIF-1α activation and stabilization (Zhang et al. 2015). Consequently, glycolysis is increased and OXPHOS pathway is inhibited via augmenting NADH dehydrogenase ubiquinone 1 alpha subcomplex, 4-like 2 (NDUFA4L2) in multiple cancer cells (Tello et al. 2011). Furthermore, important role of caveolin-1 (Cav-1) has been found to be responsible for promoting glycolysis in breast cancer associated CAFs and mesenchymal stromal cells via HIF-1α and NF-κB activation (Avagliano et al. 2018). Loss of Cav-1 increases lactate production in CAFs for aiding mitochondrial respiration of the anabolic cancer cells. Glycolysis promotion has been associated with downregulation of mitochondrial transcription factor A (TFAM), which negatively regulates Cav-1 thus supporting lactate production by CAFs and promoting tumor growth (Avagliano et al. 2018). The absence of Cav-1 favors expression of pyruvate kinase-2 (PKM2) in the stromal CAFs that promotes OXPHOS pathway in the associated breast cancer cells (Avagliano et al. 2018). Cumulatively, the continuous supply of nutrients and metabolites by the CAFs essentially drive the proliferation of the cancer cells and enables them to fulfil their energy demands.

Endothelial Cells (ECs)

Within their niche, the tumor cells encounter reduced oxygen concentration and scarcity of nutrient supply, which are compensated by the newly formed blood vessels and lymphatics through a process called angiogenesis (Lau and Vander Heiden 2020). These blood vessels and lymphatics are lined by endothelial cells (ECs) which constitute the stromal component of the TME. Normally, the blood vessels are arranged in multiple layers, whereas within the TME they are arranged in an irregular network. The tumor ECs are highly proliferative but produce fragile junctions during networking, which promotes leakiness and intravasation of cancer cells. Also, tumor ECs are highly transcriptionally active due to their increased metabolic demands, in contrast to the normal ECs, which remain quiescent for prolonged periods. Interestingly, these quiescent ECs can switch to their proliferative or migratory behavior in response to pro-angiogenic cues, thereby generating new sprouts. Tumor ECs are highly dependent on glucose metabolism as evident from enhanced mRNA expression of the glycolytic enzyme PFKFB3 (Bando et al. 2005). Suppressing PFKFB3 activity inhibits cancer cell invasion, migration and intravasation without affecting tumor growth. Xu et al. showed that PFKFB3 deletion reduces vessel perfusion in tumors (Xu et al. 2014). Further, tumor implanted mice with deficient PFKFB3 showed reduced growth. In the absence of PFKFB3, Protein kinase B remains downregulated and lactate accumulation is reduced which suppresses migration and proliferation of ECs. Pharmacological inhibition of PFKFB3 also represses proliferation and causes death of ECs, resulting in tumor vessel disintegration (Conradi et al. 2017). These observations demonstrate the significance of the glycolytic enzyme PFKFB3 in neo-vascularization of the tumor cells within the TME.

ECs can be grouped into three main categories namely, the tip cells, the stalk cells and the phalanx cells (Lau and Vander Heiden 2020). The migratory tip cells localized at the leading edge of the blood vessel guides the highly proliferative stalk cells and quiescent phalanx cells to form its lining during angiogenesis (Lau and Vander Heiden 2020). Each of these EC types have different proliferative and biosynthetic requirements, hence, they exhibit varying metabolic demands. ECs mostly depend on glycolysis for ATP generation, which is enabled by increased PFKFB3 activity, and is induced by VEGF (De Bock et al. 2013). Silencing of PFKFB3 activity results in significant loss in glycolytic flux, resulting from the reduced level of fructose 2,6-bisphosphate (De Bock et al. 2013). However, the metabolic demand of ECs is also supported by the OXPHOS pathway, specifically complex III that supports EC proliferation during angiogenesis (Diebold et al. 2019). In contrast, the phalanx cells have a lower glycolysis requirement than that of the proliferating ECs (De Bock et al. 2013). Proliferation of ECs is induced by cancer cells which help to increase the glucose uptake in ECs via elevating GLUT1 expression (Lau and Vander Heiden 2020).

There is increasing evidence that ECs influence the metabolic programs of cancer cells in various ways. Recent studies in multiple breast cancer cell types suggested that lymphatic endothelial cells (LECS) exhibit different metabolic profiles when co-cultured with one or more cell lines (Acevedo-Acevedo et al. 2020) in particular, LECs exhibit reduced OXPHOS and elevated glycolytic flux. In breast cancer cells, LECs exhibit decreased NADPH levels as a result of increased NADPH utilization which indicates the dependence of LECs on glycolysis. It is also suggested that co-culture of LECs with breast cancer cells induce lymphangiogenic signaling in these cells which may promote pro-metastatic behavior of the cancer cells. Furthermore, highly expressed lactate metabolism markers, MCTs and LDHs, have also been reported in LECs (Acevedo-Acevedo et al. 2020). Furthermore, it is reported that increased import of lactate by the ECs affects their cell-signaling pathways, such as induction of autocrine IL8/NFkB activity or HIF-1α stabilization and increased VEGF production (Acevedo-Acevedo et al. 2020). Thus, it is believed that lactate exhibits its pro-angiogenic role in ECs within a tumor microenvironment. It is reported that the tumor vasculature in colorectal adenocarcinoma expresses low level of MCT1 (Koukourakis et al. 2006), despite the fact that high levels of lactate in these cells can act as a pro-angiogenic factor for the surrounding ECs. Notably, the tumor associated ECs, have upregulated expression of LDHB enzyme, which facilitates the exogenous lactate uptake (Van Beijnum et al. 2006). Hence, presence of lactate dependent ECs in the vicinity of lactate-exporting tumor cells, depicts a switch from glucose to lactate utilization. Additionally, the endothelial cells significantly contribute to the survival of tumor cells, beyond inducing angiogenesis in a glucose-depleted environment.

Tumor Associated Macrophages (TAMs)
Macrophages constitute approximately half of the tumor mass within the tumor microenvironment and are considered to be an important component of the immune cell infiltrate (Vitale et al. 2019). In general, macrophages can be grouped into two types of activation states, termed M1 (classically activated) and M2 (alternatively activated) (Biswas and Mantovani 2010), that have different metabolic phenotypes and distinct influences on cancer cells. The M1 phenotype commences in the presence of cues such as interferon-γ (IFNγ) and Toll-like receptor (TLR) ligands that induce polarization to a pro-inflammatory state. The M1 phenotype is characterized by the expression of pro-inflammatory cytokines IL-12, TNF, inducible nitric oxide synthase (iNOS), reactive oxygen species (ROS), and microbicidal and tumoricidal activity. On the contrary, the M2 state is generated via induction of anti-inflammatory cytokines IL-4, IL-5, IL-10, IL-13 which is further associated with tissue restructuring and tumor progression. The M2 macrophages are characterized by specific chemokine receptors (CCL17, CCL18, and CCL22), a IL-12^{lo}IL-10^{hi}IL-1 decoyR^{hi}IL-1RA^{hi} phenotype and a high level of phagocytic activity (Biswas and Mantovani 2010). TAMs are often associated with poor prognosis in many cancers and their role has been documented in immunosuppression, tumor growth, invasion and metastasis (Biswas and Mantovani 2010). TAMs respond to TME-derived signals and display dynamic behavior regarding their functional phenotypes, which consequently dictates their fate as pro- or anti-inflammatory immune cells (Murray and Wynn 2011). Importantly, the M1 phenotype favors predominantly the use of glycolytic pathway (Vitale et al. 2019). The M1 macrophage enrichment is characterized by lowered OXPHOS activity which is associated with pro-inflammatory and anti-tumoral response (Kim and Bae 2018). In contrast to its dimeric form in the M1-phenotype, the increased activity of tetrameric PKM2 suppresses the M1 macrophages and stimulates M2 macrophages (Palsson-Mcdermott et al. 2015). M2 macrophages show higher oxidative phosphorylation in response to PGC1-dependent mitochondrial biogenesis (Huang et al. 2014b).

There are growing evidence that suggest how metabolically reprogrammed TAMs are associated with the functioning of the cancer cells. Reprogrammed TAMs can regulate tumor growth and survival by releasing altered cytokines and angiogenic factors (Condeelis and Pollard 2006). The immune-metabolic interplay between cancer cells and TAMs establishes an immunosuppressive landscape in the TME via chemokines, cytokines, metabolites and growth factors such as IL-6, TNF, CCL5 and CCL18 (Andrejeva and Rathmell 2017). The aforementioned immunogenic factors induce lactate dehydrogenase A (LDHA), pyruvate kinase M1/2 (PKM1/2), PDK1, G6PD, and GLUT1 (Andrejeva and Rathmell 2017). Also, IL-6 promotes glycolysis as regulated by phosphoinositide dependent protein kinase 1 (PDPK1) dependent phosphoglycerate kinase (PGK) phosphorylation (tyrosine 243) in tumor cells, leading to proliferation of human glioblastoma multiform

(Zhang et al. 2018). Moreover, PGK1 phosphorylation can be correlated significantly with malignancy and prognosis of human glioblastoma.

TME favors the recruitment of macrophages where, hypoxia-induced angiogenic factor VEGF is greatly responsible for the recruitment of TAMs (Puthenveetil and Dubey 2020). Interestingly, M2 macrophages remain highly populated in the oxygen-deficient region within the tumor, where the cancer progression rate is higher (Puthenveetil and Dubey 2020). Under such circumstances, TAMs promote HIF-1α expression which acts as a pre-requisite for glycolytic fermentation. Several glycolysis pathway enzymes namely PDK1, PGK1, GLUT1, glucokinase (GCK) and PKM2 are upregulated via HIF-1α overexpression (Puthenveetil and Dubey 2020). Furthermore, lactate generated from cancer cells stabilizes HIF-1α expression and is also responsible for the polarization of M1 to M2 phenotype, which is characterized by upregulated CD206, arginase-1 and VEGF levels (Colegio et al. 2014). Several studies indicate how cancer cells derived lactate contributes to the reprogramming of macrophages in the TME. Thyroid cancer (TC) derived lactate activates macrophages that promotes AKT1/mTOR pathway leading to elevated cytokine production and an increase in glycolysis (Arts et al. 2016). Interestingly, there is an abundance of histone methylation (H3K4me3) on the promoters of cytokine genes IL6 and TNF-α, which leads to their activation. Moreover, lactate promotes the release of anti-inflammatory cytokine IL-10, thereby eliciting its immunosuppressive role (Puthenveetil and Dubey 2020).

Metabolic reprogramming of macrophages promotes their pro-tumoral function (Penny et al. 2016). The prominent glycolytic flux in TAMs has been shown to support pro-metastatic behaviour in pancreatic ductal adenocarcinoma (PDCA) (Penny et al. 2016). Also, TAMs are highly associated with tumor vasculature formation, promotion of extravasation and EMT (Penny et al. 2016). In addition, there are transcription factors, NF-κB, HIF-1α and STAT3 that play essential roles in regulating metabolic pathways of macrophages and their polarization which in turn promotes their pro-tumoral role (Puthenveetil and Dubey 2020). NF-κB induces transcription of the pro-inflammatory gene iNOS that produces nitric oxide (Mailloux and Willmore 2014). Increased iNOS production stabilizes aerobic glycolysis in M1 macrophages via nitrosylation of HIF-1α (Li et al. 2007). Importantly, HIF-1α activation results in PD-1 ligand mediated immunosuppression (Noman et al. 2014). PD-1 ligand expression by TAMs favors a pro-tumoral role whereas, its inhibition leads to increased phagocytosis and reduced tumor growth (Gordon et al. 2017). This suggests a switch from pro- to anti-tumoral function of HIF-1α in cancer.

On the contrary, M2-activation is supported by the role of STAT3 and STAT6 leading to immunosuppression and tumor progression (Yan et al. 2016). Low ATP and glucose starved condition favor M2-like cells, which are also activated by metabolic enzyme LDHA (Assmann and Finlay 2016). TAMs have also been shown to induce tumorigenicity and support tumor vascularization in breast cancer cells (Gómez et al. 2020). It had been suggested that TAM-derived TGF-β binds to receptors of breast cancer cells and suppresses STAT1 which decreases succinate dehydrogenase (SDH) (Gómez et al. 2020). Consequently, the decreased level of succinate stabilizes HIF-1α that reprograms the tumor cell metabolism towards glycolytic flux and stimulates PD-L1 mediated immunosuppression. Thus, reprogramming metabolism in TAMs provide important insights to understand their role in the TME towards cancer progression.

Role of Oncogenes and Tumor Suppressors in Metabolic Reprogramming in Cancer

The balance between the activity of oncogene and tumor suppressor genes is critical for carcinogenesis. Simultaneous oncogene activation and tumor suppressor inactivation result in abnormal cell proliferation, leading to development of aggressive cancer. Several classes of proteins account for the oncogenic function which include growth factors: Insulin-like growth factor (IGF-1), epidermal growth factor (EGF); receptor tyrosine kinases such as epidermal growth factor receptor (EGFR); proto-oncogenes: c-Src (Src), Ras, spleen tyrosine kinase (Syk) and transcription factors: c-Myc (Myc) (Levine and Puzio-Kuter 2010). EGF-mediated fructose 1,6-bisphosphate accumulation activates EGFR in triple-negative breast cancer cells ($ER^-PR^-HER2^-$) that results in higher glucose uptake and increased lactate production, elevating the glycolytic flux. Such metabolic events are fueled by an elevated level of hexokinase and PKM2 enzymes (Lim et al. 2016). Altogether, these series of events result in breast cancer cell proliferation. Mutant EGFR is well known for its oncogenic role in lung adenocarcinoma. An association of mutated EGFR has been shown to augment aerobic glycolysis and pentose phosphate pathway (PPP) in lung cancer, which exhibits elevated GLUT3 expression and lactate production (Makinoshima et al. 2014). Pentose phosphate pathway (PPP) branches from glycolytic intermediate glucose-6-phosphate (G6P) and generates NADPH and ribose 5-phosphate which are required for antioxidant defenses and synthesis of nucleotides, fatty acids and non-essential amino acids (Ge et al. 2020). EGFR-induced Ras is known to activate PI3K/AKT pathway which leads to increased aerobic glycolysis in glioblastoma cells (Elstrom et al. 2004). The Myc proto-oncogene remains highly expressed in many cancers and is known to enhance the level of glucose transporter enzymes, lactate dehydrogenase, pyruvate dehydrogenase lipoamide kinase isozyme 1 and glycolysis pathway enzymes (Cairns et al. 2011). Ectopic Myc expression in cancer stimulates aerobic glycolysis or OXPHOS pathway to enrich the fuel source and substrates for cancer cell growth and proliferation within the TME (Cairns et al. 2011). Besides the afore-mentioned oncogenes, another important cancer linked kinase is PI3K which is a central regulator of metabolism in both transformed as well as non-transformed cells. In cancer, PI3K/AKT pathway remains highly deregulated on account of an inactivating mutation in PI3K (Jones and Thompson 2009). Elevated PI3K signaling positively regulates metabolic transformation via increasing glucose uptake and directing glycolysis through stimulation of hexokinase and phosphofructokinase (PFK) (Jones and Thompson 2009). Glioblastoma cells exhibiting constitutive Akt expression display high glucose consumption thereby, augmenting aerobic glycolysis (Elstrom et al. 2004). Taken together, there are many ways through which oncogene activation contributes to metabolic reprogramming in cancer.

Deregulation of metabolism in cancer cells is also a consequence of mutations in the tumor suppressors. The regulation of PI3K pathway is achieved by rigorous control of tumor suppressor PTEN (Wong et al. 2010). PTEN is a phosphatase that degrades the phosphoinositide downstream of PI3K and its gene remains deleted or mutated in different cancers such as prostate, breast and brain cancers (Li et al. 1997). Loss of function of PTEN is associated with mutations (missense and nonsense), chromosome deletions (homozygous/heterozygous deletion, frameshift, in-frame deletion, and truncation) and epigenetic modification such as PTEN promoter hypermethylation (Molinari and Frattini 2014). Interestingly, PTEN gain of function has been shown in human glioma, where truncated C-terminal region of PTEN (Δ51) associates with increased NFκB activity in astrocyte tumors (Fernández et al. 2014). This gain-of-function of PTENΔ51 stimulates IGF-1 synthesis which in turn leads to Akt and NFκB activity in an autocrine loop, which ultimately results in aberrant cell growth (Fernández et al. 2014). Loss of function of PTEN leads to PI3K mediated AKT1 activation, which increases glycolysis in tumor cells resulting in an increased ATP/AMP ratio (Fan et al. 2010). In fact, AKT1 upregulates the expression as well as translocation of glucose transporters to the plasma membrane and stimulates phosphorylation of PFK-2 activating glycolysis (Edinger and Thompson 2002). Moreover, increased AKT1 signaling is responsible for regulating a cluster of transcription factors by inhibiting forkhead box sub-family O (FOXO) which accelerates the glycolytic capacity of cancer cells (Khatri et al. 2010). Such studies prove that the role of tumor suppressors in cancer represent an interesting aspect of metabolic regulation.

The tumor suppressor p53 plays a key role in suppressing glycolysis, in addition to its involvement in the DNA damage response and apoptosis (Vousden and Ryan 2009). Enhanced glycolysis in cancer cells is promoted via an increased rate of glucose transport mediated by various GLUT receptors. Wild-type p53 reduces glycolysis by downregulating the transcription of GLUT1 and 4 and inhibits the expression of GLUT3 by downregulating the activity of NF-κB (Kawauchi et al. 2008). In addition, activation of p53-responsive transcription factor RRAD blocks GLUT1 translocation to the plasma membrane thereby suppressing glycolysis (Liu et al. 2019). Conversely, in cancer cells mutant p53 prefers aerobic glycolysis which is favored by increased translocation of GLUT1 to the plasma membrane owing to increased glucose import (Eriksson et al. 2017; Zhang et al. 2013). Also, p53 mutation results in decreased expression of SCO2 (Cytochrome c oxidase assembly) and GLS2 (Glutaminase 2) and increased level of GLUT1 and 4, which promotes aerobic glycolysis in place of OXPHOS (Zhang et al. 2013).

p53 also modulates metabolic enzymes in multiple ways, ranging from destabilizing HK2 mRNA (Lei Wang et al. 2014) to downregulating protein level of phosphoglycerate mutase 1 (PGAM1) which lead to shutdown of glycolysis (Kondoh et al. 2005). In addition, p53 controls the transcription of an E3 ubiquitin ligase Parkin that leads to polyubiquitination mediated degradation of HIF-1α, thereby suppressing glycolysis (Liu et al. 2017). p53 also transcriptionally induces the expression of an enzyme TIGAR (TP-53 induced glycolysis and apoptosis regulator) that has been shown to decrease the glycolytic activity of fructose-2,6-bisphosphate (Bensaad et al. 2006). Another target of p53 is G6PD, which is repressed to inhibit glucose consumption via suppression of the pentose phosphate pathway (PPP) (Jiang et al. 2011). Transportation of the glycolytic product lactate in cancer cells is also inhibited by p53 through reduced transcription of MCT1 (Boidot et al. 2012). Finally, p53 negatively regulates mTORC1 via the Sestrin1/2/AMPK axis which leads to inhibition of glycolysis (Budanov and Karin 2008). In summary, there are multiple ways through which p53 suppresses glycolysis, and all of those are reversed upon p53 mutation or epigenetic silencing in cancer, leading to the enhanced glycolysis that sustains cancer cell proliferation.

Mutant p53 is well known for its gain of function (GOF) activities and remains stabilized in cancer cells (Muller and Vousden 2013). An important role of GOF mutation in p53 has been evidenced in metabolic reprogramming. p53 gain of function mutation is known to promote GLUT1 translocation through the small GTPase RhoA/Rock signaling axis, thereby supporting the proliferation and survival of breast and lung cancer cells (Zhang et al. 2013). Furthermore, mutant p53 drives glycolysis in hepatoma cells through upregulation of HK2 expression and elevates Mtorc1-dependent phosphorylation of PKM2 enzyme in human pancreatic adenocarcinoma cells (Liu et al. 2019). The importance of p53 in regulating cancer metabolic programming is evident from its role in KRAS-mutated pancreatic cancer (Morris et al. 2019). Here, reactivation of p53 increases α-KG level using glucose-derived pyruvate. Increased level of α-KG results in TET (histone demethylase) mediated 5-hydroxymethylcytosine (5hmC) formation, which is related to tumor cell differentiation. These events lead to upregulation of a series of genes including Dynlt3, Kif3c, Nat2, Per2, etc. This is how p53 reprograms cancer cell metabolism by affecting chromatin and gene expression, suppressing the malignant progression.

Other tumor suppressors also play a role in increasing glycolytic flux in cancer cells. One such candidate is liver kinase B1 (LKB1), an upstream kinase required for AMPK activation. It remains mutated in Peutz–Jeghers syndrome designated by benign oral and gastrointestinal lesions with a tendency to develop multiple types of malignancies (Cairns et al. 2011). Mutant LKB1 has been shown to down regulate AMPK signaling along with suppressing mTOR inhibition, which contributes to tumorigenesis (Shackelford and Shaw 2009). In this manner, LKB1 mutation ensures that mTOR and HIF1 activation essentially supports glycolytic metabolism. Therefore, the evidence and observations discussed so far indicate that oncogenes and tumor suppressors mediated metabolic regulation can be an effective way of therapy in cancer.

Hormone-Regulated Metabolic Alterations

Beside the regulation of metabolic pathways in cancer cells by the TME, oncogenes and tumor suppressors, several hormones play significant roles in modulating the metabolic signaling hub of cancer cells. Altered glucose metabolism in cancer cells is often associated with specific hormonal activities, which is responsible for increasing the expression of metabolic enzymes or receptor proteins that function to elevate the level of glycolysis and support tumor cell proliferation. In response to glucose abundance, insulin is secreted from pancreatic beta cells, whereas the insulin-like peptide, IGF-1 is predominantly synthesized in the liver by growth hormone signaling pathways (Bowers et al. 2015). Insulin has to travel via circulation to reach tumor cells while IGF1 is being produced in cancer cells to serve as an alternative system as it develops a connection between cells via autocrine, paracrine and endocrine signaling (Clemmons 2007). Precisely, IGF-1 binds to IGF-1R receptors and activates the PI3K/Akt/mTOR and MAPK/ERK downstream signaling pathways which eventually promote cell proliferation (Bowers et al. 2015). The downstream kinase of mTOR, S6K1, phosphorylates ER-alpha on serine 167 and causes transcriptional activation of estrogen-responsive genes to coordinate cell proliferation (Yamnik and Holz 2010). Interestingly, IGF1 and IGF-1R are also estrogen signaling targets and operate autocrine signaling in breast cancer cell (Christopoulos et al. 2015). Thus, estrogen and insulin signaling function synergistically to activate the glycolytic pathway in luminal breast cancer. In addition, a role of insulin has been documented in cancer where it upregulates expression of the glycolytic enzyme PKM2 via PI3K/mTOR-mediated HIF1α induction (Iqbal et al. 2013). This drives aerobic glycolysis in the liver, lung and prostate cancer cells. On the contrary, insulin leads to reduced PKM2 activity which results in accumulation of NADPH, ultimately switching glycolytic flux towards macromolecule synthesis and increased growth of cancer cells (Iqbal et al. 2013).

Another hormone regulated cancer is ovarian cancer, which is regulated by the presence of follicle-stimulating hormone (FSH) and acquires malignant properties that are controlled by signaling crosstalk (Chen et al. 2016). A novel oncogene ACTL6A, has been shown to be overexpressed in ovarian cancer, which in turn is regulated by FSH-induced PI3K/AKT pathway (Zhang et al. 2019). Notably, ACTL6A overexpression positively correlates with proliferation, clonal growth and migration of cancer cells. Also, ACTL6A silencing leads to reduced glucose consumption, lactate production, pyruvate synthesis and PGK1 downregulation in ovarian cancer (Zhang et al. 2019). It has the ability to modulate FSH-driven glycolysis. Several reports indicate significant involvement of androgen receptor (AR) signaling in promoting the Warburg effect in prostate cancer. AR is known to enhance glucose uptake and utilization in various PCa cell models, consistent with increased expression of GLUT1, GLUT3, GLUT10 and GLUT12 receptors (Massie et al. 2011). Moreover, AR induces several glycolytic genes such as HK1, HK2, PFK, PFKP, ENO1, PFKFB2 and LDHA (Audet-Walsh et al. 2017; Massie et al. 2011). Studies in patient-derived xenograft PCa models show elevated level of aerobic glycolysis in the case of AR-dependent castration resistance (Zacharias et al. 2019). Altogether, AR plays crucial role in supporting Warburg effect by upregulating key glycolytic genes and glucose uptake.

Carbohydrate Metabolism Plays a Key Role in Acquisition of Different Cancer Hallmarks

Tumorigenesis is a multi-step process, based on acquisition of six different hallmarks by cancer cells as reviewed by Hanahan and Weinberg (2000). The six acquired capabilities that mediate the transformation of cells, are self-sufficiency in growth signals, insensitivity to anti-growth signals, evading apoptosis, limitless replicative potential, sustained angiogenesis and tissue invasion and metastasis (Hanahan and Weinberg 2000). Subsequently, two additional cancer hallmarks have emerged – reprogramming of energy metabolism and evading immune destruction (Hanahan and Weinberg 2011). Cellular metabolism is exploited by these different “hallmark” processes, as they demand high energy to sustain tumorigenesis. Recent evidence highlights a hybrid metabolic state where both glycolysis and OXPHOS orchestrate metabolic plasticity in cancer (Jia et al. 2019). Additionally, another concept has grown that typical “glycolytic” type cancer cells have low OXPHOS activity, whereas the “OXPHOS” type of cancer cells rely mainly on mitochondrial respiration (Jia et al. 2018). Therefore, rewiring in different pathways of carbohydrate metabolism is necessary to support the cellular transformation.

Proliferation

Besides augmenting a faster supply of ATP, the advantages of high glycolytic rate include diversion of glycolytic intermediates into various anabolic pathways such as nucleotide, amino acid and lipid synthesis which in turn helps in generating macromolecules and organelles to support cancer cell proliferation (De Berardinis and Chandel 2016). Recent evidence highlight the importance of mitochondrial metabolism in the context of cancer cell proliferation by efficiently producing ATP, TCA cycle intermediates and maintaining redox balance (De Berardinis and Chandel 2016).

Bioenergetics

Cancer cells use different nutrients, glucose, glutamine, fatty acids to meet their bioenergetic demand for maintaining high proliferation rate. The well-known bioenergetic pathway on which they rely, is aerobic glycolysis but many types of cancer cells additionally depend on mitochondrial respiration (Porporato et al. 2018). It has been observed that the malignant phenotypes in ovarian cancer stem cells (CSCs) are supported by the mitochondrial TCA cycle and oxidative phosphorylation pathways while glioblastoma, osteosarcoma and breast cancer stem cells are predominantly dependent on glycolysis (Porporato et al. 2018). Survival and growth of tumors require both oncogenic growth signals and energy sensing signals. The PI3K-AKT-mTOR axis is the principal growth signaling pathway in many cancer types and is tightly integrated with carbohydrate metabolism (Engelman 2009). PI3K-AKT signaling promotes glycolysis, leading to enhanced ATP production. On the contrary the serine-threonine kinase AMPK senses energy stress which is relayed to LKB1-AMPK signaling when there is high AMP/ATP ratio (Chaube et al. 2015), to enhance ATP production. However, bioenergetics is supported by combinatorial performance of both glycolysis and the TCA cycle. In addition to pyruvate, amino acids and fatty acids also provide substrates to the TCA cycle for mitochondrial ATP synthesis. For example, beta oxidation of fatty acids produces acetyl Co-A and two reducing equivalents NADH and FADH2 which get converted to ATP via the electron transport chain (De Berardinis and Chandel 2016). Therefore, this coordinated bioenergetics maintains a continuous supply of ATP during proliferation.

Macromolecule Biosynthesis

Carbohydrate metabolism impinges on nucleotide, protein and lipid metabolic pathways to satisfy their anabolic demands for cell growth and proliferation. Various metabolic intermediates that are produced during glycolysis and the TCA cycle serve in synthesis of these macromolecules. Fatty acid synthesis is crucial for cell division because it is essential for the formation of cellular membranes. Furthermore, the tumor cell signaling pathways require fatty acid synthesis. The pentose phosphate pathway branches from the glycolytic intermediate, glucose-6-phosphate which forms ribose-5-phosphate and NADPH that are successively utilized for nucleotide biosynthesis (De Berardinis and Chandel 2016). The TCA cycle contributes oxaloacetate and glutamate, the former being trans-aminated to aspartate and is utilized for synthesis of purine and pyrimidine bases (De Berardinis and Chandel 2016). Furthermore, amino acid precursor molecules are supplied by the TCA cycle intermediates. Metabolism and synthesis of glutamine, proline, alanine and aspartate are regulated by mitochondrial metabolism. Acetyl-CoA is also a major source of fatty acid which is a TCA cycle intermediate. Clearly carbohydrate metabolism plays a key role in promoting cancer cell growth by fueling these biosynthetic pathways.

Redox Balance

Reactive oxygen species (ROS) induces various oncogenic signaling transduction cascades including MAPK (Weinberg et al. 2010), EGFR (Liou et al. 2016) and PI3K-AKT signaling pathways (Irani et al. 1997) which lead to high proliferation rate. The master transcription factors, HIF1α (Jung et al. 2008) and NF-kB (Morgan and Liu 2011), are intricately connected to energy synthetic pathways, and also get activated by ROS. For that, cancer cells exhibit spatially localized mitochondrial and cytosolic NADPH oxidase (NOX)- dependent Reactive Oxygen Species (ROS) production. An extra-mitochondrial source of ROS production is mediated by cytosolic NOX family proteins which catalyzes transfer of electrons from cytosolic NADPH to generate ROS (Block and Gorin 2012). In response to this escalated generation of ROS, antioxidant mechanisms are quite abundant in cancer cells that prevent stress mediated cell death by maintaining cellular redox homeostasis (Locasale 2013).

Angiogenesis

Angiogenesis is a process of sprouting blood vessels within the tumor to fulfil the increased need of nutrients and oxygen supply as their proliferation rate is high. Recent investigations have established the role of glycolysis in augmenting proliferation and migration of endothelial cells during angiogenesis (Cantelmo et al. 2016). The branching of endothelial cell constituted blood vessels is mediated by the pro-angiogenic stimulator VEGF through the VEGF-Delta-Notch signaling pathway (Arima et al. 2011). During angiogenesis, endothelial cells promote glycolysis over OXPHOS as more energy and biomass is required by different subsets of endothelial cells, such as the stalk and migratory tip cells (De Bock et al. 2013). In this way, endothelial cells avoid the risk of facing oxidative stress by ROS, which is generated from OXPHOS (Eelen et al. 2015). GLUT1, LDHA and PFKFB3 expression are induced by VEGF signaling when angiogenesis is initiated, inducing a high glycolytic flux (De Bock et al. 2013; Parra-Bonilla et al. 2010; Peters et al. 2009). PFKFB3 is a kinase that catalyzes generation of fructose 2,6-bisphosphate which mediates allosteric activation of PFK1, a glycolytic rate-limiting enzyme. As such, VEGF plays a key role in inducing glycolysis to promote angiogenesis.

Invasion and Metastasis

As the tumor size grows in size, metastasis ensues, which is a multi-step process, involving local invasion of tumor cells and their dissemination into nearby blood vessels (intravasation) followed by extravasation of the circulating tumor cells into distant organs, and their colonization to form metastases (Lambert et al. 2017). Tumor cells experience phenotypic transitions as they metastasize. They undergo epithelial to mesenchymal transition (EMT) prior to entering circulation, and undergo mesenchymal to epithelial transition (MET) to enable their colonization. Below, we describe how metabolic reprogramming influences different stages of metastasis to facilitate the formation of secondary tumors as represented in Fig. 1.2.

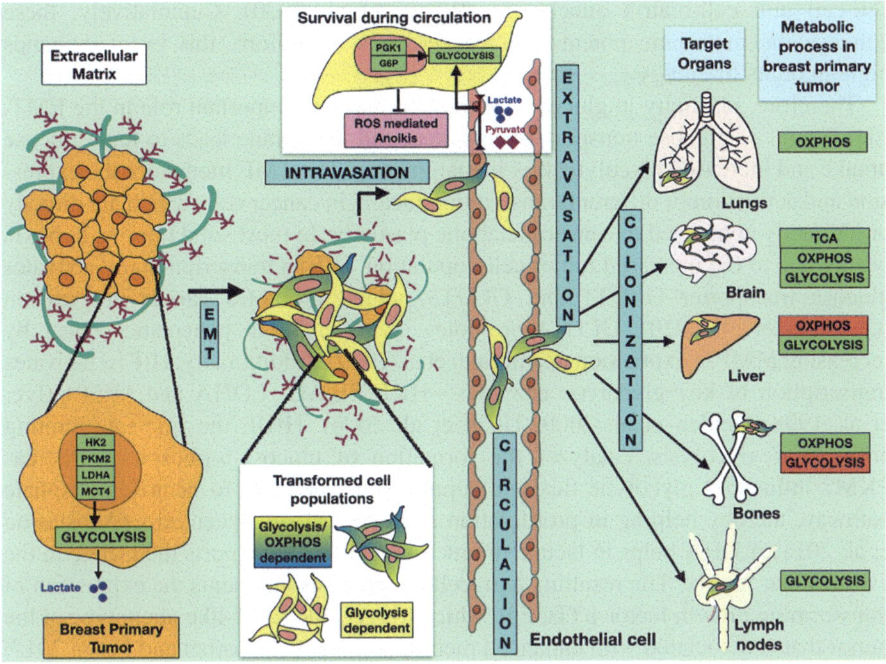

Fig. 1.2 Metabolic plasticity determining the metastatic potential of the breast cancer cells. The primary tumors have elevated expression of different glycolytic enzymes such as HKII, PKM2, LDHA and MCT4 that drives glycolysis to provide energy for the Epithelial to Mesenchymal Transition (EMT). Additionally, acidic pH due to excess lactate production by glycolysis, promotes this phenotypic transition. During EMT, separate transformed cell populations that are either dependent solely on glycolysis or a combination of glycolysis/OXPHOS pathways have been identified. Subsequently, the transformed cell population undergo intravasation to the circulatory system. The circulating tumor cells acquire glycolysis/OXPHOS hybrid phenotypes to get protection from different stresses and ROS mediated damage. Additionally, lactate and pyruvate serve as nutrients for them. Subsequently, extravasation leads to site-specific colonization to form secondary tumors. Different metabolic pathways determine the localization of the cancer cells. The target organs and the fate dictating pathways are as following. Lungs- OXPHOS; Brain- TCA, OXPHOS, Glycolysis, Liver- Glycolysis up regulated, OXPHOS down regulated; Bones- OXPHOS up regulated, glycolysis down regulated; Lymph nodes- glycolysis

Epithelial to Mesenchymal Transition (EMT)

EMT is the initial trans- differentiation event in metastasis which is acquisition of mesenchymal features from epithelial cells. This phenomenon is characterized by the loss of epithelial markers E-cadherin, cytokeratin and increment of mesenchymal markers N-cadherin and vimentin. This process entails major change in cytoskeletal organization and cell shape to promote cell motility and invasion. Additionally, there are distinct mechanisms such as alteration in expression of cell adhesion molecules (e.g., E-cadherin), disruption of basement membrane through activation of matrix metalloproteases (MMP-2, MMP-3, MMP-9) which directly results in loss of cell-cell and cell-matrix attachment (Ribatti et al. 2020). Cumulatively, these enhance the metastatic potential of cancer cells. Therefore, this event demands high amounts of energy.

Therefore, plasticity in glucose metabolism plays an important role in the EMT. Enrichment of glucose transporters on cancer cell membrane leads to high glucose uptake and increased glycolytic flux in malignancy. The TME modulates the expression and activation of different transcription factors in cancer cells which are directly or indirectly connected to cancer metabolic plasticity. In most solid tumors, hypoxia occurs due to the increased tumor cell population. HIF1α transcriptionally activates glucose transporter GLUT1 and GLUT3, which promotes cancer progression (Casciano et al. 2020). GLUT1 promotes invasiveness in pancreatic cancer by increasing MMP-2 expression (Zambrano et al. 2019). Additionally, HIF1α activates transcription of key glycolytic enzymes—HKII, PKM2, LDHA and MCT4 (Iyer et al. 1998; Heiden et al. 2009; Ullah et al. 2006). HKII, the first rate-limiting enzyme of glycolysis, catalyzes the formation of glucose-6-phosphate whereas PKM2 influences glycolytic flux and supplies intermediates to pentose phosphate pathway, thereby helping in proliferation and survival of cancer cells (Anastasiou et al. 2011). LDHA helps in lactate production and MCT4 exports the lactate to the extracellular matrix. The resulting extracellular acidosis augments the expression of transforming growth factor-β (TGFβ), which triggers the EMT-like phenotype of the tumor that is associated with enhanced metastatic potential (Corbet and Feron 2017; Riemann et al. 2019). Another EMT regulating transcription factor c-Myc, in cooperation with HIF1α induces the transcription of PDK1, thereby inhibiting mitochondrial metabolism to avoid ROS mediated damage (Kim et al. 2007). Several upstream signaling pathways of metabolic regulation, including PI3K-Akt, get activated which induces expression of MMP9 and contributes to invasion (Huang et al. 2019; Liu et al. 2020). The EMT process requires this rewiring of metabolism and the EMT driving factors regulate the expression of different metabolic genes. In basal-like breast cancer, SNAIL1 represses the transcription of Fructose 1,6-bisphosphatase-1 (FBP1), a gluconeogenic gene which catalyzes the formation of fructose-6-phosphate from fructose-1,6-bisphosphate, leading to high glycolytic flux (Dong et al. 2013). Another EMT marker, ZEB1 activates GLUT3, and correlates with low survivability in non-small cell lung carcinoma (Masin et al. 2014). Consequently, carbohydrate metabolic pathways are regulated differentially in the context of EMT, partly due to the altered expression of glycolysis and TCA cycle genes. Under hypoxic conditions, Hexokinase 2 expression is very high in squamous cell carcinoma, driving the tumor cells towards EMT (Chen et al. 2018b). Pyruvate dehydrogenase kinase-1 (PDK-1) phosphorylates the pyruvate dehydrogenase complex to inhibit its function. Low PDK1 levels divert pyruvate to the TCA cycle and is considered a poor prognostic marker in lung cancer (Sun et al. 2014). The last key enzyme of glycolysis, pyruvate kinase, catalyzes the transfer of a phosphoryl group from phosphor enol pyruvate to ADP. This leads to the generation of pyruvate and ATP. Expression of PKM2 is high in colorectal cancer and pancreatic ductal carcinoma, which favors glycolysis to enhance metastatic potential (Hamabe et al. 2014; Li et al. 2016). Enolase-1 (ENO-1) converts

2-phosphoglycerate to phospho-enol pyruvate and plays an oncogenic role promoting EMT progression in endometrial cancer (Zhao et al. 2015). The MET process is also promoted by altered metabolism. Silencing of phosphor glucose isomerase (PGI) results in MET in human lung fibrocarcinoma cells (Funasaka et al. 2007). The conversion of fructose 1,6 bisphosphate to glyceraldehyde-3 phosphate and di-hydroxy acetone phosphate is mediated by Aldolase A. Aldolase A functions to increase invasiveness via E-cadherin - EGFR signaling in bladder cancer (Li et al. 2019). Although glycolysis is the major metabolic pathway in cancer, evidence suggests that mitochondrial energy pathways such as TCA and OXPHOS also fuel the high proliferation and metastatic properties of tumor cells. A mechanistic link in between mutation in the genes encoding the TCA cycle enzymes FH, SDH, IDH and cancer metastasis has been investigated (Wei et al. 2020). Mutations in these enzymes result in accumulation of oncometabolites such as fumarate, succinate and 2-Hydroxy glutarate (2-HG) which are the pivotal regulators of metastatic cascade as they contribute to epigenetic changes (Wei et al. 2020). There is also evidence that the hybrid glycolysis/OXPHOS phenotype which supports metabolic plasticity of cancer cells also promotes EMT, given that the metastatic breast cancer cells 4T1 exhibit both glycolytic and OXPHOS activities (Dupuy et al. 2015). Therefore, as discussed, a symbiotic association between EMT and metabolic processes empowers the development of secondary tumors.

Intravasation and Circulating Tumor Cells

EMT is followed by intravasation of tumor cells, which is the invasion of cancer cells through the basement membrane into blood or lymphatic vessels and those cells are called circulating tumor cells (CTC). CTCs must combat shear force and different environmental stresses (Wei et al. 2020). For that, the CTC subpopulation utilizes distinctive metabolic pathways which are essential for their survival. PGK1/G6PD dual marker expression in the CTC subpopulation in breast cancer tissue indicates that they have both high glycolysis and pentose phosphate pathway (Chen et al. 2020). Another study found increased mitochondrial biogenesis and respiration in 4T1 derived CTCs, compared to primary tumors and lung metastasis of 4T1 cells (Jia et al. 2018). Significantly, the glycolytic activity in these 4T1-derived CTCs remains unaltered, indicating a hybrid phenotype to resist stress effects (Jia et al. 2018). Furthermore, CTCs develop different mechanisms to minimize ROS mediated cell death and to promote metastasis as mitochondrial respiration occurs (Park et al. 2020).

Extravasation and Colonization in the Metastatic Niche

Metabolic heterogeneity in tumor cells determines their metastatic potential in terms of local and distant site-specific metastasis. In this context, breast cancer cell populations engage OXPHOS in case of bone and lung metastasis or glycolysis in case of liver metastasis, indicating a role of metabolic heterogeneity in site-selective/organ-specific metastatic ability (Wang et al. 2021). Additionally, being a master regulator of glycolysis, the HIF1α-PDK1 network has been shown to promote colonization of breast cancer cells in the liver (Dupuy et al. 2015). Glycolysis and gluconeogenesis are two reversible pathways. Cancer cells depend mostly on glycolysis to obtain an adequate energy supply rather than relying on gluconeogenesis. On the contrary, the expression of FBP2, a rate-limiting gluconeogenic enzyme is upregulated in brain metastatic cells originating from breast cancer to sustain and survive by activating gluconeogenesis for their nutrient supply. Silencing of FBP2 significantly compromises the proliferation and survival of the metastatic cells (Chen et al. 2015). However, metabolic pre-conditioning of the metastatic niche has also been proposed. Secreted exosomes from the primary tumor appear to build a suitable microenvironment by modulating the metabolic status of the secondary site, thereby favoring homing of migrating cells (Peinado et al. 2017).

Immune Escape

Programming the tumor microenvironment into an immune-suppressive state has been postulated as an inordinate challenge to the host anti-tumor immunity (Hanahan and Weinberg 2011). Recent advances on check point blocking immunotherapy which deactivates the immunosuppressive receptors cytotoxic T lymphocyte associated antigen (CTLA-4) and programmed death 1 (PD-1) on tumor infiltrating lymphocytes (TIL), restores the anti-tumor immunity in triple negative breast cancer context (Keenan and Tolaney 2020). Additionally, an emerging field of immunometabolism bridges the metabolism of tumor cell and immune cell as an efficient immunotherapeutic strategy. Immunometabolism focuses on the altered metabolic fate of immune cells as a consequence of nutrient scarcity and hypoxic condition in the tumor microenvironment. Therefore, this is a remarkable way by which cancer cells can escape immune surveillance and divide in an uncontrolled manner. To cater to the extra energy demand, the tumor cells competitively feed on the nutrients available in the TME depriving the immune cells (Chang et al. 2015). Naive T cells rely on OXPHOS, whereas upon activation, there is a switch towards glycolysis and anabolic pathways (Gerriets and Rathmell 2012; Wang et al. 2011). Intriguingly, glucose deprivation in TME can upregulate PD-1 expression, leading to "functional exhaustion" of T cell (Chang et al. 2013). Thus, the metabolic fate of T cells influenced by TME, dictates its anti-tumor function. Indeed, increased glycolysis in both tumor and immune cells prompts lactate secretion in TME which lowers its pH. Low pH can stabilize HIF1α even in the presence of adequate oxygen, resulting in enhanced production of VEGF, which stimulates the population of myeloid derived suppressor cells (MDSCs) in TME (Gabrilovich and Nagaraj 2009), that has a potent immune-suppressive activity. Expansion of MDSC causes upregulated expression of immunosuppressive factors which confer tumor progression (Norian et al. 2009).

Escaping Cell Death and Acquiring Resistance

Evading cell death is a crucial hallmark of cancer cells. Cancer cells engage in different intrinsic and extrinsic mechanisms to evade cell death programs through apoptosis, thereby enabling high proliferation rate and survival in stressful conditions, and also acquiring resistance against therapies. Carbohydrate metabolism has both direct and indirect effects on apoptosis regulation. As cancer cells harness altered metabolic features, this influences cell death signaling pathways. Due to massive growth and high proliferation, ROS production is quite common in cancer cells which activates many pro-tumorigenic signaling. However, mitochondria-generated ROS contributes to increased oxidative stress in cancer cells. High levels of ROS can induce cancer cell death by initiating intrinsic apoptosis in the mitochondria or death receptor mediated extrinsic apoptotic signaling (Ichijo et al. 1997). So, they develop different mechanisms to keep high oxidative stress in check. The Warburg effect is pronounced in cancer cells, and its intermediates contribute to the pentose phosphate pathway, which is important in supplying of ROS scavenger glutathione, thus protecting cancer cells from extensive oxidative stress (Patra and Hay 2014), thereby bypassing mitochondria-dependent apoptosis. Hypoxic conditions occur in the microenvironment as the tumor cell density increases. HIF1-α transcriptionally regulates expression of different glycolytic genes in hypoxia (Semenza 2010a, b, 2012). In many types of cancers, HIF1-α is overexpressed (Semenza 2010a, b). Moreover, HIF1-α is an important mediator of anti-apoptosis (Kotch et al. 1999) where HIF1-α mediates resistance to apoptosis under hypoxic condition by regulating the glucose uptake and expression of glycolytic genes (Dong and Wang 2004). PI3K/Akt signaling regulates both metabolism and apoptosis in parallel and co-ordinates these inter-related processes (Robey and Hay 2006). Activated Akt promotes glycolysis through localizing Hexokinase II on the mitochondrial membrane, and thus it prevents the release of mitochondrial cytochrome C (Majewski et al. 2004), directing the cancer cells towards apoptosis resistance. The tumor suppressor p53 transcriptionally induces cell cycle arrest and apoptosis under stress conditions (Vousden and Prives 2009). p53 promotes many apoptosis-inducing genes such as BAX, APAF-1, NOXA, and PUMA, etc. (Bourdon et al. 2003). p53 in turn activates mitochondrial respiration while suppressing glycolysis. The expression of glucose uptake and glycolysis regulatory genes such as GLUT1, GLUT4 (Schwartzenberg-Bar-Yoseph et al. 2004) and Hexokinase II (Wang et al. 2014) are negatively regulated by p53 whereas TP53-inducible Glycolysis and Apoptosis Regulator (TIGAR) expression is positively regulated (Bensaad et al. 2006). Also, p53 regulates basal expression of Apoptosis inducing factor (AIF) and Synthesis of Cytochrome C oxidase 2 (SCO2), thereby facilitating OXPHOS. In the majority of cancer cells, p53 is mutated or functionally or epigenetically inactivated, causing the cancer cells to become highly glycolytic (Muller and Vousden 2013; Soussi and Wiman 2015).

Metabolic adaptations of cancer cells occur as a mode of developing drug resistance and that is an important area to study to develop effective treatment of therapy-resistant tumors. For example, increased glucose uptake and enhanced glycolysis mediate intrinsic or acquired chemoresistance in different tumor cells, such as ovarian and breast (Desbats et al. 2020). Therapy resistant tumors show different metabolic phenotypes in comparison to untreated tumors (Desbats et al. 2020). Therefore, we need a better understanding of how standard cancer therapy changes metabolism of the cancer cells to be able to develop novel therapeutics to combat this resistance. Notwithstanding, metabolic plasticity clearly plays a key role in acquiring different cancer hallmarks.

Epigenetic Regulation in Reprogramming Carbohydrate Metabolism in Cancer Cells and Tumor Microenvironment

In addition to genomic mutations, epigenetic modifications on chromatin such as DNA methylation, post translational modifications of histone and non-histone proteins and non-coding RNA mediated changes, play a central role in regulating metabolic reprogramming in cancer cells either by (i) directly mediating changes in expression of metabolic enzymes or by (ii) indirectly influencing their upstream signaling pathways.

Epigenetic Regulation of Metabolic Genes and Their Regulators Impacting Cancer Cells

DNA Methylation

DNA methylation, of cytosine residues at CpG sites, within gene promoters, augments transcriptional silencing (Easwaran et al. 2014). DNA methyltransferase (DNMT) enzymes catalyze DNA methylation by the transferring methyl group from S-adenosyl methionine (SAM). DNMT1 methylates hemi-methylated CpGs during DNA synthesis and is known as the maintenance methyltransferase whereas DNMT3a and DNMT3b are *de novo* methyltransferases (Li et al. 1992).

The major rate-limiting enzyme of glycolysis, Hexokinase 2 is regulated by DNA hypomethylation of its promoter which results in its high expression. Thus, there is an acquisition of glycolytic phenotype, via hypomethylation of HK2 in hepatoma as opposed to normal hepatic cells (Goel et al. 2003) which is shown in Fig. 1.3. Similar is in the case of Glioblastoma (Wolf et al. 2011). Gluconeogenesis, a reverse pathway of glycolysis, is inhibited in cancer. Fructose 1,6-bis phosphatase (FBP1), a gluconeogenic rate-limiting enzyme is transcriptionally silenced by DNMT1 in basal-like breast cancer (Dong et al. 2013). In addition, promoter hypermethylation of FBP1 is also seen in multiple cancers including liver, colon, gastric and lung cancer (Liu and Zhang 2018), suggesting increased glycolysis which re-establishes the Warburg effect. In cancer cells, different transporters are up regulated to facilitate

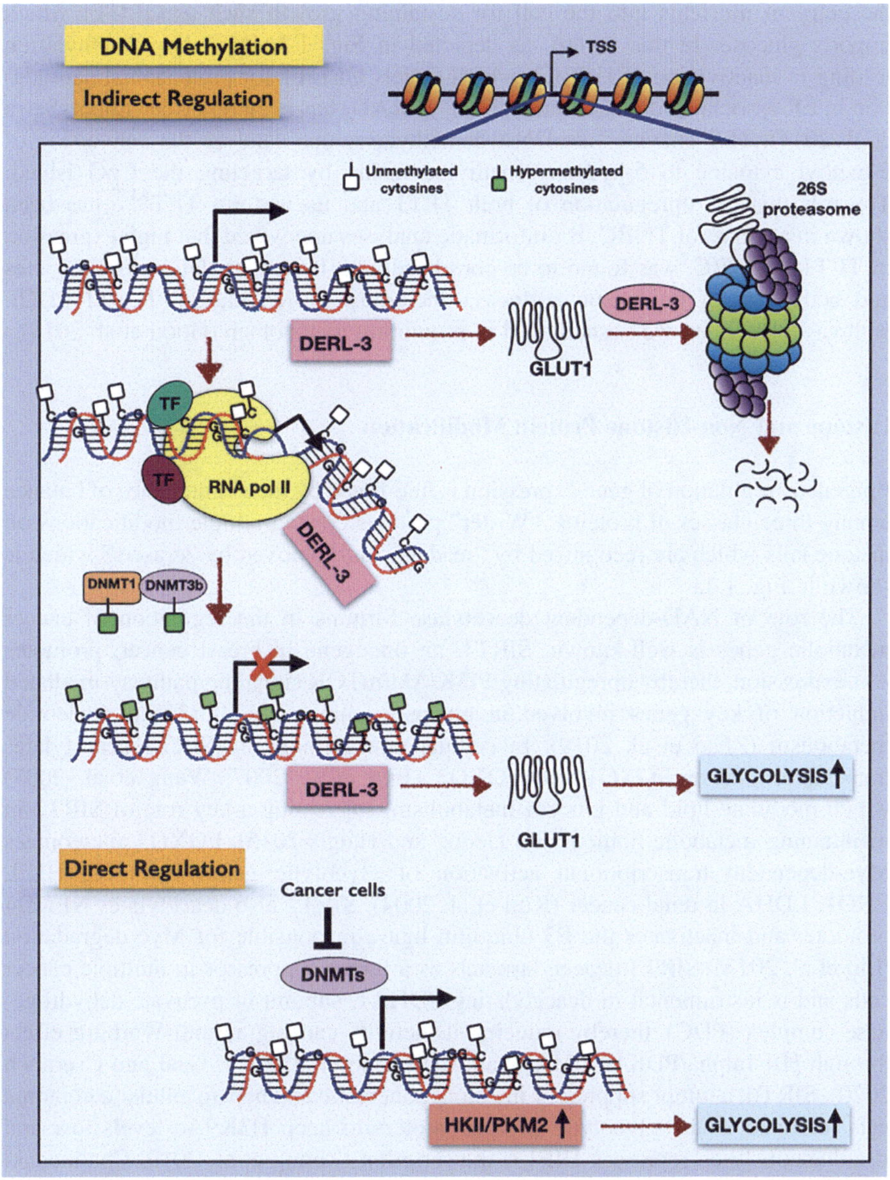

Fig. 1.3 DNA methylation regulates cancer cell metabolic pathways. Carbohydrate metabolic pathways are regulated by DNA methylation of (i) factors that indirectly regulate glucose metabolism or (ii) metabolic enzymes directly. ERAD associated E3 ubiquitin ligase DERL3 causes ubiquitin mediated degradation of GLUT1 in normal cells while in transformed cell, hypermethylation of DERL3 promoter by DNMT1 and DNMT3b results in upregulation of GLUT1. This indicates excessive glucose import and high rate of glycolysis in cancer cells. In cancer cells, DNMTs get inhibited so that promoter methylation of different glycolytic enzyme encoding genes such as HKII and PKM2 can be skipped, thus promoting glycolysis

the entry of nutrients into the cell for sustaining growth such as GLUT1 which imports glucose. In this regard, as depicted in Fig. 1.3, DNA hypermethylation leading to inactivation of DERL3, which targets GLUT1 for proteasomal degradation in ER associated degradation pathway (ERAD) has been reported (Lopez-Serra et al. 2014). TET enzymes are DNA demethylases that catalyze the conversion of 5-methyl cytosine to 5-hydroxy methyl cytosine by targeting the CpG island. Towards this, an upregulation of both TET1 and its isoform TET^{ALT} has been shown in a subset of TNBC. Bioinformatic analyses uncovered that high expression of TET1 in TNBC, was found to be correlated with hypomethylation of CpG sites and activation of oncogenic pathways including PI3K-Akt, EGFR and PDGF pathways which are well established in promoting metabolism (Good et al. 2018).

Histone and Non-Histone Protein Modification

Epigenetic regulation of gene expression is fine-tuned by the maintenance of balance among three classes of proteins. "Writer" proteins create multiple modifications on histone tails which are recognized by "readers" and removed by "erasers" which is shown in Fig. 1.4a.

The role of NAD-dependent deacetylase Sirtuins in the regulation of cancer metabolic genes is well-known. SIRT1, an oncogene in breast cancer, promotes Akt expression, thereby upregulating PI3K/Akt/mTOR signaling pathway-mediated induction of key genes involved in glucose, amino acid, lipid and nucleotide metabolism (Zhao et al. 2019). In cellular stress condition, SIRT2 deacetylates transcription factor FOXO1 and FOXO3 (Jing et al. 2007; Wang et al. 2007) which modulate lipid and glucose metabolism, suggesting a key role of SIRT2 in maintaining metabolic homeostasis (Jeong and Haigis 2015). FOXO3 antagonizes Myc-dependent transcriptional activation of glycolytic genes including HKII, ENO1, LDHA in renal cancer (Kim et al. 2004). SIRT2 also deacetylates NEDD4 promoter and inactivates the E3 ubiquitin ligase responsible for Myc degradation (Liu et al. 2013). SIRT3 deacetylase acts as a tumor suppressor in multiple cancer cells and is instrumental in deacetylating PDHA1, subunit of pyruvate dehydrogenase complex (PDC), thereby inducing its activity causing an anti-Warburg effect through HIF1alpha/PDK1/PDHA1 pathway (Zhao et al. 2019; Gaál and Csernoch 2020). SIRT6, a tumor suppressor in colon, pancreatic and hepatocellular carcinoma gets recruited to the glycolytic gene promoters to keep H3K9Ac levels low and simultaneously co-represses HIF1-α transcription (Zhong et al. 2010; Chalkiadaki and Guarente 2015). These studies establish that histone deacetylases play seminal role in regulating carbohydrate metabolism through multiple mechanisms.

Epigenetic regulation of HIF1-α transcription impacts the expression of several downstream targets involved in carbohydrate metabolism, including GLUT1, HKII, PDK1, PFK1 and LDHA (Semenza 2010a, b; Huang et al. 2017). The functional crosstalk between the epigenetic regulators fine-tunes the carbohydrate metabolic pathways. Towards this cooperative regulation of HIF1-α by SETD8/HDAC2/PHD1 complex has been well documented. SETD8, is a H4K20

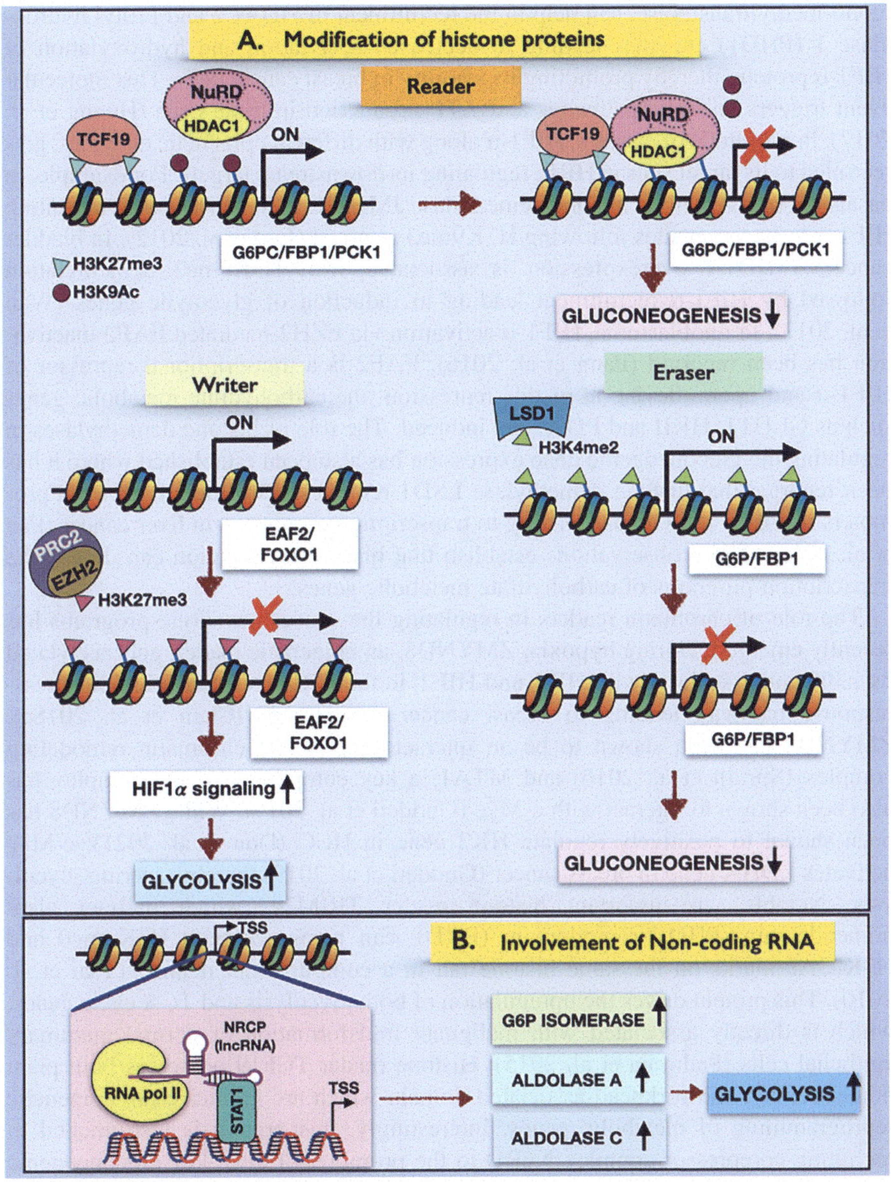

Fig. 1.4 Regulation of carbohydrate metabolic pathways by alteration of histone modification landscape and noncoding RNAs in cancer. (**a**) Histone modification is steered by "Reader," "Writer," and "Eraser" proteins to epigenetically program the metabolic genes. Histone reader TCF19 recognizes H3K4me3 mark onto gluconeogenic gene promoters and recruits NuRD complex that harbors HDAC1. Removal of H3K9Ac marks by the HDAC1 leads to transcriptional silencing of gluconeogenic genes. PRC2 complex harbors methyltransferase (writer) enzyme EZH2, which promotes H3K27me3 mark leading to a repression of EAF2/FOXO1 genes. This leads to an upregulation of HIF1-α signaling promoting glycolysis. An Eraser, histone demethylase LSD1, removes H3K4me2 mark from the gluconeogenic genes G6P/FBP1 thereby transcriptionally silencing them. (**b**) Long non-coding RNA NRCP, serves as an intermediate binding partner

mono-methyltransferase, can help in the recruitment of HDAC2 and Prolyl hydroxylase 1 (PHD1) enzymes leading to decreased acetylation and hydroxylation of HIF1-α protein, thereby promoting its stability in breast cancer cells. This molecular event triggers anabolic pathways and ATP production in these cells (Huang et al. 2017). In multitude of cancers, HIF1-α along with different epigenetic enzymes, gets recruited to its target sites in HRE, regulating its downstream targets. For example, in basal-like breast cancer, histone demethylase JMJD2C is instrumental in recruiting HIF1-α to its target sites following H3K9me3 removal (Luo et al. 2012). In bladder cancer, JMJD1A overexpression is associated with H3K9me2 demethylation followed by HIF1-α recruitment leading to induction of glycolytic genes (Wan et al. 2017). In glioblastoma, HIF1-α activation via EZH2-mediated EAF2 inactivation has been reported (Pang et al. 2016). EAF2 is a transcriptional repressor of HIF1-α and upon alleviation of this repression, the carbohydrate metabolic genes such as GLUT1, HKII and PDK1 get induced. The role of histone demethylases in regulating the gluconeogenic gene expression has also been established where it has been reported that histone demethylase LSD1 removes H3K4me2 mark from promoters of FBP1 and G6Pase leading to transcription repression in liver cancer (Pan et al. 2013). These observations establish that histone methylation can dictate the transcription programs of carbohydrate metabolic genes.

The role of chromatin readers in regulating the cancer metabolic programs has recently emerged. During hypoxia, ZMYND8, an epigenetic reader, gets acetylated by p300 and associates with HIF1 and HIF2, in the HRE sites activating the stress-response pathway leading to breast cancer progression (Chen et al. 2018c). ZMYND8 has been shown to be an interactor of NuRD chromatin remodeling complex (Spruijt et al. 2016) and MTA1, a key component of the complex has also been shown to interact with c-Myc (Guddeti et al. 2019). While ZMYND8 has been shown to positively regulate HK2 gene in HCC (Dou et al. 2021), c-Myc activates LDHA gene in breast cancer (Guddeti et al. 2019) thereby steering glycolysis. Notably, an important histone reader TRIM24, which harbors plant homeodomain (PHD)/bromodomain (BRD), can recognize both H3K4me0 and H3K23Ac marks on the same histone tail in a combinatorial manner (Tsai et al. 2010). This protein drives the upregulation of both glycolysis and TCA cycle genes, which is directly associated with malignant transformation of normal mammary epithelial cells (Pathiraja et al. 2015). Histone reader TCF19 possesses both plant homeodomain and forkhead-associated domain which are implicated in epigenetic reprogramming of metabolic genes. Interestingly, this protein is instrumental in recruiting corepressor complex NuRD to the promoter of three key gluconeogenic genes, G6PC, FBP1 and PCK1, in a high glucose condition, to mediate their transcriptional silencing in hepatocellular carcinoma (HCC) cells (Sen et al. 2017). Moreover, TCF19 has recently been reported to co-regulate the transcription of

---

Fig. 1.4 (continued) between STAT1 and RNA polymerase. Upregulation of NRCP mediates STAT1 driven glycolytic gene (Glucose-6-phosphate isomerase, Aldolase A and Aldolase C) expression

TIGAR and SCO2 genes in association with p53, which impact both glycolysis and OXPHOS pathway in response to short-term or prolonged high glucose-mediated stress situations in HCC (Mondal et al. 2021).

The chromatin remodeling complexes also play important role in regulating metabolic gene expression programs, impacting their epigenetic landscape in indirect ways. In advanced bladder cancer which is highly aggressive, BRM, the ATPase subunit of SWI/SNF is down regulated, leading to an activation of PKM2 and PRKAA (coding for AMPK1α subunit of AMPK) genes (Stachowiak et al. 2020). These studies establish the role of chromatin remodelers and histone/non-histone modifying enzymes in cancer metabolism through distinct mechanisms.

Non-Coding RNAs

Based on their size, non-coding RNAs can be classified as: 21-25 nucleotide long micro-RNA and 200nt or more as long non-coding RNA. Micro-RNA forms a stable complex with RNA induced silencing complex known as miRISC, which functions in two ways—(i) It recognizes target mRNA sequences for transcriptional silencing by degrading them, (ii) It can also regulate translational repression (Valencia-Sanchez et al. 2006). Long noncoding RNAs can regulate gene transcription through three mechanisms- interaction and recruitment of chromatin modifying enzymes, interaction with ribonucleoprotein and enhancer function (Bhat et al. 2016). Regulation of metabolic gene transcription by non-coding RNA is described in Fig. 1.4b. Recent studies have demonstrated the role of different non-coding RNA in regulating cancer metabolic reprogramming (summarized in Table 1.1).

Reliance of Tumor Microenvironment Sustenance on Epigenetic Regulation

Epigenetic regulation inevitably governs the heterogeneous and ever-changing tumor microenvironment. This influences various kinds of interactions between tumor cell and its microenvironment, facilitating the adaptation of tumor cells to different stresses, supporting growth of stromal cells (tumor promoting) and suppressing development of anti-tumor immune cells. Epigenetic regulations which dictate gene expressions in the stromal components of TME include DNA methylation, histone post-translational modifications and expression of non-coding RNAs (Marks et al. 2016). For example, DNA methylation regulates the activation of quiescent fibroblasts associated with non-small cell lung cancer via downregulating SMAD3 gene expression that results in hyper-responsiveness to TGF-β signals (Vizoso et al. 2015). Another evidence shows removal of histone methyltransferase EZH2 and corresponding H3K27me3 mark from ECM modifying

Table 1.1 Role of non-coding RNAs in regulation of carbohydrate metabolism in different cancer

Cancer types	Non-coding RNAs		Oncogene/ Tumor suppressor	Target genes/ Regulators	Molecular Function in cancer cell	Metabolic status in cancer cell	References
Gastric cancer	*micro-RNA*	miR-129-5p	Tumor suppressor	SLC2A3	Modulate PI3K-Akt and MAPK signalling	Glucose uptake↑ Glycolysis↑	Chen et al. (2018a)
Breast cancer		miR-155	Oncogene	FOXO3a, PI3KR1	Up regulate PI3R1-FOXO3a-cMyc axis.	Glycolysis↑	Kim et al. (2018)
Colorectal cancer		miR-26a	Oncogene	PDHX	–	OXPHOS↓ Glycolysis↑	Chan et al. (2015)
		miR-181a	Oncogene	PTEN	Phospho-Akt signalling activated	Lactate production↑	
Hepatocellular cancer		miR-122	Tumor suppressor	SMARD1, MAP3K3, CAT1	Promote mitochondrial metabolic gene expression	Glycolysis↑	
Lung cancer		miR-1, miR-206	Tumor suppressor	G6PD, GPD2, TKT, 6PGD,	–	Glycolysis↑ PPP, TCA Cycle↓	
Bladder cancer	*Long non-coding RNA*	UCA1	Oncogene	HKII, miR-143	STAT3 induced gene expression	Glycolysis↑	Ghafouri-Fard et al. (2020)
Hepatocellular carcinoma		MITA1	Oncogene	LKB1-AMPK	Energy stress and glucose starvation induce MTA1 expression	–	
		HOTAIR	Oncogene	HKII, HIF1 alpha	HIF1alpha regulated genes activation	Glycolysis↑	
Breast cancer		BCAR4	Oncogene	YAP	Hedgehog signalling pathway activation	Glycolysis↑	
		SNHG7	Oncogene	LDHA	c-Myc induced genes activation	Glycolysis↑	
Epithelial Ovarian cancer		NRCP	Oncogene	STAT1	STAT1 induced gene activation	Glycolysis↑	
Non-small cell lung carcinoma		NEF	Tumor suppressor	GLUT1	Reduced glucose uptake	Glycolysis↓	

proteoglycanase ADAMTSI gene promoter in CAFs, which is associated with lymph node metastasis in breast cancer patients (Marks et al. 2016).

Tumor cell and its microenvironment are connected in a bidirectional way where metabolites serve as a bridge between them. In this context, recent studies have shown that in CAFs, hypoxia induces hypomethylation in HIF-1α gene promoter which results in heightened expression of HIF-1α, and consequently an activation of PKM and LDHA and a repression of FBP1. This indicates a pro-glycolytic phenotype in CAFs which metabolically supports proliferation and progression of breast cancer cells (Becker et al. 2020). In another way, metabolite tracing revealed that the lactate secreted by pancreatic ductal adenocarcinoma (PDAC) cells cause excess production of α-ketoglutarate in neighboring mesenchymal stem cells (MSCs), which in turn activates the DNA demethylase TET. This results in increased cytosine hydroxymethylation in the promoter of a chemokine receptor gene CXCR4 which leads to differentiation of MSCs to CAFs (Bhagat et al. 2019). Thus, the metabolic symbiosis, modulated by different epigenetic regulators, mediate the sustenance of both tumor cell and its environment.

Metabolites Dictating Epigenetic Landscape Impact Cancer Progression

Various studies have shown that epigenetic regulation is influenced by the metabolites produced in the cancer cells. In particular, the metabolism-induced epigenetic alteration can be achieved by (1) modulating the homeostatic level of metabolites that act as epigenetic substrates or cofactors (SAM, acetyl coA, NAD^+,α-KG), (2) synthesis of oncometabolites that alter the functions of epigenetic enzymes (fumarate, succinate, 2-hydroxyglutarate) and (3) direct nuclear translocation of metabolic enzymes responsible for production of epigenetic co-factors (IDH3-α, PFKFB4, PKM2) (Boukouris et al. 2016; Zhao et al. 2016). A list of various metabolites along with their roles is represented in Table 1.2.

Metabolites Serving as Coenzymes or Cofactors

Availability of the coenzymes, cofactors, and allosteric modulators alter the complex epigenetic landscape of chromatin, dynamically regulated by the histone modifying enzymes. These factors determine the metabolic enzyme activity via influencing the kinetic parameters (Km, Vmax) as well as their allosteric properties (Reid et al. 2017). These metabolites are produced in homeostatic level in normal cells and can be differentially regulated to meet the excess energy demand of the cancer cells. Some of the well-known metabolites for epigenetic reprogramming include S-adenosyl methionine (SAM), Acetyl-coenzyme A (Acetyl-CoA), nicotinamide

Table 1.2 List of metabolites in epigenetic regulation of cancer

Metabolites in reprogramming	Target enzymes or proteins	Epigenetic modification	Cancer involved	Fate of cancer cells	References
Epigenetic coenzymes/cofactors					
SAM	c-Myc, H-Ras	DNA hypermethylation	Gastric cancer, colon cancer	Inhibit tumor growth	Crispo et al. (2019)
	RASSF1, SOCS2	DNA hypermethylation	Hepatocellular cancer	Promotes tumor growth	Crispo et al. (2019
	NNMT	Histone hypomethylation	Breast cancer, colon cancer, brain cancer, etc.	Invasion, migration, proliferation, Promotes tumor growth	Thomas et al. (2013), Tang et al. (2011)
Acetyl CoA	ATP-citrate lyase	Histone acetylation	Glioma, prostate cancer	Promotes tumor growth	Wellen et al. (2009), Lee et al. (2014)
NAD+	Sirtuins	Histone deacetylation	Liver cancer, lung cancer, breast cancer	Reduced survival, worse prognosis metastasis,	Etchegaray and Mostoslavsky (2016)
α-KG	TET, UTX6 demethylase	Histone demethylation	Melanoma	Suppresses dedifferentiation	Pan et al. (2016)
α-KG	IDH1	DNA and Histone demethylation	Glioblastoma	Suppresses dedifferentiation, promotes tumor growth	Pan et al. (2016)
Oncometabolites					
Fumarate	TET, α-KGDD (PHDs)	Suppresses DNA and histone demethylation	Renal cancer	EMT, invasion, metastasis, cancer progression	Sciacovelli et al. (2016), Xiao et al. (2012)
Succinate	TET, α-KGDD, KDMs	Suppresses DNA and histone demethylation	Paraganglioma	Dedifferentiation, invasion	Letouzé et al. (2013)
2-HG	2-(OG)-dependent dioxygenase	Suppresses DNA and histone demethylation	Glioma, cholangiosarcoma	Promotes tumor growth	Crispo et al. (2019)
	TET, Demethylases	DNA hypermethylation	Renal carcinoma	Promotes proliferation	Shim et al. (2014)
B-hydroxy butyrate	HDACs	Histone acetylation	Breast cancer	Metastasis, Promotes Tumor growth	Shimazu et al. (2013)

Nuclear metabolites or enzymes					
IDH3-α	SHMT	DNA demethylation	Glioblastoma	Cancer progression	May et al. (2019)
PKM2	HATs	Acetylation, methylation	Glioblastoma, breast cancer	Cell cycle progression, tumorigenesis	Desai et al. (2014)
PFKFB4	SRC-3	Histone acetylation	Breast cancer	Metastasis	Dasgupta et al. (2018)

adenine dinucleotide (NAD+), α-ketoglutarate (α-KG), etc. which are associated with glycolysis and TCA cycle as elucidated in Text Box 1.1.

SAM Regulates DNA and Histone Methylation

DNA and histone methylation are essential epigenetic modifications catalyzed by DNA methyltransferase (DNMT) and histone methyltransferase (HMT) respectively, which require SAM as a common methyl donor (Crispo et al. 2019). SAM is released from one-carbon metabolism pathway using methionine and ATP as substrates. Moreover, SAM is generated from serine metabolism, which is linked to a glycolysis intermediate 3-phosphoglycerate responsible for amino acids biosynthesis (Locasale 2013). Conversion of SAM into S-adenosyl-homocysteine (SAH) provides an inhibitory signal to DNMTs (Fig. 1a in Box 1.1) and HMTs hence, dictating their enzymatic activities (Crispo et al. 2019). SAM/SAH ratio is critical in the context of transcriptional regulation by chromatin modification. Increased concentration of SAM hypermethylates the CpG islands thereby, repressing genes involved in cancer cell proliferation and metastasis (Crispo et al. 2019). In gastric and colon cancer, SAM promotes hypermethylation of promoters of oncogenes c-Myc and H-Ras which leads to inhibition of cancer growth (Crispo et al. 2019). In addition, regression of osteosarcoma depends on the level of SAM which regulates ERK1/2 and STAT-3 pathway (Crispo et al. 2019). SAM downregulates ERK1/2 and STAT3 phosphorylation thus, leading to cell cycle inhibition and apoptosis of osteosarcoma cells. A rare genetic condition of GNMT (glycine methyltransferase) deficiency is caused due to disorder in methionine metabolism, which is characterized by elevated level of SAM that ultimately turns on the oncogene activation (Wong et al. 2017). In knockout GNMT mouse, SAM-induced methylation of tumor suppressors RASSF1 and SOCS2 has been reported to promote hepatocellular carcinoma (Martínez-Chantar et al. 2008). SAM is utilized for 1-methylnicotinamide (1-MNA) production by Nicotinamide N-methyl-transferase (NNMT) which remain overexpressed in breast (Wong et al. 2017), colon (Thomas et al. 2013) and brain cancer (Tang et al. 2011), that lowers the cellular pool of SAM and reduces methylation at H3K4, H3K9, H3K27, H4K20 (Ulanovskaya et al. 2013), thereby resulting in aggressive phenotype of cancer. On the contrary, decreased SAM/SAH ratio in methionine deficient environment is responsible for reducing global methylation of histone lysines like H3K4Me3 and consequent silencing of cell-cycle- and cancer-pathway- related genes (Dai et al. 2018). Thus, role of SAM in DNA and histone methylation has diverse effects on the epigenetic landscape which are cell type dependent.

Acetyl Co-A Contributes to Histone Acetylation

Histone acetylation and deacetylation is subjected to the activities of several metabolites that regulate histone acetyltransferases (HATs) and deacetylases (HDACs) activities, affecting several gene transcriptions in cancer. Enzymatic activity of HATs depends greatly on the availability of the cellular metabolite Acetyl-CoA that functions as an allosteric modulator and acts as an acetyl group donor for HATs (Fig. 1b in Box 1.1). The various pool of Acetyl-CoA in the nucleus, cytosol and mitochondria are furnished from pyruvate, acetate, fatty acids and amino acids. Altered level of acetyl-CoA influences level of histone acetylation and gene expression promoting cancer growth. Lee et al. reported that glycolysis derived Acetyl-CoA regulates global histone acetylation in cancer (Lee et al. 2014). Moreover, in prostate cancer and human gliomas histone acetylation remains under the regulation of ATP-citrate lyase (ACL), mediated by Akt activity. pAkt activity remains positively correlated with increased H4ac level in glioma whereas, in prostate tumors positive correlation is present with increased levels of H3K18ac, H3K9ac, and H4K12ac (Lee et al. 2014). Importantly, Acetyl-CoA dictates the kinetic behaviour and binding efficiency of epigenetic erasers and writers in acetylation reaction. Since HAT exhibits similar affinity for Acetyl-CoA and CoA (potent HAT inhibitor), the ratio of Acetyl-CoA to CoA is crucial for its activity (Pietrocola et al. 2015). Oncogene-mediated reprogramming of cancer cells results in increased glycolytic flux that supplies pyruvate for Acetyl-CoA production, thereby upregulating genes involved in cell cycle progression, growth, migration (Wellen et al. 2009; Lee et al. 2014). Furthermore, cancer cells are found to overexpress Acetyl-CoA synthase enzymes (ATP citrate-lyase and acetyl-CoA synthase short-chain family member 1) which increase acetyl-CoA and thereby enables histone acetylation and metabolic gene upregulation such as HK2, PFK-1, LDHA (Etchegaray and Mostoslavsky 2016). ACL contributes to the nuclear pool of acetyl-CoA and remains frequently overexpressed in cancer (Wellen et al. 2009). Colon cancer cell line HCT116 showed decreased histone acetylation upon silencing of ACL. Moreover, increased level of ACL also has been demonstrated in lung cancer to favor histone acetylation and expression of glycolytic enzymes in cancer cells (Migita et al. 2008). Level of nuclear acetyl-CoA is increased in presence of mutant $KRAS^{G12D}$ in pancreatic adenocarcinoma which results in increased H3 and H4 acetylation (Lee et al. 2014). In addition, Myc oncogene has been shown to elevate acetyl-CoA by upregulating glycolysis related enzymes. Role of Myc in epigenetic regulation is evident from Myc-mediated HAT-GCN5 expression which leads to multiple sites acetylation of H4 N-terminal (Knoepfler et al. 2006). Hence, from the above-discussed evidence, it is clear that acetyl-coA is an important metabolic regulator which contributes to the regulation of downstream genes in cancer.

NAD^+ and α-KG Regulates Histone Deacetylation and Demethylation

Histone acetylation and deacetylation are affected by the altered metabolic state of the cancer cells. It is evident from literature that reduced pH increases histone deacetylation, and acetate ions are co-exported with protons in the extracellular space which maintains the optimum intracellular pH (McBrian et al. 2013). The alkaline environment of cells favors the glycolytic flux by cancer cells, as a result of which histone deacetylation is inhibited. HDACs catalyzing histone deacetylation reactions can be classified into four types (Xu et al. 2007) of which class III (Sirtuin family) is NAD^+ dependent (Etchegaray and Mostoslavsky 2016). NAD+/NADH ratio is highly linked to the energy level of the cells. High energy requiring cancer cells promote aerobic glycolysis, resulting in decreased NAD+/NADH ratio which prevents deacetylase function of Sirtuins (Fig. 1b in Box 1.1) (Etchegaray and Mostoslavsky 2016). Consequently, histone hyperacetylation is enhanced which upregulates aberrant gene expression in tumor.

A TCA cycle metabolite α-ketoglutarate (α-KG) serves as a cofactor of 2-oxoglutarate-dependent dioxygenases (2-OGDO), which promotes hydroxylation and demethylation of proteins and nucleic acid (McDonough et al. 2010). This renders α-KG as an important cofactor during demethylation as observed for Ten-eleven translocation hydroxylases (TETs) (Fig. 1a in Box 1.1) and Jumonji C domain-containing lysine demethylases (JmjC-KDMs or JHDM). These enzymes perform DNA and histone demethylation, and their activities are dependent on the availability of α-KG, which is produced from isocitrate, by isocitrate dehydrogenase isoforms in mitochondria as well as cytosol (Crispo et al. 2019). Most importantly, glucose maintains enriched pool of α-KG and leads to H3K27me3, H3K9me3, H4K20me3, and TET-dependent DNA demethylation (Crispo et al. 2019). Moreover, these methylations often regulate pluripotency or stem cells characteristics (Carey et al. 2015). In addition, H3K27me is highly expressed in tumors exhibiting low α-KG level. UTX6 demethylase is most responsive to the presence of α-KG in melanoma (Pan et al. 2016). Profound expression of H3K27 methylation can also be linked to dedifferentiation of cancer cells (Pan et al. 2016). In solid tumors such as glioblastoma, chondrosarcoma, and cholangiocarcinoma, low level of α-KG helps to maintain DNA and histone hypermethylation, that can be linked to dedifferentiation and drug resistance (Pan et al. 2016; Calvert et al. 2017). Besides, metabolic rewiring also modulates specific subsets of genes to contribute to tumorigenesis. H3K4me3 levels are maintained by Trithorax complex whereas, H3K27me2/3 and H3K9me3 are maintained by polycomb-group complex 2 (PRC2) (Schuettengruber et al. 2017). However, both the regulations respond differently to the level of the cofactor, α-KG within the cell (Hwang et al. 2016). All these regulations by NAD^+ and α-KG imply their significant roles in metabolic reprogramming in the context of deacetylation and demethylation.

Box 1. Chemical reactions representing epigenetic modifications catalyzed by specific enzymes

(a) Methylation of histone and DNA requires SAM that acts as a methyl group donor. SAM is converted into SAH which acts as an inhibitory signal to DNA methyltransferases, which catalyze the methylation reaction. SAM/SAH ratio is critical in context of transcriptional regulation by chromatin modification. TET catalyzes oxidation of 5-methyl cytosine into hydroxymethyl cytosine leading to demethylation reaction. Here, α-KG acts as a cofactor for hydroxy-mediated lysine demethylation, which releases succinate as a reaction product. (b) Acetyl CoA acts as a donor of acetyl group for reactions catalyzed by HAT, which mediates acetylation of histone proteins. Deacetylation reactions are catalyzed by HDACs of which Sirtuins belongs to HDAC III family. SIrtuins require NAD^+ as the cofactor that accepts the acetyl group from acetylated lysine residue and gets converted into O-acetyl ADP ribose nicotinamide.

Fig. 1a

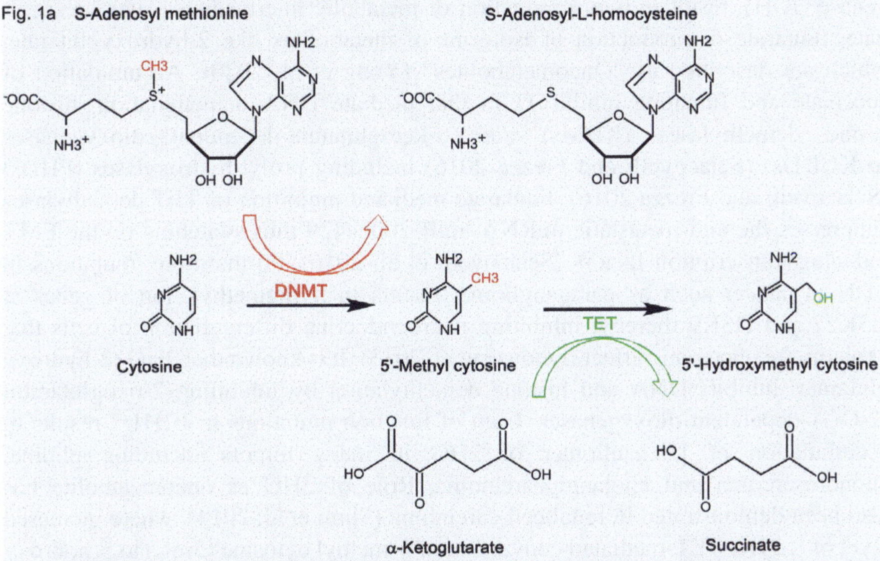

Fig. 1a. Epigenetic enzymes catalyzing reactions using SAM and α-KG as metabolites. SAM: S-Adenosyl methionine; SAH: S-Adenosyl-homocysteine; TET: Ten-eleven transloctaion hydroxylases; α-KG: α-Ketoglutarate

Fig. 1b

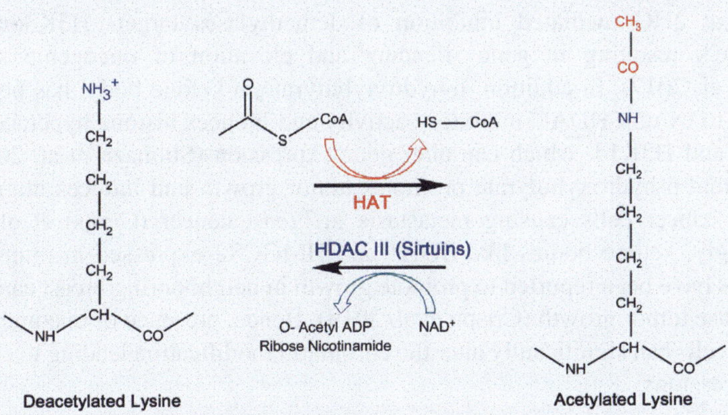

Fig. 1b. Epigenetic enzymes HAT and HDAC III catalyzing histone acetylation and deacetylation reactions using acetyl CoA and NAD^+. HAT: Histone acetyltransferase; HDAC: Histone deacetylase

Oncometabolites Regulating Metabolic Enzymes

Alteration in metabolic profile of cancer cells is achieved by two ways; one is through "reprogramming" wherein the conventional metabolic signaling pathways remain upregulated or downregulated. The other way is via "rewiring" which describes development of new metabolic characteristics caused by mutations, which are absent in normal cells (De Berardinis and Chandel 2016). Mutations in fumarate hydratase (FH), succinate dehydrogenase (SDH), and isocitrate dehydrogenase (IDH), result in over-production of metabolic intermediates such as succinate, fumarate or production of isoforms of metabolites like 2-hydroxyglutarate, which are described as "Oncometabolites" (Yong et al. 2020). Accumulation of succinate and fumarate inhibit TETs that mediate DNA demethylation, histone lysine demethylases (KDMs) and α-ketoglutarate-dependent dioxygenases (α-KGDDs) (Sciacovelli and Frezza 2016) including prolyl-hydroxylases (PHDs) (Sciacovelli and Frezza 2016). Fumarate mediated inhibition of TET demethylases suppresses the anti-metastatic miRNA, miR-200ba429 thus switching on the EMT inducing transcription factors (Sciacovelli et al. 2016). Furthermore, mutations in SDH in cancer such as paraganglioma, results in hypermethylation of genes at H3K27 and H3K9 thereby, inhibiting neuroendocrine differentiation of cells that explains the oncogenic effect (Letouzé et al. 2013). It is known that 2HG (2-hydroxy glutarate) inhibits DNA and histone demethylation by inhibiting 2-oxoglutaratre (2-OG)–dependent dioxygenases. Gain of function mutations in IDH1/2 results in accumulation of D-enantiomer of 2HG in many tumors including glioma, chondrosarcoma and cholangiocarcinoma. Role of 2HG as oncometabolite has also been demonstrated in renal cell carcinoma (Shim et al. 2014), where increased level of L-2HG TET-mediated conversion of 5-methyl cytosine (5mC) to 5-hodroxy methyl cytosine. It is shown that L-2-hydroxyglutarate dehydrogenase (L2HGDG) remains deficient which would otherwise remove excess 2HG. Janke et al. has shown that 2HG mediated inhibition of demethylases targets H3K36me2 and H3K36me3, resulting in gene silencing and elevation of oncogenic signaling (Janke et al. 2017). In addition, β-hydroxybutyrate, a ketone body, has been demonstrated to exhibit HDAC inhibitory activity and induces histone hyperacetylation at H3K9 and H3K14, which can alter gene expression (Shimazu et al. 2013).It is reported that β-hydroxybutyrate promotes tumor growth and induces migration of epithelial cancer cells causing metastasis in breast cancer (Crispo et al. 2019). Interestingly, ketone bodies like BDH1 and HMGCS2 expressed in immortalized fibroblasts have been reported to promote growth in neighbouring breast cancer cells and increase tumor growth (Crispo et al. 2019). Hence, presence of oncometabolites in cancer cells can significantly alter the chromatin modification leading to change in gene expressions.

Nuclear Metabolites Acting As a Source of Epigenetic Co-Factors

A remarkable aspect of epigenetic regulation in cancer cells is that there are certain metabolites or metabolic enzymes that translocate to the nucleus and modulate transcription of adjacent genes. Sub-cellular localization of metabolic products assists chromatin modifiers to add or remove different modifications on histone and non-histone proteins. Such molecular event is exemplified by the methyl donor SAM which remains located in the nucleus of cancer cells, although biosynthesized in the cytosol (Reytor et al. 2009). The nuclear isoform of MAT (methionine adenosyl transferase) generates SAM to contribute to methyl transferase activity and regulates methylation events. Here, MATII-α (a nuclear isoform) forms a complex SESAME (serine-responsive SAM-containing metabolic enzyme) with chromatin and transcription associated factors like Mafk, involved in differentiation of erythroid cells that ensures regulation of specific chromatin sites (Li et al. 2015; Katoh et al. 2011). The isoform MATII-α has been observed in murine cancer cells and interacts with H3K9 methyltransferase SETDB1 (Kera et al. 2013). Another such metabolite is Acetyl-CoA which normally shuttles between cytosol and mitochondria and can diffuse freely through nuclear pores. However, there are nuclear enzymes producing acetyl-CoA including ACSS2 (Acetyl-CoA synthetase short-chain family member 2) (Takahashi et al. 2006; Li et al. 2017), ACLY (ATP citrate-lyase) (Sivanand et al. 2017), PDC (pyruvate dehydrogenase complex) (Sutendra et al. 2014) and CAT (carnitine acetyltransferase) (Sebastián and Mostoslavsky 2017), which regulate transcriptional program depending upon conditions such as nutrition, disease, and microenvironment.

The nuclear PKM2 has been found to be highly expressed in different cancers such as bladder, breast, lung, kidney, and glioblastoma (Christofk et al. 2008; Desai et al. 2014). Nuclear localized PKM2 functions in cooperation with nuclear pyruvate dehydrogenase complex (PDC) thereby, maintaining the activity of histone acetyltransferases via local Acetyl-CoA production (Matsuda et al. 2016). In addition, PKM2 may regulate histone methylation, where Pyk1 (PKM2 homolog in yeast), as a part of SESAME complex, participate in H3K4me3-H3pT11 crosstalk (Li et al. 2015). Here, SESAME has been found to interact with Set1 H3K4 methyltransferase complex and get recruited to target genes, that results in phosphorylation of H3T11. Another glycolytic enzyme with nuclear associated role is PFKFB4 that acts on steroid receptor coactivator-3 (SRC-3), which consists of nuclear receptor domains and exhibits histone-acetyltransferase activity. PFKFB mediated phosphorylation of SRC-3 induces expression of glycolytic genes towards purine synthesis, which is an essential factor for metastatic progression in breast cancer (Dasgupta et al. 2018). Epigenetic modification is greatly dependent on the availability of NAD+ whose supply is assured by the enzyme LDH. Activity of SIRT1 is dependent upon NAD+ in the nucleus which is ensured by nuclear translocation of LDHA during ROS-mediated stress generation (Crispo et al. 2019), where LDHA orchestrates deacetylation activity of SIRT1.

Few TCA cycle related enzymes have been reported to play special roles in epigenetic regulation, while being in the nucleus. Fumarate hydratase is known to inhibit lysine demethylase 2B (KDM2B) via local fumarate synthesis (Jiang et al. 2015). In addition, fumarate production in nucleus suppresses the functions of epigenetic erasers thereby, leading to tumorigenesis. Also, IDH3α being a key enzyme of TCA cycle has shown its presence in cytoplasm as well as nuclear periphery in S-phase arrested cells (May et al. 2019). It interacts with serine hydroxyl methyltransferase (SHMT) and regulates pyrimidine synthesis and DNA methylation. In glioma cells, the cytosolic isoform IDH1 is localized in nucleus which impacts histone and DNA methylation. Although the role of IDH1 needs further investigation, the enzyme seems to prove its important contribution in epigenetic modulation.

Therapeutic Strategies Targeting Epigenetics-Metabolism Crosstalk in Cancer

Crosstalk between epigenetics and metabolic programs has been considered in developing some innovative therapeutic strategies among which the drugs targeting the epigenetic regulators and metabolite inhibitors are the prime focus.

Epi-Drugs

DNMT Inhibitors Two DNMT inhibitors 5-Aza-cytidine and 5-Aza-2′-deoxycytidine have been clinically approved by the US Food and Drug Administration (FDA) for treating myelodysplastic syndrome and acute myeloid leukemia (AML) (Bohl et al. 2018; Döhner et al. 2017) and are also being tested in many solid cancers (Lee et al. 2018; Singal et al. 2015). Both are cytosine analogues, which non-selectively suppress the functions of DNMT1, DNMT3A and DNMT3B (Rodríguez-Paredes and Esteller 2011). Although the role of these metabolic inhibitors in cancer is largely unknown, they can regulate a wide array of genes which lead to pleiotropic effect. There are evidence showing that oncometabolite-driven methylation can be reverted by DNMTi. DNA hypermethylation is observed in IDH1/2 mutant cancer which is sensitive to 5-Aza-cytidine and 5-Aza-2′-deoxycytidine. In IDH1 mutant glioma xenograft model, tumor was regressed when treated with 5-Aza-cytidine whereas 5-Aza-2′-deoxycytidine effectively inhibited IDH mutant glioma cell growth both in vivo and in vitro. In a further study, it has been shown that treatment of 5-Aza-2′-deoxycytidine is more effective in inducing differentiation of IDH-mutant glioma cells in comparison to mutant IDH inhibitors (Borodovsky et al. 2013; Turcan et al. 2013).

HDAC Inhibitors

a. *Class I and Class II HDAC*: HDAC inhibitors promote histone acetylation and reverse gene silencing. Emerging evidence suggest their impact on cancer metabolism. In lung cancer cells, Sodium Butyrate (NaBu) and TSA reverts glycolysis and cause transition to mitochondrial metabolic pathways (Amoêdo et al. 2011) and similar effect was scored in breast cancer cell (Rodrigues et al. 2015). Treatment with HDAC inhibitor NaBu or TSA significantly reduces glucose uptake, glycolytic flux and lactate production in colorectal cancer cells (Alcarraz-Vizán et al. 2010). Interestingly, it has been observed that the expression of GLUT1 receptor and activity of glycolytic enzyme HKII are abrogated upon treatment with HDAC inhibitors Vorinostat and Sodium Valproate in multiple myeloma cells.
b. *Class III HDAC or Sirtuins*: SIRT6, an important tumor suppressor protein, gets activated by free fatty acids (Feldman et al. 2013) and opposes glycolysis (Sebastián et al. 2012). Besides this, SIRT1 inhibitor MHY2245 exerts anti-tumor activity by obstructing PKM2/mTOR pathway (Tae et al. 2020). Future investigations are required to decipher the role of modulators of Sirtuins in cancer metabolism.

Non-Coding RNAs As Therapeutic Target

As the noncoding RNAs play an important regulatory role in cancer metabolism, targeting them (miRNA or lncRNA) has been considered to be a promising therapeutic approach. Currently, two approaches are being used for regulating miRNA expression—(i) restoring miRNA by supplying synthetic mimics and (ii) silencing the overexpressed miRNA by using anti-sense RNA (Li and Rana 2014). For example, miR-143 targets 3′UTR of HK2 gene as well as functions as a tumor suppressor. Ectopic expression of miR-143 in colon cancer cells reverses glycolytic phenotype and inhibits cancer growth (Gregersen et al. 2012). On the contrary, anti-miR-21 targeting miR-21, a PTEN suppressor, restores PTEN expression in hepatocellular carcinoma which reduces phosphorylation of downstream targets of PTEN, FAK and Akt, that are key mediators of cancer metabolism (Meng et al. 2007). Multiple lncRNAs target different steps in carbohydrate metabolism. Most notably, lncRNA directly or indirectly targets the major metabolic regulator c-Myc. Therefore, advances in promoting delivery and cellular efficacy will make it a good therapeutic option.

Tumor Metabolite Inhibitors

Glycolytic Inhibitors High glycolytic flux enhances histone acetylation which is mediated by increased Acetyl-CoA and citrate. Thus, by inhibiting glycolysis with different glycolytic inhibitors such as 2-Deoxy glucose (2-DG), a glucose analogue and 3-bromo pyruvate, a pyruvate analogue, modulate histone acetylation level (Liu et al. 2015; Moussaieff et al. 2015). Treatment with 2-DG competitively inhibits glucose-6-P production and reduces Acetyl CoA level, thereby causes global decrease in H3 and H4 acetylation in several cancer cell lines. This event is associated with increased sensitivity to DNA damaging agents (Liu et al. 2015).

TCA Cycle Inhibitors: Mutant Isocitrate Dehydrogenase Inhibitors Mutant IDH1 and IDH2 are well known in modulating epigenetic landscape in gliomas and leukaemia. These mutated enzymes produce a potential 2-Hydroxy glutarate (2-HG). $IDH1^{R132H}$ inhibitor ACI-5198 blocks 2-HG production and induces demethylation of H3K9me3 and augments gene expression that are associated with differentiation of glioma (Rohle et al. 2013). AG-120, AG-881, ML309 and GSK-321 have been identified as other novel $IDH1^{R132H}$ inhibitors (Wong et al. 2017). In leukaemia, $IDH2^{R140Q}$ inhibitor AG-221 leads to reduction of 2-HG and promotes survival (Yen et al. 2017). Another $IDH2^{R140Q}$ inhibitor, AG-6780, leads to demethylation of DNA and histones, causing tumor differentiation (Wang et al. 2013).

Methionine Metabolism Inhibitor: SAH Hydrolase Inhibitor DNMT and HMT activities are largely dependent on SAM/SAH ratio of methionine cycle. TCA cycle intermediate oxaloacetate is converted to aspartate which serves as a precursor of methionine and other amino acid synthesis. SAM and SAH are intermediates of methionine metabolism. Interestingly, 3-Deazaneplanocin-A (DZNep) is an inhibitor of S-adenosylhomocysteine hydrolase which catalyzes SAH hydrolysis. Therefore, methionine cycle is disrupted and there is a reduction in S-adenosyl methionine (SAM) level which is essential for the activity of DNMT and HMT (Glazer et al. 1986). A synergistic effect has been shown in leukaemia when DZNep is combined with DNMT inhibitor 5-Aza, through reactivation of silenced genes by histone and DNA methylation (Momparler et al. 2014).

Conclusion: Challenges and Future Perspectives

"Metabolic rewiring" has been established as a defined "cancer hallmark" which supports the core demands of the highly proliferative cancer cells with the changing tumor microenvironment. This metabolic adaptation leads to an altered genomic landscape which is responsible for growth, survival and fulfilment of the energy demands of the cancer cells. On the contrary, the epigenetic aberrations impact the pathways affecting the hallmarks of cancer by differentially regulating the oncogenes and tumor suppressor genes that promote neoplastic transformation. Taken together there exists an interplay between epigenetics and metabolism that plays a fundamental role in the manifestation of cancer.

While exploring the therapeutic options for cancer treatment, it may be suggested that epigenetic readers are emerging as one of the potent targets, amongst which the bromodomain and extra-terminal inhibitors (BETi) targeting the bromodomain functions are in clinical trial. Besides, these selective targeting of the epigenetic regulators (writer, reader and eraser) by small molecule modulators or synthetic RNA interference-based approaches have a great therapeutic potential. Importantly, epigenetics and metabolism are dynamically interconnected in cancer and serve as an important platform for therapeutic intervention through three independent approaches, (i) designing epi-drugs to target metabolic programs of cancer cells, (ii) targeting the metabolic pathways through the intermediates that in turn alter the epigenetic landscape and (iii) targeting epigenetic regulators and metabolites which act synergistically. Notably, the epigenetic and metabolic heterogeneity which contribute to therapeutic vulnerabilities and poor clinical outcome, impose great challenges in cancer treatment. However, emerging field of studies targeting the interface of epigenetics and metabolism is being explored as a potential therapeutic option.

Conventionally, total metabolic profiling in different cancers gives an overall understanding of the alteration in metabolite levels and its connection to the disease etiology. Improving the meta-analysis tools for such purpose could be very useful in interpreting the metabolic complexity of different cancers. Further, futuristic precision medicines would embark upon individual metabolome and epigenome that will allow selection of patient cohorts for specific therapies. Also, the challenges of conventional therapeutic resistance could be revisited through a new holistic outlook of targeting interface of epigenetics and metabolism. In fact, the involvement of nutrition status is of seminal importance not only for the cancer cells, but also for the supporting cells constituting the TME. Thus, an emergence of the combination therapy exploiting both epigenetic as well as metabolic paradigm may contribute towards better therapeutic development. Significantly, one such combinatorial treatment using mutant IDH1/2 inhibitors with DNMT inhibitors has been used in different cancers. Moreover, inclusion of other therapeutic approach such as targeting TME components (employing angiogenesis inhibitors or immune check point inhibitors) along with the conventional chemotherapy may also be an effective strategy to combat cancer. Site-specific drug delivery imposes another impediment to cancer therapies. To counteract this, nanoparticle-based drug carrier systems and artificial intelligence (AI)-based approaches for site-specific drug delivery are highly prospective strategies in cancer therapy. Thus, targeting the epigenetics-metabolism axis in cancer has tremendous potential in designing innovative and effective therapeutic strategies.

Acknowledgment This work was supported in part by research grants received from Basic and Applied Research in Biophysics and Material Science (RSI 4002) from the Department of Atomic Energy; Science & Engineering Research Board (SERB)(CRG/2018/000985), Department of Science & Technology; SwarnaJayanti Fellowship (DST/SJF/LSA-02/2017-18), Department of Science and Technology; Department of Biotechnology (BT/PR28920/MED/122/176/2018); S. Ramachandran National Bioscience Award for Career Development 2019 (BT/HRD-NBA-NWB/38/2019-20), Department of Biotechnology, Govt. of India to CD.

References

Acevedo-Acevedo S, Millar DC, Simmons AD, Favreau P, Cobra PF, Skala M, Palecek SP (2020) Metabolomics revealed the influence of breast cancer on lymphatic endothelial cell metabolism, metabolic crosstalk, and lymphangiogenic signaling in co-culture. Sci Rep 10(1):1–15. https://doi.org/10.1038/s41598-020-76394-7

Alcarraz-Vizán G, Boren J, Lee WNP, Cascante M (2010) Histone deacetylase inhibition results in a common metabolic profile associated with HT29 differentiation. Metabolomics 6(2):229–237. https://doi.org/10.1007/s11306-009-0192-0

Amoêdo ND, Rodrigues MF, Pezzuto P, Galina A, da Costa RM, de Almeida FCL, El-Bacha T, David Rumjanek F (2011) Energy metabolism in H460 lung cancer cells: effects of histone deacetylase inhibitors. PLoS One 6(7). https://doi.org/10.1371/journal.pone.0022264

Anastasiou D, Poulogiannis G, Asara JM, Boxer MB, Jiang JK, Shen M, Bellinger G, Sasaki AT, Locasale JW, Auld DS, Thomas CJ, Vander Heiden MG, Cantley LC (2011) Inhibition of pyruvate kinase M2 by reactive oxygen species contributes to cellular antioxidant responses. Science 334(6060):1278–1283. https://doi.org/10.1126/science.1211485

Andrejeva G, Rathmell JC (2017) Similarities and distinctions of cancer and immune metabolism in inflammation and tumors. Cell Metabolism 26(1):49–70. https://doi.org/10.1016/j.cmet.2017.06.004

Arima S, Nishiyama K, Ko T, Arima Y, Hakozaki Y, Sugihara K, Koseki H, Uchijima Y, Kurihara Y, Kurihara H (2011) Angiogenic morphogenesis driven by dynamic and heterogeneous collective endothelial cell movement. Development 138(21):4763–4776. https://doi.org/10.1242/dev.068023

Arts RJW, Plantinga TS, Tuit S, Ulas T, Heinhuis B, Tesselaar M, Sloot Y, Adema GJ, Joosten LAB, Smit JWA, Netea MG, Schultze JL, Netea-Maier RT (2016) Transcriptional and metabolic reprogramming induce an inflammatory phenotype in non-medullary thyroid carcinoma-induced macrophages. OncoImmunology 5(12). https://doi.org/10.1080/2162402X.2016.1229725

Assmann N, Finlay DK (2016) Metabolic regulation of immune responses: therapeutic opportunities. J Clin Investig 126(6):2031–2039. https://doi.org/10.1172/JCI83005

Audet-Walsh É, Dufour CR, Yee T, Zouanat FZ, Yan M, Kalloghlian G, Vernier M, Caron M, Bourque G, Scarlata E, Hamel L, Brimo F, Aprikian AG, Lapointe J, Chevalier S, Giguère V (2017) Nuclear mTOR acts as a transcriptional integrator of the androgen signaling pathway in prostate cancer. Genes Dev 31(12):1228–1242. https://doi.org/10.1101/gad.299958.117

Avagliano A, Granato G, Ruocco MR, Romano V, Belviso I, Carfora A, Montagnani S, Arcucci A (2018) Metabolic reprogramming of cancer associated fibroblasts: the slavery of stromal fibroblasts. BioMed Res Int. https://doi.org/10.1155/2018/6075403

Baltazar F, Pinheiro C, Morais-Santos F, Azevedo-Silva J, Queirós O, Preto A, Casal M (2014) Monocarboxylate transporters as targets and mediators in cancer therapy response. Histol Histopathol. 29(12):1511–1524

Bando H, Atsumi T, Nishio T, Niwa H, Mishima S, Shimizu C, Yoshioka N, Bucala R, Koike T (2005) Phosphorylation of the 6-phosphofructo-2-kinase/fructose 2,6-bisphosphatase/PFKFB3 family of glycolytic regulators in human cancer. Clin Cancer Res 11(16):5784–5792. https://doi.org/10.1158/1078-0432.CCR-05-0149

Becker LM, O'Connell JT, Vo AP, Cain MP, Tampe D, Bizarro L, Sugimoto H, McGow AK, Asara JM, Lovisa S, McAndrews KM, Zielinski R, Lorenzi PL, Zeisberg M, Raza S, LeBleu VS, Kalluri R (2020) Epigenetic reprogramming of cancer-associated fibroblasts deregulates glucose metabolism and facilitates progression of breast cancer. Cell Rep 31(9):107701. https://doi.org/10.1016/J.CELREP.2020.107701

Bensaad K, Tsuruta A, Selak MA, Vidal MNC, Nakano K, Bartrons R, Gottlieb E, Vousden KH (2006) TIGAR, a p53-inducible regulator of glycolysis and apoptosis. Cell 126(1):107–120. https://doi.org/10.1016/j.cell.2006.05.036

Bhagat TD, Von Ahrens D, Dawlaty M, Zou Y, Baddour J, Achreja A, Zhao H, Yang L, Patel B, Kwak C, Choudhary GS, Gordon-Mitchell S, Aluri S, Bhattacharyya S, Sahu S, Bhagat P, Yu Y, Bartenstein M, Giricz O, Suzuki M, Sohal D, Gupta S, Guerrero PA (2019) Lactate-mediated epigenetic reprogramming regulates formation of human pancreatic cancer-associated fibroblasts. eLife 8. https://doi.org/10.7554/ELIFE.50663

Bhat SA, Ahmad SM, Mumtaz PT, Malik AA, Dar MA, Urwat U, Shah RA, Ganai NA (2016) Long non-coding RNAs: mechanism of action and functional utility. Non-coding RNA Res 1(1): 43–50. https://doi.org/10.1016/j.ncrna.2016.11.002

Biswas SK, Mantovani A (2010) Macrophage plasticity and interaction with lymphocyte subsets: cancer as a paradigm. Nat Immunol 11(10):889–896. https://doi.org/10.1038/ni.1937

Block K, Gorin Y (2012) Aiding and abetting roles of NOX oxidases in cellular transformation. Nat Rev Cancer. 12(9):627–637. https://doi.org/10.1038/nrc3339

Bohl SR, Bullinger L, Rücker FG (2018) Epigenetic therapy: azacytidine and decitabine in acute myeloid leukemia. Expert Rev Hematol 11(5):361–371. https://doi.org/10.1080/17474086.2018.1453802

Boidot R, Veǵran F, Meulle A, Le Breton A, Dessy C, Sonveaux P, Lizard-Nacol S, Feron O (2012) Regulation of monocarboxylate transporter MCT1 expression by p53 mediates inward and outward lactate fluxes in tumors. Cancer Res 72(4):939–948. https://doi.org/10.1158/0008-5472.CAN-11-2474

Bonuccelli G, Avnet S, Grisendi G, Salerno M, Granchi D, Dominici M, Kusuzaki K, Baldini N (2014) Role of mesenchymal stem cells in osteosarcoma and metabolic reprogramming of tumor cells. Oncotarget 5(17):7575–7588. https://doi.org/10.18632/oncotarget.2243

Borodovsky A, Salmasi V, Turcan S, Fabius AWM, Baia GS, Eberhart CG, Weingart JD, Gallia GL, Baylin SB, Chan TA, Riggins GJ (2013) 5-azacytidine reduces methylation, promotes differentiation and induces tumor regression in a patient-derived IDH1 mutant glioma xenograft. Oncotarget 4(10):1737–1747. https://doi.org/10.18632/oncotarget.1408

Boukouris AE, Zervopoulos SD, Michelakis ED (2016) Metabolic enzymes moonlighting in the nucleus: metabolic regulation of gene transcription. Trends Biochem Sci 41(8):712–730. https://doi.org/10.1016/j.tibs.2016.05.013

Bourdon JC, De Laurenzi V, Melino G, Lane D (2003) p53: 25 years of research and more questions to answer. Cell Death Differ 10(4):397–399. https://doi.org/10.1038/sj.cdd.4401243

Bowers LW, Rossi EL, O'Flanagan CH, De Graffenried LA, Hursting SD (2015) The role of the insulin/IGF system in cancer: lessons learned from clinical trials and the energy balance-cancer link. Front Endocrinol 6. https://doi.org/10.3389/fendo.2015.00077

Brunelle JK, Bell EL, Quesada NM, Vercauteren K, Tiranti V, Zeviani M, Scarpulla RC, Chandel NS (2005) Oxygen sensing requires mitochondrial ROS but not oxidative phosphorylation. Cell Metab 1(6):409–414. https://doi.org/10.1016/j.cmet.2005.05.002

Budanov AV, Karin M (2008) p53 target genes Sestrin1 and Sestrin2 connect genotoxic stress and mTOR signaling. Cell 134(3):451–460. https://doi.org/10.1016/j.cell.2008.06.028

Cairns RA, Harris IS, Mak TW (2011) Regulation of cancer cell metabolism. Nat Rev Cancer 11(2): 85–95. https://doi.org/10.1038/nrc2981

Calvert AE, Chalastanis A, Wu Y, Hurley LA, Kouri FM, Bi Y, Kachman M, May JL, Bartom E, Hua Y, Mishra RK, Schiltz GE, Dubrovskyi O, Mazar AP, Peter ME, Zheng H, James CD, Burant CF, Chandel NS et al (2017) Cancer-associated IDH1 promotes growth and resistance to targeted therapies in the absence of mutation. Cell Rep 19(9):1858–1873. https://doi.org/10.1016/j.celrep.2017.05.014

Cantelmo AR, Conradi LC, Brajic A, Goveia J, Kalucka J, Pircher A, Chaturvedi P, Hol J, Thienpont B, Teuwen LA, Schoors S, Boeckx B, Vriens J, Kuchnio A, Veys K, Cruys B, Finotto L, Treps L, Stav-Noraas TE et al (2016) Inhibition of the glycolytic activator PFKFB3 in endothelium induces tumor vessel normalization, impairs metastasis, and improves chemotherapy. Cancer Cell 30(6):968–985. https://doi.org/10.1016/j.ccell.2016.10.006

Carey BW, Finley LWS, Cross JR, Allis CD, Thompson CB (2015) Intracellular α-ketoglutarate maintains the pluripotency of embryonic stem cells. Nature 518(7539):413–416. https://doi.org/10.1038/nature13981

Carling D (2017) AMPK signalling in health and disease. Curr Opin Cell Biol 45:31–37. https://doi.org/10.1016/j.ceb.2017.01.005

Carmeliet P, Jain RK (2000) Angiogenesis in cancer and other diseases. Nature 407(6801): 249–257. https://doi.org/10.1038/35025220

Casciano JC, Perry C, Cohen-Nowak AJ, Miller KD, Vande Voorde J, Zhang Q, Chalmers S, Sandison ME, Liu Q, Hedley A, McBryan T, Tang HY, Gorman N, Beer T, Speicher DW, Adams PD, Liu X, Schlegel R, McCarron JG et al (2020) MYC regulates fatty acid metabolism through a multigenic program in claudin-low triple negative breast cancer. Br J Cancer. https://doi.org/10.1038/s41416-019-0711-3

Chalkiadaki A, Guarente L (2015) The multifaceted functions of sirtuins in cancer. Nat Rev Cancer 15(10):608–624. https://doi.org/10.1038/nrc3985

Chan B, Manley J, Lee J, Singh SR (2015) The emerging roles of microRNAs in cancer metabolism. Cancer Lett 356(2):301–308. https://doi.org/10.1016/j.canlet.2014.10.011

Chang CH, Curtis JD, Maggi LB, Faubert B, Villarino AV, O'Sullivan D, Huang SCC, Van Der Windt GJW, Blagih J, Qiu J, Weber JD, Pearce EJ, Jones RG, Pearce EL (2013) Posttranscriptional control of T cell effector function by aerobic glycolysis. Cell 153(6):1239. https://doi.org/10.1016/j.cell.2013.05.016

Chang CH, Qiu J, O'Sullivan D, Buck MD, Noguchi T, Curtis JD, Chen Q, Gindin M, Gubin MM, Van Der Windt GJW, Tonc E, Schreiber RD, Pearce EJ, Pearce EL (2015) Metabolic competition in the tumor microenvironment is a driver of cancer progression. Cell 162(6):1229–1241. https://doi.org/10.1016/j.cell.2015.08.016

Chaube B, Malvi P, Singh SV, Mohammad N, Viollet B, Bhat MK (2015) AMPK maintains energy homeostasis and survival in cancer cells via regulating p38/PGC-1α-mediated mitochondrial biogenesis. Cell Death Dis 1(1):1. https://doi.org/10.1038/cddiscovery.2015.63

Chen J, Lee HJ, Wu X, Huo L, Kim SJ, Xu L, Wang Y, He J, Bollu LR, Gao G, Su F, Briggs J, Liu X, Melman T, Asara JM, Fidler IJ, Cantley LC, Locasale JW, Weihua Z (2015) Gain of glucose-independent growth upon metastasis of breast cancer cells to the brain. Cancer Res 75(3):554–565. https://doi.org/10.1158/0008-5472.CAN-14-2268

Chen J, Bai M, Ning C, Xie B, Zhang J, Liao H, Xiong J, Tao X, Yan D, Xi X, Chen X, Yu Y, Bast RC, Zhang Z, Feng Y, Zheng W (2016) Gankyrin facilitates follicle-stimulating hormone-driven ovarian cancer cell proliferation through the PI3K/AKT/HIF-1α/cyclin D1 pathway. Oncogene 35(19):2506–2517. https://doi.org/10.1038/onc.2015.316

Chen D, Wang H, Chen J, Li Z, Li S, Hu Z, Huang S, Zhao Y, He X (2018a) MicroRNA-129-5p regulates glycolysis and cell proliferation by targeting the glucose transporter SLC2A3 in gastric cancer cells. Front Pharmacol 9(May):502. https://doi.org/10.3389/fphar.2018.00502

Chen G, Zhang Y, Liang J, Li W, Zhu Y, Zhang M, Wang C, Hou J (2018b) Deregulation of hexokinase II is associated with glycolysis, autophagy, and the epithelial-mesenchymal transition in tongue squamous cell carcinoma under hypoxia. Biomed Res Int. https://doi.org/10.1155/2018/8480762

Chen Y, Zhang B, Bao L, Jin L, Yang M, Peng Y, Kumar A, Wang JE, Wang C, Zou X, Xing C, Wang Y, Luo W (2018c) ZMYND8 acetylation mediates HIF-dependent breast cancer progression and metastasis. J Clin Investig 128(5):1937–1955. https://doi.org/10.1172/JCI95089

Chen J, Ye C, Dong J, Cao S, Hu Y, Situ B, Xi X, Qin S, Xu J, Cai Z, Zheng L, Wang Q (2020) Metabolic classification of circulating tumor cells as a biomarker for metastasis and prognosis in breast cancer. J Transl Med 18(1):59. https://doi.org/10.1186/s12967-020-02237-8

Christofk HR, Vander Heiden MG, Harris MH, Ramanathan A, Gerszten RE, Wei R, Fleming MD, Schreiber SL, Cantley LC (2008) The M2 splice isoform of pyruvate kinase is important for cancer metabolism and tumour growth. Nature 452(7184):230–233. https://doi.org/10.1038/nature06734

Christopoulos PF, Msaouel P, Koutsilieris M (2015) The role of the insulin-like growth factor-1 system in breast cancer. Mol Cancer 14. https://doi.org/10.1186/s12943-015-0291-7

Clemmons DR (2007) Modifying IGF1 activity: an approach to treat endocrine disorders, atherosclerosis and cancer. Nat Rev Drug Discov 6(10):821–833. https://doi.org/10.1038/nrd2359

Colegio OR, Chu NQ, Szabo AL, Chu T, Rhebergen AM, Jairam V, Cyrus N, Brokowski CE, Eisenbarth SC, Phillips GM, Cline GW, Phillips AJ, Medzhitov R (2014) Functional polarization of tumour-associated macrophages by tumour-derived lactic acid. Nature 513(7519): 559–563. https://doi.org/10.1038/nature13490

Condeelis J, Pollard JW (2006) Macrophages: obligate partners for tumor cell migration, invasion, and metastasis. Cell 124(2):263–266. https://doi.org/10.1016/j.cell.2006.01.007

Conradi LC, Brajic A, Cantelmo AR, Bouché A, Kalucka J, Pircher A, Brüning U, Teuwen LA, Vinckier S, Ghesquière B, Dewerchin M, Carmeliet P (2017) Tumor vessel disintegration by maximum tolerable PFKFB3 blockade. Angiogenesis 20(4):599–613. https://doi.org/10.1007/s10456-017-9573-6

Corbet C, Feron O (2017) Tumour acidosis: from the passenger to the driver's seat. Nat Rev Cancer 17(10):577–593. https://doi.org/10.1038/nrc.2017.77

Crispo F, Condelli V, Lepore S, Notarangelo T, Sgambato A, Esposito F, Maddalena F, Landriscina M (2019) Metabolic dysregulations and epigenetics: a bidirectional interplay that drives tumor progression. Cells 8. https://doi.org/10.3390/cells8080798

Cui XG, Han ZT, He SH, Da Wu X, Chen TR, Shao CH, Chen DL, Su N, Chen YM, Wang T, Wang J, Song DW, Yan WJ, Yang XH, Liu T, Wei HF, Xiao J (2017) HIF1/2α mediates hypoxia-induced LDHA expression in human pancreatic cancer cells. Oncotarget 8(15): 24840–24852. https://doi.org/10.18632/oncotarget.15266

Dai Z, Mentch SJ, Gao X, Nichenametla SN, Locasale JW (2018) Methionine metabolism influences genomic architecture and gene expression through H3K4me3 peak width. Nat Commun 9(1). https://doi.org/10.1038/s41467-018-04426-y

Dasgupta S, Rajapakshe K, Zhu B, Nikolai BC, Yi P, Putluri N, Choi JM, Jung SY, Coarfa C, Westbrook TF, Zhang XHF, Foulds CE, Tsai SY, Tsai MJ, O'Malley BW (2018) Metabolic enzyme PFKFB4 activates transcriptional coactivator SRC-3 to drive breast cancer. Nature 556(7700):249–254. https://doi.org/10.1038/s41586-018-0018-1

De Berardinis RJ, Chandel NS (2016) Fundamentals of cancer metabolism. Sci Adv 2(5):e1600200. https://doi.org/10.1126/sciadv.1600200

De Bock K, Georgiadou M, Schoors S, Kuchnio A, Wong BW, Cantelmo AR, Quaegebeur A, Ghesquière B, Cauwenberghs S, Eelen G, Phng LK, Betz I, Tembuyser B, Brepoels K, Welti J, Geudens I, Segura I, Cruys B, Bifari F et al (2013) Role of PFKFB3-driven glycolysis in vessel sprouting. Cell 154(3):651–663. https://doi.org/10.1016/j.cell.2013.06.037

DeBerardinis RJ, Lum JJ, Hatzivassiliou G, Thompson CB (2008) The biology of cancer: metabolic reprogramming fuels cell growth and proliferation. Cell Metab 7(1):11–20. https://doi.org/10.1016/j.cmet.2007.10.002

Desai S, Ding M, Wang B, Lu Z, Zhao Q, Shaw K, Alfred Yung WK, Weinstein JN, Tan M, Yao J (2014) Tissue-specific isoform switch and DNA hypomethylation of the pyruvate kinase PKM gene in human cancers. Oncotarget 5(18):8202–8210. https://doi.org/10.18632/oncotarget.1159

Desbats MA, Giacomini I, Prayer-Galetti T, Montopoli M (2020) Metabolic plasticity in chemotherapy resistance. Front Oncol. 10:281. https://doi.org/10.3389/fonc.2020.00281

Diebold LP, Gil HJ, Gao P, Martinez CA, Weinberg SE, Chandel NS (2019) Mitochondrial complex III is necessary for endothelial cell proliferation during angiogenesis. Nat Metab 1(1):158–171. https://doi.org/10.1038/s42255-018-0011-x

Döhner H, Estey E, Grimwade D, Amadori S, Appelbaum FR, Büchner T, Dombret H, Ebert BL, Fenaux P, Larson RA, Levine RL, Lo-Coco F, Naoe T, Niederwieser D, Ossenkoppele GJ, Sanz M, Sierra J, Tallman MS, Tien HF et al (2017) Diagnosis and management of AML in adults: 2017 ELN recommendations from an international expert panel. Blood 129(4):424–447. https://doi.org/10.1182/blood-2016-08-733196

Dong Z, Wang J (2004) Hypoxia selection of death-resistant cells: a role for Bcl-XL. J Biol Chem 279(10):9215–9221. https://doi.org/10.1074/jbc.M312225200

Dong C, Yuan T, Wu Y, Wang Y, Fan TWM, Miriyala S, Lin Y, Yao J, Shi J, Kang T, Lorkiewicz P, St Clair D, Hung MC, Evers BM, Zhou BP (2013) Loss of FBP1 by snail-mediated repression provides metabolic advantages in basal-like breast cancer. Cancer Cell 23(3):316–331. https://doi.org/10.1016/j.ccr.2013.01.022

Dou C, Mo H, Chen T, Liu J, Zeng Y, Li S, Guo C, Zhang C (2021) ZMYND8 promotes the growth and metastasis of hepatocellular carcinoma by promoting HK2-mediated glycolysis. Pathol Res Pract 219:153345. https://doi.org/10.1016/j.prp.2021.153345

Dupuy F, Tabariès S, Andrzejewski S, Dong Z, Blagih J, Annis MG, Omeroglu A, Gao D, Leung S, Amir E, Clemons M, Aguilar-Mahecha A, Basik M, Vincent EE, St. Pierre J, Jones RG, Siegel PM (2015) PDK1-dependent metabolic reprogramming dictates metastatic potential in breast cancer. Cell Metab 22(4):577–589. https://doi.org/10.1016/j.cmet.2015.08.007

Eales KL, Hollinshead KER, Tennant DA (2016) Hypoxia and metabolic adaptation of cancer cells. Oncogenesis 5(1):e190–e190. https://doi.org/10.1038/oncsis.2015.50

Easwaran H, Tsai HC, Baylin SB (2014) Cancer epigenetics: tumor heterogeneity, plasticity of stem-like states, and drug resistance. Mol Cell 54(5):716–727. https://doi.org/10.1016/j.molcel.2014.05.015

Edinger AL, Thompson CB (2002) Akt maintains cell size and survival by increasing mTOR-dependent nutrient uptake. Mol Biol Cell 13(7):2276–2288. https://doi.org/10.1091/mbc.01-12-0584

Eelen G, De Zeeuw P, Simons M, Carmeliet P (2015) Endothelial cell metabolism in normal and diseased vasculature. Circul Res 116(7):1231–1244. https://doi.org/10.1161/CIRCRESAHA.116.302855

Elstrom RL, Bauer DE, Buzzai M, Karnauskas R, Harris MH, Plas DR, Zhuang H, Cinalli RM, Alavi A, Rudin CM, Thompson CB (2004) Akt stimulates aerobic glycolysis in cancer cells. Cancer Res 64(11):3892–3899. https://doi.org/10.1158/0008-5472.CAN-03-2904

Engelman JA (2009) Targeting PI3K signalling in cancer: opportunities, challenges and limitations. Nat Rev Cancer 9(8):550–562. https://doi.org/10.1038/nrc2664

Eriksson M, Ambroise G, Ouchida AT, Lima Queiroz A, Smith D, Gimenez-Cassina A, Iwanicki MP, Muller PA, Norberg E, Vakifahmetoglu-Norberg H (2017) Effect of mutant p53 proteins on glycolysis and mitochondrial metabolism. Mol Cell Biol 37(24). https://doi.org/10.1128/mcb.00328-17

Etchegaray JP, Mostoslavsky R (2016) Interplay between metabolism and epigenetics: a nuclear adaptation to environmental changes. Mol Cell 62(5):695–711. https://doi.org/10.1016/j.molcel.2016.05.029

Fan Y, Dickman KG, Zong WX (2010) Akt and c-Myc differentially activate cellular metabolic programs and prime cells to bioenergetic inhibition. J Biol Chem 285(10):7324–7333. https://doi.org/10.1074/jbc.M109.035584

Feldman JL, Baeza J, Denu JM (2013) Activation of the protein deacetylase SIRT6 by long-chain fatty acids and widespread deacylation by Mammalian Sirtuins. J Biol Chem 288(43):31350–31356. https://doi.org/10.1074/jbc.C113.511261

Fernández S, Genis L, Torres-Alemán I (2014) A phosphatase-independent gain-of-function mutation in PTEN triggers aberrant cell growth in astrocytes through an autocrine IGF-1 loop. Oncogene 33(32):4114–4122. https://doi.org/10.1038/ONC.2013.376

Fiaschi T, Marini A, Giannoni E, Taddei ML, Gandellini P, De Donatis A, Lanciotti M, Serni S, Cirri P, Chiarugi P (2012) Reciprocal metabolic reprogramming through lactate shuttle coordinately influences tumor-stroma interplay. Cancer Res 72(19):5130–5140. https://doi.org/10.1158/0008-5472.CAN-12-1949

Fiaschi T, Giannoni E, Taddei ML, Cirri P, Marini A, Pintus G, Nativi C, Richichi B, Scozzafava A, Carta F, Torre E, Supuran CT, Chiarugi P (2013) Carbonic anhydrase IX from cancer-associated fibroblasts drives epithelial-mesenchymal transition in prostate carcinoma cells. Cell Cycle 12(11):1791–1801. https://doi.org/10.4161/cc.24902

Funasaka T, Hu A, Yanagawa T, Hogan V, Raz A (2007) Down-regulation of phosphoglucose isomerase/autocrine motility factor results in mesenchymal-to-epithelial transition of human lung fibrosarcoma cells. Cancer Res 67(9):4236–4243. https://doi.org/10.1158/0008-5472.CAN-06-3935

Gaál Z, Csernoch L (2020) Impact of sirtuin enzymes on the altered metabolic phenotype of malignantly transformed cells. Front Oncol. 10:45. https://doi.org/10.3389/fonc.2020.00045

Gabrilovich DI, Nagaraj S (2009) Myeloid-derived suppressor cells as regulators of the immune system. Nat Rev Immunol. 9(3):162–174. https://doi.org/10.1038/nri2506

Ge T, Yang J, Zhou S, Wang Y, Li Y, Tong X (2020) The role of the pentose phosphate pathway in diabetes and cancer. Front Endocrinol 11. https://doi.org/10.3389/FENDO.2020.00365

Gerriets VA, Rathmell JC (2012) Metabolic pathways in T cell fate and function. Trends Immunol 33(4):168–173. https://doi.org/10.1016/j.it.2012.01.010

Ghafouri-Fard S, Shoorei H, Taheri M (2020) The role of long non-coding rnas in cancer metabolism: a concise review. Front Oncol. 10:555825. https://doi.org/10.3389/fonc.2020.555825

Glazer RI, Hartman KD, Knode MC, Richard MM, Chiang PK, Tseng CKH, Marquez VE (1986) 3-deazaneplanocin: a new and potent inhibitor of S-adenosylhomocysteine hydrolase and its effects on human promyelocytic leukemia cell line HL-60. Biochem Biophys Res Commun 135(2):688–694. https://doi.org/10.1016/0006-291X(86)90048-3

Goel A, Mathupala SP, Pedersen PL (2003) Glucose metabolism in cancer: evidence that demethylation events play a role in activating type II hexokinase gene expression. J Biol Chem 278(17):15333–15340. https://doi.org/10.1074/jbc.M300608200

Gómez V, Eykyn TR, Mustapha R, Flores-Borja F, Male V, Barber PR, Patsialou A, Green R, Panagaki F, Li CW, Fruhwirth GO, Ros S, Brindle KM, Ng T (2020) Breast cancer-associated macrophages promote tumorigenesis by suppressing succinate dehydrogenase in tumor cells. Sci Signal 13(652). https://doi.org/10.1126/scisignal.aax4585

Good CR, Panjarian S, Kelly AD, Madzo J, Patel B, Jelinek J, Issa JPJ (2018) TET1-Mediated hypomethylation activates oncogenic signaling in triple-negative breast cancer. Cancer Res 78(15):4126–4137. https://doi.org/10.1158/0008-5472.CAN-17-2082

Gordan JD, Simon MC (2007) Hypoxia-inducible factors: central regulators of the tumor phenotype. Curr Opin Genetics Dev. 17(1):71–77. https://doi.org/10.1016/j.gde.2006.12.006

Gordon SR, Maute RL, Dulken BW, Hutter G, George BM, McCracken MN, Gupta R, Tsai JM, Sinha R, Corey D, Ring AM, Connolly AJ, Weissman IL (2017) PD-1 expression by tumour-associated macrophages inhibits phagocytosis and tumour immunity. Nature 545(7655):495–499. https://doi.org/10.1038/nature22396

Gregersen LH, Jacobsen A, Frankel LB, Wen J, Krogh A, Lund AH (2012) MicroRNA-143 down-regulates Hexokinase 2 in colon cancer cells. BMC Cancer 12(1):232. https://doi.org/10.1186/1471-2407-12-232

Guddeti RK, Bali P, Karyala P, Pakala SB (2019) MTA1 coregulator regulates LDHA expression and function in breast cancer. Biochem Biophys Res Commun 520(1):54–59. https://doi.org/10.1016/j.bbrc.2019.09.078

Guo Y, Xiao Z, Yang L, Gao Y, Zhu Q, Hu L, Huang D, Xu Q (2020) Hypoxia-inducible factors in hepatocellular carcinoma (review). Oncol Rep 43(1):3–15. https://doi.org/10.3892/or.2019.7397

Hamabe A, Konno M, Tanuma N, Shima H, Tsunekuni K, Kawamoto K, Nishida N, Koseki J, Mimori K, Gotoh N, Yamamoto H, Doki Y, Mori M, Ishii H (2014) Role of pyruvate kinase M2 in transcriptional regulation leading to epithelial-mesenchymal transition. Proc Natl Acad Sci U S A 111(43):15526–15531. https://doi.org/10.1073/pnas.1407717111

Hanahan D, Weinberg RA (2000) The hallmarks of cancer. Cell. https://doi.org/10.1016/S0092-8674(00)81683-9

Hanahan D, Weinberg RA (2011) Hallmarks of cancer: the next generation. Cell 144(5):646–674. https://doi.org/10.1016/j.cell.2011.02.013

Heiden MGV, Cantley LC, Thompson CB (2009) Understanding the warburg effect: the metabolic requirements of cell proliferation. Science 324(5930):1029–1033. https://doi.org/10.1126/science.1160809

Huang D, Li C, Zhang H (2014a) Hypoxia and cancer cell metabolism. Acta Biochim Biophys Sin 46(3):214–219. https://doi.org/10.1093/ABBS/GMT148

Huang SCC, Everts B, Ivanova Y, O'Sullivan D, Nascimento M, Smith AM, Beatty W, Love-Gregory L, Lam WY, O'Neill CM, Yan C, Du H, Abumrad NA, Urban JF, Artyomov MN, Pearce EL, Pearce EJ (2014b) Cell-intrinsic lysosomal lipolysis is essential for alternative activation of macrophages. Nat Immunol 15(9):846–855. https://doi.org/10.1038/ni.2956

Huang R, Yu Y, Zong X, Li X, Ma L, Zheng Q (2017) Monomethyltransferase SETD8 regulates breast cancer metabolism via stabilizing hypoxia-inducible factor 1α. Cancer Lett 390:1–10. https://doi.org/10.1016/j.canlet.2016.12.038

Huang T, Chen QF, Chang BY, Shen LJ, Li W, Wu PH, Fan WJ (2019) TFAP4 promotes hepatocellular carcinoma invasion and metastasis via activating the pi3k/akt signaling pathway. Dis Markers 2019. https://doi.org/10.1155/2019/7129214

Hwang IY, Kwak S, Lee S, Kim H, Lee SE, Kim JH, Kim YA, Jeon YK, Chung DH, Jin X, Park S, Jang H, Cho EJ, Youn HD (2016) Psat1-dependent fluctuations in α-ketoglutarate affect the timing of ESC differentiation. Cell Metab 24(3):494–501. https://doi.org/10.1016/j.cmet.2016.06.014

Ichijo H, Nishida E, Irie K, Ten Dijke P, Saitoh M, Moriguchi T, Takagi M, Matsumoto K, Miyazono K, Gotoh Y (1997) Induction of apoptosis by ASK1, a mammalian MAPKKK that activates SAPK/JNK and p38 signaling pathways. Science 275(5296):90–94. https://doi.org/10.1126/SCIENCE.275.5296.90

Iqbal MA, Siddiqui FA, Gupta V, Chattopadhyay S, Gopinath P, Kumar B, Manvati S, Chaman N, Bamezai RNK (2013) Insulin enhances metabolic capacities of cancer cells by dual regulation of glycolytic enzyme pyruvate kinase M2. Mol Cancer 12(1). https://doi.org/10.1186/1476-4598-12-72

Irani K, Xia Y, Zweier JL, Sollott SJ, Der CJ, Fearon ER, Sundaresan M, Finkel T, Goldschmidt-Clermont PJ (1997) Mitogenic signaling mediated by oxidants in ras-transformed fibroblasts. http://science.sciencemag.org/. Retrieved 3 February 2021

Iyer NV, Kotch LE, Agani F, Leung SW, Laughner E, Wenger RH, Gassmann M, Gearhart JD, Lawler AM, Yu AY, Semenza GL (1998) Cellular and developmental control of O2 homeostasis by hypoxia- inducible factor 1α. Genes Dev 12(2):149–162. https://doi.org/10.1101/gad.12.2.149

Janke R, Iavarone AT, Rine J (2017) Oncometabolite d-2-hydroxyglutarate enhances gene silencing through inhibition of specific h3k36 histone demethylases. eLife 6. https://doi.org/10.7554/eLife.22451

Jeong SM, Haigis MC (2015) Sirtuins in cancer: a balancing act between genome stability and metabolism. Mol Cells 38(9):750–758. https://doi.org/10.14348/molcells.2015.0167

Jia D, Park J, Jung K, Levine H, Kaipparettu B (2018) Elucidating the metabolic plasticity of cancer: mitochondrial reprogramming and hybrid metabolic states. Cell 7(3):21. https://doi.org/10.3390/cells7030021

Jia D, Lu M, Jung KH, Park JH, Yu L, Onuchic JN, Kaipparettu BA, Levine H (2019) Elucidating cancer metabolic plasticity by coupling gene regulation with metabolic pathways. Proc Natl Acad Sci U S A 116(9):3909–3918. https://doi.org/10.1073/pnas.1816391116

Jiang P, Du W, Wang X, Mancuso A, Gao X, Wu M, Yang X (2011) P53 regulates biosynthesis through direct inactivation of glucose-6-phosphate dehydrogenase. Nat Cell Biol 13(3):310–316. https://doi.org/10.1038/ncb2172

Jiang Y, Qian X, Shen J, Wang Y, Li X, Liu R, Xia Y, Chen Q, Peng G, Lin SY, Lu Z (2015) Local generation of fumarate promotes DNA repair through inhibition of histone H3 demethylation. Nat Cell Biol 17(9):1158–1168. https://doi.org/10.1038/ncb3209

Jing E, Gesta S, Kahn CR (2007) SIRT2 regulates adipocyte differentiation through FoxO1 acetylation/deacetylation. Cell Metab 6(2):105–114. https://doi.org/10.1016/j.cmet.2007.07.003

Jones RG, Thompson CB (2009) Tumor suppressors and cell metabolism: a recipe for cancer growth. Genes Dev 23(5):537–548. https://doi.org/10.1101/gad.1756509

Jung SN, Yang WK, Kim J, Kim HS, Kim EJ, Yun H, Park H, Kim SS, Choe W, Kang I, Ha J (2008) Reactive oxygen species stabilize hypoxia-inducible factor-1 alpha protein and stimulate transcriptional activity via AMP-activated protein kinase in DU145 human prostate cancer cells. Carcinogenesis 29(4):713–721. https://doi.org/10.1093/carcin/bgn032

Kalluri R (2003) Basement membranes: structure, assembly and role in tumour angiogenesis. Nat Rev Cancer 3(6):422–433. https://doi.org/10.1038/nrc1094

Katoh Y, Ikura T, Hoshikawa Y, Tashiro S, Ito T, Ohta M, Kera Y, Noda T, Igarashi K (2011) Methionine adenosyltransferase II serves as a transcriptional corepressor of Maf oncoprotein. Mol Cell 41(5):554–566. https://doi.org/10.1016/j.molcel.2011.02.018

Kawauchi K, Araki K, Tobiume K, Tanaka N (2008) p53 regulates glucose metabolism through an IKK-NF-κB pathway and inhibits cell transformation. Nat Cell Biol 10(5):611–618. https://doi.org/10.1038/ncb1724

Keenan TE, Tolaney SM (2020) Role of immunotherapy in triple-negative breast cancer. JNCCN 18(4):479–489. https://doi.org/10.6004/jnccn.2020.7554

Kera Y, Katoh Y, Ohta M, Matsumoto M, Takano-Yamamoto T, Igarashi K (2013) Methionine adenosyltransferase II-dependent histone H3K9 methylation at the COX-2 gene locus. J Biol Chem 288(19):13592–13601. https://doi.org/10.1074/jbc.M112.429738

Khatri S, Yepiskoposyan H, Gallo CA, Tandon P, Plas DR (2010) FOXO3a regulates glycolysis via transcriptional control of tumor suppressor TSC1. J Biol Chem 285(21):15960–15965. https://doi.org/10.1074/jbc.M110.121871

Kim J, Bae JS (2018) Metabolic regulation of macrophages in tumor microenvironment. Curr Opin Hematol 25(1):52–59. https://doi.org/10.1097/MOH.0000000000000390

Kim J, DeBerardinis RJ (2019) Mechanisms and implications of metabolic heterogeneity in cancer. Cell Metab 30(3):434–446. https://doi.org/10.1016/j.cmet.2019.08.013

Kim J-w, Zeller KI, Wang Y, Jegga AG, Aronow BJ, O'Donnell KA, Dang CV (2004) Evaluation of Myc E-Box phylogenetic footprints in glycolytic genes by chromatin immunoprecipitation assays. Mol Cell Biol 24(13):5923–5936. https://doi.org/10.1128/mcb.24.13.5923-5936.2004

Kim JW, Tchernyshyov I, Semenza GL, Dang CV (2006) HIF-1-mediated expression of pyruvate dehydrogenase kinase: a metabolic switch required for cellular adaptation to hypoxia. Cell Metab 3(3):177–185. https://doi.org/10.1016/j.cmet.2006.02.002

Kim J-w, Gao P, Liu Y-C, Semenza GL, Dang CV (2007) Hypoxia-inducible factor 1 and dysregulated c-Myc cooperatively induce vascular endothelial growth factor and metabolic switches hexokinase 2 and pyruvate dehydrogenase kinase 1. Mol Cell Biol 27(21): 7381–7393. https://doi.org/10.1128/mcb.00440-07

Kim S, Lee E, Jung J, Lee JW, Kim HJ, Kim J, Yoo HJ, Lee HJ, Chae SY, Jeon SM, Son BH, Gong G, Sharan SK, Chang S (2018) MicroRNA-155 positively regulates glucose metabolism via PIK3R1-FOXO3a-cMYC axis in breast cancer. Oncogene 37(22):2982–2991. https://doi.org/10.1038/s41388-018-0124-4

Knoepfler PS, Zhang XY, Cheng PF, Gafken PR, McMahon SB, Eisenman RN (2006) Myc influences global chromatin structure. EMBO J 25(12):2723–2734. https://doi.org/10.1038/sj.emboj.7601152

Kondoh H, Lleonart ME, Gil J, Wang J, Degan P, Peters G, Martinez D, Carnero A, Beach D (2005) Glycolytic enzymes can modulate cellular life span. Cancer Res 65(1):177–185

Kotch LE, Iyer NV, Laughner E, Semenza GL (1999) Defective vascularization of HIF-1α-null embryos is not associated with VEGF deficiency but with mesenchymal cell death. Dev Biol 209(2):254–267. https://doi.org/10.1006/dbio.1999.9253

Koukourakis MI, Giatromanolaki A, Harris AL, Sivridis E (2006) Comparison of metabolic pathways between cancer cells and stromal cells in colorectal carcinomas: a metabolic survival role for tumor-associated stroma. Cancer Res 66(2):632–637. https://doi.org/10.1158/0008-5472.CAN-05-3260

Laderoute KR, Amin K, Calaoagan JM, Knapp M, Le T, Orduna J, Foretz M, Viollet B (2006) 5′-AMP-activated protein kinase (AMPK) is induced by low-oxygen and glucose deprivation conditions found in solid-tumor microenvironments. Mol Cell Biol 26(14):5336–5347. https://doi.org/10.1128/mcb.00166-06

Lambert AW, Pattabiraman DR, Weinberg RA (2017) Emerging biological principles of metastasis. Cell 168(4):670–691. https://doi.org/10.1016/j.cell.2016.11.037

Lau AN, Vander Heiden MG (2020) Metabolism in the tumor microenvironment. Annu Rev Cancer Biol 4(1):17–40. https://doi.org/10.1146/annurev-cancerbio-030419-033333

Lee JV, Carrer A, Shah S, Snyder NW, Wei S, Venneti S, Worth AJ, Yuan ZF, Lim HW, Liu S, Jackson E, Aiello NM, Haas NB, Rebbeck TR, Judkins A, Won KJ, Chodosh LA, Garcia BA, Stanger BZ et al (2014) Akt-dependent metabolic reprogramming regulates tumor cell Histone acetylation. Cell Metab 20(2):306–319. https://doi.org/10.1016/j.cmet.2014.06.004

Lee V, Wang J, Zahurak M, Gootjes E, Verheul HM, Parkinson R, Kerner Z, Sharma A, Rosner G, De Jesus-Acosta A, Laheru D, Le DT, Oganesian A, Lilly E, Brown T, Jones P, Baylin S, Ahuja N, Azad N (2018) A phase I trial of a guadecitabine (SGI-110) and irinotecan in metastatic colorectal cancer patients previously exposed to irinotecan. Clin Cancer Res 24(24):6160–6167. https://doi.org/10.1158/1078-0432.CCR-18-0421

Letouzé E, Martinelli C, Loriot C, Burnichon N, Abermil N, Ottolenghi C, Janin M, Menara M, Nguyen AT, Benit P, Buffet A, Marcaillou C, Bertherat JÔ, Amar L, Rustin P, DeReyniès A, Gimenez-Roqueplo AP, Favier J (2013) SDH mutations establish a hypermethylator phenotype in paraganglioma. Cancer Cell 23(6):739–752. https://doi.org/10.1016/j.ccr.2013.04.018

Levine AJ, Puzio-Kuter AM (2010) The control of the metabolic switch in cancers by oncogenes and tumor suppressor genes. Science 330(6009):1340–1344. https://doi.org/10.1126/science.1193494

Li Z, Rana TM (2014) Therapeutic targeting of microRNAs: current status and future challenges. Nat Rev Drug Discov. 13(8):622–638. https://doi.org/10.1038/nrd4359

Li E, Bestor TH, Jaenisch R (1992) Targeted mutation of the DNA methyltransferase gene results in embryonic lethality. Cell 69(6):915–926. https://doi.org/10.1016/0092-8674(92)90611-F

Li J, Yen C, Liaw D, Podsypanina K, Bose S, Wang SI, Puc J, Miliaresis C, Rodgers L, McCombie R, Bigner SH, Giovanella BC, Ittmann M, Tycko B, Hibshoosh H, Wigler MH, Parsons R (1997) PTEN, a putative protein tyrosine phosphatase gene mutated in human brain, breast, and prostate cancer. Science 275(5308):1943–1947. https://doi.org/10.1126/science.275.5308.1943

Li F, Sonveaux P, Rabbani ZN, Liu S, Yan B, Huang Q, Vujaskovic Z, Dewhirst MWW, Li CY (2007) Regulation of HIF-1α stability through S-nitrosylation. Mol Cell 26(1):63–74. https://doi.org/10.1016/j.molcel.2007.02.024

Li S, Swanson SK, Gogol M, Florens L, Washburn MP, Workman JL, Suganuma T (2015) Serine and SAM responsive complex SESAME regulates histone modification crosstalk by sensing cellular metabolism. Mol Cell 60(3):408–421. https://doi.org/10.1016/j.molcel.2015.09.024

Li C, Zhao Z, Zhou Z, Liu R (2016) PKM2 promotes cell survival and invasion under metabolic stress by enhancing Warburg effect in pancreatic ductal adenocarcinoma. Dig Dis Sci 61(3):767–773. https://doi.org/10.1007/s10620-015-3931-2

Li X, Yu W, Qian X, Xia Y, Zheng Y, Lee JH, Li W, Lyu J, Rao G, Zhang X, Qian CN, Rozen SG, Jiang T, Lu Z (2017) Nucleus-translocated ACSS2 promotes gene transcription for lysosomal biogenesis and autophagy. Mol Cell 66(5):684–697.e9. https://doi.org/10.1016/j.molcel.2017.04.026

Li J, Wang F, Gao H, Huang S, Cai F, Sun J (2019) ALDOLASE A regulates invasion of bladder cancer cells via E-cadherin-EGFR signaling. J Cell Biochem 120(8):13694–13705. https://doi.org/10.1002/jcb.28642

Liberti MV, Locasale JW (2016) The Warburg effect: how does it benefit cancer cells? Trends Biochem Sci. 41(3):211–218. https://doi.org/10.1016/j.tibs.2015.12.001

Lim SO, Li CW, Xia W, Lee HH, Chang SS, Shen J, Hsu JL, Raftery D, Djukovic D, Gu H, Chang WC, Wang HL, Chen ML, Huo L, Chen CH, Wu Y, Sahin A, Hanash SM, Hortobagyi GN, Hung MC (2016) EGFR signaling enhances aerobic glycolysis in triple-negative breast cancer cells to promote tumor growth and immune escape. Cancer Res 76(5):1284–1296. https://doi.org/10.1158/0008-5472.CAN-15-2478

Liou GY, Döppler H, DelGiorno KE, Zhang L, Leitges M, Crawford HC, Murphy MP, Storz P (2016) Mutant KRas-induced mitochondrial oxidative stress in acinar cells upregulates EGFR signaling to drive formation of pancreatic precancerous lesions. Cell Rep 14(10):2325–2336. https://doi.org/10.1016/j.celrep.2016.02.029

Liu GM, Zhang YM (2018) Targeting FBPase is an emerging novel approach for cancer therapy. Cancer Cell Int. 18(1):36. https://doi.org/10.1186/s12935-018-0533-z

Liu PY, Xu N, Malyukova A, Scarlett CJ, Sun YT, Zhang XD, Ling D, Su SP, Nelson C, Chang DK, Koach J, Tee AE, Haber M, Norris MD, Toon C, Rooman I, Xue C, Cheung BB, Kumar S et al (2013) The histone deacetylase SIRT2 stabilizes Myc oncoproteins. Cell Death Differ 20(3):503–514. https://doi.org/10.1038/cdd.2012.147

Liu XS, Little JB, Yuan ZM (2015) Glycolytic metabolism influences global chromatin structure. Oncotarget 6(6):4214–4225. https://doi.org/10.18632/oncotarget.2929

Liu J, Zhang C, Zhao Y, Yue X, Wu H, Huang S, Chen J, Tomsky K, Xie H, Khella CA, Gatza ML, Xia D, Gao J, White E, Haffty BG, Hu W, Feng Z (2017) Parkin targets HIF-1α for ubiquitination and degradation to inhibit breast tumor progression. Nat Commun 8(1):1–16. https://doi.org/10.1038/s41467-017-01947-w

Liu J, Zhang C, Hu W, Feng Z (2019) Tumor suppressor p53 and metabolism. J Mol Cell Biol 11(4):284–292. https://doi.org/10.1093/jmcb/mjy070

Liu C, Peng X, Li Y, Liu S, Hou R, Zhang Y, Zuo S, Liu Z, Luo R, Li L, Fang W (2020) Positive feedback loop of FAM83A/PI3K/AKT/c-Jun induces migration, invasion and metastasis in hepatocellular carcinoma. Biomed Pharmacother 123. https://doi.org/10.1016/j.biopha.2019.109780

Locasale JW (2013) Serine, glycine and one-carbon units: cancer metabolism in full circle. Nat Rev Cancer 13(8):572–583. https://doi.org/10.1038/nrc3557

Lopez-Serra P, Marcilla M, Villanueva A, Ramos-Fernandez A, Palau A, Leal L, Wahi JE, Setien-Baranda F, Szczesna K, Moutinho C, Martinez-Cardus A, Heyn H, Sandoval J, Puertas S, Vidal A, Sanjuan X, Martinez-Balibrea E, Viñals F, Perales JC et al (2014) A DERL3-associated defect in the degradation of SLC2A1 mediates the Warburg effect. Nat Commun 5(1):1–14. https://doi.org/10.1038/ncomms4608

Luo W, Chang R, Zhong J, Pandey A, Semenza GL (2012) Histone demethylase JMJD2C is a coactivator for hypoxia-inducible factor 1 that is required for breast cancer progression. Proc Natl Acad Sci U S A 109(49):E3367–E3376. https://doi.org/10.1073/pnas.1217394109

Lyssiotis CA, Kimmelman AC (2017) Metabolic interactions in the tumor microenvironment. Trends Cell Biol 27(11):863–875. https://doi.org/10.1016/j.tcb.2017.06.003

Mailloux RJ, Willmore WG (2014) S-glutathionylation reactions in mitochondrial function and disease. Front Cell Dev Biol 2. https://doi.org/10.3389/fcell.2014.00068

Majewski N, Nogueira V, Bhaskar P, Coy PE, Skeen JE, Gottlob K, Chandel NS, Thompson CB, Robey RB, Hay N (2004) Hexokinase-mitochondria interaction mediated by Akt is required to inhibit apoptosis in the presence or absence of Bax and Bak. Mol Cell 16(5):819–830. https://doi.org/10.1016/j.molcel.2004.11.014

Makinoshima H, Takita M, Matsumoto S, Yagishita A, Owada S, Esumi H, Tsuchihara K (2014) Epidermal growth factor receptor (EGFR) signaling regulates global metabolic pathways in EGFR-mutated lung adenocarcinoma. J Biol Chem 289(30):20813–20823. https://doi.org/10.1074/jbc.M114.575464

Marks DL, Olson RL, Fernandez-Zapico ME (2016) Epigenetic control of the tumor microenvironment. Epigenomics 8(12):1671–1687. https://doi.org/10.2217/EPI-2016-0110

Martínez-Chantar ML, Vázquez-Chantada M, Ariz U, Martínez N, Varela M, Luka Z, Capdevila A, Rodríguez J, Aransay AM, Matthiesen R, Yang H, Calvisi DF, Esteller M, Fraga M, Lu SC,

Wagner C, Mato JM (2008) Loss of the glycine N-methyltransferase gene leads to steatosis and hepatocellular carcinoma in mice. Hepatology 47(4):1191–1199. https://doi.org/10.1002/hep.22159

Masin M, Vazquez J, Rossi S, Groeneveld S, Samson N, Schwalie PC, Deplancke B, Frawley LE, Gouttenoire J, Moradpour D, Oliver TG, Meylan E (2014) GLUT3 is induced during epithelial-mesenchymal transition and promotes tumor cell proliferation in non-small cell lung cancer. Cancer Metabolism 2(1):11. https://doi.org/10.1186/2049-3002-2-11

Massie CE, Lynch A, Ramos-Montoya A, Boren J, Stark R, Fazli L, Warren A, Scott H, Madhu B, Sharma N, Bon H, Zecchini V, Smith DM, Denicola GM, Mathews N, Osborne M, Hadfield J, MacArthur S, Adryan B et al (2011) The androgen receptor fuels prostate cancer by regulating central metabolism and biosynthesis. EMBO J 30(13):2719–2733. https://doi.org/10.1038/emboj.2011.158

Masson N, Ratcliffe PJ (2014) Hypoxia signaling pathways in cancer metabolism: the importance of co-selecting interconnected physiological pathways. Cancer Metab 2. https://doi.org/10.1186/2049-3002-2-3

Matsuda S, Adachi J, Ihara M, Tanuma N, Shima H, Kakizuka A, Ikura M, Ikura T, Matsuda T (2016) Nuclear pyruvate kinase M2 complex serves as a transcriptional coactivator of arylhydrocarbon receptor. Nucleic Acids Res 44(2):636–647. https://doi.org/10.1093/nar/gkv967

May JL, Kouri FM, Hurley LA, Liu J, Tommasini-Ghelfi S, Ji Y, Gao P, Calvert AE, Lee A, Chandel NS, Davuluri RV, Horbinski CM, Locasale JW, Stegh AH (2019) IDH3α regulates one-carbon metabolism in glioblastoma. Sci Adv 5(1). https://doi.org/10.1126/sciadv.aat0456

McBrian MA, Behbahan IS, Ferrari R, Su T, Huang TW, Li K, Hong CS, Christofk HR, Vogelauer M, Seligson DB, Kurdistani SK (2013) Histone acetylation regulates intracellular pH. Mol Cell 49(2):310–321. https://doi.org/10.1016/j.molcel.2012.10.025

McDonough MA, Loenarz C, Chowdhury R, Clifton IJ, Schofield CJ (2010) Structural studies on human 2-oxoglutarate dependent oxygenases. Curr Opin Struct Biol 20(6):659–672. https://doi.org/10.1016/j.sbi.2010.08.006

Meng F, Henson R, Wehbe-Janek H, Ghoshal K, Jacob ST, Patel T (2007) MicroRNA-21 regulates expression of the PTEN tumor suppressor gene in human hepatocellular cancer. Gastroenterology 133(2):647–658. https://doi.org/10.1053/j.gastro.2007.05.022

Migita T, Narita T, Nomura K, Miyagi E, Inazuka F, Matsuura M, Ushijima M, Mashima T, Seimiya H, Satoh Y, Okumura S, Nakagawa K, Ishikawa Y (2008) ATP citrate lyase: activation and therapeutic implications in non-small cell lung cancer. Cancer Res 68(20):8547–8554. https://doi.org/10.1158/0008-5472.CAN-08-1235

Moldogazieva NT, Mokhosoev IM, Terentiev AA (2020) Metabolic heterogeneity of cancer cells: an interplay between HIF-1, GLUTs, and AMPK. Cancers 12. https://doi.org/10.3390/cancers12040862

Molinari F, Frattini M (2014) Functions and regulation of the PTEN gene in colorectal cancer. Front Oncol 3. https://doi.org/10.3389/FONC.2013.00326

Momparler RL, Côté S, Momparler LF, Idaghdour Y (2014) Epigenetic therapy of acute myeloid leukemia using 5-aza-2'-deoxycytidine (decitabine) in combination with inhibitors of histone methylation and deacetylation. Clin Epigenetics 6(1). https://doi.org/10.1186/1868-7083-6-19

Mondal P, Gadad SS, Adhikari S, Ramos EI, Sen S, Prasad P, Das C (2021) TCF19 and p53 regulate transcription of TIGAR and SCO2 in HCC for mitochondrial energy metabolism and stress adaptation. FASEB J 35(9). https://doi.org/10.1096/FJ.202002486RR

Morgan MJ, Liu ZG (2011) Crosstalk of reactive oxygen species and NF-κB signaling. Cell Res 21(1):103–115. https://doi.org/10.1038/cr.2010.178

Morris JP, Yashinskie JJ, Koche R, Chandwani R, Tian S, Chen CC, Baslan T, Marinkovic ZS, Sánchez-Rivera FJ, Leach SD, Carmona-Fontaine C, Thompson CB, Finley LWS, Lowe SW (2019) α-Ketoglutarate links p53 to cell fate during tumour suppression. Nature 573(7775): 595–599. https://doi.org/10.1038/s41586-019-1577-5

Moussaieff A, Rouleau M, Kitsberg D, Cohen M, Levy G, Barasch D, Nemirovski A, Shen-Orr S, Laevsky I, Amit M, Bomze D, Elena-Herrmann B, Scherf T, Nissim-Rafinia M, Kempa S, Itskovitz-Eldor J, Meshorer E, Aberdam D, Nahmias Y (2015) Glycolysis-mediated changes in acetyl-CoA and histone acetylation control the early differentiation of embryonic stem cells. Cell Metab 21(3):392–402. https://doi.org/10.1016/j.cmet.2015.02.002

Muller PAJ, Vousden KH (2013) P53 mutations in cancer. Nat Cell Biol 15(1):2–8. https://doi.org/10.1038/ncb2641

Murray PJ, Wynn TA (2011) Protective and pathogenic functions of macrophage subsets. Nat Rev Immunol 11(11):723–737. https://doi.org/10.1038/nri3073

Noman MZ, Desantis G, Janji B, Hasmim M, Karray S, Dessen P, Bronte V, Chouaib S (2014) PD-L1 is a novel direct target of HIF-1α, and its blockade under hypoxia enhanced: MDSC-mediated T cell activation. J Exp Med 211(5):781–790. https://doi.org/10.1084/jem.20131916

Norian LA, Rodriguez PC, O'Mara LA, Zabaleta J, Ochoa AC, Cella M, Allen PM (2009) Tumor-Infiltrating regulatory dendritic cells inhibit CD8+ T cell function via L-Arginine metabolism. Cancer Res 69(7):3086–3094. https://doi.org/10.1158/0008-5472.CAN-08-2826

Palsson-Mcdermott EM, Curtis AM, Goel G, Lauterbach MAR, Sheedy FJ, Gleeson LE, Van Den Bosch MWM, Quinn SR, Domingo-Fernandez R, Johnson DGW, Jiang JK, Israelsen WJ, Keane J, Thomas C, Clish C, Vanden Heiden M, Xavier RJ, O'Neill LAJ (2015) Pyruvate kinase M2 regulates hif-1α activity and il-1β induction and is a critical determinant of the warburg effect in LPS-activated macrophages. Cell Metab 21(1):65–80. https://doi.org/10.1016/j.cmet.2014.12.005

Pan D, Mao C, Wang YX (2013) Suppression of gluconeogenic gene expression by LSD1-mediated histone demethylation. PLoS One 8(6). https://doi.org/10.1371/journal.pone.0066294

Pan M, Reid MA, Lowman XH, Kulkarni RP, Tran TQ, Liu X, Yang Y, Hernandez-Davies JE, Rosales KK, Li H, Hugo W, Song C, Xu X, Schones DE, Ann DK, Gradinaru V, Lo RS, Locasale JW, Kong M (2016) Regional glutamine deficiency in tumours promotes dedifferentiation through inhibition of histone demethylation. Nat Cell Biol 18(10):1090–1101. https://doi.org/10.1038/ncb3410

Pang B, Zheng X-R, Tian J-X, Gao T-H, Gu G-Y, Zhang R, Fu Y-B, Pang Q, Li X-G, Liu Q (2016) EZH2 promotes metabolic reprogramming in glioblastomas through epigenetic repression of EAF2-HIF1α signaling. Oncotarget 7(29):45134–45143. https://doi.org/10.18632/oncotarget.9761

Park HA, Brown SR, Kim Y (2020) Cellular mechanisms of circulating tumor cells during breast cancer metastasis. Int J Mol Sci 21(14):1–19. https://doi.org/10.3390/ijms21145040

Parra-Bonilla G, Alvarez DF, Al-Mehdi A-B, Alexeyev M, Stevens T (2010) Critical role for lactate dehydrogenase A in aerobic glycolysis that sustains pulmonary microvascular endothelial cell proliferation. Am J Phys Lung Cell Mol Phys 299(4):L513–L522. https://doi.org/10.1152/ajplung.00274.2009

Pathiraja TN, Thakkar KN, Jiang S, Stratton S, Liu Z, Gagea M, Shi X, Shah PK, Phan L, Lee MH, Andersen J, Stampfer M, Barton MC (2015) TRIM24 links glucose metabolism with transformation of human mammary epithelial cells. Oncogene 34(22):2836–2845. https://doi.org/10.1038/onc.2014.220

Patra KC, Hay N (2014) The pentose phosphate pathway and cancer. Trends Biochem Sci 39(8):347–354. https://doi.org/10.1016/j.tibs.2014.06.005

Pavlides S, Whitaker-Menezes D, Castello-Cros R, Flomenberg N, Witkiewicz AK, Frank PG, Casimiro MC, Wang C, Fortina P, Addya S, Pestell RG, Martinez-Outschoorn UE, Sotgia F, Lisanti MP (2009) The reverse Warburg effect: aerobic glycolysis in cancer associated fibroblasts and the tumor stroma. Cell Cycle 8(23):3984–4001. https://doi.org/10.4161/cc.8.23.10238

Pavlova NN, Thompson CB (2016) The emerging hallmarks of cancer metabolism. Cell Metab. 23(1):27–47. https://doi.org/10.1016/j.cmet.2015.12.006

Peinado H, Zhang H, Matei IR, Costa-Silva B, Hoshino A, Rodrigues G, Psaila B, Kaplan RN, Bromberg JF, Kang Y, Bissell MJ, Cox TR, Giaccia AJ, Erler JT, Hiratsuka S, Ghajar CM,

Lyden D (2017) Pre-metastatic niches: organ-specific homes for metastases. Nat Rev Cancer 17(5):302–317. https://doi.org/10.1038/nrc.2017.6
Penny HL, Sieow JL, Adriani G, Yeap WH, Ee SC, Luis BS, Lee B, Lee T, Mak SY, Ho YS, Lam KP, Ong CK, Huang RYJ, Ginhoux F, Rotzschke O, Kamm RD, Wong SC (2016) Warburg metabolism in tumor-conditioned macrophages promotes metastasis in human pancreatic ductal adenocarcinoma. OncoImmunology 5(8). https://doi.org/10.1080/2162402X.2016.1191731
Peppicelli S, Bianchini F, Calorini L (2014) Extracellular acidity, a "reappreciated" trait of tumor environment driving malignancy: perspectives in diagnosis and therapy. Cancer Metastasis Rev 33(2–3):823–832. https://doi.org/10.1007/s10555-014-9506-4
Peters K, Kamp G, Berz A, Unger RE, Barth S, Salamon A, Rychly J, Kirkpatrick CJ (2009) Changes in human endothelial cell energy metabolic capacities during in vitro cultivation. the role of "aerobic glycolysis" and proliferation. Cell Physiol Biochem 24(5–6):483–492. https://doi.org/10.1159/000257490
Pietrocola F, Galluzzi L, Bravo-San Pedro JM, Madeo F, Kroemer G (2015) Acetyl coenzyme A: a central metabolite and second messenger. Cell Metab 21(6):805–821. https://doi.org/10.1016/j.cmet.2015.05.014
Porporato PE, Filigheddu N, Pedro JMBS, Kroemer G, Galluzzi L (2018) Mitochondrial metabolism and cancer. Cell Res 28(3):265–280. https://doi.org/10.1038/cr.2017.155
Puthenveetil A, Dubey S (2020) Metabolic reprograming of tumor-associated macrophages. Ann Transl Med 8(16):1030–1030. https://doi.org/10.21037/atm-20-2037
Reid MA, Dai Z, Locasale JW (2017) The impact of cellular metabolism on chromatin dynamics and epigenetics. Nat Cell Biol 19(11):1298–1306. https://doi.org/10.1038/ncb3629
Reytor E, Pérez-Miguelsanz J, Alvarez L, Pérez-Sala D, Pajares MA (2009) Conformational signals in the C-terminal domain of methionine adenosyltransferase I/III determine its nucleocytoplasmic distribution. FASEB J 23(10):3347–3360. https://doi.org/10.1096/fj.09-130187
Ribatti D, Tamma R, Annese T (2020) Epithelial-mesenchymal transition in cancer: a historical overview. Transl Oncol 13(6):100773. https://doi.org/10.1016/J.TRANON.2020.100773
Riemann A, Rauschner M, Gießelmann M, Reime S, Haupt V, Thews O (2019) Extracellular acidosis modulates the expression of epithelial-mesenchymal transition (EMT) markers and adhesion of epithelial and tumor cells. Neoplasia (United States) 21(5):450–458. https://doi.org/10.1016/j.neo.2019.03.004
Robey RB, Hay N (2006) Mitochondrial hexokinases, novel mediators of the antiapoptotic effects of growth factors and Akt. Oncogene 25(34):4683–4696. https://doi.org/10.1038/sj.onc.1209595
Rodrigues MF, Carvalho É, Pezzuto P, Rumjanek FD, Amoêdo ND (2015) Reciprocal modulation of histone deacetylase inhibitors sodium butyrate and trichostatin A on the energy metabolism of breast cancer cells. J Cell Biochem 116(5):797–808. https://doi.org/10.1002/jcb.25036
Rodríguez-Paredes M, Esteller M (2011) A combined epigenetic therapy equals the efficacy of conventional chemotherapy in refractory advanced non-small cell lung cancer. Cancer Discov 1(7):557–559. https://doi.org/10.1158/2159-8290.CD-11-0271
Rohle D, Popovici-Muller J, Palaskas N, Turcan S, Grommes C, Campos C, Tsoi J, Clark O, Oldrini B, Komisopoulou E, Kunii K, Pedraza A, Schalm S, Silverman L, Miller A, Wang F, Yang H, Chen Y, Kernytsky A et al (2013) An inhibitor of mutant IDH1 delays growth and promotes differentiation of glioma cells. Science 340(6132):626–630. https://doi.org/10.1126/science.1236062
Santi A, Kugeratski FG, Zanivan S (2018) Cancer associated fibroblasts: the architects of stroma remodeling. Proteomics 18(5–6):1700167. https://doi.org/10.1002/pmic.201700167
Schuettengruber B, Bourbon HM, Di Croce L, Cavalli G (2017) Genome regulation by polycomb and trithorax: 70 years and counting. Cell 171(1):34–57. https://doi.org/10.1016/j.cell.2017.08.002

Schwartzenberg-Bar-Yoseph F, Armoni M, Karnieli E (2004) The tumor suppressor p53 down-regulates glucose transporters GLUT1 and GLUT4 gene expression. Cancer Res 64(7): 2627–2633. https://doi.org/10.1158/0008-5472.CAN-03-0846

Sciacovelli M, Frezza C (2016) Oncometabolites: unconventional triggers of oncogenic signalling cascades. Free Radic Biol Med. 100:175–181. https://doi.org/10.1016/j.freeradbiomed.2016.04.025

Sciacovelli M, Gonçalves E, Johnson TI, Zecchini VR, Da Costa ASH, Gaude E, Drubbel AV, Theobald SJ, Abbo SR, Tran MGB, Rajeeve V, Cardaci S, Foster S, Yun H, Cutillas P, Warren A, Gnanapragasam V, Gottlieb E, Franze K et al (2016) Fumarate is an epigenetic modifier that elicits epithelial-to-mesenchymal transition. Nature 537(7621):544–547. https://doi.org/10.1038/nature19353

Sebastián C, Mostoslavsky R (2017) The various metabolic sources of histone acetylation. Trends Endocrinol Metab 28(2):85–87. https://doi.org/10.1016/j.tem.2016.11.001

Sebastián C, Zwaans BMM, Silberman DM, Gymrek M, Goren A, Zhong L, Ram O, Truelove J, Guimaraes AR, Toiber D, Cosentino C, Greenson JK, MacDonald AI, McGlynn L, Maxwell F, Edwards J, Giacosa S, Guccione E, Weissleder R et al (2012) The histone deacetylase SIRT6 Is a tumor suppressor that controls cancer metabolism. Cell 151(6):1185–1199. https://doi.org/10.1016/j.cell.2012.10.047

Semenza GL (2010a) Defining the role of hypoxia-inducible factor 1 in cancer biology and therapeutics. Oncogene 29(5):625–634. https://doi.org/10.1038/onc.2009.441

Semenza GL (2010b) HIF-1: upstream and downstream of cancer metabolism. Curr Opin Genetics Dev 20(1):51–56. https://doi.org/10.1016/j.gde.2009.10.009

Semenza GL (2012) Hypoxia-inducible factors in physiology and medicine. Cell 148(3):399–408. https://doi.org/10.1016/j.cell.2012.01.021

Semenza GL, Roth PH, Fang HM, Wang GL (1994) Transcriptional regulation of genes encoding glycolytic enzymes by hypoxia-inducible factor 1. J Biol Chem 269(38):23757–23763

Sen S, Sanyal S, Srivastava DK, Dasgupta D, Roy S, Das C (2017) Transcription factor 19 interacts with histone 3 lysine 4 trimethylation and controls gluconeogenesis via the nucleosome-remodeling-deacetylase complex. J Biol Chem 292(50):20362–20378. https://doi.org/10.1074/jbc.M117.786863

Shackelford DB, Shaw RJ (2009) The LKB1-AMPK pathway: metabolism and growth control in tumour suppression. Nat Rev Cancer 9(8):563–575. https://doi.org/10.1038/nrc2676

Shim EH, Livi CB, Rakheja D, Tan J, Benson D, Parekh V, Kho EY, Ghosh AP, Kirkman R, Velu S, Dutta S, Chenna B, Rea SL, Mishur RJ, Li Q, Johnson-Pais TL, Guo L, Bae S, Wei S et al (2014) L-2-hydroxyglutarate: an epigenetic modifier and putative oncometabolite in renal cancer. Cancer Discov 4(11):1290–1298. https://doi.org/10.1158/2159-8290.CD-13-0696

Shimazu T, Hirschey MD, Newman J, He W, Shirakawa K, Le Moan N, Grueter CA, Lim H, Saunders LR, Stevens RD, Newgard CB, Farese RV, De Cabo R, Ulrich S, Akassoglou K, Verdin E (2013) Suppression of oxidative stress by β-hydroxybutyrate, an endogenous histone deacetylase inhibitor. Science 339(6116):211–214. https://doi.org/10.1126/science.1227166

Singal R, Ramachandran K, Gordian E, Quintero C, Zhao W, Reis IM (2015) Phase I/II study of azacitidine, docetaxel, and prednisone in patients with metastatic castration-resistant prostate cancer previously treated with docetaxel-based therapy. Clin Genitourin Cancer 13(1):22–31. https://doi.org/10.1016/j.clgc.2014.07.008

Sivanand S, Rhoades S, Jiang Q, Lee JV, Benci J, Zhang J, Yuan S, Viney I, Zhao S, Carrer A, Bennett MJ, Minn AJ, Weljie AM, Greenberg RA, Wellen KE (2017) Nuclear acetyl-CoA production by ACLY promotes homologous recombination. Mol Cell 67(2):252–265.e6. https://doi.org/10.1016/j.molcel.2017.06.008

Soussi T, Wiman KG (2015) TP53: an oncogene in disguise. Cell Death and Differ. 22(8): 1239–1249. https://doi.org/10.1038/cdd.2015.53

Spruijt CG, Luijsterburg MS, Menafra R, Lindeboom RGH, Jansen PWTC, Edupuganti RR, Baltissen MP, Wiegant WW, Voelker-Albert MC, Matarese F, Mensinga A, Poser I, Vos HR, Stunnenberg HG, van Attikum H, Vermeulen M (2016) ZMYND8 co-localizes with NuRD on target genes and regulates poly(ADP-ribose)-dependent recruitment of GATAD2A/NuRD to sites of DNA damage. Cell Rep 17(3):783–798. https://doi.org/10.1016/j.celrep.2016.09.037

Stachowiak M, Szymanski M, Ornoch A, Jancewicz I, Rusetska N, Chrzan A, Demkow T, Siedlecki JA, Sarnowski TJ, Sarnowska E (2020) SWI/SNF chromatin remodeling complex and glucose metabolism are deregulated in advanced bladder cancer. IUBMB Life 72(6): 1175–1188. https://doi.org/10.1002/iub.2254

Sun Y, Daemen A, Hatzivassiliou G, Arnott D, Wilson C, Zhuang G, Gao M, Liu P, Boudreau A, Johnson L, Settleman J (2014) Metabolic and transcriptional profiling reveals pyruvate dehydrogenase kinase 4 as a mediator of epithelial-mesenchymal transition and drug resistance in tumor cells. Cancer Metab 2(1):20. https://doi.org/10.1186/2049-3002-2-20

Sutendra G, Kinnaird A, Dromparis P, Paulin R, Stenson TH, Haromy A, Hashimoto K, Zhang N, Flaim E, Michelakis ED (2014) A nuclear pyruvate dehydrogenase complex is important for the generation of acetyl-CoA and histone acetylation. Cell 158(1):84–97. https://doi.org/10.1016/j.cell.2014.04.046

Tae IH, Son JY, Lee SH, Ahn MY, Yoon K, Yoon S, Moon HR, Kim HS (2020) A new sirt1 inhibitor, MHY2245, induces autophagy and inhibits energy metabolism via PKM2/MTOR pathway in human ovarian cancer cells. Int J Biol Sci 16(11):1901–1916. https://doi.org/10.7150/ijbs.44343

Takahashi H, McCaffery JM, Irizarry RA, Boeke JD (2006) Nucleocytosolic acetyl-coenzyme a synthetase is required for histone acetylation and global transcription. Mol Cell 23(2):207–217. https://doi.org/10.1016/j.molcel.2006.05.040

Tang S-W, Yang T-C, Lin W-C, Chang W-H, Wang C-C, Lai M-K, Lin J-Y (2011) Nicotinamide N -methyltransferase induces cellular invasion through activating matrix metalloproteinase-2 expression in clear cell renal cell carcinoma cells. Carcinogenesis 32(2):138–145. https://doi.org/10.1093/carcin/bgq225

Tello D, Balsa E, Acosta-Iborra B, Fuertes-Yebra E, Elorza A, Ordóñez Á, Corral-Escariz M, Soro I, López-Bernardo E, Perales-Clemente E, Martínez-Ruiz A, Enríquez JA, Aragonés J, Cadenas S, Landázuri MO (2011) Induction of the mitochondrial NDUFA4L2 protein by HIF-1α decreases oxygen consumption by inhibiting complex i activity. Cell Metab 14(6): 768–779. https://doi.org/10.1016/j.cmet.2011.10.008

Thomas MG, Saldanha M, Mistry RJ, Dexter DT, Ramsden DB, Parsons RB (2013) Nicotinamide N-methyltransferase expression in SH-SY5Y neuroblastoma and N27 mesencephalic neurones induces changes in cell morphology via ephrin-B2 and Akt signalling. Cell Death Dis 4(6):e669. https://doi.org/10.1038/cddis.2013.200

Tsai W-W, Wang Z, Yiu TT, Akdemir KC, Xia W, Winter S, Tsai C-Y, Shi X, Schwarzer D, Plunkett W, Aronow B, Gozani O, Fischle W, Hung M-C, Patel DJ, Barton MC (2010) TRIM24 links a noncanonical histone signature to breast cancer. Nature 468(7326):927. https://doi.org/10.1038/NATURE09542

Turcan S, Fabius AWM, Borodovsky A, Pedraza A, Brennan C, Huse J, Viale A, Riggins GJ, Chan TA (2013) Efficient induction of differentiation and growth inhibition in IDH1 mutant glioma cells by the DNMT inhibitor Decitabine. Oncotarget 4(10):1729–1736. https://doi.org/10.18632/oncotarget.1412

Ulanovskaya OA, Zuhl AM, Cravatt BF (2013) NNMT promotes epigenetic remodeling in cancer by creating a metabolic methylation sink. Nat Chem Biol 9(5):300–306. https://doi.org/10.1038/nchembio.1204

Ullah MS, Davies AJ, Halestrap AP (2006) The plasma membrane lactate transporter MCT4, but not MCT1, is up-regulated by hypoxia through a HIF-1α-dependent mechanism. J Biol Chem 281(14):9030–9037. https://doi.org/10.1074/jbc.M511397200

Valencia-Sanchez MA, Liu J, Hannon GJ, Parker R (2006) Control of translation and mRNA degradation by miRNAs and siRNAs. Genes Dev 20(5):515–524. https://doi.org/10.1101/gad.1399806

Van Beijnum JR, Dings RP, Van Der Linden E, Zwaans BMM, Ramaekers FCS, Mayo KH, Griffioen AW (2006) Gene expression of tumor angiogenesis dissected: specific targeting of colon cancer angiogenic vasculature. Blood 108(7):2339–2348. https://doi.org/10.1182/blood-2006-02-004291

Vitale I, Manic G, Coussens LM, Kroemer G, Galluzzi L (2019) Macrophages and metabolism in the tumor microenvironment. Cell Metab 30(1):36–50. https://doi.org/10.1016/j.cmet.2019.06.001

Vizoso M, Marta Puig F, Carmona J, Maqueda M, Velásquez A, Gómez A, Labernadie A, Lugo R, Gabasa M, Rigat-Brugarolas LG, Trepat X, Ramírez J, Moran S, Vidal E, Reguart N, Perera A, Esteller M, Alcaraz J (2015) Aberrant DNA methylation in non-small cell lung cancer-associated fibroblasts. Carcinogenesis 36(12):1453–1463. https://doi.org/10.1093/CARCIN/BGV146

Vousden KH, Prives C (2009) Blinded by the light: the growing complexity of p53. Cell 137(3):413–431. https://doi.org/10.1016/j.cell.2009.04.037

Vousden KH, Ryan KM (2009) P53 and metabolism. Nat Rev Cancer 9(10):691–700. https://doi.org/10.1038/nrc2715

Wan W, Peng K, Li M, Qin L, Tong Z, Yan J, Shen B, Yu C (2017) Histone demethylase JMJD1A promotes urinary bladder cancer progression by enhancing glycolysis through coactivation of hypoxia inducible factor 1α. Oncogene 36(27):3868–3877. https://doi.org/10.1038/onc.2017.13

Wang F, Nguyen M, Qin FXF, Tong Q (2007) SIRT2 deacetylates FOXO3a in response to oxidative stress and caloric restriction. Aging Cell 6(4):505–514. https://doi.org/10.1111/j.1474-9726.2007.00304.x

Wang R, Dillon CP, Shi LZ, Milasta S, Carter R, Finkelstein D, McCormick LL, Fitzgerald P, Chi H, Munger J, Green DR (2011) The transcription factor Myc controls metabolic reprogramming upon T lymphocyte activation. Immunity 35(6):871–882. https://doi.org/10.1016/j.immuni.2011.09.021

Wang F, Travins J, DeLaBarre B, Penard-Lacronique V, Schalm S, Hansen E, Straley K, Kernytsky A, Liu W, Gliser C, Yang H, Gross S, Artin E, Saada V, Mylonas E, Quivoron C, Popovici-Muller J, Saunders JO, Salituro FG et al (2013) Targeted inhibition of mutant IDH2 in leukemia cells induces cellular differentiation. Science 340(6132):622–626. https://doi.org/10.1126/science.1234769

Wang L, Xiong H, Wu F, Zhang Y, Wang J, Zhao L, Guo X, Chang LJ, Zhang Y, You MJ, Koochekpour S, Saleem M, Huang H, Lu J, Deng Y (2014) Hexokinase 2-mediated Warburg effect is required for PTEN- and p53-deficiency-driven prostate cancer growth. Cell Rep 8(5):1461–1474. https://doi.org/10.1016/j.celrep.2014.07.053

Wang L, Zhang S, Wang X (2021) The metabolic mechanisms of breast cancer metastasis. Front Oncol 10. https://doi.org/10.3389/fonc.2020.602416

Warburg O (1956) On the origin of cancer cells. Science 123(3191):309–314. https://doi.org/10.1126/science.123.3191.309

Wei Q, Qian Y, Yu J, Wong CC (2020) Metabolic rewiring in the promotion of cancer metastasis: mechanisms and therapeutic implications. Oncogene 39:6139–6156. https://doi.org/10.1038/s41388-020-01432-7

Weinberg F, Hamanaka R, Wheaton WW, Weinberg S, Joseph J, Lopez M, Kalyanaraman B, Mutlu GM, Budinger GRS, Chandel NS (2010) Mitochondrial metabolism and ROS generation are essential for Kras-mediated tumorigenicity. Proc Natl Acad Sci U S A 107(19):8788–8793. https://doi.org/10.1073/pnas.1003428107

Wellen KE, Hatzivassiliou G, Sachdeva UM, Bui TV, Cross JR, Thompson CB (2009) ATP-citrate lyase links cellular metabolism to histone acetylation. Science 324(5930):1076–1080. https://doi.org/10.1126/science.1164097

Whitaker-Menezes D, Martinez-Outschoorn UE, Lin Z, Ertel A, Flomenberg N, Witkiewicz AK et al (2011) Evidence for a stromal-epithelial "lactate shuttle" in human tumors: MCT4 is a marker of oxidative stress in cancer-associated fibroblasts. Cell Cycle 10(11):1772–1783. https://doi.org/10.4161/cc.10.11.15659

Wolf A, Agnihotri S, Munoz D, Guha A (2011) Developmental profile and regulation of the glycolytic enzyme hexokinase 2 in normal brain and glioblastoma multiforme. Neurobiol Dis 44(1):84–91. https://doi.org/10.1016/j.nbd.2011.06.007

Wong KK, Engelman JA, Cantley LC (2010) Targeting the PI3K signaling pathway in cancer. Curr Opin Genetics Dev. 20(1):87–90. https://doi.org/10.1016/j.gde.2009.11.002

Wong CC, Qian Y, Yu J (2017) Interplay between epigenetics and metabolism in oncogenesis: mechanisms and therapeutic approaches. Oncogene 36(24):3359–3374. https://doi.org/10.1038/onc.2016.485

Xiao M, Yang H, Xu W, Ma S, Lin H, Zhu H, Liu L, Liu Y, Yang C, Xu Y, Zhao S, Ye D, Xiong Y, Guan KL (2012) Inhibition of α-KG-dependent histone and DNA demethylases by fumarate and succinate that are accumulated in mutations of FH and SDH tumor suppressors. Genes Dev 26(12):1326–1338. https://doi.org/10.1101/gad.191056.112

Xu WS, Parmigiani RB, Marks PA (2007) Histone deacetylase inhibitors: molecular mechanisms of action. Oncogene 26(37):5541–5552. https://doi.org/10.1038/sj.onc.1210620

Xu Y, An X, Guo X, Habtetsion TG, Wang Y, Xu X, Kandala S, Li Q, Li H, Zhang C, Caldwell RB, Fulton DJ, Su Y, Hoda MN, Zhou G, Wu C, Huo Y (2014) Endothelial PFKFB3 plays a critical role in angiogenesis. Arterioscler Thromb Vasc Biol 34(6):1231–1239. https://doi.org/10.1161/ATVBAHA.113.303041

Yamnik RL, Holz MK (2010) mTOR/S6K1 and MAPK/RSK signaling pathways coordinately regulate estrogen receptor α serine 167 phosphorylation. FEBS Lett 584(1):124–128. https://doi.org/10.1016/j.febslet.2009.11.041

Yan D, Wang HW, Bowman RL, Joyce JA (2016) STAT3 and STAT6 signaling pathways synergize to promote cathepsin secretion from macrophages via IRE1α activation. Cell Rep 16(11):2914–2927. https://doi.org/10.1016/j.celrep.2016.08.035

Yang H, Shu Z, Jiang Y, Mao W, Pang L, Redwood A, Jeter-Jones SL, Jennings NB, Ornelas A, Zhou J, Rodriguez-Aguayo C, Bartholomeusz G, Iles LKR, Zacharias NM, Millward SW, Lopez-Berestein G, Le XF, Ahmed AA, Piwnica-Worms H et al (2019) 6-phosphofructo-2-kinase/fructose-2,6-biphosphatase-2 regulates TP53-dependent paclitaxel sensitivity in ovarian and breast cancers. Clin Cancer Res 25(18):5702–5716. https://doi.org/10.1158/1078-0432.CCR-18-3448

Yen K, Travins J, Wang F, David MD, Artin E, Straley K, Padyana A, Gross S, Delabarre B, Tobin E, Chen Y, Nagaraja R, Choe S, Jin L, Konteatis Z, Cianchetta G, Saunders JO, Salituro FG, Quivoron C et al (2017) AG-221, a first-in-class therapy targeting acute myeloid leukemia harboring oncogenic IDH2 mutations. Cancer Discov 7(5):478–493. https://doi.org/10.1158/2159-8290.CD-16-1034

Yi W, Clark PM, Mason DE, Keenan MC, Hill C, Goddard WA, Peters EC, Driggers EM, Hsieh-Wilson LC (2012) Phosphofructokinase 1 glycosylation regulates cell growth and metabolism. Science 337(6097):975–980. https://doi.org/10.1126/science.1222278

Yong C, Stewart GD, Frezza C (2020) Oncometabolites in renal cancer. Nat Rev Nephrol. 16(3): 156–172. https://doi.org/10.1038/s41581-019-0210-z

Yu B, Shah A, Wang B, Rajaram N, Wang Q, Ramanujam N, Palmer GM, Dewhirst MW (2014) Measuring tumor cycling hypoxia and angiogenesis using a side-firing fiber optic probe. J Biophotonics 7(7):552–564. https://doi.org/10.1002/JBIO.201200187

Yu T, Yang G, Hou Y, Tang X, Wu C, Wu XA, Guo L, Zhu Q, Luo H, Du YE, Wen S, Xu L, Yin J, Tu G, Liu M (2017) Cytoplasmic GPER translocation in cancer-associated fibroblasts mediates cAMP/PKA/CREB/glycolytic axis to confer tumor cells with multidrug resistance. Oncogene 36(15):2131–2145. https://doi.org/10.1038/onc.2016.370

Zacharias N, Lee J, Ramachandran S, Shanmugavelandy S, McHenry J, Dutta P, Millward S, Gammon S, Efstathiou E, Troncoso P, Frigo DE, Piwnica-Worms D, Logothetis CJ, Maity SN, Titus MA, Bhattacharya P (2019) Androgen receptor signaling in castration-resistant prostate cancer alters hyperpolarized pyruvate to lactate conversion and lactate levels in vivo. Mol Imaging Biol 21(1):86–94. https://doi.org/10.1007/s11307-018-1199-6

Zambrano A, Molt M, Uribe E, Salas M (2019) Glut 1 in cancer cells and the inhibitory action of resveratrol as a potential therapeutic strategy. Int J Mol Sci 20(13):3374. https://doi.org/10.3390/ijms20133374

Zhang C, Liu J, Liang Y, Wu R, Zhao Y, Hong X, Lin M, Yu H, Liu L, Levine AJ, Hu W, Feng Z (2013) Tumour-associated mutant p53 drives the Warburg effect. Nat Commun 4:2935. https://doi.org/10.1038/ncomms3935

Zhang D, Wang Y, Shi Z, Liu J, Sun P, Hou X, Zhang J, Zhao S, Zhou BP, Mi J (2015) Metabolic reprogramming of cancer-associated fibroblasts by IDH3α downregulation. Cell Rep 10(8): 1335–1348. https://doi.org/10.1016/j.celrep.2015.02.006

Zhang Y, Yu G, Chu H, Wang X, Xiong L, Cai G, Liu R, Gao H, Tao B, Li W, Li G, Liang J, Yang W (2018) Macrophage-associated PGK1 phosphorylation promotes aerobic glycolysis and tumorigenesis. Mol Cell 71(2):201–215.e7. https://doi.org/10.1016/j.molcel.2018.06.023

Zhang J, Zhang J, Wei Y, Li Q, Wang Q (2019) ACTL6A regulates follicle-stimulating hormone-driven glycolysis in ovarian cancer cells via PGK1. Cell Death Dis 10(11). https://doi.org/10.1038/s41419-019-2050-y

Zhao M, Fang W, Wang Y, Guo S, Shu L, Wang L, Chen Y, Fu Q, Liu Y, Hua S, Fan Y, Liu Y, Deng X, Luo R, Mei Z, Jiang Q, Liu Z (2015) Enolase-1 is a therapeutic target in endometrial carcinoma. Oncotarget 6(17):15610

Zhao Z, Wang L, Di LJ (2016) Compartmentation of metabolites in regulating epigenomes of cancer. Mol Med 22:349–360. https://doi.org/10.2119/molmed.2016.00051

Zhao E, Hou J, Ke X, Abbas MN, Kausar S, Zhang L, Cui H (2019) The roles of sirtuin family proteins in cancer progression. Cancers 11(12):1949. https://doi.org/10.3390/cancers11121949

Zhong L, D'Urso A, Toiber D, Sebastian C, Henry RE, Vadysirisack DD, Guimaraes A, Marinelli B, Wikstrom JD, Nir T, Clish CB, Vaitheesvaran B, Iliopoulos O, Kurland I, Dor Y, Weissleder R, Shirihai OS, Ellisen LW, Espinosa JM, Mostoslavsky R (2010) The histone deacetylase Sirt6 regulates glucose homeostasis via Hif1α. Cell 140(2):280–293. https://doi.org/10.1016/j.cell.2009.12.041

Chapter 2
Iron in Cancer Progression: Does BACH1 Promote Metastasis by Altering Iron Homeostasis?

Kazuhiko Igarashi, Hironari Nishizawa, and Mitsuyo Matsumoto

Abstract The transcription factor BACH1, which is regulated by direct binding of prosthetic group heme, promotes epithelial-mesenchymal transition (EMT) and drives metastasis of diverse types of cancer cells. De-regulated target genes of BACH1 in cancer cells include those for glycolysis, oxidative phosphorylation, epithelial cell adhesion, and mesodermal cell motility. In addition, the canonical target genes of BACH1 include genes for the regulation of iron homeostasis. Importantly, cancer cells are addicted to iron. We summarize known functions of BACH1 in cancer and discuss how BACH1 may affect iron homeostasis in cancer cells to support their progression by increasing mobile iron within cells. The dependency on BACH1 for cancer progression may also confer upon cancer cells susceptibility to iron-dependent cell death ferroptosis. Finally, we discuss that the human transcription factors provide research opportunities for better understanding of cancer cell properties.

Keywords BACH1 · Cancer · Epigenetics · Epithelial-Mesenchymal transition · Iron · Metastasis · Transcription factor

Introduction

Iron is an essential metal for almost all life forms, including humans, but it is also highly toxic. Of the approximately 19,000 protein genes encoded in the human genome, it has been estimated that there are 398 protein genes whose products bind iron as heme (iron-coordinated prosthetic group), iron-sulfur clusters or as it is (Andreini et al. 2018). These iron-binding proteins include many of the metabolism-related proteins such as mitochondrial electron-transfer system enzymes, oxidoreductases, oxygen-transport proteins and demethylase enzymes (Andreini et al. 2018) (Fig. 2.1). For example, many of the histone demethylases

K. Igarashi (✉) · H. Nishizawa · M. Matsumoto
Department of Biochemistry, Tohoku University Graduate School of Medicine, Sendai, Japan
e-mail: igarashi@med.tohoku.ac.jp

T. K. Kundu, C. Das (eds.), *Metabolism and Epigenetic Regulation: Implications in Cancer*, Subcellular Biochemistry 100,
https://doi.org/10.1007/978-3-031-07634-3_2

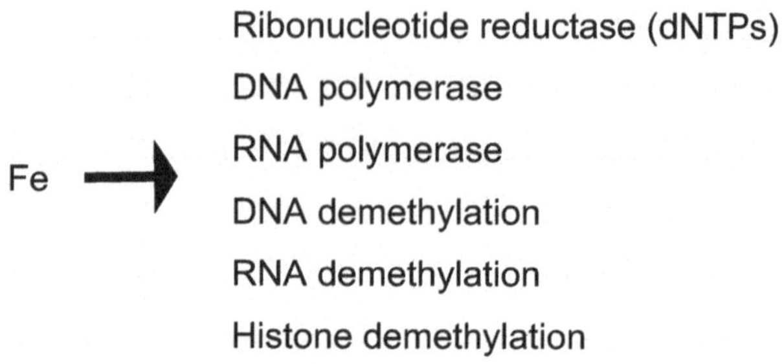

Fig. 2.1 The roles of iron in cell biochemistry. Fe-dependent reactions important for normal and cancer cells are depicted

and DNA demethylases involved in epigenetic regulation use iron as a cofactor, and their activity is greatly influenced by iron availability (Blanquart et al. 2019). Therefore, iron is involved in a wide variety of reactions, from metabolic activities to gene expression. In addition, there are a number of iron-binding proteins whose functions have yet to be elucidated (Andreini et al. 2018). On the other hand, free reduced iron that does not bind to proteins (labile iron or mobile iron in equilibrium with protein binding, hereafter referred to as mobile iron) generates hydroxyl radical (·OH) through the Fenton reaction, which leads to the oxidation of biomolecules, resulting in decreased cell function and iron-dependent cell death (ferroptosis) (Stockwell et al. 2017, 2020; Toyokuni et al. 2020) (Fig. 2.2a). Therefore, maintaining adequate iron levels and transporting it into and out of cells is literally a matter of life and death for cells and organisms. Iron utilization is not always maintained at a constant level but is greatly enhanced during cell proliferation and differentiation. In addition, changes in iron metabolism are known to be associated with a variety of pathological conditions. In particular, it is known that epithelial-mesenchymal transition (EMT), which is important for cancer metastasis, is dependent on mobile iron (Lane et al. 2014). However, the full picture of how iron dynamics are regulated in cancer cells and how iron alters the properties of cancer cells is still unknown. Cancer cells with EMT have increased sensitivity to ferroptosis (Lee et al. 2020b), and it is expected that cancer treatment strategies utilizing the mechanism of ferroptosis will be developed. However, it is still unclear why EMT-induced cells are more susceptible to ferroptosis in the first place. In this chapter, we will summarize the regulation of iron homeostasis in cancer cells and how it can be connected to cancer cell properties including increased sensitivity to ferroptosis, based on the functions of the transcription factor BACH1. We finally discuss the importance to investigate functions of many, understudied transcription factors in the context of cancer biology.

Iron Is High in Cancer Cells

In certain types of cancer cells, iron has been reported to increase. For example, prostate cancer cells secure iron by altering iron metabolism (Vela 2018). Colorectal cancer cells also increase their iron content by restricting iron export out of the cells

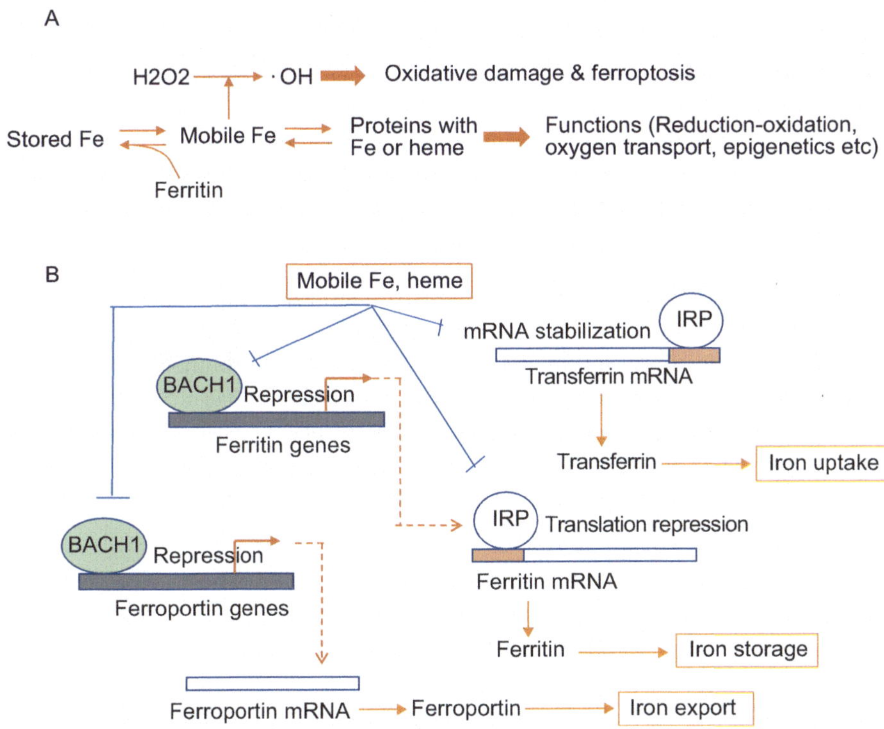

Fig. 2.2 Iron homeostasis and ferroptosis. (**a**) Essential roles of iron and the toxicity of mobile iron, leading to ferroptosis. (**b**) Cellular iron homeostasis regulated by BACH1 at the level of gene transcription and iron regulatory protein (IRP) at the level of post-transcription. BACH1 and IRP increase mobile iron whereas an increase in mobile iron and/or heme executes a negative feedback regulation

(Schwartz et al. 2021) or increase iron uptake (Xue et al. 2016), promoting mitochondrial function and proliferation. Breast cancer cells increase their iron content by using CD44-dependent uptake of iron-hyaluronate complex (Müller et al. 2020). These and other observations have established that iron is increased in cancer cells. Furthermore, epidemiological studies have shown that increased amount of systemic iron or iron intake was associated with increased risks of esophageal cancer, colorectal cancer, and lung cancer (Torti et al. 2018).

The Regulation of Iron in Cancer Cells

Basically, cancer cells utilize the regulatory mechanisms for iron homeostasis in normal cells (Fig. 2.2b), but the overall setting point of iron appears elevated. The mechanism of iron homeostasis in mammals has been studied mainly through post-transcriptional regulation. Iron Regulatory Proteins (IRPs) bind to the untranslated regions of ferritin and transferrin receptor mRNAs and regulate the expression of target proteins in response to mobile iron levels (Iwai 2019), as shown in Fig. 2.2b. For example, when the amount of mobile iron in the cell decreases, IRP binds to the target mRNAs, repressing the translation of ferritin mRNA and stabilizing transferrin mRNA. As ferritin sequesters iron from utilization and suppresses its toxicity, iron stores decrease as ferritin decreases. Transferrin receptor expression promotes iron uptake from extracellular sources. Together, these responses increase the amount of mobile iron in the cell and maintains homeostasis.

However, cancer cells modify the system for iron homeostasis by multiple ways so that iron content is increased. The activity of ferroportin is inhibited by hepcidin in colorectal cancer cells, leading to accumulation of iron (Schwartz et al. 2021). Breast cancer cells utilize lipocalin-2 to uptake iron from environment where iron is relatively scarce (e.g., cerebrospinal fluid) (Chi et al. 2020). CD44-dependent uptake of iron-hyaluronate complex appears also important in breast cancer cells (Müller et al. 2020). One of the most significant findings is that the expression of ferritin is reduced in cancer cells undergoing epithelial-mesenchymal transition (EMT) (Cook and Vanderhyden 2020).

Possible Roles for Iron in Cancer Cells

The most well-known function of iron in cancer cells is as a co-factor of ribonucleotide reductase which is located at the center of nucleotide metabolism and DNA synthesis (Fig. 2.1). Furthermore, DNA and RNA polymerases also utilize iron as a co-factor (Andreini et al. 2018). Therefore, the increases in iron in cancer cells are expected to support their proliferation.

Considering that many of the DNA and histone demethylases are also iron-dependent, the increases in iron may promote epigenetic alterations of cancer cells. In breast cancer cells, increased uptake of iron leads to PHF8-mediated erasing of dimethylation of histone H3 lysine 9 (H3K9me2), leading to increased expression of genes involved in EMT and metastasis (Müller et al. 2020). This observation suggests that iron is indeed rate-limiting in the epigenetic regulation by demethylases.

While their affinities for iron have not been studied, many studies have shown critical roles of iron-dependent histone demethylases in cancer cells, including KDM5B (Xhabija and Kidder 2019), KDM4 (Lee et al. 2020a) and KDM6A (Tran et al. 2020). Consistent with their roles in cancer cells, iron chelation has been shown to inhibit demethylases and to confer anti-tumor activities (Khodaverdian et al. 2019). Therefore, iron appears to promote epigenetic alterations to promote cancer cell proliferation and/or progression (Fig. 2.1). It should be noted that, while epigenetic modifications are often considered as static, recent findings support the notion that writing and erasing of epigenetic modifications are under equilibrium (Parry et al. 2021). Therefore, continuous supply of iron may be important to maintain pathological epigenetic status of cancer cells.

BACH1 Promotes Progression of Diverse Types of Cancers

The transcriptional repressor BACH1 belongs to the basic-leucine zipper (bZip) family of transcription factors (Igarashi et al. 2021). BACH1 and its related factor BACH2 are unique in that a prosthetic group heme directly binds to them and inhibits their activities as transcription factors. Heme inhibits the DNA binding activity of BACH1 and BACH2 and induces their nuclear export, polyubiquitination and proteasome-mediated degradation (Igarashi and Watanabe-Matsui 2014). Therefore, heme induces the expression of BACH-repressed genes. BACH1 is expressed in a wide range of tissues and cells whereas BACH2 is expressed in hematopoietic and immune cells and regulates development of erythroid and lymphoid cells, and responses of B cells and T cells to antigen and cytokine signals (Igarashi et al. 2017). Some of the BACH2 function in hematopoietic and immune cells are redundant with that of BACH1 (Itoh-Nakadai et al. 2014, 2017; Kato et al. 2018). Importantly, BACH1 promotes progression and metastasis of diverse types of cancers including breast, lung, pancreatic, colorectal and ovarian cancers by forming several gene regulatory network (GRN) with its target genes (Fig. 2.3).

1. *Breast cancer*

   Two lines of investigation reported that BACH1 promotes bone metastasis of triple negative cancer cells in mice implantation model. Rosner and colleagues investigated downstream target mRNA of Let-7, a microRNA known to function as a tumor suppressor (Yun et al. 2011). BACH1 mRNA was found to be one of the targets of Let-7, leading to a model in which Let-7 restricts tumor progression by reducing BACH1 mRNA (Yun et al. 2011). Kang and their colleagues (Liang et al. 2012) used gene expression profiling of breast cancer cell lines that show higher metastatic activity to identify key transcription factors for metastasis. They identified BACH1 as a potential regulator of cancer progression along with SMAD4 and HIF1. In both studies, authors showed a critical function of BACH1 to promote bone metastasis by using transplantation of human cancer cells with genetic manipulations of BACH1. Consistent with these observations, several studies reported that knockdown of BACH1 in human breast cancer cell line decreases migration in vitro (Davudian et al. 2016; Mansoori et al. 2019). BACH1 expression is positively correlated with the expression of genes related to the cytoskeleton and extracellular matrix in triple negative breast cancers, providing a set of prognostic markers of triple negative breast cancer patients (Lee et al. 2014). Another study has recently confirmed BACH1 as a prognostic marker for triple-negative breast cancer, with patients with high BACH1 expression exhibiting shorter overall survival and disease-free survival (Ou et al. 2019). Direct target genes of BACH1 in breast cancer include many of those involved in the oxidative phosphorylation, and thereby BACH1 represses the oxidative-phosphorylation in cancer cells (Lee et al. 2019). This is consistent with the properties of metastatic triple-negative breast cancer cells, which often show a glycolytic pattern of gene expression (Kim et al. 2013).
2. *Non-small cell lung cancer*

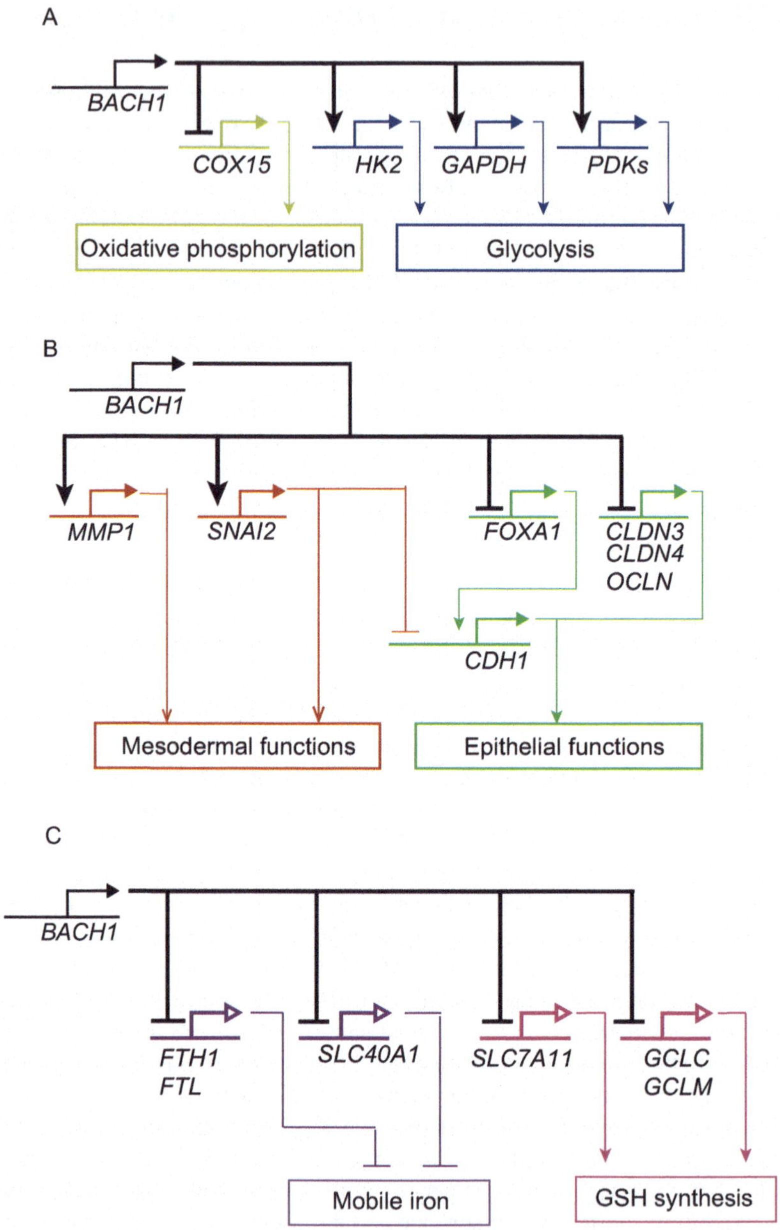

Fig. 2.3 BACH1 gene regulatory networks in cancer progression. (**a**) BACH1 promotes oxidative glycolysis. (**b**) BACH1 promotes EMT via multiple target genes. BACH1 reduces epithelial cell adhesion directly and indirectly via repressing FOXA1 and enhances cell mobility and invasion by activating some of the mesodermal genes. (**c**) BACH1 increases mobile iron and reduces GSH in cancer cells

BACH1 promotes metastasis in non-small cell lung cancer (NSCLC). Approximately one third of NSCLC possesses somatic mutations of NFE2L2 gene or its negative regulator KEAP1 gene, resulting in hyperactivation of NRF2 (Cancer Genome Atlas Research Network 2012). NRF2 induces the expression of heme oxygenase-1 (HO-1) encoded by HMOX1 (Harada et al. 2008), leading to a reduction of intracellular heme. Since heme promotes degradation of BACH1, this leads to increased protein amount of BACH1 and its binding to its target genes, including HMOX1, in NSCLC cells (Wiel et al. 2019). BACH1 promotes cell migration, invasion and metastasis of NSCLC (Lignitto et al. 2019; Wei et al. 2020; Wiel et al. 2019), but its direct target genes in these processes in NSCLC are not clear. BACH1 may stimulate metastasis by promoting glycolysis (Wiel et al. 2019). Importantly, while BACH1 and NRF2 bind to similar DNA sequences, these proteins regulate not only overlapping sets of genes but also unique genes in NSCLC cells (Wiel et al. 2019). Therefore, BACH1 and NRF2 may promote NSCLC through distinct mechanisms by regulating genes specific to one or the other.

3. *Pancreatic cancer*

   BACH1 promotes metastasis and is correlated with patient prognosis in pancreatic ductal adenocarcinoma (PDAC). We reported that BACH1 knockdown reduced proliferation of PDAC cells in vitro, especially when the initial cell density is lower, and caused the re-expression of epithelial genes, suggesting that BACH1 promotes EMT (Sato et al. 2020). When transplanted into immunodeficient mice, PDAC cells with BACH1 ablation showed less metastasis compared with control cells. Consistent with these findings, patients with higher expression of BACH1 showed poorer prognosis than those with lower BACH1 expression. BACH1 directly binds to a set of genes important for the epithelial structure and represses their expression in PDAC cells (see below) (Sato et al. 2020).

4. *Colorectal cancer*

   Colorectal cancer patients with higher expression of BACH1 show poorer prognosis (Zhu et al. 2018). Invasion and migration activities of colon cancer cell line in vitro are reduced upon BACH1 knockdown (Davudian et al. 2016; Zhu et al. 2018). When overexpressed in colon cancer cells, BACH1 enhanced tumor growth in transplanted mice (Zhu et al. 2018). Analysis of MYC-dependent super-enhancers (SE) of colorectal cancer cells led to the identification of the HOXB8 gene as a factor for maintaining malignant features of cancer cells including invasion and migration in vitro and tumor formation and metastasis in mice xenograft model (Ying et al. 2020). HOXB8 promotes metastasis by inducing the expression of BACH1. Furthermore, BACH1 appears to bind HOXB8 protein to induce the expression of their shared target genes (Ying et al. 2020). However, direct target genes co-regulated by HOXB8 and BACH1 remain elusive.

5. *Ovarian cancer*

   BACH1 expression is elevated in ovarian cancer, compared with normal ovarian tissues. While higher expression of BACH1 is associated with poorer prognosis (Han et al. 2019), the hazard ratio appears very small. Similar to experiments in other cancer types, BACH1 knockdown in ovarian cancer cells reduces migration in vitro whereas its overexpression increases migration (Han et al. 2019; Shajari et al. 2018). Similar to pancreatic cancer cells, BACH1 knockdown induces the expression of E-cadherin whereas its overexpression reduced E-cadherin expression and reduced metastasis in xenograft mouse model (Han et al. 2019).

6. *Mechanisms for the promotion of metastasis*

   Several mechanisms have been reported for the promotion of metastasis by BACH1. BACH1 enhances aerobic glycolysis, which is an increased glycolysis in spite of oxygen availability and a hallmark of cancer cells, in breast and lung cancer cells. BACH1 represses the expression of genes involved in oxidative phosphorylation in mitochondria, including COX15 which encodes for the cytochrome c oxidase assembly factor, in breast cancer cells (Lee et al. 2019). BACH1 induces the expression of genes for glycolysis, including GAPDH and HK2 which encode for glyceraldehyde-3-phosphate dehydrogenase and hexokinase-2, respectively, in lung cancer cells (Wiel et al. 2019). Repression and induction of these genes are expected to lead to aerobic glycolysis (Fig. 2.3a).

Furthermore, BACH1 regulates the expression of genes directly involved in EMT (Fig. 2.3b). In PDAC cells, BACH1 directly represses the expression of genes that encode for proteins involved in cell-cell adhesion, including a set of claudin and occludin genes (Sato et al. 2020). Since the protein products of these genes mediate the formation of tight junction, BACH1 reduces cell-cell adhesion. BACH1 further represses the expression of epithelial cadherin (CDH1) but this regulation appears indirect (Sato et al. 2020). BACH1 directly represses the expression of FOXA1, a transcription factor important for CDH1 expression. These direct and indirect repression of genes for cell-cell adhesion by BACH1 leads to a reduction in epithelial phenotype (Sato et al. 2020). In addition, BACH1 directly activates the expression of genes involved in mesenchymal phenotype. Target genes in this category include SNAI1 (Han et al. 2019; Sato et al. 2020), which is a transcription factor that represses epithelial genes, and MMP1 (Liang et al. 2012; Yun et al. 2011), which is a collagenase and is important for invasion. By regulating a cohort of genes involved in EMT, BACH1 promotes metastatic properties of cancer cells.

In addition, BACH1 further regulates other responses to promote cancer progression. BACH1 promotes angiogenesis in ovarian cancer by inducing the expression of VEGFC (Cohen et al. 2020). However, there are several other reports showing that BACH1 rather restricts angiogenesis (Jiang et al. 2015, 2017, 2020; Yusoff et al. 2020). Therefore, the regulation of angiogenesis by BACH1 may depend on contexts. BACH1 is known to repress the expression of tumor suppressor genes by recruiting DNA methyltransferase DNMT3b in melanoma and colorectal cancer cells (Fang et al. 2016; Fang et al. 2014).

Does BACH1 Regulate Iron in Cancer Cells?

1. *Iron-related target genes of BACH1*
   In addition to these roles for BACH1 in cancer progression, the fact that many of the target genes of BACH1 are related to iron metabolism (Igarashi et al. 2021) strongly suggest that BACH1 alters iron homeostasis in cancer cells (Fig. 2.2b and 2.3c). Ferritin is composed of heavy and light chain subunits. BACH1 represses both of the genes for ferritin subunits (FTH1 for heavy and FTL for light chain subunits) (Igarashi et al. 2021; Nishizawa et al. 2020). BACH1 also represses the expression of ferroportin (SLC40A1), which exports iron out of cells. In aggregate, BACH1 is expected to increase mobile iron within cancer cells by repressing the expression of these iron-related genes. BACH1 may set the homeostatic level of mobile iron in cancer cells higher than that in normal cells. If this is the case, BACH1 promotes cancer by increasing the activities of proteins that are dependent on iron, ranging from nucleotide metabolism to epigenetic alterations.

   Another interesting possibility to consider is the effect of BACH1 on ferroptosis, to which many of cancer cells show enhanced sensitivity (Stockwell et al. 2017). Ferroptosis is cell death caused by damages to cellular and organelle membrane. In this process, iron-catalyzed lipid peroxidation is the central reaction that causes cell death. BACH1 drives ferroptosis in mouse fibroblasts and cardiomyocytes by not only increasing mobile iron but also decreasing the synthesis of glutathione, one of the critical metabolites for reducing oxidized molecules including lipid peroxides. BACH1 represses the expression of genes SLC7A11, GCLC, and GCLM (Nishizawa et al. 2020) (Fig. 2.3c). SLC7A11 imports cystine which is one of the substrates for glutathione synthesis, whereas GCLC and GCLM are the subunits of glutamate-cysteine ligase, the rate limiting enzyme of the glutathione synthesis pathway. Therefore, BACH1 increases the level of iron-catalyzed production of hydroxyl radical and reduces the production of glutathione, leading to ferroptosis. We surmise that BACH1 not only promotes metastasis of cancer cells but also drives their sensitivity to ferroptosis (Igarashi et al. 2021; Nishizawa et al. 2021).
2. *Pathological and clinical implications*
   The connection of BACH1 with iron and glutathione metabolism suggest a new venue of research. First, identification of critical, iron-related genes among the BACH1 target genes will be required to understand how iron is increased in cancer cells and how iron drives the progression of cancer cells. If it becomes clear that iron-dependent proteins contribute to one or the other steps of metastasis, they will provide new therapeutic strategies, including iron-targeted therapy. Second, metastatic cancer cells driven by BACH1 are expected to show higher sensitivity to ferroptosis. Therefore, the expression of BACH1 and its target genes can be used to stratify cancer cells into ferroptosis-sensitive and resistant types. When drugs for inducing ferroptosis become available, such a stratification may be helpful for personalized precision medicine. Recently,

ferroptosis has been shown to confer population effects: cells undergoing ferroptosis secrete molecules that induce ferroptosis in surrounding cells (Nishizawa et al. 2021; Riegman et al. 2020). Such cell-cell communication mediated by ferroptosis may be important for understanding the responses of cancer cell populations to therapeutic interventions.

Transcription Factor-Based Research into Cancer Cell Properties

Human genome encodes 1,639 known or likely transcription factors (Lambert et al. 2018). A PubMed search of the human transcription factors reveals that the distribution of research papers in cancer biology is heavily skewed toward a small subset of the transcription factors whose importance in cancer is very well established: top 10 transcription factors occupy 38% of those literatures on cancer and a transcription factor whereas roughly half of the human transcription factors are with less than five publications related to cancer (Fig. 2.4). Less than half of them (706) have been

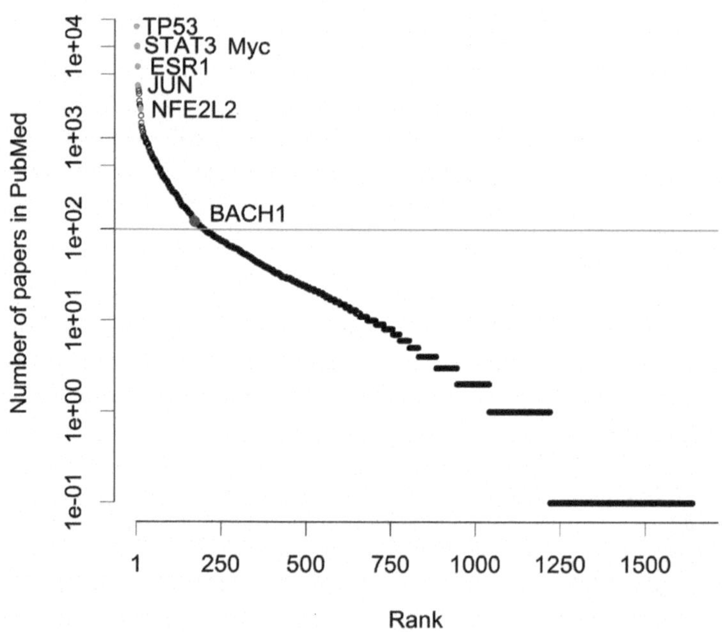

Fig. 2.4 Cancer-related publications on human transcription factors. Human transcription factors are ranked by numbers of papers in PubMed describing each factor in cancer biology as of January 12, 2021. The basic search string was "transcription factor [Title/Abstract] AND transcription AND (cancer OR leukemia OR lymphoma OR sarcoma)." Arbitrary modifications were added in case of homonymy or synonymy. The numbers of papers per transcription factors are plotted on a log scale. Those without any paper were set at 0.1 for depiction. Some of the transcription factors are shown with color dots. The green line indicates 100 papers. BACH1 is currently ranked 173rd with 123 papers

reported in more than ten papers in the context of cancer biology and gene expression. Considering that many of these transcription factors are likely to contribute to the homeostasis and/or stress responses of normal and cancer cells, fundamental knowledge on the expression of cancer cell properties must be far from enough. We need to delve into functions of these transcription factors in cancer biology more intensively. While this may sound laborious and challenging, a library of sgRNA targeting these factors combined with monitoring in vivo tumor growth and/or metastasis is likely to make such studies feasible. A library of mammalian expression plasmids for 1,588 mouse transcription factors (Yahagi and Takeuchi 2021) may be exploited. Technologies such as Perturb-Seq (Dixit et al. 2016) may aid in unraveling the complexity of transcription factor functions in the populations of cancer cells trying to survive and expand in the body. With increasing knowledge about transcription factor functions and the development of platforms that enable their targeted delivery, we can expect more transcription factor-based therapies against cancers to become available in the forthcoming years. With more knowledge on human transcription factors and their GRNs, we will be able to understand the cancer cell properties as dynamic GRNs and to dodge the dreadful fate they impose.

Acknowledgment We are indebted to stimulating discussion with and contribution from laboratory members. Investigation by the authors have been supported by Grants-in-Aid for Scientific Research from the Japan Society for the Promotion of Science (20KK0176 and 18H04021 to KI, 20K16296 and 19K23738 to HN, and 19K07680 and 16K07108 to MM), Grant-in-Aid for Joint Research by Young Researchers from Tohoku University Graduate School of Medicine (to H.N.), Gonryo Medical Foundation (to H.N.), and Takeda Science Foundation (to H.N.).

References

Andreini C, Putignano V, Rosato A, Banc L (2018) The human iron-proteome. Metallomics Integr Biometal Sci 10:1223–1231

Blanquart C, Linot C, Cartron PF, Tomaselli D, Mai A, Bertrand P (2019) Epigenetic metalloenzymes. Curr Med Chem 26:2748–2785

Cancer Genome Atlas Research Network (2012) Comprehensive genomic characterization of squamous cell lung cancers. Nature 489:519–525

Chi Y, Remsik J, Kiseliovas V, Derderian C, Sener U, Alghader M, Saadeh F, Nikishina K, Bale T, Iacobuzio-Donahue C et al (2020) Cancer cells deploy lipocalin-2 to collect limiting iron in leptomeningeal metastasis. Science 369:276–282

Cohen B, Tempelhof H, Raz T, Oren R, Nicenboim J, Bochner F, Even R, Jelinski A, Eilam R, Ben-Dor S et al (2020) BACH family members regulate angiogenesis and lymphangiogenesis by modulating VEGFC expression. Life Sci Alliance 3:e202000666

Cook DP, Vanderhyden BC (2020) Context specificity of the EMT transcriptional response. Nat Commun 11:2142

Davudian S, Shajari N, Kazemi T, Mansoori B, Salehi S, Mohammadi A, Shanehbandi D, Shahgoli VK, Asadi M, Baradaran B (2016) BACH1 silencing by siRNA inhibits migration of HT-29 colon cancer cells through reduction of metastasis-related genes. Biomed Pharmacother 84:191–198

Dixit A, Parnas O, Li B, Chen J, Fulco CP, Jerby-Arnon L, Marjanovic ND, Dionne D, Burks T, Raychowdhury R et al (2016) Perturb-Seq: dissecting molecular circuits with scalable single-cell RNA profiling of pooled genetic screens. Cell 167:1853–1866.e1817

Fang M, Ou J, Hutchinson L, Green MR (2014) The BRAF oncoprotein functions through the transcriptional repressor MAFG to mediate the CpG Island Methylator phenotype. Mol Cell 55: 904–915

Fang M, Hutchinson L, Deng A, Green MR (2016) Common BRAF(V600E)-directed pathway mediates widespread epigenetic silencing in colorectal cancer and melanoma. Proc Natl Acad Sci U S A 113:1250–1255

Han W, Zhang Y, Niu C, Guo J, Li J, Wei X, Jia M, Zhi X, Yao L, Meng D (2019) BTB and CNC homology 1 (Bach1) promotes human ovarian cancer cell metastasis by HMGA2-mediated epithelial-mesenchymal transition. Cancer Lett 445:45–56

Harada H, Sugimoto R, Watanabe A, Taketani S, Okada K, Warabi E, Siow R, Itoh K, Yamamoto M, Ishii T (2008) Differential roles for Nrf2 and AP-1 in upregulation of HO-1 expression by arsenite in murine embryonic fibroblasts. Free Radic Res 42:297–304

Igarashi K, Watanabe-Matsui M (2014) Wearing red for signaling: the heme-bach axis in heme metabolism, oxidative stress response and iron immunology. Tohoku J Exp Med 232:229–253

Igarashi K, Kurosaki T, Roychoudhuri R (2017) BACH transcription factors in innate and adaptive immunity. Nat Rev Immunol 17:437–450

Igarashi K, Nishizawa H, Saiki Y, Matsumoto M (2021) The transcription factor BACH1 at the crossroads of cancer biology: from epithelial-mesenchymal transition to ferroptosis. J Biol Chem 297:101032

Itoh-Nakadai A, Hikota R, Muto A, Kometani K, Watanabe-Matsui M, Sato Y, Kobayashi M, Nakamura A, Miura Y, Yano Y et al (2014) The transcription repressors Bach2 and Bach1 promote B cell development by repressing the myeloid program. Nat Immunol 15:1171–1180

Itoh-Nakadai A, Matsumoto M, Kato H, Sasaki J, Uehara Y, Sato Y, Ebina-Shibuya R, Morooka M, Funayama R, Nakayama K et al (2017) A Bach2-Cebp gene regulatory network for the commitment of multipotent hematopoietic progenitors. Cell Rep 18:2401–2414

Iwai K (2019) Regulation of cellular iron metabolism: iron-dependent degradation of IRP by SCF (FBXL5) ubiquitin ligase. Free Radic Biol Med 133:64–68

Jiang L, Yin M, Wei X, Liu J, Wang X, Niu C, Kang X, Xu J, Zhou Z, Sun S et al (2015) Bach1 represses Wnt/β-catenin signaling and angiogenesis. Circ Res 117:364–375

Jiang L, Yin M, Xu J, Jia M, Sun S, Wang X, Zhang J, Meng D (2017) The transcription factor Bach1 suppresses the developmental angiogenesis of zebrafish. Oxidative Med Cell Longev 2017:2143875

Jiang L, Jia M, Wei X, Guo J, Hao S, Mei A, Zhi X, Wang X, Li Q, Jin J et al (2020) Bach1-induced suppression of angiogenesis is dependent on the BTB domain. EBioMedicine 51:102617

Kato H, Itoh-Nakadai A, Matsumoto M, Ishii Y, Watanabe-Matsui M, Ikeda M, Ebina-Shibuya R, Sato Y, Kobayashi M, Nishizawa H et al (2018) Infection perturbs Bach2- and Bach1-dependent erythroid lineage 'choice' to cause anemia. Nat Immunol 19:1059–1070

Khodaverdian V, Tapadar S, MacDonald IA, Xu Y, Ho PY, Bridges A, Rajpurohit P, Sanghani BA, Fan Y, Thangaraju M et al (2019) Deferiprone: pan-selective histone lysine demethylase inhibition activity and structure activity relationship study. Sci Rep 9:4802

Kim S, Kim DH, Jung WH, Koo JS (2013) Metabolic phenotypes in triple-negative breast cancer. Tumour Biol 34:1699–1712

Lambert SA, Jolma A, Campitelli LF, Das PK, Yin Y, Albu M, Chen X, Taipale J, Hughes TR, Weirauch MT (2018) The human transcription factors. Cell 172:650–665

Lane DJ, Mills TM, Shafie NH, Merlot AM, Saleh Moussa R, Kalinowski DS, Kovacevic Z, Richardson DR (2014) Expanding horizons in iron chelation and the treatment of cancer: role of iron in the regulation of ER stress and the epithelial-mesenchymal transition. Biochim Biophys Acta 1845:166–181

Lee J, Lee J, Farquhar KS, Yun J, Frankenberger CA, Bevilacqua E, Yeung K, Kim EJ, Balazsi G, Rosner MR (2014) Network of mutually repressive metastasis regulators can promote cell heterogeneity and metastatic transitions. Proc Natl Acad Sci U S A 111:E364–E373

Lee J, Yesilkanal AE, Wynne JP, Frankenberger C, Liu J, Yan J, Elbaz M, Rabe DC, Rustandy FD, Tiwari P et al (2019) Effective breast cancer combination therapy targeting BACH1 and mitochondrial metabolism. Nature 568:254–258

Lee DH, Kim GW, Jeon YH, Yoo J, Lee SW, Kwon SH (2020a) Advances in histone demethylase KDM4 as cancer therapeutic targets. FASEB J 34:3461–3484

Lee J, You JH, Kim MS, Roh JL (2020b) Epigenetic reprogramming of epithelial-mesenchymal transition promotes ferroptosis of head and neck cancer. Redox Biol 37:101697

Liang Y, Wu H, Lei R, Chong RA, Wei Y, Lu X, Tagkopoulos I, Kung SY, Yang Q, Hu G et al (2012) Transcriptional network analysis identifies BACH1 as a master regulator of breast cancer bone metastasis. J Biol Chem 287:33533–33544

Lignitto L, LeBoeuf SE, Homer H, Jiang S, Askenazi M, Karakousi TR, Pass HI, Bhutkar AJ, Tsirigos A, Ueberheide B et al (2019) Nrf2 activation promotes lung cancer metastasis by inhibiting the degradation of Bach1. Cell 178:316–329.e318

Mansoori B, Mohammadi A, Asadzadeh Z, Shirjang S, Minouei M, Abedi Gaballu F, Shajari N, Kazemi T, Gjerstorff MF, Duijf PHG et al (2019) HMGA2 and Bach-1 cooperate to promote breast cancer cell malignancy. J Cell Physiol 234:17714–17726

Müller S, Sindikubwabo F, Cañeque T, Lafon A, Versini A, Lombard B, Loew D, Wu TD, Ginestier C, Charafe-Jauffret E et al (2020) CD44 regulates epigenetic plasticity by mediating iron endocytosis. Nat Chem 12:929–938

Nishizawa H, Matsumoto M, Shindo T, Saigusa D, Kato H, Suzuki K, Sato M, Ishii Y, Shimokawa H, Igarashi K (2020) Ferroptosis is controlled by the coordinated transcriptional regulation of glutathione and labile iron metabolism by the transcription factor BACH1. J Biol Chem 295:69–82

Nishizawa H, Matsumoto M, Chen G, Ishii Y, Tada K, Onodera M, Kato H, Muto A, Tanaka K, Igarashi K (2021) Lipid peroxidation and the subsequent cell death transmitting from ferroptotic cells to neighboring cells. Cell Death Dis 12:332

Ou X, Gao G, Bazhabayi M, Zhang K, Liu F, Xiao X (2019) MALAT1 and BACH1 are prognostic biomarkers for triple-negative breast cancer. J Cancer Res Ther 15:1597–1602

Parry A, Rulands S, Reik W (2021) Active turnover of DNA methylation during cell fate decisions. Nat Rev Genet 22:59–66

Riegman M, Sagie L, Galed C, Levin T, Steinberg N, Dixon SJ, Wiesner U, Bradbury MS, Niethammer P, Zaritsky A et al (2020) Ferroptosis occurs through an osmotic mechanism and propagates independently of cell rupture. Nat Cell Biol 22:1042–1048

Sato M, Matsumoto M, Saiki Y, Alam M, Nishizawa H, Rokugo M, Brydun A, Yamada S, Kaneko MK, Funayama R et al (2020) BACH1 promotes pancreatic cancer metastasis by repressing epithelial genes and enhancing epithelial-mesenchymal transition. Cancer Res 80:1279–1292

Schwartz AJ, Goyert JW, Solanki S, Kerk SA, Chen B, Castillo C, Hsu PP, Do BT, Singhal R, Dame MK et al (2021) Hepcidin sequesters iron to sustain nucleotide metabolism and mitochondrial function in colorectal cancer epithelial cells. Nat Metab 3:969–982

Shajari N, Davudian S, Kazemi T, Mansoori B, Salehi S, Khaze Shahgoli V, Shanehbandi D, Mohammadi A, Duijf PHG, Baradaran B (2018) Silencing of BACH1 inhibits invasion and migration of prostate cancer cells by altering metastasis-related gene expression. Artif Cells Nanomed Biotechnol 46:1495–1504

Stockwell BR, Friedmann Angeli JP, Bayir H, Bush AI, Conrad M, Dixon SJ, Fulda S, Gascón S, Hatzios SK, Kagan VE et al (2017) Ferroptosis: a regulated cell death nexus linking metabolism, redox biology, and disease. Cell 171:273–285

Stockwell BR, Jiang X, Gu W (2020) Emerging mechanisms and disease relevance of ferroptosis. Trends Cell Biol 30:478–490

Torti SV, Manz DH, Paul BT, Blanchette-Farra N, Torti FM (2018) Iron and cancer. Annu Rev Nutr 38:97–125

Toyokuni S, Yanatori I, Kong Y, Zheng H, Motooka Y, Jiang L (2020) Ferroptosis at the crossroads of infection, aging and cancer. Cancer Sci 111:2665–2671

Tran N, Broun A, Ge K (2020) Lysine demethylase KDM6A in differentiation, development, and cancer. Mol Cell Biol 40

Vela D (2018) Iron metabolism in prostate cancer; from basic science to new therapeutic strategies. Front Oncol 8:547

Wei S, Liu J, Li X, Liu X (2020) LncRNA MIR17HG inhibits non-small cell lung cancer by upregulating miR-142-3p to downregulate Bach-1. BMC Pulm Med 20:78

Wiel C, Le Gal K, Ibrahim MX, Jahangir CA, Kashif M, Yao H, Ziegler DV, Xu X, Ghosh T, Mondal T et al (2019) BACH1 stabilization by antioxidants stimulates lung cancer metastasis. Cell 178:330–345.e322

Xhabija B, Kidder BL (2019) KDM5B is a master regulator of the H3K4-methylome in stem cells, development and cancer. Semin Cancer Biol 57:79–85

Xue X, Ramakrishnan SK, Weisz K, Triner D, Xie L, Attili D, Pant A, Győrffy B, Zhan M, Carter-Su C et al (2016) Iron uptake via DMT1 integrates cell cycle with JAK-STAT3 signaling to promote colorectal tumorigenesis. Cell Metab 24:447–461

Yahagi N, Takeuchi Y (2021) Genome-wide screening of upstream transcription factors using an expression library [version 1; peer review: awaiting peer review]. F1000Research 10:51

Ying Y, Wang Y, Huang X, Sun Y, Zhang J, Li M, Zeng J, Wang M, Xiao W, Zhong L et al (2020) Oncogenic HOXB8 is driven by MYC-regulated super-enhancer and potentiates colorectal cancer invasiveness via BACH1. Oncogene 39:1004–1017

Yun J, Frankenberger CA, Kuo WL, Boelens MC, Eves EM, Cheng N, Liang H, Li WH, Ishwaran H, Minn AJ et al (2011) Signalling pathway for RKIP and Let-7 regulates and predicts metastatic breast cancer. EMBO J 30:4500–4514

Yusoff FM, Maruhashi T, Kawano KI, Nakashima A, Chayama K, Tashiro S, Igarashi K, Higashi Y (2020) Bach1 plays an important role in angiogenesis through regulation of oxidative stress. Microvasc Res 134:104126

Zhu GD, Liu F, OuYang S, Zhou R, Jiang FN, Zhang B, Liao WJ (2018) BACH1 promotes the progression of human colorectal cancer through BACH1/CXCR4 pathway. Biochem Biophys Res Commun 499:120–127

Chapter 3
Regulation of Lipid Metabolism Under Stress and Its Role in Cancer

Rimsha Munir and Nousheen Zaidi

Abstract Within the tumor microenvironment, cancer cells are often exposed to oxygen and nutrient deficiency, leading to various changes in their lipid composition and metabolism. These alterations have important therapeutic implications as they affect the cancer cells' survival, membrane dynamics, and therapy response. This chapter provides an overview of recent insights into the regulation of lipid metabolism in cancer cells under metabolic stress. We discuss how this metabolic adaptation helps cancer cells thrive in a harsh tumor microenvironment.

Keywords Cancer · De Novo Fatty acid synthesis · Fatty acid synthase · Hypoxia · Lipid metabolism · Lipidomic profiles · Metabolic stress · Nutrient deprivation

Introduction

Cancer cells within solid tumors that reside far off the functional blood vessels are exposed to a metabolically challenging environment, where oxygen and blood supply is scarce. This poor perfusion also reduces the availability of serum-derived factors/nutrients to the cancer cells. It has been indicated that oxygen- and nutrient-deprivation induces metabolic stress, which affects metabolic profiles of cancer cells (Schug et al. 2015; Daniëls et al. 2014; Ackerman and Simon 2014).

R. Munir
Cancer Biology Lab, Institute of Microbiology & Molecular Genetics, University of the Punjab, Lahore, Pakistan

Hormone Lab Lahore, Lahore, Pakistan

N. Zaidi (✉)
Cancer Biology Lab, Institute of Microbiology & Molecular Genetics, University of the Punjab, Lahore, Pakistan

Cancer Research Centre, University of the Punjab, Lahore, Pakistan
e-mail: nosheen.mmg@pu.edu.pk

T. K. Kundu, C. Das (eds.), *Metabolism and Epigenetic Regulation: Implications in Cancer*, Subcellular Biochemistry 100,
https://doi.org/10.1007/978-3-031-07634-3_3

Deregulated lipid metabolism in cancer cells has shown to be strongly associated with different malignant processes (Beloribi-Djefaflia et al. 2016; Mullen et al. 2016; Röhrig and Schulze 2016). Rapidly proliferating cancer cells require a constant supply of fatty acids for membrane biogenesis and protein modifications. Elevated de novo fatty acid synthesis is one of the hallmarks of cancer (Menendez and Lupu 2007; Brusselmans et al. 2007; Sebti 2005; Kuemmerle et al. 2011; Zaidi et al. 2013). Fatty acid synthase (FASN)—a rate-limiting enzyme in the de novo fatty acid synthesis pathway—has been extensively shown to promote cancer progression (Röhrig and Schulze 2016). Although endogenous lipogenesis has historically been considered the principal source of fatty acids in cancer cells, lipolytic phenotypes are also widely recognized (reviewed in Zaidi et al. 2013). For example, it has been reported that in addition to the markers of lipogenesis (FASN), different cancer cells also express markers of lipolysis (lipoprotein lipase, LPL) and exogenous fatty acid uptake (CD36) (Kuemmerle et al. 2011).

Interestingly, cells transformed by oncogenic HRAS (HRAS-G12V) display elevated lipid-uptake. In contrast, cells transformed by constitutively active AKT (myristoylated AKT) show increased de novo synthesis (Kamphorst et al. 2013). These results demonstrate that genetic background plays a significant role in determining cancer cells' relative dependence on lipid-uptake vs. endogenous synthesis. However, crosstalk between tumor cells and the microenvironment may also affect cancer cells' lipid acquisition mode. Under metabolic stress, cancer cells are shown to modify balances of de novo lipid synthesis, lipid uptake, and degradation (Röhrig and Schulze 2016; Kamphorst et al. 2013). This metabolic flexibility to switch between de novo lipid synthesis and lipid-uptake/degradation could be particularly important for cancer cells exposed to temporal fluctuations in oxygen and nutrient accessibility within tumor tissues.

This section summarizes the evidence on metabolic stress-induced alterations in lipid profiles of cancer cells. We discuss the regulation of lipid metabolism under oxygen- and/or nutrient-deprivation and explore the impact of metabolic stress on cancer cells' lipidomic profiles. Finally, we compare the relevant findings from in vitro and in vivo models of human cancer.

Impact of Metabolic Stress on Lipid Metabolism in Cancer Cells

Effect of Hypoxia on Lipid Metabolism in Cancer Cells

Multiple in vitro studies have reported the impact of metabolic stress—induced by oxygen and/or nutrient deprivation—on lipid metabolism in cancer cells. The effects of hypoxia have been more extensively explored (Table 3.1). Hypoxia is considered a hallmark of many human cancers. Hypoxic stress in the tumor microenvironment results from high cell proliferation rates (that consume oxygen) and aberrant blood

Table 3.1 Effect of hypoxia on lipid metabolism in cancer cells cultivated in 2D cell culture system

Methodology		Findings				
Cell culture conditions	Cell lines studied	Effect on lipid metabolism/ associated gene expression and cells proliferation	Pathway/ function of studied gene	Effects on lipid/lipidomic profiles	Effects of silencing/ inhibition of corresponding genes	References
O_2 Concentration GasPak induced hypoxia. **Serum/Supplementations** 10% FBS **Incubation time** 48 hours	**Breast Cancer cell lines** MX1 MCF7 MDA-MB157	**Under Hypoxia** → ↑**FASN** expression → ↑ **SREBP**, **p-SREBP**, **HIF1**, **p-Akt** and expression in MCF7.	 **FA synthesis** **TFs**		**Under Hypoxia** → ⊣ **HIF1** (By YC-1) blocks the expression of FASN and SREBP. → ⊣ **Akt** (By LY294002) blocks the expression of **FASN**, **SREBP**, **p-Akt** and **HIF1**.	Furuta et al. (2008)
O_2 Concentration Hypoxia: 1% O_2 Normoxia: 20 % O_2 **Serum/Supplementations** 10% FBS **Incubation time** 48 hours.	**Liver cancer cell lines** Hep3B SK-Hep-1 HepG2 **Prostate cancer cell lines** PC3M	**Under Hypoxia** → ↑**FASN** expression → ↑**LPIN1** expression → ↓**MCAD & LCAD** expression	 **TG synthesis** **FA beta-oxidation**	**Under Hypoxia** → ↑TGs levels	**Under Hypoxia** → ⊥**HIF-1α** and **HIF-2α** results in ↓ TGs levels in Hep3B cells.	Huang et al. (2014)
O_2 Concentration Normoxia: 21% O_2 Hypoxia: 2% O_2 **Serum/Supplementations** 10% FBS **Incubation time** 12 hours.	**Liver Cancer cell lines** HepG2	**Under Hypoxia** → ↑cells apoptosis in high seeding density cultures. → ↓**FASN** expression → ↓**SREBP** expression → FASN expression is correlated with cell death.				Jung et al. (2012)

(continued)

Table 3.1 (continued)

Methodology		Findings				
Cell culture conditions	Cell lines studied	Effect on lipid metabolism/ associated gene expression and cells proliferation	Pathway/ function of studied gene	Effects on lipid/lipidomic profiles	Effects of silencing/ inhibition of corresponding genes	References
O_2 Concentration Normoxia:20% O_2 Hypoxia: <0.5% O_2 **Serum Concentration** 10% FCS **Incubation time** 24 hours.	**Brain Cancer cell lines** U87 U251	**Under Hypoxia** → ↓**ACACA, ACACB, FASN** expression → ↓**HMGCR** expression	 **FA synthesis** **Mevalonate pathway**			Lewis et al. (2015)
O_2 Concentration Hypoxia: 1% O_2 Normoxia: 21 % O_2 **Serum Concentration** 10% FBS **Incubation time** 24 hours.	**Colorectal cancer cell lines** HCT-116 HCT-116 *hif1α-/-* SW1222 DLD1	**Under Hypoxia** → ↓ **ACC1** expression → ↔**FASN** expression → ↑ **SCD-1** expression → ↑**SREBP-1** expression **Under Hypoxia (hif1α-/- cells)** → ↓ **ACAT1** expression → ↑ **SREBP-1** irrespective of O_2 concentration.	 **FA synthesis** **FA desaturation** **Cholesterol Metabolism** **TF**	**Under Hypoxia (hif1α+/+ cells)** → ↑ FAs levels → ↑ TAG levels, → ↓ Desaturation index → ↑ Glycerol → ↓glycerophosphate → ↑↑TAG levels → ↑ MAG, Cho, Pcho and PC levels		Valli et al. (2015)

O_2 Concentration Normoxia :21% O_2 Hypoxia: 1% O_2 **Serum/Supplementations** 10% DFBS 90 μMU-^{13}C-acetate **Incubation time** 48 to 72 hours.	**Breast cancer cell lines** MDA-MB-468 BT-474	**Under Hypoxia** Acetate served as the main carbon source for Acetyl-Co A synthesis.			→ ┤**ACSS2** causes cells to utilize glucose and glutamine for Acetyl-Co A synthesis instead of acetate.	Bulusu et al. (2017)
O_2 Concentration Normoxia :21% O_2 Hypoxia: 1% O_2 **Serum/Supplementations** 10% FBS 2.5mM acetate **Incubation time** 12 hours.	**Liver cancer cell lines** HepG2 PLC-8024 **Breast cancer cell lines** SkBr3 **Prostate cancer cell lines** DU 145	**Under Hypoxia + Acetate supplementation**		**Under Hypoxia + Acetate supplementation** → ↑Acetate uptake → ╨**FASN** results in ↓Acetate derived *de novo* FA synthesis.	**Under Hypoxia + Acetate supplementation** → HepG2/SkBr3 cells were more sensitive to FASN inhibitor C75	Gao et al. (2016)
		→ ↑ **ACACA** RNA expression				
		→ ↑**FASN** RNA expression				
		→ ↑**ACSS2** RNA expression.	**Acetate derived acetyl Co-A synthesis**			
		→ Acetate fuels *de novo* FA synthesis which promotes the cancer cell survival.				
O_2 Concentration Normoxia :21% O_2 Hypoxia: 0.1% O_2 **Serum/Supplementations**	**Breast cancer cell lines** MCF7 MDA-MB-231	**Under Hypoxia**		**Under Hypoxia** → ↑LDs → Treatment of DMOG activates HIF-1α and restores the LD levels. → ↑TGs	**Under Hypoxia** → ┤**FABP3**, **FABP7**, and **ADFP** results in ↓LD → ┤**HIF-1α** ↓LD	Bensaad et al. (2014)
		→ ↑**FABP3** RNA expression	**Lipoprotein Metabolism**			
		→ ↑**FABP7** RNA expression	**Lipoprotein Metabolism**			

(continued)

Table 3.1 (continued)

Methodology		Findings				
Cell culture conditions	Cell lines studied	Effect on lipid metabolism/ associated gene expression and cells proliferation	Pathway/ function of studied gene	Effects on lipid/lipidomic profiles	Effects of silencing/ inhibition of corresponding genes	References
10% FCS **Incubation time** 48 hours.	**Brain cancer cell lines** U87 T98G	→ ↑**ADFP** RNA expression	**LD synthesis**	→ ↓ TGs species with 3 double bonds in MCF7 cells. → ↑ TGs species with 3 double bonds in U87 cells three → ↓Pyruvate dependent *de novo* FA synthesis → ↑ in Palmitate, Stearate and Linoleate levels.		
O_2 Concentration Hypoxia: 1% O_2 Normoxia: 20 % O_2 **Serum/Supplementations** 10% DFBS **Incubation time** 72 hours.	**Cancer cell lines** MDA-MB-468 (Breast Cancer) HeLa (Cervical Cancer) A549 (Lung Cancer)	**Under Hypoxia** Cancer cells utilized glutamine as chief carbon source for synthesis of Acetyl-Co A.		**Under Hypoxia** → ↑ Exogenous FA levels.	**Under Hypoxia** → ↓**SCD-1** activity.	Kamphorst et al. (2013)

O_2 Concentration Hypoxia: 0.5% O_2 Normoxia: 20 % O_2 **Serum/Supplementations** 10% FBS+ 200 μM Oleate/0.2% BSA complex. **Incubation time** 24 hours.	**Renal cancer cell lines** ACHN Caki-1 **Colorectal cancer cell lines** DLD-1 HCT116	**Under Hypoxia** → ↔ **ATGL** RNA expression. → ↔ **CGI58** RNA expression → ↔ **G0S2** RNA expression.	 **Lipolysis** **TF** **TF**	**Under Hypoxia** → ↓ Intracellular lipolysis. → ↑TG levels.		Zhang et al. (2017)
O_2 Concentration Hypoxia: 1% O_2 Normoxia: 21 % O_2 Cells and media were placed in low oxygen overnight. **Serum/Supplementations** 10% DFBS **Incubation time** 48 hours.	**Cancer cell lines** MDA-MB-468 (Breast Cancer) HeLa (Cervical Cancer) A549 (Lung Cancer)	**Under Hypoxia** Cancer cells utilized glutamine and acetate for the synthesis of Acetyl-Co A.				Kamphorst et al. (2014)

(continued)

Table 3.1 (continued)

Methodology		Findings				References
Cell culture conditions	Cell lines studied	Effect on lipid metabolism/associated gene expression and cells proliferation	Pathway/function of studied gene	Effects on lipid/lipidomic profiles	Effects of silencing/inhibition of corresponding genes	
O_2 Concentration Hypoxia: 1-3% O_2 Normoxia: 20 % O_2 **Serum/Supplementations** 10% DFBS **Incubation time** 48 hours.	**Cancer cell lines** A549 (Lung Cancer) H460(Lung Cancer) MDA231 (Breast Cancer) SK-MEL-5 (Skin Cancer) A431(Skin Cancer) HCT-116 (Colorectal Cancer)	**Under Hypoxia** Cancer cells used glutamine as a precursor for lipid synthesis pathways by reductive carboxylation.				Metallo et al. (2012)

O_2 Concentration Hypoxia: 1% O_2 Normoxia: 20 % O_2 **Serum/Supple-mentations** 10% FCS **Incubation time** 48 hours.	**Cancer cell lines** Huh7(Liver Cancer) HeLaM (Cervical Cancer) **Bronchial smooth muscle cells** Hbsm	**Under Hypoxia** → ↑**LIP1N1** RNA expression → **HIF1-α** & **HIF2-α** expression is correlated with **LIP1N1** RNA expression.		**Under Hypoxia** → ↑ TGs levels that are correlated with **LIP1N1** RNA expression.	**Under Hypoxia** →┤**LIP1N1** decreases the TGs accumulation.	Mylonis et al. (2012)
O_2 Concentration Hypoxia: 1% O_2 Normoxia: 21 % O_2 **Serum/Supple-mentations** 10% FBS *For EVs isolation cells were cultured in exome-depleted FBS **Incubation time** 1-72 hours.	**Prostate Cancer cell lines** LNCaP DU145	**Under Hypoxia** → ↓Cell proliferation → ↑ **FASN** expression → ↑**ACC** expression → ↑**ACLY** expression → ↑**SCD-1** expression **After Reoxygenation** → ↑Cell proliferation		**Under Hypoxia** → ↑%age of saturated FA in EV-PL as compared to parental cells. → ↑ Saturation index of membrane PL → ↑ TGs levels in EVs → ↔TGs levels in cells → ↓ Stearic acid and Oleic acid in EV-TGs as compared to parental cells. → ↑ Stearic, Palmitic and linoleic acid in EV. → ↑Myristic, Palmitic, stearic, linoleic and arachidonic acid in cells. **After Reoxygenation** → ↓Lipid accumulation		Schlaepfer et al. (2015)

(continued)

Table 3.1 (continued)

Methodology		Findings				
Cell culture conditions	Cell lines studied	Effect on lipid metabolism/ associated gene expression and cells proliferation	Pathway/ function of studied gene	Effects on lipid/lipidomic profiles	Effects of silencing/ inhibition of corresponding genes	References
O_2 Concentration Hypoxia was induced by adding 200 μM cobalt chloride **Serum/Supplementations** 10% FBS *For EVs isolation cells were cultured in exome-depleted FBS **Incubation time** 3 hours *Cells were serum starved before incubation under hypoxic	**Cervical Cancer cell lines**			**Under Hypoxia** → Principal component analysis reveal clear-cut separation between lipidomic profiles of hypoxic and normoxic cells. → ↔PL with ≥3 double bonds in acyl chains → ↔PL with <3 double bonds → ↓Mono/di-unsaturated PC → ↑Polyunsaturated PC → ↑PL with polyunsaturated acyl chains → ↓PL with mono/di-unsaturated acyl chains		Yu et al. (2014)

Key abbreviations: **ACACA**, acetyl-CoA carboxylase 1; **ACACB,** acetyl-CoA carboxylase 2; **ACSS2,** acyl Co-A synthetase-2; **ADFP,** adipose differentiation protein; **ATGL,** adipose triglyceride lipase; **Cho,** choline; **CS-FBS,** charcoal striped fetal bovine serum; **DFBS** , dialyzed fetal bovine serum; **DMOG,** dimethyloxalylglycine; **FA,** fatty acids; **FABP3,** fatty acid binding protein 3; **FABP7,** fatty acid binding protein 7; **FFA,** free fatty acids; **FASN,** fatty acid synthase; **FBS,** fetal bovine serum; **FCS,** fetal calf serum; **HIF-1α,** hypoxia-inducible factor 1-alpha; **HIF-2α,** hypoxia-inducible factor 2α; **HMGCR,** 3-hydroxy-3-methylglutaryl-CoA reductase; **LD,** lipid droplet; **LCAD,** long-Chain Specific acyl-CoA dehydrogenase; **MAG,** monoacylglycerol ; **MCAD,** Medium-chain acyl-CoA dehydrogenase; **MUFA,** monounsaturated fatty acids; **PE,** phosphatidylethanolamines; **PC,** Phosphatidylcholines; **Pcho,** propargyl-choline; **PI,** phosphatidylinositol; **PS,** phosphatidylserine; **PUFA,** polyunsaturated fatty acids; **SCD,** stearoyl-CoA Desaturase; **SFA,** saturated fatty acids; **SMEM,** spinner minimum essential medium; **SREBP,** sterol regulatory element-binding proteins; **TG,** triglycerides

Symbols: ↓, down regulated; ↑, up regulated; ↑↑, strongly up regulated; ↔, no effect; ×, low effect; ┤, gene knock down by siRNA; ╨, gene knock out; ╣, pharmacological gene inhibition

vessel development. Hypoxia promotes aggressive malignancy and is associated with poor prognosis in a wide range of cancer types (Hanahan and Weinberg 2011; Masson and Ratcliffe 2014; Rankin and Giaccia 2016). Oxygen-deprivation has been shown to induce hypoxia-inducible factors (HIFs) that mediate multiple protective mechanisms. These mechanisms help maintain oxygen homeostasis by reducing oxidative metabolism and oxygen consumption (Gordan and Simon 2007; Papandreou et al. 2006; Rankin and Giaccia 2008; Semenza 2010). HIF-1α is shown to be a key regulator of multiple cancer metabolism pathways, including glycolysis, glycogenesis, Tricarboxylic acid cycle (TCA), nucleotides/amino acids/leptin metabolism, lipid metabolism (Denko 2008; Harris 2002; Semenza 2003, 2010).

Most of the works on hypoxia-induced effects on lipid metabolism in cancer cells studied the expression of FASN (Table 3.1 and Fig. 3.1) (reviewed in Munir et al. 2019). To date, the effect of hypoxia on FASN expression appears to be inconsistent. For example, under hypoxia, the expression of FASN is reported to be significantly up-regulated in breast cancer cells (Furuta et al. 2008), down-regulated in HepG2 liver cancer cells (Jung et al. 2012), and unaffected in HCT116 colorectal cancer cells (Valli et al. 2015). Expression of multiple other genes involved in de novo lipid synthesis is also differentially regulated in various cancer cell lines (Table 3.1 and Fig. 3.1). Sterol regulatory-element binding protein (SREBP)-1 is the major transcriptional regulator of most enzymes involved in de novo lipid synthesis, including FASN. Furuta. E et al. (Furuta et al. 2008) reported that hypoxia-induced up-regulation in FASN expression in breast cancer cell lines is due to the activation of the Akt and HIF1 followed by up-regulation of SREBP-1. Conversely, the studies that reported a hypoxia-induced decrease in FASN expression showed that under such conditions SREBP-1 expression is also decreased or unaffected (Jung et al. 2012; Lewis et al. 2015).

These observations indicate that the hypoxic regulation of de novo lipid synthesis is cell-type specific, as the major difference among these contradictory reports was in the selection of cell line models. More meticulous scrutiny of the literature showed that these studies also applied different cell culture conditions. For example, some studies applied chemical induction of hypoxia (e.g., $CoCl_2$ method) (Yu et al. 2014). In contrast, others induced hypoxia by regulating O_2 levels in the incubators (Schug et al. 2015). Moreover, differences in severity (or O_2 Concentration) and/or duration of hypoxia were also noted (Table 3.1). These differences might also have caused inconsistencies in the literature. A few studies also used additional media supplementations that may have also caused altered lipid profiles in corresponding cancer cells. For instance, under hypoxic conditions acetate supplementation was shown to induce increased expression of FASN and acyl-CoA synthetase short-chain family member 2 (ACSS2)—a cytosolic enzyme that produces acetyl-CoA using acetate as a substrate (Gao et al. 2016). Under hypoxic conditions, acetate was shown to serve as the primary source of acetyl-CoA synthesis (Bulusu et al. 2017; Gao et al. 2016) that fuels de novo lipogenesis and promotes cell survival (Gao et al. 2016). However, this observation is not limited to the studies that used acetate supplementations. Multiple other works that cultivated the cells under regular complete growth-media

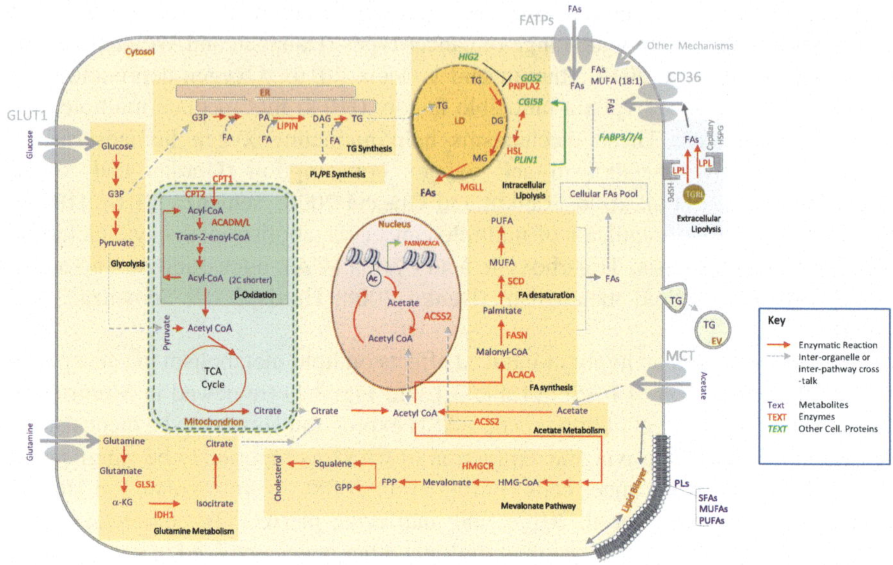

Fig. 3.1 Overview of the major lipid metabolism pathways shown to be affected by metabolic stress (Munir et al. 2019). The figure highlights all the key lipid metabolism pathways activated in a cancer cells. Major pathways are shown as boxes without outlines. The systematic names of these pathways are given at the *bottom-right corner* of each box. Abbreviations: *ACACA* acetyl-CoA carboxylase 1, *ACACB* acetyl-CoA carboxylase 2, *ACSS2* acyl Co-A synthetase-2, *ADFP* adipose differentiation protein, *ATGL* adipose triglyceride lipase, *FA* fatty acids, *FABP3* fatty acid binding protein 3, *FABP7* fatty acid binding protein 7, *FFA* free fatty acids, *FASN* fatty acid synthase, *H* hypoxia, *HIF-1α* hypoxia-inducible factor 1-alpha, *HIF-2α* hypoxia-inducible factor 2α, *HIG2* hypoxia-inducible gene 2 protein, *HMGCR* 3-hydroxy-3-methylglutaryl-CoA reductase, *LD* lipid droplet, *LCAD* long-chain specific acyl-CoA dehydrogenase, *MAG* monoacylglycerol, *MCAD* medium-chain acyl-CoA dehydrogenase, *MUFA* monounsaturated fatty acids, *PBMCs* peripheral blood mononuclear cells, *PC* phosphatidylcholines, *PE* phosphatidylethanolamines, *Pcho* propargyl-choline, *PI* phosphatidylinositol, *PS* phosphatidylserine, *PUFA* polyunsaturated fatty acids, *SCD* stearoyl-CoA desaturase, *SFA* saturated fatty acids, *SREBP* sterol regulatory element-binding proteins, *TG* triglycerides

have also shown that hypoxic cancer cells do utilize glutamine or acetate as a chief carbon source for the synthesis of acetyl-Co A (Kamphorst et al. 2013, 2014; Metallo et al. 2012).

It has also been argued that hypoxia prevents the use of glucose for fatty acid synthesis, therefore under hypoxia, cancer cells are more dependent on lipid-uptake than on de novo synthesis (Röhrig and Schulze 2016). Studies indicate that under hypoxia, cancer cells switch from lipid synthesis to uptake (Kamphorst et al. 2013; Bensaad et al. 2014), particularly of lipid species containing monounsaturated acyl chains (Kamphorst et al. 2013). Another study reported that hypoxia increases lipid uptake in breast cancer and glioblastoma cells by inducing the expression of fatty acid-binding protein 3 (FABP3) and FABP7 (Bensaad et al. 2014)—that are involved in the uptake and subcellular trafficking of fatty acids. It has also been recognized that cancer cells increase their HIF-1-mediated triglyceride/lipid-droplet (LD) accumulation in response to oxygen deprivation (Bensaad et al. 2014). It was shown that hypoxia facilitates lipids storage in lipid droplets through induction of perilipin 2 (PLIN2) (Bensaad et al. 2014). Upon reoxygenation, cells used these stored lipids for energy production and antioxidant defense (Bensaad et al. 2014). This increased the ability of the cells to survive the surge in oxidative stress associated with a sudden increase in oxygen availability. In addition to the hypoxia-induced increase in fatty acid uptake, cancer cells are shown to down-regulate intracellular lipolysis (Zhang et al. 2017) that may also contribute to LD accumulation. In various human colorectal cancer and renal cell carcinoma cell lines, the baseline intracellular lipolysis is mediated by adipose triglyceride lipase (ATGL) (Zhang et al. 2017) that is significantly reduced under hypoxia (Zhang et al. 2017). However, the expression of ATGL remains unaffected; instead, ATGL was shown to be inhibited by hypoxia-inducible gene 2 (HIG2), a HIF-1 target (Zhang et al. 2017).

Effect of Nutrient and Lipid Deprivation on Lipid Metabolism in Cancer Cells

Poor blood perfusion in tumors also reduces the availability of serum-derived factors. A low-serum environment also affects lipid metabolism in cancer cells (Fig. 3.1). Under such conditions, cancer cells were shown to be more reliant on de novo lipid synthesis (Schug et al. 2015; Peck et al. 2016). Expression of Stearoyl-CoA desaturase (SCD1)—an enzyme that introduces a double bond into certain saturated fatty acyl-CoAs producing monounsaturated fatty acids (MUFA)—is also shown to be up-regulated under conditions of reduced availability of serum-derived factors (Peck et al. 2016). Depletion of SCD resulted in a substantial reduction in cell number when cancer cells were grown in low serum while having little or no effect on non-cancer cell lines (Peck et al. 2016).

This increased dependence of cancer cells on de novo lipid synthesis (DNL) under a low-serum environment is most likely caused by diminished lipid availability in such conditions. Tumor cells are speculated to be exposed to lipid gradients, similar to those described for glucose and oxygen (Gatenby and Gillies 2004). A very limited number of works have reported the impact specifically of lipid-deprivation on lipid metabolism in cancer cells (Table 3.2). For that, cancer cells were cultivated under media containing sera that were specifically depleted of lipid or lipoproteins. It has been shown that cancer cells differentially activate and thrive on de novo lipid synthesis pathways in a low-lipid environment (Daniëls et al. 2014). Both lipogenic and non-lipogenic cancer cells were shown to raise their lipogenic activity in these conditions, albeit to a different degree. Cells that attained the highest lipogenic activity under these conditions were best able to cope with lipid reduction in terms of proliferative capacity. Supplementation of the medium with very low-density lipoproteins, free fatty acids, and cholesterol reversed this activation,

Table 3.2 Effect of low serum culture conditions on lipid metabolism in cancer cells cultivated in 2D cell culture system

Methodology		Findings				
Cell culture conditions	Cell lines studied	Effect on lipid metabolism/associated gene expression and cells proliferation	Pathway/ Function of studied gene	Effects on lipid/lipidomic profiles	Effects of silencing /inhibition of various lipid metabolism associated enzymes	References
Serum Concentration 1% vs. 10% FCS **Incubation time** 24 hours	**Breast cancer cell lines** BT20 BT549 MDA-MB-231 MDA-MB-468 MCF7 T47D SKBr3 **Breast epithelial cell line** MCF10 A				**Under LS conditions** → Cancer cells displayed increased sensitivity to **FASN** inhibitors (C75 & AZ22) in terms of proliferation rates.	Schug et al. (2015)
Serum Concentration 1% vs. 10% FCS **Incubation time** 72 hours	**Breast cancer cell lines** MDA-MB-468 MDA-MB-231 BT549 BT20	**Under LS conditions** → ↑ *De novo* fatty acid synthesis		**Under LS conditions** → Membrane phosphoglycerides are mainly composed of mono-unsaturated acyl chains. → ↑MUFA levels in DU145 cells.	**Under LS conditions** → ┤**SCD** results in ↓cell number → ┤**SCD** results in ↑MUFA & PUFAs levels and ↓SFAs levels. However, oleic acid supplementation restores the relative proportions of FA classes. → ┤**SCD** reduces the mono-	Peck et al. (2016)

	SKBr3 T47D MCF7 **Breast epithelial cell line** MCF10 A **Prostate cancer cell lines** LNCAP PC3 DU145 **Prostate epithelial cell lines** RWPE1				unsaturated acyl chains in PE, PI and PS. Supplementation of oleic acid restores the concentrations of mono-unsaturated species in cancer cells.	

(continued)

Table 3.2 (continued)

Methodology		Findings				
Cell culture conditions	Cell lines studied	Effect on lipid metabolism/associated gene expression and cells proliferation	Pathway/ Function of studied gene	Effects on lipid/lipidomic profiles	Effects of silencing /inhibition of various lipid metabolism associated enzymes	References
Serum Concentration 2% vs. 10% FBS **Incubation time** 48 hours	**Cancer Cell lines** A549 (Lung Cancer) FaDu (Pharynx Cancer) H1299 (Lung Cancer)	**Under LS conditions** → ↑ **SCD-1** RNA Expression → ↔**SCD-5** RNA Expression	 **FA Desaturation** **FA Desaturation**		**Under LS conditions** → ┤ **SCD-1** induced apoptosis → ┤ **SCD-5** inhibition did not induce apoptosis. → ╣**SCD-1** (with A939572) reduced the proliferation of cell. Supplementation of media with oleic acid reversed the effect of A939572.	Roongta et al. (2011)

Key abbreviations: **ACACA**, acetyl-CoA carboxylase 1; **ACACB,** acetyl-CoA carboxylase 2; **ACSS2,** acyl Co-A synthetase-2; **ADFP,** adipose differentiation protein; **ATGL,** adipose triglyceride lipase; **Cho,** choline; **CS-FBS,** charcoal striped fetal bovine serum; **DFBS** , dialyzed fetal bovine serum; **DMOG,** dimethyloxalylglycine; **FA,** fatty acids; **FABP3,** fatty acid binding protein 3; **FABP7,** fatty acid binding protein 7; **FFA,** free fatty acids; **FASN,** fatty acid synthase; **FBS,** fetal bovine serum; **FCS,** fetal calf serum; **HIF-1α,** hypoxia-inducible factor 1-alpha; **HIF-2α,** hypoxia-inducible factor 2α; **HMGCR,** 3-hydroxy-3-methylglutaryl-CoA reductase; **LD,** lipid droplet; **LCAD,** long-Chain Specific acyl-CoA dehydrogenase; **MAG,** monoacylglycerol ; **MCAD,** Medium-chain acyl-CoA dehydrogenase; **MUFA,** monounsaturated fatty acids; **PE,** phosphatidylethanolamines; **PC,** Phosphatidylcholines; **Pcho,** propargyl-choline; **PI,** phosphatidylinositol; **PS,** phosphatidylserine; **PUFA,** polyunsaturated fatty acids; **SCD,** stearoyl-CoA Desaturase; **SFA,** saturated fatty acids; **SMEM,** spinner minimum essential medium; **SREBP,** sterol regulatory element-binding proteins; **TG,** triglycerides

Symbols: ↓, down regulated; ↑, up regulated; ↑↑, strongly up regulated; ↔, no effect; ×, low effect; ┤, gene knock down by siRNA; ╨, gene knock out; ╣, pharmacological gene inhibition

indicating that the mere lack of lipids is sufficient to activate de novo lipogenesis in cancer cells. Consequently, cancer cells grown in lipid-reduced conditions became more dependent on de novo lipid synthesis pathways and are more sensitive to inhibitors of lipogenic pathways, such as Soraphen A and Simvastatin (Daniëls et al. 2014). Increased expression of ACSS2 is also observed in the low-lipid environment (Schug et al. 2015; Daniëls et al. 2014).

Combinatorial Effect of Hypoxia and Nutrient Deprivation on Lipid Metabolism in Cancer Cells

All the aforementioned data focuses on the impact of either oxygen- or nutrient-deprivation on lipid metabolism in cancer cells. Few studies have demonstrated the effect of hypoxia in combination with nutrient-deprivation (Table 3.3). It has been speculated that these cell culture conditions more closely mimic the in vivo conditions. Schug et al. studied in detail the synergistic effects of hypoxia and nutrient-deprivation on lipid metabolism in cancer cells (Schug et al. 2015). They used custom-formulated serum-like tissue culture medium (SMEM) containing concentrations of metabolites more physiologically relevant to human plasma than what is typically found in standard growth mediums. They reported that under low-oxygen and lipid-depleted conditions cancer cells display up-regulated de novo fatty acid synthesis. It was also shown that acetate provides major carbon source for de novo fatty acid synthesis in an ACSS2-dependent manner. These results signify a critical role for acetate consumption in the production of lipid biomass within the harsh tumor microenvironment. As mentioned above hypoxic cancer cells display increase in lipid-uptake however it might not be a very efficient mode of lipid acquisition if the exogenous lipids are also short in supply. Under such conditions, cancer cells may switch back to de novo synthesis but now fully depend on glutamine or acetate as alternative substrates (Röhrig and Schulze 2016). A previous study has shown multiple cancer cell lines that display decreased expression of genes involves in de novo lipogenesis under hypoxia, whereas when hypoxia was induced in combination with low serum the expression of these genes was increased (Lewis et al. 2015).

Impact of Metabolic Stress on Lipidomic Profiles in Cancer Cells

As discussed above certain types of cancer cells cultivated under metabolic stress display elevated de novo FA synthesis (Furuta et al. 2008; Gao et al. 2016; Peck et al. 2016; Huang et al. 2014), while some types of cancer cells display increased lipid-uptake (Kamphorst et al. 2013; Bensaad et al. 2014). One of the studies showed that cancer cells specifically scavenge lipid species containing monounsaturated acyl

Table 3.3 Effect of hypoxia+nutrient deprivation on lipid metabolism in cancer cells cultivated in 2D cell culture system

Methodology		Findings				
Cell culture conditions	**Cell lines studied**	**Effect on lipid metabolism/ associated gene expression and cells proliferation**	**Pathway/Function of studied gene**	**Effects on lipid/lipidomic profiles**	**Effects of silencing /inhibition of various lipid metabolism associated enzymes**	**References**
O_2 Concentration Hypoxia: 0.1% O_2 Normoxia: 21% O_2 **Serum/Supplementations** SMEM media. **Incubation time** 72 hours	**Breast Cancer cell lines** MDA-MB-468 **Prostate Cancer cell lines** DU145	**Under hypoxia+SMEM** → Cancer cells primarily depend upon the *de novo* fatty acid synthesis. → ↓Glucose utilization by TCA cycle.		↑PC and PE with shorter FA acyl chains.		(Schug et al. 2015)
O_2 Concentration Hypoxia: 0.1% O_2 Normoxia: 21% O_2 **Serum Concentration** 10% FCS 1% FCS **Incubation time** 72 hours	**Breast Cancer cell lines** BT474 BT474c1 **Prostate Cancer cell lines** DU145	**Under hypoxia+LS** → Cancer cells used acetate as chief carbon source. → ↑ ACSS2 expression both at mRNA and protein level.	Acetate derived acetyl Co-A synthesis	Not Evaluated	**Under Hypoxia+LS** Cancer cells more sensitive to ⊣ ACSS2.	(Schug et al. 2015)

Key: Abbreviations: *ACACA* acetyl-CoA carboxylase 1, *ACACB* acetyl-CoA carboxylase 2, *ACSS2* acyl Co-A synthetase-2, *ADFP* adipose differentiation protein, *ATGL* adipose triglyceride lipase, *Cho* choline, *CS-FBS* charcoal striped fetal bovine serum, *DFBS* dialyzed fetal bovine serum, *DMOG* dimethyloxalylglycine, *FA* fatty acids, *FABP3* fatty acid binding protein 3, *FABP7* fatty acid binding protein 7, *FFA* free fatty acids, *FASN* fatty acid synthase, *FBS* fetal bovine serum, *FCS* fetal calf serum, *HIF-1α* hypoxia-inducible factor 1-alpha, *HIF-2α* hypoxia-inducible factor 2α, *HMGCR* 3-hydroxy-3-methylglutaryl-CoA reductase, *LD* lipid droplet, *LCAD* long-chain specific acyl-CoA dehydrogenase, *MAG* monoacylglycerol, *MCAD* medium-chain acyl-CoA dehydrogenase, *MUFA* monounsaturated fatty acids, *PE* phosphatidylethanolamines, *PC* phosphatidylcholines, *Pcho* propargyl-choline, *PI* phosphatidylinositol, *PS* phosphatidylserine, *PUFA* polyunsaturated fatty acids, *SCD* stearoyl-CoA desaturase, *SFA* saturated fatty acids, *SMEM* spinner minimum essential medium, *SREBP* sterol regulatory element-binding proteins, *TG* triglycerides

Symbols: ↓, down regulated, ↑, up regulated, ↑↑, strongly up regulated, ↔, no effect, ×, low effect, ⊣, gene knock down by siRNA, ⫫, gene knock out, ⫣, pharmacological gene inhibition

Bensaad et al. (2014), Bulusu et al. (2017), Lewis et al. (2015), Schug et al. (2015)

chains (Kamphorst et al. 2013). This lipolytic vs. lipogenic acquisition of fatty acids may significantly affect the lipidomic profiles of cancer cells. The mammalian cells have a limited ability to synthesize polyunsaturated fatty acids de novo, as they lack the Δ12 desaturase. Therefore, enhanced de novo lipogenesis enriches the cancer cell membranes with saturated and/or monounsaturated fatty acids (Rysman et al. 2010). As these fatty acids are less prone to lipid peroxidation than polyunsaturated acyl chains, de novo fatty acid synthesis was proposed to make cancer cells more resistant to oxidative stress-induced cell death (Rysman et al. 2010). Moreover, as saturated lipids pack more densely their increased levels alter lateral and transverse membrane dynamics that may limit the uptake of drugs, making the cancer cells more resistant to therapy (Rysman et al. 2010). Hence, the balance between lipogenic and lipolytic acquisition of fatty acids may have important therapeutic implications.

There are multiple works that have reported impact of metabolic stress on lipid-load in cancer cells (Table 3.3 and Fig. 3.1). Cultured cells are shown to generate and increase the number of lipid droplets (LDs) in two different situations from a physiological perspective: when maintained in medium containing excess free fatty acids and/or lipoproteins or under nutrient- or oxygen-deprivation. Some of the previous works on stress-induced accumulation of LDs utilized complete nutrient-deprivation (Cabodevilla et al. 2013). This severe stress leading to cell death induces the formation of LDs (Hapala et al. 2011; Murphy and Vance 1999) and this LD accumulation was considered to be a hallmark of apoptosis (Boren and Brindle 2012).

Only few previous reports have studied the effects of metabolic stress on lipidomic profiles of cancer cells (Valli et al. 2015; Bensaad et al. 2014; Schlaepfer et al. 2015; Yu et al. 2014). Moreover, these studies had a limited scope and were only focused on stress-induced changes in phospholipid (Yu et al. 2014) or triglyceride profiles (Bensaad et al. 2014) of cancer cells. For instance, it was shown that breast cancer cells cultivated under oxygen- and nutrient-deprivation display modified phosphatidylcholine (PC) and phosphatidylethanolamine (PE) profiles mainly characterized by the presence of shorter and more saturated acyl chains (Schug et al. 2015)—indicative of enhanced de novo lipogenesis. Another study indicated that hypoxic cancer cells (Hela cells) display decreases in mono- and di-unsaturated phospholipid (PL) species and increase in polyunsaturated PL species (Yu et al. 2014).

The triglyceride (TG) profiles of cancer cells are also shown to be affected under hypoxic conditions. Schlaepfer et al. (2015) studied the effects of hypoxia on total TG-content and TG-subspecies-profiles in LNCaP prostate cancer cells and extracellular vesicles (EVs) released by these cells. They reported that total TG-content in hypoxic cells was only marginally increased, whereas TG-content in EVs from these cells was significantly elevated. Interestingly, TG subspecies profile was significantly altered in cells as well as in EVs. In cells, myristic, palmitic and stearic fatty acids (all saturated fatty acids) were significantly increased under hypoxia. Of note, essential linoleic acid and its derivative arachidonic acid—that are strongly associated with inflammation and tumor growth—were both significantly increased in TGs of hypoxic LNCaP cells. Significant increases in palmitic, stearic and linoleic acids were also observed in EVs released from these cells. Another study reported that under hypoxic conditions cellular levels of triglycerides with three double bonds significantly decreased in MCF7 breast cancer cells (Bensaad et al. 2014), but significantly increased in U87 glioblastoma cells (Bensaad et al. 2014).

The studies on the impact of metabolic stress on lipidomic profiles of cancer cells also used diverse cell line models and experimental methodologies. In additions to that these studies applied differential data normalization strategies. These differences might also cause inconsistencies in the data from various reports. It has been suggested that averaging the lipidomics data for broader lipid subgroups/categories increases the robustness of analyses and provides a useful measure against noise in the observed data. Interestingly, previous data indicate that hypoxia-induced changes in PL profiles of cancer cells are mainly evident when the effects of individual PL species are compared. However, when the average values of phospholipid species with acyl chains containing $\geq$3 double bonds were compared with that containing $<$3 double bonds no significant difference was observed between cells under normoxia and hypoxia. Another study, which reported hypoxia-induced shift in the PC profiles of MDA-MB breast cancer cells, also analyzed individual PC sub-species and presented the data. This literature review indicates that more systematically designed studies are required to understand the interplay between tumor microenvironment and lipidomic profiles of cancer cells.

Lipid/Lipidomic Profiles in Tumor Tissues and Tumor Spheroids

The data discussed above is mainly gathered from the studies that used 2D cell culture system. There is a growing consensus among the cancer biologists that cells growing in 2D cell culture inadequately reflect tumor growth because they lack features of the three-dimensional (3D) microenvironment that profoundly affects cell signaling, proliferation, differentiation, migration in vivo (Daster et al. 2017; Desoize and Jardillier 2000; Sutherland 1988; Yamada and Cukierman 2007). Multicellular tumor spheroid (MCTS) or 3D cell culture model is regarded as a more stringent and representative model of in vivo tumor growth. For studying tumor metabolism or metabolic stress this system is particularly important because like in vivo tumors, MCTs also develop hypoxic cores. Hence, MCTs are ideal models for studying effects of varying oxygen levels on tumor metabolism. However, only few studies have used this model and tumor tissues for studying lipid metabolism in cancer cells (Tables 3.4 and 3.5). Bensaad et al. compared lipid metabolism in cells at the periphery and cells in the hypoxic core of the tumor spheroid (Bensaad et al. 2014). They also studied the impact of hypoxia on lipid metabolism in 2D cell culture system. It observed that in both 2D and 3D systems hypoxic cells (U87, brain cancer cells) display significantly elevated expression of

Table 3.4 Effect of metabolic stress conditions on lipid metabolism in cancer cells cultivated in 3D cell culture

Origin of Tumor Spheroids	Specific cell culture conditions/Methodology	Findings				References
		Effects on MCTS growth or expression of lipid metabolism associated genes	Pathway/Function of studied gene	Effects on lipid/lipidomic profiles	Effects of silencing /inhibition of various lipid metabolism associated enzymes	
Breast cancer cell lines MDA-MB-468 **Prostate cancer cell lines** PC3 DU145	**3D cell culture** 1% vs. 10% Serum			Not Evaluated	**Under LS conditions** → ⊣ FASN (by AZ22) was more effective against MCTS growth.	(Schug et al. 2015)
Breast cancer cell lines T47D MDA-MB-468 **Prostate cancer cell lines** DU145	**3D cell culture** 10% serum Comparison between 2D(10% FCS) vs 3D(10% FCS)			→ ↑poly-unsaturated DG in MCTs of T47D cells → ↑ Mono-unsaturated PI and LPA.	→ ⊣ SCD results in ↓spheroid size. → ⊣ SCD results in ↓ MUFAs & PUFAs in MCTs of DU145 cells.	(Peck et al. 2016)
Liver cancer cell lines HepG2	**3D cell culture** 10% complete serum vs. 10% LR serum	**Under low-lipid conditions** → ↓Growth of tumor spheroids.		→ Not Evaluated		(Daniëls et al. 2014)
Brain cancer cell lines U87 cells	**3D cell culture** 10% serum	→ ↑ **FABP3, FABP7 & ADFP** protein expression in hypoxic core of tumor spheroids.	Lipoprotein Metabolism	→ ↑ LD levels in hypoxic core of tumor spheroids.		(Bensaad et al. 2014)
Brain cancer cell lines U87 cells	**3D cell culture** 10% serum			Not Evaluated	→ ⊣ SREBP1 results in ↓spheroid diameter.	(Lewis et al. 2015)

(continued)

Table 3.4 (continued)

Key: Abbreviations: *ACACA* acetyl-CoA carboxylase 1, *ACACB* acetyl-CoA carboxylase 2, *ACSS2* acyl Co-A synthetase-2, *ADFP* adipose differentiation protein, *ATGL* adipose triglyceride lipase, *Cho* choline, *CS-FBS* charcoal striped fetal bovine serum, *DFBS* dialyzed fetal bovine serum, *DMOG* dimethyloxalylglycine, *FA* fatty acids, *FABP3* fatty acid binding protein 3, *FABP7* fatty acid binding protein 7, *FFA* free fatty acids, *FASN* fatty acid synthase, *FBS* fetal bovine serum, *FCS* fetal calf serum, *HIF-1α* hypoxia-inducible factor 1-alpha, *HIF-2α* hypoxia-inducible factor 2α, *HMGCR* 3-hydroxy-3-methylglutaryl-CoA reductase, *LD* lipid droplet, *LCAD* long-chain specific acyl-CoA dehydrogenase, *MAG* monoacylglycerol, *MCAD* medium-chain acyl-CoA dehydrogenase, *MUFA* monounsaturated fatty acids, *PE* phosphatidylethanolamines, *PC* phosphatidylcholines, *Pcho* propargyl-choline, *PI* phosphatidylinositol, *PS* phosphatidylserine, *PUFA* polyunsaturated fatty acids, *SCD* stearoyl-CoA desaturase, *SFA* saturated fatty acids, *SMEM* spinner minimum essential medium, *SREBP* sterol regulatory element-binding proteins, *TG* triglycerides

Symbols: ↓, down regulated, ↑, up regulated, ↑↑, strongly up regulated, ↔, no effect, ×, low effect, ⊣, gene knock down by siRNA, ╨, gene knock out, ╡, pharmacological gene inhibition

Bensaad et al. (2014), Daniëls et al. (2014), Lewis et al. (2015), Peck et al. (2016), Schug et al. (2015)

Table 3.5 Effect of metabolic stress on lipid metabolism of tumors in vivo

Cancer Type/Xenograft Model	Applied Metabolic Stress condition	Findings		References
		Effect on gene expression/ tumor growth	Effect on lipid metabolism	
<u>Breast cancer xenograft</u> BT474 BT474 c 1 DU145 MDA-MB-468		⊣ FASN (AZ62) inhibits the tumor growth of BT474 xenografts. → ↑ ACSS2 expression in hypoxic regions of tumors. → ⊣ ACSS2 inhibits the tumor growth.	↑FAs precursor acetyl-Co A & PC precursor CDP choline in of BT474 xenografts.	(Schug et al. 2015)
Genetically engineered mouse model of breast cancer MMTV-PyMT		→ ↑ ACSS2 expression in hypoxic regions of tumors.		(Bulusu et al. 2017)
<u>Orthotopic Breast Cancer Xenograft</u> BT474 <u>Orthotopic Prostate Cancer Xenograft</u> DU145		→ ⊥SCD causes reduced growth of DU145 tumors.	→ ↑Unsaturated DG and PC species in BT474 tumors.	(Peck et al. 2016)
GA16 primary xenograft derived from gastric cancer patient		→ ⊣ SCD-1(A939572) inhibits the tumor growth.		(Roongta et al. 2011)
<u>Prostate Cancer Xenograft</u> PC-3		→ ⊣ ACAT (Avasimibe &Sandoz 58-035) inhibits the growth of PC-3 tumors. → ⊣ DGAT (A922500) ↔growth of PC-3 tumors. → ↓LDLR in avasimibe treated tumors.	→ ↓CE in avasimibe treated tumors.	(Yue et al. 2014)

(continued)

Table 3.5 (continued)

Hif1a+/_ mice (HET *Hif1a*+/+ mice(WT)	5 days of intermittent hypoxia (IH) VS. intermittent air control (IA)	**Under IH conditions** → ↑SREBP-1, SCAP and SCD -1expression. → ↔SREBP-2 expression.	**Under IH conditions** → ↑serum total cholesterol, HDLC, PL, TG and FFA levels both in HET and WT mice. → ↔ Serum LDLC levels. → ↔Liver cholesterol and PL levels.	(Li et al. 2006)
Brain cancer xenografts U87		→ ╨FABP3 and FABP7 inhibit the growth of tumors.	→ ╨FABP3 and FABP7 inhibit LD accumulation in tumors.	(Bensaad et al. 2014)
Liver cancer xenografts Hep3B		→ ╨LCAD enhanced the tumor growth. → ╨MCAD ↔ tumor growth.	→ ╨ LCAD and MCAD induced lipid accumulation.	(Huang et al. 2014)
iBMK cells with myrAkt iBMK cells with H-Ras^{V12G}		→ ⊣ SCD-1(CAY10566) effects the growth of Akt driven tumors.		(Kamphors t et al. 2013)
Colorectal cancer xenografts ATGL KO, HIG2 KO, and HIG2/ATGL dKO HCT116		→ ╨HIG2 delayed tumor growth.	→ ↑Neutral lipid accumulation in HIG 2 KO tumors. → ↔ Lipid accumulation in ATGL KO tumors.	(Zhang et al. 2017)
Mice carrying HT1080 flank tumors		→ ╨ACSS2 and HIF-2α inhibits tumor growth. → ╨HIF-1α↔ tumor growth.		(Chen et al. 2015)

Key: Abbreviations: *ACACA* acetyl-CoA carboxylase 1, *ACACB* acetyl-CoA carboxylase 2, *ACSS2* acyl Co-A synthetase-2, *ADFP* adipose differentiation protein, *ATGL* adipose triglyceride lipase, *Cho* choline, *CS-FBS* charcoal striped fetal bovine serum, *DFBS* dialyzed fetal bovine serum, *DMOG* dimethyloxalylglycine, *FA* fatty acids, *FABP3* fatty acid binding protein 3, *FABP7* fatty acid binding protein 7, *FFA* free fatty acids, *FASN* fatty acid synthase, *FBS* fetal bovine serum, *FCS* fetal calf serum, *HIF-1α* hypoxia-inducible factor 1-alpha, *HIF-2α* hypoxia-inducible factor 2α, *HMGCR* 3-hydroxy-3-methylglutaryl-CoA reductase, *LD* lipid droplet, *LCAD* long-chain specific acyl-CoA dehydrogenase, *MAG* monoacylglycerol, *MCAD* medium-chain acyl-CoA dehydrogenase, *MUFA* monounsaturated fatty acids, *PE* phosphatidylethanolamines, *PC* phosphatidylcholines, *Pcho* propargyl-choline, *PI* phosphatidylinositol, *PS* phosphatidylserine, *PUFA* polyunsaturated fatty acids, *SCD* stearoyl-CoA desaturase, *SFA* saturated fatty acids, *SMEM* spinner minimum essential medium, *SREBP* sterol regulatory element-binding proteins, *TG* triglycerides

Symbols: ↓, down regulated, ↑, up regulated, ↑↑, strongly up regulated, ↔, no effect, ×, low effect, ⊣, gene knock down by siRNA, ⫫, gene knock out, ⫤, pharmacological gene inhibition

Bensaad et al. (2014), Bulusu et al. (2017), Chen et al. (2015), Huang et al. (2014), Kamphorst et al. (2013), Li et al. (2006), Peck et al. (2016), Roongta et al. (2011), Schug et al. (2015), Yue et al. (2014), Zhang et al. (2017)

FABP3 and FABP7 with concomitant increase in LDs (Bensaad et al. 2014)—indicative of increased lipid-uptake. However, the effect was more pronounced in 3D system. Moreover, the effects of blocking fatty acid desaturation—via SCD-1 inhibition—were also heightened in 3D system (Peck et al. 2016). The lipidomic profiles of cells grown in 3D cell culture were shown to be different than that in 2D cell culture system. MCTs of T47D cells showed significant increase in poly-unsaturated DG and monounsaturated PI and LPA (Peck et al. 2016).

The impact of lipid-deprivation on cell proliferation is also more pronounced in tumor spheroids than in cells grown in monolayers (Daniëls et al. 2014). MCTs generated from breast or prostate cancer cell lines were more sensitive to FASN inhibition under low-serum environment than under normal growth conditions (Peck et al. 2016).

Targeting Lipid Metabolism for Cancer Therapy

Since the discovery of tumor-associated deregulations in lipid metabolism, several research studies have focused on assessing the therapeutic significance of the related pathways. Much attention has been given to the lipid acquisition pathways because it was noted that cancer cells require a steady supply of lipids for survival and disease progression.

One of the widely investigated pathways was the de novo cholesterol synthesis pathway. Particularly, the impact of statins, which targets the rate-limiting HMG-CoA reductase enzyme, on cancer cell proliferation and survival was studied. Stains are clinically utilized as cholesterol-lowering drugs. The data from some epidemiological studies showed that statins are associated with a reduced risk of cancer development. Later studies indicated that statin use is also associated with reduced mortality or recurrence in many cancers (Mallik and Chowdhury 2018). However, the meta-analysis of randomized trials in cancer showed no significant effect of adding statins to therapy on progression-free or overall survival (Farooqi et al. 2018; Emilsson et al. 2018). Recent studies emphasize a need for more extensive, randomized trials (Butler et al. 2020).

The de novo fatty acid synthesis pathway, which was thought to be the most significant mode for fatty acid acquisition in cancer cells, gained much attention as a therapeutic target. FASN was the most widely studied target among all the modulators of deregulated lipid metabolism in cancer cells. Several inhibitors of FASN have been identified and tested for their ability to kill cancer cells. The so-called next-generation FASN inhibitors—including C93, IPI-9119, and TVB-2640—display higher specificity for FASN with limited off-target effects (Koundouros and Poulogiannis 2020). However, TVB-2640 is the only clinically available inhibitor (Butler et al. 2020). It has also entered the phase II clinical trials for HER2-positive breast cancer combined with paclitaxel and trastuzumab (Heuer et al. 2017).

Recent advances indicated that apart from the de novo synthesis, other interlinked pathways also regulate fatty acid metabolism and homeostasis in cancer cells. These pathways include extracellular fatty acids uptake, fatty acid oxidation, fatty acid activation, intracellular lipolysis, triglyceride synthesis, phospholipid modification, and synthesis. It was reported that these pathways might also play distinctive pro-oncogenic roles under varying metabolic conditions and during various stages of tumor development (Munir et al. 2019, 2021). Researchers are investigating the therapeutic targeting of these other pathways and have identified several agents or inhibitors against various targets within these pathways. However, most of these inhibitors are still at the preclinical stage (Butler et al. 2020).

Conclusions & Future Perspectives

Deregulations in lipid metabolism pathways of cancer cells are widely reported. Multiple studies have indicated that these alterations are modulated by various cancer cell-intrinsic processes. However, emerging evidence suggests that these modifications are also mediated by crosstalk between tumor microenvironment and metabolic circuits within cancer cells. The tumor microenvironment is mostly hypoxic. Few studies indicate that cancer cells may display increased FA synthesis under hypoxia. However, most recent studies suggest that hypoxic cancer cells switch from endogenous synthesis to increased FA-uptake because glucose-based acetyl-CoA generation is inhibited (Fig. 3.2) (reviewed in Munir et al. 2019). Alternatively, certain cancer cells, to compensate for this down-regulation, switch to other carbon sources such as glutamine or acetate. Still, FA desaturation by SCD1 requires oxygen and is therefore deactivated in hypoxic conditions, rendering cancer cells dependent on exogenous unsaturated FAs. In addition to increased MUFA-uptake, intracellular lipolysis and FA oxidation are down-regulated under hypoxia, which results in elevated LD/TG accumulation. Increased TG synthesis and decreased lipolysis are both suggested to be strategies adapted by cancer cells to evade lipotoxicity in hypoxia. Nutrient/Lipid deprivation renders cancer cells completely dependent on endogenous FA synthesis and desaturation. When hypoxia is induced in combination with nutrient/lipid deprivation exogenous lipids are also short in supply. Hence, they switch back to de novo FA synthesis but now fully depend on glutamine or acetate as alternative substrates. Accordingly, therapeutic targeting of de novo FA synthesis in cancer cells would be most effective under conditions that limit metabolic flexibility or, possibly, by concurrently targeting several routes of lipid acquisition.

With the advent of better molecular techniques, including mass spectrometry imaging, to spatially visualize lipids in intact tissues, it will be possible to address several of the outstanding questions, to better correlate altered lipid profiles to changing microenvironments and to better explore the potential of interfering with lipid metabolism as an antineoplastic approach.

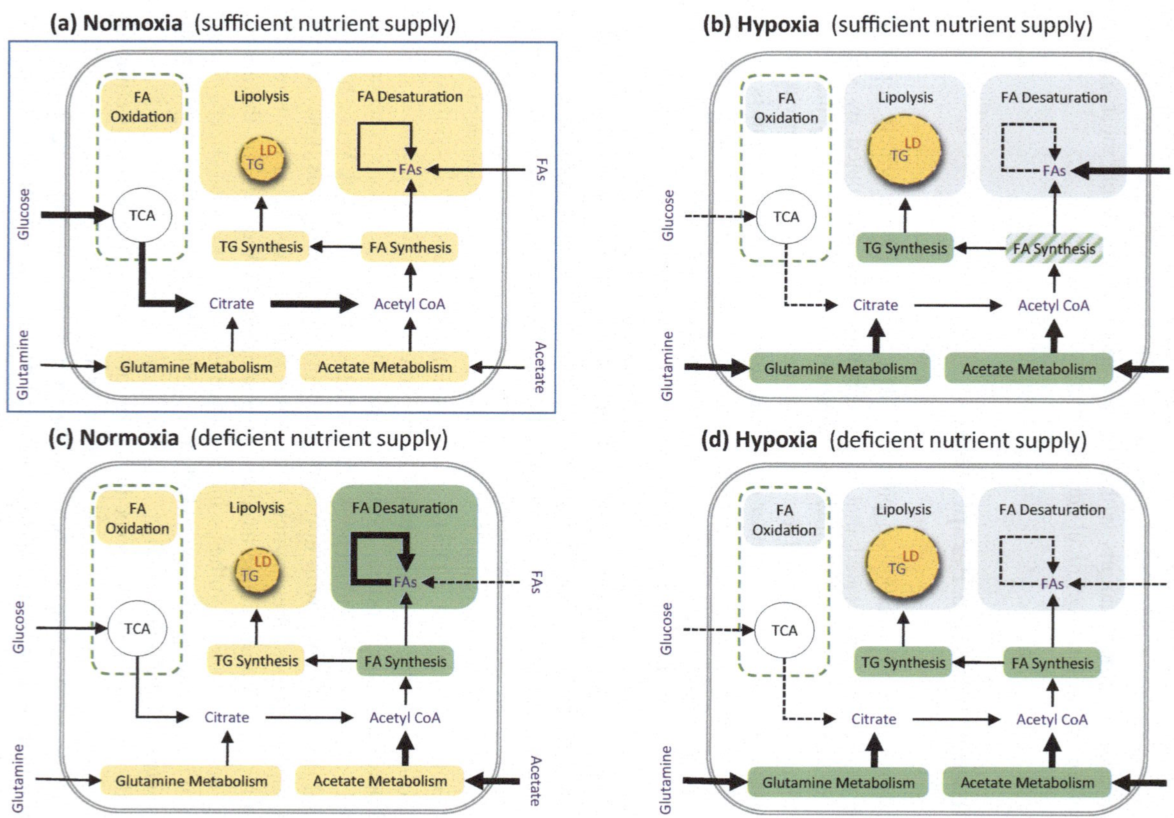

Fig. 3.2 Effect of oxygen and/or nutrient-deprivation on fatty acid metabolism in cancer cells (Munir et al. 2019). (**a**) Cancer cells under sufficient nutrients and oxygen supply mainly use glucose-derived acetyl-CoA for de novo fatty acid (FA) synthesis to support rapid cell proliferation. They can also acquire FAs from

the environment. (**b**) Hypoxic cancer cells differentially regulate FA synthesis depending on various environmental factors particularly, on nutrient availability. Hypoxia inhibits entry of glucose-derived pyruvate into TCA cycle. Hence, cells either switch to alternative carbon sources (i.e., glutamine/acetate) for FA synthesis or increase FA-uptake. FA desaturation is impaired by oxygen-deprivation therefore uptake of unsaturated FAs is particularly enhanced. FAs are rapidly incorporated into cellular triglycerides (TG) that enhances TG and Lipid droplet (LD) accumulation. (**c**) Under nutrient/lipid restriction cancer cells mainly rely on endogenous FA synthesis and desaturation. (**d**) When hypoxia is induced in combination with nutrient/lipid-deprivation cell cannot acquire FA from the environment. Hence, they switch back to de novo FA synthesis but now fully depend on glutamine or acetate as alternative substrates. Figures (**b–d**) are constructed in comparison to the normoxic state depicted in (**a**). Line thickness represents level of flux through pathway. Dashed lines indicate biosynthetic pathways that are inactive due to lack of substrates or cofactors. Color of pathway boxes in (**b–d**) represent up-regulation (green), down-regulation (grey) or differential-regulation (green/grey) of the corresponding pathway in comparison to normoxic state (with sufficient nutrient supply) (**a**)

Acknowledgments Work from the author's laboratory was supported by Alexander von Humboldt Foundation and Higher Education Commission of Pakistan (Project # 2505/R&D/11-2670).

References

Ackerman D, Simon MC (2014) Hypoxia, lipids, and cancer: surviving the harsh tumor microenvironment. Trends Cell Biol 24(8):472–478

Beloribi-Djefaflia S, Vasseur S, Guillaumond F (2016) Lipid metabolic reprogramming in cancer cells. Oncogenesis 5(1):e189

Bensaad K, Favaro E, Lewis CA, Peck B, Lord S, Collins JM, Pinnick KE, Wigfield S, Buffa FM, Li J-L (2014) Fatty acid uptake and lipid storage induced by HIF-1α contribute to cell growth and survival after hypoxia-reoxygenation. Cell Rep 9(1):349–365

Boren J, Brindle KM (2012) Apoptosis-induced mitochondrial dysfunction causes cytoplasmic lipid droplet formation. Cell Death Differ 19(9):1561–1570. https://doi.org/10.1038/cdd.2012.34

Brusselmans K, Timmermans L, Van de Sande T, Van Veldhoven PP, Guan G, Shechter I, Claessens F, Verhoeven G, Swinnen JV (2007) Squalene synthase, a determinant of raft-associated cholesterol and modulator of cancer cell proliferation. J Biol Chem 282(26):18777–18785

Bulusu V, Tumanov S, Michalopoulou E, van den Broek NJ, MacKay G, Nixon C, Dhayade S, Schug ZT, Voorde JV, Blyth K (2017) Acetate recapturing by nuclear acetyl-CoA synthetase 2 prevents loss of histone acetylation during oxygen and serum limitation. Cell Rep 18(3):647–658

Butler LM, Perone Y, Dehairs J, Lupien LE, de Laat V, Talebi A, Loda M, Kinlaw WB, Swinnen JV (2020) Lipids and cancer: emerging roles in pathogenesis, diagnosis and therapeutic intervention. Adv Drug Deliv Rev 159:245–293. https://doi.org/10.1016/j.addr.2020.07.013

Cabodevilla AG, Sanchez-Caballero L, Nintou E, Boiadjieva VG, Picatoste F, Gubern A, Claro E (2013) Cell survival during complete nutrient deprivation depends on lipid droplet-fueled beta-oxidation of fatty acids. J Biol Chem 288(39):27777–27788. https://doi.org/10.1074/jbc.M113.466656

Chen R, Xu M, Nagati JS, Hogg RT, Das A, Gerard RD, Garcia JA (2015) The acetate/ACSS2 switch regulates HIF-2 stress signaling in the tumor cell microenvironment. PLoS One 10(2):e0116515

Daniëls VW, Smans K, Royaux I, Chypre M, Swinnen JV, Zaidi N (2014) Cancer cells differentially activate and thrive on de novo lipid synthesis pathways in a low-lipid environment. PLoS One 9(9):e106913

Daster S, Amatruda N, Calabrese D, Ivanek R, Turrini E, Droeser RA, Zajac P, Fimognari C, Spagnoli GC, Iezzi G, Mele V, Muraro MG (2017) Induction of hypoxia and necrosis in multicellular tumor spheroids is associated with resistance to chemotherapy treatment. Oncotarget 8(1):1725–1736. https://doi.org/10.18632/oncotarget.13857

Denko NC (2008) Hypoxia, HIF1 and glucose metabolism in the solid tumour. Nat Rev Cancer 8(9):705–713. https://doi.org/10.1038/nrc2468

Desoize B, Jardillier J (2000) Multicellular resistance: a paradigm for clinical resistance? Crit Rev Oncol Hematol 36(2–3):193–207

Emilsson L, Garcia-Albeniz X, Logan RW, Caniglia EC, Kalager M, Hernan MA (2018) Examining bias in studies of statin treatment and survival in patients with cancer. JAMA Oncol 4(1):63–70. https://doi.org/10.1001/jamaoncol.2017.2752

Farooqi MAM, Malhotra N, Mukherjee SD, Sanger S, Dhesy-Thind SK, Ellis P, Leong DP (2018) Statin therapy in the treatment of active cancer: a systematic review and meta-analysis of

randomized controlled trials. PLoS One 13(12):e0209486. https://doi.org/10.1371/journal.pone.0209486
Furuta E, Pai SK, Zhan R, Bandyopadhyay S, Watabe M, Mo Y-Y, Hirota S, Hosobe S, Tsukada T, Miura K (2008) Fatty acid synthase gene is up-regulated by hypoxia via activation of Akt and sterol regulatory element binding protein-1. Cancer Res 68(4):1003–1011
Gao X, Lin S-H, Ren F, Li J-T, Chen J-J, Yao C-B, Yang H-B, Jiang S-X, Yan G-Q, Wang D (2016) Acetate functions as an epigenetic metabolite to promote lipid synthesis under hypoxia. Nat Commun. 7:11960
Gatenby RA, Gillies RJ (2004) Why do cancers have high aerobic glycolysis? Nat Rev Cancer 4(11):891–899. https://doi.org/10.1038/nrc1478
Gordan JD, Simon MC (2007) Hypoxia-inducible factors: central regulators of the tumor phenotype. Curr Opin Genet Dev 17(1):71–77. https://doi.org/10.1016/j.gde.2006.12.006
Hanahan D, Weinberg RA (2011) Hallmarks of cancer: the next generation. Cell 144(5):646–674. https://doi.org/10.1016/j.cell.2011.02.013
Hapala I, Marza E, Ferreira T (2011) Is fat so bad? Modulation of endoplasmic reticulum stress by lipid droplet formation. Biol Cell 103(6):271–285. https://doi.org/10.1042/BC20100144
Harris AL (2002) Hypoxia: a key regulatory factor in tumour growth. Nat Rev Cancer 2(1):38–47. https://doi.org/10.1038/nrc704
Heuer TS, Ventura R, Mordec K, Lai J, Fridlib M, Buckley D, Kemble G (2017) FASN inhibition and taxane treatment combine to enhance anti-tumor efficacy in diverse xenograft tumor models through disruption of tubulin palmitoylation and microtubule organization and FASN inhibition-mediated effects on oncogenic signaling and gene expression. EBioMedicine 16: 51–62
Huang D, Li T, Li X, Zhang L, Sun L, He X, Zhong X, Jia D, Song L, Semenza GL (2014) HIF-1-mediated suppression of acyl-CoA dehydrogenases and fatty acid oxidation is critical for cancer progression. Cell Rep 8(6):1930–1942
Jung SY, Jeon HK, Choi JS, Kim YJ (2012) Reduced expression of FASN through SREBP-1 down-regulation is responsible for hypoxic cell death in HepG2 cells. J Cell Biochem 113(12): 3730–3739
Kamphorst JJ, Cross JR, Fan J, De Stanchina E, Mathew R, White EP, Thompson CB, Rabinowitz JD (2013) Hypoxic and Ras-transformed cells support growth by scavenging unsaturated fatty acids from lysophospholipids. Proc Natl Acad Sci 110(22):8882–8887
Kamphorst JJ, Chung MK, Fan J, Rabinowitz JD (2014) Quantitative analysis of acetyl-CoA production in hypoxic cancer cells reveals substantial contribution from acetate. Cancer Metab 2(1):23
Koundouros N, Poulogiannis G (2020) Reprogramming of fatty acid metabolism in cancer. Br J Cancer 122(1):4–22. https://doi.org/10.1038/s41416-019-0650-z
Kuemmerle NB, Rysman E, Lombardo PS, Flanagan AJ, Lipe BC, Wells WA, Pettus JR, Froehlich HM, Memoli VA, Morganelli PM (2011) Lipoprotein lipase links dietary fat to solid tumor cell proliferation. Mol Cancer Ther 10:427
Lewis C, Brault C, Peck B, Bensaad K, Griffiths B, Mitter R, Chakravarty P, East P, Dankworth B, Alibhai D (2015) SREBP maintains lipid biosynthesis and viability of cancer cells under lipid- and oxygen-deprived conditions and defines a gene signature associated with poor survival in glioblastoma multiforme. Oncogene 34(40):5128–5140
Li J, Bosch-Marce M, Nanayakkara A, Savransky V, Fried SK, Semenza GL, Polotsky VY (2006) Altered metabolic responses to intermittent hypoxia in mice with partial deficiency of hypoxia-inducible factor-1α. Physiol Genomics 25(3):450–457
Mallik R, Chowdhury TA (2018) Metformin in cancer. Diabetes Res Clin Pract 143:409–419. https://doi.org/10.1016/j.diabres.2018.05.023
Masson N, Ratcliffe PJ (2014) Hypoxia signaling pathways in cancer metabolism: the importance of co-selecting interconnected physiological pathways. Cancer Metab 2(1):3. https://doi.org/10.1186/2049-3002-2-3

Menendez JA, Lupu R (2007) Fatty acid synthase and the lipogenic phenotype in cancer pathogenesis. Nat Rev Cancer 7(10):763

Metallo CM, Gameiro PA, Bell EL, Mattaini KR, Yang J, Hiller K, Jewell CM, Johnson ZR, Irvine DJ, Guarente L (2012) Reductive glutamine metabolism by IDH1 mediates lipogenesis under hypoxia. Nature 481(7381):380–384

Mullen PJ, Yu R, Longo J, Archer MC, Penn LZ (2016) The interplay between cell signalling and the mevalonate pathway in cancer. Nat Rev Cancer 16(11):718

Munir R, Lisec J, Swinnen JV, Zaidi N (2019) Lipid metabolism in cancer cells under metabolic stress. Br J Cancer 120(12):1090–1098

Munir R, Lisec J, Jaeger C, Zaidi N (2021) Abundance, fatty acid composition and saturation index of neutral lipids in colorectal cancer cell lines. Acta Biochim Pol 68(1):115–118

Murphy DJ, Vance J (1999) Mechanisms of lipid-body formation. Trends Biochem Sci 24(3): 109–115

Mylonis I, Sembongi H, Befani C, Liakos P, Siniossoglou S, Simos G (2012) Hypoxia causes triglyceride accumulation by HIF-1-mediated stimulation of lipin 1 expression. J Cell Sci 125(14):3485–3493

Papandreou I, Cairns RA, Fontana L, Lim AL, Denko NC (2006) HIF-1 mediates adaptation to hypoxia by actively downregulating mitochondrial oxygen consumption. Cell Metab 3(3): 187–197. https://doi.org/10.1016/j.cmet.2006.01.012

Peck B, Schug ZT, Zhang Q, Dankworth B, Jones DT, Smethurst E, Patel R, Mason S, Jiang M, Saunders R (2016) Inhibition of fatty acid desaturation is detrimental to cancer cell survival in metabolically compromised environments. Cancer Metab 4(1):6

Rankin EB, Giaccia AJ (2008) The role of hypoxia-inducible factors in tumorigenesis. Cell Death Differ 15(4):678–685. https://doi.org/10.1038/cdd.2008.21

Rankin EB, Giaccia AJ (2016) Hypoxic control of metastasis. Science (New York, NY) 352(6282): 175–180. https://doi.org/10.1126/science.aaf4405

Röhrig F, Schulze A (2016) The multifaceted roles of fatty acid synthesis in cancer. Nat Rev Cancer 16(11):732–749

Roongta UV, Pabalan JG, Wang X, Ryseck R-P, Fargnoli J, Henley BJ, Yang W-P, Zhu J, Madireddi MT, Lawrence RM (2011) Cancer cell dependence on unsaturated fatty acids implicates stearoyl-CoA desaturase as a target for cancer therapy. Mol Cancer Res 9(11): 1551–1561

Rysman E, Brusselmans K, Scheys K, Timmermans L, Derua R, Munck S, Van Veldhoven PP, Waltregny D, Daniels VW, Machiels J, Vanderhoydonc F, Smans K, Waelkens E, Verhoeven G, Swinnen JV (2010) De novo lipogenesis protects cancer cells from free radicals and chemotherapeutics by promoting membrane lipid saturation. Cancer Res 70(20): 8117–8126. https://doi.org/10.1158/0008-5472.CAN-09-3871

Schlaepfer IR, Nambiar DK, Ramteke A, Kumar R, Dhar D, Agarwal C, Bergman B, Graner M, Maroni P, Singh RP (2015) Hypoxia induces triglycerides accumulation in prostate cancer cells and extracellular vesicles supporting growth and invasiveness following reoxygenation. Oncotarget 6(26):22836

Schug ZT, Peck B, Jones DT, Zhang Q, Grosskurth S, Alam IS, Goodwin LM, Smethurst E, Mason S, Blyth K (2015) Acetyl-CoA synthetase 2 promotes acetate utilization and maintains cancer cell growth under metabolic stress. Cancer Cell 27(1):57–71

Sebti SM (2005) Protein farnesylation: implications for normal physiology, malignant transformation, and cancer therapy. Cancer Cell 7(4):297–300

Semenza GL (2003) Targeting HIF-1 for cancer therapy. Nat Rev Cancer 3(10):721–732. https://doi.org/10.1038/nrc1187

Semenza GL (2010) Defining the role of hypoxia-inducible factor 1 in cancer biology and therapeutics. Oncogene 29(5):625

Sutherland RM (1988) Cell and environment interactions in tumor microregions: the multicell spheroid model. Science 240(4849):177–184

Valli A, Rodriguez M, Moutsianas L, Fischer R, Fedele V, Huang H-L, Van Stiphout R, Jones D, Mccarthy M, Vinaxia M (2015) Hypoxia induces a lipogenic cancer cell phenotype via HIF1-α-dependent and-independent pathways. Oncotarget 6(4):1920

Yamada KM, Cukierman E (2007) Modeling tissue morphogenesis and cancer in 3D. Cell 130(4): 601–610. https://doi.org/10.1016/j.cell.2007.08.006

Yu Y, Vidalino L, Anesi A, Macchi P, Guella G (2014) A lipidomics investigation of the induced hypoxia stress on HeLa cells by using MS and NMR techniques. Mol BioSyst 10(4):878–890

Yue S, Li J, Lee S-Y, Lee HJ, Shao T, Song B, Cheng L, Masterson TA, Liu X, Ratliff TL (2014) Cholesteryl ester accumulation induced by PTEN loss and PI3K/AKT activation underlies human prostate cancer aggressiveness. Cell Metab 19(3):393–406

Zaidi N, Lupien L, Kuemmerle NB, Kinlaw WB, Swinnen JV, Smans K (2013) Lipogenesis and lipolysis: the pathways exploited by the cancer cells to acquire fatty acids. Prog Lipid Res 52(4): 585–589

Zhang X, Saarinen AM, Hitosugi T, Wang Z, Wang L, Ho TH, Liu J (2017) Inhibition of intracellular lipolysis promotes human cancer cell adaptation to hypoxia. eLife 6. https://doi.org/10.7554/eLife.31132

Chapter 4
Role of the Histone Acetyl Transferase MOF and the Histone Deacetylase Sirtuins in Regulation of H4K16ac During DNA Damage Repair and Metabolic Programming: Implications in Cancer and Aging

Tej K Pandita, Clayton R. Hunt, Vipin Singh, Santanu Adhikary, Shruti Pandita, Siddhartha Roy, Kenneth Ramos, and Chandrima Das

Abstract The accurate repair of genomic damage mediated by ionizing radiation (IR), chemo- or radiomimetic drugs, or other exogenous agents, is necessary for maintenance of genome integrity, preservation of cellular viability and prevention of oncogenic transformation. Eukaryotes have conserved mechanisms designed to perceive and repair the damaged DNA quite efficiently. Among the different types of DNA damage, double strand breaks (DSB) are the most detrimental. The cellular DNA DSB response is a hierarchical signaling network that integrates damage sensing and repair with chromatin structural changes that involve a range of pre-existing and induced covalent modifications. Recent studies have revealed that

---

Clayton R. Hunt, Vipin Singh and Santanu Adhikary contributed equally with all other contributors.

T. K. Pandita (✉)
The Houston Methodist Research Institute, Houston, TX, USA

Department of Cellular and Molecular Biology, Baylor College of Medicine, Houston, TX, USA

Center for Genomics and Precision Medicine, Texas A&M College of Medicine, Houston, TX, USA
e-mail: tpandita@tamu.edu

C. R. Hunt
The Houston Methodist Research Institute, Houston, TX, USA

V. Singh
Biophysics and Structural Genomics Division, Saha Institute of Nuclear Physics, Kolkata, India

Homi Bhaba National Institute, Mumbai, India

S. Adhikary · C. Das
Biophysics and Structural Genomics Division, Saha Institute of Nuclear Physics, Kolkata, India

T. K. Kundu, C. Das (eds.), *Metabolism and Epigenetic Regulation: Implications in Cancer*, Subcellular Biochemistry 100,
https://doi.org/10.1007/978-3-031-07634-3_4

pre-existing histone modifications are important contributors within this signaling/ repair network. This chapter discusses the role of a critical histone acetyl transferase (HAT) known as MOF (males absent on the first) and the histone deacetylases (HDACs) Sirtuins on histone H4K16 acetylation (H4K16ac) and DNA damage repair. We also discuss the role of this important histone modification in light of metabolic rewiring and its role in regulating human pathophysiologic states.

Keywords DNA damage response · Histone H4K16ac · MOF · Chromatin

Introduction

A diverse range of epigenetic modifications on the eukaryotic chromatin template are essential for transcription regulation. Recent studies have shown that these modifications, particularly in the core histone component of chromatin, are also involved in the DNA damage response (DDR) and repair, indicating a direct connection between transcription and DNA damage repair. Understanding the mechanisms involved in sensing and repair of damaged DNA arising from ionizing radiation as well as chemo- or radio-mimetic drugs can open up new vistas in cancer treatments through the identification of new potential therapeutic intervention points.

The DNA repair field has several challenges in studying chromatin modification effects including the lack of methods to specifically determine how DNA double strand breaks (DSBs) induced by heterotypic agents (chemotherapeutic drugs or ionizing radiation) act within a defined chromatin context. Further, it becomes difficult to conceive how specific alterations in the chromatin landscape can impact the recruitment of repair factors to DNA lesions. Depending upon the chromosomal site, the repair mechanism could be orchestrated to a different extent, for example active or gene-rich regions have facilitated repair mechanisms as opposed to condensed or gene-poor regions (Horikoshi et al. 2019; Clouaire et al. 2018). The chromatin at active or inactive sites likely harbors very different histone modifications, suggesting that these epigenetic marks are important determinants for choosing the pathways for DNA damage repair. The recent development of new

Structural Biology and Bioinformatics Division, CSIR-Indian Institute of Chemical Biology, Kolkata, India
e-mail: chandrima.das@saha.ac.in

S. Pandita
Department of Internal Medicine, Division of Hematology, Oncology and Cellular Therapy, Saint Louis University School of Medicine, St. Louis, MO, USA

S. Roy
Biophysics and Structural Genomics Division, Saha Institute of Nuclear Physics, Kolkata, India

K. Ramos
Center for Genomics and Precision Medicine, Texas A&M College of Medicine, Houston, TX, USA

technologies for generating site-specific DSBs at targeted chromatin loci that can be further subjected to detailed modification analysis, shows major scientific advances impacting future clinical interventions.

MOF

Allfrey and co-workers first reported histone acetylation in 1964 (Allfrey et al. 1964), and subsequent studies identified additional histone and non-histone substrates (Choudhary et al. 2009). The requisite modification enzymes are categorized as lysine acetyltransferases (KATs) (Allis et al. 2007) and have been shown to play essential roles in both normal and malignant development (Sun et al. 2015; Singh et al. 2020a). In fact, histone and non-histone protein acetylation have been shown to regulate normal blood cell development (Mishima et al. 2011; Aerbajinai et al. 2007; Bruserud et al. 2006). Analysis of interaction networks during hematopoiesis identified multiple chromatin factor complexes including NuA4/P300/CBP/HBO1 (Huang et al. 2013). Zhang and coworkers have reported selective binding of the PHD finger of MLL4 to histone H4K16ac, thus linking the MLL4 and MOF (Zhang et al. 2019).

Lysine acetyltransferases fall into five families based on homology and mechanistic features. Members of the largest KAT subdivision, the MYST family (MOZ, MOF, Ybf2, Sas2 and Tip60) all harbor a highly conserved MYST domain composed of a C2HC-type zinc finger and a canonical acetyl coenzyme A binding motif (Yang and Ullah 2007). MOF [K (lysine) acetyltransferase 8, KAT8] is the best-characterized MYST-family protein and mediates histone H4 lysine 16 acetylation (H4K16ac) (Huang et al. 2013; Zhang et al. 2019; Yang and Ullah 2007) and is essential for murine embryogenesis (Smith et al. 2005; Gupta et al. 2005, 2008). Another HAT, Tip60, has also been reported to acetylate H4K16 (Tang et al. 2013).

Males absent on the first (MOF), was initially identified as a component of the dosage compensation complex (DCC) in *Drosophila melanogaster* (Belote and Lucchesi 1980). It is also a critical factor of the male-specific lethal (MSL) complex in flies which balances out X chromosome gene expression between males (XY) and females (XX) (Hilfiker et al. 1994, 1997; Bone and Kuroda 1996). MOF binding and acetylation of male X chromosome histone H4 lysine 16 (H4K16ac), correlates with enhanced transcription (Hilfiker et al. 1994, 1997; Bone and Kuroda 1996), while MOF mutations result in H4K16ac loss and subsequent male fly death (Rea et al. 2007). Additionally, MOF is reported to have protein propionylation activity, though the biological significance of this is not clear (Han et al. 2018).

A genome wide H4K16ac chromatin immunoprecipitation study in HEK293 cells identified and mapped 25,893 associated DNA regions (average length of 692 nucleotides), 82% of which were located within genes as discussed by Horikoshi and co-workers (Horikoshi et al. 2013). This suggests that MOF targeted transcription has important consequences when a selective molecular cue dictates the expression of a subset of genes in specific cell types (Horikoshi et al. 2013). Stimuli like DNA damage can modulate MOF function to activate ATM autophosphorylation (Gupta et al. 2005), with the resultant H4K16ac levels impacting physiological phenomena such as the DDR and oncogenesis (Gupta et al. 2008; Li et al. 2010). The DNA damage response of MOF is highly conserved in mammals and *Drosophila* as it was found that ionizing radiation exposure reduces survival of *Drosophila mof* mutants and enhanced chromosome aberrations upon *mof* knockdown in Drosophila SL-2 embryonic stem cells (Bhadra et al. 2012).

Among the types of DNA damage induced by physical or chemical agents, DSBs are the most harmful lesions, and their repair is critical for cellular survival. Regulation of DSB repair, both spatially and temporally, is dependent on histone modifiers and specific histone modifications (Williamson et al. 2012; Widlak et al. 2006; Verger and Crossley 2004; Vidanes et al. 2005). Gupta and coworkers reported that MOF interacts with ATM and plays a critical role in regulating MOF phosphorylation by ATM in response to the ionizing radiation (Gupta et al. 2005). In addition, MOF also suppresses replicative stress, as its depletion decreases replication fork speed and increases replication fork stalling (Singh et al. 2018).

Mammalian MOF influences the cellular DNA damage response (DDR) and the DSB repair pathways (Gupta et al. 2014) through its function as a regulator of chromatin structure. H4K16ac is a unique histone modification that regulates higher order chromatin folding which can indirectly affect the access of non-histone proteins to DNA (Shogren-Knaak et al. 2006). Histone H4K16ac levels prior to irradiation correlate with the DDR (Gupta et al. 2005; Sharma et al. 2010). MOF also promotes both DNA repair pathways, nonhomologous end joining (NHEJ) as well as homologous recombination (HR) (Sharma et al. 2010). Li et al showed that MOF depletion in mouse embryonic fibroblasts reduced MDC1, 53BP1, and BRCA1 containing repair foci, likely due to changes in chromatin structure (Li et al. 2010). Consistent with these studies, Gupta and coworkers reported that ATM-dependent phosphorylation of MOF at Thr392 (Gupta et al. 2008) was essential for DNA repair by homologous recombination in S- and G2-phases of cell cycle, but dispensable during G1-phase (Singh et al. 2018). MOF function itself is dependent upon ATM which phosphorylates MOF at T392/Mutation of MOF (T392A) reduces BRCA1 accumulation at DSBs, while 53BP1 persists at sites during the S/G2 phases of the cell cycle (Gupta et al. 2014). These studies revealed that MOF influences ATM functions during the DDR (Gupta et al. 2005), while MOF is phosphorylated post-irradiation by ATM on T392, an important modification for DSB repair by homologous recombination (HR) (Fig. 4.1) (Gupta et al. 2008, 2014).

SIRTUINS

Sirtuins are a family of enzymatic proteins with a common, highly conserved domain that carries out NAD+ dependent deacetylation (Imai et al. 2000). Sirtuins have been detected in almost all organisms, from animals to bacteria and plants (Sinclair and

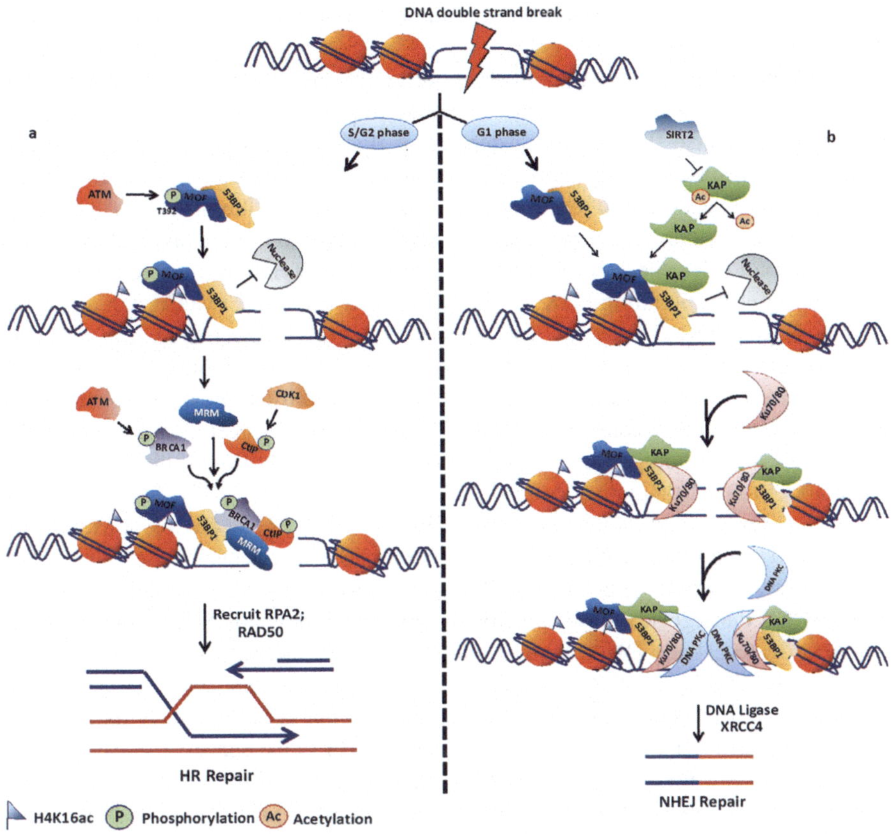

Fig. 4.1 Role of MOF and SIRTUINS in DNA damage repair. (**a**) In *HR (Homologous recombination)* repair, ATM phosphorylates MOF and facilitates recruitment of MRN complex, phosphorylated BRCA1 and CTIP at the double strand break (DSB) sites. This further helps in recruitment of RAD50 to execute the HR process. (**b**) In *NHEJ (Non-homologous end joining)* repair, SIRT2 deacetylates KAP, which assist in its recruitment at the DSB site along with MOF and 53BP1. KU70/80 along with DNA PKC gets recruited at the DSB site, which further assists DNA ligase to complete the repair process

Guarente 2006), where they are critical for health and survival due to their functional roles in aging, metabolism, DNA repair, tumorigenesis and stress tolerance (Sinclair and Guarente 2006; Meng et al. 2020; Rajman et al. 2018; Vazquez et al. 2017). Sirtuin was discovered initially in the budding yeast *Saccharomyces cerevisiae* as Sir2 (Silent Information Regulator) from where the word "Sirtuin" was coined (Guarente 1999). The mammalian homologs of the yeast *Sir2* gene encode SIRT1-SIRT7, having conserved catalytic domains but variations at the N and C termini. This N and C tail variation is a key factor in determining enzyme activity, binding target and cellular localization (Mei et al. 2016). In mammals, sirtuins can be localized to the nuclear (SIRT1, 6 and 7), mitochondrial (SIRT3-5) or cytosolic (SIRT2) compartments (Kiran et al. 2013; Tanno et al. 2012; Lombard et al. 2007), though pathological or other stressful conditions may affect localization (Eldridge et al. 2020).

The DDR can be initiated due to several factors including oxidative stress, UV light, genotoxic stress, etc. Against the resulting DNA damage, cells mount a predictable response that is highly dependent upon signaling and repair proteins which are themselves highly regulated functionally by post-translational modification, particularly acetylation reactions (Dantuma and van Attikum 2016). At an early stage in the DNA damage response, a global deacetylation event takes place in the cell. In human cells, there is rapid and reversible deacetylation of selective histone acetylation marks including H3K9ac, H3K56ac and H4K16ac (McCord et al. 2009; Tjeertes et al. 2009). These acetylation marks have been associated with transcription activation. Moreover, the H4K16ac mark can reportedly inhibit Mg^{2+}-induced higher order chromatin folding, and thereby altering structure, in the absence of histone H1 (Shogren-Knaak et al. 2006). Consequently, H4K16ac is an especially promising candidate for detailed examination of the contribution of histone acetylation marks to DDR regulatory control.

Of the seven mammalian Sirtuins, SIRT1 and SIRT2 are responsible for H4K16ac deacetylation with SIRT1 described as the key player in regulating the DDR via two different means. First, SIRT1 acts as a deacetylase of proteins involved in the DNA damage response and repair pathways. Second, SIRT1 deacetylates histones located at the DNA damage site (Ding et al. 2016; O'Hagan et al. 2011; Jeong et al. 2007). H4K16ac is likely a major target candidate, since SIRT1 knockdown led to persistent H4K16 acetylation (Hida et al. 2007). A more direct role in chromatin repression is also evident since SIRT1 also deacetylates suppressor of variegation 3–9 homologue 1 (SUV39H1), a histone methyltransferase that when activated induces production of the repressor markers H3K9me3, H4K20me3 and H3K79me2 (Vaquero et al. 2007). Consistent with these data, $SIRT1^{-/-}$ mice display increased H3K9 acetylation and decreased K9me3 (Wang et al. 2008). Genotoxic stress can induce SirT1 relocalization from a repetitive DNA loci to a DSB position, reorganizing the chromatin structure with resulting transcriptional changes that are associated with age related modifications (Oberdoerffer et al. 2008). SIRT1 was also found to function synergistically with ATM in a complex, where the recruitment of SIRT1 to the DNA breaks is dependent on ATM, where the former stabilizes ATM via regulating its autophosphorylation (Dobbin et al. 2013). SIRT1 binds to and deacetylates NBS1 which is then a substrate for phosphorylation by ATM at Ser343. Phosphorylated NBS1 forms the MRN complex, following its association with MRE11 and RAD50, which plays a critical role in DNA damage detection and the subsequent repair response (Hughes et al. 2011; Yuan and Seto 2007). In addition, SIRT1 induces NHEJ and inhibits HR by stabilizing the KAP1 interaction with 53BP1 after it is phosphorylated by ATM at Ser824 (Fig. 4.1) (Lin et al. 2015). KAP1 is a well-known regulator of genomic stability which induces heterochromatin formation and silences genes (Lin et al. 2015).

Induction of oxidative DNA damage in colorectal and colon cancer cells by hydrogen peroxide (H_2O_2) causes MSH2-MSH6 to form a heterodimer which then strongly associates with chromatin and is instrumental in recruiting PCNA and DNA methyltransferase 1 (DNMT1) to DNA damage sites (Ding et al. 2016). Recruitment of SirT1 at DSBs along with EZH2, DNMT1 and DNMT3 lead to repressive histone alterations such as hypoacetyl H4K16 or trimethylation of H3K9 and H3K27 and thus produces chromatin compaction at the damaged site. In this regard, Ding and his coworkers have suggested that DNMT1 functions as a scaffold protein that brings SIRT1, EZH2 and DNMT3B to the DNA damage site (Ding et al. 2016). In a study by Zhong and co-workers, SirT1 deacetylated not only H4K16ac but also hMOF thus affecting hMOF recruitment leading to decreased expression of important DNA DSB repair genes including BRCA2, RAD50 and FANCA (Zhong et al. 2017a).

While SIRT2 is mainly distributed in the cytosol, it causes H4K16ac deacetylation during the mitotic stage of cell cycle (Vaquero et al. 2006). Further studies by Vaquero and co-workers have shown that SIRT2 dependent H4K16 deacetylation facilitates H4K20 methylation and tumor suppression (Serrano et al. 2013). Initially SIRT2 plays a crucial role in the mitotic phase of cell cycle progression causing M-phase exit (Dryden et al. 2003) and to deacetylate CDH1 and CDC20, co-activators of APC/C that regulate APC/C activity (Kim et al. 2011). Recently SIRT2's role in genome stability and the DDR was linked to its function as a histone methyltransferase (HMT) that regulates monomethyl deposition at H4K20, a critically important function in genome stability and DNA repair (Serrano et al. 2013).

Histone H4 Lysine 16 Acetylation (H4K16ac) in DNA Damage Repair

The initial report indicating a role for histone acetyltransferase in DDR was the observation that loss of acetyltransferase activity in TIP60 mutant cells was defective in DNA DSB repair (Ikura et al. 2000). The acetylation of histone tails has been proposed as critically important to create the loosely compacted chromatin structure favorable for DNA repair machinery access (Tamburini and Tyler 2005; Hunt et al. 2013). Post DNA repair, histone deacetylases are important to restore the original native chromatin structure.

In eukaryotes, reversible acetylation occurs on histone H4 lysines (K) at 5, 8, 12 and 16 in the amino-terminal tail (Allfrey et al. 1968). Hyperacetylation of histone H4 could lead to unfolding of the nucleosomal fiber (Tse et al. 1998). Studies revealed that histone H4K16ac regulates the chromatin structure by modulating the association of the chromatin fiber with other histone and non-histone proteins (Shogren-Knaak et al. 2006). Zhang and co-workers have suggested that acetylation of H4K16 could reduce the inter-nucleosomal interactions facilitating the binding of the non-histone proteins to the acidic patch of histone H4 (Zhang et al. 2017). Bird and co-workers reported that Esa1 (Essential SAS2-related Acetyltransferase) mediated H4K16ac was essential for DNA repair in yeast (Bird et al. 2002) and have also proposed that a similar mechanism should operate in mammals. Gupta and coworkers established that upon DNA damage, hMOF is responsible for ATM activation (Gupta et al. 2005) and that the role of MOF in the DDR is conserved in mammals and *Drosophila* (Bhadra et al. 2012; Sharma et al. 2010). MOF is the major HAT acetylating histone H4 at K16 (Smith et al. 2000, 2005; Gupta et al. 2005, 2008; Akhtar and Becker 2000; Neal et al. 2000; Taipale et al. 2005). Our recent studies revealed that pre-existing H4K16ac marks correlate with the enrichment of HR related proteins and HR frequency in gene-rich regions (Horikoshi et al. 2019). Artificially increasing H4K16ac in gene-poor regions neither increased recruitment of HR proteins nor HR per se (Horikoshi et al. 2019). It has been reported that MOF interacts with several other proteins described previously (Sharma et al. 2010) that will provide significant information about the regulation of HR in gene-rich regions.

SIRTUINS Influence on Metabolic Regulation and Cancer

Activation of the DNA damage response initiates metabolic changes that include an increase in nucleotide synthesis levels as well as anabolic glucose metabolism and a reduction in glutamine anaplerosis (Sayin et al. 2017). The metabolic context is the sum total of all chemical reactions taking place inside the cells and organism to maintain viability. It has been clear for the last two decades that metabolism can be a key regulator for growth abnormalities of malignant cells. In this regard SIRTUINs influence metabolism and cell growth. For example, SIRT2 regulates the multifaceted oncoprotein c-MYC, an important regulator of cell transformation (Morrish et al. 2008) as an early response gene that activates many downstream signaling pathways. The MYC family oncoproteins c-MYC, n-MYC and l-MYC are usually upregulated during cancer and contribute to the malignant phenotype (Gustafson and Weiss 2010). The n-MYC oncoprotein is usually 20–25% overexpressed in neuroblastoma (Dzieran et al. 2018), whereas 50% c-MYC overexpression is detected in human primary pancreatic tumors (Mahlamaki et al. 2002). A study by Liu and co-workers, found that in neuroblastoma and pancreatic cancer cells, the oncoproteins n-MYC and c-MYC, respectively, upregulates SIRT2 protein levels via a post-transcriptional mechanism (Liu et al. 2013). The increased SIRT2 protein levels then work to stabilize MYC proteins and enhance cancer cell proliferation (Liu et al. 2013). SIRT2 ensures MYC protein stability by directly binding to the *NEDD4* gene core promoter region to inhibit transcription via deacetylation of H4K16ac (Fig. 4.2) (Liu et al. 2013). Inhibition of SIRT2 with AC-93253 and Salermide upregulated NEDD4, a well-known ubiquitin protein ligase which suppresses cancer cell progression by ubiquitinating and degrading MYC oncoprotein. Once *NEDD4* gene expression is downregulated, MYC protein levels simultaneously increase and begin impacting regulatory pathways, including cellular metabolism (Liu et al. 2013). A major impact of c-MYC is the regulation of glycolysis (Miller et al. 2012). Lactate dehydrogenase A is responsible for converting pyruvate to lactate and is one of the twenty putative targets of c-MYC (Miller et al. 2012).

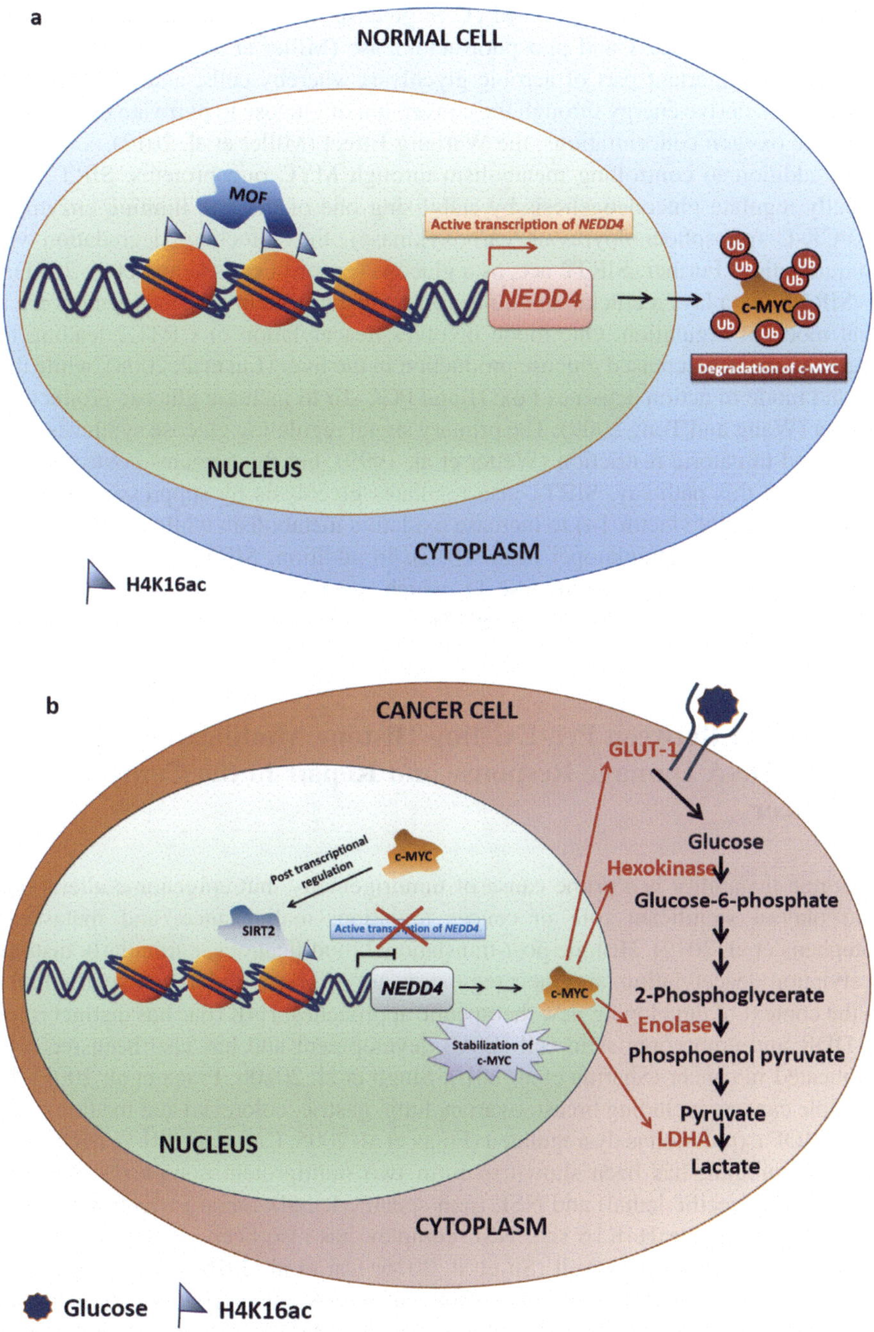

Fig. 4.2 Role of H4K16 acetylation in regulating glucose metabolism. (**a**) MOF mediated H4K16 acetylation activates NEDD4, which further polyubiquitinates c-MYC, thereby repressing the c-MYC target genes implicated in glycolysis. (**b**) During oncogenesis, SIRT2 mediated deacetylation of H4K16ac represses NEDD4, which thereby stabilizes c-MYC and activate its target genes, thus inducing aerobic glycolysis

Additional regulatory targets of c-MYC in glucose metabolism includes GLUT-1, hexokinase-2, enolase-1 and phosphofructokinase (Miller et al. 2012). These proteins are an important part of aerobic glycolysis whereby cells, and preferentially cancer cells, derive energy through the conversion of glucose to pyruvate even under adequate oxygen concentrations, the Warburg Effect (Miller et al. 2012).

In addition to controlling metabolism through MYC oncoproteins, SIRT2 can directly regulate gluconeogenesis by stabilizing one of the rate limiting enzymes, PEPCK-C (phosphoenolpyruvate carboxykinase), thus blocking degradation via ubiquitination. Further, SIRT2 acts as a glucose sensor in cells (Jiang et al. 2011).

SIRT1 also plays a crucial, but quite controversial, role in gluconeogenesis via a dual mode of regulation. One mode involves deacetylation of CRTC2 leading to degradation and decreased glucose production in the liver (Liu et al. 2008), while the second mode of action activates FoxO1 and PGC-1α to increase glucose production in liver (Wang and Tong 2009). The primary signal regulating glucose synthesis may be related to caloric restriction (Wetter et al. 1999), but this remains controversial. Along with this pathway, SIRT1 also regulates glycolysis by suppressing HIF-1α (hypoxia inducible factor 1α) to increase oxidative metabolism while decreasing the rate of glycolysis (Houtkooper et al. 2012). In addition, SIRT1 also deacetylates PGAM-1(phosphoglycerate mutase-1) which decreases catalytic activity thus inhibiting glycolysis (Hallows et al. 2012).

Connection Between Pre-Existing Histone Modifications and the DNA Damage Response and Repair in the Context of Cancer

Genomic instability is a prime cause of tumorigenicity, but epigenome alterations also play a significant role in cancer initiation, maintenance, and metastasis (Stephens et al. 2012). Histone post-translational modifications, particularly histone acetylation/deacetylation, and their regulators are of prime importance, particularly in the context of developing new therapeutic approaches. H4K16ac has distinct roles in DDR and embryonic stem cell (ESC) development and has also been recently implicated in cancer (Sharma et al. 2010; Singh et al. 2020b; Fraga et al. 2005). In multiple cancers including breast, ovarian, lung, gastric, colorectal and medulloblastoma MOF expression is dysregulated (Fraga et al. 2005; Cao et al. 2014). MOF, as a catalytic subunit, has been shown to form two multiprotein complexes, namely, MSL (male-specific lethal) and NSL (non-specific lethal), among which both complexes can acetylate H4K16 site. NSL complex has also been shown to acetylate H4K5 and H4K8 sites as well (Su et al. 2016; Cai et al. 2010). These regulatory aspects of both complexes introduce significant complexity to MOF functionality. In breast and hepatocellular carcinoma, higher levels of MOF are associated with better survival (Cao et al. 2014; Zhu et al. 2015; Zhang et al. 2014). In breast cancer it was reported that hMOF plays a key regulatory role in suppression of epithelial-mesenchymal transition (EMT). Lu and co-workers have shown that with increasing MOF activity through its autoacetylation at K274 (Sun et al. 2011) in mouse embryonic fibroblast (MEF) and in the MDAMB-231 cell line, MOF causes LSD1 (Lysine-specific demethylase 1) acetylation. LSD1 represses gene transcription by causing mono- or di-methylation of histone H3 at Lysine 4 (Shi et al. 2004). LSD1 acetylation decreases binding at the epithelial gene promoters resulting in increased E-cadherin gene expression and suppression of EMT (Fig. 4.3a) (Luo et al. 2016). Interestingly, DNA demethylase Ten Eleven Translocation 1 (TET1) protein can influence hMOF auto-acetylation and affect its DNA repair function (Zhong et al. 2017b). TET1 forms complex with SIN3A and hMOF facilitating its recruitment to chromatin (Zhong et al. 2017b). Further, hMOF autoacetylation is prevented by TET1 thereby promoting its enzymatic activity leading to H4K16ac to regulate expression of downstream DNA repair genes such as *RAD50*, *BRCA1*, *RAD51* and *TP53BP1* which can act as tumor suppressors (Zhong et al. 2017b).

In contrast, MOF also acts as a tumor promoter and has a significant role in tumor initiation in patients with gastric cancer, thyroid cancer, and bladder cancer (Cao et al. 2014; Zhu et al. 2021). Enhanced expression of MOF was also found in 50 percent of NSCLC (non-small cell lung cancer) patients (Su et al. 2016). MOF promotes cellular proliferation of thyroid cancer cells by increasing the expression of TNK2 which phosphorylates AKT leading to activation of the PI3K/AKT pathway (Li et al. 2020). It was also observed that in bladder cancer T24 cells, the commonly used chemotherapeutic drug GEM inhibits both endogenous and exogenous MOF protein expression, thereby downregulating histone H4 acetylation at Lys5, Lys8, and Lys16 sites (Zhu et al. 2021).

In the context of hematopoietic lineage commitment, MOF has been demonstrated to play pivotal role in regulating the transcription factor network priming the cell towards erythroid lineage (Pessoa Rodrigues et al. 2020). In the transcription factor network RUNX1 and GFI1B are two critical factors; while MOF binds to the Runx1 promoter, the latter regulates GFI1B expression (Pessoa Rodrigues et al. 2020). GFI1B on the other hand acts as a positive regulator of MOF and thus the network constitutes a feedforward pathway for erythroid lineage commitment (Pessoa Rodrigues et al. 2020), which can go awry in hematological malignancies. Loss of H4K16ac due to MOF loss is also associated with a poorer prognosis for hematological malignancies and colorectal adenocarcinomas (Fraga et al. 2005). MOF is overexpressed in kidney tumors and associated with chromosomal mis-segregation and cell proliferation (Singh et al. 2020b). A more subtle MOF regulatory mechanism is observed in endometrial cancer where hMOF expression is regulated by estrogen in a concentration dependent manner, with lower estrogen concentrations (i.e. 10^{-9}–10^{-8} mol/L) enhancing hMOF expression leading to tumor cell proliferation, invasion, and migration while higher concentrations ($>10^{-7}$ mol/L) inhibiting hMOF expression and the above mentioned biological effects (Qi et al. 2020). Low estrogen levels have been shown to inhibit apoptosis while higher concentrations induce a response regulated via hMOF through PI3K/Akt signaling pathways (Fig. 4.3b) (Qi et al. 2020). Aside from this, MOF activity has been shown to play a crucial role in ER alpha signaling pathway in

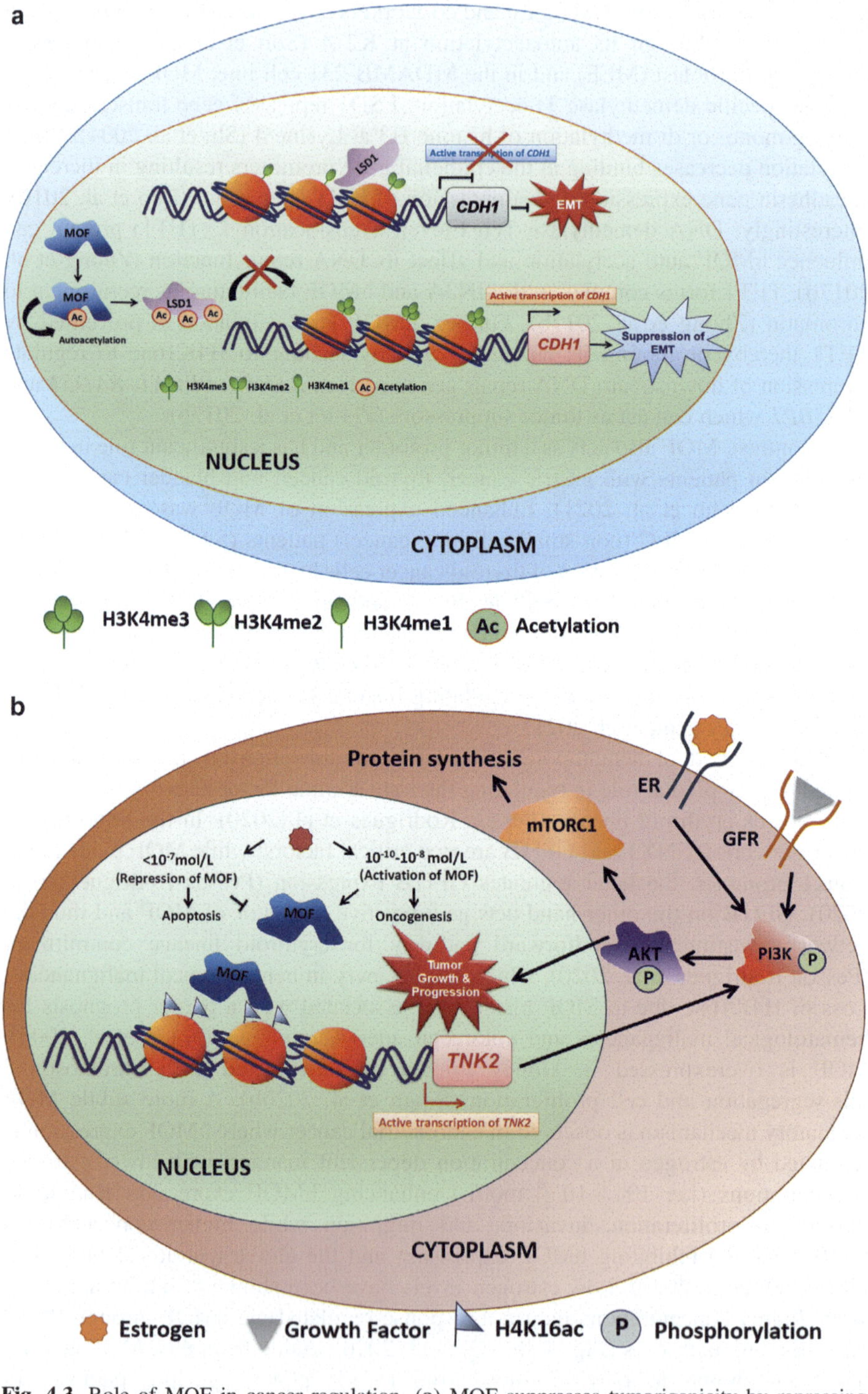

Fig. 4.3 Role of MOF in cancer regulation. (**a**) MOF suppresses tumorigenicity by repressing EMT. MOF is autoacetylated, which further acetylates LSD1. Acetylated LSD1 fails to get recruited

c

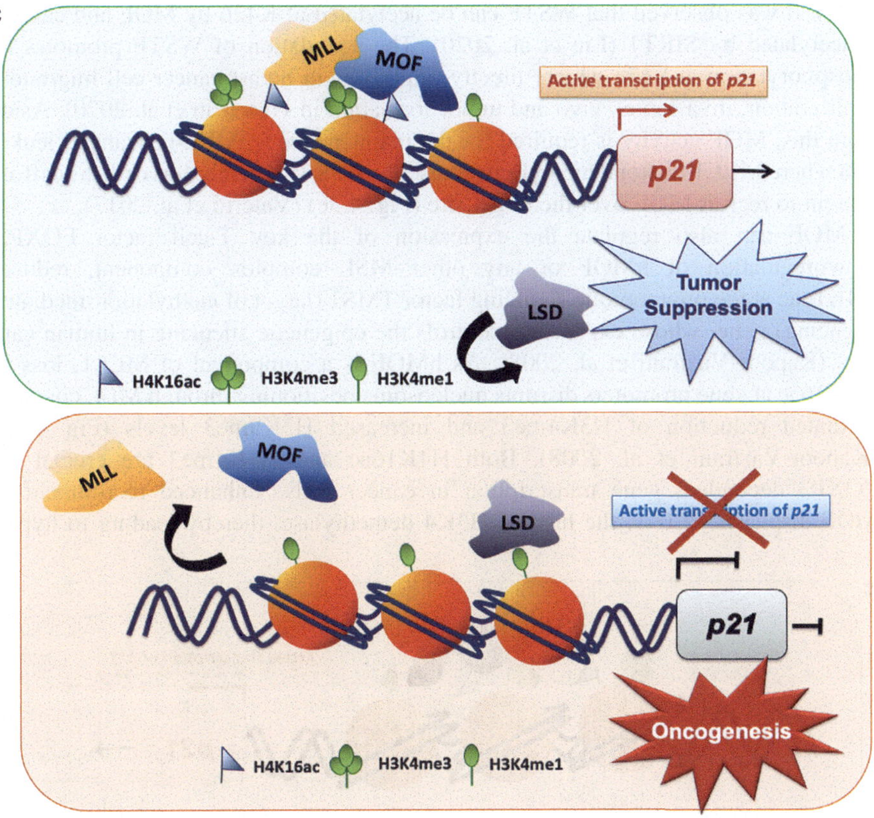

Fig. 4.3 (continued) at *CDH1* gene promoter, thus inducing CDH1 expression and suppressing EMT (epithelial to mesenchymal transition). Absence of MOF promotes LSD1 recruitment at *CDH1* promoter. (**b**) Estrogen regulates MOF expression during oncogenesis. MOF expression is regulated by estrogen concentration as $<10^{-7}$ mol/L of estrogen inhibits MOF expression thereby inducing apoptosis. Low estrogen concentration between 10^{-8} and 10^{-10} mol/L induces MOF expression, thereby hyper H4K16 acetylation, promoting oncogenesis by inducing *TNK2* expression. Estrogen receptor (ER), Growth factor receptor (GFR) and TNK2 phosphorylates PI3K, which further phosphorylates AKT, thus inducing mTORC1 and thereby oncogenesis. (**c**) Role of MOF in tumor suppression through multiple mechanisms. i. MOF and MLL gets are recruited at gene promoter of tumor suppressor p21 and to activates its transcription. ii. Absence of MOF leads to displacement of MLL by LSD, thereby repressing p21 expression and promoting oncogenesis

Hepatocellular Carcinoma (HCC) (Wei et al. 2021). ER alpha functions as a tumor suppressor in the context of HCC (Wei et al. 2021). Interestingly, it has been shown that hMOF interacts with ER alpha and acetylates the protein (Wei et al. 2021). This prevents its polyubiquitination-mediated degradation and stabilizes the protein thereby turning on ER alpha target genes where MOF can also act as coactivator (Wei et al. 2021). In another study in luminal breast cancer, Williams syndrome transcription factor (WSTF) was shown to be an activator of ER alpha (Liu et al.

2020). It was observed that WSTF can be acetylated at K426 by MOF and can get deacetylated by SIRT1 (Liu et al. 2020). The acetylation of WSTF promotes its phosphorylation at S158 which directly implicates in breast cancer cell migration, proliferation, invasion ex vivo and tumor formation in vivo (Liu et al. 2020). Aside from this, MOF activity is required for the maintenance of MLL-rearranged leukemia where MLL1 (histone methyltransferase for H3K79) merely serves as a scaffold protein to recruit MOF over the target site (Fig. 4.3c) (Valerio et al. 2017).

MOF can also regulate the expression of the key T-cell factor FOXP3. Downregulation of hMOF or any other MSL complex component, reduces H4K16ac at the proapoptotic signaling factor TMS1 (*t*arget of *m*ethylation-mediated *s*ilencing) gene, whose expression controls the epigenetic silencing in human cancers (Kapoor-Vazirani et al. 2008). As hMOF is a component of MLL1, loss of H4K16ac at gene promoters disrupts nucleosome positioning through MLL complex mediated reduction of H3K4me3 and increased H3K9me3 levels (Fig. 4.3c) (Kapoor-Vazirani et al. 2008). Both H4K16ac and H3K4me3 are crucial for FOXP3 dependent gene transcription in cancer cells. Enhanced recruitment of MOF displaces PLU-1, the histone H3K4 demethylase, thereby leading to hyper-

c

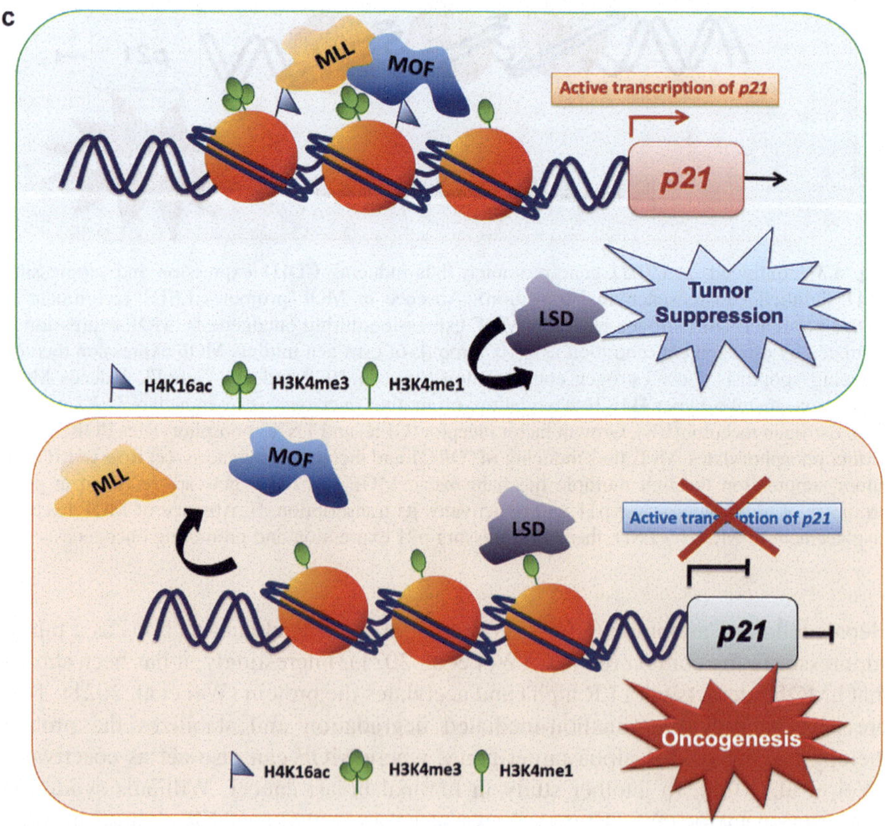

Fig. 4.3 (continued)

enrichment of H4K16ac and H3K4me3 at several FOXP3-associated genes (Katoh et al. 2011). Somatic as well as targeted mutations of FOXP3 in several cancers, including prostate cancer also perturb the nuclear localization of MOF (Katoh et al. 2011).

Another MYST family histone acetyl transferase, TIP60, has also been shown to play an important role in cancer. In gastric cancer, TIP60 expression decreases as the tumor size in patients increases, leading to low H4K16ac levels in gastric cancers with lymph node metastasis (Sakuraba et al. 2011). Expression of TIP60 may also play an important role in malignant glioblastomas as it positively regulates EGFR and through it the downstream proteins AKT and ERK1/2 involved in glioblastoma development (Dong et al. 2019).

Opposing the tumorigenic functions of acetyl transferases, deacetylases such as SIRT1 can have tumor suppressive roles as exemplified by inhibition of NF-kB signaling in gastric cancer (Yang et al. 2013). Although SIRT1/SIRT2 mediated H4K16ac deacetylation has established role in tumorigenicity, but the mitochondrial Sirtuin SIRT4 has not yet been explored in this context. Most recently, SIRT4 mediated deacetylation of H4K16 was found to exerts a tumor suppressive activity in breast cancer cells via modulation of SIRT1 expression and a repressive effect on glutamine metabolism (Du et al. 2020).

The multiple different modes of cancer mediated alterations in H4K16 acetylation levels highlights the importance of this histone mark. Different acetylation enzymes (hMOF or TIP60) and deacetylase SIRTUINS can regulate H4K16ac levels either positively or negatively impacting cell proliferation depending upon the cancer cell type. The H4K16 histone mark, therefore, has been identified as a key epigenetic regulator in various types of cancer.

Recruitment of Repair Proteins at DSBs Correlates with H4K16ac Status

Horikoshi and coworkers reported that there was no significant alteration of H4K16ac levels upon DSB induction in gene-rich or -poor regions (Horikoshi et al. 2019). However, while there was no difference in DSB repair by NHEJ with regard to H4K16ac chromatin status, there was relatively higher DSB repair by the HR pathway in H4K16ac-enriched chromatin regions. Higher H4K16ac levels are associated with gene-rich as opposed to gene-poor regions (Horikoshi et al. 2019). Surprisingly, increasing local H4K16ac levels by ectopic expression of a nuclease-dead dCas9-hMOFfusion protein (dCas9-hMOF) in gene-poor regions does not increase HR repair frequency (Horikoshi et al. 2019). The differences in HR frequency can be correlated with altered recruitment of HR-related repair factors at individual gene-rich or gene-poor sites (Horikoshi et al. 2019). Examined at greater detail, no significant differences in phosphorylated H2AX (γ-H2AX) were detected at the DSB sites, but KU80 and 53BP1 signals were high at DSBs in gene-poor regions with reduced H4K16ac, whereas RAD51, BRCA1, and RPA2 levels were high in gene-rich regions with enhanced H4K16ac (Horikoshi et al. 2019). G1-arrested cells had increased levels of KU80 and 53BP1 at the DSBs in gene/H4K16ac poor regions, whereas RAD51, BRCA1, and RPA2 levels are higher at gene/H4K16ac-rich sites in S-phase cells thus impacting the pathway choice for DSB repair. It was also observed that recruitment of HR proteins to DSB sites during DNA resection is assisted by SMARCAD1, an ATP dependent DNA helicase, which was recruited to DNA DSBs (Chakraborty et al. 2018). Interestingly, SMARCAD1 has proposed roles in nucleosome displacement, and this function is exerted in an ATM-dependent manner during the DNA resection process leading to HR (Chakraborty et al. 2018; Chen et al. 2012). Furthermore, the analysis of RNA polymerase ll recruitment to DSBs shows significant enrichment in gene-rich regions as compared with gene-poor regions, a correlation between DSB repair by HR, H4K16ac levels, and transcriptional (Horikoshi et al. 2019).

MOF Suppresses DNA Replicative Stress by Facilitating Resolution of Stalled Replication Forks

DNA replication leads to precise duplication of the genome and accurate chromosome segregation to the daughter cells. Significantly, inhibition of DNA replication fork progression, fork stability, or accurate fork re-initiation after replication stalling impact genome stability (Mazouzi et al. 2014; Zeman and Cimprich 2014). This event, termed replication stress, is induced by factors related to DNA Polymerase stalling, topoisomerase trapping, sensitivity to fragile sites, R-loop formation, oncogenic stress or stabilization of DNA secondary structures like G quadruplexes. If such impediments are not properly removed, chromosomal instability or oncogene activation are possible (Halazonetis et al. 2008).

Aberrant DNA replication can lead to replication fork stalling and fork collapse. Stalled forks accumulate single stranded DNA (ssDNA), which can ultimately be converted into DSBs after the fork collapses (Blow et al. 2011; Branzei and Foiani 2007). Thus, replicative stress can trigger the cellular DDR, a response that is dependent upon the surrounding chromatin environment.

Hydroxyurea (Hu) treatment of cells depletes the available nucleotide pool for the DNA polymerases required for efficient DNA replication, thereby stalling the replication fork. MOF depletion significantly increases chromosomal aberrations and reduces the survival of HU treated cells (Singh et al. 2018). This suggests that MOF not only plays a crucial role in maintenance of genome stability during DNA replication, but also has an important role under conditions of replicative stress (Singh et al. 2018).

Singh and co-workers reported that cellular DNA replication, under normal growth or during replicative stress conditions, is affected by MOF (Singh et al. 2018). Evaluation of replication fork dynamics at single-molecule resolution by the DNA fiber assay revealed defective replication fork progression upon MOF depletion which retarded cell growth (Han et al. 2018; Meng et al. 2020). Upon encountering DNA template obstacles, the replication forks stall and ultimately collapse leading to accumulation of DNA DSBs. The role of MOF in replication fork dynamics in MOF-depleted cells was studied after HU or aphidicolin (APH) treatment using the DNA fiber analysis (Singh et al. 2018). Interestingly, the DNA fiber tracts could be of three distinct types, continuing elongation forks, stalled forks and forks fired with new DNA replication origin (Singh et al. 2018). In absence of MOF, HU treatment could enhance the frequencies of stalled forks with new DNA replication origins ~four fold, as compared to the control cells. Further, the MOF-depleted cells showed a significant reduction in the fork speed (Singh et al. 2018). Overall, these results indicate that MOF plays a crucial role in the recovery of the replication fork, once movement is unblocked (Singh et al. 2018). Notably, in the absence of MOF the HU-mediated DNA damage sites revealed an increased persistence of γ-H2AX foci. This clearly identified a defect in processing of the DNA damage. Furthermore, MOF depletion resulted in a significant reduction in Mre11 foci, a marker of the early stage of DNA damage response (Singh et al. 2018). This suggests that MOF might be a crucial factor during early events of DNA end resection. MOF depletion also led to decreased recruitment of other downstream HR proteins, including RPA and RAD51 to the DNA damage sites (Singh et al. 2018). Significantly, the repair defects observed upon MOF depletion were not unique to HU-induced fork collapse, but could also be seen upon treatment with Cisplatin, a DNA cross-linking agent. Furthermore, absence of MOF led to a significant reduction in RIF1 and 53BP1, the classical NHEJ-related proteins, suggesting that MOF can also influence the NHEJ pathway (Gupta et al. 2014; Sharma et al. 2010). Whether these effects are due to reduced chromatin accessibility in the absence of H4K16ac or more direct effects remains to be determined.

Role of H4K16ac in Aging

The aging process is an intrinsically complex process associated with lifespan. Aging can involve increased individual vulnerability to physiological and environmental risk due to gradual declines in functional capacities at the cellular level (Veldurthy et al. 2016; Johnson et al. 1999). Over the last few decades scientists have tried to uncover the root causes and cumulative factors contributing to aging. In this regard, efforts have been made to understand the effect of genetic modulations and environmental factors on aging in twin studies. These studies have indicated that only around 25% of the lifespan variation during aging was associated with genetic factors (Skytthe et al. 2003; Ljungquist et al. 1998; Herskind et al. 1996; Iachine et al. 2006). In exploring other factors, Vaupel and co-workers analyzed non-heritable regulators. These factors, including parental economic status or educational background as well as other similar aspects, accounted for 25% of lifespan, whereas life circumstances such as medical and psychologic conditions contributed 50% (Vaupel et al. 1998). Scientists are now trying to explain the complex aging

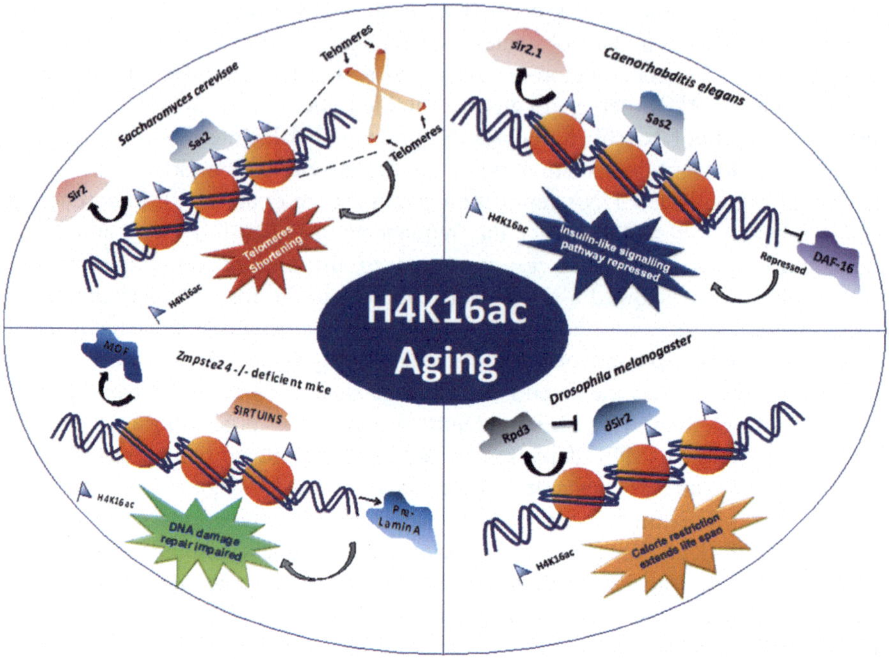

Fig. 4.4 H4K16 acetylation in aging. H4K16 acetylation role in aging plays an important role in *Saccharomyces cerevisiae*, *Caenorhabditis elegans*, *Drosophila melanogaster* and *Mus musculus* in the aging context through multiple mechanisms involving telomere maintenance, insulin-like signaling, calorie restrictions and DNA damage repair pathways

process by considering the interactions between genetic and environmental factors, meaning that epigenetic modulations have become an important consideration in regulating gene expression without changing the DNA sequence (Probst et al. 2009).

Multiple lines of evidence have shown that epigenetic alterations may be a hallmark for aging (Huidobro et al. 2013). In mammals, for example, the early onset of cellular senescence correlated to declining H4K16ac levels. Zhou and co-workers have shown that in the HGPS mouse model for premature aging (zmpste24-deficient mice), mis-positioning of MOF due to unprocessed prelamin A accumulation, causes hypoacetylation of target genes and early senescence (Krishnan et al. 2011). Ectopically overexpressed hMOF or HDAC inhibition increased H4K16ac and delayed early senescence in Zmpste24$^{-/-}$ MEFs in culture (Krishnan et al. 2011).

H4K16ac plays a contradictory role in telomere maintenance in yeast where during physiologic aging global H4K16ac levels were found to increase. As in newly born budding yeast, it was observed that Sir2, along with Sas2, maintains low levels of H4K16ac at telomeres and sub-telomeric regions (Dang et al. 2009). As the cells age, there was a decline in the level of Sir2 and a simultaneously increase in Sas2 activity which lead to increased H4K16ac and loss of histones in the telomeric region of X element (Dang et al. 2009) (Fig. 4.4). However, the contradiction regarding low H4K16ac causing senescence in mammals and high H4K16ac causing aging in yeast might be explained by tissue type specific requirement for both HATs and HDACs and their expression which help to perform acetylation and deacetylation of histones involved in regulating the aging process.

A comparison of H4K16ac levels in embryonic stem cells and neuronal progenitor cells (NPC), indicated genome wide changes during differentiation. In undifferentiated embryonic stem cells, H4K16ac levels were much greater than in NPCs and other differentiated cells, suggesting a possible function in stem cell differentiation or pluripotent growth requirements (Taylor et al. 2013: Mujoo et al. 2017). Alternatively, this may simply reflect a reduced amount of heterochromatin in embryonic stem cells.

Conclusion

From the regulation of various processes including transcription, cell cycle progression, chromatin organization, nuclear structure maintenance, DDR and apoptosis, H4K16ac has emerged as a critical site to control these key processes and maintain cellular homeostasis. This chapter summarizes recent studies and discoveries regarding the importance of H4K16ac epigenetic mark and how its function is regulated by using MOF and SIRTUINS. Within the context of DDR, hMOF-mediated acetylation of histone H4 at K16 regulates the functional association between the chromatin fiber and the non-histone chromatin proteins thereby loosening the higher order chromatin folding and critically impacting DNA repair and transcription. MOF mediated regulation of HR upon phosphorylation by ATM becomes critical by helping in the recruitment of the MRN complex, phosphorylation of BRCA1 and CTIP at the DSB sites, further helping with the recruitment of RAD50 to execute the HR process. However, similar regulation is not observed for NHEJ, where SIRT2 has been found to play a critical role by deacetylating KAP, to assist with its recruitment to the DSB site along with MOF and 53BP1. Apart from their function in NHEJ, a conflicting mode of regulation between SIRT2 and MOF was observed in the regulation of metabolism, where SIRT2 stabilizes c-MYC to fulfill the increased metabolic demands of the cells, while MOF tries to shut it down.

In comparison to their mode of regulation in NHEJ, the functional regulation of MOF and SIRT2 in metabolism leads us to wonder if any competitive interactions are playing any role in NHEJ or the differences are exerted at the level of target regulation. Thus, in this regard the future challenge includes efforts to understand how MOF and SIRT2 functions are being regulated and whether or not their function and recruitment is dependent upon external stimuli or is it simply expression dependent for different histone targeted or non-histone targeted proteins. Moreover, increasing evidence suggests a role for MOF in cancer where different models of regulation may facilitate or inhibit cancer outcomes. Although these different models of regulation work in different types of cancers, a detailed study of their mechanism by which MOF mediates tumorigenicity is needed to understand their functional signaling cascade and future development of cancer therapeutic strategies. In addition, it will be important to identify other downstream non-histone targets for MOF that may be important to recognize its activity and specificity of these processes.

Acknowledgments This work was supported by grants GM109768 and RO1 CA129537 (T.K.P.). This work was supported in part by research grants received from Basic and Applied Research in Biophysics and Material Science (RSI 4002) from the Department of Atomic Energy; Science & Engineering Research Board (SERB)(CRG/2018/000985), Department of Science & Technology; SwarnaJayanti Fellowship (DST/SJF/LSA-02/2017-18), Department of Science and Technology; Department of Biotechnology (BT/PR28920/MED/122/176/2018); S. Ramachandran National Bioscience Award for Career Development 2019 (BT/HRD-NBA-NWB/38/2019-20), Department of Biotechnology, Govt. of India to CD. Department of Science and Technology (EMR/2016/006233), Govt. of India to SR. SA thanks the Council of Scientific and Industrial Research (CSIR), Govt. of India, for fellowship support. Dr. Ramos is the recipient of a GURI award from the State of Texas, USA.

Conflict of Interest The authors declare no competing interests.

References

Aerbajinai W, Zhu J, Gao Z, Chin K, Rodgers GP (2007) Thalidomide induces gamma-globin gene expression through increased reactive oxygen species-mediated p38 MAPK signaling and histone H4 acetylation in adult erythropoiesis. Blood 110:2864–2871

Akhtar A, Becker PB (2000) Activation of transcription through histone H4 acetylation by MOF, an acetyltransferase essential for dosage compensation in Drosophila. Mol Cell 5:367–375

Allfrey VG, Faulkner R, Mirsky AE (1964) Acetylation and methylation of histones and their possible role in the regulation of Rna synthesis. Proc Natl Acad Sci U S A 51:786–794

Allfrey VG, Pogo BG, Littau VC, Gershey EL, Mirsky AE (1968) Histone acetylation in insect chromosomes. Science 159:314–316

Allis CD, Berger SL, Cote J, Dent S, Jenuwien T, Kouzarides T, Pillus L, Reinberg D, Shi Y, Shiekhattar R et al (2007) New nomenclature for chromatin-modifying enzymes. Cell 131:633–636

Belote JM, Lucchesi JC (1980) Male-specific lethal mutations of Drosophila melanogaster. Genetics 96:165–186

Bhadra MP, Horikoshi N, Pushpavallipvalli SN, Sarkar A, Bag I, Krishnan A, Lucchesi JC, Kumar R, Yang Q, Pandita RK et al (2012) The role of MOF in the ionizing radiation response is conserved in Drosophila melanogaster. Chromosoma 121:79–90

Bird AW, Yu DY, Pray-Grant MG, Qiu Q, Harmon KE, Megee PC, Grant PA, Smith MM, Christman MF (2002) Acetylation of histone H4 by Esa1 is required for DNA double-strand break repair. Nature 419:411–415

Blow JJ, Ge XQ, Jackson DA (2011) How dormant origins promote complete genome replication. Trends Biochem Sci 36:405–414

Bone JR, Kuroda MI (1996) Dosage compensation regulatory proteins and the evolution of sex chromosomes in Drosophila. Genetics 144:705–713

Branzei D, Foiani M (2007) Interplay of replication checkpoints and repair proteins at stalled replication forks. DNA Repair (Amst) 6:994–1003

Bruserud O, Stapnes C, Tronstad KJ, Ryningen A, Anensen N, Gjertsen BT (2006) Protein lysine acetylation in normal and leukaemic haematopoiesis: HDACs as possible therapeutic targets in adult AML. Expert Opin Ther Targets 10:51–68

Cai Y, Jin J, Swanson SK, Cole MD, Choi SH, Florens L, Washburn MP, Conaway JW, Conaway RC (2010) Subunit composition and substrate specificity of a MOF-containing histone acetyltransferase distinct from the male-specific lethal (MSL) complex. J Biol Chem 285: 4268–4272

Cao L, Zhu L, Yang J, Su J, Ni J, Du Y, Liu D, Wang Y, Wang F, Jin J et al (2014) Correlation of low expression of hMOF with clinicopathological features of colorectal carcinoma, gastric cancer and renal cell carcinoma. Int J Oncol 44:1207–1214

Chakraborty S, Pandita R, Hambarde S, Mattoo AR, Charaka V, Ahmad KM, Iyer SP, Hunt RA, Pandita TK (2018) SMARCAD1 phosphorylation and ubiquitination are required for resection during DNA double-strand break repair. iScience 2:123–135

Chen X, Cui D, Papusha A, Zhang X, Chu CD, Tang J, Chen K, Pan X, Ira G (2012) The Fun30 nucleosome remodeller promotes resection of DNA double-strand break ends. Nature 489:576–580

Choudhary C, Kumar C, Gnad F, Nielsen ML, Rehman M, Walther TC, Olsen JV, Mann M (2009) Lysine acetylation targets protein complexes and co-regulates major cellular functions. Science 325:834–840

Clouaire T, Rocher V, Lashgari A, Arnould C, Aguirrebengoa M, Biernacka A, Skrzypczak M, Aymard F, Fongang B, Dojer N et al (2018) Comprehensive mapping of histone modifications at DNA double-strand breaks deciphers repair pathway chromatin signatures. Mol Cell. 72(2): 250–262.e6

Dang W, Steffen KK, Perry R, Dorsey JA, Johnson FB, Shilatifard A, Kaeberlein M, Kennedy BK, Berger SL (2009) Histone H4 lysine 16 acetylation regulates cellular lifespan. Nature 459:802–807

Dantuma NP, van Attikum H (2016) Spatiotemporal regulation of posttranslational modifications in the DNA damage response. EMBO J 35:6–23

Ding N, Bonham EM, Hannon BE, Amick TR, Baylin SB, O'Hagan HM (2016) Mismatch repair proteins recruit DNA methyltransferase 1 to sites of oxidative DNA damage. J Mol Cell Biol 8: 244–254

Dobbin MM, Madabhushi R, Pan L, Chen Y, Kim D, Gao J, Ahanonu B, Pao PC, Qiu Y, Zhao Y et al (2013) SIRT1 collaborates with ATM and HDAC1 to maintain genomic stability in neurons. Nat Neurosci 16:1008–1015

Dong Z, Zou J, Li J, Pang Y, Liu Y, Deng C, Chen F, Cui H (2019) MYST1/KAT8 contributes to tumor progression by activating EGFR signaling in glioblastoma cells. Cancer Med 8:7793–7808

Dryden SC, Nahhas FA, Nowak JE, Goustin AS, Tainsky MA (2003) Role for human SIRT2 NAD-dependent deacetylase activity in control of mitotic exit in the cell cycle. Mol Cell Biol 23:3173–3185

Du L, Liu X, Ren Y, Li J, Li P, Jiao Q, Meng P, Wang F, Wang Y, Wang YS et al (2020) Loss of SIRT4 promotes the self-renewal of breast cancer stem cells. Theranostics 10:9458–9476

Dzieran J, Rodriguez Garcia A, Westermark UK, Henley AB, Eyre Sanchez E, Trager C, Johansson HJ, Lehtio J, Arsenian-Henriksson M (2018) MYCN-amplified neuroblastoma maintains an aggressive and undifferentiated phenotype by deregulation of estrogen and NGF signaling. Proc Natl Acad Sci U S A 115:E1229–E1238

Eldridge MJG, Pereira JM, Impens F, Hamon MA (2020) Active nuclear import of the deacetylase Sirtuin-2 is controlled by its C-terminus and importins. Sci Rep 10:2034

Fraga MF, Ballestar E, Villar-Garea A, Boix-Chornet M, Espada J, Schotta G, Bonaldi T, Haydon C, Ropero S, Petrie K et al (2005) Loss of acetylation at Lys16 and trimethylation at Lys20 of histone H4 is a common hallmark of human cancer. Nat Genet 37:391–400

Guarente L (1999) Diverse and dynamic functions of the Sir silencing complex. Nat Genet 23:281–285

Gupta A, Sharma GG, Young CS, Agarwal M, Smith ER, Paull TT, Lucchesi JC, Khanna KK, Ludwig T, Pandita TK (2005) Involvement of human MOF in ATM function. Mol Cell Biol 25: 5292–5305

Gupta A, Guerin-Peyrou TG, Sharma GG, Park C, Agarwal M, Ganju RK, Pandita S, Choi K, Sukumar S, Pandita RK et al (2008) The mammalian ortholog of Drosophila MOF that acetylates histone H4 lysine 16 is essential for embryogenesis and oncogenesis. Mol Cell Biol 28:397–409

Gupta A, Hunt CR, Hegde ML, Chakraborty S, Chakraborty S, Udayakumar D, Horikoshi N, Singh M, Ramnarain DB, Hittelman WN et al (2014) MOF phosphorylation by ATM regulates 53BP1-mediated double-strand break repair pathway choice. Cell Rep 8:177–189

Gustafson WC, Weiss WA (2010) Myc proteins as therapeutic targets. Oncogene 29:1249–1259

Halazonetis TD, Gorgoulis VG, Bartek J (2008) An oncogene-induced DNA damage model for cancer development. Science 319:1352–1355

Hallows WC, Yu W, Denu JM (2012) Regulation of glycolytic enzyme phosphoglycerate mutase-1 by Sirt1 protein-mediated deacetylation. J Biol Chem 287:3850–3858

Han Z, Wu H, Kim S, Yang X, Li Q, Huang H, Cai H, Bartlett MG, Dong A, Zeng H et al (2018) Revealing the protein propionylation activity of the histone acetyltransferase MOF (males absent on the first). J Biol Chem 293:3410–3420

Herskind AM, McGue M, Holm NV, Sorensen TI, Harvald B, Vaupel JW (1996) The heritability of human longevity: a population-based study of 2872 Danish twin pairs born 1870-1900. Hum Genet 97:319–323

Hida Y, Kubo Y, Murao K, Arase S (2007) Strong expression of a longevity-related protein, SIRT1, in Bowen's disease. Arch Dermatol Res 299:103–106

Hilfiker A, Yang Y, Hayes DH, Beard CA, Manning JE, Lucchesi JC (1994) Dosage compensation in Drosophila: the X-chromosomal binding of MSL-1 and MLE is dependent on Sxl activity. EMBO J 13:3542–3550

Hilfiker A, Hilfiker-Kleiner D, Pannuti A, Lucchesi JC (1997) mof, a putative acetyl transferase gene related to the Tip60 and MOZ human genes and to the SAS genes of yeast, is required for dosage compensation in Drosophila. EMBO J 16:2054–2060

Horikoshi N, Kumar P, Sharma GG, Chen M, Hunt CR, Westover K, Chowdhury S, Pandita TK (2013) Genome-wide distribution of histone H4 lysine 16 acetylation sites and their relationship to gene expression. Genome Integr 4:3

Horikoshi N, Sharma D, Leonard F, Pandita RK, Charaka VK, Hambarde S, Horikoshi NT, Gaur Khaitan P, Chakraborty S, Cote J et al (2019) Pre-existing H4K16ac levels in euchromatin drive DNA repair by homologous recombination in S-phase. Commun Biol 2:253

Houtkooper RH, Pirinen E, Auwerx J (2012) Sirtuins as regulators of metabolism and healthspan. Nat Rev Mol Cell Biol 13:225–238

Huang HT, Kathrein KL, Barton A, Gitlin Z, Huang YH, Ward TP, Hofmann O, Dibiase A, Song A, Tyekucheva S et al (2013) A network of epigenetic regulators guides developmental haematopoiesis in vivo. Nat Cell Biol 15:1516–1525

Hughes KJ, Meares GP, Hansen PA, Corbett JA (2011) FoxO1 and SIRT1 regulate beta-cell responses to nitric oxide. J Biol Chem 286:8338–8348

Huidobro C, Fernandez AF, Fraga MF (2013) Aging epigenetics: causes and consequences. Mol Asp Med 34:765–781

Hunt CR, Ramnarain D, Horikoshi N, Iyengar P, Pandita RK, Shay JW, Pandita TK (2013) Histone modifications and DNA double-strand break repair after exposure to ionizing radiations. Radiat Res 179:383–392

Hjelmborg JB, Iachine I, Skytthe A, Vaupel JW, McGue M, Koskenvuo M, Kaprio J, Pedersen NL, Christensen K (2006) Genetic influence on human lifespan and longevity. Hum Genet 119:312–321

Ikura T, Ogryzko VV, Grigoriev M, Groisman R, Wang J, Horikoshi M, Scully R, Qin J, Nakatani Y (2000) Involvement of the TIP60 histone acetylase complex in DNA repair and apoptosis. Cell 102:463–473

Imai S, Armstrong CM, Kaeberlein M, Guarente L (2000) Transcriptional silencing and longevity protein Sir2 is an NAD-dependent histone deacetylase. Nature 403:795–800

Jeong J, Juhn K, Lee H, Kim SH, Min BH, Lee KM, Cho MH, Park GH, Lee KH (2007) SIRT1 promotes DNA repair activity and deacetylation of Ku70. Exp Mol Med 39:8–13

Jiang W, Wang S, Xiao M, Lin Y, Zhou L, Lei Q, Xiong Y, Guan KL, Zhao S (2011) Acetylation regulates gluconeogenesis by promoting PEPCK1 degradation via recruiting the UBR5 ubiquitin ligase. Mol Cell 43:33–44

Johnson FB, Sinclair DA, Guarente L (1999) Molecular biology of aging. Cell 96:291–302

Kapoor-Vazirani P, Kagey JD, Powell DR, Vertino PM (2008) Role of hMOF-dependent histone H4 lysine 16 acetylation in the maintenance of TMS1/ASC gene activity. Cancer Res 68:6810–6821

Katoh H, Qin ZS, Liu R, Wang L, Li W, Li X, Wu L, Du Z, Lyons R, Liu CG et al (2011) FOXP3 orchestrates H4K16 acetylation and H3K4 trimethylation for activation of multiple genes by recruiting MOF and causing displacement of PLU-1. Mol Cell 44:770–784

Kim HS, Vassilopoulos A, Wang RH, Lahusen T, Xiao Z, Xu X, Li C, Veenstra TD, Li B, Yu H et al (2011) SIRT2 maintains genome integrity and suppresses tumorigenesis through regulating APC/C activity. Cancer Cell 20:487–499

Kiran S, Chatterjee N, Singh S, Kaul SC, Wadhwa R, Ramakrishna G (2013) Intracellular distribution of human SIRT7 and mapping of the nuclear/nucleolar localization signal. FEBS J 280:3451–3466

Krishnan V, Chow MZ, Wang Z, Zhang L, Liu B, Liu X, Zhou Z (2011) Histone H4 lysine 16 hypoacetylation is associated with defective DNA repair and premature senescence in Zmpste24-deficient mice. Proc Natl Acad Sci U S A 108:12325–12330

Li X, Corsa CA, Pan PW, Wu L, Ferguson D, Yu X, Min J, Dou Y (2010) MOF and H4 K16 acetylation play important roles in DNA damage repair by modulating recruitment of DNA damage repair protein Mdc1. Mol Cell Biol 30:5335–5347

Li D, Yang Y, Chen B, Guo X, Gao S, Wang M, Duan M, Li X (2020) MOF regulates TNK2 transcription expression to promote cell proliferation in thyroid cancer. Front Pharmacol 11: 607605

Lin YH, Yuan J, Pei H, Liu T, Ann DK, Lou Z (2015) KAP1 deacetylation by SIRT1 promotes non-homologous end-joining repair. PLoS One 10:e0123935

Liu Y, Dentin R, Chen D, Hedrick S, Ravnskjaer K, Schenk S, Milne J, Meyers DJ, Cole P, Yates J 3rd et al (2008) A fasting inducible switch modulates gluconeogenesis via activator/coactivator exchange. Nature 456:269–273

Liu PY, Xu N, Malyukova A, Scarlett CJ, Sun YT, Zhang XD, Ling D, Su SP, Nelson C, Chang DK et al (2013) The histone deacetylase SIRT2 stabilizes Myc oncoproteins. Cell Death Differ 20: 503–514

Liu Y, Zhang YY, Wang SQ, Li M, Long YH, Li YF, Liu YK, Li YH, Wang YQ, Mi JS et al (2020) WSTF acetylation by MOF promotes WSTF activities and oncogenic functions. Oncogene 39: 5056–5067

Ljungquist B, Berg S, Lanke J, McClearn GE, Pedersen NL (1998) The effect of genetic factors for longevity: a comparison of identical and fraternal twins in the Swedish Twin Registry. J Gerontol A Biol Sci Med Sci 53:M441–M446

Lombard DB, Alt FW, Cheng HL, Bunkenborg J, Streeper RS, Mostoslavsky R, Kim J, Yancopoulos G, Valenzuela D, Murphy A et al (2007) Mammalian Sir2 homolog SIRT3 regulates global mitochondrial lysine acetylation. Mol Cell Biol 27:8807–8814

Luo H, Shenoy AK, Li X, Jin Y, Jin L, Cai Q, Tang M, Liu Y, Chen H, Reisman D et al (2016) MOF acetylates the histone demethylase LSD1 to suppress epithelial-to-mesenchymal transition. Cell Rep 15:2665–2678

Mahlamaki EH, Barlund M, Tanner M, Gorunova L, Hoglund M, Karhu R, Kallioniemi A (2002) Frequent amplification of 8q24, 11q, 17q, and 20q-specific genes in pancreatic cancer. Genes Chromosomes Cancer 35:353–358

Mazouzi A, Velimezi G, Loizou JI (2014) DNA replication stress: causes, resolution and disease. Exp Cell Res 329:85–93

McCord RA, Michishita E, Hong T, Berber E, Boxer LD, Kusumoto R, Guan S, Shi X, Gozani O, Burlingame AL et al (2009) SIRT6 stabilizes DNA-dependent protein kinase at chromatin for DNA double-strand break repair. Aging (Albany NY) 1:109–121

Mei Z, Zhang X, Yi J, Huang J, He J, Tao Y (2016) Sirtuins in metabolism, DNA repair and cancer. J Exp Clin Cancer Res 35:182

Meng F, Qian M, Peng B, Peng L, Wang X, Zheng K, Liu Z, Tang X, Zhang S, Sun S et al (2020) Synergy between SIRT1 and SIRT6 helps recognize DNA breaks and potentiates the DNA damage response and repair in humans and mice. Elife 9:e55828

Miller DM, Thomas SD, Islam A, Muench D, Sedoris K (2012) c-Myc and cancer metabolism. Clin Cancer Res 18:5546–5553

Mishima Y, Miyagi S, Saraya A, Negishi M, Endoh M, Endo TA, Toyoda T, Shinga J, Katsumoto T, Chiba T et al (2011) The Hbo1-Brd1/Brpf2 complex is responsible for global acetylation of H3K14 and required for fetal liver erythropoiesis. Blood 118:2443–2453

Morrish F, Neretti N, Sedivy JM, Hockenbery DM (2008) The oncogene c-Myc coordinates regulation of metabolic networks to enable rapid cell cycle entry. Cell Cycle 7:1054–1066

Mujoo K, Pandita RK, Tiwari A, Charaka V, Chakraborty S, Singh DK, Hambarde S, Hittelman WN, Horikoshi N, Hunt CR, Khanna KK, Kots AY, Brian Butler E, Murad F, Pandita TK (2017) Differentiation of human induced pluripotent or embryonic stem cells decreases the DNA damage repair by homologous recombination. Stem Cell Rep 9(5):1660–1674. https://doi.org/10.1016/j.stemcr.2017.10.002

Neal KC, Pannuti A, Smith ER, Lucchesi JC (2000) A new human member of the MYST family of histone acetyl transferases with high sequence similarity to Drosophila MOF. Biochim Biophys Acta 1490:170–174

Oberdoerffer P, Michan S, McVay M, Mostoslavsky R, Vann J, Park SK, Hartlerode A, Stegmuller J, Hafner A, Loerch P et al (2008) SIRT1 redistribution on chromatin promotes genomic stability but alters gene expression during aging. Cell 135:907–918

O'Hagan HM, Wang W, Sen S, Destefano Shields C, Lee SS, Zhang YW, Clements EG, Cai Y, Van Neste L, Easwaran H et al (2011) Oxidative damage targets complexes containing DNA methyltransferases, SIRT1, and polycomb members to promoter CpG Islands. Cancer Cell 20: 606–619

Pessoa Rodrigues C, Herman JS, Herquel B, Valsecchi CIK, Stehle T, Grun D, Akhtar A (2020) Temporal expression of MOF acetyltransferase primes transcription factor networks for erythroid fate. Sci Adv 6:eaaz4815

Probst AV, Dunleavy E, Almouzni G (2009) Epigenetic inheritance during the cell cycle. Nat Rev Mol Cell Biol 10:192–206

Qi Y, Tan M, Zheng M, Jin S, Wang H, Liu J, Wang P, Nie X, Gao L, Lin B (2020) Estrogen/estrogen receptor promotes the proliferation of endometrial carcinoma cells by enhancing hMOF expression. Jpn J Clin Oncol 50:241–253

Rajman L, Chwalek K, Sinclair DA (2018) Therapeutic potential of NAD-boosting molecules: the in vivo evidence. Cell Metab 27:529–547

Rea S, Xouri G, Akhtar A (2007) Males absent on the first (MOF): from flies to humans. Oncogene 26:5385–5394

Sakuraba K, Yokomizo K, Shirahata A, Goto T, Saito M, Ishibashi K, Kigawa G, Nemoto H, Hibi K (2011) TIP60 as a potential marker for the malignancy of gastric cancer. Anticancer Res 31: 77–79

Sayin VI, LeBoeuf SE, Singh SX, Davidson SM, Biancur D, Guzelhan BS, Alvarez SW, Wu WL, Karakousi TR, Zavitsanou AM et al (2017) Activation of the NRF2 antioxidant program generates an imbalance in central carbon metabolism in cancer. Elife 6:e28083

Serrano L, Martinez-Redondo P, Marazuela-Duque A, Vazquez BN, Dooley SJ, Voigt P, Beck DB, Kane-Goldsmith N, Tong Q, Rabanal RM et al (2013) The tumor suppressor SirT2 regulates cell cycle progression and genome stability by modulating the mitotic deposition of H4K20 methylation. Genes Dev 27:639–653

Sharma GG, So S, Gupta A, Kumar R, Cayrou C, Avvakumov N, Bhadra U, Pandita RK, Porteus MH, Chen DJ et al (2010) MOF and histone H4 acetylation at lysine 16 are critical for DNA damage response and double-strand break repair. Mol Cell Biol 30:3582–3595

Shi Y, Lan F, Matson C, Mulligan P, Whetstine JR, Cole PA, Casero RA, Shi Y (2004) Histone demethylation mediated by the nuclear amine oxidase homolog LSD1. Cell 119:941–953

Shogren-Knaak M, Ishii H, Sun JM, Pazin MJ, Davie JR, Peterson CL (2006) Histone H4-K16 acetylation controls chromatin structure and protein interactions. Science 311:844–847

Sinclair DA, Guarente L (2006) Unlocking the secrets of longevity genes. Sci Am 294(48–51): 54–47

Singh DK, Pandita RK, Singh M, Chakraborty S, Hambarde S, Ramnarain D, Charaka V, Ahmed KM, Hunt CR, Pandita TK (2018) MOF suppresses replication stress and contributes to resolution of stalled replication forks. Mol Cell Biol 38:e00484–e00417

Singh M, Bacolla A, Chaudhary S, Hunt CR, Pandita S, Chauhan R, Gupta A, Tainer JA, Pandita TK (2020a) Histone acetyl transferase MOF orchestrates outcomes at the crossroad of oncogenesis, DNA damage response, proliferation, and stem cell development. Mol Cell Biol. 40(18):e00232–e00220

Singh M, Bacolla A, Chaudhary S, Hunt CR, Pandita S, Chauhan R, Gupta A, Tainer JA, Pandita TK (2020b) Histone acetyltransferase MOF orchestrates outcomes at the crossroad of oncogenesis, DNA damage response, proliferation, and stem cell development. Mol Cell Biol. 40(18): e00232–e00220

Skytthe A, Pedersen NL, Kaprio J, Stazi MA, Hjelmborg JV, Iachine I, Vaupel JW, Christensen K (2003) Longevity studies in GenomEUtwin. Twin Res 6:448–454

Smith ER, Pannuti A, Gu W, Steurnagel A, Cook RG, Allis CD, Lucchesi JC (2000) The drosophila MSL complex acetylates histone H4 at lysine 16, a chromatin modification linked to dosage compensation. Mol Cell Biol 20:312–318

Smith ER, Cayrou C, Huang R, Lane WS, Cote J, Lucchesi JC (2005) A human protein complex homologous to the Drosophila MSL complex is responsible for the majority of histone H4 acetylation at lysine 16. Mol Cell Biol 25:9175–9188

Stephens PJ, Tarpey PS, Davies H, Van Loo P, Greenman C, Wedge DC, Nik-Zainal S, Martin S, Varela I, Bignell GR et al (2012) The landscape of cancer genes and mutational processes in breast cancer. Nature 486:400–404

Su J, Wang F, Cai Y, Jin J (2016) The functional analysis of histone acetyltransferase MOF in tumorigenesis. Int J Mol Sci 17:99

Sun B, Guo S, Tang Q, Li C, Zeng R, Xiong Z, Zhong C, Ding J (2011) Regulation of the histone acetyltransferase activity of hMOF via autoacetylation of Lys274. Cell Res 21:1262–1266

Sun XJ, Man N, Tan Y, Nimer SD, Wang L (2015) The role of histone acetyltransferases in Normal and malignant hematopoiesis. Front Oncol 5:108

Taipale M, Rea S, Richter K, Vilar A, Lichter P, Imhof A, Akhtar A (2005) hMOF histone acetyltransferase is required for histone H4 lysine 16 acetylation in mammalian cells. Mol Cell Biol 25:6798–6810

Tamburini BA, Tyler JK (2005) Localized histone acetylation and deacetylation triggered by the homologous recombination pathway of double-strand DNA repair. Mol Cell Biol 25:4903–4913

Tang J, Cho NW, Cui G, Manion EM, Shanbhag NM, Botuyan MV, Mer G, Greenberg RA (2013) Acetylation limits 53BP1 association with damaged chromatin to promote homologous recombination. Nat Struct Mol Biol 20:317–325

Tanno M, Kuno A, Horio Y, Miura T (2012) Emerging beneficial roles of sirtuins in heart failure. Basic Res Cardiol 107:273

Taylor GC, Eskeland R, Hekimoglu-Balkan B, Pradeepa MM, Bickmore WA (2013) H4K16 acetylation marks active genes and enhancers of embryonic stem cells, but does not alter chromatin compaction. Genome Res 23:2053–2065

Tjeertes JV, Miller KM, Jackson SP (2009) Screen for DNA-damage-responsive histone modifications identifies H3K9Ac and H3K56Ac in human cells. EMBO J 28:1878–1889
Tse C, Sera T, Wolffe AP, Hansen JC (1998) Disruption of higher-order folding by core histone acetylation dramatically enhances transcription of nucleosomal arrays by RNA polymerase III. Mol Cell Biol 18:4629–4638
Valerio DG, Xu H, Chen CW, Hoshii T, Eisold ME, Delaney C, Cusan M, Deshpande AJ, Huang CH, Lujambio A et al (2017) Histone acetyltransferase activity of MOF is required for MLL-AF9 leukemogenesis. Cancer Res 77:1753–1762
Vaquero A, Scher MB, Lee DH, Sutton A, Cheng HL, Alt FW, Serrano L, Sternglanz R, Reinberg D (2006) SirT2 is a histone deacetylase with preference for histone H4 Lys 16 during mitosis. Genes Dev 20:1256–1261
Vaquero A, Scher M, Erdjument-Bromage H, Tempst P, Serrano L, Reinberg D (2007) SIRT1 regulates the histone methyl-transferase SUV39H1 during heterochromatin formation. Nature 450:440–444
Vaupel JW, Carey JR, Christensen K, Johnson TE, Yashin AI, Holm NV, Iachine IA, Kannisto V, Khazaeli AA, Liedo P et al (1998) Biodemographic trajectories of longevity. Science 280:855–860
Vazquez BN, Thackray JK, Serrano L (2017) Sirtuins and DNA damage repair: SIRT7 comes to play. Nucleus 8:107–115
Veldurthy V, Wei R, Oz L, Dhawan P, Jeon YH, Christakos S (2016) Vitamin D, calcium homeostasis and aging. Bone Res 4:16041
Verger A, Crossley M (2004) Chromatin modifiers in transcription and DNA repair. Cell Mol Life Sci 61:2154–2162
Vidanes GM, Bonilla CY, Toczyski DP (2005) Complicated tails: histone modifications and the DNA damage response. Cell 121:973–976
Wang F, Tong Q (2009) SIRT2 suppresses adipocyte differentiation by deacetylating FOXO1 and enhancing FOXO1's repressive interaction with PPARgamma. Mol Biol Cell 20:801–808
Wang RH, Sengupta K, Li C, Kim HS, Cao L, Xiao C, Kim S, Xu X, Zheng Y, Chilton B et al (2008) Impaired DNA damage response, genome instability, and tumorigenesis in SIRT1 mutant mice. Cancer Cell 14:312–323
Wei S, Liu W, Sun N, Wu Y, Song H, Wang C, Wang S, Zou R, Lin L, Zeng K et al (2021) MOF upregulates the estrogen receptor alpha signaling pathway by its acetylase activity in hepatocellular carcinoma. Cancer Sci 112:1865–1877
Wetter TJ, Gazdag AC, Dean DJ, Cartee GD (1999) Effect of calorie restriction on in vivo glucose metabolism by individual tissues in rats. Am J Phys 276:E728–E738
Widlak P, Pietrowska M, Lanuszewska J (2006) The role of chromatin proteins in DNA damage recognition and repair. Histochem Cell Biol 125:119–126
Williamson EA, Wray JW, Bansal P, Hromas R (2012) Overview for the histone codes for DNA repair. Prog Mol Biol Transl Sci 110:207–227
Yang XJ, Ullah M (2007) MOZ and MORF, two large MYSTic HATs in normal and cancer stem cells. Oncogene 26:5408–5419
Yang Q, Wang B, Gao W, Huang S, Liu Z, Li W, Jia J (2013) SIRT1 is downregulated in gastric cancer and leads to G1-phase arrest via NF-kappaB/cyclin D1 signaling. Mol Cancer Res 11: 1497–1507
Yuan Z, Seto E (2007) A functional link between SIRT1 deacetylase and NBS1 in DNA damage response. Cell Cycle 6:2869–2871
Zeman MK, Cimprich KA (2014) Causes and consequences of replication stress. Nat Cell Biol 16: 2–9
Zhang J, Liu H, Pan H, Yang Y, Huang G, Yang Y, Zhou WP, Pan ZY (2014) The histone acetyltransferase hMOF suppresses hepatocellular carcinoma growth. Biochem Biophys Res Commun 452:575–580
Zhang R, Erler J, Langowski J (2017) Histone acetylation regulates chromatin accessibility: role of H4K16 in inter-nucleosome interaction. Biophys J 112:450–459

Zhang Y, Jang Y, Lee JE, Ahn J, Xu L, Holden MR, Cornett EM, Krajewski K, Klein BJ, Wang SP et al (2019) Selective binding of the PHD6 finger of MLL4 to histone H4K16ac links MLL4 and MOF. Nat Commun 10:2314

Zhong J, Ji L, Chen H, Li X, Zhang J, Wang X, Wu W, Xu Y, Huang F, Cai W et al (2017a) Acetylation of hMOF modulates H4K16ac to regulate DNA repair genes in response to oxidative stress. Int J Biol Sci 13:923–934

Zhong J, Li X, Cai W, Wang Y, Dong S, Yang J, Zhang J, Wu N, Li Y, Mao F et al (2017b) TET1 modulates H4K16 acetylation by controlling auto-acetylation of hMOF to affect gene regulation and DNA repair function. Nucleic Acids Res 45:672–684

Zhu L, Yang J, Zhao L, Yu X, Wang L, Wang F, Cai Y, Jin J (2015) Expression of hMOF, but not HDAC4, is responsible for the global histone H4K16 acetylation in gastric carcinoma. Int J Oncol 46:2535–2545

Zhu H, Wang Y, Wei T, Zhao X, Li F, Li Y, Wang F, Cai Y, Jin J (2021) KAT8/MOF-mediated anti-cancer mechanism of gemcitabine in human bladder cancer cells. Biomol Ther (Seoul) 29: 184–194

Chapter 5
Autophagy in Cancer: A Metabolic Perspective

Sweta Sikder, Atanu Mondal, Chandrima Das, and Tapas K. Kundu

Abstract Autophagy is an intracellular catabolic degradative process in which damaged cellular organelles, unwanted proteins and different cytoplasmic components get recycled to maintain cellular homeostasis or metabolic balance. During autophagy, a double membrane vesicle is formed to engulf these cytosolic materials and fuse to lysosomes wherein the entire cargo degrades to be used again. Because of this unique recycling ability of cells, autophagy is a universal stress response mechanism. Dysregulation of autophagy leads to several diseases, including cancer, neurodegeneration and microbial infection. Thus, autophagy machineries have become targets for therapeutics. This chapter provides an overview of the paradoxical role of autophagy in tumorigenesis in the perspective of metabolism.

Keywords Autophagy · Cancer · Histone modifications · Metabolism

S. Sikder
Transcription and Disease Laboratory, Molecular Biology and Genetics Unit, Jawaharlal Nehru Centre for Advanced Scientific Research, Bangalore, India

A. Mondal · C. Das
Biophysics and Structural Genomics Division, Saha Institute of Nuclear Physics, Kolkata, India

Homi Bhaba National Institute, Mumbai, India

T. K. Kundu (✉)
Transcription and Disease Laboratory, Molecular Biology and Genetics Unit, Jawaharlal Nehru Centre for Advanced Scientific Research, Bangalore, India

Division of Cancer Biology, CSIR-Central Drug Research Institute, Lucknow, India
e-mail: tapas@jncasr.ac.in

T. K. Kundu, C. Das (eds.), *Metabolism and Epigenetic Regulation: Implications in Cancer*, Subcellular Biochemistry 100,
https://doi.org/10.1007/978-3-031-07634-3_5

Introduction

Upon exposure to any intracellular or extracellular stress signal, cells safeguard themselves by altering their gene expression which affect either the catabolic or anabolic pathways. Catabolism of cellular materials is mediated majorly by two pathways, the proteasomal degradation pathway and the autophagic pathway. While the proteasome pathway is used majorly for protein degradation, the autophagy pathway degrades other macromolecules and damaged organelles as well. The term "autophagy" encompasses various types of processes, such as micro-autophagy, macro-autophagy, and chaperone-mediated autophagy (Codogno and Meijer 2005). It is an evolutionary conserved mechanism in eukaryotic cells that engulfs cell's wastage, such as damaged organelles, macromolecules, through the formation of a double membraned vesicle called autophagosome. In mammalian cells, maturation of autophagosomes requires its fusion with endocytic compartments and lysosomes. A large number of genes involved in the process of autophagy usually referred to as ATG (Tsukada and Ohsumi 1993) (Table 5.1) were identified through genetic screens in yeast. There are some core ATG genes which are involved in autophagosome formation in both selective and non-selective autophagy summarized in the table below (Mizushima et al. 2011; Nakatogawa 2020; Levine and Kroemer 2019). The selectivity is conferred to the process by specific factors which act as a bridge between the cargo and autophagosome. Selective autophagy involves capture and degradation of specific cargo such as mitochondria (known as mitophagy), peroxisomes (pexophagy), ribosomes (ribophagy), protein aggregates (aggrephagy), pathogens (xenophagy), part of the nucleus (nucleophagy), endoplasmic reticulum (reticulophagy).

In the mammalian system, the core autophagy pathway commences with the formation of an isolation membrane called phagophore. The phagophore comprises of at least five molecular components: (1) the Atg1/unc-51-like kinase (ULK) complex; (2) two transmembrane proteins, Atg9 and vacuole membrane protein 1 (VMP1); (3) the Beclin 1/ class III phosphatidylinositol 3-kinase (PI3K) complex; (4) two ubiquitin-like protein (Atg12 and Atg8/LC3) conjugation systems; and (5) proteins that mediate fusion between autophagosomes and lysosomes (Yang and Klionsky 2010) (Fig. 5.1). This isolation membrane can originate from diverse sources including the endoplasmic reticulum, the outer mitochondrial membrane or the plasma membrane. Comparable to the Atg1 protein of yeast, the ULK proteins (ULK1 and ULK2) play a key role in autophagy induction in mammalian cells. The ULK proteins function downstream of the mammalian target of rapamycin (mTOR) complex 1 (mTORC1), which is constituted of mTOR, Raptor, mLST8/GßL, Deptor and PRAS40. During nutrient rich conditions, mTORC1 is active and possesses kinase activity, forming a multiprotein complex that contains ULK1, Atg13, FIP200 and Atg101. During starvation conditions, mTORC1 is inactivated leading to its dissociation from the ULK complex (Rabanal-Ruiz et al. 2017). This leads to dephosphorylation and activation of ULK1 (or ULK2) and Atg13, which are mTORC1 targets (Qin et al. 2010). ULK1/2 can also be positively regulated by

Table 5.1 Autophagy genes and their functions: list of autophagy genes with their human ortholog are listed with each of their functions. Adapted from Tanida et al. (2004), Murrow and Debnath (2013)

Yeast gene	Human ortholog	Function
ATG1	ULK1, ULK2	Serine-threonine kinase; induction/membrane recycling
ATG2	C14orf103	Peripheral membrane protein; retrograde ATG9 transport (from PAS)
ATG3	hATG3	E2-like; ATG8 conjugation
ATG4	hATG4A (autophagin-2), hATG4B (autophagin-1), hATG4C(autophagin-3), hATG4 (autophagin-4)	Cysteine protease; assists in the cleavage of ATG8
ATG5	hATG5	Conjugated to ATG12; involved in formation/expansion of autophagosomes
ATG6	BECN1	Helps in PI3P synthesis; formation/expansion; often related to tumorigenesis
ATG7	hATG7	An E1-like enzyme for Atg12 and Atg 8 homologues
ATG8	MAP1LC3A, MAP1LC3B, GABARAP, GABARAPL1	Ubiquitin-like protein conjugated to PE; formation/expansion of autophagosomes
ATG9	hATG9A, hATG9B	Transmembrane protein shuttling between PAS and peripheral membranous structures
ATG10	hATG10	E2-like enzyme; ATG5-ATG12 conjugation
ATG11	Unknown	Cargo receptor/adaptor and assists in anterograde (to PAS) ATG9 transport
ATG12	hATG12	Ubiquitin-like protein conjugated to ATG5; formation/expansion
ATG13	Unknown	Modulates ATG1 activity; formation/expansion
ATG14	Unknown	PI3P synthesis; formation/expansion
ATG15		Lipase-like; degradation of vesicle within vacuole
ATG16	ATG16.1 ATG16.2	Forms homodimers and associates with ATG5-ATG12 complex; formation/expansion
ATG17	Unknown	Modulates ATG1 activity; regulate size of autophagosome
ATG18	WIPI1, WIPI2	PI(3,5)P2 binding protein involved in vacuole membrane recycling; PI3P binding protein involved in retrograde ATG9 recycling

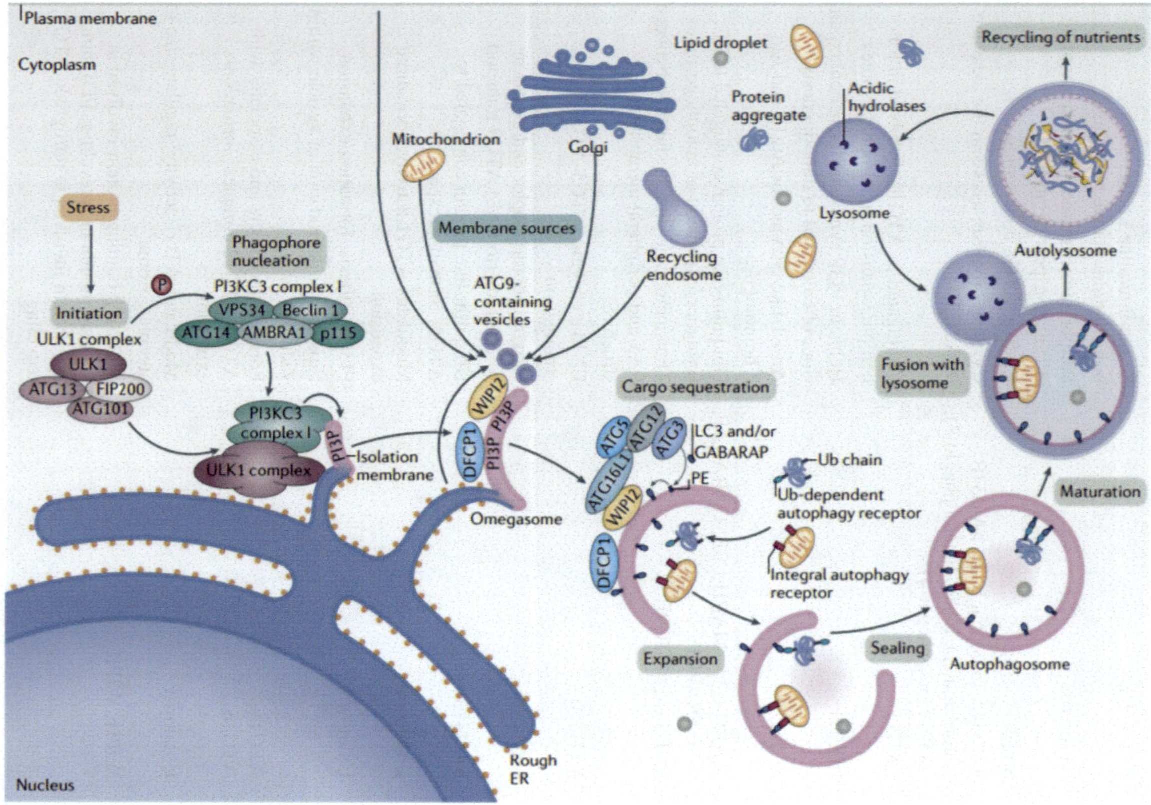

Fig. 5.1 The autophagic process. Upon nutrient or growth factor deprived conditions, AMPK is activated with inhibition of mTOR leading to activation of ULK, which phosphorylates Beclin-1, leading to VPS34 activation and phagophore formation. ULK forms a complex with FIP200 and ATG13, whereas VPS34 function requires a regulatory subunit, VPS15 (p150), and Beclin-1, which further mediates the association of other regulatory factors such as AMBRA, ATG14, UVRAG, and BIF- 1. Multiple ATG proteins such as ATG5 and ATG7 constitute two “ubiquitin-like conjugation systems” that catalyze the formation of

phosphatidylethanolamine (PE)-conjugated LC3 (LC3-II) and direct its proper incorporation into the phagophore membrane, where it serves as docking site of adaptor proteins (and bound cargos). The closure of an elongated phagophore marks the formation of a mature autophagosome, which eventually fuses with a lysosome, leading to cargo degradation and recycling of nutrients and metabolites. Ub, ubiquitin. Source: Dikic and Elazar (2018)

directly interacting with several subunits of the energy-sensing kinase, AMP-activated protein kinase (AMPK) (Alers et al. 2012). Although the exact molecular mechanism of how ULK1/2 activates downstream components of the autophagic machinery still remains elusive, it is shown to phosphorylate Ambra1, a component of the Beclin 1/Class III PI3K complex. The mTORC1 mediated regulation of autophagy is a tightly coordinated circuitry consisting of various feedback loops, which leads to a on or off state that ultimately causes induction or inhibition of autophagy by ULK1/2 (Hill et al. 2019). The Beclin 1 core complex is another major effector protein complex which induces autophagy. It includes Beclin 1, Vps15, Vps34 and likely, Ambra1 (He and Levine 2010). This complex allosterically activates the class III PI3K Vps34 to generate phosphatidylinositol-3-phosphate (PI3P). Generation of PI3P assists in the recruitment of other effectors such as the double FYVE domain-containing protein 1 (DFCP1) and WD-repeat protein interacting with phosphoinositides (WIPI) family proteins leading to the initiation of vesicle nucleation or autophagosome formation. Depending upon its interacting partners Beclin1 functions either as an autophagy inducer or repressor. Atg14/Atg14L/Barkor, (Beclin 1-associated autophagy-related key regulator) is a positive regulator of autophagy induction which is crucial for PI3K activity. UVRAG (UV radiation resistance associated gene) has been found to compete with Atg14 to bind to Beclin 1 and may positively regulate PI3K activity in a cell type-specific manner. Further, its interactions with class CVps/HOPS complexes augments autophagosome fusion with the late endosome/lysosome. Rubicon (RUN domain protein as Beclin 1 interacting and cysteine-rich containing), also adversely regulates autophagy (as well as endocytic trafficking) upon its interaction with Beclin 1/PI3K complexes (He and Levine 2010). Thus, the multiprotein Beclin1 complex through its different interacting partners directly affects the autophagy process. Cycling of the other core autophagy protein Atg9 between specific subcellular components might be crucial for providing lipids to the phagophore membrane. This cycling process of Atg9 requires the involvement of other proteins such as Atg1/ULK1, Vps34 or UVRAG/ Bif-1-containing Beclin 1 complex (Kim et al. 2011; Goodwin et al. 2017; Karanasios et al. 2016; Hill et al. 2018).

Two distinct ubiquitin-like conjugation systems functions as a part of the vesicle elongation process. The first pathway- involves Atg12 to be covalently conjugated to Atg5, with the help of Atg7 (E1-like enzyme) and Atg10 (E2-like enzyme). Subsequently, upon association with Atg16, a multimeric complex of Atg12-Atg5-Atg16 is formed. This multimeric complex possesses E3 ligase activity for LC3 (Yang and Klionsky 2010). The second pathway starts when phosphatidylethanolamine (PE) gets conjugated with the glycine residue of yeast Atg8 (mammalian LC3) by the consecutive action of Atg4 (protease), Atg7 (E1-like enzyme) and Atg3 (E2-like enzyme). This conjugation transforms the soluble form of LC3 (named LC3-I) to the autophagic vesicle-associated form, LC3-II (Tanida et al. 2004; Yang and Klionsky 2010). This lipidated form of LC3 (LC3-II) remains associates with the autophagosome membrane and can be used as a potent marker for the detection of cellular autophagy (Barth et al. 2010). Several proteins in the mammalian cells harbor a LC3-interacting region (LIR) that directly interacts with LC3 (along with its paralogs) and is instrumental in targeting ubiquitinated proteins, damaged mitochondria, or other organelles to the autophagic machinery. The best studied among them are p62 (also known as sequestosome1, SQSTM1) and NBR1 (Neighbor of BRCA1), both of which recognize ubiquitinated proteins as well as BNIP3L, which binds to mitochondrial membranes (Larsen et al. 2010; Lazarou et al. 2015). The fine regulation of these autophagy adaptors is not yet comprehended; however, they hold an essential key to understand how specific stress stimuli trigger selective autophagic pathway.

Autophagy: A Mechanism of Cellular Defense

The process of autophagy provides an adaptive response to deliver energy and nutrients to cells during starvation conditions. In yeast, deprivation of nutrients, or adverse changes in extracellular environment promotes the activation of autophagy which in turn digests different cytosolic contents through specific pathways depending upon the extracellular or intracellular condition (Murrow and Debnath 2013). The degradation process converts the macromolecules into their respective micromolecules (amino acids, monosaccharides, fatty acids etc.) which can be reused by the cell (Poillet-Perez et al. 2018; Karsli-Uzunbas et al. 2014). The cellular biosynthetic capacity and ATP levels are maintained during periods of stress through the autophagic pathway. It recycles the amino acids for de novo protein synthesis and can regulates cellular metabolic pathways (like TCA cycle) there by maintaining cellular ATP levels. Following starvation, autophagy has also been reported to be upregulated in almost all tissues in mice. Mice lacking essential core autophagy genes ATGs (like Atg5 or Atg7) have reduced level of plasma and tissue amino acids concentration, which leads to neonatal death (Kuma et al. 2004). Depletion of autophagic genes like Atg7 by RNA interference (RNAi) on mammalian cell-based model or inhibition of autophagy through administration of small molecule inhibitors such as chloroquine (CQ) or 3-methyadenine (3-MA) promotes cell death. This signifies that autophagy mediated catabolic function might play important cytoprotective role (Liu et al. 2020; Yue et al. 2021; Elswood et al. 2020). The self-eating process does not protect cells indefinitely but rather it helps the cells to cope during adverse situations. Thus, autophagy acts as a salvage mechanism which helps in the sustenance of cells by recycling basic cellular components to sustain core metabolic processes during stressful conditions.

The stress-induced autophagy is regulated primarily at two critical nodes; by the AMP activated protein kinase (AMPK) and mammalian target of rapamycin (mTOR). Here the AMPK signaling cascade is described (Fig. 5.2). AMPK acts as a sensing kinase responding to low intracellular energy conditions through the levels of AMP to ATP. Under conditions of high AMP/ATP ratio, AMPK phosphorylates

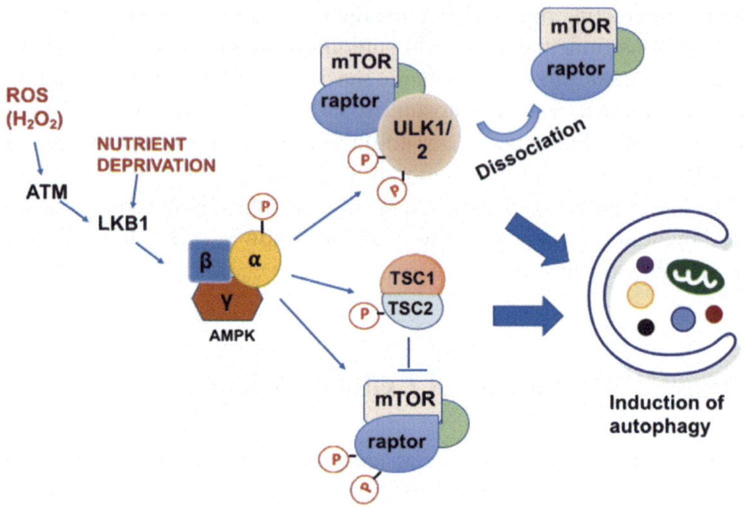

Fig. 5.2 AMPK signaling pathway in autophagy: AMPK activates autophagy at various steps by directly phosphorylating different protein substrates involved in its initiation phases. Upon conditions of nutrient stress or through Reactive oxygen species (ROS) AMPK is activated. Upon nutrient rich conditions or in the absence of oxidative stress, there is an inactivation of AMPK and activation of mTORC1. Subsequently, Ulk1 is phosphorylated at Ser 757 by activated mTORC1 which prevents the interaction between Ulk1 and AMPK. Upon activation of AMPK, there is phosphorylation of TSC2 and Raptor leading to mTORC1 inhibition, when cellular energy sources are depleted. This leads to a reduction in Ulk1 Ser 757 phosphorylation promoting its association with AMPK and consequent phosphorylation at Ser 317 and Ser 777 residues of ULK1. AMPK-mediated Phosphorylated Ulk1 is active which subsequently initiates autophagy. Figure adapted from Alers et al. (2012), Qin et al. (2010)

ULK1 thereby activating it and subsequently inhibiting mTOR complex 1 (mTORC1) via phosphorylation of Raptor (Herzig and Shaw 2017; Jia et al. 2018; Kim and Lee 2014; Kim et al. 2011). Thus, both AMPK and mTOR can regulates autophagy and subsequently can control cellular metabolic machinery. The Hypoxia inducible factor 1 (HIF-1) signaling is another major pathway which regulates autophagy under the conditions of hypoxia (Lu et al. 2018). HIF-1 directly affects the transcription of the BH3-only protein BNIP3, which is essential for induction of autophagy. It binds directly to autophagy thus alleviating the inhibitory interaction between Bcl-2 and Beclin 1 (the mammalian ortholog of yeast Atg6) (You et al. 2015; Zhou et al. 2009). Interestingly, p53 was reported to target different kinds of autophagy related genes. While p53 family member proteins such as p63 and p73 may also contribute in regulating the autophagic gene network, p53 plays the main regulatory role within this family (Sharif et al. 2019; Napoli and Flores 2013; He et al. 2013; Wangyang et al. 2017). During DNA damage, p53 mediated autophagy do not cause cell cycle arrest rather it promotes global transcriptional inhibition and enhance p53 mediated cell death. Various studies also support this connection between transcriptional network of autophagic genes and p53 leading to cell death (Wang et al. 2017a; Mrakovcic and Fröhlich 2018; Lee and Oh 2021; Liu et al. 2017). Interestingly, p53, induced by various cellular stresses plays dual roles in autophagy regulation. Cytosolic localization of p53 was reported to suppress autophagy although the detailed molecular mechanisms and the fine balance between pro−/anti-autophagic function of p53 remains to be elucidated (Levine and Abrams 2008).

Epigenetic Regulation of Autophagy

As discussed in the earlier sections, macroautophagy is a major pathway of self-digestion which is critical for maintaining cellular homeostasis and longevity. Investigating the regulation of autophagy has been garnering attention among biologists as it might unravel many unanswered questions regarding its selectivity and misregulation in various pathogenic conditions. Autophagy is described as a cytosolic recycling event which requires several core and effector proteins for its proper functioning. However, recent research advances reveal that the nucleus might play an important role in its regulation through its diverse histone post translational modifications (Füllgrabe et al. 2014). Whether an autophagic event lead to cell death/survival is dependent upon diverse epigenetic modifiers such as miRNAs or histone modifications in association with transcription factors (Yang and Liang 2015). For a long period of time, the understanding about the epigenetic regulation of autophagy was limited to the cytosolic interaction of HATs (histone acetyltransferases) and HDACs (histone deacetylases) with different proteins involved in autophagic pathway. Recent studies however reveal their transcription regulatory role upon nuclear import (Füllgrabe et al. 2014). The effect of histone modifications in the adaptive response of autophagy acts in three tiers. During a rapid autophagic response, in addition to the post-translational modification of cytosolic proteins a battery of transcription factors upregulates the expression of autophagy protein encoding genes through the active histone marks (H3R17me2, H3K9Ac, Shin et al. 2016). However, after prolonged exposure to autophagic stimuli, there is a global change observed in the histone modification pattern helping it to act as a negative regulatory feedback loop. Recently, it has been shown that coactivator-associated arginine methyltransferase 1 (CARM1) or protein arginine methyltransferase 4 (PRMT4) which asymmetrically methylate arginine residues in histone H3 and several non-histone proteins, under glucose starvation, methylates pontin, a chromatin remodeling factor. The methylated pontin interact with FOXO3a and recruit histone acetyltransferase Tip60, resulting in the hyperacetylation of histone H4 in the enhancers of autophagy genes thereby activating their expression (Fig. 5.3). The unique CARM1-Pontin-FOXO3a signaling axis establishes a new paradigm in the regulation of autophagy (Yu et al. 2020).

The global downregulation of active histone marks prevents the transcription factor mediated autophagic gene activation, thus acting as a mechanism to prevent lethality in cells caused by sustained high level of autophagy.

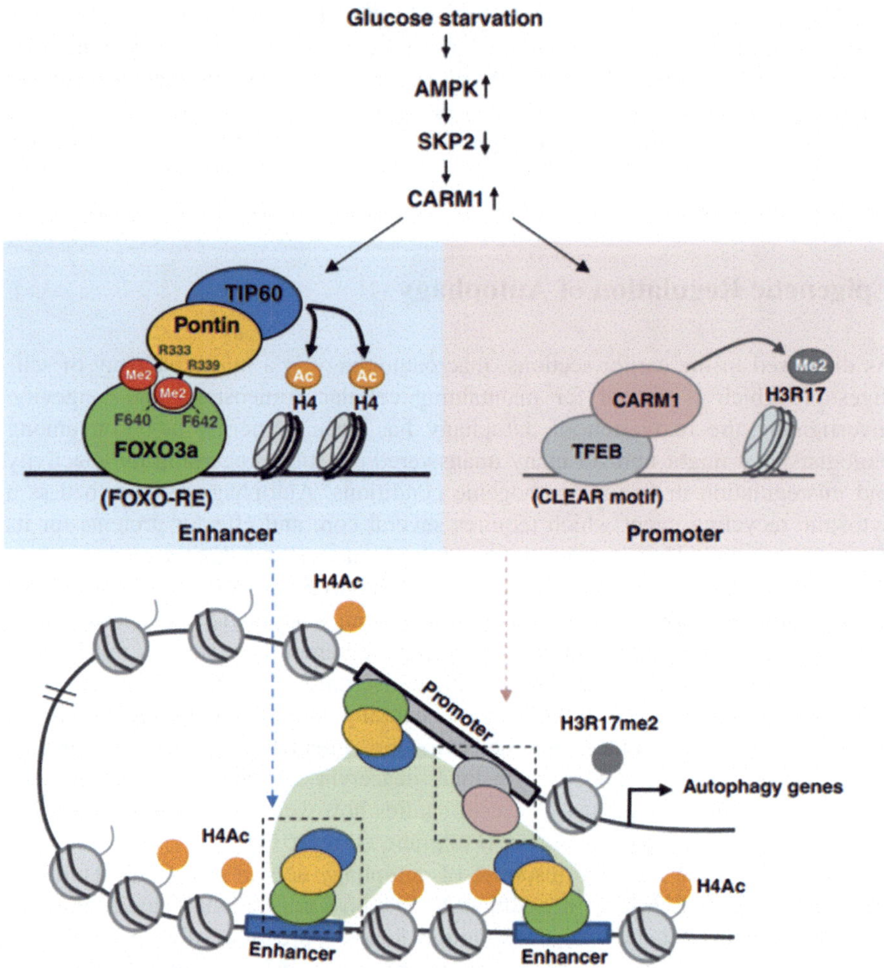

Fig. 5.3 CARM1–Pontin–FOXO3a signaling axis in the activation Autophagy gene: Upon glucose starvation, activated CARM1 methylates Pontin. The methyl binding domain of FOXO3a bind with the methylated pontin, which recruit acetyltransferase Tip60.The recruited Tip60 acetylate histone H4 that recruit mediator complex and activate the transcription. (Source: Yu et al. (2020))

The global chromatin silencing by histone modifications (H4K16Ac) (Füllgrabe et al. 2014), and also by miRNAs might also help in energy conservation during times of prolonged starvation. Thus, chromatin dynamics modulated by the interplay of several histone and non-histone factors articulately regulate the process of autophagy thus critically maintaining the cellular homeostasis and heath (Tables 5.2 and 5.3).

Table 5.2 Regulation of autophagy at transcriptional level. Different transcription factors regulate autophagy through the differential expression of the target genes (Füllgrabe et al. 2014; Hu et al. 2016; You et al. 2015)

Transcription factors		
Name	Core autophagy genes regulated at the transcriptional level	Effect on autophagy
ATF4	ATG5, BH3-only LC3 and ULK1	Enhanced autophagy
ATF5	mTOR	Suppressed autophagy
β-Catenin	SQSTM1	Suppressed autophagy
C/EBPβ	BNIP3, LC3 and ULK1	Enhanced autophagy
CHOP	ATG5 and LC3	Enhanced autophagy
E2F1	ATG5, BNIP3, LC3 and ULK1	Enhanced autophagy
FOXO1	ATG5, ATG12, ATG14, BECN1, BNIP3, LC3, and VPS34	Enhanced autophagy
FOXO3	ATG5, BH3-only LC3 and ULK1	Enhanced autophagy or suppressed autophagy
GATA1	LC3	Enhanced autophagy
HIF1	BNIP3	Enhanced autophagy
JUN	BECN1 and LC3	Enhanced autophagy
NF-κB	BCL2, BECN1, BNIP3 and SQSTM1	Enhanced autophagy or suppressed autophagy
p53	ATG2, ATG4, ATG7, ATG10, BCL2, BH3-only, ULK1 and UVRAG	In the cytosol: suppressed autophagy; In the nucleus: enhanced autophagy
p63	ATG3, ATG4, ATG5, ATG7, ATG9, ATG10, BECN1, LC3 and ULK1	Enhanced autophagy
p73	ATG5, ATG7 and UVRAG	Enhanced autophagy
SOX2	ATG10	Enhanced autophagy
SREBP2	LC3, ATG4B and ATG4D	Enhanced autophagy
STAT1	ATG12 and BECN1	Suppressed autophagy
STAT3	ATG3, BCL2 and BNIP3	Suppressed autophagy
TFEB	ATG4, ATG9, BCL2, LC3, SQSTM1, UVRAG and WIPI	Enhanced autophagy
ZKSCAN3	LC3, ULK1 and WIP1	Suppressed autophagy

Autophagy in Cancer

Autophagy being a master pro-survival mechanism against various stresses such as starvation, accumulation of damaged and unwanted intracellular protein and organelle has been implicated in various physiological processes such as development, anti-aging, elimination of microorganisms, cell death, tumor suppression, and antigen presentation (Mizushima et al. 2020). In case of mammalian system, autophagy is involved either in pathogenesis such as chronic viral and bacterial infection or in different human diseases such as atherosclerosis, cancer and even neurodegenerative diseases (Levine and Kroemer 2008; Racanelli et al. 2018). Detailed molecular mechanism of autophagic pathway and its epigenetic regulation has been studied

Table 5.3 Epigenetic regulation of autophagy. Table summarizing the effect of different histone modifications as well as the role of miRNAs in the process of autophagy (Baek and Kim 2017; Bánréti et al. 2013; Easwaran and Baylin 2010; Wei et al. 2015; Füllgrabe et al. 2014)

Histone modifications		
Histone mark	Epigenetic enzyme	Effect on autophagy
H3R17me2	CARM1	Enhanced autophagy
H3K27me3	EZH2	Suppressed autophagy
H3K9me2	G9a	Suppressed autophagy
H4K16Ac	hMOF	Suppressed autophagy
H3K56Ac	EP300/KAT3B/p300 and KAT2A/GCN5	Suppressed autophagy
H4K20me	SETD8 and SUV420	Enhanced autophagy
H2BK120ub	RNF20–RNF40	Suppressed autophagy

miRNA	
Autophagy target genes	miRNA
RB1CC1	miR-10a
ULK1	miR-20a and miR-106b
ULK2	miR-885-3p
ATG9	miR-34a
ATG14	miR-195
BCL2	miR-15, miR-16, miR-21, miR-24-2, miR-195, miR-196b, miR-204, miR-205 and miR-365-2
BECN1	miR-30a, miR-30d, miR-376b and miR-519a
UVRAG	miR-374a and miR-630
ATG4	miR-101 and miR-376b
ATG5	miR-30a, miR-30d, miR-180a and miR-374a
ATG7	miR-17 and miR-375
ATG10	miR-519a
ATG12	miR-30d and miR-630
ATG16	miR-519a and miR-885-3p
BNIP3	miR-145 and miR-210
LC3	miR-204
SQSTM1	miR-17, miR-20, miR-93 and miR-106
ATG2	miR-30d and miR-130a
ATG9	miR-34a

extensively and recently the relationship between pathogenesis and autophagy in various diseases is gaining importance (Kundu and Thompson 2008). Autophagy acts as a two-way sword during tumorigenesis. In general, it can suppress as well as promote oncogenesis, depending upon the stage of cancer (Fig. 5.4). During initial cancer stage, autophagy protects cells from further DNA damage by eliminating aggregated proteins and damage cellular organelles that may cause genomic instability (Mathew et al. 2007; Bhutia et al. 2013). Even in some cancer types where apoptotic cell death is impaired, autophagy may initiate cell death machinery. On the other hand, autophagy can also promote tumor growth by protecting it from

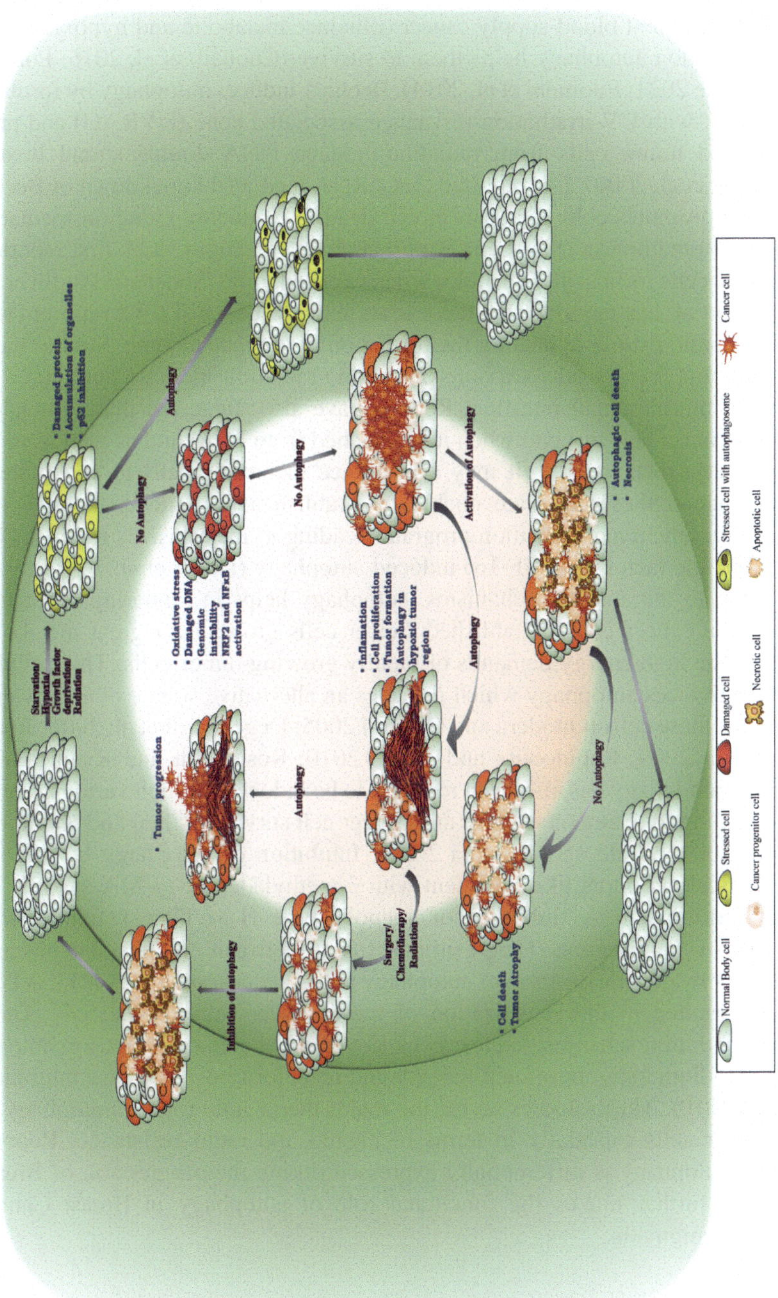

Fig. 5.4 Autophagy in regulation of cancer and its therapeutic aspect. Autophagy may suppress as well as promote tumor formation. Different stress situations can be combatted through activation of autophagic pathways. Cellular instabilities may cause tumorigenesis which can be antagonized by activating autophagic pathways. On the other hand, autophagy in growing tumor cells may support its proliferation which may be treated by chemotherapy followed by autophagic inhibition

inflammation and necrosis (Gu et al. 2016; Yunyan et al. 2015). Due to a high proliferation rate, cancer cells require higher amount of nutrient and oxygen. Often because of insufficient blood supply cancer cells face metabolic and hypoxic stress and in this situation autophagy helps them to survive (Chouaib et al. 2016; Poillet-Perez and White 2021; Filomeni et al. 2014). Beclin 1 induces autophagy by forming a core complex with UV-irradiation-resistance-associated gene (UVRAG) and protect established tumor cells from radiation-induced DNA double strand breaks (DSBs) (Liang et al. 2006). It is reported that siRNA mediated knockdown of Beclin 1 or UVRAG promotes colorectal cancer cell death by inducing radiation-mediated DNA double strand breaks (Shoshana Paglin et al. 2001). Paglin et al. first reported that neoplastic cells form acidic vacuoles when exposed to IR (Shoshana Paglin et al. 2001). Liang et al. has reported that there is an increase LC3-II/LC3-I ratio in Hela cells upon radiation stress indicating the formation of autophagosome. Studies have also reported that 7 days post expoure of gamma-radiation increase autophagy in crypt cells of murine small intestine. Reports have also suggested that autophagy may support cancer cell survival when it is detached from ECM and as well as in its colonization at a distance site. It may also induce metastatic cells to enter into a dormancy stage if these cells are unable to establish a new niche. Notably the epithelial to mesenchymal transition programs leading to metastasis is regulated by hypoxia inducible factor-1α (HIF-1α) induced autophagy (Hoare et al. 2011).

In addition to the above mechanisms, autophagy helps in rendering resistance property to the cancer cells. Established cancer cells grow under great metabolic stress due to the high energy demands of rapidly growing tumor cells. This cellular environment induces autophagy which provides an alternative source of energy and metabolites to these cells (Onodera and Ohsumi 2005; Degenhardt et al. 2006; Jones and Thompson 2009; Rabinowitz and White 2010; Rosenfeldt and Ryan 2009). Furthermore, autophagy has also been reported to be induced through various cancer therapies, leading to chemo resistance and cancer cell survival (Chen and Karantza-Wadsworth 2009; White and DiPaola 2009). Inhibition of autophagy by various pharmacological inhibitors like treatment with 3-methyladenine (3-MA, a PI3K III inhibitor), Bafilomycin A (inhibitor of vacuolar-type H+-ATPase), chloroquine (CQ) or hydroxychloroquine (HCQ) (impair fusion between autophagosomes and lysosomes), or by genetic deletion of autophagy related genes, such as ATG5, ATG6 and ATG7 resulted in the sensitization of these cancer cells (Table 5.4). Thus, autophagy inhibition sensitized cancer cells to various therapies such as genotoxic chemo- and radiotherapy, hormonal therapy and receptor tyrosine kinase inhibition (Chen et al. 2010). These studies reveal the major therapeutic role of autophagy in treating cancer cells especially in terms of chemo and radio resistance. Reports suggest that autophagy is differentially expressed during the progression of Breast cancer which further makes the functional role of autophagy in Breast Cancer manifestation enigmatic.

Table 5.4 Autophagy inhibitors in different cancers. Preclinical studies with autophagy small molecular inhibitors as well as inhibition of genes by siRNA treatment for cancer treatment. Inhibition of autophagy in these cancer cells results in rapid cell death (Choi 2012; Hoare et al. 2011; Li et al. 2013; Maycotte and Thorburn 2014)

Cancer type	Treatment	Cell death by autophagy inhibition		
		Early step	Late step	siRNAs
Breast cancer	γ-radiation			Beclin 1, ATG3, ATG4b, ATG4c, ATG5, ATG12
	Tamoxifen	3-MA		ATG5, Beclin 1, ATG7
	Bortezomib			LC3, HDAC6, ATF4
	Trastuzumab	3-MA	Baf	LC3
	Sulforaphane		Baf	
Colorectal cancer	Bortezomib			Beclin1
	γ-radiation			Beclin 1, ATG5, ATG7, UVRAG
	5-FU	3-MA		ATG7
	Sulforaphane	3-MA		
Glioma	Temozolomide		Baf	
	4-HPR	3-MA	Baf	
	γ-radiation	3-MA	Baf	Beclin 1, ATG5
	Imatinib		Baf	
Multiple myeloma	Bortezomib		Baf	
Prostate cancer	Androgen deprivation ADI-PEG20 (arginine deaminase)	3-MA		Beclin1
	Saracatinib (Src kinase)	3-MA	CQ	
Skin cancer	Cisplatin	3-MA		ATG5
Lymphoma	Tamoxifen		CQ	ATG5
Gastric cancer	Quercetin		CQ	ATG5, Beclin1

Autophagy and Cancer Therapy

In normal cells autophagy play a cytoprotective in different mild stressful conditions. Similarly, it may also promote resistance towards death of proliferative cancer cells under nutrient limitation, low oxygen availability and different other stresses (Lum et al. 2005; Degenhardt et al. 2006; Altman et al. 2011). Thus, autophagy may help in the better survival of cancer cells. However, the importance of autophagy in cancer is complicated, even confusing and controversial. Several studies have shown its effect towards cancer cell death (Mathew et al. 2007; Qu et al. 2003; Yue et al. 2003), whereas other literature has established its function in promoting cancer cell proliferation (Fung et al. 2008; Yang et al. 2011). It may also increase resistance against chemotherapy (Ma et al. 2014; Park et al. 2013; Sui et al. 2014) and ultimately may prevent apoptosis of cancer cells. In many studies, inhibition of autophagy is reported to enhance the effect of various anticancer therapy (Mancias and Kimmelman 2011). Recent studies also demonstrated about the enhancement of cancer cell death by induction of autophagy (Amaravadi et al. 2011; White and DiPaola 2009). So, autophagy may have diverse function in the regulation of cancer cells in a stage-specific manner.

Enhancement of Effectiveness of Anticancer Therapies by Inhibiting Autophagy

Upon cancer therapy, both normal and cancer cells show elevated level of autophagosome formation, but the amount is higher in case of cancer cells. Thus, it has a cytoprotective role in some cases which suggests that suppression of autophagy may promote apoptosis of cancer cell in response to chemotherapy (Amaravadi et al. 2007; Bellodi et al. 2009; Sheen et al. 2011). It is reported that siRNA mediated suppression of different autophagic genes such as *ATG5, ATG6, ATG7* along with bafilomycin A1 treatment increase the effectiveness of different anticancer drugs (Chen et al. 2010). Different drugs such as CQ or HCQ that suppress autophagosome maturation is reported to increase the cytotoxic effect of anticancer drugs and induce cancer cell death (Amaravadi et al. 2007; Boya et al. 2005; Liang et al. 2014; Maclean et al. 2008). Specific blocker of PI3K pathway with HQC is reported to enhance the effectiveness of chemotherapeutic drugs (Carew et al. 2007; Wu et al. 2010). Clinical trial is going on to understand how patients would be benefited with it. In case of breast cancer treatment, 3-methyladenine (3-MA), an autophagy (PI3K) inhibitor, and siRNA mediated inhibition of ATG5, ATG7, Beclin1 cumulatively enhance the effectiveness of anticancer drug tamoxifen (Schoenlein et al. 2009). (Table 5.4) There are several such other studies are being carried out. However, specific combination of autophagy inhibitor that would be effective in different cancer subtype is yet elusive.

Enhancement of Effectiveness of Anticancer Therapies by Promoting Autophagy

Many early growing tumors, especially solid tumor, show defect in autophagy and are reported to show chemotherapeutic drug resistance and possess high apoptotic threshold. In this kind of tumor types, induction of autophagy may increase the rate of cell death either by apoptosis or by mitophagy. Earlier in vitro studies reported that EB1089 (an analogue of vitamin D), arsenic trioxide can promote autophagy induced apoptotic cell death (Høyer-Hansen et al. 2005; Kanzawa et al. 2003).

Cancer cells, those lack pro-apoptotic genes such as BAX, BAC and caspases, are resistant to apoptotic cell death. In this kind of cancer, treatment with specific chemotherapeutic agents such as dexamethasone, etoposide and fenretinide, induces autophagy mediated cell death in vitro (Fazi et al. 2008; Grandér et al. 2009; Laane et al. 2009). Most of the reported autophagy inducing chemotherapeutic drugs blocks the activity of mTOR, thereby indirectly or directly activates AMPK. For example metformin (most common anti-diabetic drug) is reported to activate AMPK directly and induces autophagy by inhibition of mTOR or by phosphorylation of ULK1 by activated AMPK in melanoma and cervical cancer (Tomic et al. 2011; Xiao et al. 2012). Rapamycin is studied to stabilize raptor mTOR-complex by binding with FKBP12 and thus suppress mTOR activity and resulting in autophagy induction (Kim et al. 2002). It was also reported to suppress proliferation of neuroblastoma cells through induction of autophagy mediated cell cycle arrest and suppress tumor growth in murine sarcoma by autophagy mediated depletion of cancer stem cells (Lin et al. 2018; Shi et al. 2019). Recently different histone deacetylase inhibitors (HDACi-s) including SAHA, OSU-HDAC-42 are reported to be effective in autophagy induced cell death in chondrosarcoma cell line (Yamamoto et al. 2008). Studies on hepatocellular carcinoma (HCC) shows that SAHA and OSU-HDAC-42 induce autophagic cell death when treated with 3-MA (Liu et al. 2010).

As autophagy has both pro and anti-cancer effect, we have to understand the cancer type more accurately to answer the question whether to induce or suppress autophagy. Although many research are going on to implicate autophagy in cancer therapy, further research needs to be done to understand the proper molecular mechanisms and side effects of using it in human patients.

Autophagy and Metabolism Crosstalk

Autophagy is a tightly regulated cellular response to a myriad of stresses that regulates the cellular homeostasis. It may promote three distinct types of lysosome mediated degradation machinery, such as: (1) Formation of autophagosome, a double membrane cytosolic structure that digest biomolecules and damaged cellular organelles, known as macroautophagy; (2) Even lysosome can also engulf cytosolic macromolecules and micromolecules directly by a process called microautophagy; and (3) Autophagy mediated by different chaperone (Glick et al. 2010; Parzych and Klionsky 2014).These various types of autophagy pathways, are tightly coordinated inside a cell for fulfilling their energy demands and meeting up their needs for the synthesis of the essential biomolecules for vital processes (e.g., growth and proliferation) aided by the extracellular stimuli and carbon source (such as amino acids and glucose) availability. If there is an insufficient nutrient availability to sustain the different cellular processes (such as protein synthesis) or a deprivation of ATP pools in the cells to cater to the cellular energy requirement, autophagy pathway can get activated to rapidly disintegrate the available resources and reuse the generated pool of biomolecules. Several metabolic scenarios that can trigger autophagy and are summarized below

(i) *AMP/ATP ratio*

AMP-activated Protein Kinase (AMPK) can sense the available amount of ATP. The different stressed conditions in cells result in elevated levels of AMP and ADP that in turn activate AMPK. Upon activation, AMPK promotes different catabolic pathways such as glycolysis and fatty acid oxidation and suppresses various anabolic reactions such as lipid and protein synthesis and subsequently may trigger autophagy (Hardie and Pan 2002; Na et al. 2018; Zhang et al. 2019). Although AMPK gets activated in AMP and ADP rich environment, a minimal amount of ATP is required to act as a phosphate group donor for the activation of AMPK to induce autophagy (Filomeni et al. 2015). Hence rapid induction of AMP-ADP/ATP ratio higher than a threshold level may suppress AMPK mediated autophagy and promotes cell death (Hardie 2014; Roach 2011; Xiao et al. 2011). mTORC1 inhibits ULK1, AMBRA1, ATG14 and transcription factor EB (TFEB) by phosphorylation and subsequently suppresses autophagy (Nazio et al. 2013; Kim et al. 2011). While there is higher AMP/ATP ratio in cell, AMPK gets activated and phosphorylates mTORC1 component raptor at S722 and S792 residue and promotes its binding with 14-3-3 that ultimately leads to its suppression (Gwinn et al. 2008). Apart from this, mTORC1 can also be suppressed when activated AMPK phosphorylates TSC2 (inhibitor of mTORC1) at S1387 residue (Inoki et al. 2006). Upon inactivation of mTORC1, AMPK phosphorylates ULK1 at its S317 and S777 residue to activate it which in turn (Kim et al. 2011) positively regulates atg13 and FIP200. Subsequently, they promote the recruitment of microtubule associated PI3K complex mediated by AMBRA1 resulting in the formation of autophagic vesicles from mitochondria and endoplasmic reticulum (Kim et al. 2011). Hexokinase II (HKII) can also suppress the activity of mTORC1 by binding with it when cellular level of glucose-6-phosphate decrease during low glucose condition (Roberts et al. 2014). Under such situations, phosphorylation of BECN1 on S93 and S96 residue by AMPK increases the class III PI3K activity of VPS34. Activated AMPK can also phosphorylate VPS34 on T163 and S165 residues. Thus, activation of VPS34 leads to the suppression of the endosome-Golgi trafficking mediated non-autophagic function (Kim et al. 2011). AMPK promotes the transactivation of nicotinamide phosphoribosyltransferase (NAMPT) encoding gene in a FOXO1 dependent manner (Cantó et al. 2009). The enzyme NAMPT is involved in NAD^+ production. Thus, AMPK may activate SIRT 1 indirectly.

(ii) *NADH/NAD$^+$ ratio*

Upon low nutrient availability, different metabolic process become compromised that lead to the decrease of NADH/NAD$^+$ ratio. In this high NAD$^+$ condition, various Sirtuins get activated and resulting in Sirtuin-mediated autophagic activation (Ou et al. 2014). It is reported that artificial supply of nicotinamide or nicotinamide riboside increase cellular NAD level that in turn inhibits the activity of NAD$^+$ dependent enzymes like PARP1. It also increases cellular Sirtuin level by increasing NAD$^+$ concentration (Gibson and Kraus 2012; Houtkooper et al. 2012). In a study, it is also demonstrated that SIRT1 may have a role in autophagy as it is directly involved with many autophagic genes such as ATG5, ATG7, ATG12 and LC3 (Lee et al. 2008).

(iii) *Low acetyl CoA*

Low nutrient availability decreases the level of acetyl CoA in cytosol as well as in mitochondria. Acetyl CoA acts as an acetyl group donor for the activity of different histone acetyl transferases (HATs) and thus may promote autophagy (Mariño et al. 2014). It is also reported that during starvation, hyperactivated AKT1-mediated activation of ATP Citrate Lyase (ACLY) increases cellular acetyl CoA level and thus inhibits autophagy (Lee et al. 2014). Alteration of cellular acetyl CoA level in different physiological conditions from fed to fasting state, affects the activity of E1A-binding protein p300 (EP300). EP300 may inhibit different autophagic proteins such as ATG5, ATG7, ATG12 and LC3. Thus, it plays important role in regulating autophagic machinery (Lee and Finkel 2009). In one hand EP300 negatively regulates the activity of α-tubulin acetyltransferase (ATAT), also known as MEC17, on the other hand activated AMPK promotes ATAT1 activity. Even AMPK can phosphorylate EP300 at S89 residue and causes its repression (Yang et al. 2001). Thus, AMPK mediated EP300 suppressing situation, MEC17 stimulates autophagy by acetylating α-tubulin (Mackeh et al. 2014).

(iv) *High level of ROS*

Cellular ROS level increases during different stressful conditions. Elevated level of ROS stimulates autophagic machinery in several ways. Higher amount of ROS damages mitochondria. ROS mediated activation of HIF-1 promotes the expression Bcl-2 family member proteins BNIP3 and BNIP3-like (BNIP3L) that stimulates the removal of damaged mitochondria by an autophagic process known as mitophagy. High level of ROS also damages DNA and as a result activation of ATM occur. This in turn activates TSC2 and suppresses mTORC1 for the successive promotion of autophagy. Even oxidation of different autophagic protein such as ATG4 by ROS may augment the process of autophagy (Mackeh et al. 2014; Filomeni et al. 2015).

Epigenetic Regulation of Metabolic Pathways and Its Implication in Autophagy

(i) *Epigenetic regulation of autophagy by AMPK signaling*

AMPK can phosphorylates nutrient transporter proteins, such as glucose transporter 1 (GLUT1), and metabolic enzymes such as adipose triglyceride lipase (ATGL) and acetyl-CoA carboxylase 1 (ACC1) and thus can be considered as an important player for the regulation of cellular bioenergetics (Fullerton et al. 2013). Activated AMPK can also regulate acetyl CoA level. AMPK inhibits the formation of malonyl-CoA from acetyl CoA by phosphorylating ACC1 (Fullerton et al. 2013). It can also phosphorylate and activate Acetyl CoA Synthetase 2 (ACSS2) to promote its translocation to the nucleus where it interacts with TFEB and increases acetyl CoA level in a locus-specific manner. Thus, it promotes the activity of HATs that finally may lead to the induction of autophagy and lysosomal biogenesis (Li et al. 2017; Zhang et al. 2018). Further, AMPK can directly phosphorylate p300 resulting in its activation and consequently can lead to acetylation-mediated transcriptional activation of target genes, some of which are components of the autophagy machinery (Lee and Finkel 2009; Lim et al. 2012).

AMPK can also influence the activity of HDACs by phosphorylating their cofactors directly or by interfering with substrate availability in response to metabolic activity. Phosphorylation of HDAC4 and HDAC5 by AMPK promotes its cytosolic translocation where they interact with 14-3-3 proteins resulting in activation of autophagic genes (Chen et al. 2015). Inhibition of HDAC5 and dissociation of HDAC4 and HDAC5 from 14-3-3 protein leads to their nuclear import and causes suppression of autophagic genes (Peixoto et al. 2012).

AMPK can activate SIRT1 by increasing NAD^+/NADH ratio or by phosphorylating GAPDH. Phosphorylated GAPDH then translocate into the nucleus and may interact with SIRT1, releasing the sequestration of SIRT1 from its repressor (Chang et al. 2015). Then SIRT1 mediated deacetylation of *Atgs* and FOXO1 induce autophagic gene expression (Cantó et al. 2009) and it also deacetylates LC3 that promotes its cytosolic transport from nucleus to form autophagosome. During nutrient rich condition, BRD4 binds to autophagy and lysosomal gene promoter and recruits G9a methyltransferase, to suppress autophagy and lysosomal gene expression. Under nutrient deprived condition, activated SIRT1 deacetylates histones that lead to the release of BRD4 from autophagy and lysosomal genes promoter resulting in transcriptional activation of autophagy (Sakamaki et al. 2017).

Activated AMPK phosphorylates EZH2 and disrupt PRC2 complex and reduces H3K27me3 mark which was reported to downregulate the expression of TSC2 and cause autophagy induction via inhibition of mTORC1 (Wan et al.

2018; Wei et al. 2015). Upon nutrient deprivation, activated AMPK increases the level of fumarate by phosphorylation-mediated inhibition of fumarase enzyme. High level of fumarate suppresses KDM2A activity and decreases H3K36me2 level, thus it inhibits mTOR via PI3K/AKT pathway resulting in autophagosome formation (Lu et al. 2019; Tanaka et al. 2015; Wang et al. 2017b). Low nutrient availability also promotes AMPK mediated phosphorylation of FOXO3 and translocation of CARM1 into the nucleus. Then phosphorylated FOXO1 mediated transcriptional suppression of SKP2 and interaction of CARM1 with TFEB inside the nucleus lead to the expression of lysosomal and autophagy related genes (Shin et al. 2016). Cystic fibrosis transmembrane conductance regulator (CFTR) induces LC3, Atg12 and Beclin1 expression thus may promote autophagosome formation (Yang et al. 2018). DNMT1 suppress the expression of CFTR by methylating its promoter. Activated AMPK suppress DNA methylation by phosphorylating DNMT1 and thus it promotes the expression of CFTR mediated autophagosome formation (Marin et al. 2017).

Activation of AMPK by glucose starvation promotes its translocation to the nucleus where it increases serine 36 phosphorylation of H2B on chromatin that in turn increases the transcription of Atg genes in colon cancer cells (Bungard et al. 2010).

Box

Metabolic autophagy and Quality control autophagy: When through the autophagic recycling process nutrients are generated for cell growth, it is termed as metabolic autophagy. Through the quality control autophagy cells maintain the clean environment by degrading damaged as well as surplus organelles, infectious bodies and unwanted aggregates. The metabolic autophagy is directly in the cellular metabolism.

(continued)

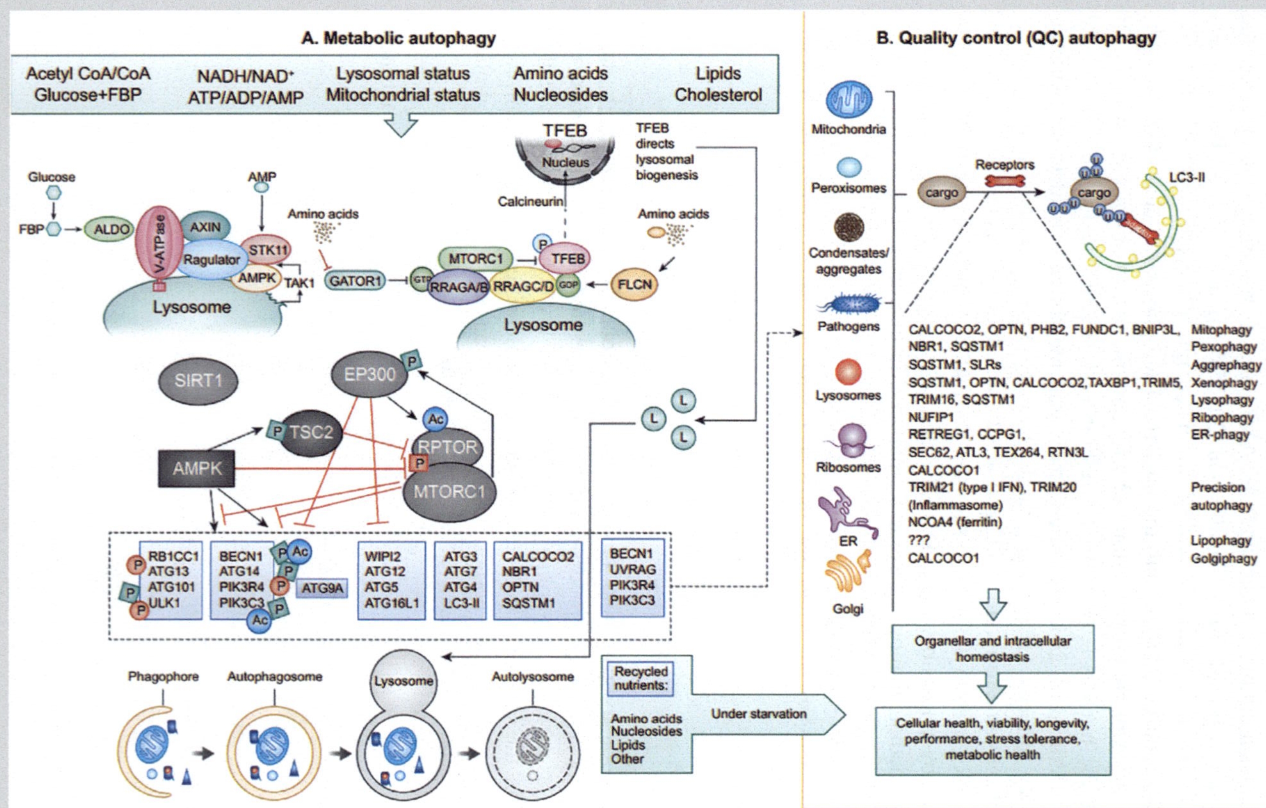

Metabolic autophagy and quality control autophagy: Metabolic autophagy helps to maintain the level and balance of nutrients where AMPK and mTOR play central role. Adapted from (Deretic and Kroemer 2021)

(ii) *Epigenetic regulation of autophagy by mTOR signaling*

Under insufficient nutrient availability, mTOR gets inactivated. Thus, mTOR mediated inhibition of autophagy is averted. During nutrient-rich conditions, EZH2 suppress TSC2 transcription (mTORC1 inhibitor) by increasing H3K27me3 mark in its promoter region that lead to the activation of mTORC1 and suppression of autophagy (Wei et al. 2015). In a study it is demonstrated that starvation causes mTORC1 inhibition and subsequently decreases of H3K4me2 and increases of H3K4me3 mark (Lee et al. 2017). Upon nutrient stress, regulation of mTORC1 activity by Set2 mediated H3K36me have direct connection to transcriptional fidelity although the mechanism is not well understood yet (McDaniel et al. 2017).

In yeast upon nutrient deprivation, there is a dynamic regulation of mTORC1-PP2A signaling and mTORC2-AKT signaling that phosphorylates TFIID subunit 12 (transcription initiation factor) of SAGA complex (Laboucarié et al. 2017). In mammals, induction of autophagy cause GCN5 (histone acetyltransferase) inhibition, which results in decrease TFEB acetylation and promotes autophagic and lysosomal gene expression (Wang et al. 2020). In yeast, GCN5 is known to inhibit autophagy by acetylating Atg7 (Zhang et al. 2017). Further, p300 acetylates some important autophagy related proteins including Beclin1/VPS34, LC3, Atg5, Atg7, and Atg12 and thus inhibiting autophagy (Sun et al. 2015; Lee and Finkel 2009). Activated mTOR inhibits autophagy by phosphorylation mediated activation of p300 (Wan et al. 2017). A study reported that rapamycin treatment cause mTORC1 inhibition that led to the decrease of hMOF level and enhanced deacetylation of H4K16ac, resulting in decrease of Atg gene expression (Füllgrabe et al. 2013). During nutrient affluence, mTOR enhances nuclear import of Foxk1 that in turn recruits the Sin3-HDAC1/2 complexes and reduces the acetylation of H4 resulting in suppression of essential autophagy genes expression (Bowman et al. 2014).

Conclusion/Future Perspectives

The self-eating recycling process autophagy, not only sense the metabolic changes but also maintain it through highly orchestrated mechanisms. Our understanding of role of autophagy in cancer is yet to be comprehensive. However, considering the fact that during malignancy dramatic alteration of metabolism is a hallmark, autophagy must be playing critical role in cancer initiation, progression and metastasis. Beyond Warburg effect, that is, increased level of glycolysis: altered balance of glucose fermentation and oxidative phosphorylation, the tumor suppressor p53 and oncogene c-myc regulate several metabolic pathways in cancer. The tumorigenesis has also been linked to mutation or overexpression of metabolic enzyme genes such as succinate dehydrogenase, fumarate hydratase, pyruvate kinase, and isocitrate dehydrogenase. The glucose metabolism, oncogene function and these metabolic enzymes are also essential part of autophagy machineries. How these components are functioning in the dynamic network and incidental opposing role of autophagy in cancer are being elucidated. But the rules of the jungle are yet to be understood.

Acknowledgments We acknowledge JNCASR, Department of Biotechnology, Government of India for the financial support. TKK is a recipient of Sir J.C. Bose National Fellowship. SS was a CSIR Senior Research Fellow. CD acknowledges financial support from Basic and Applied Research in Biophysics and Material Science (RSI 4002) from the Department of Atomic Energy. We also want to thank Kruthi HT for her assistance in figure and manuscript preparation.

References

Alers S, Löffler AS et al (2012) Role of AMPK-mTOR-Ulk1/2 in the regulation of autophagy: cross talk, shortcuts, and feedbacks. Mol Cell Biol 32(1):2–11

Altman BJ, Jacobs SR et al (2011) Autophagy is essential to suppress cell stress and to allow BCR-Abl-mediated leukemogenesis. Oncogene 30(16):1855–1867

Amaravadi RK, Yu D et al (2007) Autophagy inhibition enhances therapy-induced apoptosis in a Myc-induced model of lymphoma. J Clin Invest 117(2):326–336

Amaravadi RK, Lippincott-Schwartz J et al (2011) Principles and current strategies for targeting autophagy for cancer treatment. Clin Cancer Res 17(4):654–666

Baek SH, Kim KI (2017) Epigenetic control of autophagy: nuclear events gain more attention. Mol Cell 65(5):781–785

Bánréti A, Sass M et al (2013) The emerging role of acetylation in the regulation of autophagy. Autophagy 9(6):819–829

Barth S, Glick D et al (2010) Autophagy: assays and artifacts. J Pathol 221(2):117–124

Bellodi C, Lidonnici MR et al (2009) Targeting autophagy potentiates tyrosine kinase inhibitor-induced cell death in Philadelphia chromosome-positive cells, including primary CML stem cells. J Clin Invest 119(5):1109–1123

Bhutia SK, Mukhopadhyay S et al (2013) Autophagy: cancer's friend or foe? Adv Cancer Res. 118: 61–95

Bowman CJ, Ayer DE et al (2014) Foxk proteins repress the initiation of starvation-induced atrophy and autophagy programs. Nat Cell Biol 16(12):1202–1214

Boya P, González-Polo RA et al (2005) Inhibition of macroautophagy triggers apoptosis. Mol Cell Biol 25(3):1025–1040

Bungard D, Fuerth BJ et al (2010) Signaling kinase AMPK activates stress-promoted transcription via histone H2B phosphorylation. Science 329(5996):1201–1205

Cantó C, Gerhart-Hines Z et al (2009) AMPK regulates energy expenditure by modulating NAD+ metabolism and SIRT1 activity. Nature 458(7241):1056–1060

Carew JS, Nawrocki ST et al (2007) Targeting autophagy augments the anticancer activity of the histone deacetylase inhibitor SAHA to overcome Bcr-Abl-mediated drug resistance. Blood 110(1):313–322

Chang C, Su H et al (2015) AMPK-dependent phosphorylation of GAPDH triggers Sirt1 activation and is necessary for autophagy upon glucose starvation. Mol Cell 60(6):930–940

Chen N, Karantza-Wadsworth V (2009) Role and regulation of autophagy in cancer. Biochim Biophys Acta 1793(9):1516–1523

Chen S, Rehman SK et al (2010) Autophagy is a therapeutic target in anticancer drug resistance. Biochim Biophys Acta 1806(2):220–229

Chen S, Yin C et al (2015) AMPK-HDAC5 pathway facilitates nuclear accumulation of HIF-1α and functional activation of HIF-1 by deacetylating Hsp70 in the cytosol. Cell Cycle 14(15): 2520–2536

Choi KS (2012) Autophagy and cancer. Exp Mol Med 44(2):109–120

Chouaib S, Noman MZ et al (2016) Hypoxic stress: obstacles and opportunities for innovative immunotherapy of cancer. Oncogene 36(4):439–445
Codogno P, Meijer AJ (2005) Autophagy and signaling: their role in cell survival and cell death. Cell Death Differ 12(Suppl 2):1509–1518
Degenhardt K, Mathew R et al (2006) Autophagy promotes tumor cell survival and restricts necrosis, inflammation, and tumorigenesis. Cancer Cell 10(1):51–64
Deretic V, Kroemer G (2021) Autophagy in metabolism and quality control: opposing, complementary or interlinked functions? Autophagy 2021:1–10
Dikic I, Elazar Z (2018) Mechanism and medical implications of mammalian autophagy. Nat Rev Mol Cell Biol 19(6):349–364
Easwaran HP, Baylin SB (2010) Role of nuclear architecture in epigenetic alterations in cancer. Cold Spring Harb Symp Quant Biol 75:507–515
Elswood J, Pearson SJ et al (2020) Autophagy regulates functional differentiation of mammary epithelial cells. Autophagy 17(2):420–438
Fazi B, Bursch W et al (2008) Fenretinide induces autophagic cell death in caspase-defective breast cancer cells. Autophagy 4(4):435–441
Filomeni G, De Zio D et al (2014) Oxidative stress and autophagy: the clash between damage and metabolic needs. Cell Death Differ 22(3):377–388
Filomeni G, De Zio D et al (2015) Oxidative stress and autophagy: the clash between damage and metabolic needs. Cell Death Differ 22(3):377–388
Fullerton MD, Galic S et al (2013) Single phosphorylation sites in Acc1 and Acc2 regulate lipid homeostasis and the insulin-sensitizing effects of metformin. Nat Med 19(12):1649–1654
Füllgrabe J, Lynch-Day MA et al (2013) The histone H4 lysine 16 acetyltransferase hMOF regulates the outcome of autophagy. Nature 500(7463):468–471
Füllgrabe J, Klionsky DJ et al (2014) The return of the nucleus: transcriptional and epigenetic control of autophagy. Nat Rev Mol Cell Biol 15(1):65–74
Fung C, Lock R et al (2008) Induction of autophagy during extracellular matrix detachment promotes cell survival. Mol Biol Cell 19(3):797–806
Gibson BA, Kraus WL (2012) New insights into the molecular and cellular functions of poly (ADP-ribose) and PARPs. Nat Rev Mol Cell Biol 13(7):411–424
Glick D, Barth S et al (2010) Autophagy: cellular and molecular mechanisms. J Pathol 221(1):3–12
Goodwin JM, Dowdle WE et al (2017) Autophagy-independent lysosomal targeting regulated by ULK1/2-FIP200 and ATG9. Cell Rep 20(10):2341–2356
Grandér D, Kharaziha P et al (2009) Autophagy as the main means of cytotoxicity by glucocorticoids in hematological malignancies. Autophagy 5(8):1198–1200
Gu Y, Li P et al (2016) Autophagy-related prognostic signature for breast cancer. Mol Carcinog 55(3):292–299
Gwinn DM, Shackelford DB et al (2008) AMPK phosphorylation of raptor mediates a metabolic checkpoint. Mol Cell 30(2):214–226
Hardie DG (2014) AMPK—sensing energy while talking to other signaling pathways. Cell Metab 20(6):939–952
Hardie DG, Pan DA (2002) Regulation of fatty acid synthesis and oxidation by the AMP-activated protein kinase. Biochem Soc Trans 30(Pt 6):1064–1070
He C, Levine B (2010) The Beclin 1 interactome. Curr Opin Cell Biol 22(2):140–149
He Z, Liu H et al (2013) p73 regulates autophagy and hepatocellular lipid metabolism through a transcriptional activation of the ATG5 gene. Cell Death Differ 20(10):1415–1424
Herzig S, Shaw RJ (2017) AMPK: guardian of metabolism and mitochondrial homeostasis. Nat Rev Mol Cell Biol 19(2):121–135
Hill SM, Wrobel L et al (2018) Post-translational modifications of Beclin 1 provide multiple strategies for autophagy regulation. Cell Death Differ 26(4):617–629
Hill SM, Wrobel L et al (2019) Post-translational modifications of Beclin 1 provide multiple strategies for autophagy regulation. Cell Death Differ 26(4):617–629

Hoare M, Young AR et al (2011) Autophagy in cancer: having your cake and eating it. Semin Cancer Biol 21(6):397–404

Houtkooper RH, Pirinen E et al (2012) Sirtuins as regulators of metabolism and healthspan. Nat Rev Mol Cell Biol 13(4):225–238

Høyer-Hansen M, Bastholm L et al (2005) Vitamin D analog EB1089 triggers dramatic lysosomal changes and Beclin 1-mediated autophagic cell death. Cell Death Differ 12(10):1297–1309

Hu L, Wang H et al (2016) Crosstalk between autophagy and intracellular radiation response (review). Int J Oncol 49(6):2217–2226

Inoki K, Ouyang H et al (2006) TSC2 integrates Wnt and energy signals via a coordinated phosphorylation by AMPK and GSK3 to regulate cell growth. Cell 126(5):955–968

Jia J, Abudu YP et al (2018) Galectins control MTOR and AMPK in response to lysosomal damage to induce autophagy. Autophagy 15(1):169–171

Jones RG, Thompson CB (2009) Tumor suppressors and cell metabolism: a recipe for cancer growth. Genes Dev 23(5):537–548

Kanzawa T, Kondo Y et al (2003) Induction of autophagic cell death in malignant glioma cells by arsenic trioxide. Cancer Res 63(9):2103–2108

Karanasios E, Walker SA et al (2016) Autophagy initiation by ULK complex assembly on ER tubulovesicular regions marked by ATG9 vesicles. Nat Commun 7(1):12420

Karsli-Uzunbas G, Guo JY et al (2014) Autophagy is required for glucose homeostasis and lung tumor maintenance. Cancer Discov 4(8):914–927

Kim KH, Lee M-S (2014) Autophagy—a key player in cellular and body metabolism. Nat Rev Endocrinol 10(6):322–337

Kim DH, Sarbassov DD et al (2002) mTOR interacts with raptor to form a nutrient-sensitive complex that signals to the cell growth machinery. Cell 110(2):163–175

Kim J, Kundu M et al (2011) AMPK and mTOR regulate autophagy through direct phosphorylation of Ulk1. Nat Cell Biol 13(2):132–141

Kuma A, Hatano M et al (2004) The role of autophagy during the early neonatal starvation period. Nature 432(7020):1032–1036

Kundu M, Thompson CB (2008) Autophagy: basic principles and relevance to disease. Annu Rev Pathol 3(1):427–455

Laane E, Tamm KP et al (2009) Cell death induced by dexamethasone in lymphoid leukemia is mediated through initiation of autophagy. Cell Death Differ 16(7):1018–1029

Laboucarié T, Detilleux D et al (2017) TORC1 and TORC2 converge to regulate the SAGA co-activator in response to nutrient availability. EMBO Rep 18(12):2197–2218

Larsen KB, Lamark T et al (2010) A reporter cell system to monitor autophagy based on p62/SQSTM1. Autophagy 6(6):784–793

Lazarou M, Sliter DA et al (2015) The ubiquitin kinase PINK1 recruits autophagy receptors to induce mitophagy. Nature 524(7565):309–314

Lee IH, Finkel T (2009) Regulation of autophagy by the p300 acetyltransferase. J Biol Chem 284(10):6322–6328

Lee H-Y, Oh S-H (2021) Autophagy-mediated cytoplasmic accumulation of p53 leads to apoptosis through DRAM-BAX in cadmium-exposed human proximal tubular cells. Biochem Biophys Res Commun 534:128–133

Lee IH, Cao L et al (2008) A role for the NAD-dependent deacetylase Sirt1 in the regulation of autophagy. Proc Natl Acad Sci U S A 105(9):3374–3379

Lee JV, Carrer A et al (2014) Akt-dependent metabolic reprogramming regulates tumor cell histone acetylation. Cell Metab 20(2):306–319

Lee D, An J et al (2017) SHPRH regulates rRNA transcription by recognizing the histone code in an mTOR-dependent manner. Proc Natl Acad Sci U S A 114(17):E3424–E3433

Levine B, Abrams J (2008) p53: the Janus of autophagy? Nat Cell Biol 10(6):637–639

Levine B, Kroemer G (2008) Autophagy in the pathogenesis of disease. Cell 132(1):27–42

Levine B, Kroemer G (2019) Biological functions of autophagy genes: a disease perspective. Cell 176(1–2):11–42

Li X, Xu HL et al (2013) Autophagy modulation as a target for anticancer drug discovery. Acta Pharmacol Sin 34(5):612–624
Li X, Yu W et al (2017) Nucleus-translocated ACSS2 promotes gene transcription for lysosomal biogenesis and autophagy. Mol Cell 66(5):684–697.e689
Liang C, Feng P et al (2006) Autophagic and tumour suppressor activity of a novel Beclin1-binding protein UVRAG. Nat Cell Biol 8(7):688–698
Liang X, Tang J et al (2014) Suppression of autophagy by chloroquine sensitizes 5-fluorouracil-mediated cell death in gallbladder carcinoma cells. Cell Biosci 4(1):10
Lim J-Y, Oh M-A et al (2012) AMP-activated protein kinase inhibits TGF-β-induced fibrogenic responses of hepatic stellate cells by targeting transcriptional coactivator p300. J Cell Physiol 227(3):1081–1089
Lin X, Han L et al (2018) Rapamycin inhibits proliferation and induces autophagy in human neuroblastoma cells. Biosci Rep 38(6)
Liu YL, Yang PM et al (2010) Autophagy potentiates the anti-cancer effects of the histone deacetylase inhibitors in hepatocellular carcinoma. Autophagy 6(8):1057–1065
Liu Q, Fang Q et al (2017) Resveratrol-mediated apoptosis in renal cell carcinoma via the p53/AMP-activated protein kinase/mammalian target of rapamycin autophagy signaling pathway. Mol Med Rep
Liu X, Meng L et al (2020) Regulation of FN1 degradation by the p62/SQSTM1-dependent autophagy–lysosome pathway in HNSCC. Int J Oral Sci 12(1)
Lu N, Li X et al (2018) HIF-1α/Beclin1-mediated autophagy is involved in neuroprotection induced by hypoxic preconditioning. J Mol Neurosci 66(2):238–250
Lu DH, Yang J et al (2019) Lysine demethylase 2A promotes the progression of ovarian cancer by regulating the PI3K pathway and reversing epithelial-mesenchymal transition. Oncol Rep 41(2): 917–927
Lum JJ, Bauer DE et al (2005) Growth factor regulation of autophagy and cell survival in the absence of apoptosis. Cell 120(2):237–248
Ma XH, Piao SF et al (2014) Targeting ER stress-induced autophagy overcomes BRAF inhibitor resistance in melanoma. J Clin Invest 124(3):1406–1417
Mackeh R, Lorin S et al (2014) Reactive oxygen species, AMP-activated protein kinase, and the transcription cofactor p300 regulate α-tubulin acetyltransferase-1 (αTAT-1/MEC-17)-dependent microtubule hyperacetylation during cell stress. J Biol Chem 289(17):11816–11828
Maclean KH, Dorsey FC et al (2008) Targeting lysosomal degradation induces p53-dependent cell death and prevents cancer in mouse models of lymphomagenesis. J Clin Invest 118(1):79–88
Mancias JD, Kimmelman AC (2011) Targeting autophagy addiction in cancer. Oncotarget 2(12): 1302–1306
Marin TL, Gongol B et al (2017) AMPK promotes mitochondrial biogenesis and function by phosphorylating the epigenetic factors DNMT1, RBBP7, and HAT1. Sci Signal 10(464)
Mariño G, Pietrocola F et al (2014) Regulation of autophagy by cytosolic acetyl-coenzyme A. Mol Cell 53(5):710–725
Mathew R, Kongara S et al (2007) Autophagy suppresses tumor progression by limiting chromosomal instability. Genes Dev 21(11):1367–1381
Maycotte P, Thorburn A (2014) Targeting autophagy in breast cancer. World J Clin Oncol 5(3): 224–240
McDaniel SL, Hepperla AJ et al (2017) H3K36 methylation regulates nutrient stress response in Saccharomyces cerevisiae by enforcing transcriptional fidelity. Cell Rep 19(11):2371–2382
Mizushima N, Yoshimori T et al (2011) The role of Atg proteins in autophagosome formation. Annu Rev Cell Dev Biol 27(1):107–132
Mizushima N, Longo DL et al (2020) Autophagy in human diseases. N Engl J Med 383(16): 1564–1576
Mrakovcic M, Fröhlich L (2018) p53-mediated molecular control of autophagy in tumor cells. Biomol Ther 8(2):14

Murrow L, Debnath J (2013) Autophagy as a stress-response and quality-control mechanism: implications for cell injury and human disease. Annu Rev Pathol 8:105–137
Na RS, Ma C et al (2018) Itraconazole attenuates hepatic gluconeogenesis and promotes glucose uptake by regulating AMPK pathway. Exp Ther Med 15(2):2165–2171
Nakatogawa H (2020) Mechanisms governing autophagosome biogenesis. Nat Rev Mol Cell Biol 21(8):439–458
Napoli M, Flores ER (2013) The family that eats together stays together: new p53 family transcriptional targets in autophagy. Genes Dev 27(9):971–974
Nazio F, Strappazzon F et al (2013) mTOR inhibits autophagy by controlling ULK1 ubiquitylation, self-association and function through AMBRA1 and TRAF6. Nat Cell Biol 15(4):406–416
Onodera J, Ohsumi Y (2005) Autophagy is required for maintenance of amino acid levels and protein synthesis under nitrogen starvation. J Biol Chem 280(36):31582–31586
Ou X, Lee MR et al (2014) SIRT1 positively regulates autophagy and mitochondria function in embryonic stem cells under oxidative stress. Stem Cells 32(5):1183–1194
Park JM, Huang S et al (2013) Prognostic impact of Beclin 1, p62/sequestosome 1 and LC3 protein expression in colon carcinomas from patients receiving 5-fluorouracil as adjuvant chemotherapy. Cancer Biol Ther 14(2):100–107
Parzych KR, Klionsky DJ (2014) An overview of autophagy: morphology, mechanism, and regulation. Antioxid Redox Signal 20(3):460–473
Peixoto P, Castronovo V et al (2012) HDAC5 is required for maintenance of pericentric heterochromatin, and controls cell-cycle progression and survival of human cancer cells. Cell Death Differ 19(7):1239–1252
Poillet-Perez L, White E (2021) Role of tumor and host autophagy in cancer metabolism. Genes Dev 35:23–24
Poillet-Perez L, Xie X et al (2018) Autophagy maintains tumour growth through circulating arginine. Nature 563(7732):569–573
Qin L, Wang Z et al (2010) ER stress negatively regulates AKT/TSC/mTOR pathway to enhance autophagy. Autophagy 6(2):239–247
Qu X, Yu J et al (2003) Promotion of tumorigenesis by heterozygous disruption of the beclin 1 autophagy gene. J Clin Invest 112(12):1809–1820
Rabanal-Ruiz Y, Otten EG et al (2017) mTORC1 as the main gateway to autophagy. Essays Biochem 61(6):565–584
Rabinowitz JD, White E (2010) Autophagy and metabolism. Science 330(6009):1344–1348
Racanelli AC, Kikkers SA et al (2018) Autophagy and inflammation in chronic respiratory disease. Autophagy 14(2):221–232
Roach PJ (2011) AMPK -> ULK1 -> autophagy. Mol Cell Biol 31(15):3082–3084
Roberts DJ, Tan-Sah VP et al (2014) Hexokinase-II positively regulates glucose starvation-induced autophagy through TORC1 inhibition. Mol Cell 53(4):521–533
Rosenfeldt MT, Ryan KM (2009) The role of autophagy in tumour development and cancer therapy. Expert Rev Mol Med 11:e36
Sakamaki JI, Wilkinson S et al (2017) Bromodomain protein BRD4 is a transcriptional repressor of autophagy and lysosomal function. Mol Cell 66(4):517–532.e519
Schoenlein PV, Periyasamy-Thandavan S et al (2009) Autophagy facilitates the progression of ERalpha-positive breast cancer cells to antiestrogen resistance. Autophagy 5(3):400–403
Sharif T, Martell E et al (2019) Regulation of the proline regulatory axis and autophagy modulates stemness in TP73/p73 deficient cancer stem-like cells. Autophagy 15(5):934–936
Sheen JH, Zoncu R et al (2011) Defective regulation of autophagy upon leucine deprivation reveals a targetable liability of human melanoma cells in vitro and in vivo. Cancer Cell 19(5):613–628
Shi H, Zhang L et al (2019) Rapamycin may inhibit murine S180 sarcoma growth by regulating the pathways associated with autophagy and cancer stem cells. J Cancer Res Ther 15(2):398–403
Shin HJ, Kim H et al (2016) AMPK-SKP2-CARM1 signalling cascade in transcriptional regulation of autophagy. Nature 534(7608):553–557

Shoshana Paglin TH, Delohery T, Hackett N, McMahill M, Sphicas E, Domingo D, Yahalom J (2001) A novel response of cancer cells to radiation involves autophagy and formation of acidic vesicles. Cancer Res 61(2):439–444

Sui X, Kong N et al (2014) JNK confers 5-fluorouracil resistance in p53-deficient and mutant p53-expressing colon cancer cells by inducing survival autophagy. Sci Rep 4:4694

Sun T, Li X et al (2015) Acetylation of Beclin 1 inhibits autophagosome maturation and promotes tumour growth. Nat Commun 6:7215

Tanaka Y, Yano H et al (2015) Mild glucose starvation induces KDM2A-mediated H3K36me2 demethylation through AMPK to reduce rRNA transcription and cell proliferation. Mol Cell Biol 35(24):4170–4184

Tanida I, Ueno T et al (2004) LC3 conjugation system in mammalian autophagy. Int J Biochem Cell Biol 36(12):2503–2518

Tomic T, Botton T et al (2011) Metformin inhibits melanoma development through autophagy and apoptosis mechanisms. Cell Death Dis 2(9):e199

Tsukada M, Ohsumi Y (1993) Isolation and characterization of autophagy-defective mutants of Saccharomyces cerevisiae. FEBS Lett 333(1–2):169–174

Wan W, You Z et al (2017) mTORC1 phosphorylates acetyltransferase p300 to regulate autophagy and lipogenesis. Mol Cell 68(2):323–335.e326

Wan L, Xu K et al (2018) Phosphorylation of EZH2 by AMPK suppresses PRC2 methyltransferase activity and oncogenic function. Mol Cell 69(2):279–291.e275

Wang S, Wang K et al (2017a) Overaccumulation of p53-mediated autophagy protects against betulinic acid-induced apoptotic cell death in colorectal cancer cells. Cell Death Dis 8(10): e3087–e3087

Wang T, Yu Q et al (2017b) O-GlcNAcylation of fumarase maintains tumour growth under glucose deficiency. Nat Cell Biol 19(7):833–843

Wang Y, Huang Y et al (2020) Acetyltransferase GCN5 regulates autophagy and lysosome biogenesis by targeting TFEB. EMBO Rep 21(1):e48335

Wangyang YZ, Xin, Liu G-w, Li D-y, Feng Y-b, Guo T-y, Ma C, Wang T (2017) Upregulation of P63 inhibits chondrocyte autophagy thereby enhancing the malignant progression of osteoarthritis. Pharmazie 72(6):361–364(364)

Wei FZ, Cao Z et al (2015) Epigenetic regulation of autophagy by the methyltransferase EZH2 through an MTOR-dependent pathway. Autophagy 11(12):2309–2322

White E, DiPaola RS (2009) The double-edged sword of autophagy modulation in cancer. Clin Cancer Res 15(17):5308–5316

Wu Z, Chang PC et al (2010) Autophagy blockade sensitizes prostate cancer cells towards Src family kinase inhibitors. Genes Cancer 1(1):40–49

Xiao B, Sanders MJ et al (2011) Structure of mammalian AMPK and its regulation by ADP. Nature 472(7342):230–233

Xiao X, He Q et al (2012) Metformin impairs the growth of liver kinase B1-intact cervical cancer cells. Gynecol Oncol 127(1):249–255

Yamamoto S, Tanaka K et al (2008) Suberoylanilide hydroxamic acid (SAHA) induces apoptosis or autophagy-associated cell death in chondrosarcoma cell lines. Anticancer Res 28(3a):1585–1591

Yang Z, Klionsky DJ (2010) Mammalian autophagy: core molecular machinery and signaling regulation. Curr Opin Cell Biol 22(2):124–131

Yang Y, Liang C (2015) MicroRNAs: an emerging player in autophagy. ScienceOpen Res 2015

Yang W, Hong YH et al (2001) Regulation of transcription by AMP-activated protein kinase: phosphorylation of p300 blocks its interaction with nuclear receptors. J Biol Chem 276(42): 38341–38344

Yang S, Wang X et al (2011) Pancreatic cancers require autophagy for tumor growth. Genes Dev 25(7):717–729

Yang A, Jiao Y et al (2018) Homocysteine activates autophagy by inhibition of CFTR expression via interaction between DNA methylation and H3K27me3 in mouse liver. Cell Death Dis 9(2): 169

You L, Wang Z et al (2015) The role of STAT3 in autophagy. Autophagy 11(5):729–739

Yu YS, Shin HR et al (2020) Pontin arginine methylation by CARM1 is crucial for epigenetic regulation of autophagy. Nat Commun 11(1):6297

Yue Z, Jin S et al (2003) Beclin 1, an autophagy gene essential for early embryonic development, is a haploinsufficient tumor suppressor. Proc Natl Acad Sci U S A 100(25):15077–15082

Yue J-Y, Wang Y-J et al (2021) Silencing of ATG2 and ATG7 promotes programmed cell death in wheat via inhibition of autophagy under salt stress. Ecotoxicol Environ Saf 225:112761

Yunyan G, Zhang M, Peng F, Fang L, Zhang Y, Liang H, Zhou W, Ao L, Guo Z (2015) The BRCA1/2-directed miRNA signature predicts a good prognosis in ovarian cancer patients with wild-type BRCA1/2. Oncotarget 6:2397–2406

Zhang S, Liang M et al (2017) Phototrophy and starvation-based induction of autophagy upon removal of Gcn5-catalyzed acetylation of Atg7 in Magnaporthe oryzae. Autophagy 13(8): 1318–1330

Zhang J, Wang J et al (2018) Importance of TFEB acetylation in control of its transcriptional activity and lysosomal function in response to histone deacetylase inhibitors. Autophagy 14(6): 1043–1059

Zhang Y, Chen J et al (2019) Involvement of AMPK activation in the inhibition of hepatic gluconeogenesis by Ficus carica leaf extract in diabetic mice and HepG2 cells. Biomed Pharmacother 109:188–194

Zhou J, Zhang W et al (2009) PPARgamma activation induces autophagy in breast cancer cells. Int J Biochem Cell Biol 41(11):2334–2342

Part II
Epigenetic Regulation of Cellular Metabolic Pathways

Chapter 6
Long Non-coding RNAs, Lnc$^{(ing)}$ RNA Metabolism to Cancer Biology

Sourav Dey, Arushi Misra, and R. Selvi Bharathavikru

Abstract The eukaryotic genome is represented by a vast proportion of non-coding regions, which in recent years have been attributed to have important functional roles in gene regulation through a myriad of processes, ranging from proper localization, correct folding and, most importantly, spatial and temporally regulated expression of genes. One of the major contributing factors in these processes is ribonucleic acid (RNA) metabolism, which comprises the RNA-nucleoprotein (RNP) complexes that interact with and instruct the genome to function. Long non-coding RNAs are an integral component of several RNPs, and herein we provide an overview of the understanding of the long non-coding RNAs, their characteristics, their function and their balancing act as dual modulators in cancer manifestation and progression.

Keywords Non-coding RNA · LncRNA · Cancer · Signalling · MALAT1 · NEAT1 · Homeostasis

Introduction

The regulation of gene expression is an intricately controlled process through a coordinated interplay between the structural organization and functional manifestation of the eukaryotic genome, both at the level of transcription and with post-transcriptional regulation. Epigenetic modifications have been known to regulate gene expression. The cascade of events from cellular cues to the actual effects is controlled at distinct steps by epigenetic modifications. The crosstalk between epigenetics and RNA regulatory processes is also being studied now, with more evidence linking ribonucleic acid (RNA) as a regulator of epigenetic processes (Holoch and Moazed 2015). In recent years, with the advent of better sequencing

---

S. Dey · A. Misra · R. Selvi Bharathavikru (✉)
RNP Biology Laboratory, Department of Biological Sciences, Indian Institute of Science Education and Research (IISER)-Berhampur, Transit Campus, Govt ITI Building, Engineering School Junction, Berhampur, Ganjam, Odisha, India
e-mail: brselvi@iiserbpr.ac.in

T. K. Kundu, C. Das (eds.), *Metabolism and Epigenetic Regulation: Implications in Cancer*, Subcellular Biochemistry 100,
https://doi.org/10.1007/978-3-031-07634-3_6

methodologies supported by technological advances in cellular and molecular characterization, the erstwhile 'junk' (non-coding) region of the genome is now demonstrated to be an important modulator of how and when genes are expressed. The ENCODE project, after careful curation, projected an estimate of about 80% of the non-coding genome to be interacting with regulatory proteins and thus enabling functional outcomes (ENCODE 2012). The non-coding region of the genome is transcribed to non-coding RNA (ncRNA), which includes both small and long non-coding RNAs.

In this context, we discuss the role of an important class of non-coding RNAs, the long non-coding RNAs (LncRNAs), in modulating RNA metabolism, resulting in the perturbation of signalling pathways and leading to cancer manifestation.

Origin and Development

Long non-coding RNAs (LncRNA) are longer than 200 nucleotides and were initially discovered in the transcripts (of mouse origin). These can be detected either in the cytoplasm or in the nucleus of the cells where they are expressed. LncRNAs are observed to have a broad variation in their length, structure and function. Computational studies have shown that less than 5% of LncRNA are derived from the RNA transcripts that have the potential to be translated into proteins.

Long non-coding RNAs are derived through different modes, including the following:

- Protein-coding genes with structural damage can transform into LncRNAs.
- Abreast non-transcribed regions can generate LncRNAs through chromosomal rearrangements.
- The retro-transposition of a copy of a non-coding gene can form an LncRNA.
- Tandem duplication events can lead to the development of LncRNAs with adjacent repeats.
- Inserting a transposon can also generate a functional LncRNA.
- LncRNAs are generated from the mitochondrial genome.

Based on the location where they originate, they can be classified as intergenic LncRNA (lincRNA), intronic LncRNA (sense/anti-sense), exonic LncRNA and overlapping LncRNA (Kung et al. 2013). Although there has been extensive research on LncRNA expression in disease development, only little is known about the origin and development of LncRNAs, which are conserved across some mammalian lineages. A timeline of the various investigations related to identification and characterization is summarized in Fig. 6.1.

Briefly, the functional aspect of an unstable RNA intermediate that carries information from genes to ribosomes for protein synthesis was described by Brenner et al. (1961). Further, a model for the regulation of gene expression in eukaryotes where ncRNAs act as regulatory intermediates for relaying signals was postulated in 1969 (Britten and Davidson 1969). Introns were then discovered as ncRNA elements

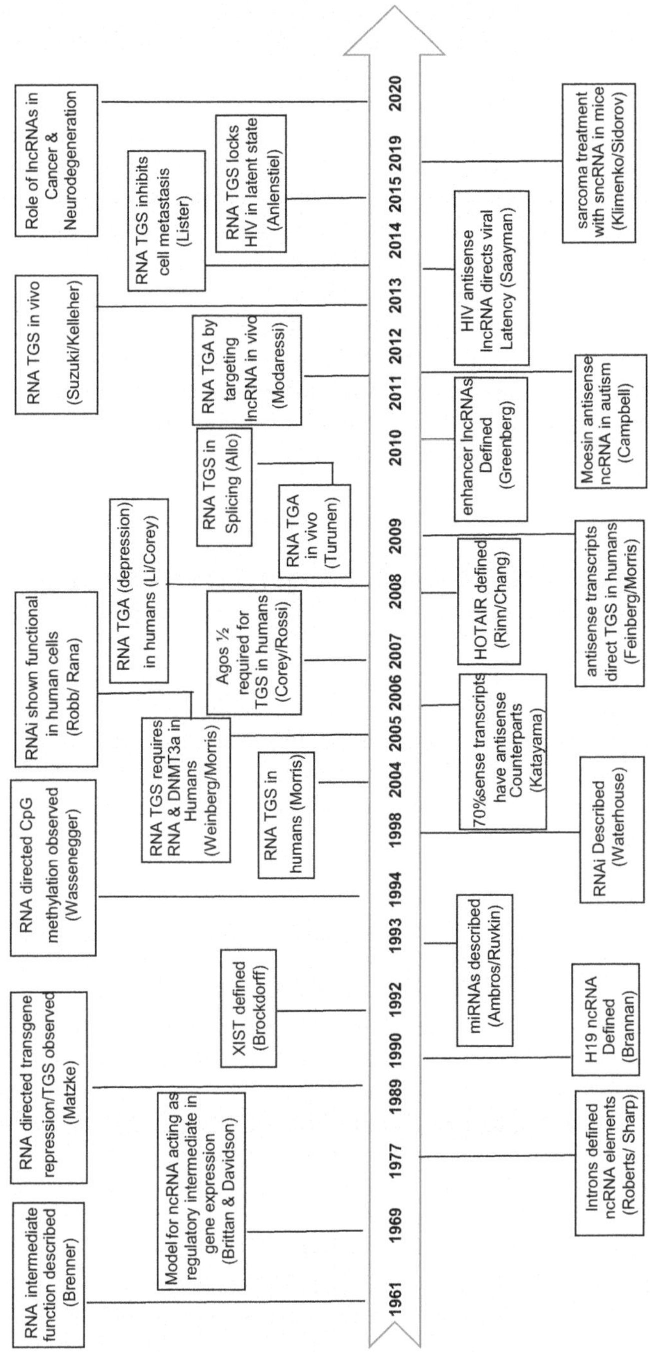

Fig. 6.1 The timeline associated with long non-coding RNA research, starting from discovery, identification, characterization and therapeutic targeting

(Berget et al. 1977; Chow et al. 1977), but in a few years, RNA-directed transgene repression/transcriptional gene silencing (TGS) was observed (Matzke et al. 1989), which was further clarified with the discovery of gene-specific regulatory roles of LncRNA with the discovery of H19 LncRNA involved in epigenetic regulation (Brannan et al. 1990) and XIST (Brockdorff et al. 1992). This was followed by the description of microRNAs (miRNAs) (Lee et al. 1993). Within the next few years, RNA interference, the process through which these ncRNAs interfere with transcription, was described (Waterhouse et al. 1998). With further knowledge of the processing events, transgenic silencing was reported in humans (Morris et al. 2004). In 2005, it was found that RNA interference is functional in the nucleus of human cells (Robb et al. 2005) and that AGO1/2 is required for transgene silencing (Rand et al. 2005). In 2010, a new class of LncRNA was described, known as "enhancer LncRNA" (Kim et al. 2010). Although enhancer RNAs are different from LncRNAs, these two species of non-coding RNAs share similar roles in the activation of mRNAs. In vivo transcriptional gene silencing via non-coding RNAs, such as small interfering RNAs (siRNAs) and miRNAs, was a new advancement (Suzuki et al. 2013), which led to the observation that RNA TGS inhibits cell metastasis (Lister et al. 2015). This machinery was then studied in relation to the human immunodeficiency virus (HIV), through which it was found that HIV anti-sense LncRNA directs viral latency (Saayman et al. 2014) and RNA TGS locks HIV in the latent state (Ahlenstiel et al. 2015). After the recent characterization of Bloodlinc (trans-acting super-enhancer RNA), an enhancer RNA also functioning as an LncRNA was characterized (Alvarez-Dominguez et al. 2017). In recent years, LncRNA has been one of the major targets of molecular studies since its functional role in not only cancer but also different kinds of diseases and syndromes, like Alzheimer's disease, has been observed and intensely researched. The most interesting aspect of LncRNA, which makes it all the more coveted, is the dual role it played in different tissues.

Concepts and Facts

LncRNAs are a massive, unexplored class of non-coding RNAs. According to the NONCODE database (http://www.noncode.org/), 96308 LncRNA genes have been identified in the human genome (Liu et al. 2005). Moreover, studies on the sequencing of the human genome in 2001 showed that the majority of the genome except 1.2% is non-coding, though it was later discovered that the genome is pervasively transcribed into numerous ncRNAs with their controversial functions. According to lnc2Cancer (http://www.bio-bigdata.net/lnc2cancer), a human LncRNA database, a total of 165 cancer sub-types have been found to be associated with 1614 different LncRNAs. Therefore, the sequences, structures and functions of all experimentally validated functional LncRNA must be manually collected. EVLncRNA (https://www.sdklab-biophysics-dzu.net/EVLncRNA2/) is one such database that was released in 2016, and more recently, its newer version with more recent (until May 2020) articles has been released (Zhou et al. 2018, 2021).

Functional Roles of LncRNA

LncRNAs were earlier considered to be transcriptional noise but later were demonstrated to be functional molecules, and just like proteins, their role in cellular functions has also been shown. These functions are manifested through various mechanisms, such as cell proliferation and apoptosis, cell development and differentiation, X-chromosome inactivation and genomic imprinting (Lee and Bartolomei 2013; Akhade et al. 2017). Since they can mediate their effects in a cis or trans manner by directly binding to deoxyribonucleic acid (DNA), RNA or proteins, they have also been found to play important roles in human diseases like coronary artery disease, amyotrophic lateral sclerosis, Alzheimer's disease as well as cancer with either an oncogenic or a tumour-suppressive role. But the mechanism by which LncRNA governs these disease processes remains mysterious and is being investigated. Thus, understanding LncRNA is of utmost importance in the study of human diseases and their biology.

Major Long Non-coding RNAs

There are about 300 annotated LncRNAs, of which approximately 180 have been characterized in humans. A brief overview of a representative subset is presented in Table 6.1.

Regulatory Roles

LncRNAs have been ascribed dual regulatory roles in gene expression, where they can either activate or repress gene expression through the modulation of transcription factor availability, chromatin accessibility, RNA interaction with RNA-binding proteins (RBPs) and signalling networks.

Association with RBPs

About 7% of all the proteins are RBPs that regulate the functioning and fate of a broad range of transcripts, thereby contributing to cellular homeostasis. Recent proteomics and transcriptomics studies have revealed that LncRNAs are often

Table 6.1 Representative subset of lncRNAs and their associated functions

LncRNA	Function	Expression	Role of lncRNA	Mechanism of action	Associated cancer
H19	mRNA decay	Up-regulated	Inhibits apoptosis and promotes cell proliferation	Involves hedgehog signalling	Osteocarcinoma
	Pre-miRNA processing				
HOTAIR	Pre-miRNA processing	Up-regulated	Promotes cell proliferation	Invasion and secretion of MMP2 and MMP9	Various
MALAT1	Alternative splicing	Up-regulated	Promotes cell proliferation and metastasis	Activates P13K/AKT pathway	Various
	Nuclear export			Involvement of RhoA/ ROCK pathway	
NEAT1	Alternative splicing	Up-regulated	Induces DNA damage response	Associates with inflammasomes	Various
	Pre-miRNA processing		Assembling of paraspeckles		
LINC00161	mRNA decay	Up-regulated	Increases cisplatin-induced apoptosis	Sponges endogenous mir-645	Hepatocellular carcinoma
	mRNA translation		Reverses cisplatin resistance	Upregulates *IFIT2*	
MEG3	mRNA translation	Down-regulated	Associated with poor overall cell survival	Positively regulates *p53* expression	Colorectal cancer
HIF2PUT	mRNA translation	Down-regulated	Promotes cell proliferation and migration	Positively correlated with *HIF-2α* expression	Osteocarcinoma
			Decreases the percentage of CD133-expressing cells		
			Impairs the osteocarcinoma-stem-sphere-forming ability of MG63 cells		
MF12	Alternative splicing	Up-regulated	Promotes cell proliferation and inhibits apoptosis	Upregulates *FOXP4* expression	Osteocarcinoma
NEAT1	mRNA translation	Down-regulated	Inhibits cellular growth		Leukaemia and multiple myeloma

included in RBP-based complexes. mRNA decay is an essential regulatory mechanism in gene expression. Long ncRNAs regulate mRNA decay by interacting with the RBPs and thereby modulate important cellular functions and pathological events. NORAD (non-coding RNA activated by DNA damage) is one such example of LncRNA, and it is highly conserved and abundantly expressed in the mammalian tissues as a result of tissue damage. This LncRNA functions as a multivalent binding platform for the PUMILIO (PUM) family of RBPs (Lee et al. 2016).

DNA Damage Response (DDR)

Our cells are equipped with impressive DDR systems for repairing different DNA lesion types. Tumour suppressor protein p53 can modulate all DNA damage responses in either a transcription-dependent or -independent manner. NEAT1, an LncRNA that is an indispensable structural component of paraspeckles, has been identified and validated as a p53 target. These studies have also shown the impairment of DDR efficiency in NEAT1 depleted cells (Adriaens et al. 2016; Huarte et al. 2010). In 2015, it was reported by Blume et al. that the induction of NEAT1 could possibly play a role in the DNA damage pathway in chronic lymphocytic leukaemia (CLL), which further demonstrated that once DNA damage occurs, NEAT1 is induced in the presence of functional p53 but not in CLL-carrying p53 mutation, and once it is induced, it leads to cell death (Blume et al. 2015). Adriaens et al. also demonstrated that treatment with Nutlin-3, which is a p53 inducer, induces NEAT1 expression levels, and its silencing leads to the accumulation of DNA damage, which was confirmed by increased levels of YH2AX (histone H2AX) as well as enhancement in the phosphorylated fraction of KAP1 (KRAB-associated protein 1), which is an ATM substrate (ataxia-telangiectasia mutated kinase) (Adriaens et al. 2016).

Maintenance of Chromatin States

Chromatin is a nucleoprotein complex that is a very dynamic entity in the cell which governs gene expression. The chromatin states are defined by the proteins that are part of it and the various epigenetic factors that interact with chromatin. XIST LncRNA is a pioneering example under this category (Loda and Heard 2019). Several LncRNAs have been shown to be a part of chromatin, both active and inactive chromatin. They act as scaffolds providing a binding site for the activator and/or repressor complexes and, through cis interactions, can regulate gene expression. A well-known example under this category is the LncRNA HOTAIR (Khalil et al. 2009). HOTAIR and chromatin remodelling complexes, such as polycomb-repressive complexes 1 and 2 (PRC1 and PRC2) (Rinn et al. 2007), mediate the mono-ubiquitinylation of lysine 119 of histone 2A (H2AK119ub) (Wang et al. 2004) and the di- and tri-methylation of histone H3 on lysine 27 (H3K27me2 and

H3K27me3) (Cao and Zhang 2004), respectively, as well as lysine-specific demethylase 1 (LSD1)/CoREST, which demethylates mono- and di-methylated histone 3 at lysine 4 (H3K4) and G9a histone methyltransferase, which catalyzes histone 3 lysine 9 di- and tri-methylation (H3K9me2 and H3K9me3). This RNA-protein interaction between HOTAIR and PRC2 is essential for the proper localization of PRC2 across the HOXD cluster. In fact, the deletion of HOTAIR results in the mis-localization of PRC2 from the HOXD cluster, in trans, and concomitantly activates HOXD genes (Shi et al. 2004). Thus, LncRNA can act as recruiters of epigenetic regulators to the chromatin and also as tethering junctions holding the chromatin in a particular state.

Transcription

The role of LncRNA in the context of transcription regulation is multifaceted. Intergenic LncRNAs have been divided into two major categories on the basis of chromatin marks at their promoters: those emanating from enhancer regions or those transcribed from promoter-like LncRNA loci (Marques et al. 2013). Studies have shown that LncRNA can act as scaffolds, recruiting co-activators and transcription complexes to facilitate transcription (Hung et al. 2011). On the other hand, they have also been shown to repress transcription by directly blocking access, acting as decoys or competing with the transcription factor itself (Willingham et al. 2005). Some studies have also shown the role of LncRNA directly in Pol-II-mediated transcription as well as in acting as co-activators. Examples of some of these LncRNAs that modulate transcription through direct or indirect means include Gas5, SRA and NRON (Lanz et al. 1999; Schnell et al. 2004). H19 is one of the highly expressed LncRNAs in cancer, and it mediates (1) SP1-TGFBR2 (TF-gene) interaction through let-7b and miR-200b, (2) ETS1-TGFBR2 (TF- gene) interaction through miR-29a and miR-200b and (3) STAT3-KLF11 (TF-gene) interaction through miR-17. Such regulatory triplets can be used to predict the potential function of LncRNAs and miRNAs in cancer. This critical regulation could play a pivotal role in understanding the molecular mechanisms of cancer initiation and progression and could serve as biomarkers for cancer diagnosis, drug development and therapeutic strategy development (Li et al. 2020a).

Post-transcription

RNA processing is composed of several distinct steps involving several RBPs. After transcription, these RBPs facilitate the overall turnover of the RNA and its stability, degradation and localization. These events together determine the dynamic and steady-state levels of mRNAs. LncRNAs have been shown to participate, along with RBPs, in modulating at least some of these post-transcriptional mechanisms. It is notable that one of the most characterized LncRNAs, NEAT1, is localized to paraspeckles and is an essential structural component of this membrane-less organelle. MALAT1 has been shown to localize with nuclear speckles. These organelles have important roles in RNA processing events (Tripathi et al. 2010; Tsuiji et al. 2011). SRSF1 (SR splicing factor 1) can interact with MALAT1 through its RRM domain (RNA recognition motifs). It is required for the efficient localization of SRSF1 as well as several other splicing factors to nuclear speckles. SR protein interactions with MALAT1 in the nuclear speckles alter their phosphorylation status, thereby leading to alterations in alternative splicing patterns (Tripathi et al. 2010). Apart from the facilitation of correct localization, LncRNAs are also known to act as sponges, thus intervening with miRNA and mRNA interactions.

Post-translational

Although evidence supporting the role of LncRNA in post-translational regulation is limited, there are recent studies that show that LncRNA-p21 can associate with polysome-bound β-catenin (Ebralidze et al. 2008; Yoon et al. 2012). There have also been other studies showing LncRNA interaction with polysomes and suggesting that LncRNA are capable of fine-tuning gene expression through all stages of transcription and translation. LncRNA can also interact directly with proteins and regulate their stability by preventing protein ubiquitination and degradation. Liu et al. found that MT1JP acts as a crucial factor in cellular transformation through the modulation of p53 translation through the binding and stabilization of the RNA-binding protein TIAR (Liu et al. 2016). LoNA, a nucleolar-specific LncRNA, has been detected to reduce the production of rRNA ribosome biosynthesis by binding to nucleolin through its 5′ region and sequestering it, thereby suppressing rRNA transcription and diminishing fibrillarin activity by recruiting 3′ end, thereby reducing rRNA methylation (Li et al. 2018a).

LncRNA, Dual Regulators of Signalling Pathways in Cancer

The functional activity of long non-coding RNAs is not limited to the physiological events in cellular processes; it has a very important and regulatory role in several signalling pathways which are directly or indirectly involved in cancer or tumourigenesis. LncRNA functions as a regulatory component by interacting with either proteins, such as receptors and transcription factors, or RNA and lipids. LncRNAs are known to modulate the signalling transduction by directly associating with receptors, several protein kinases, transcription factors and various downstream signalling molecules. They can enhance phosphorylation events or down-regulate the signalling pathways by altering the enzymatic activity of kinases.

To date, various researchers have investigated the role of LncRNA in cancer progression and have identified this to occur by altering signalling pathways. The association of LncRNA Lnc-DC with STAT3 prevents SHP1-mediated dephosphorylation in JAK-STAT signalling (Wang et al. 2014). Understanding the functional activity and regulatory role of LncRNA in signalling pathways should create the direction for therapeutic targets in cancer treatment.

LncRNAs Are Involved in Different Steps of the Signalling Cascade

Multiple signalling cascades are activated by different extra and intracellular signals that regulate cellular activites (Lin and Yang 2018):

- Receptor dimerization (homo- or hetero-dimer) could be regulated by LncRNA.
- The recruitment of protein kinases to receptors could be enhanced because of the association of LncRNA with activation/inactivation.
- Sometimes LncRNA itself acts as a second messenger in signalling cascades.
- By associating with kinases, LncRNA modifies effector molecules through post-translational mechanisms in a kinase-dependent manner.
- LncRNA regulates the entry of downstream signalling components, especially transcription factors' entry into the nucleus by associating and modulating the transcription factor binding to cognate sites.

Routes Through Which LncRNAs Regulate Signalling Pathways

The involvement of LncRNA in the signalling cascade seems to occur in six functional ways, as represented in Fig. 6.2 (Lin and Yang 2018):

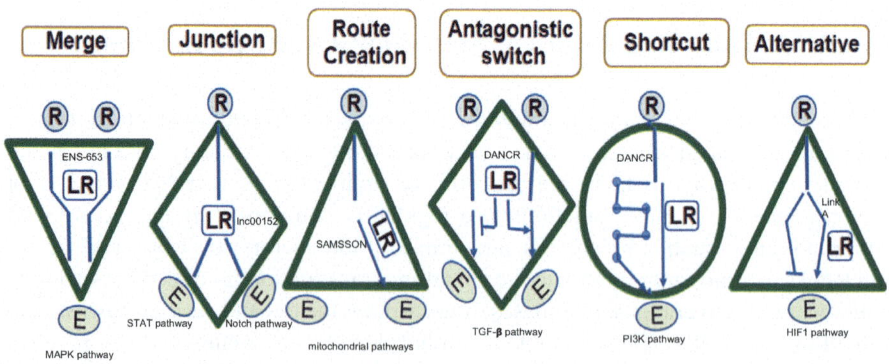

Fig. 6.2 Schematic representation of the road map of how LncRNAs modulate signalling pathways

1. *Merger:* two different kinases that are activated by two closely placed signalling pathways are connected by LncRNA and activate similar effectors, leading to common cellular effects. Because of this, any ligand-receptor activation leads to a cellular effect. For example, LncRNA ENS-653 associates with ERK1/2 and ERK 5, thus activating the MAPK signalling cascade (Song et al. 2019).
2. *Junction:* sometimes LncRNA couples with receptors in a way that can create a junction where multiple kinases can associate with it during receptor-ligand activation; thus, different signalling pathways are initiated in a kinase-dependent manner. An example of this kind is lnc00152, which is induced through STAT2, which acts as a junctional LncRNA, which can lead to the activation of either STAT-mediated signalling pathways or Notch signalling pathway through the modulation of the microRNA network (Binder et al. 2020; Bian et al. 2017).
3. *Route creation:* LncRNA forms a bridge between specific receptor-activated kinases with more than one effector. It can drive a signalling pathway towards one effector at a time, depending on the cellular environment and the specific ligand. As a result, specific receptor-ligand activation can move forward to different signalling pathways in response to cellular context. An example of this is LncRNA SAMSSON, which upon interaction with p32 is localized to the mitochondria and thus regulated through metabolic cues (Goding 2016).
4. *Antagonist switch:* LncRNA may associate with two different kinases, which are activated by their respective receptors, and regulate the signalling in a way that when one signalling is on, the other will be kept off, as observed in DANCR non-coding RNA, which inhibits the TGF-β pathway during the early stage of cancer manifestation, whereas it activates the β-catenin signalling pathway for cancer progression (Pan et al. 2020).
5. *Shortcut:* in some signalling cascades, there is more than one kinase activated by another in a sequential event, ultimately resulting in effector activation. The association of LncRNA in this type of signalling can bypass all intermediate kinases and directly activate effector response. An example is DANCR lincRNA, which activates the PI3K/Akt signalling pathway through RXRA-mediated phosphorylation by GSKβ and thus facilitates the rapid activation of cell signalling events (Tang et al. 2018).
6. *Alternative:* the disruption of some signalling pathways could be continued by the association of LncRNA in an alternative way, where LncRNA sometimes recruits an intermediate component or itself acts as a mediator, as seen in the case of BRK, which is a non-receptor tyrosine kinase, which interacts with Link A RNA and is recruited to the EGFR tyrosine kinase, and together they facilitate BRK activation and HIF 1 pathway stabilization (Lin et al. 2016).

Signalling Pathways Regulated by LncRNA: Building Blocks in Cancer Biology

Emerging research has shown that LncRNA is involved in the regulation of several signalling events. It can be associated with most of the intermediate signalling components (reviewed in Camacho et al. 2018). This mode of LncRNA involvement in signalling pathways has been reported to promote cellular growth, metastasis and angiogenesis in cancer, as depicted in Fig. 6.3. For example, LncRNA OIP5-AS1 promotes epithelial-mesenchymal transition (EMT) in laryngeal squamous cell carcinoma, and LncRNA NEAT2 (a highly conserved LncRNA) promotes metastasis in lung cancer. Some of the major signalling pathways that are regulated by LncRNA are as follows.

LncRNA and Wnt Signalling

Wnt signalling is conserved from embryo to adult and plays a crucial role in physiology. Cellular proliferation, growth and differentiation are driven by this signalling during embryogenesis, whereas in adults, it is involved in maintaining tissue homeostasis. There are three major WNT-signalling pathways: (i) canonical (β-catenin dependent), (ii) non-canonical (WNT/PCP) and (iii) WNT/Ca2+ signalling. Among them, β-catenin-mediated signalling has been more extensively studied. In the absence of Wnt, the cytoplasmic level of β-catenin is decreased by ubiquitin-mediated proteasomal degradation. Two kinases GSK3/CK1 are bound to β-catenin with the help of scaffold proteins Axin and adenomatous polyposis coli (APC). Multiple phosphorylations of β-catenin on serine and threonine residues by kinases lead to degradation. In the nucleus, the transcription of downstream genes is kept off by the Groucho protein. When Wnt binds to its receptors, β-catenin is stabilized and enters the nucleus, and along with co-activator TCF, it starts the transcription of several growth-related Wnt target genes.

Recently, it has been shown that LncRNA interacts with a variety of signalling pathways, and Wnt signalling is one of them. The perturbation of Wnt signalling by LncRNA in different cancers is summarized below. LncRNA is directly involved in the regulation of the intermediate component of Wnt signalling (for example β-catenin) and sometimes also suppresses the signalling, resulting in cancer progression. In colorectal cancer, LncRNA CCAT-2 helps in the recruitment of TCF3 (a co-activator of Wnt-targeted gene transcription) to the promoter, directly associating with it, resulting in the activation of Wnt signalling. Another LncRNA RBM-AS1 acts as a transcriptional activator of Wnt-targeted genes by binding with β-catenin and recruiting it to the promoter (Javed et al. 2020). In renal cancer, cell carcinoma LncRNA CRND modifies the PI3K/AKT/GSK3-beta signalling pathway in an unknown manner, resulting in the up-regulation of the β-catenin level in cells, which leads to the activation of Wnt signalling (Yang et al. 2018). In hepato-cellular carcinoma, LncRNA TCF7 recruits the SWI/SNF chromatin remodelling complex to the promoter of the TCF co-activator of Wnt-targeted gene expression, leading to the activation of Wnt signalling (Wang et al. 2015). LncRNA CASC11 stabilizes hnRNP-K (a poly-C binding protein family), which protects β-catenin from GSK3 by interacting with Axin and GSK3, leading to the up-regulation of β-catenin signalling (Hu et al. 2018). In non-small cell lung cancer (NSCLC) cells, LncRNA AK126698 (Fu et al. 2016) and, in colorectal cancer,

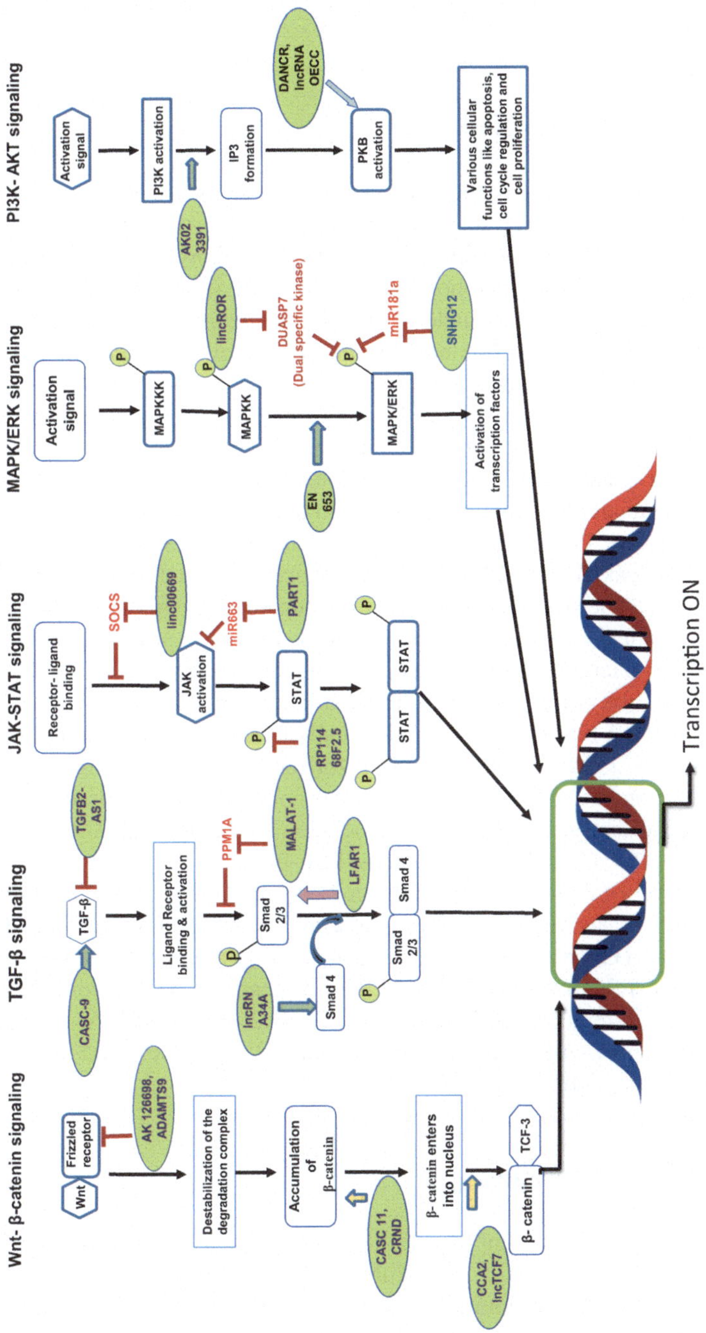

Fig. 6.3 The different signalling cascades and their effect on the regulation of gene expression are represented. Highlighted in green are the different LncRNAs, and their modulatory effects on cancer signalling networks are shown

LncRNA ADAMTS9-A1 (Li et al. 2020a, b) both suppress Wnt signalling by targeting and inhibiting Wnt receptor Frizzled 8 and down-regulating β-catenin, respectively.

LncRNA and TGF-Beta Signalling

The TGF-beta/Smad signalling pathway works through the TGF-β super-family ligands (TGF β-1, 2,3,5; BMP, activin), which bind to the type II receptor, allow type I dimerization with the type II receptor and are phosphorylated. Following this, the activated type I receptor phosphorylates Smad2 and Smad3, which form complexes with Smad4 and enter the nucleus, thus regulating gene expression. Cellular growth, differentiation, cellular homeostasis and several other cellular activities are maintained by TGF-β signalling.

Recent research has revealed that LncRNA can cause the up- and down-regulation of TGF-β signalling, which is summarized below. LncRNA can recruit phosphatases as a result of the LncRNA-mediated dephosphorylation of the intermediate components of the signalling cascade. In hepatic cells, LncRNA MALAT1 recruits a phosphatase, PPM1A, to the phosphorylated Smad2/Smad3, together with other proteins, such as SETD2. PPM1A dephosphorylates Smad2/Smad3 so there is a reduction of Smad4 activation, leading to the termination of signalling (Zhang et al. 2020). LncRNA interferes with the recruitment of transcription factors to the promoters of the TGF-β target gene through epigenetic modification. LncRNA TGFB2-AS1 keeps the transcription of TGF-β targeted genes turned off through H3K12 methylation. LncRNA binds to the 3′ terminal end of the EED adaptor of polycomb-repressive complex 2 (PRC2) and recruits it for methylation (Papoutsoglou et al. 2019). The involvement of LncRNA CASC9 has an important role in the development of colorectal cancer, wherein the association of LncRNA CASC9 with its binding protein CPSF3 stabilizes the ligands TGF-β2 and TRET, thus resulting in the hyperactivation of signalling, resulting in more accumulation of Smad3 phosphorylation (Luo et al. 2019). LncRNA acts as an activator of almost all the components of the TGF-beta signalling cascade, causing the progression of cancer. In hepatocellular carcinoma, LncRNA MALAT1 up-regulates TGF-β concentration, NORAD, KR19 up-regulate TGFBR1 and TGFBR2; LncRNA34A up-regulate Smad4; KRT19, LncRNA LFAR1 up-regulate Smad2/3 complex, and all of these up-regulate TGF-β signalling (Han et al. 2020).

LncRNA and JAK-STAT Pathway

JAK-STAT signalling is driven by cytokine receptors as well as by hormone signalling. JAK (Janus kinase) is part of the receptor, activated upon the binding of cytokines to its specific receptor. The activated JAK kinase then phosphorylates the cytosolic part of the receptor, which is the binding site of the SH2 domain containing the protein STAT (signal transducer and activator of transcription); the bound STAT is also phosphorylated by JAK, leading to the dimerization of STAT, which enters into the nucleus, regulating key processes like cell division, cell growth, differentiation and cell migration.

The normal activation and inactivation of this signalling are maintained through the dephosphorylation of receptors and JAK by SHP1 phosphatase and through ubiquitin-mediated destruction by the SOCS (suppressor of cytokine signalling) protein. LncRNA binds and inactivates such inhibitors. LncRNA LINC00669 in nasopharyngeal cell carcinoma (NPC) associates with the SOCS protein and blocks its ubiquitin ligase activity so that STAT1 levels increase, resulting in longer activation of JAK-STAT signalling and nasopharyngeal carcinoma progression (Qing et al. 2020).

LncRNA can up-regulate the component of JAK-STAT signalling cascade from ligand to intermediate components, leading to an increase in the phosphorylation event of kinases and STAT. LncRNA also acts as a double negative gate, which is the repressor of JAK-STAT signalling. In NSCLC, LncRNA PART1 acts as a repressor of microRNA miR-635, which is a repressor of JAK1 and JAK3 gene transcription. The sponging mechanism of PART1 accelerates JAK/STAT stabilization and phosphorylation, ultimately causing cancer progression (Zhu et al. 2019). A few LncRNAs act as a negative regulator of JAK-STAT signalling by down-regulating STAT. In colorectal cancer, LncRNA RP11-468F2.5, binds to STAT5 and STAT6, resulting in the inhibition of JAK/STAT-mediated cell proliferation and promoting apoptosis (Jiang et al. 2019).

LncRNA and PI3/AKT Pathway

Several cytokines and RTK signalling events activate and recruit PI3K (phosphatidyl-inositol-3 kinase). PI3K uses phospholipids (phosphatidylinositol) as a substrate and generates PI-3 phosphates (PI (3,4) P2 and PI (3,4,5) P3)', which act as docking sites of other proteins, such as protein kinase-B or Akt. LncRNA up-regulates the expression of signalling components, such as protein kinase B, which promotes cellular signalling by phosphorylation events. LncRNA DANCR promotes the expression of the PKB gene through the phosphorylation of RXRA (thyroxine receptor) at 49/78 serine residues (Tang et al. 2018). In lung cancer, LncRNA OECC has been shown to associate with Akt genes, stabilizing the mRNA and promoting gene expression (Zou et al. 2019). In gastric cancer, LncRNA AK023391 accelerates the PI3K/Akt pathway through an increase in phosphorylated PI3K levels (Huang et al. 2017).

LncRNA and MAPK/ERK Signalling

Another important signalling pathway required for cellular proliferation, gene expression, differentiation, mitosis and cell survival is MAPK (mitogen-activated protein kinases)/ERK (extracellular-signal-regulated kinase) signalling, also known as RAS-RAF-MAPK signalling. MAPKs are serine threonine kinases that phosphorylate several proteins including transcription factors. This phosphorylation event acts as an on/off switch for signalling.

In the case of MAPK signalling also, LncRNA plays a role in both up- and down-regulation. LncRNA can stabilize and help in maintaining the phosphorylated intermediate components of signalling cascades through an increase in the phosphorylated MEK1/2 and C-RAF levels, which activates p38 activation through phosphorylation (Liu et al. 2019; Wang et al. 2018a, b). LncRNA ENS-653 associates with ERK1/2 and ERK5 (an intermediate components of the MAPK signalling cascade) and promotes phosphorylation, leading to the progression of MAPK signalling in papillary thyroid cancer (Song et al. 2019). In breast cancer, LncRNA linc-ROR associates with dual-specific kinase DUASP7, which is a negative regulator of ERK phosphorylation, and blocks its activity, thus resulting in the up-regulation of p-ERK (Peng et al. 2017). Some LncRNAs function as endogenous RNAs, which compete with other miRNAs that are repressors of MAPK signalling, ultimately resulting in the up-regulation of MAPK signalling. In NSCLC, miRNA miR-181a blocks the phosphorylation of MAPK1 and MAP2K1. LncRNA SNHG12 inhibits miR-181a and up-regulates the phosphorylated MAPK1 and MAP 2K1, leading to the activation of signalling (Wang et al. 2017).

LncRNA as Tumour Suppressors

LncRNA partakes in gene regulation both in a cis and in a trans manner. Mature transcripts of LncRNA have been found to regulate the expression of nearby or far-off genes in a sequence-specific manner by regulating the structure and organization of their chromatin, by directly binding to their promoter or enhancer regions or by aiding the development of complexes that support or mask active transcription (Johnson et al. 2012; Kopp and Mendell 2018). A number of LncRNAs are delineated as mediators of epigenetic silencing because it has been found that they recruit complexes like polycomb-repressive complex 2 (PRC2), which is responsible for the tri-methylation of lysine at the 27th position in histone 3 (H3K27me3), referred to as repressive modification, which silences the transcription of the genetic locus (Tsai et al. 2010; Hirata et al. 2015). Thus, LncRNAs not only promote or activate signalling pathways, but they are also a part of repressive complexes, acting as negative regulators of cancer progression. A few examples are listed below.

GAS5

Growth arrest-specific 5, a non-coding gene, sequesters miR-21, thus functioning as an endogenous sponge, and as a result, reciprocal repression of both miR-21 and

GAS5 occurs. This interaction occurs on the putative miR-21 binding site at exon 4 of GAS5 and RISC (RNA-induced silencing complex) (Zhang et al. 2013). GAS5 also inhibits TNBC (triple-negative breast cancer) cell invasion through miR-196a-5p/forkhead box protein O1(FOXO)/PI3K (phosphatidylinositol 3-kinase)/AKT axis, which initiates the competitive binding of GAS5 to miR-196a-5p (Li et al. 2018b).

MT1JP

Reported as a tumour suppressor, MT1JP (metallothionein 1J pseudogene) modulates the p53 protein expression level, thereby regulating p53-related signalling pathways (Liu et al. 2016). Reduced MT1JP expression leads to significant lymphatic metastasis and an advanced TNM (tumour node metastasis) stage for gastric cancer patients. In vitro and in vivo studies also show the suppressive role of MT1JP in the migration and invasion of gastric cancer (Xu et al. 2018a; Zhang et al. 2018).

LET

As the name suggests, low expression in tumour (LET) LncRNA is usually down-regulated in most cancer types. It is negatively regulated by hypoxia-induced deacetylase 3 (HDAC3), thus promoting cellular invasion by nuclear factor 90 (NF90) stabilization, which is an outcome of its down-regulation (Sun et al. 2015). In nasopharyngeal carcinoma cells, its repression has been linked to the EZH2-mediated H3K27 histone tri-methylation of the LET promoter, which leads to the inhibition of cell proliferation and apoptosis (Yang et al. 2013).

Some LncRNAs involved in inhibiting distant metastases are listed below.

MALAT1

Metastasis-associated lung adenocarcinoma transcript 1 is both a tumour-promoting and a tumour-suppressing LncRNA. Its escalated expression is associated with relapse and metastatic progression in breast cancer; on the other hand, a few recent studies provide insight into its metastasis-suppressing activity. Mechanistically, it prevents the association of prometastatis transcription factor TEA domain family member 2 (TEAD) with its co-activator YAP (Yes-associated protein) (Kim et al. 2018). Kim et al. reported that the somatic knockout of MALAT1 specifically induced metastasis. Kwok et al. suggested that the PTEN-miRNA-MALAT1 axis may promote tumourigenesis and reported the first evidence proving that MALAT1 possesses novel tumour-suppressive capabilities in breast cancer (Kwok et al. 2018).

MEG3

Known as maternally expressed gene 3, this LncRNA belongs to the DLK1-MEG3 imprinted locus. MEG3 expression inhibition has been found in various types of human tumours and tumour cell lines, including those derived from the brain, bladder, bone marrow, breast, cervix, colon, liver, lung, meninges and prostate (Zhou et al. 2012; Braconi et al. 2011; Wang et al. 2012). MEG 3 expression is lost or significantly decreased in 25% of neuroblastomas (Astuti et al. 2005), 81% of hepatocellular cancers (Braconi et al. 2011) and 82% of gliomas. Recently, in meningiomas, also MEG3 expression was examined, and it was found that the loss of MEG3 expression is strongly linked to tumour grade. It was detected in one of 11 grade II, four of nine grade I and none of seven grade III meningiomas (Zhang et al. 2010).

XIST

X-inactive specific transcript (XIST) is considered to be a potential tumour suppressor in cancer (Xing et al. 2018). A pathway screening based gene set enrichment analysis (GSEA) study has shown that XIST activates moesin (MSN) on the X chromosome resulting in the activation of c-Met signalling pathway. The loss of c-Met signalling decreases the transmigration abilities of breast cancer cells and suppresses breast-to-brain metastatic abilities in vivo (Xu et al. 2018b).

LncRNAs as Therapeutic Targets

The oncogenic and suppressive functions of LncRNA have been shown to have a deterministic role in cancer (Fig. 6.4). Due to their involvement in various cellular processes and their functional outcome, LncRNAs are considered to be an attractive therapeutic target, especially against cancer. Their restricted appearance and ability to interact with proteins, lipids as well as RNAs make them amenable to being

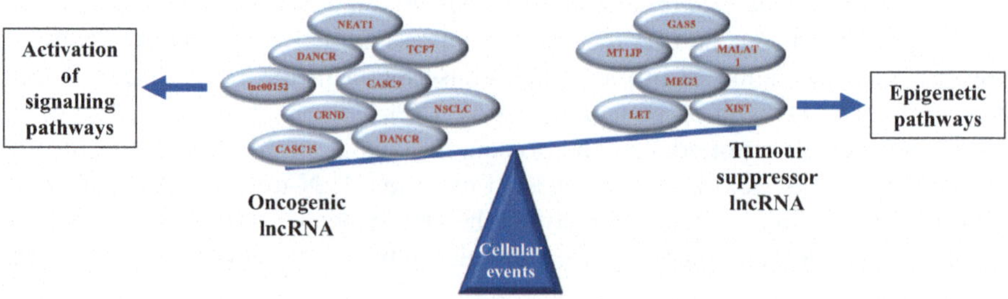

Fig. 6.4 The following figure represents the involvement of LncRNA as a dual regulator in cancer. Different oncogenic and tumour suppressor LncRNAs are shown in highlighted oval shape

targeted by existing anti-cancer drugs. In papillary thyroid carcinoma (PTC), LncRNA BANCR regulates EMT through the up- and down-regulation of E-cadherin and N-cadherin and regulates the RAF/MEK/ERK pathway by increasing the phosphorylation event of intermediates. BANCR inhibitors can be used as a therapeutic agent in PTC as well as various disorders that are causally related to the pathway (Wang et al. 2018a, b).

LncRNA acts as a dual regulator in cancer, and the regulation mechanism in most cases occurs through genetic regulation and/or epigenetic regulation. So targeting LncRNA at all the following levels can be considered from a therapeutic perspective.

Nucleic Acid Modulators of LncRNA

Small interfering RNAs (siRNAs) form a complex with RISC (RNA-induced silencing complex) and bind to the target region and degrade them. siRNA can be used to target LncRNA as a therapeutic strategy. In esophageal squamous cell carcinoma, the knockdown of LncRNA CASC9 by siRNA down-regulates its metastasis (Liang et al. 2018). Targeting LncRNA MALAT1 through siRNA is very effective in metastasis suppression in human bladder cancer and lung adenocarcinoma (Matsui and Corey 2017). A single stranded oligonucleotide of 15–25 bases which is complementary to the RNA, known as anti-sense oligonucleotides (ASO), is synthesized. It can bind to the RNA and degrade it by recruiting RNase H. Nowadays, the use of ASOs to target LncRNA is showing very effective results. In breast cancer, use of MALAT1 ASO has shown a reduction of metastasis and slower tumour formation (Arun et al. 2016). Prokaryotic organisms used the CRISPR system as an immune defence system to degrade foreign genetic materials. CRISPR with CAS9 nuclease binds the target side and degrades it. Now, this system is used widely as a genome editing tool in mammalian cells and can be applied to target LncRNA. Recently, it has been shown that metastasis can be controlled by the CRISPR/Cas9-mediated inhibition of LncRNA LOC389641 in pancreatic duct carcinoma (Goyal et al. 2017). In several signalling pathways, the regulatory effect of LncRNA has been shown to occur through the sponging of miRNA. The interaction of LncRNA and miRNA regulates EMT in various cancers; in lung cancer, miRNA217 interacts with MALAT1 and triggers its decay, causing a decrease of EMT (Lu et al. 2015). Hence, targeting this crosstalk of LncRNA and miRNA would be an important therapeutic target.

Small Molecule Modulators of LncRNA

LncRNA interaction with the target is governed by structural adaptations. This can be targeted by various small molecules to disrupt the function of LncRNA. The sponging of miRNA-186 by LncRNA HULC causes the progression of Ewing sarcoma. The small molecule YK-4-279 inhibits such sponging activity of LncRNA HULC (Mercatelli et al. 2020). Several plant-derived natural compounds are found to be effective regulators of LncRNA. Some studies have shown that LncRNA MALAT1, NEAT11, H19 and PVT1 are affected by natural compounds. In multiple myeloma cells, NEAT1 promotes cell proliferation, migration and invasion. The treatment of resveratrol causes the down-regulation of NEAT activity (Geng et al. 2018).

Future Prospects

In this overview, we have provided a glimpse of LncRNAs and their modulatory role in cancer. The reach of LncRNA in modulating other diseases is equally important. LncRNAs exist along with several RBPs in several multi-protein complexes, and their role in phase transition is yet to be conclusively determined. LncRNAs are thus at the crossroads of several signalling pathways and have the ability to modulate gene expression accordingly. It is now clear that these non-coding RNAs, although are inert in the sense of translation, have far reaching effects in development and disease through their ability to act as dual modulators. Moreover, understanding their role at the molecular level in cell-fate specification and pluripotency mechanisms would play a critical role in defining the stem cell basis of cancers as well as de-differentiation events that occur during tumourigenesis and thus help in connecting the different hallmarks of cancer biology.

Acknowledgments Work in the authors' laboratory is supported by funding from Science and Engineering Research Board, Department of Science and Technology, Department of Biotechnology, and Indian Institute of Science Education and Research, Berhampur, MHRD, Govt. of India. SD is a UGC-JRF.

References

Adriaens C, Standaert L, Barra J, Latil M, Verfaillie A, Kalev P, Boeckx B, Wijnhoven P.W, Radaelli E, Vermi W, et al. (2016) p53 induces formation of NEAT1 LncRNA-containing paraspeckles that modulate replication stress response and chemosensitivity. Nat Med 22: 861–868

Ahlenstiel CL, Suzuki K, Marks K, Symonds GP, Kelleher AD (2015) Controlling HIV-1: non-coding RNA gene therapy approaches to a functional cure. Front Immunol 6:474

Akhade VS, Pal D, Kanduri C (2017) Long noncoding RNA: genome organization and mechanism of action. Adv Exp Med Biol 1008:47–74

Alvarez-Dominguez JR, Knoll M, Gromatzky AA, Lodish HF (2017) The super-enhancer- derived aLncRNA-EC7/Bloodlinc potentiates red blood cell development in trans. Cell Rep 19: 2503–2514

Arun G, Diermeier S, Akerman M, Chang K, Wilkinson JE, Hearn S, Kim Y, MacLeod R, Krainer AR, Norton L et al (2016) Differentiation of mammary tumors and reduction in metastasis upon Malat1 LncRNA loss. Genes Dev 30:34–51

Astuti D, Latif F, Wagner K, Gentle D, Cooper WN, Catchpoole D, Grundy R, Ferguson-Smith AC, Maher ER (2005) Epigenetic alteration at the DLK1-GTL2 imprinted domain in human neoplasia: analysis of neuroblastoma, phaeochromocytoma and Wilms' tumour. Br J Cancer. 92:1574–1580

Berget SM, Moore C, Sharp PA (1977) Spliced segments at the 5 terminus of adenovirus 2 late mRNA. Proc Natl Acad Sci U S A 74:3171–3175

Bian Z, Zhang J, Li M, Feng Y, Yao S, Song M, Qi X, Fei B, Yin Y, Hua D, Huang Z (2017) Long non-coding RNA LINC00152 promotes cell proliferation, metastasis, and confers 5-FU resistance in colorectal cancer by inhibiting miR-139-5p. Oncogenesis 6:395

Binder S, Zipfel I, Friedrich M, Riedel D, Ende S, Kämpf C, Wiedemann K, Buschmann T, Puppel S. H, Reiche K, et al. (2020) Master and servant: LINC00152 - a STAT3-induced long noncoding RNA regulates STAT3 in a positive feedback in human multiple myeloma. BMC Med Genom. 13(1): 22

Blume CJ, Hotz-Wagenblatt A, Hullein J, Sellner L, Jethwa A, Stolz T, Slabicki M, Lee K, Sharathchandra A, Benner A et al (2015) p53-dependent non-coding RNA networks in chronic lymphocytic leukemia. Leukemia 29:2015–2023

Braconi C, Henry JC, Kogure T, Schmittgen T, Patel T (2011) The role of microRNAs in human liver cancers. Sem Oncol 38:752–763

Brannan CI, Dees EC, Ingram RS, Tilghman SM (1990) The product of the H19 gene may function as an RNA. Mol Cell Biol 10:28–36

Brenner S, Jacob F, Meselson M (1961) An unstable intermediate carrying information from genes to ribosomes for protein synthesis. Nature 190:576–581

Britten RJ, Davidson EH (1969) Gene regulation for higher cells: a theory. Science 165:349–357

Brockdorff N, Ashworth A, Kay GF, McCabe VM, Norris DP, Cooper PJ, Swift S, Rastan S (1992) The product of the mouse Xist gene is a 15 kb inactive X-specific transcript containing no conserved ORF and located in the nucleus. Cell 71:515–526

Camacho CV, Choudhari R, Gadad SS (2018) Long noncoding RNAs and cancer, an overview. Steroids 133:93–95

Cao R, Zhang Y (2004) The functions of E(Z)/EZH2-mediated methylation of lysine 27 in histone H3. Curr Opin Genet Dev 14:155–164

Chow LT, Gelinas RE, Broker TR, Roberts RJ (1977) An amazing sequence arrangement at the 5 ends of adenovirus 2 messenger RNA. Cell 12:1–8

Ebralidze AK, Guibal FC, Steidl U, Zhang P, Lee S, Bartholdy B, Jorda MA, Petkova V, Rosenbauer F, Huang G, Dayaram T (2008) PU. 1 expression is modulated by the balance of functional sense and antisense RNAs regulated by a shared cis-regulatory element. Genes Dev 22:2085–2092

ENCODE Project Consortium (2012) An integrated encyclopedia of DNA elements in the human genome. Nature 489:57–74

Fu X, Li H, Liu C, Hu B, Li T, Wang Y (2016) Long noncoding RNA AK126698 inhibits proliferation and migration of non-small cell lung cancer cells by targeting Frizzled-8 and suppressing Wnt/β-catenin signaling pathway. Onco Targets Ther 9:3815–3827

Geng W, Guo X, Zhang L, Ma Y, Wang L, Liu Z, Ji H, Xiong Y (2018) Resveratrol inhibits proliferation, migration and invasion of multiple myeloma cells via NEAT1-mediated Wnt/β-catenin signaling pathway. Biomed Pharmacother 107:484–494

Goding CR (2016) Targeting the lncRNA SAMMSON reveals metabolic vulnerability in melanoma. Cancer Cell 29:619–621

Goyal A, Myacheva K, Groß M, Klingenberg M, Arque DB, Diederichs S (2017) Challenges of CRISPR/Cas9 applications for long non-coding RNA genes. Nucleic Acids Res 45:1–13

Han M, Liao Z, Liu F, Chen X, Zhang B (2020) Modulation of the TGF-beta signaling pathway by long noncoding RNA in hepatocellular carcinoma. Biomark Res 8:1–11

Hirata H, Hinoda Y, Shahryari V, Deng G, Nakajima K, Tabatabai Z. L, Ishii N, Dahiya, R. (2015) Long noncoding RNA MALAT1 promotes aggressive renal cell carcinoma through Ezh2 and interacts with miR-205. Cancer Res 75:1322–1331

Holoch D, Moazed D (2015) RNA-mediated epigenetic regulation of gene expression. Nat Rev Genet 16:71–84

Hu XY, Hou PF, Li TT, Quan HY, Li ML, Lin T, Liu JJ, Bai J, Zheng JN (2018) The roles of Wnt/β-catenin signaling pathway related LncRNA in cancer. Int J Biol Sci 14:2003–2011

Huang Y, Zhang J, Hou L, Wang G, Liu H, Zhang R, Chen X, Zhu J (2017) LncRNA AK023391 promotes tumorigenesis and invasion of gastric cancer through activation of the PI3K/Akt signaling pathway. J Exp Clin Cancer Res 36:1–14

Huarte M, Guttman M, Feldser D, Garber M, Koziol MJ, Kenzelmann-Broz D, Khalil AM, Zuk O, Amit I, Rabani M et al (2010) A large intergenic noncoding RNA induced by p53 mediates global gene repression in the p53 response. Cell 142:409–419

Hung T, Wang Y, Lin MF, Koegel AK, Kotake Y, Grant GD, Horlings HM, Shah N, Umbricht C, Wang P, Wang Y (2011) Extensive and coordinated transcription of noncoding RNAs within cell-cycle promoters. Nat Genet 43:621–629

Javed Z, Khan K, Sadia H, Raza S, Salehi B, Sharifi-Rad J, Cho WC (2020) LncRNA and Wnt signaling in colorectal cancer. Cancer Cell Int 20:1–10

Jiang L, Zhao XH, Mao YL, Wang JF, Zheng HJ, You QS (2019) Long non-coding RNA RP11-48E2.5 curtails colorectal cancer cell proliferation and stimulate apoptosis via JAK/STAT signaling pathway targeting STAT5 and STAT6. J Exp Clin Cancer Res 38:1–16

Johnson R, Derrien T, Bussotti G, Tanzer A, Djebali S, Tilgner H, Guernec G, Martin D, Merkel A, Knowles DG et al (2012) The GENCODE v7 catalog of human long noncoding RNAs: analysis of their gene structure, evolution, and expression. Genome Res 22:1775–1789

Khalil AM, Guttman M, Huarte M, Garber M, Raj A, Rivea Morales D, Thomas K, Presser A, Bernstein BE, van Oudenaarden A et al (2009) Many human large intergenic noncoding RNAs associate with chromatin-modifying complexes and affect gene expression. Proc Natl Acad Sci U S A 106:11667–11672

Kim TK, Hemberg M, Gray JM, Costa AM, Bear DM, Wu J, Harmin DA, Laptewicz M, Barbara-Haley K, Kuersten S et al (2010) Widespread transcription at neuronal activity-regulated enhancers. Nature 465:182–187

Kim J, Piao HL, Kim BJ, Yao F, Han Z, Wang Y, Xiao Z, Siverly AN, Lawhon SE, Ton BN et al (2018) Long noncoding RNA MALAT1 suppresses breast cancer metastasis. Nat Genet 50: 1705–1715

Kopp F, Mendell JT (2018) Functional classification and experimental dissection of long noncoding RNAs. Cell 172:393–407

Kung JT, Colognori D, Lee JT (2013) Long noncoding RNAs: past, present, and future. Genetics 193:651–669

Kwok ZH, Roche V, Chew XH, Fadieieva A, Tay Y (2018) A non-canonical tumor suppressive role for the long non-coding RNA MALAT1 in colon and breast cancers. Int J Cancer 143:668–678

Lanz RB, McKenna NJ, Onate SA, Albrecht U, Wong J, Tsai SY, Tsai MJ, O'Malley BW (1999) A steroid receptor coactivator, SRA, functions as an RNA and is present in an SRC-1 complex. Cell 97:17–27

Lee JT, Bartolomei MS (2013) X-inactivation, imprinting, and long noncoding RNAs in health and disease. Cell 152:1308–1323

Lee RC, Feinbaum RL, Ambros V (1993) The C. elegans heterochronic gene lin-4 encodes small RNAs with antisense complementarity to lin-14. Cell 75:843–854

Lee S, Kopp F, Chang TC, Sataluri A, Chen B, Sivakumar S, Yu H, Xie Y, Mendell JT (2016) Noncoding RNA NORAD regulates genomic stability by sequestering PUMILIO proteins. Cell 164:69–80

Li D, Zhang J, Wang M, Li X, Gong H, Tang H, Chen L, Wan L, Liu Q (2018a) Activity dependent LoNA regulates translation by coordinating rRNA transcription and methylation. Nat Commun 9:1726

Li S, Zhou J, Wang Z, Wang P, Gao X, Wang Y (2018b) Long noncoding RNA GAS5 suppresses triple negative breast cancer progression through inhibition of proliferation and invasion by competitively binding miR-196a-5p. Biomed Pharmacother 104:451–457

Li A, Mallik S, Luo H, Jia P, Lee DF, Zhao Z (2020a) H19, a long non-coding RNA, mediates transcription factors and target genes through interference of microRNAs in pan-cancer. Mol Ther Nucl Acids 21:180–191

Li N, Li J, Mi Q, Xie Y, Li P, Wang L, Binang H, Wang Q, Wang Y, Chen Y, Wang Y, Mao H, Du L, Wang C (2020b) Long non-coding RNA ADAMTS9-AS1 suppresses colorectal cancer by inhibiting Wnt/beta catenin signaling pathway and is a potential diagnostic marker. J Cell Mol Med 24:11318–11329

Liang Y, Chen X, Wu Y, Li J, Zhang S, Wang K, Guan X, Yang K, Bai Y (2018) LncRNA CASC9 promotes esophageal squamous cell carcinoma metastasis through upregulating LAMC2 expression by interacting with the CREB-binding protein. Cell Death Diff 25:1980–1995

Lin C, Yang L (2018) Long noncoding RNA in cancer: wiring signaling circuitry. Trends Cell Biol 28:287–301

Lin A, Li C, Xing Z, Hu Q, Liang K, Han L, Wang C, Hawke D. H, Wang S, Zhang Y, et al. (2016) The LINK-A lncRNA activates normoxic HIF1α signalling in triple-negative breast cancer. Nat Cell Biol 18:213–224

Lister NC, Clemson M, Morris KV (2015) RNA-directed epigenetic silencing of Periostin inhibits cell motility. R Soc Open Sci 2:140545

Liu C, Bai B, Skogerbø G, Cai L, Deng W, Zhang Y, Bu D, Zhao Y, Chen R (2005) NONCODE: an integrated knowledge database of non-coding RNAs. Nucleic Acids Res 33:D112–D115

Liu L, Yue H, Liu Q, Yuan J, Li J, Wei G, Chen X, Lu Y, Guo M, Luo J, Chen R (2016) LncRNA MT1JP functions as a tumor suppressor by interacting with TIAR to modulate the p53 pathway. Oncotarget 7:15787–15800

Liu B, Zhao H, Zhang L, Shi X (2019) Silencing of long-non-coding RNA ANCR suppresses the migration and invasion of osteosarcoma cells by activating the p38 MAPK signalling pathway. BMC Cancer 19:1–9

Loda A, Heard E (2019) Xist RNA in action: past, present, and future. PLoS Genet. 15:e1008333

Lu L, Luo F, Liu Y, Liu X, Shi L, Lu X, Liu Q (2015) Posttranscriptional silencing of the LncRNA MALAT1 by miR-217 inhibits the epithelial-mesenchymal transition via enhancer of zeste homolog 2 in the malignant transformation of HBE cells induced by cigarette smoke extract. Toxicol Appl Pharmacol 289:276–285

Luo K, Geng J, Zhang Q, Xu Y, Zhou X, Huang Z, Shi KQ, Pan C, Wu J (2019) LncRNA CASC9 interacts with CPSF3 to regulate TGF-beta signaling in colorectal cancer. J Exp Clin Cancer Res 38:1–16

Marques AC, Hughes J, Graham B et al (2013) Chromatin signatures at transcriptional start sites separate two equally populated yet distinct classes of intergenic long noncoding RNAs. Genome Biol 14:R131

Matsui M, Corey RD (2017) Noncoding RNAs as drug targets. Nat Rev Drug Discov 16:167–179

Matzke MA, Primig M, Trnovsky J, Matzke AJM (1989) Reversible methylation and inactivation of marker genes in sequentially transformed tobacco plants. EMBO J 8:643–649

Mercatelli N, Fortini D, Palombo R, Paronetto, M. P. (2020) Small molecule inhibition of Ewing sarcoma cell growth via targeting the long non coding RNA HULC. Cancer Lett 469:111–123

Morris KV, Chan SW, Jacobsen SE, Looney DJ (2004) Small interfering RNA-induced transcriptional gene silencing in human cells. Science 305:1289–1292

Pan L, Xiao X, Zhao Y, Yin L, Fu M, Zhang X, Jiang P (2020) The functional roles of long noncoding RNA DANCR in human cancers. J Cancer 11:6970–6981

Papoutsoglou P, Tsubakihara Y, Caja L, Morén A, Pallis P, Ameur A, Heldin C.H, Moustakas A. (2019) The TGFB2-AS1 LncRNA regulates TGF-beta signaling by modulating corepressor activity. Cell Rep 28:3182–3198

Peng WX, Huang JG, Yang L, Gong AH, Mo YY (2017) Linc-RoR promotes MAPK/ERK signaling and confers estrogen-independent growth of breast cancer. Mol Cancer 16:1–11

Qing X, Tan GL, Liu HW, Li W, Ai JG, Xiong SS, Yang MQ, Wang TS (2020) LINC00669 insulate the JAK/STAT suppressor SOCS1 to promote nasopharyngeal cancer cell proliferation and invasion. J Exp Clin Cancer Res. 39:1–16

Rand TA, Petersen S, Du F, Wang X (2005) Argonaute2 cleaves the anti-guide strand of siRNA during RISC activation. Cell 123:621–629
Rinn JL, Kertesz M, Wang JK, Squazzo SL, Xu X, Brugmann SA, Goodnough LH, Helms JA, Farnham PJ, Segal E, Chang HY (2007) Functional demarcation of active and silent chromatin domains in human HOX loci by noncoding RNAs. Cell. 129:1311–1323
Robb GB, Brown KM, Khurana J, Rana TM (2005) Specific and potent RNAi in the nucleus of human cells. Nat Struct Mol Biol. 12:133–137
Saayman S, Ackley A, Turner AW, Famiglietti M, Bosque A, Clemson M, Planelles V, Morris KV (2014) An HIV-encoded antisense long noncoding RNA epigenetically regulates viral transcription. Mol Therapy 22:1164–1175
Schnell JR, Dyson HJ, Wright PE (2004) Structure, dynamics, and catalytic function of dihydrofolate reductase. Annu Rev Biophys Biomol Struct 33:119–140
Shi Y, Lan F, Matson C, Mulligan P, Whetstine JR, Cole PA, Casero RA, Shi Y (2004) Histone demethylation mediated by the nuclear amine oxidase homolog LSD1. Cell 119:941–953
Song B, Li R, Zuo Z, Tan J, Liu L, Ding D, Lu Y, Hou D (2019) LncRNA ENST00000539653 acts as an oncogenic factor via MAPK signalling in papillary thyroid cancer. BMC Cancer. 19:2–12
Sun Q, Liu H, Li L, Zhang S, Liu K, Liu Y, Yang C (2015) Long noncoding RNA-LET, which is repressed by EZH2, inhibits cell proliferation and induces apoptosis of nasopharyngeal carcinoma cell. Med oncol. 32:226
Suzuki K, Hattori S, Marks K, Ahlenstiel C, Maeda Y, Ishida T, Millington M, Boyd M, Symonds G, Cooper D. A, et al. (2013) Promoter targeting shRNA suppresses HIV-1 infection in vivo through transcriptional gene silencing. Mol Ther Nucleic Acids. 2(12):e137.
Tang J, Zhong G, Zhang H, Yu B, Wei F, Luo L, Kang Y, Wu J, Jiang J, Li Y et al (2018) LncRNA DANCR upregulates PI3K/AK signaling through activating serine phosphorylation of RXRA. Cell Death Dis. 9:1–12
Tripathi V, Ellis J.D, Shen Z, Song D.Y, Pan Q, Watt A.T, Freier S.M, Bennett C.F, Sharma A, Bubulya P.A, Blencowe B.J. (2010) The nuclear-retained noncoding RNA MALAT1 regulates alternative splicing by modulating SR splicing factor phosphorylation. Mol Cell. 39:925-938.
Tsai MC, Manor O, Wan Y, Mosammaparast N, Wang JK, Lan F, Shi Y, Segal E, Chang HY (2010) Long noncoding RNA as modular scaffold of histone modification complexes. Science 329:689–693
Tsuiji H, Yoshimoto R, Hasegawa Y, Furuno M, Yoshida M, Nakagawa S (2011) Competition between a noncoding exon and introns: Gomafu contains tandem UACUAAC repeats and associates with splicing factor-1. Genes Cells. 16:479–490
Wang H, Wang L, Erdjument-Bromage H, Vidal M, Tempst P, Jones R.S, Zhang Y. (2004) Role of histone H2A ubiquitination in Polycomb silencing. Nature. 431: 873–878.
Wang P, Ren Z, Sun P (2012) Overexpression of the long non-coding RNA MEG3 impairs in vitro glioma cell proliferation. J Cell Biochem 113(6):1868–1874. https://doi.org/10.1002/jcb.24055
Wang P, Xue Y, Han Y, Lin L, Wu C, Xu S, Jiang Z, Xu J, Liu Q, Cao X (2014) The STAT3-binding long noncoding RNA lnc-DC controls human dendritic cell differentiation. Science. 344:310–313
Wang Y, He L, Du Y, Zhu P, Huang G, Luo J, Yan X, Ye B, Li C, Xia P et al (2015) The long noncoding RNA LncTCF7 promote self-renewal of human liver cancer stem cells through activation of Wnt signaling. Cell Stem Cell. 16:413–425
Wang P, Chen D, Ma H, Li Y (2017) LncRNA SNHG12 contributes to multidrug resistance through activating the MAPK/Slug pathway by sponging miR-181a in non-small cell lung cancer. Oncotarget. 8:1–16
Wang Y, Lin X, Fu X, Yan W, Lin F, Kuang P, Luo Y, Lin E, Hong X, Wu G (2018a) Long non-coding RNA BANCR regulates cancer stem cell markers in papillary thyroid cancer via the RAF/MEK/ERK signaling pathway. Oncol Rep. 40:859–866
Wang Y, Gu J, Lin X, Yan W, Yang W, Wu G (2018b) LncRNA BANCR promotes EMT in PTC via the Raf/MEK/ERK signaling pathway. Oncol Lett 15:5865–5870

Waterhouse PM, Graham MW, Wang MB (1998) Virus resistance and gene silencing in plants can be induced by simultaneous expression of sense and antisense RNA. Proc Natl Acad Sci USA 95:13959–13964

Willingham AT, Orth AP, Batalov S, Peters EC, Wen BG, Aza-Blanc P, Hogenesch JB, Schultz PG (2005) A strategy for probing the function of noncoding RNAs finds a repressor of NFAT. Science. 309:1570–1573

Xing F, Liu Y, Wu SY, Wu K, Sharma S, Mo YY, Feng J, Sanders S, Jin G, Singh R et al (2018) Loss of XIST in breast cancer activates MSN-c-Met and reprograms microglia via exosomal miRNA to promote brain metastasis. Cancer Res. 78:4316–4330

Xu Y, Zhang G, Zou C, Zhang H, Gong Z, Wang W, Ma G, Jiang P, Zhang W (2018a) LncRNA MT1JP suppresses gastric cancer cell proliferation and migration through MT1JP/MiR-214-3p/RUNX3 Axis. Cell Physiol Biochem. 46:2445–2459

Xu R, Zhu X, Chen F, Huang C, Ai K, Wu H, Zhang L, Zhao X (2018b) LncRNA XIST/miR-200c regulates the stemness properties and tumourigenicity of human bladder cancer stem cell-like cells. Cancer Cell Int. 18:41

Yang F, Huo XS, Yuan SX, Zhang L, Zhou WP, Wang F, Sun SH (2013) Repression of the long noncoding RNA-LET by histone deacetylase 3 contributes to hypoxia-mediated metastasis. Mol Cell. 49:1083–1096

Yang G, Shen T, Yi X, Zhang Z, Tang C, Wang L, Zhou Y, Zhou W (2018) Crosstalk between long non-coding RNA and Wnt/beta-catenin signaling in cancer. J Cell Mol Med. 22:2062–2070

Yoon J.H, Abdelmohsen K, Srikantan S, Yang X, Martindale J.L, De S, Huarte M, Zhan M, Becker K.G, Gorospe M. (2012) LincRNA-p21 suppresses target mRNA translation. Mol Cell. 47:648-655.

Zhang X, Gejman R, Mahta A, Zhong Y, Rice KA, Zhou Y, Cheunsuchon P, Louis DN, Klibanski A (2010) Maternally expressed gene 3, an imprinted noncoding RNA gene, is associated with meningioma pathogenesis and progression. Cancer Res 70(6):2350–2358. https://doi.org/10.1158/0008-5472.CAN-09-3885. Epub 2010 Feb 23

Zhang Z, Zhu Z, Watabe K, Zhang X, Bai C, Xu M, Wu F, Mo Y. Y. (2013) Negative regulation of LncRNA GAS5 by miR-21. Cell Death Diff. 20:1558–1568.

Zhang G, Li S, Lu J, Ge Y, Wang Q, Ma G, Zhao Q, Wu D, Gong W, Du M et al (2018) LncRNA MT1JP functions as a ceRNA in regulating FBXW7 through competitively binding to miR-92a-3p in gastric cancer. Mol Cancer. 17:87

Zhang J, Han C, Song K, Chen W, Ungerleider N, Yao L, Ma W, Wu T (2020) The long noncoding RNA MALAT1 regulates TGF-beta signaling through formation of a LncRNA-protein complex with Smads, SETD2 and PPM1A in hepatic cells. PLos One. 1371:1–16

Zhou Y, Zhang X, Klibanski A (2012) MEG3 noncoding RNA: a tumor suppressor. J Mol Endocrinol. 48:R45–R53

Zhou B, Zhao H, Yu J, Guo C, Dou X, Song F, Hu G, Cao Z, Qu Y, Yang Y, Zhou Y, Wang J (2018) EVLncRNA: a manually curated database for long non-coding RNAs validated by low-throughput experiments. Nucleic Acids Res. 46:D100–D105

Zhou B, Ji B, Liu K, Hu G, Wang F, Chen Q, Yu R, Huang P, Ren J, Guo C et al (2021) EVLncRNA 2.0: an updated database of manually curated functional long non-coding RNAs validated by low-throughput experiments. Nucleic Acids Res. 49:D86–D91

Zhu D, Yu Y, Wang W, Wu K, Liu D, Yang Y, Zhang C, Qi Y, Zhao S (2019) Long noncoding RNA PART1 promotes progression of non-small cell lung cancer cells via JAK-STAT signaling pathway. Cancer Med. 8:6064–6481

Zou Y, Zhang B, Mao Y, Zhang H, Hong W (2019) Long noncoding RNA OECC promotes cell proliferation and metastasis through the PI3K/Akt/mTOR signaling pathway in human lung cancer. Oncol lett. 18:3017–3024

Chapter 7
Modulation of DNA/RNA Methylation Signaling Mediating Metabolic Homeostasis in Cancer

Pallawi Choubey, Harshdeep Kaur, and Kushagra Bansal

Abstract Nucleic acid methylation is a fundamental epigenetic mechanism that impinges upon several cellular attributes, including metabolism and energy production. The dysregulation of deoxyribonucleic acid (DNA)/ribonucleic acid (RNA) methylation can lead to metabolic rewiring in the cell, which in turn facilitates tumor development. Here, we review the current knowledge on the interplay between DNA/RNA methylation and metabolic programs in cancer cells. We also discuss the mechanistic role of these pathways in tumor development and progression.

Keywords DNA methylation · RNA methylation · CpG · 5mC · m6A · Warburg effect · Tumor metabolism · Cancer

Introduction

The chemical modification of two fundamental biomolecules, deoxyribonucleic acid (DNA) and ribonucleic acid (RNA), is central to many biological processes. Methylation is one such chemical modification that is common in a wide variety of organisms—from bacteria to mammals. In the language of chemistry, methylation is a form of alkylation reaction that involves the replacement of a hydrogen atom with a methyl group. DNA/RNA methylation has been shown to be an important regulatory step in various biological processes, such as embryogenesis (Frank et al. 1991; Razin and Cedar 1993), X-chromosome inactivation (Mohandas et al. 1981; Hellman and Chess 2007), phase separation (Ries et al. 2019), the monoallelic expression of imprinted genes (Li et al. 1993; Kaneda et al. 2004), and cellular

Pallawi Choubey and Harshdeep Kaur contributed equally to this work.

P. Choubey · H. Kaur · K. Bansal (✉)
Molecular Biology and Genetics Unit (MBGU), Jawaharlal Nehru Centre for Advanced Scientific Research (JNCASR), Jakkur, Bangalore, India
e-mail: bansalk@jncasr.ac.in

T. K. Kundu, C. Das (eds.), *Metabolism and Epigenetic Regulation: Implications in Cancer*, Subcellular Biochemistry 100,
https://doi.org/10.1007/978-3-031-07634-3_7

differentiation (Jackson et al. 2004; Lee et al. 2014). Given the regulatory role of nucleic acid methylation in animal physiology, dysregulation at any step in "the life cycle of methylation" can lead to a plethora of clinical disorders, including tumor development.

Cancer is not a single disease but has usually been viewed as a group of abnormalities characterized by the uncontrolled proliferation of cells (Hanahan and Weinberg 2011). Initial research considered the accumulation of genetic mutations as one of the major causes of neoplasia. However, recent discoveries have established that epigenetic modifications harmonize the genetic events in contributing to the complex process of neoplastic development and progression (Baylin and Jones 2011; Sandoval and Esteller 2012). The limitless replicative potential of tumor cells is accompanied by a high demand for energy and building-block molecules (Heiden et al. 2009), leading to the reprogramming of fundamental metabolic pathways. Recent evidence suggests that epigenetic alterations, such as changes in DNA/RNA methylation, support reprogrammed metabolic activities of cancer cells and that these alterations contribute to the acquisition and maintenance of malignant properties by neoplastic tissues.

In this chapter, we provide a conceptual framework for the link between alterations in DNA/RNA methylation and metabolic reprogramming in tumor cells and the contribution of these mechanisms to the process of carcinogenesis. These concepts will serve as a paradigm for developing novel strategies to therapeutically treat various human cancers.

Nucleic Acid Methylation

DNA Methylation

DNA methylation is a post-replicative modification achieved by the covalent transfer of a methyl group to the C-5 position of the cytosine pyrimidine ring, predominantly at palindromic CpG dinucleotides, leading to the generation of 5-methylcytosine (5mC) (Bird 2002) (Fig. 7.1). Experiments in the late 1970s proved that the presence of 5mC mostly in the context of symmetrical CpG dinucleotides is an elegant mechanism in which hemimethylated CpG sites are recognized after DNA replication and are methylated in daughter strands to support the semi-conservative copying of methylation pattern (Holliday and Pugh 1975; Riggs 1975; Bird 1978). Although CpG is the primary site for cytosine methylation in mammalian DNA, low levels of methylation at sites other than CpG (CpA, CpT, and CpC) have also been reported in some biological contexts.

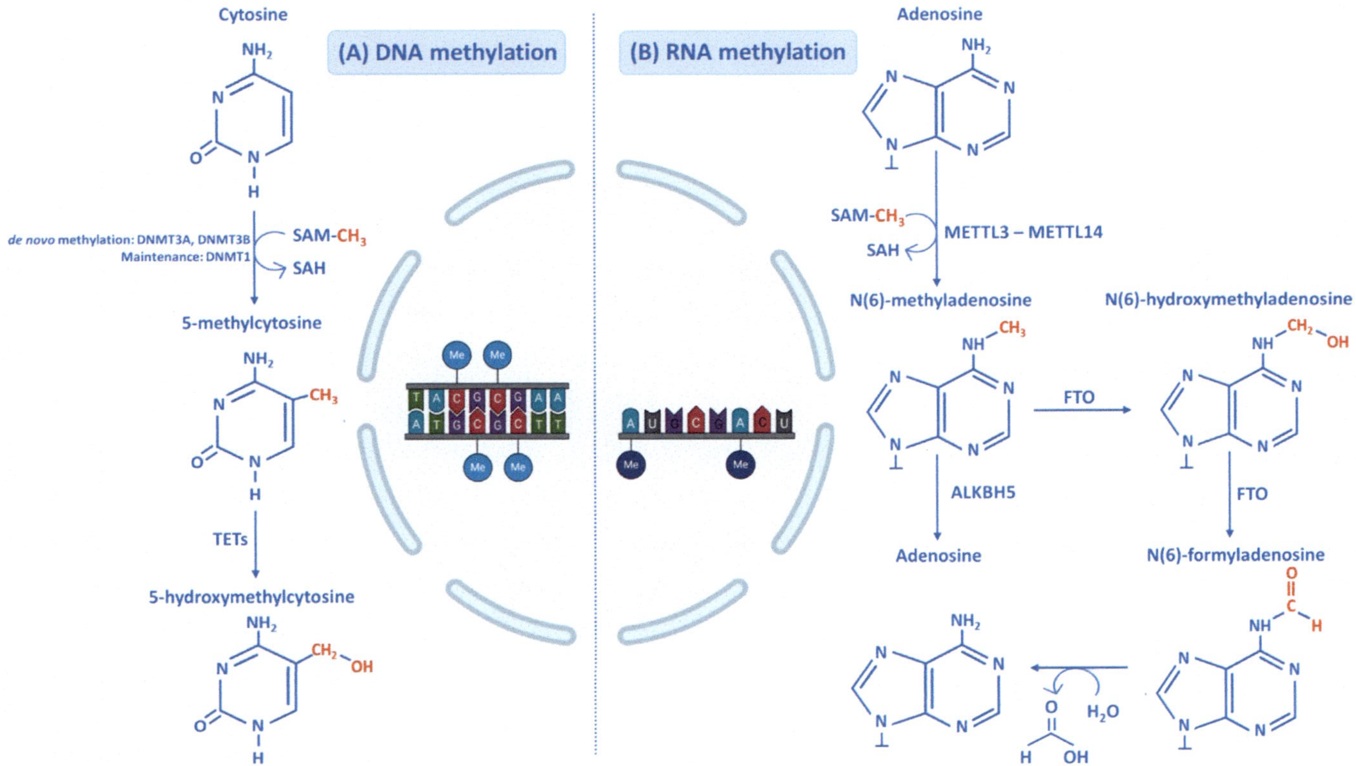

Fig. 7.1 Nucleic acid methylation. (**a**) DNA methylation: the transfer of methyl group from SAM to the C-5 position of the cytosine pyrimidine ring, resulting in the production of 5-methylcytosine. The methylation reaction is catalyzed by DNMTs. 5-methylcytosine is demethylated to 5-hydroxymethylcytosine by the TET family of enzymes. (**b**) RNA methylation: adenosine of RNA polymerase II-derived transcripts is methylated at the N(6) position by the methyltransferase

Fig. 7.1 (continued) complex (MTC), composed of METTL3-METTL14 heterodimer, to produce N(6)-methyladenosine. The methylation product is demethylated directly to adenosine by ALKBH5 and through intermediates, such as N(6)-hydroxymethyladenosine and N(6)-formyladenosine, by FTO. DNMT: DNA methyltransferase; SAM: S-adenosyl methionine; SAH: S-adenosyl homocysteine; TET: ten-eleven translocation; ALKBH5: AlkB homolog 5; FTO: fat mass and obesity-associated protein

The Writers and Erasers of DNA Methylation

"The life cycle of DNA methylation" proceeds through three phases: establishment (or the so-called *de novo* DNA methylation), maintenance, and demethylation. The methylation reaction is carried out by enzymes known as DNA methyltransferases (DNMTs). These enzymes use a similar catalytic mechanism involving base flipping to deliver the target base within a DNA molecule to the catalytic pocket of the enzyme and utilize S-adenosyl methionine (SAM) as the methyl donor (Klimasauskas et al. 1994). SAM is converted into S-adenosyl homocysteine (SAH) after the transfer of the methyl group, which is hydrolyzed by SAH hydrolase. SAH hydrolase is essential for maintaining methylation homeostasis as SAH is a potent inhibitor of DNMTs (Deguchi and Barchas 1971).

The human genome carries five DNMT-encoding genes: DNMT1, DNMT2, DNMT3A, DNMT3B, and DNMT3L. There are three canonical DNMTs: DNMT1, DNMT3A, and DNMT3B (Bestor et al. 1988; Bestor 1992; Okano et al. 1998a; Okano et al. 1999) (Fig. 7.1). In comparison, DNMT2 and DNMT3L are noncanonical DNMTs as they do not have catalytic DNMT activity (Okano et al. 1998b; Yoder and Bestor 1998; Aapola et al. 2000; Lyko 2018).

DNMT3A and DNMT3B are the two major *de novo* DNA methylation enzymes (Okano et al. 1998a, 1999). Structurally, these enzymes contain two chromatin-reading domains—Pro-Trp-Trp-Pro (PWWP) domain (Qiu et al. 2002; Ge et al. 2004) and ATRX–DNMT3–DNMT3L (ADD) domain- at N-terminus (Aapola et al. 2000), and a highly conserved catalytic methyltransferase (MTase) domain in the C-terminus (Cheng and Blumenthal 2008) (Fig. 7.2). These enzymes have been shown to exhibit a preference for purine bases at the 5′ end of the CpG sites and pyrimidines at the 3′ end (Jurkowska et al. 2011). Apart from methylating cytosine residues in CpG islands, they can also methylate in a non-CpG context (Aoki et al. 2001; Gowher and Jeltsch 2001). The methylation pattern, after its establishment, is maintained by DNMT1, a methylation maintenance enzyme, to reproduce the parental DNA methylation pattern in the newly synthesized DNA daughter strand (Bashtrykov et al. 2012).

DNMT3L, a catalytically inactive DNMT variant lacking the N-terminal part of the regulatory domain, is an essential cofactor for the activity of DNMT3 enzymes (Aapola et al. 2000; Chédin et al. 2002) and functions by promoting the affinity of DNMT3 enzymes for DNA (Jia et al. 2007; Ooi et al. 2007) (Fig. 7.2).

In contrast to other DMNTs, DNMT2 was shown to be a tRNA methyltransferase that methylates a small set of tRNAs at a single, specific site (cytosine 38 (C38)) close to the anticodon (Goll et al. 2006). The function of C38 tRNA methylation has been linked to the fidelity of protein translation (Shanmugam et al. 2015).

Mammalian 5mC can be reversed to its unmodified state in two major ways: passive and active. First, the lack of DNA methylation maintenance can lead to the dilution of 5mC during DNA replication, a process known as passive DNA demethylation. Second, active DNA demethylation is achieved by the action of the ten-eleven translocation (TET) family of enzymes, which antagonize DNMT family

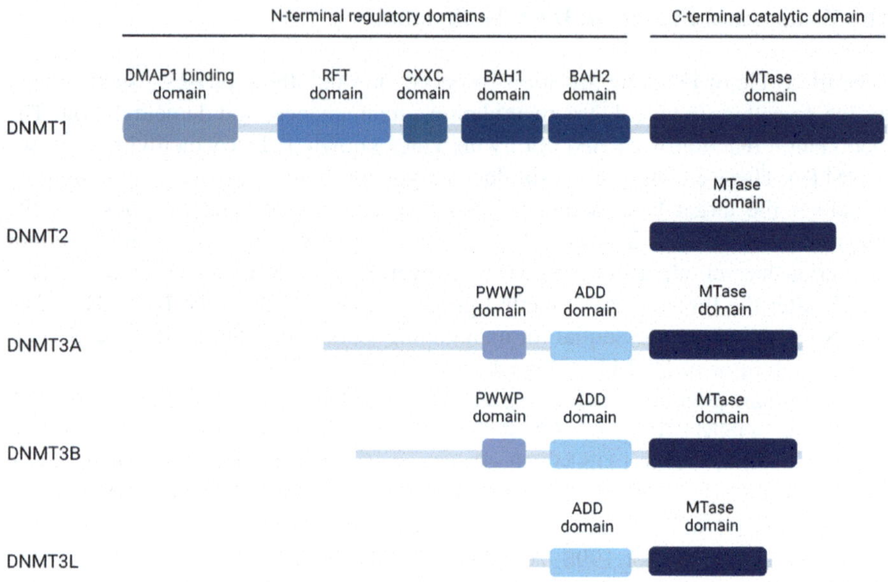

Fig. 7.2 The schematic representation of the structural and functional domains of the DNMT family of enzymes. DNMT: DNA methyltransferase; DMAP1: DNMT1-associated protein 1; RFT: replication foci targeting; CXXC: Cys-X-X-Cys; BAH: bromo-adjacent homology; PWWP: Pro-Trp-Trp-Pro; ADD: ATRX–DNMT3–DNMT3L; MTase: methyltransferase

members to prevent the hypermethylation of DNA (Wu and Zhang 2017) (Fig. 7.1). In mammals, the TET family consists of three paralogues (TET1, TET2, and TET3). The TET family of enzymes are methylcytosine dioxygenases, which progressively oxidize 5mC to 5-hydroxymethylcytosine (5hmC) (Tahiliani et al. 2009), 5-formylcytosine (5fC), and 5-carboxylcytosine (5caC) (He et al. 2011; Ito et al. 2011) to promote the reversal of DNA methylation (Pastor et al. 2013). The oxidation reaction is followed by the replication-dependent dilution of oxidized 5mC or thymine DNA glycosylase (TDG)-mediated excision of 5fC and 5caC coupled with the activity of the base excision repair (BER) pathway (Cortázar et al. 2011; Cortellino et al. 2011).

Cellular Function of DNA Methylation

CpG DNA methylation is widespread across the mammalian genome, where 70–80% of cytosine residues are methylated. The methylation of CpG residues is a fundamental mechanism of gene repression (Moore et al. 2013). Interestingly, active genes also carry methylated CpG in the transcribed region (intragenic DNA), and methylation regulates alternative splicing by recruiting methyl-CpG-binding protein 2 (MeCP2) (Young et al. 2005; Maunakea et al. 2013). The exception to this pattern is CpG-rich regions, also known as CpG islands, in the promoter regions of active genes, which were shown to be characteristically unmethylated when compared to coding regions (Bird et al. 1985; Deaton and Bird 2011).

Recent studies have identified genomic regions located near classical CpG islands, named CpG shores, and these regions are related to cancer-specific methylation changes (Irizarry et al. 2009). Interestingly, the hypermethylation of tumor suppressor genes is observed in neoplastic tissues, leading to the silencing of tumor suppressor genes (Ohtani-Fujita et al. 1993; Rhee et al. 2002).

Non-CpG methylation, functionally associated with the regulation of gene expression in neuronal (Lister et al. 2013) and embryonic stem cells (Lister et al. 2009; Laurent et al. 2010), has also been shown to mediate cancer pathogenesis. Recent data suggest that non-CpG methylation of tumor suppressor genes, such as p53 and Notch3, can play a role in lung and breast carcinoma, respectively (Kouidou et al. 2005; Xiao et al. 2020). The regulation of non-CpG methylation also contributes to tumor progression by promoting HIF1 expression, as evidenced in breast cancer (Li et al. 2019a).

RNA Methylation

RNA methylation was first reported as early as the late 1950s (Adler et al. 1958), and to date, more than 70 RNA methylations are known across various RNA species. Some well-known RNA methylations are N(6)-methyladenosine (m6A) (Dominissini et al. 2012), N(6),2'-O-dimethyladenosine (m6Am) (Keith et al. 1978; Linder et al. 2015), N(1)-methyladenosine (m1A) (Dominissini et al. 2016), 5-methylcytosine (m5C) (Squires et al. 2012), and 7-methyl guanosine (m7G) (Zhang et al. 2019a). Out of these RNA methylations, m6A represents one of the best-studied and the most abundant RNA methylation detected on RNA species with an average of one to three m6A modifications per RNA transcript (Fig. 7.1). Cytosolic mRNA molecules do not undergo m6A methylation, making m6A solely an imprint of nuclear processes (Zaccara et al. 2019).

The Writers and Erasers of m6A RNA Methylation

N(6)-adenosine methylation in mRNA and other RNA polymerase II-derived transcripts is mainly catalyzed by a multisubunit writer or methyltransferase complex (MTC). MTC has METTL3 as an enzymatic component and METTL14 as an allosteric activator, which is essential for binding the target RNA substrates and deciding the methylation site (Bokar et al. 1997; Liu et al. 2014) (Fig. 7.1). In addition, ZCCHC4 (Ma et al. 2019) and METTL5–TRMT112 (Van Tran et al. 2019) complex act as N(6)-adenosine methyltransferase for 28S and 18S rRNA, respectively, while U6 small nuclear RNA is methylated by METTL16 (Pendleton et al. 2017).

RNA demethylases (erasers), fat mass and obesity-associated protein (FTO) (Jia et al. 2011) and AlkB homolog 5 (ALKBH5) (Zheng et al. 2013), remove the methyl group from m6A residues in RNA. FTO is the first demethylase of m6A discovered, which catalyzes the oxidation of m6A to create an intermediate N(6)-hydroxymethyladenosine (hm6A) and further oxidizes hm6A to N(6)-formyladenosine (Fu et al. 2013) (Fig. 7.1). Recent findings suggest that FTO can also demethylate m6Am, both in mRNA and in small nuclear RNA (snRNA) (Mauer et al. 2017, 2019). The second RNA demethylase, ALKBH5, unlike FTO, demethylates m6A by directly removing the methyl group from the methylated adenosine (Fig. 7.1). Unlike FTO, ALKBH5 is specific to m6A and shows no activity towards m6Am (Zou et al. 2016).

Cellular Function of m6A RNA Methylation

The adenosyl-(6)-methylation of RNA is involved in a wide variety of fundamental cellular processes (Zhou et al. 2020). One of the best-known and significant function of m6A, identified in the late 1970s, is to cause mRNA instability (Sommer et al. 1978). The depletion of m6A methyltransferase METTL3 is associated with an increase in the half-life of mRNAs (Ke et al. 2017). However, there are reports where m6A increased the stability of certain mRNAs (Fry et al. 2017; Huang et al. 2018; Visvanathan et al. 2018). Thus, m6A may regulate mRNA stability in a context-dependent manner (He and He 2021).

Immunofluorescence-based studies have demonstrated the colocalization of m6A writers and erasers with nuclear speckles, a well-known site for splicing factors to accumulate, suggesting a potential role of m6A in mRNA splicing (Zheng et al. 2013; Zhao et al. 2014). Furthermore, m6A has been shown to impact 3′-UTR length, indirectly affecting translation (Ke et al. 2017). In addition, mRNAs containing m6A in their 5' UTR can be translated in a cap-independent manner (Meyer et al. 2015).

m6A RNA methylation is also known to profoundly affect various attributes of human malignancies (Sun et al. 2019; Wang et al. 2020b; Han et al. 2020b; Chen et al. 2019). Aberrant m6A modification, often via the dysregulated action of writers, readers, and erasers, promotes cancer development by stabilizing tumor-promoting RNAs (Zhang et al. 2019b; Shen et al. 2020) and facilitating the translation of these RNAs (Vu et al. 2017). In addition, RNA methylation influences the fate of tumors by regulating anti-tumor immunity (Lou et al. 2021; Zhang et al. 2021).

Metabolic Reprogramming of Cancer Cells

Cancer cells are high-energy demanding cells (Heiden et al. 2009), and to sustain their growth even under stress conditions, they implement various metabolic strategies. One such strategy is the “switch” from oxidative phosphorylation (OXPHOS) to glycolysis. This phenomenon was first discovered by Otto Warburg in the 1920s, when he found that the metabolic profile of cancer cells is significantly different from nontransformed cells (Warburg 1925). Otto Warburg observed that cancer cells catabolize large quantities of glucose via glycolysis to generate lactate instead of channelizing the metabolites into the tricarboxylic acid (TCA) cycle and OXPHOS, even in oxygen-rich conditions (Warburg 1924, 1956). This reprogrammed metabolism has now been established as a core hallmark of cancer (Hanahan and Weinberg 2011). Aerobic glycolysis, also named as Warburg effect, is bioenergetically inferior as it yields a lower fraction of adenosine triphosphate (ATP) pool in comparison to OXPHOS; however, it still has the capacity to produce more energy than OXPHOS by generating ATP at a faster rate (Pfeiffer et al. 2001). However, existing data suggest that cancer cells do not reprogram basic metabolic pathways toward aerobic glycolysis to generate ATP faster, but this diversion serves several other advantages to the tumor cells (Zu and Guppy 2004). The enhanced uptake of glucose and metabolic reprogramming to elevate glycolytic flux assist tumor cells in adapting to fluctuating oxygen levels, supply metabolites to biosynthetic pathways, and acidify the nearby microenvironment to promote tumor invasion (Kroemer and Pouyssegur 2008; Heiden et al. 2009; Estrella et al. 2013). Moreover, averting OXPHOS helps the tumor cells to minimize reactive oxygen species (ROS) production, which otherwise could cause devastating damage to the cells (Liberti and Locasale 2016; Locasale and Cantley 2011). Therefore, the paradox of the Warburg effect works in favor of cancer cells.

Apart from glucose, other sources of energy used by tumor cells are glutamine and branched-chain amino acids (BCAAs) (Goto et al. 1977; Reitzer et al. 1979; Tönjes et al. 2013; Pavlova and Thompson 2016). A major fraction of them is diverted to replenish the TCA cycle, which supplies biosynthetic precursors for proteins, lipids, and glutathione biosynthesis (DeBerardinis et al. 2007; Meng et al. 2010). A shift to glutaminolysis is an adaptive mechanism exploited by tumor cells to sustain deregulated proliferation. It helps the cells to meet increased nitrogen demand for nucleotide synthesis (Dang 2010). In addition, glutaminolysis produces acetyl-coenzyme A (acetyl-CoA), which is a precursor for the lipid biosynthesis pathway. This is critical for tumor progression as aerobic glycolysis does not allow the entry of pyruvate in the TCA cycle for the production of acetyl-CoA (Warburg et al. 1927; Weinhouse et al. 1956). For energy storage and membrane biogenesis, lipids play an important role. Enzymes involved in lipogenesis are, indeed, found to be upregulated in several tumors (Abramson 2011).

Mutations in metabolic genes, changes in their epigenetic landscape, or post-translational modifications have long been associated with predisposition to cancer and its progression (Novak 2004; Vogelstein et al. 2013). An intricate two-way crosstalk exists between the epigenome and the metabolome of cancer cells. Metabolic reprogramming has a fundamental role in the epigenetic regulation of gene expression. It provides essential cofactors required for epigenetic modifications. For example, SAM, a methyl donor in DNA and RNA methylation reaction, is a product of methionine metabolism. Reciprocally, epigenetic dysfunction can modify the expression of genes encoding metabolic enzymes (Gupta et al. 2014; Johnson et al. 2015).

Nucleic Acid Methylation and Tumor Metabolism

DNA Methylation and Tumor Metabolism

It is widely reported that DNA methylation impacts cancer-related metabolic pathways. DNA methylation can modulate the expression of genes encoding metabolic enzymes, either directly or indirectly, resulting in reprogrammed metabolism in cancer cells. Direct modulation involves the dysregulation of transcription by methylation/demethylation at the promoters of genes encoding metabolic pathway proteins, and indirect methods include the deregulation of oncogene-driven pathways, like mTOR, Akt, HIF1 etc., which have downstream metabolic targets (Wong et al. 2017; Li et al. 2020b). Several neoplastic traits are supported by reprogrammed metabolism, which emanates from a change in the methylome landscape. Figure 7.3 summarizes the DNA methylation-induced metabolic changes discussed in this chapter.

Glucose Metabolism

A number of studies have pointed towards a link between DNA methylation and glucose metabolism in cancer cells. To maintain metabolic homeostasis, cancer cells have hypomethylated promoters of certain genes involved in glycolysis. Hexokinase 2 (HK2), a rate-limiting enzyme that acts as a Warburg effect mediator, has hypomethylated CpG island shore in its promoter in hepatocellular carcinoma (HCC) (Lee et al. 2016). Unlike normal hepatocytes, where the expression of this enzyme is at the basal level, hepatomas have marked overexpression of HK2, favoring glycolytic flux to meet their metabolic requirements (DeWaal et al. 2018). Similar findings are noted in glioblastoma multiforme (GBM) (Wolf et al. 2011) and ovarian cancer (Lu et al. 2019). Paradoxically, DNA methyltransferases can also promote the expression of HK2 by downregulating factors that negatively control the expression of HK2. For example, the overexpression of DNMT3A in ovarian cancer cells reduces the expression of a small noncoding RNA, miR-145, known to inhibit the Warburg effect by targeting HK2 (Zhang et al. 2018).

Pyruvate kinase isoform M2 (PKM2) is another rate-limiting glycolytic enzyme that promotes the uptake of glucose and production of lactate, apart from its moonlighting function, like the transcriptional regulation of HIF1 and other tumor-promoting factors (Luo et al. 2011; Anastasiou et al. 2012; Gao et al. 2012; Yang and Lu 2013; Crispo et al. 2019). PKM2 gene is hypomethylated at sites within intron 1 near its promoter, which explains its overexpression in several tumors than their respective normal differentiated tissues (Desai et al. 2014; Yen et al. 2016). PKM2

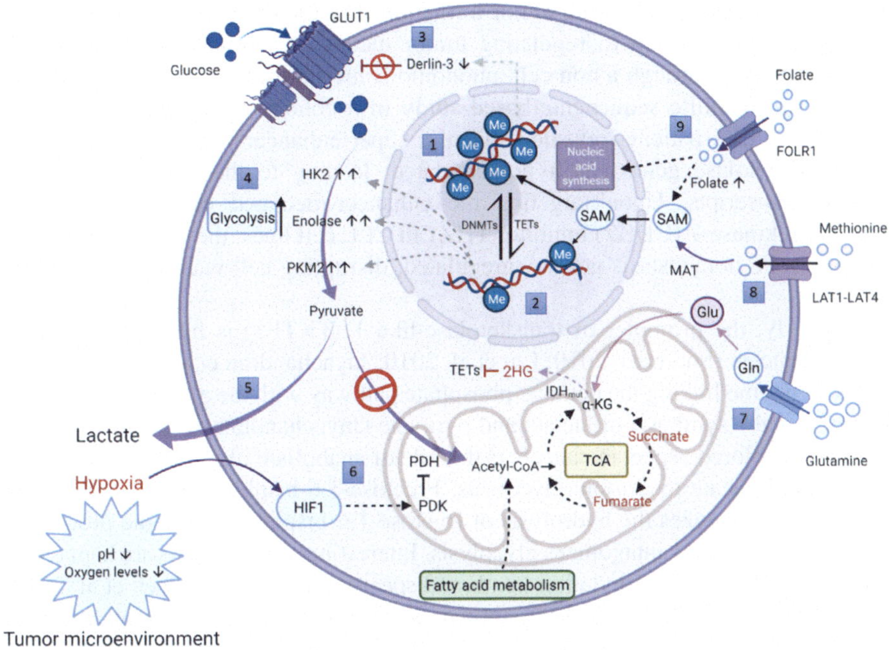

Fig. 7.3 DNA methylation-induced metabolic reprogramming of cancer cells. (1) Hypermethylated DNA. (2) Hypomethylated DNA. (3) Cancer cells silence Derlin-3 through DNA hypermethylation and upregulate the expression of glucose transporter GLUT1, facilitating elevated glucose uptake to meet high energy demands. (4) Some glycolytic enzymes, such as HK2, ENO2, and PKM2, have hypomethylated promoters, resulting in their upregulation, favoring glycolytic flux and high energy production. (5) Pyruvate is anaerobically converted to lactate, creating an acidic tumor microenvironment. (6) The negative regulators of HIF1 have hypermethylated promoters and fail to maintain the HIF1 levels in check. Elevated HIF1 activates PDK, which consequentially phosphorylates PDH to halt the conversion of pyruvate to acetyl-CoA, promoting the Warburg effect. (7) The TCA cycle is fortified through fatty acid and amino acid metabolism, with glutamine serving as the source of α-KG. In IDH mutant cancers, the levels of oncometabolite 2HG are high, leading to the competitive inhibition of DNA demethylases—TETs. (8) Upregulated LAT1 and LAT4 increase the uptake of methionine, which is further converted to SAM by MAT, supplying more methyl groups for establishing methylation marks. (9) The expression of folate receptor FOLR1 is DNA methylation status dependent. FOLR1 mediates the entry of folate into the cell, which serves as a one-carbon donor to SAM and nucleic acid biosynthesis. GLUT: glucose transporter; HK2: hexokinase 2; ENO2: enolase 2; PKM2: pyruvate kinase isoform M2; HIF1: hypoxia-inducible factor 1; PDK: pyruvate dehydrogenase kinase; PDH: pyruvate dehydrogenase; α-KG: α-ketoglutarate; IDH_{mut}: isocitrate dehydrogenase mutant; 2HG: 2-hydroxyglutarate; TET: ten-eleven translocation; LAT: L-type (leucine-preferring) amino acid transporter; SAM: S-adenosyl methionine; MAT: methionine adenosyltransferase; FOLR1: folate receptor 1; Gln: glutamine; Glu: glutamate; TCA: tricarboxylic acid cycle

overexpression is associated with poor prognosis in various cancer types, including head and neck cancers, gastric cancer, as well as squamous cell carcinomas (Desai et al. 2014; Jiayuan et al. 2016). Surprisingly, PKM2 knockout mice have been shown to spontaneously develop hepatocarcinoma, along with systemic metabolic changes, instead of attenuating tumor formation (Dayton et al. 2016). These results suggest that in addition to regulating tumor metabolism, PKM2 plays a role in preventing HCC through a non-cell-autonomous mechanism.

A recent bisulfite sequencing-based study in a cohort of chronic lymphocytic leukemia (CLL) patients revealed that the super-enhancer corresponding to the glycolytic enolase gene was hypomethylated, leading to the overexpression of enolase transcripts. Upon targeting the enhancer-mediated effect with cyclin-dependent kinase-7 (CDK7) inhibitor THZ1 in CLL cell lines, the genes responsible for metabolic homeostasis got downregulated, disrupting cell viability (Shull et al. 2016).

Similarly, the promoter of transketolase-like 1 (TKTL1) is hypomethylated in various cancers (Sun et al. 2010; Liu et al. 2010; Jayachandran et al. 2016). TKTL1 is a key enzyme linking the pentose-phosphate pathway with the glycolytic pathway. It regulates the synthesis of lactate and pyruvate (Jayachandran et al. 2016).

A less explored aspect in cancers is the role of anabolism of glucose in switching the metabolic state to aerobic glycolysis. Fructose-1,6-bisphosphatase (FBP) is an enzyme that catalyzes the hydrolysis of fructose-1,6-bisphosphate in the process of gluconeogenesis and antagonizes glycolysis. Interestingly, FBP levels are kept under check by DNA hypermethylation in a broad spectrum of cancers (Chen et al. 2011; Li et al. 2013). DNMT1 has been shown to form a complex with Snail and G9a proteins, which methylates the promoter of FBP in basal-like breast cancer (Dong et al. 2013).

Furthermore, an example of indirect regulation of glycolytic genes in cancer is the hypermethylation of the Derlin-3 promoter, which leads to the overexpression of glucose transporter 1 (GLUT1). To promote metabolic homeostasis, cancer cells upregulate the expression of high-affinity GLUTs not just by increasing their synthesis but also through the epigenetic silencing of the gene responsible for its degradation—Derlin-3 (Lopez-Serra et al. 2014). This significantly increases glucose uptake and its catalysis to final metabolites—lactate and pyruvate—which leads to acidification of the microenvironment, favoring tumor proliferation and invasion.

Another indirect impact of DNA methylation on the Warburg effect is the transcriptional silencing of tumor suppressor genes. Several tumor suppressor genes regulate signaling pathways that control tumor metabolism. The PI3K/Akt/mTOR pathway, together with MYC and HIF1 transcription factors, has been implicated in glycolytic and other tumor-related metabolic pathways (Hoxhaj and Manning 2020). Conversely, several tumor suppressor genes, including PTEN (Soria et al. 2002), VHL (Herman et al. 1994), and PHD (Rawluszko et al. 2013), which negatively regulate this pathway, are epigenetically silenced by DNA methylation in the CpG island of their respective promoters.

The TCA cycle or Krebs cycle is the central hub for many metabolic pathways. It is part of glucose metabolism, whereby glucose is catabolized to produce pyruvate, which is fed into the TCA cycle as acetyl-CoA. The role of TCA cycle in metabolic homeostasis in tumor cells has largely been ignored due to the wide acceptance of the notion that tumor cells primarily depend on aerobic glycolysis, in comparison to the TCA cycle, for metabolic demands. Tumor cells, indeed, shunt glucose away from the TCA cycle but turn to other fuel sources, like amino acids and fatty acids, to feed into the cycle.

In differentiated cells, the entry of pyruvate into the TCA cycle is mediated by pyruvate dehydrogenase (PDH), which converts pyruvate into acetyl-CoA in mitochondria. Under the hypoxic tumor microenvironment, HIF1 levels are elevated, which can activate PDH kinase (PDK), an enzyme that inhibits acetyl-CoA synthesis by phosphorylating PDH (Kim et al. 2006). Moreover, genes encoding negative regulators of HIF1 level and function, such as WW-domain-containing oxidoreductase (Wwox) and CBP/p300-interacting transactivator with ED-rich carboxy-terminal domain 4 (CITED4), have hypermethylated promoters, augmenting HIF1 and thereby enhancing the Warburg effect by exempting pyruvate from entering the TCA cycle (Huang et al. 2011; Ekizoglu et al. 2015). On a similar line, the loss of DNMT3A, as observed in primary and malignant renal cell carcinoma, leads to the unscheduled epigenetic activation of HIF2, causing the cancer cells to traverse the hypoxic barrier (Lachance et al. 2014). HIF2 is also shown to regulate the expression of GLUT1, which is indicative of another indirect link between DNMTs and glycolysis (Li et al. 2009).

Specific mutations in a TCA cycle enzyme, isocitrate dehydrogenase (IDH1/2), are common in malignancies like glioma (Parsons et al. 2008), chondrosarcoma, and other solid tumors (Cairns and Mak 2013). IDH converts isocitrate into α-ketoglutarate but impedes its further reduction to 2-hydroxyglutarate (2HG), a structural analog and oncometabolite that is a competitive inhibitor of the TET family of DNA demethylases (Dang et al. 2009; Xu et al. 2011). Therefore, IDH mutant cancers have a global hypermethylation profile and impaired cell differentiation (Figueroa et al. 2010; Sasaki et al. 2012).

Other TCA cycle enzymes, like citrate synthase (CS), succinate dehydrogenase (SDH), and fumarate hydratase (FH), are often mutated or dysregulated in a wide variety of cancers (Wong et al. 2017). SDH catalyzes the conversion of succinate to fumarate. The accumulation of succinate is believed to promote the stabilization of HIF1, thus contributing to carcinogenesis (Selak et al. 2005). Recently, a genome-wide methylation analysis highlighted that the SDH gene is methylated in a subset of neuroendocrine tumors (Bernardo-Castiñeira et al. 2018). This explains the absence of SDH protein in cancer cells that lack SDH mutation. The CpG island in the promoter of fumarate hydratase (FH) is often unmethylated in renal tumors (Dulaimi et al. 2004). It has been hypothesized that the accumulation of fumarate in the nucleus influences the epigenetic landscape by inhibiting the activity of epigenetic erasers. Like 2HG, fumarate and succinate are also oncometabolites that inhibit TETs and other demethylases, resulting in genome-wide histone and DNA methylation (Xiao et al. 2012). Both these molecules also induce epithelial-to-mesenchymal transition and disrupt DNA repair (Ortiz-Masiá et al. 2020; Sciacovelli et al. 2016).

In summary, these characteristic epigenetic changes in glucose metabolism regulators contribute to metabolic homeostasis in a lot of human malignancies.

Lipid Metabolism

Owing to the importance of lipids in building up plasma membrane, functioning as signaling molecules and as a vital source of energy, cancer cells amend lipid metabolic pathways to their own benefit. The key lipid metabolism and related pathways altered across various cancer types include PPAR signaling, fatty acid degradation, arachidonic acid metabolism, and cholesterol metabolism pathways (Fernández et al. 2020). A recent study used multiomics data to show that some of the lipid metabolism genes that were differentially expressed in cancer cells compared to their differentiated counterpart cells were regulated by DNA methylation (Hao et al. 2019).

Apolipoprotein A1 (APOA1), an HDL component that transports cholesterol to the liver through the bloodstream, has a hypermethylated promoter region in various cancers (Wang et al. 2016; Hao et al. 2019). Interestingly, hepatitis B virus (HBV) infection, known to promote HCC, induces the upregulation of DNMT1, DNMT3A, and DNMT3B with a concomitant decrease in APOA1 expression (Wang et al. 2016; Li et al. 2015). APOA1 has antioxidant and anti-inflammatory properties; therefore, it is maintained under a suppressed state in cancer cells to favor their progression (Su et al. 2010; Zamanian-Daryoush et al. 2013). Another enzyme, phospholipase A2 (PLA2), which hydrolyzes lecithin to lysophospholipid releasing arachidonic acid, is upregulated in breast carcinoma and pan-gynecological cancers, and this phenomenon is associated with promoter hypomethylation (Hao et al. 2019). In contrast, PLA2 has hypermethylated promoter region in several renal cancers (Hao et al. 2019). This disparity in methylation patterns can be explained by the tissue-specific expression landscape of lipid metabolism pathway regulators.

MeCP2, a known reader of DNA methylation, is reported to be frequently amplified in several human malignancies, such as ovarian, cervical, liver, and uterine cancers (Neupane et al. 2016). Interestingly, MeCP2 null male mice display increased expression of rate-limiting enzymes involved in *de novo* lipogenesis and cholesterol synthesis in the liver. The increased expression of lipid metabolism genes in MeCP2 null mice leads to fatty liver disease, commonly associated with an increased risk of hepatocellular malignancies (Kyle et al. 2016; Dhamija et al. 2019).

Rewired lipid metabolism has also been shown to favor metastasis. CD36, a receptor for fatty acid internalization, is often upregulated in metastasis-initiating cells and has been shown to be solely responsible for the initiation of metastasis in oral squamous cell carcinoma (Pascual et al. 2017; Sakurai et al. 2020). CD36 amplification is also correlated with metastasis in melanoma (Nath and Chan 2016). Interestingly, a recent study demonstrated that the deletion of DNMT3A increased the number of carcinogen-induced epidermal tumors in mice, and it correlated with the upregulation of a unique set of genes related to lipid metabolism, including CD36 in these tumors. These results suggest that DNA methylation pathways control the predisposition of epidermal stem cells to generate metastatic epidermal tumors via the regulation of lipid metabolism (Rinaldi et al. 2017).

Amino Acid Metabolism

Recent studies have shed light on the various anaplerotic roles of amino acids in supporting cancer cell homeostasis. Though tumor metabolism largely focuses on aerobic glycolysis for energy, amino acids like glutamine and branched amino acids (valine, leucine, and isoleucine) contribute substantially to bioenergetics and biosynthetic pathways. These amino acids serve as carbon and nitrogen donors for nucleotide synthesis (Zhang et al. 2008; Shuvalov et al. 2017).

The dysregulation of amino acid transporters is one of the hallmarks of cancer (Lopes et al. 2021; Pavlova and Thompson 2016). DNMT1, DNMT3B, and MeCP2 indirectly regulate the expression of a high-affinity glutamine transporter, ASCT2. miR-137, a tumor suppressor microRNA, silences ASCT2 by binding to its 3'UTR. Therefore, ASCT2 expression is inversely correlated with miR-137 in multiple cancers. DNMTs and MeCP2 cooperate to promote the methylation of the miR-137 promoter, conversely reactivating the expression of ASCT2 in tumor cells (Dong et al. 2017). The expression of an amino acid antiporter L-type (leucine-preferring) amino acid transporter 1 (LAT1), which mediates the influx of BCAAs and bulky amino acids into the cell in exchange for the efflux of intracellular essential amino acids, is elevated in oral cancer due to hypomethylation of SLC7A5, the gene that encodes LAT1 (Mukhopadhyay et al. 2020). The overexpression of LAT1, along with LAT4 (SLC43A2), increases the uptake of methionine, which generates the methyl provider SAM in the presence of methionine adenosyltransferase and ATP (Fuchs and Bode 2005; Haase et al. 2007). The deregulation of SAM levels has a major impact on the methylome of various cancers (Pascale et al. 2018; Mahmood et al. 2018).

In recent years, there is growing evidence that several amino acid metabolic enzymes are upregulated in cancers. A cytosolic enzyme, branched-chain amino acid transaminase 1 (BCAT1), that is involved in the catalysis of branched-chain amino acids to their corresponding α-keto acids and glutamate, is overexpressed in many solid tumors (Tönjes et al. 2013; Ananieva et al. 2016; Cuyàs et al. 2016; Mayers et al. 2016; Dey et al. 2017). However, IDH mutant gliomas have low BCAT1 expression, corresponding to the hypermethylation of its promoter (Tönjes et al. 2013). Another possible metabolic link between BCAT1 and IDH could be that BCAT1 uses α-ketoglutarate for BCAA transamination, but in IDH mutants, α-ketoglutarate is unavailable. Some reports state that the overexpression of BCAT1 correlates well with enhanced cancer growth. BCAT1-null non-small cell lung carcinoma (NSCLC) cells, when xenografted in mice, displayed impaired ability to form a tumor (Mayers et al. 2016). This indicates that BCAT1 is necessary for the proliferation of NSCLC. Recently, another study demonstrated that in pancreatic ductal adenocarcinoma (PDAC), BCAT1 levels are normal but mitochondrial BCAA transaminase (BCAT2) is overexpressed (Li et al. 2020a). Thus, not all tumors have elevated BCAT1 (Mayers et al. 2016). Interestingly, cell-free methylated BCAT1 DNA is usually found circulating in the blood of colorectal cancer patients, proving to be a good biomarker (Symonds et al. 2016).

A recent study analyzed global DNA methylation status in a cisplatin-resistant bladder cancer cell line and demonstrated the hypermethylation of the promoters of genes associated with arginine and proline metabolism, such as glutamate dehydrogenase 1 (GLUD1), arginase 2 (ARG2), glutamate dehydrogenase 2 (GLUD2), argininosuccinate synthase 1 (ASS1), and spermidine/spermine N(1)-acetyltransferase (SAT1). The hypermethylation of promoters of these metabolic genes strongly correlated with their decreased gene expression levels. Interestingly, arginine depletion or SAT1 overexpression restored the cisplatin sensitivity of the bladder cancer cell line, suggesting that decreased expression of amino acid metabolic genes may play a critical role in the chemoresistance of bladder cancer (Austin et al. 2018).

Nucleotide Metabolism

Given the high multiplication rate of cancer cells, nucleotide biosynthesis is an absolute necessity for DNA and RNA production. Cells have two metabolic ways to synthesize nucleotides: *de novo* and salvage. Cancer cells can adopt either of the two, but some cancers prefer one over the other. For example, the papillary subtype of breast cancer displays increased *de novo* nucleotide biosynthesis in comparison to the epithelial-mesenchymal transition (EMT) subtype of breast cancer (Ogrodzinski et al. 2020). Folate acts as a one-carbon donor in the biosynthesis of methyl groups and nucleic acids (Tibbetts and Appling 2010). It is proposed that the expression of folate receptor gene FOLR1 is regulated through DNA methylation (Chung-Tsen and Dolnick 1993; Saikawa et al. 1995; Elwood et al. 1997). In colorectal cancer, FOLR1 has hypermethylated CpG sites in the vicinity of its translation start codon and hypomethylated CpG sites at the distal region of the promoter (Farkas et al. 2015). Similarly, the 5' regulatory region of FOLR1 shows a trend towards hypomethylation in urothelial carcinoma (Erichsen et al. 2018). FOLR1 mRNA expression is also suggested to correlate with the aggressiveness of cancer (Chen et al. 2012; Siu et al. 2012). On the other side, it has been shown that folate depletion may lead to the upregulation of FOLR1 (Chung-Tsen and Dolnick 1993; Kelemen 2006), inducing global DNA hypomethylation (Liu and Ward 2010). Besides, high FOLR1 mRNA expression was reported to correlate with global DNA hypomethylation in type II ovarian cancers (Notaro et al. 2016). Not surprisingly, the inhibition of folate metabolism is used to eliminate malignant cells.

RNA Methylation and Tumor Metabolism

In the past decade, the importance of RNA methylation has been brought under the spotlight due to expanding insights into its indispensable role in tumorigenesis. Some studies have also highlighted its role in cancer metabolic reprogramming (Fry et al. 2017; Han et al. 2020a; Shen et al. 2020). It can either directly control

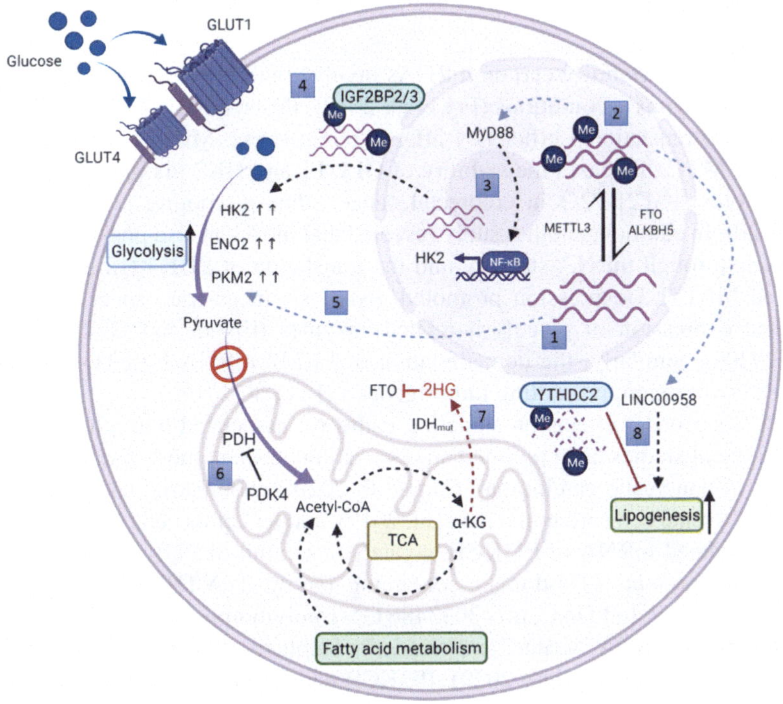

Fig. 7.4 RNA methylation promotes metabolic changes in cancer cells. (1) Unmethylated mRNA. (2) Methylated mRNA. (3) m6A methylation of MyD88 mRNA by METTL3 in cancer cells promotes NF-κB signaling, resulting in the transcriptional activation of HK2. (4) The recruitment of reader protein-IGF2BP2/3 by METTL3 increases the stability of GLUT1 and HK2 transcripts, promoting glycolysis. Elevated METTL3 expression increases the levels of ENO2 and GLUT4. (5) The upregulation of PKM2 through the demethylation of its mRNA aids in tumorigenesis. (6) Increased stability and the translation of PDK4 due to m6A methylation promote glycolysis. (7) In IDH mutant cancers, m6A demethylase FTO is competitively inhibited by 2HG. (8) The enhanced expression of lncRNA LINC00958 through m6A methylation promotes lipogenesis. m6A reader YTHDC2 decreases the stability of lipogenic mRNAs. HK2: hexokinase 2; GLUT: glucose transporter; ENO2: enolase 2; PKM2: pyruvate kinase isoform M2; PDK4: pyruvate dehydrogenase kinase 4; PDH: pyruvate dehydrogenase; α-KG: α-ketoglutarate; IDH_{mut}: isocitrate dehydrogenase mutant; 2HG: 2-hydroxyglutarate; TCA: tricarboxylic acid cycle; FTO: fat mass and obesity-associated protein; YTHDC2: YT521-B homology domain-containing 2

the translation of metabolic enzymes and nutrient transporters or indirectly regulate oncogenes and other metabolic pathways. Recent reports suggest that some of the metabolic features of tumor cells can be attributed to one of the well-studied RNA methylation, m6A. Figure 7.4 illustrates the role of RNA methylation in tumor metabolism.

Glucose Metabolism

m6A is known to stabilize certain mRNAs involved in glycolysis, like GLUT1 and MYC, under hypoxic conditions (Fry et al. 2017). Postsynthesis RNA modification helps regain translational efficiency after hypoxic stress. METTL3, an enzymatic subunit of MTC, increases the stability of GLUT1 and HK2 via the recruitment of reader protein-IGF2BP2/3 in colorectal cancer, thus promoting glycolysis (Shen et al. 2020). In addition, another study revealed that m6A can also positively regulate glycolysis through mRNA stability and the translation of PDK4 (Li et al. 2020c). Elevated METTL3 expression promoted glycolysis in gastric cancer through the increased expression of glycolysis-related enzymes (ENO2 and GLUT4) (Wang et al. 2020a). Similarly, the demethylation of PKM2 mRNA by FTO leads to its enhanced expression, promoting tumorigenesis (Li et al. 2019b).

m6A has broader effects on signaling pathways associated with glucose metabolism. Several studies have reported an association between m6A and the PI3K/Akt pathway in pancreatic carcinoma (Tian et al. 2020), endometrial cancer cells (Liu et al. 2018), renal cell carcinoma (Li et al. 2017), and leukemia cells (Vu et al. 2017). m6A-methylated mRNA of PI3KCB, a catalytic subunit of PI3K, is recognized by m6A reader protein YTH-domain-containing family 2 (YTHDF2) to negatively impact its stability. In PDAC, rs142933486 G>T polymorphism in PI3KCB disrupts its recognition by m6A "writer" and "reader" complexes, thus increasing its mRNA and protein levels (Tian et al. 2020). PI3KCB can further activate Akt to promote glycolytic flux. Interestingly, m6A attenuates Akt by methylating/demethylating its regulators, thereby fostering proliferation and tumorigenicity (Liu et al. 2018). Another glycolysis-associated signaling pathway regulated by m6A includes NF-κB. HK2 is a known direct transcriptional target of NF-κB (Londhe et al. 2018). Therefore, cancer cells install m6A methylation on MyD88 mRNA by METTL3 to promote NF-κB signaling and, hence, glycolysis (Yu et al. 2020).

Lipid Metabolism

The elevated rate of lipogenesis allows cancer cells to adjust to high metabolic demands and maintain their proliferation rate (Ray and Roy 2018). METTL3-installed m6A modification leads to the upregulation of long noncoding RNA (lncRNA) LINC00958 in HCC to facilitate lipogenesis. Key enzymes in lipogenesis were found to be dependent on LINC00958 expression (Zuo et al. 2020). Hepatocyte-specific METTL3 knockout mice have dramatically reduced levels of enzymes involved in fatty acid metabolism and the regulators of cholesterol metabolism (Xie et al. 2019). In various cancer types, the levels of m6A eraser FTO are known to be high (Kaklamani et al. 2011; Liu et al. 2017). FTO overexpression in HepG2 cells, a human liver cancer cell line, promoted lipogenesis via the upregulation of lipogenic factors (Chen et al. 2018). YT521-B homology domain-containing 2 (YTHDC2), an m6A reader, has been shown to bind to mRNA of lipogenic genes, like SREBP1c, FASN, SCD1, and ACC1, resulting in the reduced stability of these mRNAs and decreased gene expression. Interestingly, YTHDC2 expression is downregulated in fatty liver disease, a known risk factor for liver cancer (Zhou et al. 2021).

m6A also impinges on key signaling pathways that regulate lipid metabolism. AMP-activated protein kinase (AMPK) has emerged as a key regulator of lipid metabolism as the activation of AMPK suppresses fatty acid/cholesterol biosynthesis and promotes fatty acid oxidation (Ruderman and Prentki 2004; Li et al. 2014; Choi et al. 2017). Interestingly, the silencing of METTL3 in a human bladder cancer cell line, T24, led to differential expression of genes involved in the AMPK signaling pathway (Xie et al. 2020). In summary, m6A plays an important role in lipid metabolic reprogramming in cancer.

Amino Acid and Nucleotide Metabolism

Glutamine participates in reinforcing tumor growth and vitality by replenishing the TCA cycle (Heiden et al. 2009). Glutamine is converted to α-ketoglutarate upon its entry into the cell. The m6A demethylase FTO is an α-ketoglutarate-dependent dioxygenase that is competitively inhibited by 2HG produced in IDH mutants (Elkashef et al. 2017; Su et al. 2018). Therefore, the role of this axis in relation to tumorigenesis needs to be reconnoitered. The interaction between m6A and other amino acid signaling pathways still remains to be explored. Likewise, the role of RNA modifications in nucleotide metabolism is an uncharted territory.

Crosstalk Between Methylome and Metabolome: A Target for Therapeutics

Considering the vast crosstalk between methylome and metabolome in cancer cells, targeting one would affect the other. Therapies aiming at tumor metabolism might reverse the dysregulation of methylation pattern, and drugs targeting methylation regulators may modulate tumor metabolism. There are currently several inhibitors of epigenetic or metabolic enzymes in various stages of development or clinical trials. DNA methyltransferase inhibitors 5-azacytidine and 5-aza-2′-deoxycytidine have been approved by the American and European drug regulatory agencies to treat hematological malignancies (Kaminskas et al. 2005). These are cytidine analogs that are known to inhibit DNMT1, DNMT3A, and DNMT3B (Christman 2002; Oka et al. 2005). These drugs have also shown promising laboratory results in IDH mutant gliomas by reversing the hypermethylation phenotype and suppressing growth (Borodovsky et al. 2013; Turcan et al. 2013). In clinical trials, however, the therapeutic effects of DNMT inhibitors have so far been poor in solid tumors (Aparicio and Weber 2002; Braiteh et al. 2008).

Cancer-associated RNA modification machinery components can also serve as potential therapeutic targets for cancer. m6A RNA demethylase FTO requires α-ketoglutarate for its activity; hence, small molecules, like meclofenamic acid (MA) and an analog of α-ketoglutarate, R-2-hydroxyglutarate (R-2HG), act as competitive inhibitors of FTO and have shown significant therapeutic efficacy in leukemia and glioma cells (Huang et al. 2015b; Cui et al. 2017; Su et al. 2018). A recent preclinical study reported that FTO inhibitors FB23 and FB23-2 suppress the proliferation and promote the apoptosis of acute myeloid leukemia cells in vitro (Huang et al. 2019). Though this field is still in its infancy, these observations will help it propel forward and build new opportunities for cancer therapeutics.

Another strategy being considered in the field of cancer therapeutics is to target the metabolites responsible for decorating the genome with methyl groups. SAH hydrolase, an enzyme involved in methylation homeostasis maintenance, is a target in anti-tumor therapies (Glazer et al. 1986). Though DZNep, a SAH hydrolase inhibitor, can inhibit global DNA methylation, it is ineffective in reactivating silenced genes. But when used with 5-aza, it shows a synergistic effect and has been demonstrated to activate silenced genes (Momparler et al. 2014).

Some innovative cancer therapeutics involve direct targeting of tumor metabolism. The inhibition of glycolytic pathway by 2-deoxyglucose (2-DG), a glucose analog, has been demonstrated to have anti-tumor activity (Chen and Guéron 1992; Huang et al. 2015a). 2-DG is metabolized by HK2, and it competitively inhibits glucose-6-phosphate production. Similarly, 3-bromopyruvate (3-BrPA), a potent inhibitor of HK2, has been shown to trigger cancer cell death through the inhibition of tumor cell metabolism and the depletion of ATP (Chen et al. 2009; Gong et al. 2014). Lonidamine, a derivative of indazole-3-carboxylic acid, has been shown to inhibit aerobic glycolysis and impair energy metabolism in tumor cells (Floridi et al. 1981; Nath et al. 2016). Glutaminase inhibitors, like compound 968 and CB-839, have also shown efficacy in solid and hematological cancers (Wang et al. 2010; Simpson et al. 2012; Gross et al. 2014). The synthesis of oncometabolite 2HG can be controlled by drugs AG-120, AG-881, ML309, GSK321, and GSK864 (Wong et al. 2017). Additionally, AGI-6780, another mutant IDH2 inhibitor, has been shown to reverse gene expression patterns through DNA demethylation in leukemia (Wang et al. 2013; Golub et al. 2019).

Altogether, these data strongly justify the targeting of methylome/metabolome as an efficient approach for the treatment of cancer. Although many preclinical and clinical studies have validated the therapeutic use of various agents targeting either methylation or metabolism, combining metabolic and epigenetic therapies might provide an effective strategy to inhibit tumors. Table 7.1 sums up the therapies targeting methylome and metabolome in cancer cells discussed in this chapter.

Table 7.1 Methylome- and metabolome-based therapeutics for cancer. A table detailing methylome- and metabolome-based therapies for various malignancies

Inhibitor(s)	Target protein(s)	Mode of action
5-azacytidine, 5-aza-2′-deoxycytidine	DNA methyltransferases (DNMTs)	Inhibition of DNMT1, DNMT3A, and DNMT3B
Meclofenamic acid (MA), R-2-hydroxyglutarate (R-2HG), FB23 and FB23-2	Fat mass and obesity-associated protein (FTO)	Inhibition of RNA demethylase-FTO
3-deazaneplanocin A (DZNep)	S-adenosyl homocysteine (SAH) hydrolase	Indirect suppression of S-adenosyl methionine-dependent epigenetic methylations by inhibiting SAH hydrolase
2-deoxyglucose (2-DG), 3-bromopyruvate (3-BrPA)	Hexokinase	Inhibition of hexokinase and glycolytic pathway
Lonidamine	Monocarboxylate transporter/s (MCT), mitochondrial pyruvate carrier (MPC), electron transport chain complex II, and hexokinase	Inhibition of transporter/s responsible for lactate export and carrier involved in pyruvate uptake into mitochondria, leading to intracellular tumor acidification. Also known to inhibit complex II of the mitochondrial electron transport chain and hexokinase of the glycolytic pathway
Compound 968, CB-839	Glutaminase	Inhibition of glutaminase that promotes the deamination of glutamine into glutamate. Glutaminolysis generates α-KG and other TCA cycle intermediates, which in turn could impact epigenetic methylation/s
AG-120, AG-881, ML309, GSK321, GSK864, and AGI-6780	Mutant isocitrate dehydrogenase (IDH)	Inhibition of a mutated form of IDH and suppression of the formation of oncometabolite, 2-hydroxyglutarate (2HG). 2HG is a competitive inhibitor of DNA/RNA demethylases

Summary and Future Perspectives

Cancer cells attune various cellular networks to favor their growth and sustenance. They selectively activate or inhibit existing metabolic pathways to maintain a pool of biomolecules required for accelerated proliferation. A significant proportion of this metabolic reprogramming, a core hallmark of cancer, is orchestrated by epigenetic mechanisms. DNA/RNA methylation promotes the Warburg effect by regulating the expression of genes, thus regulating fundamental metabolic pathways. Evidence gathered over the last few decades reiterates the complexity of the crosstalk between methylome and tumor metabolism. We are beginning to understand the interplay between methylome and metabolome in cancer; however, there are numerous questions to be answered. For example, why are the genes involved in the same metabolic pathway regulated differentially by nucleic acid methylation? Do DNA and RNA methylation pathways play a cooperative role in tumor metabolism? Parsing the molecular mechanisms underpinning the crosstalk between methylome and metabolome would help in fully understanding the roles of these pathways in cancers and would also aid in identifying novel therapeutic targets for cancers. Moreover, most of the available data on the methylation status of metabolic genes and potential therapeutics derives from simplistic in vitro biological systems. The potential targets should be validated in vivo for better reflection of the composite nature of the carcinomas. A combinatorial approach involving epigenetic drugs and metabolic inhibitors is a promising way forward, and along with conventional cancer therapeutics, it may transform our viewpoint on the treatment of cancers.

In conclusion, we have come far in understanding the role of nucleic acid methylation in mediating metabolic homeostasis in cancer, but a long path lies ahead to fine map the regulatory pathways linking the nucleic acid methylation landscape and tumor metabolism.

Acknowledgments We thank JNCASR for intramural financial support. KB is a recipient of DBT/Wellcome Trust India Alliance Intermediate Fellowship (IA/I/19/1/504276). PC and HK are supported by research fellowships from JNCASR. Figures in this book chapter were created with BioRender.com.

References

Aapola U, Shibuya K, Scott HS, Ollila J, Vihinen M, Heino M, Shintani A, Kawasaki K, Minoshima S, Krohn K, Antonarakis SE, Shimizu N, Kudoh J, Peterson P (2000) Isolation and initial characterization of a novel zinc finger gene, DNMT3L, on 21q22.3, related to the cytosine-5-methyltransferase 3 gene family. Genomics. https://doi.org/10.1006/geno.2000.6168

Abramson HN (2011) The lipogenesis pathway as a cancer target. J Med Chem 54. https://doi.org/10.1021/jm2005805

Adler M, Weissmann B, Gutman AB (1958) Occurrence of methylated purine bases in yeast ribonucleic acid. J Biol Chem 230(2):717–723

Ananieva EA, Powell JD, Hutson SM (2016) Leucine metabolism in T cell activation: mTOR signaling and beyond. Adv Nutr 7. https://doi.org/10.3945/an.115.011221

Anastasiou D, Yu Y, Israelsen WJ, Jiang JK, Boxer MB, Hong BS, Tempel W, Dimov S, Shen M, Jha A, Yang H, Mattaini KR, Metallo CM, Fiske BP, Courtney KD, Malstrom S, Khan TM, Kung C, Skoumbourdis AP, Veith H, Southall N, Walsh MJ, Brimacombe KR, Leister W, Lunt SY, Johnson ZR, Yen KE, Kunii K, Davidson SM, Christofk HR, Austin CP, Inglese J, Harris MH, Asara JM, Stephanopoulos G, Salituro FG, Jin S, Dang L, Auld DS, Park HW, Cantley LC, Thomas CJ, Vander Heiden MG (2012) Pyruvate kinase M2 activators promote tetramer formation and suppress tumorigenesis. Nat Chem Biol 8:839–847. https://doi.org/10.1038/nchembio.1060

Aoki A, Suetake I, Miyagawa J, Fujio T, Chijiwa T, Sasaki H, Tajima S (2001) Enzymatic properties of the de novo-type mouse DNA (cytosine-5) methyltransferases. Nucleic Acids Res. https://doi.org/10.1093/nar/29.17.3506

Aparicio A, Weber JS (2002) Review of the clinical experience with 5-azacytidine and 5-aza-2-′-deoxycytidine in solid tumors. Curr Opin Investig Drugs 3(4):627–633

Austin Y, You S, Kim M, Gupta A, Park MH, Weisenberger DJ, Liang G, Kim J (2018) Rewiring of cisplatin-resistant bladder cancer cells through epigenetic regulation of genes involved in amino acid metabolism. Theranostics 8. https://doi.org/10.7150/THNO.25130

Bashtrykov P, Jankevicius G, Smarandache A, Jurkowska RZ, Ragozin S, Jeltsch A (2012) Specificity of dnmt1 for methylation of hemimethylated CpG sites resides in its catalytic domain. Chem Biol. https://doi.org/10.1016/j.chembiol.2012.03.010

Baylin SB, Jones PA (2011) A decade of exploring the cancer epigenome-biological and translational implications. Nat Rev Cancer. https://doi.org/10.1038/nrc3130

Bernardo-Castiñeira C, Valdés N, Sierra MI, Sáenz-De-Santa-María I, Bayón GF, Perez RF, Fernández AF, Fraga MF, Astudillo A, Menéndez R, Fernández B, Del Olmo M, Suarez C, Chiara MD (2018) SDHC promoter methylation, a novel pathogenic mechanism in parasympathetic paragangliomas. J Clin Endocrinol Metab 103:295–305. https://doi.org/10.1210/jc.2017-01702

Bestor TH (1992) Activation of mammalian DNA methyltransferase by cleavage of a Zn binding regulatory domain. EMBO J. https://doi.org/10.1002/j.1460-2075.1992.tb05326.x

Bestor T, Laudano A, Mattaliano R, Ingram V (1988) Cloning and sequencing of a cDNA encoding DNA methyltransferase of mouse cells. The carboxyl-terminal domain of the mammalian enzymes is related to bacterial restriction methyltransferases. J Mol Biol. https://doi.org/10.1016/0022-2836(88)90122-2

Bird AP (1978) Use of restriction enzymes to study eukaryotic DNA methylation. II. The symmetry of methylated sites supports semi-conservative copying of the methylation pattern. J Mol Biol. https://doi.org/10.1016/0022-2836(78)90243-7

Bird A (2002) DNA methylation patterns and epigenetic memory. Genes Dev. https://doi.org/10.1101/gad.947102

Bird A, Taggart M, Frommer M, Miller OJ, Macleod D (1985) A fraction of the mouse genome that is derived from islands of nonmethylated, CpG-rich DNA. Cell. https://doi.org/10.1016/0092-8674(85)90312-5

Bokar JA, Shambaugh ME, Polayes D, Matera AG, Rottman FM (1997) Purification and cDNA cloning of the AdoMet-binding subunit of the human mRNA (N6-adenosine)-methyltransferase. RNA 3(11):1233–1247

Borodovsky A, Salmasi V, Turcan S, Fabius AWM, Baia GS, Eberhart CG, Weingart JD, Gallia GL, Baylin SB, Chan TA, Riggins GJ (2013) 5-azacytidine reduces methylation, promotes differentiation and induces tumor regression in a patient-derived IDH1 mutant glioma xenograft. Oncotarget 4:1737–1747. https://doi.org/10.18632/oncotarget.1408

Braiteh F, Soriano AO, Garcia-Manero G, Hong D, Johnson MM, Silva LDP, Yang H, Alexander S, Wolff J, Kurzrock R (2008) Phase I study of epigenetic modulation with 5-azacytidine and valproic acid in patients with advanced cancers. Clin Cancer Res 14:6296–6301. https://doi.org/10.1158/1078-0432.CCR-08-1247

Cairns RA, Mak TW (2013) Oncogenic isocitrate dehydrogenase mutations: mechanisms, models, and clinical opportunities. Cancer Discov. https://doi.org/10.1158/2159-8290.CD-13-0083

Chédin F, Lieber MR, Hsieh CL (2002) The DNA methyltransferase-like protein DNMT3L stimulates de novo methylation by Dnmt3a. Proc Natl Acad Sci U S A. https://doi.org/10.1073/pnas.262443999

Chen W, Guéron M (1992) The inhibition of bovine heart hexokinase by 2-deoxy-d-glucose-6-phosphate: characterization by 31P NMR and metabolic implications. Biochimie 74:867–873. https://doi.org/10.1016/0300-9084(92)90070-U

Chen Z, Zhang H, Lu W, Huang P (2009) Role of mitochondria-associated hexokinase II in cancer cell death induced by 3-bromopyruvate. Biochim Biophys Acta Bioenerg 1787:553–560. https://doi.org/10.1016/j.bbabio.2009.03.003

Chen M, Zhang J, Li N, Qian Z, Zhu M, Li Q, Zheng J, Wang X, Shi G (2011) Promoter hypermethylation mediated downregulation of FBP1 in human hepatocellular carcinoma and colon cancer. PLoS One 6:e25564. https://doi.org/10.1371/journal.pone.0025564

Chen YL, Chang MC, Huang CY, Chiang YC, Lin HW, Chen CA, Hsieh CY, Cheng WF (2012) Serous ovarian carcinoma patients with high alpha-folate receptor had reducing survival and cytotoxic chemo-response. Mol Oncol 6:360–369. https://doi.org/10.1016/j.molonc.2011.11.010

Chen A, Chen X, Cheng S, Shu L, Yan M, Yao L, Wang B, Huang S, Zhou L, Yang Z, Liu G (2018) FTO promotes SREBP1c maturation and enhances CIDEC transcription during lipid accumulation in HepG2 cells. Biochim Biophys Acta Mol Cell Biol Lipids 1863:538–548. https://doi.org/10.1016/j.bbalip.2018.02.003

Chen XY, Zhang J, Zhu JS (2019) The role of m(6)A RNA methylation in human cancer. Mol Cancer 18(1):103. https://doi.org/10.1186/s12943-019-1033-z

Cheng X, Blumenthal RM (2008) Mammalian DNA methyltransferases: a structural perspective. Structure. https://doi.org/10.1016/j.str.2008.01.004

Choi J, Kim KJ, Koh EJ, Lee BY (2017) Gelidium elegans regulates the AMPK-PRDM16-UCP-1 pathway and has a synergistic effect with orlistat on obesity-associated features in mice fed a high-fat diet. Nutrients. https://doi.org/10.3390/nu9040342

Christman JK (2002) 5-Azacytidine and 5-aza-2′-deoxycytidine as inhibitors of DNA methylation: mechanistic studies and their implications for cancer therapy. Oncogene 21. https://doi.org/10.1038/sj.onc.1205699

Chung-Tsen H, Dolnick BJ (1993) Altered folate-binding protein mRNA stability in KB cells grown in folate-deficient medium. Biochem Pharmacol 45:2537–2545. https://doi.org/10.1016/0006-2952(93)90235-O

Cortázar D, Kunz C, Selfridge J, Lettieri T, Saito Y, MacDougall E, Wirz A, Schuermann D, Jacobs AL, Siegrist F, Steinacher R, Jiricny J, Bird A, Schär P (2011) Embryonic lethal phenotype reveals a function of TDG in maintaining epigenetic stability. Nature. https://doi.org/10.1038/nature09672

Cortellino S, Xu J, Sannai M, Moore R, Caretti E, Cigliano A, Le Coz M, Devarajan K, Wessels A, Soprano D, Abramowitz LK, Bartolomei MS, Rambow F, Bassi MR, Bruno T, Fanciulli M, Renner C, Klein-Szanto AJ, Matsumoto Y, Kobi D, Davidson I, Alberti C, Larue L, Bellacosa A (2011) Thymine DNA glycosylase is essential for active DNA demethylation by linked deamination-base excision repair. Cell. https://doi.org/10.1016/j.cell.2011.06.020

Crispo F, Condelli V, Lepore S, Notarangelo T, Sgambato A, Esposito F, Maddalena F, Landriscina M (2019) Metabolic dysregulations and epigenetics: a bidirectional interplay that drives tumor progression. Cell 8. https://doi.org/10.3390/cells8080798

Cui Q, Shi H, Ye P, Li L, Qu Q, Sun G, Sun G, Lu Z, Huang Y, Yang CG, Riggs AD, He C, Shi Y (2017) m6A RNA methylation regulates the self-renewal and tumorigenesis of glioblastoma stem cells. Cell Rep 18:2622–2634. https://doi.org/10.1016/j.celrep.2017.02.059

Cuyàs E, Fernández-Arroyo S, Alarcón T, Lupu R, Joven J, Menendez JA (2016) Germline BRCA1 mutation reprograms breast epithelial cell metabolism towards mitochondrial-dependent biosynthesis: evidence for metforminbased "starvation" strategies in BRCA1 carriers. Oncotarget 7:52974–52992. https://doi.org/10.18632/oncotarget.9732

Dang CV (2010) Glutaminolysis: supplying carbon or nitrogen or both for cancer cells? Cell Cycle 9. https://doi.org/10.4161/cc.9.19.13302

Dang L, White DW, Gross S, Bennett BD, Bittinger MA, Driggers EM, Fantin VR, Jang HG, Jin S, Keenan MC, Marks KM, Prins RM, Ward PS, Yen KE, Liau LM, Rabinowitz JD, Cantley LC, Thompson CB, Vander Heiden MG, Su SM (2009) Cancer-associated IDH1 mutations produce 2-hydroxyglutarate. Nature 462:739–744. https://doi.org/10.1038/nature08617

Dayton TL, Gocheva V, Miller KM, Israelsen WJ, Bhutkar A, Clish CB, Davidson SM, Luengo A, Bronson RT, Jacks T, Vander Heiden MG (2016) Germline loss of PKM2 promotes metabolic distress and hepatocellular carcinoma. Genes Dev. https://doi.org/10.1101/gad.278549.116

Deaton AM, Bird A (2011) CpG islands and the regulation of transcription. Genes Dev. https://doi.org/10.1101/gad.2037511

DeBerardinis RJ, Mancuso A, Daikhin E, Nissim I, Yudkoff M, Wehrli S, Thompson CB (2007) Beyond aerobic glycolysis: transformed cells can engage in glutamine metabolism that exceeds the requirement for protein and nucleotide synthesis. Proc Natl Acad Sci U S A. https://doi.org/10.1073/pnas.0709747104

Deguchi T, Barchas J (1971) Inhibition of transmethylations of biogenic amines by S-adenosylhomocysteine. enhancement of transmethylation by adenosylhomocysteinase. J Biol Chem. https://doi.org/10.1016/S0021-9258(18)62211-2

Desai S, Ding M, Wang B, Lu Z, Zhao Q, Shaw K, Alfred Yung WK, Weinstein JN, Tan M, Yao J (2014) Tissue-specific isoform switch and DNA hypomethylation of the pyruvate kinase PKM gene in human cancers. Oncotarget 5:8202–8210. https://doi.org/10.18632/oncotarget.1159

DeWaal D, Nogueira V, Terry AR, Patra KC, Jeon SM, Guzman G, Au J, Long CP, Antoniewicz MR, Hay N (2018) Hexokinase-2 depletion inhibits glycolysis and induces oxidative phosphorylation in hepatocellular carcinoma and sensitizes to metformin. Nat Commun 9. https://doi.org/10.1038/s41467-017-02733-4

Dey P, Baddour J, Muller F, Wu CC, Wang H, Liao WT, Lan Z, Chen A, Gutschner T, Kang Y, Fleming J, Satani N, Zhao D, Achreja A, Yang L, Lee J, Chang E, Genovese G, Viale A, Ying H, Draetta G, Maitra A, Wang YA, Nagrath D, Depinho RA (2017) Genomic deletion of malic enzyme 2 confers collateral lethality in pancreatic cancer. Nature 542:119–123. https://doi.org/10.1038/nature21052

Dhamija E, Paul SB, Kedia S (2019) Non-alcoholic fatty liver disease associated with hepatocellular carcinoma: an increasing concern. Indian J Med Res 149(1):9–17. https://doi.org/10.4103/ijmr.IJMR_1456_17

Dominissini D, Moshitch-Moshkovitz S, Schwartz S, Salmon-Divon M, Ungar L, Osenberg S, Cesarkas K, Jacob-Hirsch J, Amariglio N, Kupiec M, Sorek R, Rechavi G (2012) Topology of the human and mouse m6A RNA methylomes revealed by m6A-seq. Nature. https://doi.org/10.1038/nature11112

Dominissini D, Nachtergaele S, Moshitch-Moshkovitz S, Peer E, Kol N, Ben-Haim MS, Dai Q, Di Segni A, Salmon-Divon M, Clark WC, Zheng G, Pan T, Solomon O, Eyal E, Hershkovitz V, Han D, Doré LC, Amariglio N, Rechavi G, He C (2016) The dynamic N1-methyladenosine methylome in eukaryotic messenger RNA. Nature. https://doi.org/10.1038/nature16998

Dong C, Yuan T, Wu Y, Wang Y, Fan TWM, Miriyala S, Lin Y, Yao J, Shi J, Kang T, Lorkiewicz P, St Clair D, Hung MC, Evers BM, Zhou BP (2013) Loss of FBP1 by snail-mediated repression provides metabolic advantages in basal-like breast cancer. Cancer Cell 23: 316–331. https://doi.org/10.1016/j.ccr.2013.01.022

Dong J, Xiao D, Zhao Z, Ren P, Li C, Hu Y, Shi J, Su H, Wang L, Liu H, Li B, Gao P, Qing G (2017) Epigenetic silencing of microRNA-137 enhances ASCT2 expression and tumor glutamine metabolism. Oncogene 6(7):e356. https://doi.org/10.1038/oncsis.2017.59

Dulaimi E, De Caceres II, Uzzo RG, Al-Saleem T, Greenberg RE, Polascik TJ, Babb JS, Grizzle WE, Cairns P (2004) Promoter hypermethylation profile of kidney cancer. Clin Cancer Res 10: 3972–3979. https://doi.org/10.1158/1078-0432.CCR-04-0175

Ekizoglu S, Bulut P, Karaman E, Kilic E, Buyru N (2015) Epigenetic and genetic alterations affect the WWOX gene in head and neck squamous cell carcinoma. PLoS One 10. https://doi.org/10.1371/journal.pone.0115353

Elkashef SM, Lin AP, Myers J, Sill H, Jiang D, Dahia PLM, Aguiar RCT (2017) IDH mutation, competitive inhibition of FTO, and RNA methylation. Cancer Cell 31. https://doi.org/10.1016/j.ccell.2017.04.001

Elwood PC, Nachmanoff K, Saikawa Y, Page ST, Pacheco P, Roberts S, Chung KN (1997) The divergent 5′ termini of the α human folate receptor (hFR) mRNAs originate from two tissue-specific promoters and alternative splicing: characterization of the α hFR gene structure. Biochemistry. https://doi.org/10.1021/bi962070h

Erichsen L, Ghanjati F, Beermann A, Poyet C, Hermanns T, Schulz WA, Seifert HH, Wild PJ, Buser L, Kröning A, Braunstein S, Anlauf M, Jankowiak S, Hassan M, Bendhack ML, Bravo MA, Santourlidis S (2018) Aberrant methylated key genes of methyl group metabolism within the molecular etiology of urothelial carcinogenesis. Sci Rep 8:1–14. https://doi.org/10.1038/s41598-018-21932-7

Estrella V, Chen T, Lloyd M, Wojtkowiak J, Cornnell HH, Ibrahim-Hashim A, Bailey K, Balagurunathan Y, Rothberg JM, Sloane BF, Johnson J, Gatenby RA, Gillies RJ (2013) Acidity generated by the tumor microenvironment drives local invasion. Cancer Res 73:1524–1535. https://doi.org/10.1158/0008-5472.CAN-12-2796

Farkas SA, Befekadu R, Hahn-Strömberg V, Nilsson TK (2015) DNA methylation and expression of the folate transporter genes in colorectal cancer. Tumor Biol 36:5581–5590. https://doi.org/10.1007/s13277-015-3228-2

Fernández LP, Gómez de Cedrón M, Ramírez de Molina A (2020) Alterations of lipid metabolism in cancer: implications in prognosis and treatment. Front Oncol 10. https://doi.org/10.3389/fonc.2020.577420

Figueroa ME, Abdel-Wahab O, Lu C, Ward PS, Patel J, Shih A, Li Y, Bhagwat N, Vasanthakumar A, Fernandez HF, Tallman MS, Sun Z, Wolniak K, Peeters JK, Liu W, Choe SE, Fantin VR, Paietta E, Löwenberg B, Licht JD, Godley LA, Delwel R, Valk PJM, Thompson CB, Levine RL, Melnick A (2010) Leukemic IDH1 and IDH2 mutations result in a hypermethylation phenotype, disrupt TET2 function, and impair hematopoietic differentiation. Cancer Cell 18:553–567. https://doi.org/10.1016/j.ccr.2010.11.015

Floridi A, Paggi MG, Marcante ML, Silvestrini B, Caputo A, De Martino C (1981) Lonidamine, a selective inhibitor of aerobic glycolysis of murine tumor cells. J Natl Cancer Inst 66(3):497–499

Frank D, Keshet I, Shani M, Levine A, Razin A, Cedar H (1991) Demethylation of CpG islands in embryonic cells. Nature. https://doi.org/10.1038/351239a0

Fry NJ, Law BA, Ilkayeva OR, Holley CL, Mansfield KD (2017) N6-methyladenosine is required for the hypoxic stabilization of specific mRNAs. RNA. https://doi.org/10.1261/rna.061044.117

Fu Y, Jia G, Pang X, Wang RN, Wang X, Li CJ, Smemo S, Dai Q, Bailey KA, Nobrega MA, Han KL, Cui Q, He C (2013) FTO-mediated formation of N6-hydroxymethyladenosine and N 6-formyladenosine in mammalian RNA. Nat Commun. https://doi.org/10.1038/ncomms2822

Fuchs BC, Bode BP (2005) Amino acid transporters ASCT2 and LAT1 in cancer: partners in crime? Semin Cancer Biol 15. https://doi.org/10.1016/j.semcancer.2005.04.005

Gao X, Wang H, Yang JJ, Liu X, Liu ZR (2012) Pyruvate kinase M2 regulates gene transcription by acting as a protein kinase. Mol Cell 45:598–609. https://doi.org/10.1016/j.molcel.2012.01.001

Ge YZ, Pu MT, Gowher H, Wu HP, Ding JP, Jeltsch A, Xu GL (2004) Chromatin targeting of de novo DNA methyltransferases by the PWWP domain. J Biol Chem. https://doi.org/10.1074/jbc.M312296200

Glazer RI, Hartman KD, Knode MC, Richard MM, Chiang PK, Tseng CKH, Marquez VE (1986) 3-Deazaneplanocin: a new and potent inhibitor of S-adenosylhomocysteine hydrolase and its effects on human promyelocytic leukemia cell line HL-60. Biochem Biophys Res Commun 135:688–694. https://doi.org/10.1016/0006-291X(86)90048-3

Goll MG, Kirpekar F, Maggert KA, Yoder JA, Hsieh CL, Zhang X, Golic KG, Jacobsen SE, Bestor TH (2006) Methylation of tRNAAsp by the DNA methyltransferase homolog Dnmt2. Science. https://doi.org/10.1126/science.1120976

Golub D, Iyengar N, Dogra S, Wong T, Bready D, Tang K, Modrek AS, Placantonakis DG (2019) Mutant isocitrate dehydrogenase inhibitors as targeted cancer therapeutics. Front Oncol 9:417. https://doi.org/10.3389/fonc.2019.00417

Gong L, Wei Y, Yu X, Peng J, Leng X (2014) 3-Bromopyruvic acid, a hexokinase II inhibitor, is an effective antitumor agent on the hepatoma cells: in vitro and in vivo findings. Anti Cancer Agents Med Chem. https://doi.org/10.2174/1871520614666140416105309

Goto M, Shinno H, Ichihara A (1977) Isozyme patterns of branched-chain amino acid transaminase in human tissues and tumors. Gann Jpn J Cancer Res. https://doi.org/10.20772/cancersci1959.68.5_663

Gowher H, Jeltsch A (2001) Enzymatic properties of recombinant Dnmt3a DNA methyltransferase from mouse: the enzyme modifies DNA in a non-processive manner and also methylates non-CpA sites. J Mol Biol. https://doi.org/10.1006/jmbi.2001.4710

Gross MI, Demo SD, Dennison JB, Chen L, Chernov-Rogan T, Goyal B, Janes JR, Laidig GJ, Lewis ER, Li J, MacKinnon AL, Parlati F, Rodriguez MLM, Shwonek PJ, Sjogren EB, Stanton TF, Wang T, Yang J, Zhao F, Bennett MK (2014) Antitumor activity of the glutaminase inhibitor CB-839 in triple-negative breast cancer. Mol Cancer Ther 13:890–901. https://doi.org/10.1158/1535-7163.MCT-13-0870

Gupta V, Gopinath P, Iqbal M, Mazurek S, Wellen K, Bamezai R (2014) Interplay between epigenetics & cancer metabolism. Curr Pharm Des 20:1706–1714. https://doi.org/10.2174/13816128113199990536

Haase C, Bergmann R, Fuechtner F, Hoepping A, Pietzsch J (2007) L-type amino acid transporters LAT1 and LAT4 in cancer: uptake of 3-O-methyl-6-18F-fluoro-L-dopa in human adenocarcinoma and squamous cell carcinoma in vitro and in vivo. J Nucl Med 48:2063–2071. https://doi.org/10.2967/jnumed.107.043620

Han X, Wang L, Han Q (2020a) Advances in the role of m6A RNA modification in cancer metabolic reprogramming. Cell Biosci. 10. https://doi.org/10.1186/s13578-020-00479-z

Han X, Wang M, Zhao YL, Yang Y, Yang YG (2020b) RNA methylations in human cancers. Semin Cancer Biol. https://doi.org/10.1016/j.semcancer.2020.11.007

Hanahan D, Weinberg RA (2011) Hallmarks of cancer: the next generation. Cell. https://doi.org/10.1016/j.cell.2011.02.013

Hao Y, Li D, Xu Y, Ouyang J, Wang Y, Zhang Y, Li B, Xie L, Qin G (2019) Investigation of lipid metabolism dysregulation and the effects on immune microenvironments in pan-cancer using multiple omics data. BMC Bioinform 20:195. https://doi.org/10.1186/s12859-019-2734-4

He PC, He C (2021) m(6) A RNA methylation: from mechanisms to therapeutic potential. EMBO J 40(3):e105977. https://doi.org/10.15252/embj.2020105977

He YF, Li BZ, Li Z, Liu P, Wang Y, Tang Q, Ding J, Jia Y, Chen Z, Li N, Sun Y, Li X, Dai Q, Song CX, Zhang K, He C, Xu GL (2011) Tet-mediated formation of 5-carboxylcytosine and its excision by TDG in mammalian DNA. Science. https://doi.org/10.1126/science.1210944

Heiden MGV, Cantley LC, Thompson CB (2009) Understanding the warburg effect: the metabolic requirements of cell proliferation. Science 324:Science. https://doi.org/10.1126/science.1160809

Hellman A, Chess A (2007) Gene body-specific methylation on the active X chromosome. Science. https://doi.org/10.1126/science.1136352

Herman JG, Latif F, Weng Y, Lerman MI, Zbar B, Liu S, Samid D, Duan DSR, Gnarra JR, Linehan WM, Baylin SB (1994) Silencing of the VHL tumor-suppressor gene by DNA methylation in renal carcinoma. Proc Natl Acad Sci U S A 91:9700–9704. https://doi.org/10.1073/pnas.91.21.9700

Holliday R, Pugh JE (1975) DNA modification mechanisms and gene activity during development. Science. https://doi.org/10.1126/science.1111098

Hoxhaj G, Manning BD (2020) The PI3K–AKT network at the interface of oncogenic signalling and cancer metabolism. Nat Rev Cancer 20. https://doi.org/10.1038/s41568-019-0216-7

Huang KT, Takano EA, Mikeska T, Byrne DJ, Dobrovic A, Fox SB (2011) Aberrant DNA methylation but not mutation of CITED4 is associated with alteration of HIF-regulated genes in breast cancer. Breast Cancer Res Treat 130:319–329. https://doi.org/10.1007/s10549-011-1657-1

Huang C-C, Wang S-Y, Lin L-L, Wang P-W, Chen T-Y, Hsu W-M, Lin T-K, Liou C-W, Chuang J-H (2015a) Glycolytic inhibitor 2-deoxyglucose simultaneously targets cancer and endothelial cells to suppress neuroblastoma growth in mice. Dis Model Mech. https://doi.org/10.1242/dmm.021667

Huang Y, Yan J, Li Q, Li J, Gong S, Zhou H, Gan J, Jiang H, Jia GF, Luo C, Yang CG (2015b) Meclofenamic acid selectively inhibits FTO demethylation of m6A over ALKBH5. Nucleic Acids Res 43:373–384. https://doi.org/10.1093/nar/gku1276

Huang H, Weng H, Sun W, Qin X, Shi H, Wu H, Zhao BS, Mesquita A, Liu C, Yuan CL, Hu YC, Hüttelmaier S, Skibbe JR, Su R, Deng X, Dong L, Sun M, Li C, Nachtergaele S, Wang Y, Hu C, Ferchen K, Greis KD, Jiang X, Wei M, Qu L, Guan JL, He C, Yang J, Chen J (2018) Recognition of RNA N 6-methyladenosine by IGF2BP proteins enhances mRNA stability and translation. Nat Cell Biol. https://doi.org/10.1038/s41556-018-0045-z

Huang Y, Su R, Sheng Y, Dong L, Dong Z, Xu H, Ni T, Zhang ZS, Zhang T, Li C, Han L, Zhu Z, Lian F, Wei J, Deng Q, Wang Y, Wunderlich M, Gao Z, Pan G, Zhong D, Zhou H, Zhang N, Gan J, Jiang H, Mulloy JC, Qian Z, Chen J, Yang CG (2019) Small-molecule targeting of oncogenic FTO demethylase in acute myeloid leukemia. Cancer Cell 35:677–691.e610. https://doi.org/10.1016/j.ccell.2019.03.006

Irizarry RA, Ladd-Acosta C, Wen B, Wu Z, Montano C, Onyango P, Cui H, Gabo K, Rongione M, Webster M, Ji H, Potash JB, Sabunciyan S, Feinberg AP (2009) The human colon cancer methylome shows similar hypo- and hypermethylation at conserved tissue-specific CpG island shores. Nat Genet. https://doi.org/10.1038/ng.298

Ito S, Shen L, Dai Q, Wu SC, Collins LB, Swenberg JA, He C, Zhang Y (2011) Tet proteins can convert 5-methylcytosine to 5-formylcytosine and 5-carboxylcytosine. Science. https://doi.org/10.1126/science.1210597

Jackson M, Krassowska A, Gilbert N, Chevassut T, Forrester L, Ansell J, Ramsahoye B (2004) Severe global DNA hypomethylation blocks differentiation and induces histone hyperacetylation in embryonic stem cells. Mol Cell Biol. https://doi.org/10.1128/mcb.24.20.8862-8871.2004

Jayachandran A, Lo PH, Chueh AC, Prithviraj P, Molania R, Davalos-Salas M, Anaka M, Walkiewicz M, Cebon J, Behren A (2016) Transketolase-like 1 ectopic expression is associated with DNA hypomethylation and induces the Warburg effect in melanoma cells. BMC Cancer 16. https://doi.org/10.1186/s12885-016-2185-5

Jia D, Jurkowska RZ, Zhang X, Jeltsch A, Cheng X (2007) Structure of Dnmt3a bound to Dnmt3L suggests a model for de novo DNA methylation. Nature. https://doi.org/10.1038/nature06146

Jia G, Fu Y, Zhao X, Dai Q, Zheng G, Yang Y, Yi C, Lindahl T, Pan T, Yang YG, He C (2011) N6-Methyladenosine in nuclear RNA is a major substrate of the obesity-associated FTO. Nat Chem Biol. https://doi.org/10.1038/nchembio.687

Jiayuan W, Hu L, Chen M, Cao W, Chen H, He T (2016) Pyruvate kinase M2 overexpression and poor prognosis in solid tumors of digestive system: evidence from 16 cohort studies. OncoTargets Therapy 9. https://doi.org/10.2147/OTT.S106508

Johnson C, Warmoes MO, Shen X, Locasale JW (2015) Epigenetics and cancer metabolism. Cancer Lett 356. https://doi.org/10.1016/j.canlet.2013.09.043

Jurkowska RZ, Siddique AN, Jurkowski TP, Jeltsch A (2011) Approaches to enzyme and substrate design of the murine Dnmt3a DNA methyltransferase. ChemBioChem. https://doi.org/10.1002/cbic.201000673

Kaklamani V, Yi N, Sadim M, Siziopikou K, Zhang K, Xu Y, Tofilon S, Agarwal S, Pasche B, Mantzoros C (2011) The role of the fat mass and obesity associated gene (FTO) in breast cancer risk. BMC Med Genet. 12. https://doi.org/10.1186/1471-2350-12-52

Kaminskas E, Farrell AT, Wang YC, Sridhara R, Pazdur R (2005) FDA drug approval summary: azacitidine (5-azacytidine, Vidaza™) for injectable suspension. Oncologist 10:176–182. https://doi.org/10.1634/theoncologist.10-3-176

Kaneda M, Okano M, Hata K, Sado T, Tsujimoto H, Li E, Sasaki H (2004) Essential role for de novo DNA methyltransferase Dnmt3a in paternal and maternal imprinting. Nature. https://doi.org/10.1038/nature02633

Ke S, Pandya-Jones A, Saito Y, Fak JJ, Vågbø CB, Geula S, Hanna JH, Black DL, Darnell JE, Darnell RB (2017) m6A mRNA modifications are deposited in nascent pre-mRNA and are not required for splicing but do specify cytoplasmic turnover. Genes Dev. https://doi.org/10.1101/gad.301036.117

Keith JM, Ensinger MJ, Moss B (1978) HeLa cell RNA(2'-O-methyladenosine-N6-)-methyltransferase specific for the capped 5'-end of messenger RNA. J Biol Chem. https://doi.org/10.1016/s0021-9258(17)34652-5

Kelemen LE (2006) The role of folate receptor α in cancer development, progression and treatment: cause, consequence or innocent bystander? Int J Cancer 119. https://doi.org/10.1002/ijc.21712

Kim JW, Tchernyshyov I, Semenza GL, Dang CV (2006) HIF-1-mediated expression of pyruvate dehydrogenase kinase: a metabolic switch required for cellular adaptation to hypoxia. Cell Metab 3:177–185. https://doi.org/10.1016/j.cmet.2006.02.002

Klimasauskas S, Kumar S, Roberts RJ, Cheng X (1994) Hhal methyltransferase flips its target base out of the DNA helix. Cell. https://doi.org/10.1016/0092-8674(94)90342-5

Kouidou S, Agidou T, Kyrkou A, Andreou A, Katopodi T, Georgiou E, Krikelis D, Dimitriadou A, Spanos P, Tsilikas C, Destouni H, Tzimagiorgis G (2005) Non-CpG cytosine methylation of p53 exon 5 in non-small cell lung carcinoma. Lung Cancer 50(3):299–307. https://doi.org/10.1016/j.lungcan.2005.06.012

Kroemer G, Pouyssegur J (2008) Tumor cell metabolism: cancer's Achilles' heel. Cancer Cell 13. https://doi.org/10.1016/j.ccr.2008.05.005

Kyle SM, Saha PK, Brown HM, Chan LC, Justice MJ (2016) MeCP2 co-ordinates liver lipid metabolism with the NCoR1/HDAC3 corepressor complex. Hum Mol Genet 25(14): 3029–3041. https://doi.org/10.1093/hmg/ddw156

Lachance G, Uniacke J, Audas TE, Holterman CE, Franovic A, Payette J, Lee S (2014) DNMT3a epigenetic program regulates the HIF-2alpha oxygen-sensing pathway and the cellular response to hypoxia. Proc Natl Acad Sci U S A 111(21):7783–7788. https://doi.org/10.1073/pnas.1322909111

Laurent L, Wong E, Li G, Huynh T, Tsirigos A, Ong CT, Low HM, Sung KWK, Rigoutsos I, Loring J, Wei CL (2010) Dynamic changes in the human methylome during differentiation. Genome Res. https://doi.org/10.1101/gr.101907.109

Lee HJ, Hore TA, Reik W (2014) Reprogramming the methylome: erasing memory and creating diversity. Cell Stem Cell. https://doi.org/10.1016/j.stem.2014.05.008

Lee HG, Kim H, Son T, Jeong Y, Kim SU, Dong SM, Park YN, Lee JD, Lee JM, Park JH (2016) Regulation of HK2 expression through alterations in CpG methylation of the HK2 promoter during progression of hepatocellular carcinoma. Oncotarget. 7:41798–41810. https://doi.org/10.18632/oncotarget.9723

Li E, Beard C, Jaenisch R (1993) Role for DNA methylation in genomic imprinting. Nature. https://doi.org/10.1038/366362a0

Li Z, Bao S, Wu Q, Wang H, Eyler C, Sathornsumetee S, Shi Q, Cao Y, Lathia J, McLendon RE, Hjelmeland AB, Rich JN (2009) Hypoxia-inducible factors regulate tumorigenic capacity of glioma stem cells. Cancer Cell. 15(6):501–513. https://doi.org/10.1016/j.ccr.2009.03.018

Li H, Wang J, Xu H, Xing R, Pan Y, Li W, Cui J, Zhang H, Lu Y (2013) Decreased fructose-1,6-bisphosphatase-2 expression promotes glycolysis and growth in gastric cancer cells. Mol Cancer 12:110. https://doi.org/10.1186/1476-4598-12-110

Li H, Min Q, Ouyang C, Lee J, He C, Zou MH, Xie Z (2014) AMPK activation prevents excess nutrient-induced hepatic lipid accumulation by inhibiting mTORC1 signaling and endoplasmic reticulum stress response. Biochim Biophys Acta Mol basis Dis 1842:1844–1854. https://doi.org/10.1016/j.bbadis.2014.07.002

Li H, Yang F, Gao B, Yu Z, Liu X, Xie F, Zhang J (2015) Hepatitis B virus infection in hepatocellular carcinoma tissues upregulates expression of DNA methyltransferases. Int J Clin Exp Med 8(3):4175–4185

Li X, Tang J, Huang W, Wang F, Li P, Qin C, Qin Z, Zou Q, Wei J, Hua L, Yang H, Wang Z (2017) The M6A methyltransferase METTL3: acting as a tumor suppressor in renal cell carcinoma. Oncotarget. 8:96103–96116. https://doi.org/10.18632/oncotarget.21726

Li C, Xiong W, Liu X, Xiao W, Guo Y, Tan J, Li Y (2019a) Hypomethylation at non-CpG/CpG sites in the promoter of HIF-1alpha gene combined with enhanced H3K9Ac modification contribute to maintain higher HIF-1alpha expression in breast cancer. Oncogene 8(4):26. https://doi.org/10.1038/s41389-019-0135-1

Li J, Zhu L, Shi Y, Liu J, Lin L, Chen X (2019b) M6A demethylase FTO promotes hepatocellular carcinoma tumorigenesis via mediating PKM2 demethylation. Am J Transl Res. 11(9): 6084–6092

Li JT, Yin M, Wang D, Wang J, Lei MZ, Zhang Y, Liu Y, Zhang L, Zou SW, Hu LP, Zhang ZG, Wang YP, Wen WY, Lu HJ, Chen ZJ, Su D, Lei QY (2020a) BCAT2-mediated BCAA catabolism is critical for development of pancreatic ductal adenocarcinoma. Nat Cell Biol 22. https://doi.org/10.1038/s41556-019-0455-6

Li T, Mao C, Wang X, Shi Y, Tao Y (2020b) Epigenetic crosstalk between hypoxia and tumor driven by HIF regulation. J Exp Clin Cancer Res. 39. https://doi.org/10.1186/s13046-020-01733-5

Li Z, Peng Y, Li J, Chen Z, Chen F, Tu J, Lin S, Wang H (2020c) N 6-methyladenosine regulates glycolysis of cancer cells through PDK4. Nat Commun 11:1–16. https://doi.org/10.1038/s41467-020-16306-5

Liberti MV, Locasale JW (2016) The Warburg effect: how does it benefit cancer cells? Trends Biochem Sci 41. https://doi.org/10.1016/j.tibs.2015.12.001

Linder B, Grozhik AV, Olarerin-George AO, Meydan C, Mason CE, Jaffrey SR (2015) Single-nucleotide-resolution mapping of m6A and m6Am throughout the transcriptome. Nat Methods. https://doi.org/10.1038/nmeth.3453

Lister R, Pelizzola M, Dowen RH, Hawkins RD, Hon G, Tonti-Filippini J, Nery JR, Lee L, Ye Z, Ngo QM, Edsall L, Antosiewicz-Bourget J, Stewart R, Ruotti V, Millar AH, Thomson JA, Ren B, Ecker JR (2009) Human DNA methylomes at base resolution show widespread epigenomic differences. Nature. https://doi.org/10.1038/nature08514

Lister R, Mukamel EA, Nery JR, Urich M, Puddifoot CA, Johnson ND, Lucero J, Huang Y, Dwork AJ, Schultz MD, Yu M, Tonti-Filippini J, Heyn H, Hu S, Wu JC, Rao A, Esteller M, He C, Haghighi FG, Sejnowski TJ, Behrens MM, Ecker JR (2013) Global epigenomic reconfiguration during mammalian brain development. Science. https://doi.org/10.1126/science.1237905

Liu J, Ward RL (2010) Folate and one-carbon metabolism and its impact on aberrant DNA methylation in cancer. Adv Genet 71:79–121. https://doi.org/10.1016/B978-0-12-380864-6.00004-3

Liu H, Huang D, McArthur DL, Boros LG, Nissen N, Heaney AP (2010) Fructose induces transketolase flux to promote pancreatic cancer growth. Cancer Res 70:6368–6376. https://doi.org/10.1158/0008-5472.CAN-09-4615

Liu J, Yue Y, Han D, Wang X, Fu Y, Zhang L, Jia G, Yu M, Lu Z, Deng X, Dai Q, Chen W, He C (2014) A METTL3-METTL14 complex mediates mammalian nuclear RNA N6-adenosine methylation. Nat Chem Biol. https://doi.org/10.1038/nchembio.1432

Liu Y, Wang R, Zhang L, Li J, Lou K, Shi B (2017) The lipid metabolism gene FTO influences breast cancer cell energy metabolism via the PI3K/AKT signaling pathway. Oncol Lett 13: 4685–4690. https://doi.org/10.3892/ol.2017.6038

Liu J, Eckert MA, Harada BT, Liu SM, Lu Z, Yu K, Tienda SM, Chryplewicz A, Zhu AC, Yang Y, Huang JT, Chen SM, Xu ZG, Leng XH, Yu XC, Cao J, Zhang Z, Liu J, Lengyel E, He C (2018) m 6 A mRNA methylation regulates AKT activity to promote the proliferation and tumorigenicity of endometrial cancer. Nat Cell Biol. 20:1074–1083. https://doi.org/10.1038/s41556-018-0174-4

Locasale JW, Cantley LC (2011) Metabolic flux and the regulation of mammalian cell growth. Cell Metab 14. https://doi.org/10.1016/j.cmet.2011.07.014

Londhe P, Yu PY, Ijiri Y, Ladner KJ, Fenger JM, London C, Houghton PJ, Guttridge DC (2018) Classical NF-κB metabolically reprograms sarcoma cells through regulation of hexokinase 2. Front Oncol 8:104. https://doi.org/10.3389/fonc.2018.00104

Lopes C, Pereira C, Medeiros R (2021) ASCT2 and LAT1 contribution to the hallmarks of cancer: from a molecular perspective to clinical translation. Cancers (Basel) 13(2). https://doi.org/10.3390/cancers13020203

Lopez-Serra P, Marcilla M, Villanueva A, Ramos-Fernandez A, Palau A, Leal L, Wahi JE, Setien-Baranda F, Szczesna K, Moutinho C, Martinez-Cardus A, Heyn H, Sandoval J, Puertas S, Vidal A, Sanjuan X, Martinez-Balibrea E, Viñals F, Perales JC, Bramsem JB, Ørntoft TF, Andersen CL, Tabernero J, McDermott U, Boxer MB, Heiden MGV, Albar JP, Esteller M (2014) A DERL3-associated defect in the degradation of SLC2A1 mediates the Warburg effect. Nat Commun. 5. https://doi.org/10.1038/ncomms4608

Lou X, Wang JJ, Wei YQ, Sun JJ (2021) Emerging role of RNA modification N6-methyladenosine in immune evasion. Cell Death Dis 12(4):300. https://doi.org/10.1038/s41419-021-03585-z

Lu J, Wang L, Chen W, Wang Y, Zhen S, Chen H, Cheng J, Zhou Y, Li X, Zhao L (2019) miR-603 targeted hexokinase-2 to inhibit the malignancy of ovarian cancer cells. Arch Biochem Biophys 661:1–9. https://doi.org/10.1016/j.abb.2018.10.014

Luo W, Hu H, Chang R, Zhong J, Knabel M, O'Meally R, Cole RN, Pandey A, Semenza GL (2011) Pyruvate kinase M2 is a PHD3-stimulated coactivator for hypoxia-inducible factor 1. Cell. https://doi.org/10.1016/j.cell.2011.03.054

Lyko F (2018) The DNA methyltransferase family: a versatile toolkit for epigenetic regulation. Nat Rev Genet. https://doi.org/10.1038/nrg.2017.80

Ma H, Wang X, Cai J, Dai Q, Natchiar SK, Lv R, Chen K, Lu Z, Chen H, Shi YG, Lan F, Fan J, Klaholz BP, Pan T, Shi Y, He C (2019) N 6-Methyladenosine methyltransferase ZCCHC4 mediates ribosomal RNA methylation. Nat Chem Biol. https://doi.org/10.1038/s41589-018-0184-3

Mahmood N, Cheishvili D, Arakelian A, Tanvir I, Khan HA, Pépin AS, Szyf M, Rabbani SA (2018) Methyl donor S-adenosylmethionine (SAM) supplementation attenuates breast cancer growth, invasion, and metastasis in vivo; therapeutic and chemopreventive applications. Oncotarget. 9:5169–5183. https://doi.org/10.18632/oncotarget.23704

Mauer J, Luo X, Blanjoie A, Jiao X, Grozhik AV, Patil DP, Linder B, Pickering BF, Vasseur JJ, Chen Q, Gross SS, Elemento O, Debart F, Kiledjian M, Jaffrey SR (2017) Reversible methylation of m6 Am in the 5′ cap controls mRNA stability. Nature. https://doi.org/10.1038/nature21022

Mauer J, Sindelar M, Despic V, Guez T, Hawley BR, Vasseur JJ, Rentmeister A, Gross SS, Pellizzoni L, Debart F, Goodarzi H, Jaffrey SR (2019) FTO controls reversible m 6 Am RNA methylation during snRNA biogenesis. Nat Chem Biol. https://doi.org/10.1038/s41589-019-0231-8

Maunakea AK, Chepelev I, Cui K, Zhao K (2013) Intragenic DNA methylation modulates alternative splicing by recruiting MeCP2 to promote exon recognition. Cell Res. https://doi.org/10.1038/cr.2013.110

Mayers JR, Torrence ME, Danai LV, Papagiannakopoulos T, Davidson SM, Bauer MR, Lau AN, Ji BW, Dixit PD, Hosios AM, Muir A, Chin CR, Freinkman E, Jacks T, Wolpin BM, Vitkup D, Vander Heiden MG (2016) Tissue of origin dictates branched-chain amino acid metabolism in mutant Kras-driven cancers. Science 353:1161–1165. https://doi.org/10.1126/science.aaf5171

Meng M, Chen S, Lao T, Liang D, Sang N (2010) Nitrogen anabolism underlies the importance of glutaminolysis in proliferating cells. Cell Cycle. 9:3921–3932. https://doi.org/10.4161/cc.9.19.13139

Meyer KD, Patil DP, Zhou J, Zinoviev A, Skabkin MA, Elemento O, Pestova TV, Qian SB, Jaffrey SR (2015) 5′ UTR m6A promotes Cap-independent translation. Cell. https://doi.org/10.1016/j.cell.2015.10.012

Mohandas T, Sparkes RS, Shapiro LJ (1981) Reactivation of an inactive human X chromosome: evidence for X inactivation by DNA methylation. Science. https://doi.org/10.1126/science.6164095

Momparler RL, Côté S, Momparler LF, Idaghdour Y (2014) Epigenetic therapy of acute myeloid leukemia using 5-aza-2'-deoxycytidine (decitabine) in combination with inhibitors of histone methylation and deacetylation. Clin Epigenet. https://doi.org/10.1186/1868-7083-6-19

Moore LD, Le T, Fan G (2013) DNA methylation and its basic function. Neuropsychopharmacology. https://doi.org/10.1038/npp.2012.112

Mukhopadhyay S, Ghosh S, Das D, Arun P, Roy B, Biswas NK, Maitra A, Majumder PP (2020) Application of random forest and data integration identifies three dysregulated genes and enrichment of central carbon metabolism pathway in oral cancer. BMC Cancer 20:1–9. https://doi.org/10.1186/s12885-020-07709-0

Nath A, Chan C (2016) Genetic alterations in fatty acid transport and metabolism genes are associated with metastatic progression and poor prognosis of human cancers. Sci Rep 6:1–13. https://doi.org/10.1038/srep18669

Nath K, Guo L, Nancolas B, Nelson DS, Shestov AA, Lee SC, Roman J, Zhou R, Leeper DB, Halestrap AP, Blair IA, Glickson JD (2016) Mechanism of antineoplastic activity of lonidamine. Biochim Biophys Acta 1866(2):151–162. https://doi.org/10.1016/j.bbcan.2016.08.001

Neupane M, Clark AP, Landini S, Birkbak NJ, Eklund AC, Lim E, Culhane AC, Barry WT, Schumacher SE, Beroukhim R, Szallasi Z, Vidal M, Hill DE, Silver DP (2016) MECP2 is a frequently amplified oncogene with a novel epigenetic mechanism that mimics the role of activated RAS in malignancy. Cancer Discov 6(1):45–58. https://doi.org/10.1158/2159-8290.CD-15-0341

Notaro S, Reimer D, Fiegl H, Schmid G, Wiedemair A, Rössler J, Marth C, Zeimet AG (2016) Evaluation of folate receptor 1 (FOLR1) mRNA expression, its specific promoter methylation and global DNA hypomethylation in type I and type II ovarian cancers. BMC Cancer 16:589. https://doi.org/10.1186/s12885-016-2637-y

Novak K (2004) Epigenetics changes in cancer cells. MedGenMed Medscape Gen Med 6

Ogrodzinski MP, Teoh ST, Lunt SY (2020) Metabolomic profiling of mouse mammary tumor-derived cell lines reveals targeted therapy options for cancer subtypes. Cell Oncol 43:1117–1127. https://doi.org/10.1007/s13402-020-00545-1

Ohtani-Fujita N, Fujita T, Aoike A, Osifchin NE, Robbins PD, Sakai T (1993) CpG methylation inactivates the promoter activity of the human retinoblastoma tumor-suppressor gene. Oncogene. 8(4):1063–1067

Oka M, Meacham AM, Hamazaki T, Rodić N, Chang LJ, Terada N (2005) De novo DNA methyltransferases Dnmt3a and Dnmt3b primarily mediate the cytotoxic effect of 5-aza-2-′-deoxycytidine. Oncogene 24:3091–3099. https://doi.org/10.1038/sj.onc.1208540

Okano M, Xie S, Li E (1998a) Cloning and characterization of a family of novel mammalian DNA (cytosine-5) methyltransferases [1]. Nat Genet. https://doi.org/10.1038/890

Okano M, Xie S, Li E (1998b) Dnmt2 is not required for de novo and maintenance methylation of viral DNA in embryonic stem cells. Nucleic Acids Res. https://doi.org/10.1093/nar/26.11.2536

Okano M, Bell DW, Haber DA, Li E (1999) DNA methyltransferases Dnmt3a and Dnmt3b are essential for de novo methylation and mammalian development. Cell. https://doi.org/10.1016/S0092-8674(00)81656-6

Ooi SKT, Qiu C, Bernstein E, Li K, Jia D, Yang Z, Erdjument-Bromage H, Tempst P, Lin SP, Allis CD, Cheng X, Bestor TH (2007) DNMT3L connects unmethylated lysine 4 of histone H3 to de novo methylation of DNA. Nature. https://doi.org/10.1038/nature05987

Ortiz-Masiá D, Gisbert-Ferrándiz L, Bauset C, Coll S, Mamie C, Scharl M, Esplugues JV, Alós R, Navarro F, Cosín-Roger J, Barrachina MD, Calatayud S (2020) Succinate activates EMT in intestinal epithelial cells through SUCNR1: a novel protagonist in fistula development. Cell. https://doi.org/10.3390/cells9051104

Parsons DW, Jones S, Zhang X, Lin JCH, Leary RJ, Angenendt P, Mankoo P, Carter H, Siu IM, Gallia GL, Olivi A, McLendon R, Rasheed BA, Keir S, Nikolskaya T, Nikolsky Y, Busam DA, Tekleab H, Diaz LA, Hartigan J, Smith DR, Strausberg RL, Marie SKN, Shinjo SMO, Yan H, Riggins GJ, Bigner DD, Karchin R, Papadopoulos N, Parmigiani G, Vogelstein B, Velculescu

VE, Kinzler KW (2008) An integrated genomic analysis of human glioblastoma multiforme. Science 321:1807–1812. https://doi.org/10.1126/science.1164382

Pascale RM, Feo CF, Calvisi DF, Feo F (2018) Deregulation of methionine metabolism as determinant of progression and prognosis of hepatocellular carcinoma. Transl Gastroenterol Hepatol. 3. https://doi.org/10.21037/tgh.2018.06.04

Pascual G, Avgustinova A, Mejetta S, Martín M, Castellanos A, Attolini CSO, Berenguer A, Prats N, Toll A, Hueto JA, Bescós C, Di Croce L, Benitah SA (2017) Targeting metastasis-initiating cells through the fatty acid receptor CD36. Nature 541:41–45. https://doi.org/10.1038/nature20791

Pastor WA, Aravind L, Rao A (2013) TETonic shift: biological roles of TET proteins in DNA demethylation and transcription. Nat Rev Mol Cell Biol. https://doi.org/10.1038/nrm3589

Pavlova NN, Thompson CB (2016) The emerging hallmarks of cancer metabolism. Cell Metab 23. https://doi.org/10.1016/j.cmet.2015.12.006

Pendleton KE, Chen B, Liu K, Hunter OV, Xie Y, Tu BP, Conrad NK (2017) The U6 snRNA m6A methyltransferase METTL16 regulates SAM synthetase intron retention. Cell. https://doi.org/10.1016/j.cell.2017.05.003

Pfeiffer T, Schuster S, Bonhoeffer S (2001) Cooperation and competition in the evolution of ATP-producing pathways. Science 292:504–507. https://doi.org/10.1126/science.1058079

Qiu C, Sawada K, Zhang X, Cheng X (2002) The PWWP domain of mammalian DNA methyltransferase Dnmt3b defines a new family of DNA-binding folds. Nat Struct Biol. https://doi.org/10.1038/nsb759

Rawluszko AA, Bujnicka KE, Horbacka K, Krokowicz P, Jagodziński PP (2013) Expression and DNA methylation levels of prolyl hydroxylases PHD1, PHD2, PHD3 and asparaginyl hydroxylase FIH in colorectal cancer. BMC Cancer. https://doi.org/10.1186/1471-2407-13-526

Ray U, Roy SS (2018) Aberrant lipid metabolism in cancer cells – the role of oncolipid-activated signaling. FEBS J 285. https://doi.org/10.1111/febs.14281

Razin A, Cedar H (1993) DNA methylation and embryogenesis. EXS 64. https://doi.org/10.1007/978-3-0348-9118-9_15

Reitzer LJ, Wice BM, Kennell D (1979) Evidence that glutamine, not sugar, is the major energy source for cultured HeLa cells. J Biol Chem 254:2669–2676. https://doi.org/10.1016/S0021-9258(17)30124-2

Rhee I, Bachman KE, Park BH, Jair KW, Yen RWC, Schuebel KE, Cui H, Feinberg AP, Lengauer C, Kinzler KW, Baylin SB, Vogelstein B (2002) DNMT1 and DNMT3b cooperate to silence genes in human cancer cells. Nature. https://doi.org/10.1038/416552a

Ries RJ, Zaccara S, Klein P, Olarerin-George A, Namkoong S, Pickering BF, Patil DP, Kwak H, Lee JH, Jaffrey SR (2019) m6A enhances the phase separation potential of mRNA. Nature. https://doi.org/10.1038/s41586-019-1374-1

Riggs AD (1975) X inactivation, differentiation, and DNA methylation. Cytogenet Genome Res. https://doi.org/10.1159/000130315

Rinaldi L, Avgustinova A, Martín M, Datta D, Solanas G, Neus P, Benitah SA (2017) Loss of dnmt3a and dnmt3b does not affect epidermal homeostasis but promotes squamous transformation through PPAR-g. eLife 6. https://doi.org/10.7554/eLife.21697

Ruderman N, Prentki M (2004) AMP kinase and malonyl-CoA: targets for therapy of the metabolic syndrome. Nat Rev Drug Discov 3. https://doi.org/10.1038/nrd1344

Saikawa Y, Price K, Hance KW, Chen TY, Elwood PC (1995) Structural and functional analysis of the human KB Cell folate receptor gene P4 promoter: cooperation of three clustered Sp1-binding sites with initiator region for basal promoter activity. Biochemistry 34:9951–9961. https://doi.org/10.1021/bi00031a018

Sakurai K, Tomihara K, Yamazaki M, Heshiki W, Moniruzzaman R, Sekido K, Tachinami H, Ikeda A, Imaue S, Fujiwara K, Noguchi M (2020) CD36 expression on oral squamous cell carcinoma cells correlates with enhanced proliferation and migratory activity. Oral Dis 26:745–755. https://doi.org/10.1111/odi.13210

Sandoval J, Esteller M (2012) Cancer epigenomics: beyond genomics. Curr Opin Genet Dev. https://doi.org/10.1016/j.gde.2012.02.008

Sasaki M, Knobbe CB, Munger JC, Lind EF, Brenner D, Brüstle A, Harris IS, Holmes R, Wakeham A, Haight J, You-Ten A, Li WY, Schalm S, Su SM, Virtanen C, Reifenberger G, Ohashi PS, Barber DL, Figueroa ME, Melnick A, Zúñiga-Pflücker JC, Mak TW (2012) IDH1 (R132H) mutation increases murine haematopoietic progenitors and alters epigenetics. Nature 488:656–659. https://doi.org/10.1038/nature11323

Sciacovelli M, Gonçalves E, Johnson TI, Zecchini VR, Da Costa ASH, Gaude E, Drubbel AV, Theobald SJ, Abbo SR, Tran MGB, Rajeeve V, Cardaci S, Foster S, Yun H, Cutillas P, Warren A, Gnanapragasam V, Gottlieb E, Franze K, Huntly B, Maher ER, Maxwell PH, Saez-Rodriguez J, Frezza C (2016) Fumarate is an epigenetic modifier that elicits epithelial-to-mesenchymal transition. Nature. https://doi.org/10.1038/nature19353

Selak MA, Armour SM, MacKenzie ED, Boulahbel H, Watson DG, Mansfield KD, Pan Y, Simon MC, Thompson CB, Gottlieb E (2005) Succinate links TCA cycle dysfunction to oncogenesis by inhibiting HIF-α prolyl hydroxylase. Cancer Cell 7:77–85. https://doi.org/10.1016/j.ccr.2004.11.022

Shanmugam R, Fierer J, Kaiser S, Helm M, Jurkowski TP, Jeltsch A (2015) Cytosine methylation of tRNA-Asp by DNMT2 has a role in translation of proteins containing poly-Asp sequences. Cell Discov. https://doi.org/10.1038/celldisc.2015.10

Shen C, Xuan B, Yan T, Ma Y, Xu P, Tian X, Zhang X, Cao Y, Ma D, Zhu X, Zhang Y, Fang JY, Chen H, Hong J (2020) m(6)A-dependent glycolysis enhances colorectal cancer progression. Mol Cancer. 19(1):72. https://doi.org/10.1186/s12943-020-01190-w

Shull AY, Luo J, Pei L, Lee E-J, Liu J, Choi J, Awan FT, Shi H (2016) DNA hypomethylation within B-cell enhancers and super enhancers reveal a dependency on immune and metabolic mechanisms in chronic lymphocytic leukemia. Blood 128:1049–1049. https://doi.org/10.1182/blood.v128.22.1049.1049

Shuvalov O, Petukhov A, Daks A, Fedorova O, Vasileva E, Barlev NA (2017) One-carbon metabolism and nucleotide biosynthesis as attractive targets for anticancer therapy. Oncotarget 8. https://doi.org/10.18632/oncotarget.15053

Simpson NE, Tryndyak VP, Pogribna M, Beland FA, Pogribny IP (2012) Modifying metabolically sensitive histone marks by inhibiting glutamine metabolism affects gene expression and alters cancer cell phenotype. Epigenetics 7:1413–1420. https://doi.org/10.4161/epi.22713

Siu MKY, Kong DSH, Chan HY, Wong ESY, Ip PPC, Jiang LL, Ngan HYS, Le XF, Cheung ANY (2012) Paradoxical impact of two folate receptors, FRα and RFC, in ovarian cancer: effect on cell proliferation, invasion and clinical outcome. PLoS One 7. https://doi.org/10.1371/journal.pone.0047201

Sommer S, Lavi U, Darnell JE (1978) The absolute frequency of labeled N-6-methyladenosine in HeLa cell messenger RNA decreases with label time. J Mol Biol. https://doi.org/10.1016/0022-2836(78)90183-3

Soria J-C, Lee H-Y, Lee JI, Wang L, Issa J-P, Kemp BL, Liu DD, Kurie JM, Mao L, Khuri FR (2002) Lack of PTEN expression in non-small cell lung cancer could be related to promoter methylation 1. Clin Cancer Res. 8(5):1178–1184

Squires JE, Patel HR, Nousch M, Sibbritt T, Humphreys DT, Parker BJ, Suter CM, Preiss T (2012) Widespread occurrence of 5-methylcytosine in human coding and non-coding RNA. Nucleic Acids Res. https://doi.org/10.1093/nar/gks144

Su F, Kozak KR, Imaizumi S, Gao F, Amneus MW, Grijalva V, Ng C, Wagner A, Hough G, Farias-Eisner G, Anantharamaiah GM, Van Lenten BJ, Navab M, Fogelman AM, Reddy ST, Farias-Eisner R (2010) Apolipoprotein A-I (apoA-I) and apoA-I mimetic peptides inhibit tumor development in a mouse model of ovarian cancer. Proc Natl Acad Sci U S A 107:19997–20002. https://doi.org/10.1073/pnas.1009010107

Su R, Dong L, Li C, Nachtergaele S, Wunderlich M, Qing Y, Deng X, Wang Y, Weng X, Hu C, Yu M, Skibbe J, Dai Q, Zou D, Wu T, Yu K, Weng H, Huang H, Ferchen K, Qin X, Zhang B, Qi J, Sasaki AT, Plas DR, Bradner JE, Wei M, Marcucci G, Jiang X, Mulloy JC, Jin J, He C,

Chen J (2018) R-2HG exhibits anti-tumor activity by targeting FTO/m6A/MYC/CEBPA signaling. Cell 172:90–105.e123. https://doi.org/10.1016/j.cell.2017.11.031
Sun W, Liu Y, Glazer CA, Shao C, Bhan S, Demokan S, Zhao M, Rudek MA, Ha PK, Califano JA (2010) Human cancer biology TKTL1 is activated by promoter hypomethylation and contributes to head and neck squamous cell carcinoma carcinogenesis through increased aerobic glycolysis and HIF1α stabilization. Clin Cancer Res 16:857–866. https://doi.org/10.1158/1078-0432.CCR-09-2604
Sun T, Wu R, Ming L (2019) The role of m6A RNA methylation in cancer. Biomed Pharmacother 112:108613. https://doi.org/10.1016/j.biopha.2019.108613
Symonds EL, Pedersen SK, Baker RT, Murray DH, Gaur S, Cole SR, Gopalsamy G, Mangira D, LaPointe LC, Young GP (2016) A blood test for methylated BCAT1 and IKZF1 vs. a fecal immunochemical test for detection of colorectal neoplasia. Clin Transl Gastroenterol 7(e137). https://doi.org/10.1038/ctg.2015.67
Tahiliani M, Koh KP, Shen Y, Pastor WA, Bandukwala H, Brudno Y, Agarwal S, Iyer LM, Liu DR, Aravind L, Rao A (2009) Conversion of 5-methylcytosine to 5-hydroxymethylcytosine in mammalian DNA by MLL partner TET1. Science. https://doi.org/10.1126/science.1170116
Tian J, Zhu Y, Rao M, Cai Y, Lu Z, Zou D, Peng X, Ying P, Zhang M, Niu S, Li Y, Zhong R, Chang J, Miao X (2020) N 6-methyladenosine mRNA methylation of PIK3CB regulates AKT signalling to promote PTEN-deficient pancreatic cancer progression. Gut 69:2180–2192. https://doi.org/10.1136/gutjnl-2019-320179
Tibbetts AS, Appling DR (2010) Compartmentalization of mammalian folate-mediated one-carbon metabolism. Annu Rev Nutr 30:57–81. https://doi.org/10.1146/annurev.nutr.012809.104810
Tönjes M, Barbus S, Park YJ, Wang W, Schlotter M, Lindroth AM, Pleier SV, Bai AHC, Karra D, Piro RM, Felsberg J, Addington A, Lemke D, Weibrecht I, Hovestadt V, Rolli CG, Campos B, Turcan S, Sturm D, Witt H, Chan TA, Herold-Mende C, Kemkemer R, König R, Schmidt K, Hull WE, Pfister SM, Jugold M, Hutson SM, Plass C, Okun JG, Reifenberger G, Lichter P, Radlwimmer B (2013) BCAT1 promotes cell proliferation through amino acid catabolism in gliomas carrying wild-type IDH1. Nat Med 19:901–908. https://doi.org/10.1038/nm.3217
Turcan S, Fabius AWM, Borodovsky A, Pedraza A, Brennan C, Huse J, Viale A, Riggins GJ, Chan TA (2013) Efficient induction of differentiation and growth inhibition in IDH1 mutant glioma cells by the DNMT inhibitor decitabine. Oncotarget 4:1729–1736. https://doi.org/10.18632/oncotarget.1412
Van Tran N, Ernst FGM, Hawley BR, Zorbas C, Ulryck N, Hackert P, Bohnsack KE, Bohnsack MT, Jaffrey SR, Graille M, Lafontaine DLJ (2019) The human 18S rRNA m6A methyltransferase METTL5 is stabilized by TRMT112. Nucleic Acids Res. https://doi.org/10.1093/nar/gkz619
Visvanathan A, Patil V, Arora A, Hegde AS, Arivazhagan A, Santosh V, Somasundaram K (2018) Essential role of METTL3-mediated m 6 A modification in glioma stem-like cells maintenance and radioresistance. Oncogene. https://doi.org/10.1038/onc.2017.351
Vogelstein B, Papadopoulos N, Velculescu VE, Zhou S, Diaz LA, Kinzler KW (2013) Cancer genome landscapes. Science 340. https://doi.org/10.1126/science.1235122
Vu LP, Pickering BF, Cheng Y, Zaccara S, Nguyen D, Minuesa G, Chou T, Chow A, Saletore Y, Mackay M, Schulman J, Famulare C, Patel M, Klimek VM, Garrett-Bakelman FE, Melnick A, Carroll M, Mason CE, Jaffrey SR, Kharas MG (2017) The N 6-methyladenosine (m 6 A)-forming enzyme METTL3 controls myeloid differentiation of normal hematopoietic and leukemia cells. Nat Med 23:1369–1376. https://doi.org/10.1038/nm.4416
Wang JB, Erickson JW, Fuji R, Ramachandran S, Gao P, Dinavahi R, Wilson KF, Ambrosio ALB, Dias SMG, Dang CV, Cerione RA (2010) Targeting mitochondrial glutaminase activity inhibits oncogenic transformation. Cancer Cell 18:207–219. https://doi.org/10.1016/j.ccr.2010.08.009
Wang F, Travins J, DeLaBarre B, Penard-Lacronique V, Schalm S, Hansen E, Straley K, Kernytsky A, Liu W, Gliser C, Yang H, Gross S, Artin E, Saada V, Mylonas E, Quivoron C, Popovici-Muller J, Saunders JO, Salituro FG, Yan S, Murray S, Wei W, Gao Y, Dang L, Dorsch M, Agresta S, Schenkein DP, Biller SA, Su SM, de Botton S, Yen KE (2013) Targeted

inhibition of mutant IDH2 in leukemia cells induces cellular differentiation. Science (New York, NY) 340. https://doi.org/10.1126/SCIENCE.1234769

Wang Y, Hao J, Liu X, Wang H, Zeng X, Yang J, Li L, Kuang X, Zhang T (2016) The mechanism of apoliprotein A1 down-regulated by Hepatitis B virus. Lipids Health Dis 15. https://doi.org/10.1186/s12944-016-0232-5

Wang Q, Chen C, Ding Q, Zhao Y, Wang Z, Chen J, Jiang Z, Zhang Y, Xu G, Zhang J, Zhou J, Zou X, Wang S, Sun B (2020a) METTL3-mediated m 6 A modification of HDGF mRNA promotes gastric cancer progression and has prognostic significance. Gut 69:1193–1205. https://doi.org/10.1136/gutjnl-2019-319639

Wang T, Kong S, Tao M, Ju S (2020b) The potential role of RNA N6-methyladenosine in Cancer progression. Mol Cancer 19(1):88. https://doi.org/10.1186/s12943-020-01204-7

Warburg O (1924) Uber den Stoffwechsel der Carcinomzelle. Biochem Z 152:309–344

Warburg O (1925) The metabolism of carcinoma cells 1. J Cancer Res 9:148–163. https://doi.org/10.1158/jcr.1925.148

Warburg O (1956) On the origin of cancer cells. Science 123:309–314. https://doi.org/10.1126/science.123.3191.309

Warburg O, Wind F, Negelein E (1927) The metabolism of tumors in the body. J Gen Physiol 8(6):519–530. https://doi.org/10.1085/jgp.8.6.519

Weinhouse S, Warburg O, Burk D, Schade AL (1956) On respiratory impairment in cancer cells. Science 124:267–272. https://doi.org/10.1126/science.124.3215.267

Wolf A, Agnihotri S, Munoz D, Guha A (2011) Developmental profile and regulation of the glycolytic enzyme hexokinase 2 in normal brain and glioblastoma multiforme. Neurobiol Dis 44:84–91. https://doi.org/10.1016/j.nbd.2011.06.007

Wong CC, Qian Y, Yu J (2017) Interplay between epigenetics and metabolism in oncogenesis: mechanisms and therapeutic approaches. Oncogene 36. https://doi.org/10.1038/onc.2016.485

Wu X, Zhang Y (2017) TET-mediated active DNA demethylation: mechanism, function and beyond. Nat Rev Genet. https://doi.org/10.1038/nrg.2017.33

Xiao M, Yang H, Xu W, Ma S, Lin H, Zhu H, Liu L, Liu Y, Yang C, Xu Y, Zhao S, Ye D, Xiong Y, Guan KL (2012) Inhibition of α-KG-dependent histone and DNA demethylases by fumarate and succinate that are accumulated in mutations of FH and SDH tumor suppressors. Genes Dev 26:1326–1338. https://doi.org/10.1101/gad.191056.112

Xiao W, Liu X, Niu X, Li C, Guo Y, Tan J, Xiong W, Fan L, Li Y (2020) The frequency of CpG and non-CpG methylation of Notch3 gene promoter determines its expression levels in breast cancer cells. Exp Cell Res 386(2):111743. https://doi.org/10.1016/j.yexcr.2019.111743

Xie W, Ma LL, Xu YQ, Wang BH, Li SM (2019) METTL3 inhibits hepatic insulin sensitivity via N6-methyladenosine modification of Fasn mRNA and promoting fatty acid metabolism. Biochem Biophys Res Commun 518:120–126. https://doi.org/10.1016/j.bbrc.2019.08.018

Xie H, Li J, Ying Y, Yan H, Jin K, Ma X, He L, Xu X, Liu B, Wang X, Zheng X, Xie L (2020) METTL3/YTHDF2 m^6 A axis promotes tumorigenesis by degrading SETD7 and KLF4 mRNAs in bladder cancer. J Cell Mol Med 24:4092–4104. https://doi.org/10.1111/jcmm.15063

Xu W, Yang H, Liu Y, Yang Y, Wang P, Kim SH, Ito S, Yang C, Wang P, Xiao MT, Liu LX, Jiang WQ, Liu J, Zhang JY, Wang B, Frye S, Zhang Y, Xu YH, Lei QY, Guan KL, Zhao SM, Xiong Y (2011) Oncometabolite 2-hydroxyglutarate is a competitive inhibitor of α-ketoglutarate-dependent dioxygenases. Cancer Cell 19:17–30. https://doi.org/10.1016/j.ccr.2010.12.014

Yang W, Lu Z (2013) Nuclear PKM2 regulates the Warburg effect. Cell Cycle. https://doi.org/10.4161/cc.26182

Yen CY, Huang HW, Shu CW, Hou MF, Yuan SSF, Wang HR, Chang YT, Farooqi AA, Tang JY, Chang HW (2016) DNA methylation, histone acetylation and methylation of epigenetic modifications as a therapeutic approach for cancers. Cancer Lett 373. https://doi.org/10.1016/j.canlet.2016.01.036

Yoder JA, Bestor TH (1998) A candidate mammalian DNA methyltransferase related to pmt1p of fission yeast. Hum Mol Genet. https://doi.org/10.1093/hmg/7.2.279

Young JI, Hong EP, Castle JC, Crespo-Barreto J, Bowman AB, Rose MF, Kang D, Richman R, Johnson JM, Berget S, Zoghbi HY (2005) Regulation of RNA splicing by the methylation-dependent transcriptional repressor methyl-CpG binding protein 2. Proc Natl Acad Sci U S A. https://doi.org/10.1073/pnas.0507856102

Yu J, Shen L, Liu Y, Ming H, Zhu X, Chu M, Lin J (2020) The m6A methyltransferase METTL3 cooperates with demethylase ALKBH5 to regulate osteogenic differentiation through NF-κB signaling. Mol Cell Biochem 463:203–210. https://doi.org/10.1007/s11010-019-03641-5

Zaccara S, Ries RJ, Jaffrey SR (2019) Reading, writing and erasing mRNA methylation. Nat Rev Mol Cell Biol. https://doi.org/10.1038/s41580-019-0168-5

Zamanian-Daryoush M, Lindner D, Tallant TC, Wang Z, Buffa J, Klipfell E, Parker Y, Hatala D, Parsons-Wingerter P, Rayman P, Yusufishaq MSS, Fisher EA, Smith JD, Finke J, DiDonato JA, Hazen SL (2013) The cardioprotective protein apolipoprotein a1 promotes potent anti-tumorigenic effects. J Biol Chem 288:21237–21252. https://doi.org/10.1074/jbc.M113.468967

Zhang Y, Morar M, Ealick SE (2008) Structural biology of the purine biosynthetic pathway. Cell Mol Life Sci 65. https://doi.org/10.1007/s00018-008-8295-8

Zhang S, Pei M, Li Z, Li H, Liu Y, Li J (2018) Double-negative feedback interaction between DNA methyltransferase 3A and microRNA-145 in the Warburg effect of ovarian cancer cells. Cancer Sci 109(9):2734–2745. https://doi.org/10.1111/cas.13734

Zhang LS, Liu C, Ma H, Dai Q, Sun HL, Luo G, Zhang Z, Zhang L, Hu L, Dong X, He C (2019a) Transcriptome-wide mapping of internal N7-methylguanosine methylome in mammalian mRNA. Mol Cell. https://doi.org/10.1016/j.molcel.2019.03.036

Zhang Y, Kang M, Zhang B, Meng F, Song J, Kaneko H, Shimamoto F, Tang B (2019b) m(6)A modification-mediated CBX8 induction regulates stemness and chemosensitivity of colon cancer via upregulation of LGR5. Mol Cancer 18(1):185. https://doi.org/10.1186/s12943-019-1116-x

Zhang M, Song J, Yuan W, Zhang W, Sun Z (2021) Roles of RNA methylation on tumor immunity and clinical implications. Front Immunol 12:641507. https://doi.org/10.3389/fimmu.2021.641507

Zhao X, Yang Y, Sun BF, Shi Y, Yang X, Xiao W, Hao YJ, Ping XL, Chen YS, Wang WJ, Jin KX, Wang X, Huang CM, Fu Y, Ge XM, Song SH, Jeong HS, Yanagisawa H, Niu Y, Jia GF, Wu W, Tong WM, Okamoto A, He C, Danielsen JMR, Wang XJ, Yang YG (2014) FTO-dependent demethylation of N6-methyladenosine regulates mRNA splicing and is required for adipogenesis. Cell Res. https://doi.org/10.1038/cr.2014.151

Zheng G, Dahl JA, Niu Y, Fedorcsak P, Huang CM, Li CJ, Vågbø CB, Shi Y, Wang WL, Song SH, Lu Z, Bosmans RPG, Dai Q, Hao YJ, Yang X, Zhao WM, Tong WM, Wang XJ, Bogdan F, Furu K, Fu Y, Jia G, Zhao X, Liu J, Krokan HE, Klungland A, Yang YG, He C (2013) ALKBH5 is a mammalian RNA demethylase that impacts RNA metabolism and mouse fertility. Mol Cell. https://doi.org/10.1016/j.molcel.2012.10.015

Zhou Y, Kong Y, Fan W, Tao T, Xiao Q, Li N, Zhu X (2020) Principles of RNA methylation and their implications for biology and medicine. Biomed Pharmacother 131:110731. https://doi.org/10.1016/j.biopha.2020.110731

Zhou B, Liu C, Xu L, Yuan Y, Zhao J, Zhao W, Chen Y, Qiu J, Meng M, Zheng Y, Wang D, Gao X, Li X, Zhao Q, Wei X, Wu D, Zhang H, Hu C, Zhuo X, Zheng M, Wang H, Lu Y, Ma X (2021) N (6)-methyladenosine reader protein YT521-B homology domain-containing 2 suppresses liver steatosis by regulation of mRNA stability of lipogenic genes. Hepatology 73(1):91–103. https://doi.org/10.1002/hep.31220

Zou S, Toh JDW, Wong KHQ, Gao YG, Hong W, Woon ECY (2016) N 6-Methyladenosine: a conformational marker that regulates the substrate specificity of human demethylases FTO and ALKBH5. Sci Rep. https://doi.org/10.1038/srep25677

Zu XL, Guppy M (2004) Cancer metabolism: facts, fantasy, and fiction. Biochem Biophys Res Commun 313. https://doi.org/10.1016/j.bbrc.2003.11.136

Zuo X, Chen Z, Gao W, Zhang Y, Wang J, Wang J, Cao M, Cai J, Wu J, Wang X (2020) M6A-mediated upregulation of LINC00958 increases lipogenesis and acts as a nanotherapeutic target in hepatocellular carcinoma. J Hematol Oncol 13:5. https://doi.org/10.1186/s13045-019-0839-x

Chapter 8
Nutritional Epigenetics: How Metabolism Epigenetically Controls Cellular Physiology, Gene Expression and Disease

Pragyan Mishra, Shibangini Beura, Ritu Ghosh, and Rahul Modak

Abstract The regulation of gene expression is a dynamic process that is influenced by both internal and external factors. Alteration in the epigenetic profile is a key mechanism in the regulation process. Epigenetic regulators, such as enzymes and proteins involved in posttranslational modification (PTM), use different cofactors and substrates derived from dietary sources. For example, glucose metabolism provides acetyl CoA, S-adenosylmethionine (SAM), α- ketoglutarate, uridine diphosphate (UDP)-glucose, adenosine triphosphate (ATP), nicotinamide adenine dinucleotide (NAD^+), and fatty acid desaturase (FAD), which are utilized by chromatin-modifying enzymes in many intermediary metabolic pathways. Any alteration in the metabolic status of the cell results in the alteration of these metabolites, which causes dysregulation in the activity of chromatin regulators, resulting in the alteration of the epigenetic profile. Such long-term or repeated alteration of epigenetic profile can lead to several diseases, like cancer, insulin resistance and diabetes, cognitive impairment, neurodegenerative disease, and metabolic syndromes. Here we discuss the functions of key nutrients that contribute to epigenetic regulation and their role in pathophysiological conditions.

Keywords Carbon metabolism · Histone deacetylation · Dietary metabolites · Vitamins · Metabolic pathways · Metabolic disorders

List of Abbreviations

5meTHF	5-methyl-tetrahydrofolate
AceCS2	Acetyl-CoA synthase enzyme

Pragyan Mishra and Shibangini Beura contributed equally with all other contributors.

P. Mishra · S. Beura · R. Ghosh · R. Modak (✉)
Infection and Epigenetics Group, School of Biotechnology, Kalinga Institute of Industrial Technology, Bhubaneswar, Odisha, India
e-mail: rahul.modak@kiitbiotech.ac.in

T. K. Kundu, C. Das (eds.), *Metabolism and Epigenetic Regulation: Implications in Cancer*, Subcellular Biochemistry 100,
https://doi.org/10.1007/978-3-031-07634-3_8

ADIPOQ	Adiponectin
Agrp	Agouti-related neuropeptide gene
Ahcy	Adenosylhomocysteinase
AMP	Adenosine monophosphate
APOE	Apolipoprotein E
ATGL	Adipose triglyceride lipase
ATP	Adenosine triphosphate
Aβ	β-amyloid peptide
BRCA1	Breast cancer 1
CBS	Cystathionine synthase
COMT	Catechol-O-methyltransferase
CTH	Cystathione γ lyase
CVD	Cardiovascular disease
DNA	Deoxyribonucleic acid
DNMT	DNA methyltransferase
EAA	Essential amino acid
EGCG	Epigallocatechin gallate
ESCs	Embryonic stem cells
FAD	Flavin adenine dinucleotide
FATP1	Fatty acid transport protein 1
Gcn5 KAT	General control nonderepressible 5 lysine acetyltransferase
GNMT	Glycine N-methyltransferase
Hcy	Homocysteine
HDAC	Histone deacetylase
HDAC	Lysine deacetylase
HIF	Hypoxia-inducible factor
Hnf4α	Hepatocyte nuclear-4-alpha
HOX5	Homeobox-leucine zipper protein HOX5
HPUFA-LCMFA	High polyunsaturated fatty acid – low-medium chain fatty acid
JmjC domain	Jumonji C domain
JMJD	JmjC demethylase
KAT	Lysine acetyltransferase
KDM1A	Lysine-specific histone demethylase 1A
K_m	Michaelis constant
LBW	Low birth weight
LC-PUFA	Low-carbon polyunsaturated fatty acid
LEP	Leptin
LINE-1	Long interspersed nuclear element
lncRNAs	Long noncoding RNAs
LSD1	Lysine Residues
MafK	Maf bZIP transcription factor F
MAPK	Mitogen-activated protein kinase
MAT	Methionine adenosyl transferase

MGMT	O6-methylguanine-DNA methyltransferase
miRNA	Micro-RNA
MS	Methionine synthase
MSR	Methionine synthase reductase
MTHF	Methylenetetrahydrofolate
MTHFR	Methylenetetrahydrofolate reductase
MTHFR	Methylenetetrahydrofolate reductase
NADH	Nicotinamide adenine dinucleotide hydride
NADP	Nicotinamide adenine dinucleotide phosphate
NBW	Normal birth weight
ncRNAs	Noncoding RNAs
NEAA	Nonessential amino acid
PCK1	Phosphoenolpyruvate carboxykinase
PDK4	Pyruvate dehydrogenase kinase 4
PPARGC1A/PGC1α	Peroxisome proliferator activated receptor gamma coactivator 1-alpha
PPARγ	Peroxisome proliferator-activated receptor γ
PRPP	5-phosphoribosyl-1-pyrophosphate
PTM	Posttranslational modifications
RNA	Ribonucleic acid
SAGA	Spt-Ada-Gcn5 acetyltransferase
SAH	S-adenosyl homocysteine
SAM	AdoMet
SAM	S-adenosylmethionine
SHMT	Serine hydroxy methyltransferase
SIRT1	Sirtuin 1
T2D	Type 2 diabetes
TCA	Tricarboxylic acid cycle
TDH	Threonine dehydrogenase
TET	Ten-eleven translocation
THF	Tetrahydrofolate
TMAO	Trimethylamine N-oxide
UDP	Uridine diphosphate
YMC	Yeast metabolic cycle
α-KG	α-ketoglutarate

Introduction

Human gene expression across populations is influenced by dietary and lifestyle patterns. Alteration in gene expression through the consumption of nutrients, including vitamins and mineral supplements, has been demonstrated through their interaction at the epigenome level. In a eukaryotic system, gene expression is regulated via epigenetic mechanisms through changes in the chemical modifications of deoxyribonucleic acid (DNA) and proteins or by altering the accessibility of DNA due to its association with histones, nonhistone proteins, and noncoding ribonucleic acids (ncRNAs). Particularly, in response to intracellular signals, the dynamic alteration of histone posttranslational modifications (PTMs) is known to modify DNA-chromatin interactions, which can be reversed and, in turn, alter protein function, stability, and interaction with other proteins. DNA and protein methylation, serine-threonine phosphorylation, lysine acetylation, and ubiquitination are some of the most well-studied posttranslational modifications. Enzymes of intermediary metabolism are posttranslationally modified by acetylation, that acts as a regulatory mechanism and helps to adapt to the changes in metabolite flux that determines the energetic state of the cell (Hirschey et al. 2010; Newman et al. 2012).

Chromatin modifications and its associated enzymes that are involved in pathogenesis contributes to sustained deregulated metabolic phenotype. There is an increasing number of studies that associate epigenetic changes with the diagnosis and prognosis of several diseases. Metabolism refers to interconnected enzyme-catalyzed chemical reactions involved in sustained energy production and the anabolism and catabolism of complex molecules by the cell. Any alterations in energy metabolism lead to sustained stable epigenetic changes that are inherited through the germline and affect the progeny (Jang and Serra 2014). Sensing the availability of nutrients, cells integrate signals to coordinate cellular processes and metabolism through multiple mechanisms. Nutrients are involved in influencing cellular responses as they provide substrates for the posttranslational modification of proteins, which in turn affects the epigenome by regulating gene expression patterns (Gut and Verdin 2013). These intermediary substrates or metabolism products are sensed by chromatin-associated enzymes, thereby processing this information into chromatin through protein PTMs. An instance of such a phenomenon is the activation of genes driven by glucose metabolism, characterized by the increase in glucose-derived histone acetylation. Strikingly, the crucial metabolites acting as cofactors or substrates, such as acetyl-CoA, S-adenosylmethionine (SAM), α- ketoglutarate, uridine diphosphate (UDP)-glucose, adenosine triphosphate (ATP), nicotinamide adenine dinucleotide (NAD^+), and fatty acid desaturase (FAD), are utilized by chromatin-modifying enzymes in many of the intermediary metabolism core pathways. Any disturbances in the metabolic status of the cell lead to fluctuations in the concentration of these metabolites, which causes dysregulation in the activity of chromatin regulators and proteins as well as histone modifications, which link nutritional status with several disease states, for example, cancer, cognitive impairment, neurodegenerative diseases, insulin resistance and diabetes, and metabolic syndrome (Lu and Thompson 2012; Kaelin and McKnight 2013).

Many enzymes, including chromatin modifiers and metabolic pathway regulators, provide pivotal links between nutrition, metabolism, and gene regulation. In this chapter, we will be citing examples of such nutrient-derived substrates and cofactors as listed in Table 8.1 involved in programming the epigenome as well as evidence of deregulated metabolism contributing to disease. In two separate aspects, we will particularly focus on different metabolic products acting as sources or precursors of DNA methylation and histone modifications and how changes in these levels relate to several diseases.

Table 8.1 Dietary components and metabolites required for epigenetic modifications

Nutrient source/substrates and metabolites	Epigenetic modifications	References
Folic acid, Vitamin B12, Vitamin B6, Choline, Betaine — Methionine synthesis	DNA methylation Histone methylation	Zeisel (2017), Meng et al. (2018)
Methionine (SAM synthesis)	DNA methylation Histone methylation	Niculescu and Zeisel (2002)
Vitamin B2/riboflavin (FAD synthesis)	Histone demethylation	Culhane and Cole (2007), Barile et al. (2016)
Isocitrate, Glutamine, Arginine, Histidine, Proline — 2-oxoglutarate/α-ketoglutarate, O2, Fe(II)	Histone demethylation	Sánchez-Fernández et al. (2013)
Pyruvate, Cysteine, Threonine — Acetyl CoA	Histone acetylation	Wang et al. (2009), Sutendra et al. (2014)
Tryptophan, ATP — NAD+/NADP+	Histone deacetylation	Shibata (2018)
Resveratrol	Histone deacetylation	Bagul et al. (2015)
Dietary fibers (butyrate synthesis)	Histone acetylation (HDACi)	Davie (2003)

Nutritional Requirements of the Cell: Amino Acids, Vitamins, and Minerals

Amino acids act as the building blocks of proteins and key precursors for the synthesis of several physiologically important biomolecules and metabolites, such as polyamines, neurotransmitters, hormones, creatine, small peptides, nitric oxide, etc. (Wu 2009). Traditionally, amino acids are categorized as either nutritionally essential (EAA—phenylalanine, histidine, methionine, leucine, isoleucine, threonine, tryptophan, lysine, and valine) or nonessential (NEAA—tyrosine, proline, alanine, glutamate, taurine, asparagine, arginine, aspartate, serine, glycine, cysteine, and glutamine) amino acids. In recent years, evidence has been gathered of amino acids serving not only as signaling molecules but also as regulators of the protein phosphorylation cascade and gene expression. Amino acids such as methionine, serine, glycine, histidine is required for the synthesis of DNA by being part of one-carbon unit metabolism, involving the methyl cycle. Disturbances in the optimum balance and level of amino acids, along with their products, are associated with multiple disorders and diseases.

Our body cannot synthesize nutritionally essential micro- and macrominerals, because of which they need to be provided in the diet. Sodium, phosphorus, potassium, sulfur, calcium, magnesium, and chloride are the macrominerals, and iodine, zinc, molybdenum, fluoride, iron, manganese, nickel, silicon, vanadium, chromium, copper, selenium, and cobalt are the microminerals required by the body. These nutrients are important second messengers for cell signaling, act as a foundation for development, are needed for the maintenance of the skeletal system, and also regulate intracellular and extracellular osmolarities, which maintain cell and blood volume (Zimmermann 2006). Many of these micro- and macrominerals serve as cofactors for various enzymes and metalloproteins and are required in redox reaction and during electron transport as well (Burdge and Lillycrop 2010).

Apart from these inorganic nutrients, organic nutrients, i.e., vitamins, are required in small amounts for cellular biochemical reactions. Some of the vitamins are absorbed by means of portal circulation through the intestinal tract, which includes water-soluble vitamins, such as members of the vitamin B complex and vitamin C (Oommen et al. 2005). These water-soluble vitamins, including vitamin B complex members, such as folate, vitamin B_1 (thiamin), vitamin B_2 (riboflavin), vitamin B_3 (niacin), vitamin B_5 (pantothenic acid), vitamin B_6 (pyridoxal, pyridoxine, pyridoxamine), vitamin B_{12} (cobalamin), and biotin are stored in limited quantities in the body due to their water solubility, as a result of which they must be constantly supplemented in the diet (Oommen et al. 2005; Burdge and Lillycrop 2010). Other vitamins, including vitamins D, A, E, and K, are categorized as lipid-soluble vitamins. Except for vitamin D and niacin, most of these vitamins need to be provided in the diet as they cannot be synthesized by the body. In physiological processes carried out by the cell, vitamins act as cofactors for enzymes required during the metabolism of glucose, fatty acids, and amino acids; ATP production; one-carbon unit metabolism; etc. as shown in Fig. 8.1.

Metabolism and Epigenetics

Presently, there are four mechanisms that mediate epigenetic regulation: DNA methylation of cytosine residues at the 5′ position within CpG dinucleotides, chromatin modifications, histone modifications (phosphorylation, methylation, acetylation, sumoylation, ubiquitination), and mechanisms involving noncoding RNAs (microRNAs (miRNAs) or long noncoding RNAs (lncRNAs)) that require specific enzymes (protein/DNA methyltransferases and demethylases, histone lysine acetyltransferases and deacetylases, etc.) for each of these processes. In this section,

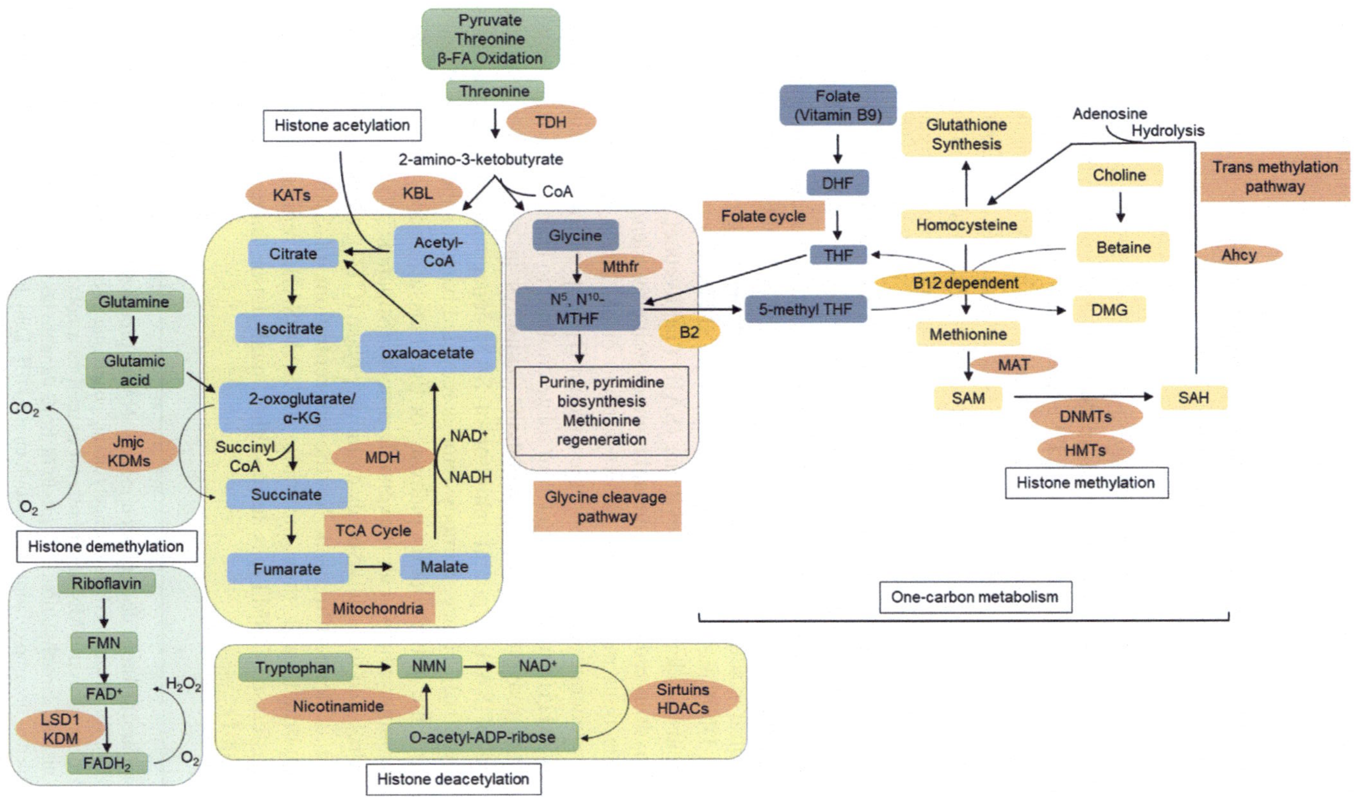

Fig. 8.1 Intermediary metabolites of metabolic pathways involved in epigenetic modifications. Metabolites participating in important biochemical pathways during cellular energy balance regulating chromatin modifications are shown in the schematic below. The TCA cycle is the central pathway that connects other pathways, and acetyl-CoA generated by β-oxidation of fatty acids (FA), glycolysis, and threonine catabolism, along with KATs, is used in histone acetylation. Micronutrient folate (vitamin B9), in one-carbon metabolism, generates SAM via methionine, which serves as a methyl donor for DNA and histone methylation. This methionine is again recycled back in the pathway from SAH (S-adenosyl homocysteine) by utilizing 5-methyl tetrahydrofolate (THF) and homocysteine.

we will be discussing the role of nutrient-derived metabolites that are essential for maintaining each of these biochemical processes and cellular plasticity.

S-Adenosyl Methionine (SAM) and the Methyl Cycle: Methylation of Histones and DNA

The methionine cycle or activated methyl cycle is centered around sulfur amino acids, methionine, and cysteine metabolism. Serving as the precursor, the essential amino acid methionine in presence of ATP is converted to S-adenosyl methionine (SAM), which acts as the main methyl donor in cells in one-carbon metabolism required for methylation of DNA and histone lysine residues (Zeisel 2006; Ly et al. 2012). Another source, a vitamin B_9 (folate) derivative, tetrahydrofolate (THF), combines with glycine, acts as a methyl group donor by being methylated on the N-5 atom (N^5-MTHF), and generates methionine (the precursor of SAM). This dependency on micronutrients and folate in one-carbon metabolism reveals the direct link between nutrition and methylation (Anderson et al. 2012). The positive charge on an adjacent sulfur atom in the S-adenosylmethionine methyl group renders the methyl group of SAM more reactive than N^5-MTHF, whose methyl group transfer potential is sufficiently low. SAM donates its methyl group to the acceptor molecule via DNA methyltransferase family of (DNMTs) enzyme catalyzed reactions that yields S-adenosylhomocysteine (SAH) (Tehlivets et al. 2013), which is then further hydrolyzed to adenosine and homocysteine by adenosylhomocysteinase (Ahcy), in the trans-methylation pathway (Mato et al. 2002). This cycle is then reversed to regenerate methionine in the vitamin-B_{12}-dependent reaction by transferring the N^5-MTHF methyl group (from folate/choline) to homocysteine. If homocysteine is not removed by other metabolic pathways in cells, it leads to the accumulation of SAH, a potent inhibitor of SAM-dependent methyltransferases (Lu and Mato 2012). Hence, homocysteine is recycled back for the regeneration of methionine via the remethylation pathway or the transsulfuration pathway to produce cystathionine, which is an irreversible process (Stipanuk 2004). Cystathionine is further hydrolyzed to cysteine by cystathione γ lyase (CTH), which constitutes the final reaction of transsulfuration. So the concentrations of SAM and SAH are critically important, and thus the ratio of SAM/SAH, regulated by glycine N-methyltransferase (GNMT), is termed the "methylation index" (Rowling and Schalinske 2001).

Fig. 8.1 (continued) Histone deacetylation requires NAD^+, produced from nicotinamide mononucleotide (NMN), for sirtuin-mediated deacetylation, generating O-acetyl-ADP ribose and nicotinamide. FAD derived from FMN and riboflavin is used as a cofactor by LSD demethylases for carrying out histone demethylation. The group of JmjC demethylases is involved in demethylation reaction coupled to decarboxylation of 2-oxoglutarate. DHF-Dihydrofolate, N^5, N^{10}-MTHF- 5, 10, methylene THF

Folate derivative tetrahydrofolate is generated by threonine catabolism in mitochondria, which produces acetyl-CoA and glycine by the action of threonine dehydrogenase (TDH), yielding intermediate molecule 2-amino-3-ketobutyrate. This short-lived intermediate, supplemented by CoA and 2-amino-3-ketobutyrate ligase, yields acetyl-CoA and glycine as the final products. Out of these, acetyl-CoA is either fed into the TCA cycle or acts as a building block for metabolites, while glycine is used for the conversion of tetrahydrofolate to N^5-MTHF, which acts as a one-carbon donor in biosynthetic reactions (nucleotide synthesis, regeneration of methionine). As N^5-MTHF is required for methionine regeneration, TDH-mediated threonine catabolism is important for maintaining intracellular levels of SAM as it may get affected if the cells are deprived of threonine (Shyh-Chang et al. 2013). Also, in mouse embryonic stem cells (ESCs), it is observed that there is a delayed deposition of H3K4me2/me3 marks on histone tails due to threonine restriction, and this epigenetic response was selective since threonine deficiency did not affect the deposition of H3K4me1, H3K27me3, H3K36me3, H3K9me3, and H3K79me3 marks on the chromatin (Shyh-Chang et al. 2013). It can be further stated that threonine catabolism generating both acetyl-CoA and SAM are necessary for histone acetylation and methylation respectively and the enzyme threonine dehydrogenase controls these two unique properties of cells.

DNA is methylated at the C5 position of cytosine in CpG dinucleotides, which requires the high-energy methyl donor SAM (Bird 2002). Both histone and DNA methylation and demethylation are carried out by several enzymes, known as methyltransferases and demethylases. Methionine adenosyltransferase enzymes (MAT) catalyze the conversion of methionine to SAM in an ATP-dependent reaction (Sakata et al. 1993). *MAT1A*, *MAT2A*, and *MAT2B* are the genes involved in AdoMet (SAM) synthesis, and MAT IIα, one of the MAT isoforms, associates with MafK, a bZip transcription factor that either activates or represses gene expression depending on its heterodimeric partner. It is shown that the MafK transcription factor associates with MAT IIα methyl transferase enzyme and directly recruits it to target genes for transcriptional repression (Katoh et al. 2011). As the synthesis of SAM requires methionine, choline, betaine, ATP, and micronutrients Mg2+ and K+, the generation of the methyl group as donor depends on the ingestion of these substances. So these reports might indicate the production of SAM in dictating the epigenetic state of cells.

Flavin Adenine Dinucleotide (FAD), 2-Oxoglutarate-Dependent Demethylases

Vitamin B2 or riboflavin generates riboflavin 5-phosphate when phosphorylated by riboflavin kinase, which is further converted to flavin adenine dinucleotide (FAD) by FAD synthetase in mitochondria. Yang Shi et al. in 2004 showed that KDM1A/LSD1 is a FAD-dependent demethylase that demethylates H3K4-methylated lysine residues (LSD), both in vitro and in vivo, establishing histone methylation as a reversible process (Shi et al. 2004). FAD acts as a cofactor for LSD demethylases to oxidize mono- and dimethylated histones as the enzyme needs a protonated lysine epsilon amino group. Histone demethylation reaction by LSD1 produces reactive oxygen species that may react with other macromolecules or adjacent DNAs and affect the transcription of genes (Perillo et al. 2008). Due to the nuclear localization of LSD1, it is sensitive to changes in the availability of FAD, which arises from other flavin-linked oxidases and dehydrogenases, including the TCA cycle and fatty acid oxidation (Hino et al. 2012). LSD1 directly binds to and represses the transcription of *PPARγ coactivator-1 α*, *FATP1*, *PDK4*, and *ATGL* genes associated with H3K4 demethylation (Hino et al. 2012).

2-oxoglutarate-dependent dioxygenases are chromatin-modifying enzymes, such as TET proteins (ten-eleven translocation) and JmjC-domain-containing histone demethylases, which also require Fe (II) and oxygen for their activity (Loenarz and Schofield 2011). Unlike LSD demethylases, JmjC histone demethylases can remove methyl groups from trimethylated arginines and lysines. 2-oxoglutarate/α-ketoglutarate (α-KG) can be produced from different sources, such as from isocitrate, by isocitrate dehydrogenase 2/3 in the mitochondrial TCA cycle. Apart from isocitrate, α-KG can also be derived from amino acids, glutamine, arginine, histidine, and proline, which are converted into glutamic acid and then transaminated to generate 2-oxoglutarate/α-KG. The 2-oxoglutarate produced from the TCA cycle through the action of the 2-oxoglutarate dehydrogenase complex gets converted into succinyl CoA and is also used for generating acetyl-CoA and pyruvate from amino acids lysine and cysteine, respectively. The JmjC group of histone demethylases have preferences for different histone methylation marks, such as the lysine demethylase, KDM5A which recognizes and exhibits demethylation of methylated H3K4 specifically. TET proteins Fe (II) and α-KG-dependent demethylases catalyze the oxidation of 5-methylcytosine (5mC) from methylated DNA (Cortellino et al. 2011). Certain 2-oxoglutarate-dependent dioxygenases, for example, EglN prolyl hydroxylases, are sensitive to fluctuations in oxygen availability, but the K_m values of oxygen are not known to TET proteins and JmjC histone methylases (Hirsila et al. 2003). However, indirectly it has been suggested that some of the JmjC demethylases can be O_2 sensitive, particularly mRNAs for several JmjC demethylase genes, such as JMJD1A, JMJD2B and JARIDC, responsible for H3K9 and H3K4 demethylation respectively are induced by transcription factor HIF and hypoxia (Chervona and Costa 2012). Further, it is not very clear if TET proteins can sense α-KG levels, but succinate and fumarate (TCA cycle intermediates) are known to inhibit TET, indicating that TET activity might be regulated by the concentration of these intermediates Xiao et al. 2012).

In a recent report, it was shown that from the tryptophan pathway, three indirect metabolites, i.e., kynurenine, 3-OH-kynurenine, and anthranilate, function as epigenetic modulators by altering the expression of hypothalamic function genes. Nutritional conditions and dietary supplementation of these metabolites resulted in elevated H3K4me3 and H2AS40 O-Glc-NAcylation levels, contributing to upregulated gene expression (Hayakawa et al. 2019). In another context, under methionine restriction conditions in mouse liver, bulk levels of H3K4me3 are globally altered due to changes in SAM concentrations (Dai et al. 2018).

Acetyl-CoA and Histone Acetylation

Ingested nutrients entering catabolic pathways produce intermediary metabolites, such as acetyl-CoA, which acts as the universal acetyl donor for acetylation used by histone acetyltransferase enzymes in eukaryotes. Acetyl-CoA is generated through the conversion of pyruvate by the mitochondrial pyruvate dehydrogenase complex through either the β-oxidation of fatty acids or the oxidation of glucose. During nutrient-deficit conditions or fasting, there is a switch to fatty acid oxidation after the depletion of glycogen stores, generating acetyl-CoA, which then fuels the TCA cycle, oxidative phosphorylation, and ATP synthesis. On the contrary, in diet-sufficient conditions, the basic energy needs of the cell are satisfied by the primary pathway for acetyl-CoA generation, i.e., glycolysis. An excess amount of acetyl-CoA generated from the pathway is exported to the cytosol from mitochondria as citrate. In the cytosol, when citrate combines with ATP and CoA, acetyl-CoA is regenerated, along with oxaloacetate, by the enzyme ATP-citrate lyase, which is required as a carbon donor for anabolic processes, such as amino acid and cholesterol synthesis and lipogenesis (Wellen and Thompson 2012). As mentioned in the earlier section, acetyl-CoA can be generated from amino acid threonine by mitochondrial TDH enzyme, and from acetate, CoA and ATP is generated by acetyl CoA synthase, along with reductive carboxylation of α-ketoglutarate. This phenomenon of anabolic or catabolic flux during the transition from feeding to fasting leads to changes in acetyl-CoA levels in mice, wherein fasting leads to increased lysine acetylation of proteins (Hebert et al. 2013).

Studies in different systems have shown that there is significant fluctuation in acetyl-CoA levels in cells. In nutrient-limiting conditions, *Saccharomyces cerevisiae* enters a robust metabolic cycle, where it requires a 4–5-h cycle of oxidative and reductive metabolism to stockpile energetically important building blocks for the oxidative phase before entering the division or reductive phase (Tu et al. 2007). There is fluctuation in acetyl-CoA levels due to the yeast metabolic cycle (YMC) between the oxidative and reductive phases, which affects the enzymatic activities of multiple lysine acetyltransferases (Berndsen and Denu 2008). At the end of the oxidative phase, the expression of genes in their growth phase depends on acetyl-CoA levels. When the direct source of acetyl-CoA, i.e., glucose, is absent, acetyl-CoA is generated through the conversion of ethanol to acetate; that is, it is further activated by acetyl-CoA synthetase. This leads to increased acetyl-CoA levels, which correlates with the upregulation of growth-related genes, along with genes encoding amino acid synthesis, ribosomal RNA and protein synthesis that are associated with histone hyperacetylation (Cai et al. 2011). The Gcn5 KAT enzyme of the SAGA complex is responsible for the acetylation of growth genes, and its deficiency abrogates H3 and H4 hyperacetylation, along with the expression of growth-related genes. Thus, in yeast, this mechanism of histone acetylation and transcriptional regulation is observed, where acetyl-CoA acts as a driver metabolite for cellular responses.

High glucose concentrations in cells increase mitochondrial acetyl-CoA levels through increased glycolysis and pyruvate generation, and deleting ATP citrate lyase decreases histone acetylation by inhibiting the conversion of citrate to acetyl-CoA. The ATP citrate lyase deficiency, does not affect acetylation of histones as it is substituted by increased levels of acetate, suggesting histone acetylation mainly depends on cytosolic pools acetyl-CoA derived from glucose (Sivanand et al. 2018).

NAD+ and Histone Deacetylation

As histone acetylation is regulated by acetyl-CoA levels, deacetylation, the reversible process, is also controlled by intracellular energy levels. Among histone deacetylases, sirtuins act as metabolic sensors, sensing the energy carrier NAD^+. Nicotinamide adenine dinucleotide (NAD) derives its nicotinate moiety from tryptophan and yields nicotinate ribonucleotide and inorganic pyrophosphate by combining with 5-phosphoribosyl-1-pyrophosphate (PRPP). AMP moiety is then transferred from ATP to nicotinate ribonucleotide to form desamido-NAD. The final step involves the transfer of ammonia produced from the amide group of glutamine to the carboxyl group of nicotinate, synthesizing NAD. A phosphorylated derivative of NAD, i.e., NADP, is generated via the NAD kinase enzyme through the transfer of a phosphoryl group from ATP to the 2′ hydroxyl group. $NADP^+$ and NAD^+ act as electron carriers by accepting electrons and getting converted into reduced forms. The intracellular NAD^+/NADH ratio is 100:1, suggesting that NAD^+ primarily functions as an electron acceptor for burning hydrocarbons.

Fluctuations in NAD^+ levels due to the nutritional state of cells affect the activity of histone deacetylase enzymes. Histone deacetylases act either as NAD^+-independent enzymes, producing deacetylated substrate and free acetate, or as NAD^+-dependent enzymes yielding, deacetylated substrate, O-acetyl-ADP-ribose and nicotinamide (Denu 2005; Feldman et al. 2012). Out of these two categories, the latter enzymes are members of the sirtuin family of histone deacetylases. The acetyl-CoA synthase enzyme (AceCS2), in an appropriate nutritional condition, is acetylated on lysine residue, which inhibits the conversion of acetate to acetyl-CoA by the enzyme, a mechanism conserved from bacteria to mammals (Hirschey et al. 2011). This process is reversed through the SIRT3-mediated deacetylation of acetyl-CoA synthase. Transcriptional coactivator PGC1α is involved in the regulation of genes that control intermediary metabolism, and SIRT1 is known to deacetylate PGC1α (Lerin et al. 2006). The acetylated form of PGC1α is inactive, whereas SIRT1-mediated deacetylation reactivates it, which shows that in both cases of AceCS2 and PGC1α, the availability of ample nutrients inhibits the activities of these proteins through acetylation (Hirschey et al. 2011). There are reports to support the fact that the expression of sirtuins is adaptable to the metabolic state of the cell. Nutrient-limiting conditions have been linked to rising NAD^+ levels, thereby leading to sirtuin activation. Alternatively, studies in yeast have shown that NAD^+ or nicotinamide levels do not change during calorie restriction, which can be associated with the activation of sirtuins (Evans et al. 2010). Hence, it is still unclear if the activity of sirtuins can be linked to changes in the intracellular levels of NAD^+; however, it is well documented that intracellular levels of acetyl-CoA influence protein acetylation.

Butyrate, a short-chain fatty acid produced from dietary fibers in the lumen of the colon, acts as a potential inhibitor of many HDACs (IC_{50}–90 μM in HT–29 cells) (Waldecker et al. 2008). It gets rapidly oxidized to acetyl-CoA in colonocytes, where it also serves as the main energy source, and there are findings to show that a fiber-rich diet reduces the chances of colon cancer and colitis in humans (Roediger 1982; Kim and Milner 2007). Increased butyrate levels (5 mM) occurring at the proximal colon epithelial wall inhibits class I/II HDAC enzymatic activity and consequently increase H3 lysine acetylation. This initiates a transcriptional regulation program that reduces proliferation. But the colon crypt cells are exposed to much lower levels of butyrate (tenfold low), which has no significant effect on the activity of HDACs in vitro (Donohoe et al. 2012). The citrate shuttle removes butyrate-derived acetyl-CoA from the mitochondria and induces ATP-citrate-lyase-dependent histone acetylation, which targets the subset of genes unrelated to high butyrate concentrations and involves genes that act as key regulators of cellular proliferation. But it is still unknown whether a diet rich in fiber results in elevated levels of butyrate than normal levels and has an anticancer effect due to the inhibition of HDACs. As an alternative energy source in low glucose or fasting conditions, ketone bodies are generated in the liver, which are consumed and reabsorbed as a carbon source by the neurons for ATP production (Maalouf et al. 2009). The ketone body β-hydroxybutyrate (β-OHB), which is similar to butyrate and associated with prolonged fasting, and calorie restriction diets also serves as an HDAC inhibitor (Maalouf et al. 2009). In one instance, β-OHB inhibits HDAC 1,3 and 4, thereby increasing the acetylation of H3K9 and K14 lysine residues and establishing chromatin arrangements for the expression of oxidative damage response key regulators in the kidney, which includes Foxo3a transcription factor and metallothionein Mt2 (Shimazu et al. 2013). This illustrates the significance of a glucose sparing energy metabolite source which can also regulate transcriptional program by acting as a signaling molecule associated with fasting.

Altered epigenetic profile is closely associated with several diseased states, such as diabetes, cancer, obesity, neurodegeneration, inflammation, dyslipidemia, and hypertension. Environmental changes involving altered nutritional states can affect epigenetics and, hence, diseases regulated by epigenetic changes. Table 8.2 gives us a brief idea about diseased state of an individual due to alterations in epigenetic modifications as a result of nutritional availability. One-carbon metabolism, which has a role in DNA and histone methylation, is influenced by diet, where the availability of nutrients (methionine, folate, choline, betaine, and vitamin B class) during pregnancy can induce epigenetically reprogrammed phenotypes in the progeny (Dolinoy et al. 2007; Sinclair et al. 2007). Again, biomarkers associated with folate deficiency (one-carbon metabolism) have been related to an increased risk of cancer (Song et al. 2012). In the upcoming section, we will basically focus on these changes in nutritional state and its associated disease phenotypes.

Table 8.2 Diet-mediated epigenetic regulation leading to diseased alterations

Diet	Epigenetic mechanism	Genes affected	Consequences	References
High-fat diet	DNA methylation	*Fgf21*, *Scd1* *Ankrd26*, *Hoxa5*	• Maternal hyperglycemia, hyperinsulinemia, hyperleptinemia	Milagro et al. (2013), Raciti et al. (2017), Yuan et al. (2018)
	Hypermethylation	*PPARG*, *HAND2*, *HOXC6*, *SORBS2*	• Reduced adipogenesis • Impaired triglyceride uptake • Insulin sensitivity	Keller et al. (2017), Perfilyev et al. (2017)
Low-protein diet	DNA methylation	*Ankrd26*, *Hoxa5*	• Altered regulation of energy balance • Increased gluconeogenesis and peroxisomal fatty acid oxidation • Alteration in cell cycle proliferation and developmental process	Bellinger et al. (2004), Lillycrop et al. (2007), Lillycrop et al. (2008), Breton et al. (2009), Coupé et al. (2009), Coupé et al. (2010), Jousse et al. (2011)
Bioactive compounds	H3 hypoacetylation	HDAC1	• Blocked expression of inflammatory modulators	Vahid et al. (2015), Liu et al. (2016)
	H3 hyperacetylation	MMP-2, TGF-β, TIMP1	• Alterations in the expression of cancer determinants	Yuan et al. (2018)
Calorie restriction	DNA methylation	IGF2/H19 PCK1 IGF2, INSIGF2, GNASAS1, MEG3, IL-10, LEP WT1, TNF-α PPARGC1A ADIPOQ, LEP	• Schizophrenia and coronary heart disease, hypercholesterolemia, hypertension • Alteration of protein abundance in the fetus	Heijmans et al. (2008), Nijland et al. (2010), Zhang et al. (2010), Milagro et al. (2013), Hjort et al. (2017)

Metabolic Regulation of Disease Through Epigenetic Route

Fetal Reprogramming

Fetal reprogramming refers to the adaptations of developing fetus to the changes reconditioned within the uterus, particularly from middle to terminal pregnancy period resulting in physical, emotional and psychological implications. These implications might have a probability of lifelong imprint. These imprints are often due to alteration of epigenetic modifications in response to environmental conditions, like nutritional status and the lifestyle of parents, which in turn can increase the risk of disease in the offspring. For example, an altered DNA methylation pattern in an obese parent due to consumption of a high-fat diet can induce cancer in the offspring (Tokunaga et al. 2013). The optimum nutritional status of an offspring depends on nutrient uptake by the mother during the developmental stages of the fetus as well as the nutritional environment to which the sperm of the father is exposed just before and during conception. Also, paternal and maternal exposures to a diet rich in high protein, fat, carbohydrate, sugar, fiber, and other dietary supplements have a substantial role in it (McCarty et al. 2009). Furthermore, the seasonal availability of food rich in dietary supplements like vitamin D, iron, etc. have a rejuvenating impact on the fetal immune function by reprogramming its epigenome. A healthy immune system helps in strong defense mechanism against diseases like cancer, CVD and autoimmune disorders. Any mismatch during fetal adjustments to maternal homeostasis and the postnatal environment might lead to abnormal heritable alterations that can cause disadvantageous effects on fetal development.

To get an in-depth insight into fetal reprogramming, we need to understand "One-carbon metabolism." It is a biochemical link between the parental environment and the epigenetic modifications bestowed upon the fetus due to the interchange of methyl groups (1C) mediated by nutrients and dietary supplements, which act as both methyl donors and acceptors in this process. Deficiency in the one carbon metabolites leading to poor health condition of the fetus, by genetic and/or non-genetic intergeneration transmission is due to probable variation during the exchange of methyl groups. Dietary methyl donors, like vitamin B-12, folic acid, zinc, and choline, when supplemented in appropriate proportions at the right stage of fetal development have beneficial long-term consequences, like the maintenance of DNA methylation patterns (Fenech and Ferguson 2001), prevention of anti-inflammatory responses (McCarty et al. 2009), proper development of fetal brain and liver (Davison et al. 2009), and promotion of intrauterine growth. The regulation of DNA methylation mediated by vitamin D availability at the maternal-fetal interface and its metabolism within the placenta itself has established a feedback loop of interaction between vitamin D and its receptor, CYP27B1, leading to the transcription of its responsive genes. Any change in this entire process is known to have obstetrical syndromes. A deficiency in LC-PUFAs (low-carbon polyunsaturated fatty acids), one of the major methyl group acceptors, can lead to mutations in angiogenic factor genes, vascular dysregulation, and abnormal placentation due to aberrant DNA methylation, which might increase the risk of cardiovascular diseases (Khot et al. 2017). Figure 8.2 explains the link between one-carbon metabolites and epigenetic reprogramming. However, fish oil supplements rich in ω-3 fatty acids have beneficial effects, like lowered blood lipid concentrations consistent over two generations due to global DNA hypomethylation. During the gestational period, if the mother is confronted with a scarcity of nutrients (undernutrition), it can lead to intrauterine growth restriction. Maternal obesity due to the abundance of nutrients (overnutrition), on the other hand, can impose detrimental effects on the developing fetus rather than prove to be beneficial. These effects can lead to long-term risks of diseases, like obesity, metabolic dysregulation, chronic inflammation, etc. through

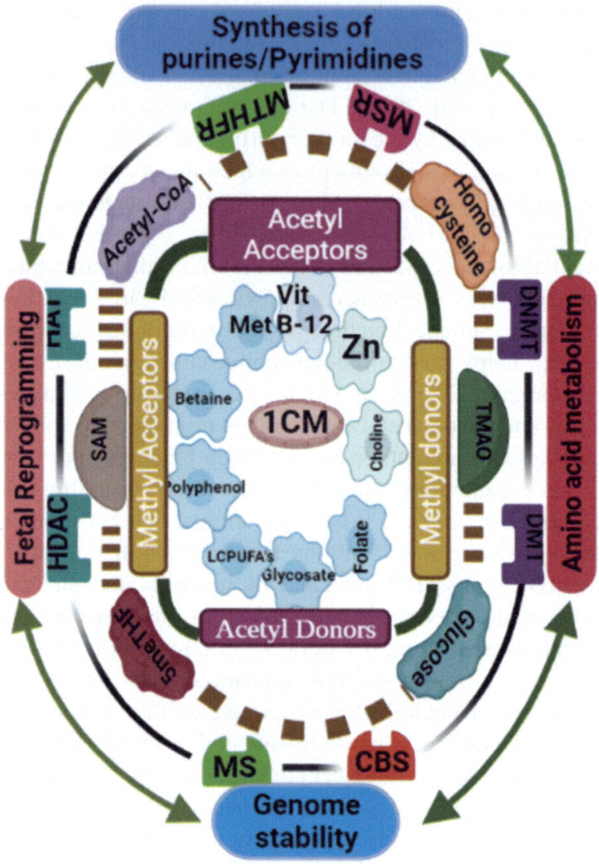

Fig. 8.2 Epigenetic reprogramming by nutrient-induced one-carbon metabolism. Metabolites of one-carbon metabolism, like vitamin B-12, betaine, choline, etc., act as acetyl group and methyl group donors and acceptors. SAM, acetyl-CoA, TMAO, etc. serve as cofactors for the interaction between these donor-acceptor groups and constituent enzymes for one-carbon metabolism, like MS (methyl synthase), CBS (cystathionine synthase), etc. The possible outcomes of the remodeling of the enzyme-substrate reaction due to the opportune availability of appropriate nutrients include genome stability, fetal reprogramming, biosynthesis of purines and pyrimidines, etc.

the regulation of certain epigenetic modifications, like H3K4 methylation (Park et al. 2008; Tosh et al. 2010), Sirt1 deacetylation (Suter et al. 2012), and IGF1 acetylation in the offspring (Fu et al. 2004). These modifications are known to have been transmitted to the subsequent generation through meiosis via epigenetic reprogramming.

Previous studies depict a correlation between one in five metabolite concentrations with a change in the methylation patterns of five out of 993 CpG islands (Knight et al. 2018). For example, 5meTHF (5-methyl-tetrahydrofolate) concentration in cord blood was found to be associated with a methylation level of MAGUL scaffold protein 2(DLG2), which is involved in neuronal development, inflammation, and hypoxia (Mazzanti et al. 2015). The metabolite concentration of maternal homocysteine, TMAO (trimethylamine N-oxide), SAM (S-adenosylmethionine), and 5meTHF was proportional to their concentrations in cord blood, implying the reflection of the former's one-carbon metabolism in the latter as a possible result of placental transfer. Furthermore, 19 out of 28 DNA methylation sites were common in maternal and fetal blood. This establishes a link between the metabolite concentration of maternal and cord blood and their relationship with DNA methylation. A balanced interplay between one-carbon metabolites and DNA methylation patterns encouraged by the right proportion of dietary folate and folic acid intake has been known to reduce any adverse effects in the developing fetus, like abnormal brain tissue development, including neural tube defects (Jacob et al. 1998; Fekete et al. 2010; Chang et al. 2011; Guéant et al. 2013; Padmanabhan and Watson 2013; Joubert et al. 2016). However, overlapping associations between the same metabolite with different CpG methylation sites in maternal and cord blood have brought the possibility of DNA methylation being regulated by multiple one-carbon metabolites (or) by the same metabolite at multiple sites.

In conclusion, efficient epigenetic reprogramming is required for proper fetal development. This can be done by quantitative determination of threshold concentration of nutrients and dietary supplements during pregnancy. This would lessen the noxious and deleterious effects on both maternal and fetal health thereby reducing the susceptibility to diseased condition.

Cardiovascular Disease (CVD)

Diet mediated reversible modification of the epigenome resulting in alterations of acetylation, methylation, phosphorylation and noncoding RNA regulation in the aging vascular cells wave a possible contributing factor toward the onset of CVD. Atherosclerosis, endothelial dysfunction, thrombosis, vascular inflammation, and fibrosis with several impaired pathways are the risk factors for CVD (1a-3). Disease progression of CVD is known to be regulated by both hypomethylation of vascular muscle cells, long interspersed nuclear element (LINE-1), CpG sites in the genes of Wnt signaling pathway (Albanese et al. 2018; Findeisen et al. 2013; Muka et al. 2016) and hypermethylation of genes involved in one-carbon metabolism, ATP production and homocysteine metabolism respectively (Tabas 2017). Genome-wide methylation studies have shown that overfeeding by high-fat diet has led to hypermethylation in the promoter regions of the genes responsible for adipocyte differentiation and lipid metabolism, thereby hindering reduced adipogenesis and impaired triglyceride uptake (Keller et al. 2016; Perfilyev et al. 2017), thus increasing the risk of CVD. In contrast to this, starvation by calorie restriction, leading to hypomethylation at the transcription start site of the SIRT1 gene, has been found to reduce the risk of CVD due to its possible effect on weight and adiposity (Weiss et al. 2016; Zullo et al. 2018). The regulation of the function of KATs (lysine acetyltransferase) and KDACs (lysine deacetylase) through the intake of dietary methyl group donors (folate, methionine, vitamin B-12) and bioactive compounds (polyphenols, curcuminoids, stilbenes) has significant effects on potential CVD determinants, like mitochondrial dysfunction, myocardial fibrosis, vascular aging, and inflammation. Nutritional intervention, like a Mediterranean diet, can alter the abnormal upregulation of host-derived miR-122 due to the high-fat diet preventing cardiovascular risks (Lombardi et al. 2020). Focusing on the dietary pattern of an individual including changes in the lifestyle and mood rather than narrowing down to the nutrients might help us have a better understanding about the interplay between nutrigenomics, epigenomics and their joint modulation of CVD leading to the design of personalized therapy.

Obesity and Type 2 Diabetes

Evidences from literature studies depict low heritability of obesity (2%) and type 2 diabetes (5–10%) despite of contrasting high disease prevalence. This suggests the probable role of epigenetic changes due to lifestyle modification especially differential nutritional status of an individual. DNA methylation, which is involved in mammalian development, gene silencing leading to protection from spurious activity inside the cell and genome stability during meiosis (Li 2018) is regulated during cellular duplication (by DNMT1) and early embryonic stage (by DNMT2 and DNMT3). Synthesis of SAM, a potential methyl donor in the process of DNA methylation, is mediated by dietary supplements like methionine, folate, choline, betaine and B2, B6, B12 vitamins, hence the establishment of methylome is vulnerable to the nutritional factors having a direct/reciprocal impact on human's health. Availability of methyl donors like folic acid, vitamin B-12, and choline in the maternal diet of Agouti mouse has been found to have suppressed the aberrant overexpression of agouti related neuropeptide gene (Agrp) thus preventing obesity and diabetes (Cropley et al. 2006). The carryover of this protective signature to the subsequent two generations validates the germline transmission of epigenetic modification subjected to changes in the nutritional environment.

Suboptimal/complete lack of protein in maternal-fraternal diet have known to cause hypermethylation of pancreatic β-cell function, hepatocyte nuclear-4-alpha (Hnf4α) (Sandovici et al. 2011) and DNA-sperm hypomethylation respectively, resulting in adiposity in offsprings and metabolic dysfunction like insulin resistance and T2D in adults. Global calorie restriction mediated alteration in DNA methylation i.e., hypermethylation at IGF2/H19 locus (Perkins et al. 2012) and hypomethylation at PCK1 promoter has led to negative regulation of insulin like growth factor-2 and gluconeogenic enzyme PCK1 in the liver. DNA methylation influencing the risk factors of obesity and T2D due to restricted calorie intake is dependent on different aspects like the amount and duration of restriction, the rate of response to restriction and the new-born weight. For example, exposing low birth weight (LBW) and normal birth weight (NBW) individuals under a 36-hour fasting period had hypermethylation of PPARGC1A (Peroxisome proliferator activated receptor gamma coactivator 1-alpha) and adiponectin (ADIPOQ) in both, but no change in the expression of leptin (LEP) in the former as compared to the later, provided all the three genes are involved in glucose and fatty acid metabolism in liver and brain (Jørgensen et al. 2015; Hjort et al. 2017). This differential response to fasting showed that the potential risk of being affected by metabolic dysfunctional disease due to altered epigenetic modifications is higher in LBW than in NBW due to the former's less flexible nature in adapting to the available nutritional environment. The high-fat dietary content of breast milk has proved to change the HPUFA-LCMFA (high polyunsaturated fatty acid–low-medium chain fatty acid) ratio, thereby altering the DNA methylation of enzymes involved in fatty acid metabolism (stearoyl-CoA desaturase-1) and dopamine-opioid regulation in the liver and brain, respectively. The later would explain the behavioural preference towards consumption of palatable food items thus summing up toward adiposity (Vucetic et al. 2010).

Increased susceptibility to obesity and T2D due to alteration in the DNA methylation patterns of respective genes like HOX5 (adipose remodeling gene) and PPARGC1A (gene involved in fatty acid metabolism) can be reversed by replacing high fat diet with normal diet. Although, diet might lead to global hypermethylation but it's the constituents, that mediate the differential regulation of genes which might serve as an intervention in the treatment of such metabolic diseases.

Alzheimer's Disease

Very little is known about the regulation of neurodegenerative diseases through bioactive-food-mediated epigenetic modification. One of the most widely studied neurodegenerative disease is the Alzheimer's disease (AD). There is a high chance of discovering nutritional intervention strategies for the treatment and therapy of the same provided the already established knowledge about diet mediated regulation of the epigenome. Prominent past evidences project the alteration in DNA methylation and histone acetylation by nutrients such as folate, B vitamins, methionine, choline, and betaine along with the efficacy of non-nutrient phytochemicals like polyphenols, glycosylates in modification of epigenetic targets but their application to delay the degeneration of neurocognitive function seems to be still in its infant stage. Folate deficiency has led to decrease in the levels of a potent methyl donor SAM, thus promoting a gradual increase in hypomethylation thereby enhancing the risk factors of Aβ formation by regulating a cascade of alterations involving DNA, proteins, neurotransmitters, hormones, and phospholipids. These aftereffects can, however, be partially reduced through folic acid supplementation (Tchantchou et al. 2006; Mastroeni et al. 2011). Prominent natural products like epigallocatechin gallate (EGCG), caffeic acid, chlorogenic acid (Fang et al. 2003; Lee and Zhu 2006), soy derived isoflavones like genistein and daidzein and glucosinolates like phenethyl isothiocyanate have known to reverse the hypermethylation of certain genes contributing to the progression of AD like apolipoprotein E (APOE) and methylenetetrahydrofolate reductase (MTHFR) (Davinelli et al. 2014).

Apart from methylation, the interplay between KATs and KDACs, regulated by nutritional availability in the disease progression of AD seems to be interesting. A decline in histone acetylation and a rise in the levels of HDAC2 and HDAC6 in the cerebrocortical and hippocampus regions of AD patients (Ding et al. 2008; Gräff et al. 2012) project a positive correlation with the pathogenesis of amyloid-beta plaque formation, leading to memory loss and severe neurocognitive dysfunctions.

These aberrant deacetylation levels can be reversed by the action of Sirt1, an HDAC inhibitor that has proved efficient in neuronal protection by suppressing β-amyloid production and inhibiting the deacetylation of major transcription factors involved in stress tolerance. Dietary food supplements, like proanthocyanins, resveratrol, quercetin, fisetin, catechins, and isoflavones, serve as enhancers of SIRT1's function, thereby reducing Aβ toxicity and impairing its adverse effects on NF-kβ and MAPK signaling pathways, thus leading to the modulation of amyloid precursor protein (APP) processing with improved memory (Yokozawa et al. 2011; Vepsäläinen et al. 2013).

The deficiency/excessive adequacy of the above-mentioned bioactive compounds, especially folate and vitamin B, which function as cofactors during one-carbon metabolism, has led to polymorphism in enzymes regulating diet-mediated epigenetic mechanisms. Methylene tetrahydrofolate reductase (MTHFR), methionine synthase (MS), methionine synthase reductase (MSR), cystathionine synthase (CBS), and serine hydroxymethyltransferase (SHMT) are some of these enzymes involved in maintaining homeostasis between Hcy-SAM levels. This balance is essential for slowing down the disease progression of AD (Troesch et al. 2016). However, long cohort experiments covering a comparatively larger population should be performed to understand the correlation between neuronal pathophysiology and their nutrition-mediated epigenetic regulation.

Cancer

Cancer is characterized by abnormal cell mass growth, leading to the formation of a tumor. Methods to combat this disease include the downregulation of tumor suppression genes and alteration of the microenvironment within and around the same decelerating abrupt cell multiplication. Recent advances in the field of chemotherapy, radiation therapy, and hormonal therapy are found to be comparatively effective, but upcoming interventions, due to nutritional epigenetics, have been found to be more intriguing. These interventions majorly include alterations in DNA methylation and histone modification within the cancerous region. The inhibition of enzyme activity in the above-mentioned mechanisms by the dietary availability of nutrients and bioactive compounds is known to have been established by either altering the substrate concentration (or) by increasing the concentration of the reaction product that can form a negative feedback loop. For example, tea catechins and bioflavonoids are known to interfere with the substrate of catechol-O-methyltransferase (COMT), whereas epigallocatechin-3-gallate (EGCG) is known to inhibit methyltransferase activity by increasing S-adenosyl homocysteine (SAH) concentration in the same reaction (Stefanska et al. 2012). Adequate intake of nutrients, like folate, methionine, cobalamin, pyridoxine, and riboflavin, is recognized to display anticancer effects through hypermethylation at the promoter sites of p16, p15, Rb, p14, BRCA1, MGMT, estrogenic receptor (ER) alpha, progesterone receptor (PR), and E-cadherin (Singhal et al. 1987), whereas nutrient deficiency has led to DNA damage and abnormal methylation, resulting in genomic instability due to aberrant hypomethylation (Bishop and Ferguson 2015), thus increasing the risk of cancers like chronic lymphatic leukemia, breast and ovarian cancer, and liver and prostate cancer (Yang et al. 2001; Bishop and Ferguson 2015; Daniel and Tollefsbol 2015). Since one-carbon metabolism is associated with three or more interlinking pathways, the deficiency of one nutrient can be compensated by the supplementation of another. For example, the low intake of methionine, pyridoxine, cobalamin, niacin, and riboflavin imposes an increased risk of breast cancer, which can be partially compensated by folate supplementation, especially in estrogen-negative premenopausal women (Maruti et al. 2009; Shrubsole et al. 2011). Evidence of increased deacetylation in many cancers influencing gene expression and kinetics inside a cell points to disruption in the acetylation-deacetylation cycle mediated by KATs and KDACs, respectively. Certain bioactive compounds, like diallyl sulfide, parsley, thyme, peppermint, basil herb, celery, curcumin, etc., help in restoring this balance by inhibiting HDAC activity, thereby reducing the invasion, differentiation, and survival of cancer cells (Andreescu et al. 2018). Treatment of cancer mediated by nutritional epigenetics has been thought to be more effective as compared to other therapies due to their dosage-dependent cytotoxicity. Also, the dosage of nutrients supplemented for the treatment is thought to be much lesser than that of the already existing therapeutic treatments for acheiving the same level of progress in reduction of disease pathogenesis.

Past evidence shows that the cofactors required for the enzymes to carry out epigenetic modification mediated disease regulation are synthesized in the mitochondria, mostly as by-products of metabolic pathways, like the TCA cycle, fatty acid metabolism, etc. Inadequate availability of the former might lead to the dysregulation of the same. Figure 8.3 explains in detail the interaction between epigenetic reprogramming, mitochondrial dysfunction mediated substrate availability, and disease progression.

Conclusion

Epigenetic regulation of a healthy cell physiology and metabolism being influenced by the environmental stimulants is well understood. However, the most intriguing and exciting part is to study the effect of nutritional availability on the modulation of epigenetic modification mediated metabolic pathways. Diet serves as a source of nutrients, like vitamin B, folate, omega-3 fatty acids, etc. which act as precursor molecules for the synthesis of cofactors like glucose, acetyl CoA, SAM, Hcy, etc. in the mitochondria of the cell to facilitate important metabolic pathways. Inappropriate exposure of proper nutrition might hamper the availability of co-substrates like NAD, FAD, MTHF, etc. for epigenetic enzymes thus leading to altered epigenome. This could serve as the causative agent of the lethal after-effects of a diseased condition that may develop later in the future. An inadequate diet may cause either scarcity (or) over-availability of nutrients. This would lead to epigenetic alteration of

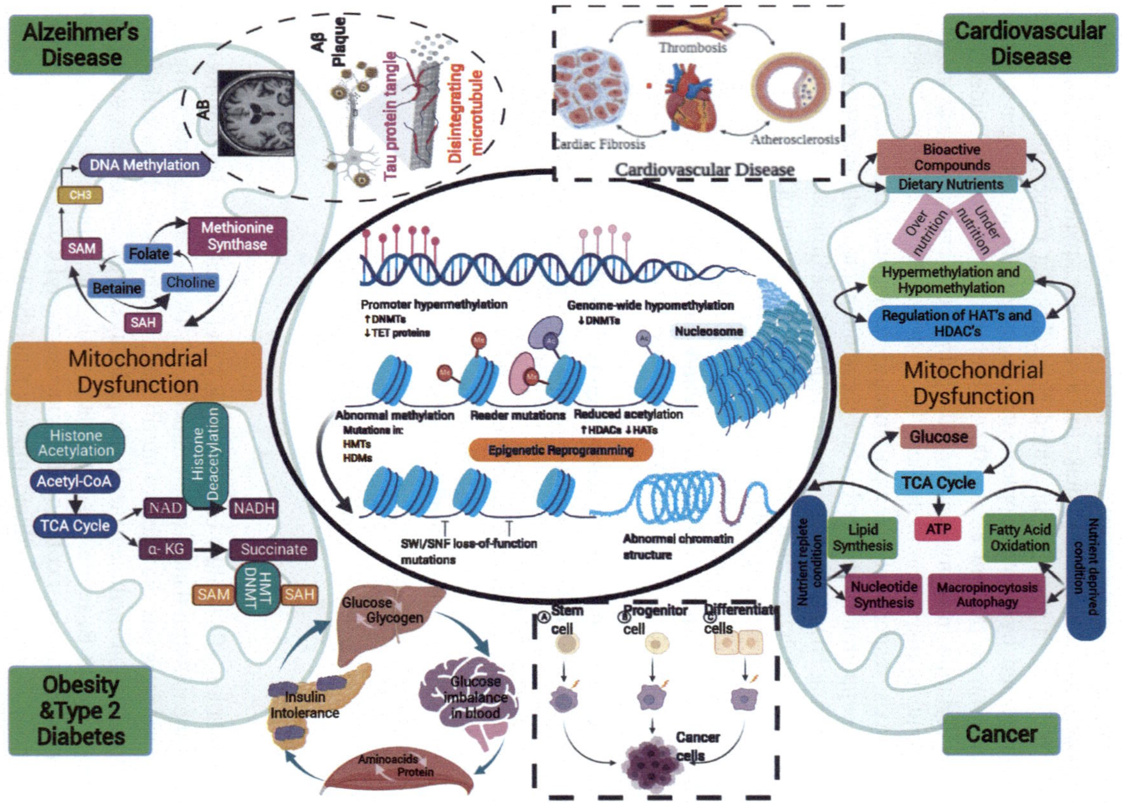

Fig. 8.3 Diet-mediated epigenetic regulation of disease. Mitochondrial dysfunction alters the availability of cofactors essential for metabolism-mediated epigenetic reprogramming, like acetyl-CoA, SAM, etc. This leads to disturbance in the regulation of epigenetic enzymes, like KATs, KDACs, HMTs (histone methyltransferases), and DNMTs (DNA methyltransferases). thereby leading to severe disease progression

biochemical processes like DNA methylation and histone modifications, which would have an adverse effect on cellular plasticity. Therefore, monitoring the time and amount of nutritional exposure to an individual is of utmost importance. However, mouse model experimentation has shown the efficacy of dietary supplementation and calorie restriction in reversing epigenetic modifications, thereby reducing the deleterious effects of nutritional scarcity and overnutrition, like abnormal disease progression in cancer and Alzheimer's. Further studies are required to understand the nutritional availability mediated imbalance in 1-carbon metabolites under variable conditions like different lifestyles, genetic backgrounds, and sex specific epigenetic and phenotypic transmissions. This would help us in understanding the correlation between nutrigenomics and epigenomics. In-depth scientific knowledge about nutritional epigenetics and its implications would help in disentanglement of genetic and epigenetic impact on complex traits. Furthermore, a better understanding of diet-mediated epigenetic reprogramming of a disease may serve as a biomarker for critical developmental stages of a disease.

Acknowledgment The research of the RM lab is funded by DAE-BRNS (Sanction No. 35/14/14/2015-BRNS/35024) and Dept. of Biotechnology (DBT) (BT/PR15263/MED/29/995/2015) Govt. of India. RM also thanks DBT, GoI, for the partial funding (BT/MED/30/SP19662/2018-reg). SB thanks the Indian Council for Medical Research (ICMR), Govt. of India, for senior research fellowship (F.No.45/07/2020-BIO/BMS).

Conflict of Interest The authors declare no competing interest.

References

Albanese I, Khan K, Barratt B, Al-Kindi H, Schwertani A (2018) Atherosclerotic calcification: Wnt is the hint. J Am Heart Assoc 7(4):e007356

Anderson OS, Sant KE, Dolinoy DC (2012) Nutrition and epigenetics: an interplay of dietary methyl donors, one-carbon metabolism and DNA methylation. J Nutr Biochem 23(8):853–859

Andreescu N, Puiu M, Niculescu M (2018) Effects of dietary nutrients on epigenetic changes in cancer. Cancer Epigenet Precis Med 2018:121–139

Bagul PK, Deepthi N, Sultana R, Banerjee SK (2015) Resveratrol ameliorates cardiac oxidative stress in diabetes through deacetylation of NFkB-p65 and histone 3. J Nutr Biochem 26(11): 1298–1307

Barile M, Giancaspero TA, Leone P, Galluccio M, Indiveri C (2016) Riboflavin transport and metabolism in humans. J Inherit Metab Dis 39(4):545–557

Bellinger L, Lilley C, Langley-Evans SC (2004) Prenatal exposure to a maternal low-protein diet programmes a preference for high-fat foods in the young adult rat. Br J Nutr 92(3):513–520

Berndsen CE, Denu JM (2008) Catalysis and substrate selection by histone/protein lysine acetyltransferases. Curr Opin Struct Biol 18(6):682–689

Bird A (2002) DNA methylation patterns and epigenetic memory. Genes Dev 16(1):6–21

Bishop KS, Ferguson LR (2015) The interaction between epigenetics, nutrition and the development of cancer. Nutrients 7(2):922–947

Breton C, Lukaszewski M-A, Risold P-Y, Enache M, Guillemot J, Riviere G, Delahaye F, Lesage J, Dutriez-Casteloot I, Laborie C (2009) Maternal prenatal undernutrition alters the response of

POMC neurons to energy status variation in adult male rat offspring. Am J Physiol Endocrinol Metab 296(3):E462–E472

Burdge GC, Lillycrop KA (2010) Nutrition, epigenetics, and developmental plasticity: implications for understanding human disease. Annu Rev Nutr 30:315–339

Cai L, Sutter BM, Li B, Tu BP (2011) Acetyl-CoA induces cell growth and proliferation by promoting the acetylation of histones at growth genes. Mol Cell 42(4):426–437

Chang H, Zhang T, Zhang Z, Bao R, Fu C, Wang Z, Bao Y, Li Y, Wu L, Zheng X (2011) Tissue-specific distribution of aberrant DNA methylation associated with maternal low-folate status in human neural tube defects. J Nutr Biochem 22(12):1172–1177

Chervona Y, Costa M (2012) The control of histone methylation and gene expression by oxidative stress, hypoxia, and metals. Free Radic Biol Med 53(5):1041–1047

Cortellino S, Xu J, Sannai M, Moore R, Caretti E, Cigliano A, Le Coz M, Devarajan K, Wessels A, Soprano D (2011) Thymine DNA glycosylase is essential for active DNA demethylation by linked deamination-base excision repair. Cell 146(1):67–79

Coupé B, Grit I, Darmaun D, Parnet P (2009) The timing of "catch-up growth" affects metabolism and appetite regulation in male rats born with intrauterine growth restriction. Am J Phys Regul Integr Comp Phys 297(3):R813–R824

Coupé B, Amarger V, Grit I, Benani A, Parnet P (2010) Nutritional programming affects hypothalamic organization and early response to leptin. Endocrinology 151(2):702–713

Cropley JE, Suter CM, Beckman KB, Martin DI (2006) Germ-line epigenetic modification of the murine Avy allele by nutritional supplementation. Proc Natl Acad Sci 103(46):17308–17312

Culhane JC, Cole PA (2007) LSD1 and the chemistry of histone demethylation. Curr Opin Chem Biol 11(5):561–568

Dai Z, Mentch SJ, Gao X, Nichenametla SN, Locasale JW (2018) Methionine metabolism influences genomic architecture and gene expression through H3K4me3 peak width. Nat Commun 9(1):1–12

Daniel M, Tollefsbol TO (2015) Epigenetic linkage of aging, cancer and nutrition. J Exp Biol 218(1):59–70

Davie JR (2003) Inhibition of histone deacetylase activity by butyrate. J Nutr 133(7):2485S–2493S

Davinelli S, Calabrese V, Zella D, Scapagnini G (2014) Epigenetic nutraceutical diets in Alzheimer's disease. J Nutr Health Aging 18(9):800–805

Davison JM, Mellott TJ, Kovacheva VP, Blusztajn JK (2009) Gestational choline supply regulates methylation of histone H3, expression of histone methyltransferases G9a (Kmt1c) and Suv39h1 (Kmt1a), and DNA methylation of their genes in rat fetal liver and brain. J Biol Chem 284(4): 1982–1989

Denu JM (2005) The Sir2 family of protein deacetylases. Curr Opin Chem Biol 9(5):431–440

Ding H, Dolan PJ, Johnson GV (2008) Histone deacetylase 6 interacts with the microtubule-associated protein tau. J Neurochem 106(5):2119–2130

Dolinoy DC, Huang D, Jirtle RL (2007) Maternal nutrient supplementation counteracts bisphenol A-induced DNA hypomethylation in early development. Proc Natl Acad Sci 104(32): 13056–13061

Donohoe DR, Collins LB, Wali A, Bigler R, Sun W, Bultman SJ (2012) The Warburg effect dictates the mechanism of butyrate-mediated histone acetylation and cell proliferation. Mol Cell 48(4):612–626

Evans C, Bogan KL, Song P, Burant CF, Kennedy RT, Brenner C (2010) NAD+ metabolite levels as a function of vitamins and calorie restriction: evidence for different mechanisms of longevity. BMC Chem Biol 10(1):1–10

Fang MZ, Wang Y, Ai N, Hou Z, Sun Y, Lu H, Welsh W, Yang CS (2003) Tea polyphenol (−)-epigallocatechin-3-gallate inhibits DNA methyltransferase and reactivates methylation-silenced genes in cancer cell lines. Cancer Res 63(22):7563–7570

Fekete K, Berti C, Cetin I, Hermoso M, Koletzko BV, Decsi T (2010) Perinatal folate supply: relevance in health outcome parameters. Matern Child Nutr 6:23–38

Feldman JL, Dittenhafer-Reed KE, Denu JM (2012) Sirtuin catalysis and regulation. J Biol Chem 287(51):42419–42427

Fenech M, Ferguson LR (2001) Vitamins/minerals and genomic stability in humans. Elsevier, New York

Findeisen HM, Kahles FK, Bruemmer D (2013) Epigenetic regulation of vascular smooth muscle cell function in atherosclerosis. Curr Atheroscler Rep 15(5):1–8

Fu Q, McKnight RA, Yu X, Wang L, Callaway CW, Lane RH (2004) Uteroplacental insufficiency induces site-specific changes in histone H3 covalent modifications and affects DNA-histone H3 positioning in day 0 IUGR rat liver. Physiol Genomics 20(1):108–116

Gräff J, Rei D, Guan J-S, Wang W-Y, Seo J, Hennig KM, Nieland TJ, Fass DM, Kao PF, Kahn M (2012) An epigenetic blockade of cognitive functions in the neurodegenerating brain. Nature 483(7388):222–226

Guéant J-L, Namour F, Guéant-Rodriguez R-M, Daval J-L (2013) Folate and fetal programming: a play in epigenomics? Trends Endocrinol Metab 24(6):279–289

Gut P, Verdin E (2013) The nexus of chromatin regulation and intermediary metabolism. Nature 502(7472):489–498

Hayakawa K, Nishitani K, Tanaka S (2019) Kynurenine, 3-OH-kynurenine, and anthranilate are nutrient metabolites that alter H3K4 trimethylation and H2AS40 O-GlcNAcylation at hypothalamus-related loci. Sci Rep 9(1):1–12

Hebert AS, Dittenhafer-Reed KE, Yu W, Bailey DJ, Selen ES, Boersma MD, Carson JJ, Tonelli M, Balloon AJ, Higbee AJ (2013) Calorie restriction and SIRT3 trigger global reprogramming of the mitochondrial protein acetylome. Mol Cell 49(1):186–199

Heijmans BT, Tobi EW, Stein AD, Putter H, Blauw GJ, Susser ES, Slagboom PE, Lumey L (2008) Persistent epigenetic differences associated with prenatal exposure to famine in humans. Proc Natl Acad Sci 105(44):17046–17049

Hino S, Sakamoto A, Nagaoka K, Anan K, Wang Y, Mimasu S, Umehara T, Yokoyama S, Kosai K-i, Nakao M (2012) FAD-dependent lysine-specific demethylase-1 regulates cellular energy expenditure. Nat Commun 3(1):1–12

Hirschey MD, Shimazu T, Goetzman E, Jing E, Schwer B, Lombard DB, Grueter CA, Harris C, Biddinger S, Ilkayeva OR (2010) SIRT3 regulates mitochondrial fatty-acid oxidation by reversible enzyme deacetylation. Nature 464(7285):121–125

Hirschey M, Shimazu T, Huang J-Y, Schwer B, Verdin E (2011) SIRT3 regulates mitochondrial protein acetylation and intermediary metabolism. Cold Spring Harbor symposia on quantitative biology. Cold Spring Harbor Laboratory Press, Cold Spring Harbor

Hirsila M, Koivunen P, Gunzler V, Kivirikko KI, Myllyharju J (2003) Characterization of the human prolyl 4-hydroxylases that modify the hypoxia-inducible factor. J Biol Chem 278(33): 30772–30780

Hjort L, Jørgensen SW, Gillberg L, Hall E, Brøns C, Frystyk J, Vaag AA, Ling C (2017) 36 h fasting of young men influences adipose tissue DNA methylation of LEP and ADIPOQ in a birth weight-dependent manner. Clin Epigenetics 9(1):1–12

Jacob RA, Gretz DM, Taylor PC, James SJ, Pogribny IP, Miller BJ, Henning SM, Swendseid ME (1998) Moderate folate depletion increases plasma homocysteine and decreases lymphocyte DNA methylation in postmenopausal women. J Nutr 128(7):1204–1212

Jang H, Serra C (2014) Nutrition, epigenetics, and diseases. Clin Nutr Res 3(1):1–8

Jørgensen SW, Bluck L, Hjort L, FÊrch K, Thankamony A, Gillberg L, Friedrichsen M, Dunger DB, Vaag AA (2015) Erratum to: metabolic response to 36 hours of fasting in Young men born small vs appropriate for gestational age. Diabetologia 58(1):204

Joubert BR, Herman T, Felix JF, Bohlin J, Ligthart S, Beckett E, Tiemeier H, Van Meurs JB, Uitterlinden AG, Hofman A (2016) Maternal plasma folate impacts differential DNA methylation in an epigenome-wide meta-analysis of newborns. Nat Commun 7(1):1–8

Jousse C, Parry L, Lambert-Langlais S, Maurin AC, Averous J, Bruhat A, Carraro V, Tost J, Letteron P, Chen P (2011) Perinatal undernutrition affects the methylation and expression of the leptin gene in adults: implication for the understanding of metabolic syndrome. FASEB J 25(9): 3271–3278

Kaelin WG Jr, McKnight SL (2013) Influence of metabolism on epigenetics and disease. Cell 153(1):56–69

Katoh Y, Ikura T, Hoshikawa Y, Tashiro S, Ito T, Ohta M, Kera Y, Noda T, Igarashi K (2011) Methionine adenosyltransferase II serves as a transcriptional corepressor of Maf oncoprotein. Mol Cell 41(5):554–566

Keller M, Hopp L, Liu X, Wohland T, Rohde K, Cancello R, Klös M, Bacos K, Kern M, Eichelmann F, ArneDietrich MR, Schön DG, Lohmann T, Dreßler M, Stumvoll M, Kov P (2016) Genome-wide DNA promoter methylation and transcriptome analysis in humanadipose tissue unravels novel candidate genes for obesity. Mol Metab. https://doi.org/10.1016/j.molmet.2016.11.003

Keller M, Hopp L, Liu X, Wohland T, Rohde K, Cancello R, Klös M, Bacos K, Kern M, Eichelmann F (2017) Genome-wide DNA promoter methylation and transcriptome analysis in human adipose tissue unravels novel candidate genes for obesity. Mol Metab 6(1):86–100

Khot VV, Chavan-Gautam P, Mehendale S, Joshi SR (2017) Variable methylation potential in preterm placenta: implication for epigenetic programming of the offspring. Reprod Sci 24(6): 891–901

Kim YS, Milner JA (2007) Dietary modulation of colon cancer risk. J Nutr 137(11):2576S–2579S

Knight AK, Park HJ, Hausman DB, Fleming JM, Bland VL, Rosa G, Kennedy EM, Caudill MA, Malysheva O, Kauwell GP (2018) Association between one-carbon metabolism indices and DNA methylation status in maternal and cord blood. Sci Rep 8(1):1–9

Lee WJ, Zhu BT (2006) Inhibition of DNA methylation by caffeic acid and chlorogenic acid, two common catechol-containing coffee polyphenols. Carcinogenesis 27(2):269–277

Lerin C, Rodgers JT, Kalume DE, Kim S-h, Pandey A, Puigserver P (2006) GCN5 acetyltransferase complex controls glucose metabolism through transcriptional repression of PGC-1α. Cell Metab 3(6):429–438

Li Y (2018) Epigenetic mechanisms link maternal diets and gut microbiome to obesity in the offspring. Front Genet 9:342

Lillycrop KA, Slater-Jefferies JL, Hanson MA, Godfrey KM, Jackson AA, Burdge GC (2007) Induction of altered epigenetic regulation of the hepatic glucocorticoid receptor in the offspring of rats fed a protein-restricted diet during pregnancy suggests that reduced DNA methyltransferase-1 expression is involved in impaired DNA methylation and changes in histone modifications. Br J Nutr 97(6):1064–1073

Lillycrop KA, Phillips ES, Torrens C, Hanson MA, Jackson AA, Burdge GC (2008) Feeding pregnant rats a protein-restricted diet persistently alters the methylation of specific cytosines in the hepatic PPARα promoter of the offspring. Br J Nutr 100(2):278–282

Liu D, Perkins JT, Hennig B (2016) EGCG prevents PCB-126-induced endothelial cell inflammation via epigenetic modifications of NF-κB target genes in human endothelial cells. J Nutr Biochem 28:164–170

Loenarz C, Schofield CJ (2011) Physiological and biochemical aspects of hydroxylations and demethylations catalyzed by human 2-oxoglutarate oxygenases. Trends Biochem Sci 36(1): 7–18

Lombardi R, Iuculano F, Pallini G, Fargion S, Fracanzani AL (2020) Nutrients, genetic factors, and their interaction in non-alcoholic fatty liver disease and cardiovascular disease. Int J Mol Sci 21(22):8761

Lu SC, Mato JM (2012) S-adenosylmethionine in liver health, injury, and cancer. Physiol Rev 92(4):1515–1542

Lu C, Thompson CB (2012) Metabolic regulation of epigenetics. Cell Metab 16(1):9–17

Ly A, Hoyt L, Crowell J, Kim Y-I (2012) Folate and DNA methylation. Antioxid Redox Signal 17(2):302–326

Maalouf M, Rho JM, Mattson MP (2009) The neuroprotective properties of calorie restriction, the ketogenic diet, and ketone bodies. Brain Res Rev 59(2):293–315

Maruti SS, Ulrich CM, White E (2009) Folate and one-carbon metabolism nutrients from supplements and diet in relation to breast cancer risk. Am J Clin Nutr 89(2):624–633

Mastroeni D, Grover A, Delvaux E, Whiteside C, Coleman PD, Rogers J (2011) Epigenetic mechanisms in Alzheimer's disease. Neurobiol Aging 32(7):1161–1180

Mato JM, Corrales FJ, Lu SC, Avila MA (2002) S-Adenosylmethionine: a control switch that regulates liver function. FASEB J 16(1):15–26

Mazzanti G, Di Sotto A, Vitalone A (2015) Hepatotoxicity of green tea: an update. Arch Toxicol 89(8):1175–1191

McCarty MF, Barroso-Aranda J, Contreras F (2009) High-dose folate and dietary purines promote scavenging of peroxynitrite-derived radicals–clinical potential in inflammatory disorders. Med Hypotheses 73(5):824–834

Meng J, Wang L, Wang J, Zhao X, Cheng J, Yu W, Jin D, Li Q, Gong Z (2018) Methionine adenosyltransferase4 mediates DNA and histone methylation. Plant Physiol 177(2):652–670

Milagro F, Mansego M, De Miguel C, Martinez J (2013) Dietary factors, epigenetic modifications and obesity outcomes: progresses and perspectives. Mol Asp Med 34(4):782–812

Muka T, Koromani F, Portilla E, O'Connor A, Bramer WM, Troup J, Chowdhury R, Dehghan A, Franco OH (2016) The role of epigenetic modifications in cardiovascular disease: a systematic review. Int J Cardiol 212:174–183

Newman JC, He W, Verdin E (2012) Mitochondrial protein acylation and intermediary metabolism: regulation by sirtuins and implications for metabolic disease. J Biol Chem 287(51): 42436–42443

Niculescu MD, Zeisel SH (2002) Diet, methyl donors and DNA methylation: interactions between dietary folate, methionine and choline. J Nutr 132(8):2333S–2335S

Nijland MJ, Mitsuya K, Li C, Ford S, McDonald TJ, Nathanielsz PW, Cox LA (2010) Epigenetic modification of fetal baboon hepatic phosphoenolpyruvate carboxykinase following exposure to moderately reduced nutrient availability. J Physiol 588(8):1349–1359

Oommen AM, Griffin JB, Sarath G, Zempleni J (2005) Roles for nutrients in epigenetic events. J Nutr Biochem 16(2):74–77

Padmanabhan N, Watson ED (2013) Lessons from the one-carbon metabolism: passing it along to the next generation. Reprod Biomed Online 27(6):637–643

Park JH, Stoffers DA, Nicholls RD, Simmons RA (2008) Development of type 2 diabetes following intrauterine growth retardation in rats is associated with progressive epigenetic silencing of Pdx1. J Clin Invest 118(6):2316–2324

Perfilyev A, Dahlman I, Gillberg L, Rosqvist F, Iggman D, Volkov P, Nilsson E, Risérus U, Ling C (2017) Impact of polyunsaturated and saturated fat overfeeding on the DNA-methylation pattern in human adipose tissue: a randomized controlled trial. Am J Clin Nutr 105(4):991–1000

Perillo B, Ombra MN, Bertoni A, Cuozzo C, Sacchetti S, Sasso A, Chiariotti L, Malorni A, Abbondanza C, Avvedimento EV (2008) DNA oxidation as triggered by H3K9me2 demethylation drives estrogen-induced gene expression. Science 319(5860):202–206

Perkins E, Murphy SK, Murtha AP, Schildkraut J, Jirtle RL, Demark-Wahnefried W, Forman MR, Kurtzberg J, Overcash F, Huang Z (2012) Insulin-like growth factor 2/H19 methylation at birth and risk of overweight and obesity in children. J Pediatr 161(1):31–39

Raciti GA, Spinelli R, Desiderio A, Longo M, Parrillo L, Nigro C, D'Esposito V, Mirra P, Fiory F, Pilone V (2017) Specific CpG hyper-methylation leads to Ankrd26 gene down-regulation in white adipose tissue of a mouse model of diet-induced obesity. Sci Rep 7(1):1–13

Roediger W (1982) Utilization of nutrients by isolated epithelial cells of the rat colon. Gastroenterology 83(2):424–429

Rowling MJ, Schalinske KL (2001) Retinoid compounds activate and induce hepatic glycine N-methyltransferase in rats. J Nutr 131(7):1914–1917

Sakata SF, Shelly L, Ruppert S, Schutz G, Chou J (1993) Cloning and expression of murine S-adenosylmethionine synthetase. J Biol Chem 268(19):13978–13986

Sánchez-Fernández EM, Tarhonskaya H, Al-Qahtani K, Hopkinson RJ, McCullagh JS, Schofield CJ, Flashman E (2013) Investigations on the oxygen dependence of a 2-oxoglutarate histone demethylase. Biochem J 449(2):491–496

Sandovici I, Smith NH, Nitert MD, Ackers-Johnson M, Uribe-Lewis S, Ito Y, Jones RH, Marquez VE, Cairns W, Tadayyon M (2011) Maternal diet and aging alter the epigenetic control of a promoter–enhancer interaction at the Hnf4a gene in rat pancreatic islets. Proc Natl Acad Sci 108(13):5449–5454

Shi Y, Lan F, Matson C, Mulligan P, Whetstine JR, Cole PA, Casero RA, Shi Y (2004) Histone demethylation mediated by the nuclear amine oxidase homolog LSD1. Cell 119(7):941–953

Shibata K (2018) Organ co-relationship in tryptophan metabolism and factors that govern the biosynthesis of nicotinamide from tryptophan. J Nutr Sci Vitaminol 64(2):90–98

Shimazu T, Hirschey MD, Newman J, He W, Shirakawa K, Le Moan N, Grueter CA, Lim H, Saunders LR, Stevens RD (2013) Suppression of oxidative stress by β-hydroxybutyrate, an endogenous histone deacetylase inhibitor. Science 339(6116):211–214

Shrubsole MJ, Shu XO, Li H-L, Cai H, Yang G, Gao Y-T, Gao J, Zheng W (2011) Dietary B vitamin and methionine intakes and breast cancer risk among Chinese women. Am J Epidemiol 173(10):1171–1182

Shyh-Chang N, Locasale JW, Lyssiotis CA, Zheng Y, Teo RY, Ratanasirintrawoot S, Zhang J, Onder T, Unternaehrer JJ, Zhu H (2013) Influence of threonine metabolism on S-adenosylmethionine and histone methylation. Science 339(6116):222–226

Sinclair KD, Allegrucci C, Singh R, Gardner DS, Sebastian S, Bispham J, Thurston A, Huntley JF, Rees WD, Maloney CA (2007) DNA methylation, insulin resistance, and blood pressure in offspring determined by maternal periconceptional B vitamin and methionine status. Proc Natl Acad Sci 104(49):19351–19356

Singhal RP, Mays-Hoopes LL, Eichhorn GL (1987) DNA methylation in aging of mice. Mech Ageing Dev 41(3):199–210

Sivanand S, Viney I, Wellen KE (2018) Spatiotemporal control of acetyl-CoA metabolism in chromatin regulation. Trends Biochem Sci 43(1):61–74

Song Y, Manson JE, Lee I-M, Cook NR, Paul L, Selhub J, Giovannucci E, Zhang SM (2012) Effect of combined folic acid, vitamin B6, and vitamin B12 on colorectal adenoma. J Natl Cancer Inst 104(20):1562–1575

Stefanska B, Karlic H, Varga F, Fabianowska-Majewska K, Haslberger A (2012) Epigenetic mechanisms in anti-cancer actions of bioactive food components–the implications in cancer prevention. Br J Pharmacol 167(2):279–297

Stipanuk MH (2004) Sulfur amino acid metabolism: pathways for production and removal of homocysteine and cysteine. Annu Rev Nutr 24:539–577

Sutendra G, Kinnaird A, Dromparis P, Paulin R, Stenson TH, Haromy A, Hashimoto K, Zhang N, Flaim E, Michelakis ED (2014) A nuclear pyruvate dehydrogenase complex is important for the generation of acetyl-CoA and histone acetylation. Cell 158(1):84–97

Suter MA, Chen A, Burdine MS, Choudhury M, Harris RA, Lane RH, Friedman JE, Grove KL, Tackett AJ, Aagaard KM (2012) A maternal high-fat diet modulates fetal SIRT1 histone and protein deacetylase activity in nonhuman primates. FASEB J 26(12):5106–5114

Tabas I (2017) 2016 Russell Ross memorial lecture in vascular biology: molecular–cellular mechanisms in the progression of atherosclerosis. Arterioscler Thromb Vasc Biol 37(2): 183–189

Tchantchou F, Graves M, Ortiz D, Chan A, Rogers E, Shea T (2006) S-adenosyl methionine: a connection between nutritional and genetic risk factors for neurodegeneration in Alzheimer's disease. J Nutr Health Aging 10(6):541

Tehlivets O, Malanovic N, Visram M, Pavkov-Keller T, Keller W (2013) S-adenosyl-L-homocysteine hydrolase and methylation disorders: yeast as a model system. Biochim Biophys Acta (BBA) 1832(1):204–215

Tokunaga M, Takahashi T, Singh RB, De Meester F, Wilson DW (2013) Nutrition and epigenetics. Med Epigenet 1(1):70–77

Tosh DN, Fu Q, Callaway CW, McKnight RA, McMillen IC, Ross MG, Lane RH, Desai M (2010) Epigenetics of programmed obesity: alteration in IUGR rat hepatic IGF1 mRNA expression and histone structure in rapid vs. delayed postnatal catch-up growth. Am J Physiol Gastrointest Liver Physiol 299(5):G1023–G1029

Troesch B, Weber P, Mohajeri MH (2016) Potential links between impaired one-carbon metabolism due to polymorphisms, inadequate B-vitamin status, and the development of Alzheimer's disease. Nutrients 8(12):803

Tu BP, Mohler RE, Liu JC, Dombek KM, Young ET, Synovec RE, McKnight SL (2007) Cyclic changes in metabolic state during the life of a yeast cell. Proc Natl Acad Sci 104(43): 16886–16891

Vahid F, Zand H, Nosrat-Mirshekarlou E, Najafi R, Hekmatdoost A (2015) The role dietary of bioactive compounds on the regulation of histone acetylases and deacetylases: a review. Gene 562(1):8–15

Vepsäläinen S, Koivisto H, Pekkarinen E, Mäkinen P, Dobson G, McDougall GJ, Stewart D, Haapasalo A, Karjalainen RO, Tanila H (2013) Anthocyanin-enriched bilberry and blackcurrant extracts modulate amyloid precursor protein processing and alleviate behavioral abnormalities in the APP/PS1 mouse model of Alzheimer's disease. J Nutr Biochem 24(1):360–370

Vucetic Z, Kimmel J, Totoki K, Hollenbeck E, Reyes TM (2010) Maternal high-fat diet alters methylation and gene expression of dopamine and opioid-related genes. Endocrinology 151(10):4756–4764

Waldecker M, Kautenburger T, Daumann H, Busch C, Schrenk D (2008) Inhibition of histone-deacetylase activity by short-chain fatty acids and some polyphenol metabolites formed in the colon. J Nutr Biochem 19(9):587–593

Wang J, Alexander P, Wu L, Hammer R, Cleaver O, McKnight SL (2009) Dependence of mouse embryonic stem cells on threonine catabolism. Science 325(5939):435–439

Weiss EP, Albert SG, Reeds DN, Kress KS, McDaniel JL, Klein S, Villareal DT (2016) Effects of matched weight loss from calorie restriction, exercise, or both on cardiovascular disease risk factors: a randomized intervention trial. Am J Clin Nutr 104(3):576–586

Wellen KE, Thompson CB (2012) A two-way street: reciprocal regulation of metabolism and signalling. Nat Rev Mol Cell Biol 13(4):270–276

Wu G (2009) Amino acids: metabolism, functions, and nutrition. Amino Acids 37(1):1–17

Xiao M, Yang H, Xu W, Ma S, Lin H, Zhu H, Liu L, Liu Y, Yang C, Xu Y (2012) Inhibition of α-KG-dependent histone and DNA demethylases by fumarate and succinate that are accumulated in mutations of FH and SDH tumor suppressors. Genes Dev 26(12):1326–1338

Yang X, Yan L, Davidson NE (2001) DNA methylation in breast cancer. Endocr Relat Cancer 8(2): 115–127

Yokozawa T, Lee YA, Cho EJ, Matsumoto K, Park CH, Shibahara N (2011) Anti-aging effects of oligomeric proanthocyanidins isolated from persimmon fruits. Drug Discov Therap 5(3): 109–118

Yuan X, Tsujimoto K, Hashimoto K, Kawahori K, Hanzawa N, Hamaguchi M, Seki T, Nawa M, Ehara T, Kitamura Y (2018) Epigenetic modulation of Fgf21 in the perinatal mouse liver ameliorates diet-induced obesity in adulthood. Nat Commun 9(1):1–16

Zeisel SH (2006) Choline: critical role during fetal development and dietary requirements in adults. Annu Rev Nutr 26:229–250

Zeisel S (2017) Choline, other methyl-donors and epigenetics. Nutrients 9(5):445

Zhang S, Rattanatray L, MacLaughlin SM, Cropley JE, Suter CM, Molloy L, Kleemann D, Walker SK, Muhlhausler BS, Morrison JL (2010) Periconceptional undernutrition in normal and overweight ewes leads to increased adrenal growth and epigenetic changes in adrenal IGF2/H19 gene in offspring. FASEB J 24(8):2772–2782

Zimmermann MB (2006) The influence of iron status on iodine utilization and thyroid function. Annu Rev Nutr 26:367–389

Zullo A, Simone E, Grimaldi M, Gagliardi M, Zullo L, Matarazzo M, Mancini F (2018) Effect of nutrient deprivation on the expression and the epigenetic signature of sirtuin genes. Nutr Metab Cardiovasc Dis 28(4):418–424

Chapter 9
Epigenetic Reprogramming of the Glucose Metabolic Pathways by the Chromatin Effectors During Cancer

Payel Mondal, Niharika Tiwary, Amrita Sengupta, Sinjini Dhang, Siddhartha Roy, and Chandrima Das

Abstract Glucose metabolism plays a vital role in regulating cellular homeostasis as it acts as the central axis for energy metabolism, alteration in which may lead to serious consequences like metabolic disorders to life-threatening diseases like cancer. Malignant cells, on the other hand, help in tumor progression through abrupt cell proliferation by adapting to the changed metabolic milieu. Metabolic intermediates also vary from normal cells to cancerous ones to help the tumor manifestation. However, metabolic reprogramming is an important phenomenon of cells through which they try to maintain the balance between normal and carcinogenic outcomes. In this process, transcription factors and chromatin modifiers play an essential role to modify the chromatin landscape of important genes related directly or indirectly to metabolism. Our chapter surmises the importance of glucose metabolism and the role of metabolic intermediates in the cell. Also, we summarize the influence of histone effectors in reprogramming the cancer cell metabolism. An interesting aspect of this chapter includes the detailed methods to detect the aberrant metabolic flux, which can be instrumental for the therapeutic regimen of cancer.

Keywords Glucose metabolism · Warburg effect · Metabolic intermediates · Epigenetic reader · Histone modification · Glycolysis · OXPHOS · TCA cycle ·

P. Mondal · C. Das (✉)
Biophysics and Structural Genomics Division, Saha Institute of Nuclear Physics, Kolkata, India

Homi Bhaba National Institute, Mumbai, India
e-mail: chandrima.das@saha.ac.in

N. Tiwary · A. Sengupta
Biophysics and Structural Genomics Division, Saha Institute of Nuclear Physics, Kolkata, India

S. Dhang · S. Roy
Structural Biology & Bio-Informatics Division, CSIR-Indian Institute of Chemical Biology, Kolkata, India

Academy of Scientific and Innovative Research (AcSIR), Ghaziabad, India

T. K. Kundu, C. Das (eds.), *Metabolism and Epigenetic Regulation: Implications in Cancer*, Subcellular Biochemistry 100,
https://doi.org/10.1007/978-3-031-07634-3_9

Gluconeogenesis · Hypoxia · Extracellular flux analysis · Techniques to measure · Hepatocellular carcinoma

Abbreviations

2-DG	2-deoxy-D-glucose
3-MG	3-O-methylglucose
53BP1	p53 binding protein 1
ACO1	Aconitase 1
Af9	ALL1-fused gene from chromosome 9
ALDOA	Aldolase A
AML	Acute Myeloid Leukaemia
APCI	Atmospheric Pressure Chemical Ionization
ATP	Adenosine Triphosphate
BARD1	BRCA1-associated RING domain protein 1
BIRC5	Baculoviral IAP repeat-containing protein 5
BRCA1	BReast CAncer gene 1
BRD	Bromodomain
CBP	CREB Binding Protein
CHD1	Chromodomain Helicase DNA binding protein 1
CHD4	Chromodomain Helicase DNA binding protein 4
CTCF	CCCTC-binding factor
D-2-HG	D-2-hydroxyglutarate
DAF2	Diaminofluorescein-2
DAF-FM	4-Amino-5-Methylamino-2′,7′-Difluorofluorescein
DCD	double Chromodomain
DCFH-DA	Dichlorodihydrofluorescein diacetate
DDR	DNA Damage Response
DNA	Deoxyribonucleic Acid
DNMT	DNA methyltransferase
DNMT1	DNA Methyltransferase 1
DOT1L	Disruptor of telomeric silencing 1-like
DPF3B	Double PHD Finger-containing protein 3b
dTMP	Thymidylate
DTNB	5,5′-dithiobis-(2-nitrobenzoic acid)
EED	Embryonic Ectoderm Development protein 1
EI	Electron Impact ionization
ENO1	α-Enolase
ERα	Estrogen Receptor α
ETC	Electron transport chain
EZH2	Enhancer of Zeste homolog 2
FAB	Fast Atom Bombardment
FAD	Flavin Adenine Dinucleotide

FBP1	Fructose-1,6 Bisphosphatase 1
FHA	Fork-Head Associated
G6P	Glucose-6-phosphatase
GAPDH	Glyceraldehyde 3-phosphate dehydrogenase
GC	Gas Chromatography
GCN5	General Control Non-derepressible 5
GDH	Glutamate Dehydrogenase
GLP	G9a-like Protein
GLS	Glutaminase
GLUT	Glucose Transporter
GPI	Glucose-6-phosphate Isomerase
GSH	Reduced Glutathione
GSSG	Oxidized Glutathione
H2O2	Hydrogen peroxide
HAT	Histone acetyltransferase
HCC	Hepatocellular Carcinoma
HDAC	Histone deacetylase
HIC1	HIC ZBTB Transcriptional Repressor 1
HIF1a	Hypoxia Inducing Factor 1a
HK2	Hexokinase 2
hL3MBTL	Human Lithal (3) Malignant Brain Tumor-like protein
HMGS-1	Hydroxymethylglutaryl-CoA synthase 1
HMT	Histone methyltransferase
HP1	Heterochromatin Protein 1
HR	Homologous recombination
HRE	Hypoxia Response Elements
HRP	Horseradish peroxidase
IDH1/2	Isocitrate dehydrogenase 1/2
IDH2	Isocitrate dehydrogenase 2
ING2	Inhibitor of Growth Protein 2
JARID1A	Jumonji/ARID domain-containing protein 1A
JC-1	Tetraethylbenzimidazolylcarbocyanine Iodide
JMJD	Jumonji C domain-containing protein
JMJD2A	Jumonji Domain-containing 2A
KISS1	KiSS-1 Metastasis Suppressor
KLR5	Kruppel-like receptor 5
KMT	Lysine methyltransferase
L-2-HG	L-2-hydroxyglutarate
LC	Liquid Chromatography
LDHA	Lactate dehydrogenase A
LDHC	Lactate Dehydrogenase C
LSD1	Lysine-Specific Demethylase 1
MALDI	Matrix-Assisted Laser Desorption Ionization
MAT	Methionine adenosyltransferase

MBT	Malignant brain tumor
MDC1	Mediator of DNA damage checkpoint protein 1
MDH	Malate dehydrogenase
MRG15	MORF-related gene 15
MS	Mass Spectrometry
MTA	Metastasis Associated Protein
NAD^+	Nicotinamide Adenine Dinucleotide
NADPH	Nicotinamide Adenine Dinucleotide Phosphate
NED	Naphthyl Ethylene Diamine Dihydrochloride
NHEJ	Non-Homologous End Joining
NMR	Nuclear Magnetic Resonance
NR	Nuclear Receptors
NuRD	Nucleosome Remodeling and Deacetylase
OGDH	Oxoglutarate dehydrogenase
OGT	O-GlcNAc transferase
OXPHOS	Oxidative Phosphorylation
PARP	Poly (ADP-ribose) polymerase
PCK1	Pyruvate Carboxykinase 1
PDK1	Pyruvate Dehydrogenase Kinase 1
PGK1	Phosphoglycerate Kinase 1
PGM1	Phosphoglycerate Mutase 1
PHD	Plant Homeodomain
PHF19	PHD Finger Protein 19
PHF20L1	PHD finger protein 20-like protein 1
PK	Pyruvate Kinase
PKM2	Pyruvate Kinase isozyme M2
PPAR	Proliferator Activated Receptor
PRC2	Polycomb repressive complexes 2
PRMT	Protein Arginine methyltransferase
RBC	Red Blood Corpuscles
RNF	Ring Finger proteins
RNS	Reactive Nitrogen Species
ROS	Reactive oxygen species
Rtt106	Regulator of Ty1 transposition protein 106
SAH	S-adenosylhomocysteine
SAM	S-Adenosyl Methionine
Sas5	Something About Silencing 5
SCO2	Cytochrome C Oxidase Assembly
SDH	Succinate dehydrogenase complex subunits A, B, C
SIM	SUMO interacting motif
SIRT	Sirtuin
SMARCAD1	SWI/SNF-Related, Matrix-Associated Actin-Dependent Regulator of Chromatin, Subfamily A, Containing DEAD/H Box 1 protein
SMN	Survival of motor neuron

SPF30	Splicing factor 30
SRA	SET and RING-associated
SSRP1	Structure Specific Recognition Protein 1
STK38	Serine/Threonine Kinase 38
SUMO	Small Ubiquitin-related Modifier
Taf14	Transcription initiation factor TFIID subunit 14
TCA	Tricarboxylic Acid
TDRD3	Tudor Domain-containing 3 protein
TET	Ten-eleven Translocation
TGM2	Transglutaminase 2
THF	Tetrahydrofolate
TIGAR	Tp53 Induced Glycolysis and Apoptosis Regulator
TMRE	Tetramethylrhodamine ethyl ester, perchlorate
TNB	2-nitro-5-thiobenzoic acid
TRIM24	Tripartite Motif-containing protein 24
TTD	tandem Tudor domain
UBC9	Ubiquitin carrier protein 9
UDP-GlcNac	UDP-N-acetylglucosamine
UDR	Ub-dependent recruitment
UHRF1	Ubiquitin-like PHD and RING finger domain-containing protein 1
WDR5	WD Repeat-containing protein 5
Yaf9	Yeast homolog of ALL1-fused gene from chromosome 9
ZMYND11	Zinc finger MYND domain-containing protein 11
ZMYND8	Zinc finger MYND domain-containing protein 8
α-KG	α-Ketoglutarate
βOHB	β-hydroxybutyrate

Introduction

The tumor microenvironment reprograms cellular pathways in favor of cancer progression through regulating metabolic pathways and nutrients acquisition (De Berardinis and Chandel 2016). The altered metabolic activity supports tumor growth during nutrient stress situations (Boroughs and Deberardinis 2015). Maintaining metabolic homeostasis in a cell is essential for its sustenance, hence deviation from a normal trend result in metabolic disorder to tumorigenesis. Longer than a decade, studies were revolving around what is the need of an inefficient bioenergetic process called the "Warburg Effect" in the context of cancer metabolism (Upadhyay et al. 2013). As the research progresses, the above-proposed phenomenon started to elicit its importance in the means of synthesizing anabolic molecules instead of just a source of energy production. "Warburg Effect" is now connected with the nucleotide, lipid, and protein metabolism and is no longer restricted only to carbohydrate metabolism. In this way, it became evident how all the metabolic pathways are connected and deregulation in any one of them can ultimately lead to tumorigenesis (Liberti and Locasale 2016). As we speak about the importance of tight regulation of metabolic functions in the cell, metabolic intermediates are also crucial for maintaining the health of the same (Frezza 2020). The formation of metabolites not only helps to synthesize other building blocks for the cell to better survive and proliferate, but it also influences several epigenetic writer proteins to modify the promoter region of important direct or indirect metabolic genes (Pavlova and Thompson 2016). Metabolites such as S-Adenosyl Methionine (SAM), Acetyl CoA, Nicotinamide Adenine Dinucleotide ($NAD^{+)}$, Flavin Adenine Dinucleotide (FAD), Tetrahydrofolate (THF) act as cofactors for different methyltransferases, acetyltransferases, deacetylases, etc., in several reactions respectively and donate their methyl, acetyl groups or help to release the same from a substrate accordingly (Wong et al. 2017). While the abundance of various metabolites in normal cell regulates different cellular functions, there exist a group of molecules called oncometabolites that are formed under aberrant situations and shift the cellular homeostasis towards tumorigenesis (Collins et al. 2017; Khatami et al. 2019). Although metabolites influence gene transcription, a core network of transcription factors and epigenetic regulators manage their spatial and temporal expression. The regulations imparted by transcription factors are well studied, but the function of epigenetic readers in the same context is being explored recently and has become the center of attraction as they potentially read the modification on gene promoter region and determine their fate contextually either towards gene synthesis or to repression through bringing several eraser proteins on the same gene loci (van Knaap and Verrijzer 2016). Thus epigenetic regulation dictates the fate of a gene and hence chromatin readers/effectors are essential to carry out the function (Bannister and Kouzarides 2011). Deciphering the role of metabolites and epigenetic factors involved in this process of regulation of cancer progression opens up a wide avenue of research on the aspect of targeted therapy for the early diagnosis and treatment of cancer. Among all the metabolic processes, carbohydrate metabolism being at the central axis has immense role in the regulation of oncogenesis. Here in this chapter, we have discussed in-depth the role of glucose metabolism and the function of epigenetic readers in the context of cancer.

Role of Glucose Metabolism in Cancer Manifestation

Normal Cells: Glycolysis and TCA Cycle Are Balanced

Glucose Metabolism is the principal source for all the major biological processes. Normal non-proliferating cells under aerobic conditions utilize glucose and through a series of metabolic reactions, convert it into pyruvate, which then enters the Tricarboxylic Acid (TCA) cycle to produce the majority of the Adenosine Triphosphate (ATP) pool (Hay 2016). So, the main energy consumption from glucose metabolic activities in normal cells happens through the pathway of mitochondrial oxidative phosphorylation (OXPHOS) that produces more ATP than glycolysis.

In the case of rapidly proliferating cells, both glycolysis and OXPHOS are upregulated but maintain a homeostatic balance to provide adequate levels of ATP (Liberti and Locasale 2016). Simultaneously these processes cause rapid biomass production and accumulation of metabolic intermediates that support the rate of proliferation. When these cells encounter hypoxic conditions, OXPHOS being an aerobic process, can no longer generate ATP (Wheaton and Chandel 2011). But to cater to the cellular proliferation rates, there is a need for a constant supply of energy that shifts the cells towards glycolysis. During this event, glucose uptake and glycolysis are highly upregulated with the concomitant release of higher amounts of lactate (Ngo et al. 2015).

Cancer Cells: Energy Production Shifts to Glycolysis—Warburg Effect

One of the important and unique features of cancer cells is to harness energy through aerobic glycolysis and for that they rapidly take up glucose from the surroundings (Han et al. 2015). Instead of harnessing energy through major catabolic pathways like OXPHOS, cancer cells follow the unconventional yet faster route of energy production through aerobic glycolysis. This is necessary for the cancer cells to rapidly proliferate and fulfil their increased biomass requirements. In the year 1920, German Scientist Otto Warburg discovered this phenomenon known as Warburg Effect (Koppenol et al. 2011). Glucose being the central macronutrient of all the metabolic processes helps in sustaining mammalian life as oxidation of its carbon bonds releases energy in the form of ATP. Glucose breaks down to lactate following glycolysis, or upon complete oxidation through respiration in mitochondria, it produces Carbon-di-oxide and ATP. Lactate is one of the essential metabolic intermediates as it provides the immediate ingredients for a tumor cell to sustain its life or a proliferating cell to divide. So these cells increase their rate of glucose uptake and lactate production and harness energy through glycolysis even in the presence of oxygen with intact and functional mitochondria. This process, known as the Warburg Effect, has been studied extensively since then in the context of cancer cell metabolism (Liberti and Locasale 2016).

Deregulation of Glucose Metabolism in Cancer: Cause and Effects

Over the past many years, extensive studies have revealed several mechanisms that affect glucose metabolism leading to metabolic disorders and other related diseases

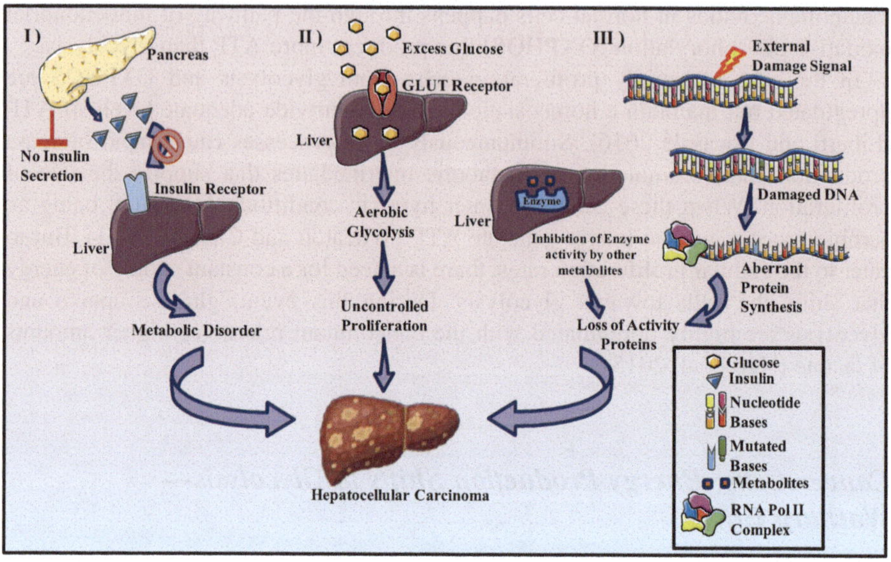

Fig. 9.1 The metabolic milieu impacts liver function culminating towards Hepatocellular carcinoma (HCC). (I) Defects in Insulin secretion or mutations in Insulin receptors can lead to metabolic disorders. (II) Excess glucose in the surrounding or defects in glucose transporters and GLUT receptors induce glycolytic pathway leading to uncontrolled proliferation. (III) Environmental factors can invoke genotoxic stress-causing mutation in important metabolic enzymes. Further, the pool of oncometabolite produced due to aberrant mutation in key metabolic genes can lead to loss of activity of other pathway proteins

(Fig. 9.1). The mechanisms can be listed as; (I) defects in secretion or transport of the hormone Insulin from the beta cells of Islets of Langerhans to hepatic tissues, (II) defects in glucose transport into the destined tissues by different glucose transporters and Glut receptors, (III) metabolic deregulation or gene mutation resulted in abnormalities in enzymes directly or indirectly associated with glucose metabolism.

Glucagon is released from the pancreas when glucose levels are very low. It helps the liver to transfer stored glycogen into glucose, which then is circulated into the bloodstream to maintain the blood glucose levels. On the other hand, high blood glucose levels cause the release of insulin. Insulin helps glucose to be taken up and used by insulin-dependent and metabolically active tissues, such as muscle or hepatic cells (Elrick et al. 1956). When there is some discrepancy in the secretion of insulin due to genetic abnormality, or impaired transport of the same due to inactive insulin receptors or insensitivity of insulin to hepatic cells, it causes hyperglycemia in blood, the outcome of which is type 1 or type 2 diabetes mellitus. Prolong exposure to high blood glucose ultimately results in hepatocellular carcinoma (Han et al. 2017) (Fig. 9.1(I)).

In hepatocytes, GLUT receptors are required for glucose uptake, which is dependent upon the glucose concentrations in the intra- and extracellular milieu, thereby regulating the transcription programs of liver cells (Thorens 2015). During metabolic deregulation or in presence of high glucose in the surrounding, this control of the receptors is lost. As a result of that, an uncontrolled uptake of glucose causes a high rate of glycolysis, and the Warburg effect predominates in the cancer cells (Fig. 9.1 (II)).

Metabolic pathways are tightly regulated at every reaction step by the function of different metabolic enzymes encoded by nuclear and mitochondrial genes. If any kind of mutation is introduced to the genes of these enzymes, that may result in the production of either non-functional enzymes or no protein at all. This in turn abrogates the fine balance in the metabolic pathways leading to the progression of cancer (Fig. 9.1(III)).

Strategy of Cancer Cells Beyond Warburg Effect for Their Better Survival

Cancer cell metabolism is highly influenced by the Warburg effect. But the initial understanding of this phenomenon was that most of the cancer cells shift to aerobic glycolysis due to impaired mitochondrial OXPHOS. However, recent studies have shown that most of the cancer cell has intact and functional mitochondria and hence regular OXPHOS (Hsu and Sabatini 2008). Current studies have shown that spontaneous tumors though depend on aerobic glycolysis for their sustenance and progression, they have healthy mitochondria. On the other hand, aerobic glycolysis can result from several internal events such as the sudden conversion of proto-oncogenes to oncogenes, significant loss of tumor suppressor proteins, mutations in mitochondrial DNA, as well as external factors like hypoxia, tissue of origin (Griguer et al. 2005; Fantin et al. 2006; Zheng 2012). So, aerobic glycolysis provides rapid energy to the cancer cells for their survival and proliferation, whereas downstream other glucose catabolic pathway like TCA cycle remains the major source of the metabolic intermediates that are utilized as building blocks for the biomass production of the cancer cells.

Glycolysis and TCA Cycle Intermediates Help Cancer Cells for Biomass Production

The glycolytic pathway provides glucose-6-phosphate that can go into the pentose phosphate pathway to synthesize pentose sugar of nucleotides. It also produces 3-phosphoglycerate that goes into the purine synthesis pathway. The synthesis of tyrosine, tryptophan, and phenylalanine is aided by phosphoenol pyruvate, another metabolite from glycolysis. On the other hand, the TCA cycle causes the release of citrate, oxaloacetate into cytoplasm that helps in the synthesis of fatty acid and sterol, and aspartate, asparagines, and pyrimidine nucleotides, respectively. Hence, any organism depends on glycolysis and the TCA cycle present at the central axis of all metabolic meshwork for both energy production and synthesis of metabolic building blocks. In the case of the cancer cells, they undergo rapid proliferation and hence require not only these building blocks for cell division but also a higher amount of energy to steer the biological processes. So, the ratio of glycolysis to OXPHOS increases in response to the proliferative frequency of the cells, oxygen imbalance, or other mutations to the pathway genes of the TCA cycle (Goetzman and Prochownik 2018).

When a normal cell reprograms towards pluripotency through the influence of metabolic alterations, there is a sudden increase in the rate of cell proliferation and abundance in biomass. This metabolic reprogramming is mostly under the regulation of chromatin structure by different epigenetic modifiers and a core transcription factor network (Zangle et al. 2013). Metabolic intermediates influence the function of these enzymatic reactions. The regulatory mechanisms of these metabolic intermediates can give the premises of the understanding of how cancer cells survive and proliferate for their own benefit. The following section details the function of metabolic intermediates in the process of epigenetic regulation on glucose metabolism and how the chromatin effectors alter the transcription landscape in accordance with the metabolic reprogramming.

Epigenetic Influences on Glucose Metabolism That Reprogram Cancer Cells Towards Survival

Energy metabolism is indispensable for cellular survival, although the metabolic intermediates and the pathways are regulated by them (He et al. 2019). Tumor cells have high proliferative potential and invasiveness that help them to metastasize to distant organs. To achieve this, tumor cells can reprogram their metabolic pathways to meet the excess energy demand which is distinctly different from a normal cell. Hence targeting metabolic intermediates might be a promising way of inhibiting tumor cells proliferation and metastasis. Instead of annihilating a tumor cell, its energy supply could be restrained by altering the stoichiometric levels of metabolic intermediates that in turn lead to longer survival of cancer patients. Several known metabolic intermediates are well studied and implicated in the epigenetic regulations as well. Few of them are listed here with their mode of action and regulations.

Metabolic Intermediates and Their Connection to Epigenetic Regulation

There have been several studies carried out to understand the connection between metabolism and epigenetics of a cell (Fanucchi et al. 2021; Fan et al. 2015; Feldman

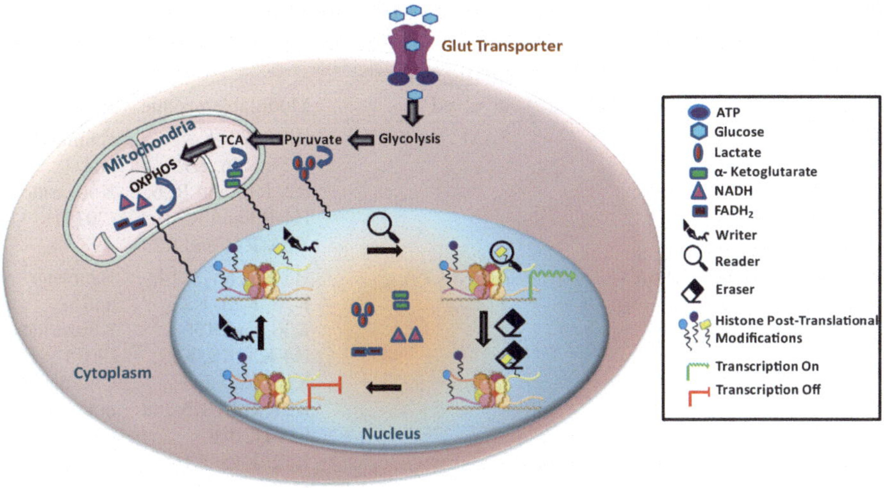

Fig. 9.2 Metabolic intermediates influence transcriptional reprogramming through reader/effector. The glucose metabolic pathway produces several intermediates such as Lactate, NADH, FADH2, α-Ketoglutarate, which directly or indirectly regulates gene transcription by influencing the function of different chromatin modifiers. While the writer or eraser proteins generate or remove those modifications, the epigenetic reader/effecters recognize them and dictate downstream signals

et al. 2012; Haws et al. 2020; Keating and El-Osta 2015; Thakur and Chen 2019). It has been found that metabolic intermediates actively participate and help the process of alteration in gene expression by directly or indirectly influencing the chromatin modifiers (Fig. 9.2). Some of the well-known common metabolites and their function have been described in the next section. Interestingly, many emerging metabolites are found to be responsible for non-canonical histone modification (Dai et al. 2020). These modifications have been summarized in Table 9.1.

S-Adenosyl Methionine (SAM)

S-Adenosyl Methionine (SAM) in general acts as the principal methyl donor in the cell for different biological reactions (Mato et al. 2013). It gets synthesized from methionine and ATP with the help of an enzyme named methionine adenosyltransferase (MAT) (Giulidori et al. 1984). It acts as a methyl group donor to lysine and arginine amino acids of proteins or nucleotide bases of DNA and RNA, or even to other small molecules (Landgraf et al. 2016). It mainly helps the different methyltransferases like lysine methyltransferase (KMTs), Histone methyltransferase (HMTs), DNA methyltransferase (DNMTs), and Protein arginine methyltransferase (PRMTs) (van Knaap and Verrijzer 2016).

Table 9.1 Metabolites influence Histone post-translational modification

Substrate for modification	Type of modification	Metabolites involved	Modifiable residue	References
Histone	Methylation	SAM	H3-K4, K9, K26, K27, K36, K79, R2, R17 and R26; H4-R3 and K20; H2A-R11 and R29	Bannister and Kouzarides (2011)
	Acetylation	Acetyl CoA	H3-K4, K9, K14, K18, K23, K27, K36, and K56; H4-K5, K8, K12, K16, K20 and K91; H2A-K5 and K9; H2B-K5, K12, K15, K16, K20, and K120	Bannister and Kouzarides (2011), Cluntun et al. (2015), Wellen et al. (2009)
	Propionylation	Propionate	H3- K23, and K56; H4- K5, K8, K12, K16, K31, K44, K77, K79, K91, and K95; H2A- K95, and K125	Chen et al. (2007), Sabari et al. (2017)
	Lactylation	Lactate	H3-K9, K14, K18, K23, K27, K56, and K79; H4-K5, K8, K12, K16, K31, K77 and K91; H2A-K11, K13 and K115; H2B-K5, K11, K15, K16, K20, K23, K43, K85, K108, K116, and K120	Zhang et al. (2019a)
	Succinylation	Succinyl-CoA	H3-K14, K23, K27, K56, K79, and K122; H4-K12, K31, K77, K79 and K91; H2A-K9, K36 and K95; H2B-K5, K34, K43, K46, K85, K108, K116, and K120	Wang et al. (2017)

(continued)

Table 9.1 (continued)

Substrate for modification	Type of modification	Metabolites involved	Modifiable residue	References
	Benzoylation	Benzoyl-CoA	H3- K4, K9, K14, K18, K23, and K27; H4- K5, K8, K12, and K16; H2A- K9, K13, and K36; H2B- K5, K11, K16, and K20; H1- K25, K33, K104, K147, and K190	Huang et al. (2018)
	Crotonylation	Crotonyl-CoA	H3-K4, K9, K18, K23, K27, K56, K79, and K122; H4-K5, K8, K12, K16, K59, K77, K79 and K91; H2A-K36, K95, K118, K119 and K125; H2B-K5, K11, K12, K15, K16, K20, K23, K35, K108, and K116	Gowans et al. (2019)
	Butyrylation	Butyrate	H3- K9, K14, K18, K23, K27, K79, K115, and K121; H4- K5, K8, K12, K16, K31, K44, K77, K79, and K91; H2A- K5, and K95; H2B- K5, K11, K12, K15, K16, K20, K23, and K46	Goudarzi et al. (2016)

(continued)

Table 9.1 (continued)

Substrate for modification	Type of modification	Metabolites involved	Modifiable residue	References
	β-hydroxybutyrylation	β-hydroxybutyrate	H3- K4, K9, K14, K18, K23, K27, K56, K79, K115, and K122; H4- K5, K8, K12, K31, K77, and K91; H2A- K5, K9, K13, K36, K95, K118, K119, and K125; H2B- K5, K11, K16, K20, K34, K85, K108, K116, and K120; H1- K31, K33, K45, K51, K62, K63, K74, K84, K89, K105, K135, and K168	Zhang et al. (2019b)
	2-hydroxyiso butyrylation	2-hydroxyisobutyrate	H3- K4, K9, K14, K18, K23, K27, K36, K56, K64, K79, and K122; H4- K5, K8, K12, K16, K31, K44, K59, K77, K79, and K91; H2A- K5, K9, K36, K74, K75, K95, and K118; H2B- K5, K12, K20, K23, K24, K34, K43, K46, K57, K85, K108, K116, and K120; H1- K22, K25, K26, K45, K51, K62, K63, K74, K80, K84, K89, K96, K109, K116, K120, K128, K135, K147, K158, K167, and K212	Dai et al. (2014)
	Malonylation	Malonate	H3- K14, K18, K23, K56, K79, and K122; H4- K8, K31, K77, K79, and K95; H2A- K95; H2B- K5, K34, K43, K108, K116, and K120; H1- K45, K62, K63, K74, K84, K89, K96, K105, and K158	Xie et al. (2012)

(continued)

Table 9.1 (continued)

Substrate for modification	Type of modification	Metabolites involved	Modifiable residue	References
	Glutarylation	Glutarate	H2B- K5, K116, and K120	Tan et al. (2014)
	Homocysteinylation	Homocystein	H3- K9, K18, K23, K27, K36, K37, K79, and K115; H4- K8, K16, K31, K44, K59, K77, and K79; H2A- K9, K15, K74, K75, K95, K99, K118, and K119; H2B- K5, K11, K12, K15, K16, K20, K23, K24, K34, K43, K46, K57, K108, K116, K120, and K125	Zhang et al. (2018)
	Serotonylation	Serotonin	H3Q5	Anastas and Shi (2019), Farrelly et al. (2019)
	Dopaminylation	Dopamine	H3Q5	Lepack et al. (n.d.)
	O-GlcNAcylation	UDP-GlcNAc	H3- S10, and T32; H3.3- T80; H4- T47; H2A- S40, and T101; H2A.X- T101, and S139; H2B- S36, T52, S55, S56, S64, S91, S112, and S123	Chen et al. (2013), Zhang et al. (2011)
	ADP-Ribosylation	NAD^+	H3-K27 and K37; H4-K16; H2A-K13, and H2B-K30	Messner et al. (2010), Messner and Hottiger (2011)
	MGO Adduction	Methylglyoxal (MGO)	H3-R8, R17, R26, R53, R63, R72, R128; H4-R23, R55, R91, R95; H2A-R35, H2A. X-R20 AND H2B-R79, R86, R92, and R99	Galligan et al. (2018), Kold-Christensen and Johannsen (2020)

Acetyl CoA

An acetyl CoA moiety is typically composed of an acetyl group (CH_3CO) attached to coenzyme A via a thioester bond (CoASH) (Alrob and Lopaschuk 2014). While glucose is converted into CO_2 and H_2O via its complete oxidation to produce ATP, Acetyl CoA is formed as an important intermediate in the process (Schug et al. 2015). The abundance of this intermediate dictates the overall energy state of the cell. Other than participating in various metabolic pathways as a precursor of different metabolites, it influences multiple proteins in their function of acetylation to epigenetically regulate gene expression (Pietrocola et al. 2015). Histone acetylation on its lysine residues is highly sensitive to the pool of acetyl CoA across the species. Being an obligate cofactor for the acetylation reaction, its abundance in the nucleo-cytosolic pool has a direct effect on the enzymatic activity of histone acetyltransferases (HATs), especially for those proteins that have a higher affinity for the acetyl CoA (Xiong and Guan 2012). Histone acetylation neutralizes the charge on the protein resulting in the unwinding of DNA from the histone (Hongs et al. 1993; Luger and Richmond 1998). Acetylation on the selective site of the nucleosome may influence the breathing dynamics and stability of the same. This in turn increases the accessibility of histone tails to different transcription factors and epigenetic modifiers, leading to more open chromatin conformation (Bowman and Poirier 2015). Not only in acetylation, Acetyl CoA is involved in other histone modifications as well, such as Histone butyrylation, crotonylation, propionylation, as these acylations use acyl-CoA moieties that may use acetyl CoA as their precursors (Schnekenburger and Diederich 2015). This in turn facilitates the transcription program and helps specific transcription factors to bind to their cognate sites (Filippakopoulos and Knapp 2014; Zeng et al. 2010). Acetyl CoA, as well as the modifying enzymes i.e., HATs and Histone deacetylases (HDACs), regulate the histone acetylation programs, which in combination with other histone post-translational modifications modulate the transcriptional landscape of chromatin (Fukuda et al. 2006).

Nicotinamide Adenine Dinucleotide (NAD^+)

Nicotinamide Adenine Dinucleotide (NAD^+) serves as a critical cofactor that carries electrons during any oxidation-reduction reactions in all living cells (Imai and Guarente 2014). NAD^+ also functions along with sirtuins (SIRTs) and modifies histone and non-histone proteins and DNA (Imai and Guarente 2014). It acts as an H^+ carrier and regulates the electron transport chain (ETC) to maintain the normal metabolic homeostasis of cells (Abeti and Duchen 2012). Recent evidence shows Lysine residues on Histones can be reversely modified with mono or poly-ADP ribose moiety produced from NAD^+ by the enzymatic reaction of poly (ADP-ribose) polymerase [PARP]. Sirtuin family proteins, namely SIRT4 (Haigis et al. 2006) and SIRT6 (Liszt et al. 2005; Mao et al. 2011), also ribosylate histones with the help of

NAD^+. Nicotinamide Adenine Dinucleotide Phosphate (NADPH) has recently been shown to bind HDAC3 and disrupt its interaction with the co-activators Ncor1/2 (nuclear receptor corepressor 1/2). There exists a competitive dynamic regulation between NADPH and D-myoinositol-(1,4,5,6)-tetrakisphosphate-a positive regulator of HDAC3 and Ncor, with NADPH having a higher affinity for HDAC3. Lower NADPH levels can, therefore, impede transcription by abrogating histone acetylation (Li et al. 2021). Another indirect role for NADPH has been implicated in the Isocitrate dehydrogenase 1/2 (IDH1/2)-catalyzed conversion of α-Ketoglutarate (α-KG) to 2-hydroxyglutarate (2-HG), with the cofactor being involved in the activity of the oncogenic enzymes (Sun et al. 2021a). In addition, the cells being exposed to reactive oxygen species (ROS), NADPH can protect against its toxicity and help in the regeneration of glutathione (GSH) (Rush et al. 1985). Impaired levels of NADPH can lead to an increased rate of oxidative stress mediated cell death. As it plays an indispensable role in DNA replication and repair and stress response pathways, a functional deregulation of NADPH has a close association with tumorigenicity which is an emerging field of research (Dong and Wang 2017; Huang et al. 2017). NADPH also plays a vital role in the regulation of lipid metabolism. High levels of fatty acid metabolism in tumor cells require large amounts of NADPH as substrates, hence inhibiting NADPH synthesis may target tumor cells.

Tetrahydrofolate (THF)

Tetrahydrofolate (THF) is the biologically active form of folic acid found in vitamin B9, a part of the vitamin B complex (Shuvalov et al. 2017). It resides in the cytoplasm, nucleus, and mitochondria and takes part in one-carbon metabolism as a carrier of those carbon units in N5 or N10 or even both N5 and N10 compounds (Shuvalov et al. 2017; Tibbetts and Appling 2010; Yang and MacKenzie 1993). THF is an important intermediate in one-carbon pathways that functions in purine and thymidylate (dTMP) biosynthesis and amino acid metabolism (Anderson et al. 2012; Field et al. 2014). Hence it has an important role in DNA replication (Shuvalov et al. 2017). During histone and DNA methylation, S-adenosylhomocysteine (SAH) is generated from SAM and is further metabolized to homocysteine. Tetrahydrofolate acts as a methyl donor and feeds into the SAM biosynthetic pathway by regenerating methionine from homocysteine. This methionine can then be conjugated to ATP by methionine adenosyltransferase to form SAM. This cycle is essential to maintain a SAM/SAH ratio that directly impacts cellular histone and DNA methylation status. In addition, SAH can act as a competitive inhibitor of SAM to downregulate the activity of histone and DNA methyltransferases (Etchegaray and Mostoslavsky 2016; Yu et al. 2018). Several studies have established the effect of anti-folate compounds in cancer development; however, the role of folate in the same context is yet not deciphered (Chen et al. 2017; Neradil et al. 2012; Wang et al. 2013).

Flavin Adenine Dinucleotide (FAD)

Flavin adenine dinucleotide (FAD) is an indispensable factor for one and two-carbon redox, nucleophilic and electrophilic reactions. It is commonly known as the nucleotide form of Vitamin riboflavin or B2 (Mishanina and Kohen 2015). As a cofactor for redox reactions, it is involved in the activities of oxidation-reduction in mitochondria, cytoplasm, and even in nuclear compartments (Molano-Arevalo et al. 2014). The total FAD synthase pool of nucleus, cytosol, and mitochondria gets influenced by the presence or absence of oxygen and nutrients and constitutes a flavin network that functions in various cellular metabolic and epigenetic events (Giancaspero et al. 2013). The ratio of FAD/$FADH_2$ directly affects the functions of lysine-specific demethylases (LSD1/2) in epigenetic regulation of several genes (Hino et al. 2012; Hou and Yu 2010; Li et al. 2007). Hence, FAD being regulated by the nutrient or oxygen availability, and intra- or extracellular environmental conditions, regulates gene expression by participating in demethylation reactions to maintain cellular homeostasis (Etchegaray and Mostoslavsky 2016).

Other Metabolites for Non-canonical Histone Modifications

Acyl-coA

In recent times, emerging metabolic precursors are the acyl-CoA molecules that are involved in histone modifications. Molecules such as crotonyl CoA (Sabari et al. 2015; Fellows et al. 2018), succinyl-CoA (Wang et al. 2017), malonyl CoA, butyryl CoA (Goudarzi et al. 2016), benzoyl-CoA (Huang et al. 2018), etc., function as the donor for the groups like crotonate, succinate, malonate, butyrate, benzoate respectively to the histone tails. Acetyl CoA plays important role in the pathways of generating these acyl-CoA molecules. Apart from these short-chain fatty acids (SCFA), other metabolic intermediates such as lactate (Zhang et al. 2019a), propionate, glutarate, β-hydroxybutyrate (Xie et al. 2016; Zhang et al. 2019a), 2-hydroxyisobutyrate, etc., play an essential role in imparting the corresponding modifications on histones.

Though it is still poorly understood, pieces of evidence suggest that the different HATs perform the enzymatic reaction of transferring these higher molecular weight moieties to their respective histone tails (Sabari et al. 2015; Wang et al. 2017). These modifications in general impart the opening of chromatin structure, aiding the transcription processes.

UDP-N-acetylglucosamine (UDP-GlcNAc)

UDP-N-acetylglucosamine (UDP-GlcNac) is produced from fructose-6-phosphate and is an indispensable precursor for the synthesis of cell wall peptidoglycan, lipopolysaccharides. Recently, it has been reported that histones also can get O-GlcNAcylated at the hydroxyl moiety of its serine/threonine residues with the help of UDP-GlcNAc (Zhang et al. 2011). The enzymatic reaction is majorly carried out by *O*-GlcNAc transferase (OGT), a member of the polycomb group proteins (Chen et al. 2013). This modification essentially helps to regulate gene transcription and might influence chromatin structure while DNA replication occurs.

Monoamines

Different metabolic processes give rise to the production of several monoamines like dopamine, serotonin, etc. These monoamines are important neurotransmitters. While serotonin is essential in the maintenance of neuronal circuits, dopamine plays important role in regulating the reward circuit of the brain. Recent evidence has emerged showing these neurotransmitters to be directly associated with the Glutamine 5 of histone H3 (H3Q5ser or H3Q5dop) with the help of the enzyme transglutaminase 2 (TGM2) (Lepack et al. 2020; Farrelly et al. 2019). There is a crosstalk between H3K4me3 and H3Q5ser as the trimethylation at lysine 4 is required for the serotonylation at glutamine 5 of histone H3.

Oncometabolites

Oncometabolites are very important in the development and progression of cancer. The specific metabolic pathways where the oncometabolite are formed and the downstream signaling cascades that they regulate represent the most potential therapeutic targets (Sciacovelli and Frezza 2016). The clinical laboratories play a critical role in identifying the specific oncometabolite production pathways and developing targeted sensitive assays to measure these aberrant metabolites and their downstream effectors. Development of these assays can aid the treatment as screening tools for initial and follow-up measures of tumorigenesis, as well as help to detect minimal residual disease and recurrence.

D-2-hydroxyglutarate (D-2-HG), L-2-hydroxyglutarate (L-2-HG), succinate, and fumarate are recognized as oncometabolites. They share a similar structure and function within the tricarboxylic acid cycle and their mechanisms of action to promote tumorigenesis is also quite similar (Khatami et al. 2019). Hence targeting these particular oncometabolites can be a potential therapeutic regimen to treat specific cancers. As these metabolites specifically are produced in cancer cells, it is easy to keep the healthy cells intact while treatment for the cancerous cell can be effective (Collins et al. 2017).

The oncometabolites 2-hydroxyglutarate, fumarate, and succinate have a profound effect on the epigenetic status of the cell. IDH1/2 mutants drive the formation of 2-HG from α-Ketoglutarate (α-KG), with the former being a potent competitive inhibitor of the latter. Thus, 2-HG accumulation leads to reduced activity of Ten-eleven Translocation (TET) proteins and Jumonji C domain-containing

(JMJDs) proteins resulting in hypermethylated DNA and histone profiles (Sun et al. 2021b). Additionally, increased methylation on CCCTC sequences blocks binding of CCCTC-binding factor (CTCF) proteins and reorganization of topologically associated domains, adding another layer of complexity to 2-HG-mediated epigenetic reprogramming. Fumarate and succinate similarly act as inhibitors of α-KG, and mutations in fumarate hydratase and succinate dehydrogenase are known to be oncogenic (Xiao et al. 2012).

Metabolites are important in imparting changes in the chromatin landscape by influencing the activities of writer or eraser proteins to create or remove different epigenetic marks on the promoter of important genes. These modifications are specifically recognized by a plethora of reader/effector proteins which can reprogram the epigenetic landscape by recruiting the writers or erasers and subsequently fine-tune the downstream functions of those genes implicated in different pathways (Musselman et al. 2012). The following section encompasses the diverse reader family and their specific domain conformations (Fig. 9.3) that aid the proteins in their specific recognition of substrates and accomplish their function (Table 9.2).

Different Epigenetic Reader Domains and Their Functions

Methylation Readers

Any protein can typically get methylated on either one of its two amino acid resides—lysine or arginine. In general, there are three different methylation states of lysine, they are either monomethylated, demethylated, or trimethylated (Theillet et al. 2012). The following section details the various modifications on either lysine or arginine and their specific readers whose structural conformation gives rise to different types of domains that recognize histone methylation.

Lysine Methylation

The most common sites for methylation comprise six lysine residues of Histone H3, i.e., K4, K9, K26, K27, K36, and K79: one residue of H4 (K20) and one of H1 (K26). There are several specific methyl–binding domains identified and characterized to date, such as–Chromodomain, double Chromodomain (DCD), Chromobarrel, Tudor, tandem Tudor domain (TTD), WD40, ankyrin, Malignant brain tumor (MBT), Plant Homeodomain (PHD).

Though the various methyl-binding domains are diverse in structural aspects, most of them bind the PTM through an aromatic cage. The complex formation occurs between the methylammonium group of the substrate and the aromatic ring present in the reader domain as well as via hydrophobic and van der walls interactions.

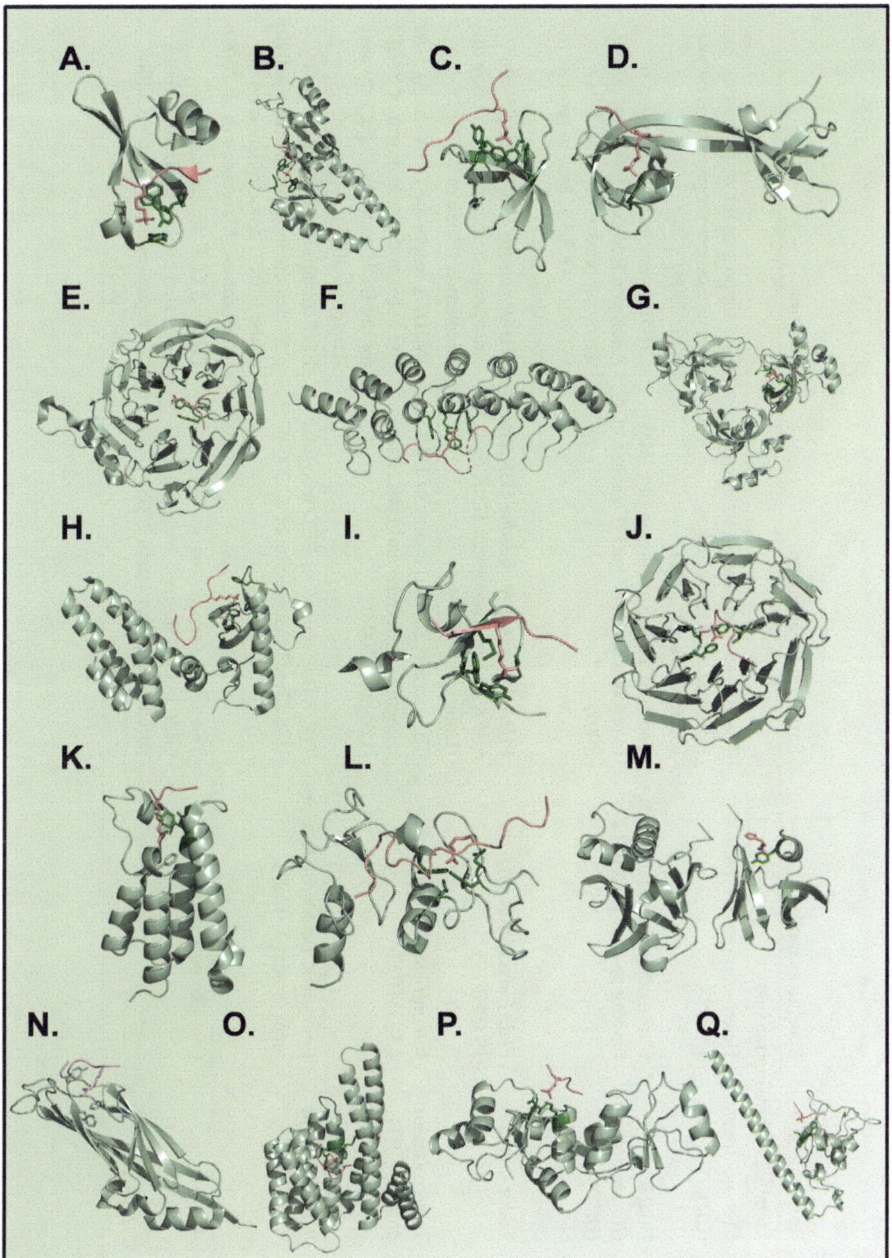

Fig. 9.3 Co-crystal structures of different chromatin readers (Green) interacting with modified histones (Pink). (**a**) HP1 (1KNE), (**b**) CHD1 (2B2W), (**c**) PHF19 (6WAU), (**d**) JMJD2A (2QQS), (**e**) EED (3IIW), (**f**) GLP (3B95), (**g**) hL3MBTL (2RHX), (**h**) ZMYND11 (4N4H), (**i**) ING2 (2G6Q), (**j**) WDR5 (2H6N), (**k**) GCN5 (1E6I), (**l**) DPF3b (2KWJ), (**m**) Rtt106 (3TW1), (**n**) AF9 (4TMP), (**o**)14-3-3 (2C1J), (**p**) MDC1 (2AZM), (**q**) Survivin (3UIG). PDB ID of each co-crystal structure is represented in parenthesis

Table 9.2 Details of different Histone modification reader domain: structure and function

Histone modification recognition	Reader domain	Specific example	Structural organization	Function
	Methylation binding module			
Lysine Methylation	Chromo domain (CD)	Hetero chromatin Protein 1 (HP1)	The HP1 chromodomain interacts with H3K9me2/3 (Jacobs and Khorasanizadeh 2002)	HP1 suppresses tumor-promoting genes to retard cancer progression (Dialynas et al. 2008)
	Double Chromodomain (DCD)	Chromodomain Helicase DNA binding protein 1 (CHD1)	The double chromodomain of CHD1 interacts with H3K4me3 (Flanagan et al. 2005)	CHD1 enhances the expression of Androgen Receptor (AR) responsive tumor suppressor genes such as FOXO1, NKX3.1, etc., in prostate tissue (Mills 2017)
	Tudor domain	PHD Finger Protein 19 (PHF19)	PHF1/19 recognizes histone H3K36Me3 and testis-specific variant H3tK27me3 (Dong et al. 2020)	PHF19 recruits PRC2 complex to embryonic stem cell genes during differentiation (Brien et al. 2012)
	Tandem Tudor domain (TTD)	Jumonji Domain containing 2A (JMJD2A)	The tandem Tudor domains of JMJD2A recognize H3K4me3 and H4K20me3 (Lee et al. 2008)	JMJD2A marks the DNA damage site and helps in the recruitment of 53BP1 for the DDR pathway to get activated (Lee et al. 2008)
	WD40 domain	Embryonic Ectoderm development protein 1 (EED)	The WD40-repeats of EED recognize H3K27me3 (Margueron et al. 2009)	EED subunit of PRC2 allosterically activates the methyltransferase activity of the H3K27 trimethylation on the nucleosomes (Margueron et al. 2009)
	Ankyrin	G9a-like protein (GLP)	The ankyrin repeat domain of G9a and GLP, bind H3K9me1 and H3K9me2 (Brent and Marmorstein 2008)	G9a and GLP are euchromatin-associated histone methyltransferase that represses transcription (Collins et al. 2008)

	Malignant brain tumor (MBT) domain	Human Lithal (3) Malignant Brain Tumor-like protein (hL3MBTL)	MBT domain of hL3MBTL1 recognizes H4K20me1 and H4K20me2 (Li et al. 2007)	hL3MBTL1acts as a tumor suppressor by silencing E2F target genes such as *MYC* and *CYCLIN E1* (Trojer et al. 2007)
	PWWP	Zinc finger MYND domain-containing protein 11 (ZMYND11)	PWWP domain of ZMYND11 recognizes histone variant H3.3K36me3 (Wen et al. 2014)	ZMYND11 regulate transcription elongation, pre-mRNA splicing and can act as tumor suppression (Ladendor et al. 2001; Wen et al. 2014)
	Plant homeodomain (PHD) finger	Inhibitor of Growth Protein 2 (ING2)	ING2 PHD finger recognizes H3K4me3 (Peña et al. 2006)	ING2 functions as a tumor suppressor (Li et al. 2011); stabilizes the mSin3a-HDAC1 complex, thereby repressing *Cyclin D1* expression in response to DNA damage (Shi et al. 2006)
Arginine Methylation	WD40 domain	WD Repeat-containing protein 5 (WDR5)	WDR5 does not read out the methylation state of histone H3K4 directly, but instead, it recognizes unmodified histone H3R2 (Ruthenburg et al. 2006; Suganuma et al. 2008)	It is important in maintaining euchromatin structure in human cells (Migliori et al. 2012)
	Acetylation binding module			
Lysine Acetylation	Bromodomain	General Control Non-derepressible 5 (GCN5)	The Gcn5p bromodomain recognizes H4K16ac (Owen et al. 2000)	The bromodomain of GCN5 enhances its HAT activity and consequent transcription activation (Mutlu and Puigserver 2021)
	Double PHD finger (DPF)	Double PHD Finger-containing protein 3b (DPF3b)	DPF3b tandem PHD finger (PHD12) recognizes H3K14ac (Zeng et al. 2010)	The DPF3b in association with the BAF complex activates transcription during heart and muscle development (Lange et al. 2008)
	Double Pleckstrin homology (PH)	Regulator of Ty1 transposition protein 106 (Rtt106)	The double PH domains of Rtt106 interact with histone H3K56ac (Su et al. 2012)	Rtt106 along with CAF1 plays important role in histone deposition during DNA replication (Clemente-Ruiz et al. 2011)

(continued)

Table 9.2 (continued)

Histone modification recognition	Reader domain	Specific example	Structural organization	Function
	YEATS domain	AF9	AF9 YEATS domain interacts with histone H3K9Ac (Li et al. 2014).	AF9-mediated H3K9ac recognition is instrumental in recruiting Dot1l to chromatin and regulate transcription (Li et al. 2014).
	Phosphorylation binding module			
Serine Phosphorylation	14-3-3	14-3-3 protein beta/alpha (14-3-3)	14-3-3 binds to histone H3S10p (Macdonald et al. 2005)	It transcriptionally regulates anti-apoptotic or mitotic genes (Winter et al. 2008)
	BRCT	Mediator of DNA damage checkpoint protein 1 (MDC1)	The BRCT domain of MDC1 recognizes the γH2A.X (H2A.X variant phosphorylated at S139) following DNA damage (Stucki et al. 2005)	MDC1 interacts with other DDR proteins and functions at the early stage of DNA damage response (Coster and Goldberg 2010)
Threonine Phosphorylation	BIR	Baculoviral IAP repeat-containing protein 5 (BIRC5 or Survivin)	Survivin binds to the histone H3T3p (Du et al. 2012)	Survivin inhibits Caspase activation, thereby leads to suppression of Apoptosis and progression of the cell cycle (Rivadeneira et al. 2015)

The specificity of reader protein is imparted by the interaction of its surrounding residues. Chromodomain, chromobarrel, Tudor, TTD, MBT, and PWWP modules are structurally related and contain a characteristic β-barrel topology. They together are called the Royal superfamily.

The smallest of all the methyl-binding domains is the chromodomain that comprises four curved β-Strands and an α-helix. In general, it recognizes trimethylated lysine residues, but there are exceptions as well (Bannister et al. 2001; Fischle et al. 2003; Min et al. 2003). Heterochromatin Protein 1 (HP1) is an example of chromodomain-containing histone methylation reader that recognizes H3K9me3 (Jacobs and Khorasanizadeh 2002; Nielsen et al. 2002) and has implications in suppressing tumor-promoting genes to retard cancer progression (Dialynas et al. 2008). Chromodomain Helicase DNA binding protein 1 (CHD1) contains double Chromodomain (DCD) that recognizes H3K4me3 (Flanagan et al. 2005; He et al. 2010) and increases the expression of Androgen receptor (AR) responsive tumor suppressor genes such as *FOXO1, NKX3.1*, etc., in prostate tissue (Mills 2017).

Other Royal family members carry five stranded β-barrel, and the methylated histone tail falls upon an open edge of the β-barrel closely near the aromatic cage of the reader domain. Single Tudor or TTD-containing reader proteins are the best examples of these Royal family members (Ballaré et al. 2012; Bian et al. 2011). PHD Finger Protein 19 (PHF19) contains the Tudor domain that recognizes H3K36me3 (Dong et al. 2020), and recruits multi-subunit Polycomb repressive complexes 2 (PRC2) to its target tumor suppressor gene promoter to selectively silence them and to form heterochromatin structure (Brien et al. 2012).

MBT, another methylation recognizing domain, reads lower methylated states of the amino acids. If there are 3MBT modules present, they form a propeller-like structure, whereas four-repeat 4MBT looks like a rhombus structure (Grimm et al. 2007; Min et al. 2007). MBT domains show less specificity towards the substrate as their mode of recognition is limited to few contacts with histone tails (Guo et al. 2009). MBT domain-containing Human Lithal (3) Malignant Brain Tumor-like protein (hL3MBTL) reads H4K20me1 (Li et al. 2007) and suppresses E2F target genes such as *MYC* and *CYCLIN E1,* showing its tumor suppressor function (Trojer et al. 2007).

On the other hand, PWWP or chromobarrel-containing proteins show greater specificity as they recognize heavily methylated (mostly tri) lysines at H3K36 (Vezzoli et al. 2010; Wu et al. 2011). A PWWP domain-containing protein Zinc finger MYND domain-containing protein 11 (ZMYND11) reads histone variant H3.3K36me3 (Wen et al. 2014) and acts as a tumor suppressor (Ladendor et al. 2001). Chromobarrel-containing protein MORF-related gene 15 (MRG15) reads H3K36me2/3 (Zhang et al. 2006) and regulates transcriptional processes and homology repair mediated DNA Damage Response (DDR) pathways (Sun et al. 2020).

PHD is a distinct methylation binder and not a member of the Royal family of methylation readers. It has a characteristic C4HC3 motif that contains two zinc ions having cross–brace topology. Most of the PHD fingers are specific readers of H3K4Me3, though few exceptions are also there. Inside the PHD finger, most of the proteins hold the trimethylated H3K4 in an aromatic cage formed through the specific arrangements of antiparallel β-strands (Li et al. 2006; Musselman and Kutateladze 2011; Peña et al. 2006; Shi et al. 2006; Wysocka et al. 2006). PHD finger-containing Jumonji/ARID domain-containing protein 1A (JARID1A) reads H3K4me3 and gets recruited there to remove the mark to prevent hematopoietic differentiation. Removal of this particular mark from the gene loci keeps the transcription happen constitutively, leading to Acute Myeloid Leukemia (AML) (Wang et al. 2009b).

Another important lysine-methylation reader module is WD40 which recognizes trimethylated substrate through an aromatic cage residing at the top of a seven-bladed β-propeller channel (Margueron et al. 2009). One of the well-known WD40 family members is Embryonic Ectoderm Development protein 1 (EED), a component of the canonical PRC2 repressor complex (Margueron et al. 2009; Xu et al. 2010).

Arginine Methylation

In the case of Arginine methylation, the omega nitrogen of the amino acid gets either monomethylated or dimethylated symmetrically (Rme2s) or asymmetrically (Rme2a). The reported arginine residues that get methylated are listed as H3R2, H3R17, H3R26, H4R3, H2AR11, and H2AR29. The information about the writer, eraser, and even the readers are limited.

A subset of Tudor domains recognizes non-histone arginine methylation but the structural information is limited. Recent studies have shown Tudor domain of Tudor Domain-containing 3 protein (TDRD3) recognizes H3R17me2a and H4R3me2a (Liu et al. 2012; Yang et al. 2010), but no information about the structural conformation of this interaction is available. However, other studies of proteins like Survival of motor neuron (SMN) and Splicing factor 30 (SPF30)-containing Tudor domains that recognize arginine methylation reveals probable structural information which suggests that similar to lysine methylation, methylated arginine also gets incorporated in the aromatic cage of the Tudor beta-barrel (Tripsianes et al. 2011). The function of this domain-containing TDRD3 protein is mainly implicated in helping to recruit a co-transcriptional activator to its cognate gene loci (Narayanan et al. 2017).

Another arginine methylation reader domain is the WD40 repeat of WD Repeat-containing protein 5 (WDR5) (Schuetz et al. 2006). It favors H3R2me2s over H3K4me0 or H3K4me3 residues (Couture et al. 2006; Han et al. 2006; Ruthenburg et al. 2006). While the initial residues of the H3 peptide make primary contact with WDR5 protein, the methylated arginine residue gets incorporated in the deep channel of beta-propeller where the guanidine moiety of the methylated residue gets trapped in between two phenylalanine rings of WD40 repeats (Suganuma et al. 2008). This interaction between the WDR5 WD40 repeats with H3R2me2s has shown importance in maintaining euchromatin structure (Migliori et al. 2012).

Acetylation Readers

Apart from being methylated, lysine residues of histones can also get acetylated. To date, the residues found to be acetylated can be listed as H2A-K5 and K9; H2B-K5, K12, K15, K16, K20, and K120; H3-K4, K9, K14, K18, K23, K27, K36, and K56; H4-K5, K8, K12, K16, K20, and K91. Acetylation in general causes the chromatin structure to be opened for transcription as it weakens the interaction between positively charged histones and negatively charged DNA. At present, three acetylation reader domains have been identified and studied.

The first identified acetyl-lysine reader domain is Bromodomain (Dhalluin et al. 1999; Mujtaba et al. 2007). Though there is not much sequence similarity across the various bromodomain-containing proteins, they harbor a highly conserved four-helix bundle structure. Its two inter-helical structures form the deep hydrophobic pocket where the acetyl-lysine interacts. While the isolated bromodomain may loosely associate with the acetylated lysine, the surrounding residues impart specification to its recognition (Filippakopoulos et al. 2012). Bromodomain-containing BRD family proteins are the most common example of this group of acetyl-readers (Cieniewicz et al. 2014; Mutlu and Puigserver 2021; Owen et al. 2000). General Control Non-derepressible 5 (GCN5) promotes human lung cancer cell growth by positively regulating the transcription of genes in cell cycle progression, gene transcription, and suppression of apoptosis (Mao et al. 2009; Mutlu and Puigserver 2021).

Another more recent finding shows double PHD finger (DPF) of Double PHD Finger-containing protein 3b (DPF3B) recognizes acetylated H3K14 residue through forming a zinc finger topology with a broader interface (Zeng et al. 2010). This typical structure gives this domain a unique scaffold-like organization that helps the protein recognize and bind the specific modification. One of the two PHD fingers of this protein binds to the first four N-terminal residues of H3K14ac, while the other PHD module traps K14ac in its binding pocket made of hydrophobic and charged residues. It regulates transcription during heart and muscle development along with the BAF complex (Lange et al. 2008). Histone binding of DPF3b plays an important regulatory role in the transcriptional programs involved in chronic lymphocytic leukemia and breast cancer, advance stages of which may lead to lymph node metastases (Hoyal et al. 2005; Theodorou et al. 2013).

The double pleckstrin homology (PH) domain-containing Regulator of Ty1 transposition protein 106 (Rtt106) is a histone chaperone that recently has been identified to be recognizing H3K56ac (Su et al. 2012). The PH domain is structurally comprised of a seven-stranded β-barrel whose one of the open ends being covered with one or two α-helixes. The acetylated lysine gets trapped in a cleft-like structure consisting of an aspartate, a tyrosine, and an arginine at the interface of the c-terminal α-helix and the β-barrel. It plays important role in histone deposition during DNA replication in association with CAF1 (Clemente-Ruiz et al. 2011). Human homolog of Rtt106 is Structure specific recognition protein 1 (SSRP1), which has been implicated in the regulation of cancer (Liao et al. 2017; Ding et al. 2016; Wu et al. 2019).

Another important acetyl reader domain is YEATS domain. It was identified in Yeast homologue of ALL1-fused gene from chromosome 9 (Yaf9) protein (Wang et al. 2009a). The name of the domain came from the initial of the first five identified protein names, i.e., Yaf9, Transcription initiation factor TFIID subunit 14 (Taf14) and Something About Silencing 5 (Sas5) of Yeast and ALL1-fused gene from chromosome 9 (Af9) and ENL of human (Schulze et al. 2009). The sequences of these proteins are also mostly conserved throughout the species (Masson et al. 2003). They share eight-stranded immunoglobulin fold. They mainly recognize H3K9Ac mark utilizing a Serine-lined aromatic cage (Li et al. 2014; Wang et al. 2009a). Taf14 recently has been identified as crotonylation binding protein as well (Andrews et al. 2016; Gowans et al. 2019). It has been implicated in the repression of energy-demanding gene expression during metabolically stressed conditions in Yeast (Gowans et al. 2019).

Phosphorylation Readers

Phosphorylation on a moiety, either Serine or Threonine, causes unwinding of DNA structures from histone due to electrostatic alteration by the bulky and negatively charged phosphate group. Phosphorylation by various kinases and recognition of them by different cytoplasmic proteins are well studied, but information on histone phosphorylation readers is very little. It is known that T3, T6, S10, T11, S28, and T45 serine/threonine residues of H3, S1 residue of H4, T120 of H2A, S14 of H2B and S139 of human histone variant H2A.X (known as γH2A.X upon phosphorylation) are susceptible to phosphorylation. These modifications are biologically connected mostly with the DNA damage response pathway or transcriptional regulation functions.

One of the well-known phospho-modification binding domains is BRCT which is formed with a four-stranded β-sheet at the core surrounded by α-helices on both the side of each β-sheet, giving it an α-β sandwich-like structure overall. In the BRCT domain, the tandem repeats are aligned such a way that they give rise to a hydrophobic interface where the γH2A.X resides (Stucki et al. 2005). The phosphorylated S139 residue is mostly recognized by the tandem BRCT domain-containing protein BRCA1, which functions widely in the DNA damage response pathway (Chen et al. 1999; Manke et al. 2003).

Another phospho-motif recognizing domain-containing protein is 14-3-3ζ. The phosphorylated H3S10 gets bound inside a fissure formed by five helices of this nine helical protein that typically is U- shaped (Macdonald et al. 2005). This recognition by 14-3-3ζ is functionally implicated during transcriptional regulations of various anti-apoptotic or mitosis-related genes (Winter et al. 2008).

The newest identified phosphorylation binding domain is BIR which recognizes H3T3ph (Du et al. 2012). A typical example is Survivin that inhibits Caspase activation, thereby leading to suppression of apoptosis (Rivadeneira et al. 2015).

The BIR domain is stabilized by a single zinc ion at the center of a small αβ structure. The phosphorylated residue falls in a small positively charged area surrounded by an electronegative surface and is anchored by several hydrogen bonds formed within the BIR domain with the help of its aspartate and glutamate residues.

Ubiquitination Readers

Apart from chemical modification of histones (such as methylation, acetylation, and phosphorylation), small protein molecules also get attached covalently to it and subsequently regulate diverse biological functions. Such modifications can be listed as ubiquitination, SUMOylation, Neddylation, etc. This covalent modification can be of different types. It can be either mono or poly ubiquitination. Again, poly ubiquitination can be straight-chain or branched-chain as well. Ubiquitination of proteins majorly leads to their degradation; however, it also acts as a quality control check point in vesicular trafficking, DNA repair, transcription regulation, cell cycle and growth factor signaling, stress response, etc. biological pathways (Karve and Cheema 2011). While most of the cellular cytoplasmic ubiquitin-binding-domain-containing proteins carry out these functions, histone ubiquitination readers are mostly involved in the function of maintaining genomic stability. There are, to date, around sixteen different modules or domains identified that recognize ubiquitinated residues. Histones mostly get mono-ubiquitinated. The most susceptible residues can be listed as H2AK13/15, K118/119, K125/127/129; H2BK120; H3K14/18/23, K23/36/37, K121/122/125; H4K91, and K31 (Mattiroli and Penengo 2021).

Reader of H2A Ubiquitination

Different ubiquitination on the same histone recruits different reader proteins. The reader of H2AK15ub was first identified to be p53 binding protein 1 (53BP1) that contains Ub-dependent recruitment (UDR) motif that binds the modification and functions in the Non-Homologous End Joining (NHEJ) pathway (Mallette et al. 2012). This recognition and binding are facilitated by H2AK20me2 that occurs in the pre-replicative DNA favoring NHEJ in the G1 phase of the cell cycle. On the other hand, the same H2AK13/15ub along with unmethylated H2AK20, which is a feature of post-replicative DNA, is recognized by BRCA1-associated RING domain protein 1 (BARD1) (Nakamura et al. 2019), the partner of BReast CAncer gene 1 (BRCA1) protein and facilitates Homologous recombination (HR) mode of DNA damage response. Hence HR is preferred in the case of the S and G2 phases of the cell cycle. This evidence shows how crosstalk between different epigenetic marks is important in determining the cellular fate (Becker et al. 2020). H2AK13/15ub also gets recognized by different DDR pathway proteins such as Ring Finger proteins RNF168, RNF169, and RAD18 and is implicated in DNA damage repairs.

The other lysine moieties of H2A that get ubiquitinated are K125/127/129. Recognition of these marks by BRCA1/BARD1 recruits SWI/SNF-Related, Matrix-Associated Actin-Dependent Regulator Of Chromatin, Subfamily A, Containing DEAD/H Box 1 protein (SMARCAD1) (Densham et al. 2016), a well-known chromatin remodeller at the damaged sites. This recruitment causes nucleosome exchange and previously ubiquitinated H2AK125/127/129, then gets removed along with 53BP1 bound to it. This results in the finishing of DNA end resection as a part of the HR repair machinery (Uckelmann et al. 2018). So, this specific recognition happens during DNA repair and causes alteration in chromatin architecture.

Reader of H2B Ubiquitination

Ubiquitination on H2B occurs typically on the K120 residue (So et al. 2019). This modification leads to involve other writer enzymes like Disruptor of telomeric silencing 1-like (DOT1L), SET1 to act to methylate their specific targets. H2BK120ub stimulates DOT1L to increase methylation on H3K79 (Worden et al. 2019; Worden and Wolberger 2019) and SET1 on H3K4 marks. This ubiquitination-methylation crosstalk is very unique in a sense as a methyltransferase DOT1L binds Ubiquitin at its Ile36 hydrophobic patch and acts on nucleosome by binding through its acidic amino acid residues. This particular crosstalk is important in the context of gene transcription, post-transcriptional RNA modification, and DNA replication (Worden et al. 2019). On the other hand, how SET1 interacts with H2BK120ub and increases H3K4me3 is yet elusive, but the multisubunit organization and H2BK120ub induced H3K4me3 create the premises of investigation in the context of DDR pathways. Thus the crosstalk between epigenetic modifications is indispensable for translating the information (Dover et al. 2002; Shema-Yaacoby et al. 2013; Sun and Allis 2002). Evidences show that ubiquitination at histone H2BK120 is a result of glucose utilization through the glycolytic pathway. It is important for glucose to be uptaken and utilized by the enzymes of glycolytic pathways for the occurrence of H2B120ub, as non-hydrolysable analogs of glucose failed to aid the ubiquitination process (Gao and Xu 2011).

Reader of H3 and H4 Ubiquitination

Readers of histone H3 and H4 ubiquitination marks are not very well studied. It has been seen that mostly the methyltransferases such as DNA Methyltransferase 1 (DNMT1) (reads H3K14/18/23), GCN5 (reads H3K23/36/37) interact with these Ub marks and functions in the maintenance of DNA methylation during DNA replication or transcriptional regulations (Zhang et al. 2017). H3K14ub, in *Schizosaccharomyces pombe*, is read by clr4 and has an important role to play during heterochromatin regulation (Oya et al. 2019). On the other hand, H4 ubiquitinations are important in the context of DNA damage and repair pathways.

For example, H4K31ub is read by Serine/Threonine Kinase 38 (STK38) that has crosstalk with H3K79me3 and regulates DDR (Qin et al. 2019, 2020).

SUMOylation Readers

SUMO (Small Ubiquitin-related MOdifier) modifications on different proteins are now being explicitly studied (Kerscher 2007). But information on histone SUMOylation, however, is limited. Few groups have shown that histone H3 and H4 get SUMOylated on its lysine residues with the help of the E2 SUMO conjugating enzyme Ubiquitin carrier protein 9 (UBC9). Histone SUMOylation readers and erasers are yet to identify. Nevertheless, the SUMO interacting motif (SIM)-containing proteins are known to be recognizing SUMOylated proteins. Three varients of SUMO proteins are there in nature: SUMO1, SUMO2, and SUMO3 that get covalently attached to different proteins to make them SUMOylated (Geiss-Friedlander and Melchior 2007). SIMs generally interact with these SUMO proteins. The SIMs are hydrophobic. They reside in the groove of the α-helix and β-strand of SUMO proteins (Kerscher 2007; Shih et al. 2007). It has been speculated that SIMs presume an intra-molecular β-sheet conformation while interacting with the β2 strand of SUMO1 protein.

Mostly the phenylalanines, valines, and leucines of SUMO1 are involved in the interaction with SIMs. SIMs have shown their important function during DNA replication, glycosylation, etc., when they involve in recognizing SUMOylated proteins either to activate or to repress the gene transcription of these processes. Available literature on SUMOylation of histones and their connection to metabolism or cancer is still not clearly understood. However, studies show SUMOylation on various non-histone proteins have different implications in metabolic alterations. It has been reported that nuclear receptors (NRs) are involved in the regulation of metabolism leading to anti-inflammatory responses. SUMOylation to specific NRs such as Proliferator Activated Receptor (PPARs) (Treuter and Venteclef 2011), Kruppel-like receptor 5 (KLR5) (Oishi et al. 2008) results in either positive or negative regulation of metabolic pathways. SUMOylation of the mevalonate pathway enzyme Hydroxymethylglutaryl-CoA synthase 1 (HMGS-1) is important for the process of aging and has tremendous potential for becoming therapeutic target in the regulation of cholesterol homeostasis (Sapir et al. 2014).

A detailed study on the histone effectors imparting their role in glucose metabolism leading to cancer has been discussed in the next segment of the chapter.

Mechanisms of Reprogramming the Metabolic Landscape Through Epigenetic Regulators

A considerable volume of literature exists on the dynamic regulation of gene expression in glucose metabolism by epigenetic writers and erasers. The role of chromatin readers in this context, on the other hand, has not yet been as well characterized. It is an interesting endeavor to explore the functions of chromatin readers in the regulation of different glucose metabolic pathways and cellular responses.

TRIM24

Mechanism of Action

Tripartite Motif-containing protein 24 (TRIM24), another moniker for Transcription Intermediary Factor 1 alpha (TIF1a), contains tandem C-terminal PHD/Bromo domains (Fong et al. 2018). The role of this protein as a mediator in several gene expression programs has been known for a while now. It is co-recruited with members of the nuclear receptor family of transcription factors such as estrogen receptor alpha, thyroid receptor, and retinoic acid receptor (Kikuchi et al. 2009). Recent studies have shed some light on the fascinating mechanism of action of this dual reader domain-containing protein. Using the PHD finger to recognize and bind to unmethylated H3K4, TRIM24 then shows a strong affinity for acetylated H3K23 via its bromodomain and subsequently upregulates its target genes (Fig. 9.4A). Lysine-Specific Demethylase 1 (LSD1) has been shown to specifically demethylate H3K4me2/1 upstream of TRIM24 binding, and CREB Binding Protein/protein 300 (CBP/p300) is a proposed HAT for H3K23 in this regulatory pathway (Appikonda et al. 2016). Estrogen receptor α (ERα) has well-known roles in glucose homeostasis, affecting glucose transport, glycolysis, and the TCA cycle (Chen et al. 2009), and is also known to recruit CBP/p300 and GCN5 to its target promoters prior to the function of TRIM24.

Glucose Uptake

MDA-MB231 (a triple-negative breast cancer type), MCF-7 (a Luminal A-type cancer), SKBR3 (an HER-2 positive cell line), and transformed human mammary epithelial cells overexpressing TRIM24 show increased uptake of glucose, with GLUT1 and GLUT4 being upregulated significantly (Pathiraja et al. 2015). In addition, GLUT3 is found to be overexpressed in another type of cancer—head and neck squamous cell carcinoma (Wang et al. 2018a).

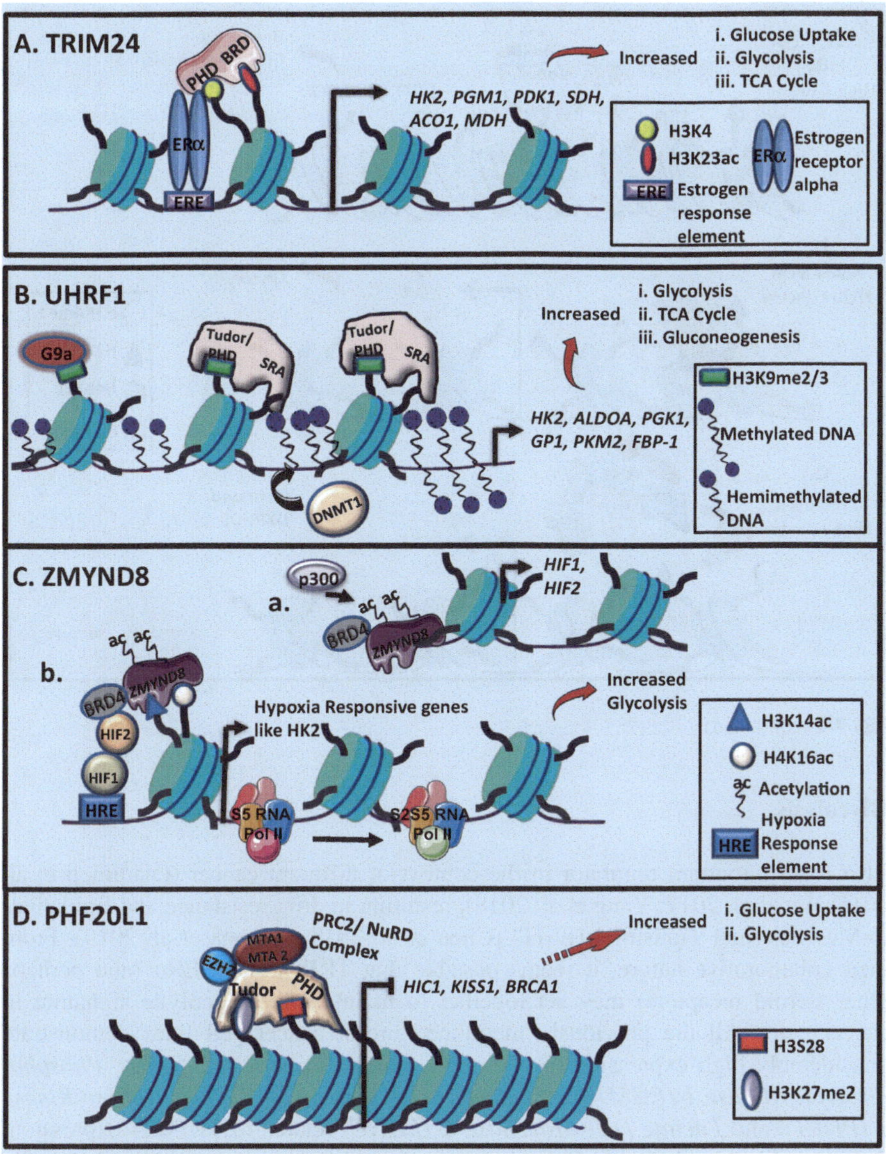

Fig. 9.4 Epigenetic readers recognize Histone marks and regulate different gene transcription to ultimately determine the fate of Glucose metabolism. (**A**) TRIM24 recognizes unmodified H3 through its PHD finger and H3K23ac by its BRD module to regulate several glucose metabolic genes to induce glucose uptake, glycolysis, and TCA cycle helping cancer progression. (**B**) UHRF1 recognizes H3K9me2/3 through its PHD/Tudor domain to bring DNMT1 at hemimethylated DNA of glucose metabolic genes to completely methylate them and increases glycolysis, TCA cycle, and Gluconeogenesis. (**C**) ZMYND8 regulates *HIF1/2* gene expression (**a**). It recognizes dual histone marks H3K14ac and H3K16ac and along with HIF1/2 protein, it regulates hypoxia-responsive genes like HK2 to increase glycolysis (**b**). (**D**) PHF20L1 recognizes unmodified H3S28 through its PHD finger and H3K27me2 by its Tudor domain to suppress important tumor suppressor genes like *HIC1, KISS1, BRCA1* that in turn would downregulate glucose uptake or glycolysis. (**E**) TCF19 through its PHD finger recognizes H3K4me3 and depending upon the time of exposure to high glucose (**a** High glucose, **b** Short-term high glucose, and **c** Prolonged high glucose) in the milieu, it regulates different genes through distinct mechanisms to ultimately dictate the glucose metabolic pathways

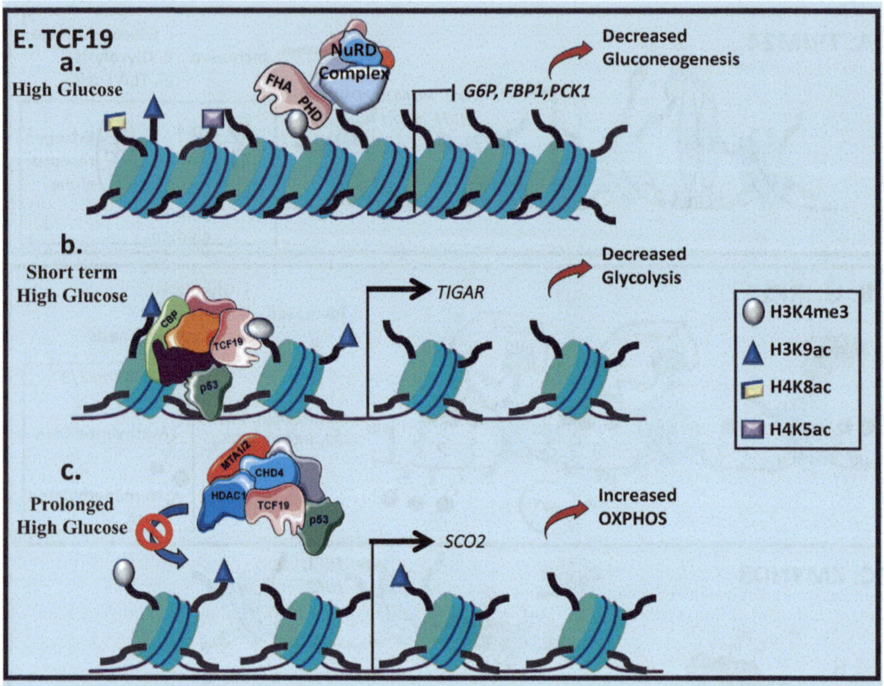

Fig. 9.4 (continued)

Glycolysis

ERα is an important regulator in the context of different cancer (Oladimeji et al. 2016; Wu et al. 2017; Yang et al. 2018), resulting in drug resistance and Epithelial-to-Mesenchymal Transition (EMT) (Chen et al. 2018a; Cuesta et al. 2019). From their collaborative nature, it seems possible that TRIM24 and ERα (and perhaps other steroid receptors) may act together to maintain the glycolytic signature in cancer cells. All the previously mentioned breast cancer cell lines demonstrate considerably high expression of glycolytic genes *Hexokinase 2 (HK2), Phosphoglycerate Mutase 1 (PGM1), α-Enolase (ENO1), Pyruvate Dehydrogenase Kinase 1 (PDK1)*, and *Lactate Dehydrogenase C (LDHC)* related to TRIM24 expression, demonstrating its broad role in transforming the tumor metabolic landscape (Pathiraja et al. 2015). A similar TRIM24-related glycolytic profile is found in head and neck squamous cell carcinoma as well (Wang et al. 2018a).

TCA Cycle

Surprisingly, and perhaps counter-intuitively, TRIM24 also seems to enhance the expression of TCA cycle genes *Succinate Dehydrogenase complex subunits A, B, C (SDH), Isocitrate Dehydrogenase 2 (IDH2), Aconitase 1 (ACO1), Oxoglutarate Dehydrogenase (OGDH)*, and *Malate Dehydrogenase (MDH)*, and cells overexpressing this reader show an enhanced oxygen consumption rate as well (Pathiraja et al. 2015). It appears that, in addition to exploiting aerobic glycolysis, certain cancer cells also maintain a high mitochondrial capacity, indicating that a combination of both energetic pathways is necessary to boost their survival.

UHRF1

Mechanism of Action

An interesting new study has implicated a significant role of the chromatin reader Ubiquitin-like PHD and RING finger domain-containing protein 1 (UHRF1) in glucose metabolism. UHRF1 is a well-known adaptor in the crosstalk between DNA and histone methylation. The protein contains N-terminal TTD, a PHD finger, and a C-Term SET and RING-associated (SRA) domain which mediate its dual interactions with both DNA and histone. Several models have been proposed for its mechanism of action but, thus far, it is understood that UHRF1 recognizes and binds to hemimethylated DNA during replication via its SRA domain and the affinity of this interaction is greatly enhanced by the recognition of H3K9me2/3 via the TTDs and/or, possibly, the PHD finger (Nady et al. 2011; Xue et al. 2019). Subsequently, the reader recruits DNMT1, which carries forward the DNA methylation imprint in daughter cells (Fig. 9.4B). DNA and histone methylation are tightly linked processes, the latter sometimes potentiating the former and vice versa. H3K9 dimethylation by methyl transferase G9a seems to be an upstream event of the recruitment of DNMT1 (Nady et al. 2011), and other histone lysine methyltransferases could be involved in local gene silencing following this event.

Gluconeogenesis

A dual inhibitor of G9a and DNMT activities, CM-272, was shown to have a profound reversal effect on epigenetically silenced genes in hepatocellular carcinoma. One of these genes was *Fructose Bisphosphatase-1 (FBP-1)*, the rate-limiting enzyme in gluconeogenesis (Bárcena-Varela et al. 2019). Since FBP-1 is an inhibitor of glycolysis, it is a potent tumor suppressor and negative regulator of the Warburg effect (Hirata et al. 2016). Consistent with this, lactate production was considerably reduced in CM-272 treated cells (Bárcena-Varela et al. 2019).

Glucose Uptake and Glycolysis

Surprisingly, GLUT1, as well as glycolytic enzymes such as *HK2, Aldolase A (ALDOA), Phosphoglycerate Kinase 1 (PGK1), Glucose-6-phosphate Isomerase (GPI)*, and *Pyruvate Kinase isozyme M2 (PKM2)* were downregulated in CM-272 treated cells (Bárcena-Varela et al. 2019). It could be as a result of the drug rescuing the expression of key repressor/s of these genes in HCC. Clearly, the G9a-UHRF1-DNMT1 axis is a key driver of metabolism-related tumorigenesis in liver cell lines.

PHF20L1

Mechanism of Action

PHD finger protein 20-like protein 1 (PHF20L1), a reader protein is known to interact with methylated non-histone proteins, has recently been shown to have a key regulatory role as a chromatin reader in glycolytic gene expression. The protein has two N-terminal Tudor domains, as well as MBT domain and PHD finger. While the in vivo interactions of the latter two have not been well studied in the context of the protein's functionality, the Tudor domain has been found to bind the H3K27me2 mark on histones. PHF20L1 binds to this mark when H3S28 is not phosphorylated and subsequently recruits the PRC2 and Nucleosome Remodeling and Deacetylase (NuRD) complexes to silence its target genes (Fig. 9.4C). This is achieved by deacetylation of H3K27 and an increase in di- and trimethylation of the same residue at gene promoters. The middle portion of the protein, characterized as a "PRC2-NuRD interacting domain," is responsible for the interaction of PHF20L1 with Enhancer of Zeste homolog 2 (EZH2) and Metastasis Associated Protein 1/2 (MTA1/MTA2) components of the aforementioned remodelers (Hou et al. 2020).

Glucose Uptake and Glycolysis

Overexpression of PHF20L1 indirectly promotes aerobic glycolysis in breast cancer cells and this effect seems to be a result of the silencing of several tumor suppressor genes, such as *HIC1*, *KISS1*, and *BRCA1*, which are involved in attenuating the Warburg effect. Interestingly, aberrant expression of PHF20L1 regulates the expression of GLUT1 as well as other important glycolytic genes such as *HK2, PKM, Glyceraldehyde 3-phosphate dehydrogenase (GAPDH), Lactate dehydrogenase A (LDHA)*, and *PGK1* (Hou et al. 2020).

Hypoxia Response

Notedly, PHF20L1 is an important transcriptional target of the Hypoxia Inducing Factor 1a (HIF1a) /MYC complex and hence, appears to have significant function in the hypoxia response program (Hou et al. 2020).

ZMYND8

ZMYND8 (Zinc finger MYND domain-containing protein 8), another PHD/Bromo domain-containing chromatin reader, has been strongly associated with the progression of various types of tumors. ZMYND8 has shown diverse functions in regulating tumorigenicity in the perspective of transcriptional reprogramming, where it has shown both activating and repressing roles (Adhikary et al. 2016; Basu et al. 2017a, b). Recent findings have established it as an important epigenetic regulator sensitizing the chemo-refractory breast carcinoma (Mukherjee et al. 2020).

Mechanism of Action and Role in Hypoxia Response

Although a direct role in the regulation of glucose metabolic genes has not yet been elucidated, a recent study found ZMYND8 to be involved in HIF-mediated gene expression (Chen et al. 2018b). HIF-activated cells show a strong preference for aerobic glycolysis, reduced oxidative phosphorylation, and increased angiogenesis in the tumor microenvironment (Choudhry and Harris 2018). The aforementioned study not only found ZMYND8 to be a transcriptional regulator of HIF-1 and HIF-2 but also that it associates with these proteins at the hypoxia response elements (HREs) of target genes, including those involved in glycolysis, to regulate their expression. ZMYND8 reads the H3K14ac and H4K16ac marks at the HREs of active genes and facilitates the switching of RNA polII from its S5-phosphorylated state to the S2-phosphorylated state, thereby allowing promoter release and subsequent transcriptional elongation (Fig. 9.4D.b). Interestingly, ZMYND8 works in concert with another chromatin reader- the bromodomain-containing BRD4, to drive HIF gene expression (Fig. 9.4D.a). This association with BRD4 is mediated by acetylation of key lysine residues on ZMYND8 by p300 (Chen et al. 2018b). Although it seems clear that ZMYND8 has a role to play in tumor metabolic reprogramming, it remains to be seen whether it is directly recruited to glycolytic gene promoters or controls the expression of other regulators which target these genes.

TCF19

Mechanism of Action

Transcription factor 19 is a chromatin reader which was recently characterized as a-type 1 diabetes-associated locus (Cheung et al. 2011). Although it was initially recognized as a growth-regulated cDNA and described as a transactivating factor in 1991 (alias SC1), very little has emerged about this protein since then. Sequence studies have revealed that TCF19 consists of PHD finger and Fork-Head Associated (FHA) domain, interspersed by a proline-rich linker region. The FHA domain may be required by TCF19 for interaction with phosphoproteins or nuclear signaling and may function in recruiting effector proteins to DNA (Krautkramer et al. 2013). Its exact function remains to be elucidated. The PHD finger, on the other hand, was recently demonstrated to be important for TCF19 binding to histones via H3K4me3. Counter-intuitively, TCF19 reads this activating mark and is instrumental in recruiting the multisubunit NuRD complex, including Chromodomain Helicase DNA binding protein 4 (CHD4), MTA1, and HDAC1, to silence its target genes (Fig. 9.4E). This recruitment is followed by a decrease in levels of H4K5ac, H4K8ac, and H3K9ac at the promoters (Sen et al. 2017)

Glucose Uptake

TCF19 plays an important role in the uptake of glucose at an optimum level to maintain a sustainable glucose concentration in the cell. It regulates *GLUT1/3* receptors to minimize glucose uptake at high glucose concentrations (Mondal et al. 2020). Abrogation of TCF19 results in loss of its regulation on *GLUT1/3* receptors and a substantial increase in cellular glucose uptake.

Glycolysis and OXPHOS

TCF19 influences GSIS depending upon the blood glucose level (Ndiaye et al. 2017). Apart from that, emerging studies reveal that TCF19 in concert with other non-histone proteins regulates glycolysis and OXPHOS to maintain cellular homeostasis through a distinct mechanism. TCF19 upon short-term high glucose stress activates *Tp53 Induced Glycolysis and Apoptosis Regulator* (*TIGAR*) to negatively regulate glycolysis resulting in cell cycle arrest and sustenance. On the other hand, upon prolonged high glucose treatment, it activates the gene *Cytochrome C Oxidase Assembly (SCO2)* by removing the NuRD complex from the promoter resulting in high OXPHOS leading to apoptosis (Mondal et al. 2021).

Gluconeogenesis

Abnormal endogenous glucose production is characteristic of diabetes. Genes involved in gluconeogenesis, such as *glucose-6-phosphatase (G6P), fructose-1,6 bisphosphatase (FBP1),* and *pyruvate carboxykinase 1 (PCK1),* are significantly downregulated by TCF19. This control is exerted by TCF19 binding at the insulin response elements at these gene promoters (Sen et al. 2017).

Chromatin readers, versatile and adaptive as they are, find different ways to pull the strings of their target genes. Glucose metabolism, whether under normal or cancer conditions, is sophisticatedly controlled by various readers either directly or indirectly. How they do, this is a rich field ready to be ploughed and exploited. Hence detection of different diseased conditions and implementation of epigenetic reader there as therapeutic target is an interesting and emerging field of study. The next section of the book chapter deals with the detail methodology of metabolic aberration leading to cancer.

Methods to Determine the Metabolic Aberrations During Glucose Metabolism due to Alteration in Transcription Programs

The diagnosis of any disease depends mainly on the detection process. The earlier the detection, the better the survival (Importance Of Early Detection Of A Disease (epainassist.com)). So, several studies have been carried out to understand the biochemical reactions that occur during these metabolic processes, followed by the techniques to selectively identify these reactions. Different colorimetric, fluorometric, mass spectrometric methods have been established to understand the differential levels of metabolites in diseased person in comparison to a healthy individual (Smith et al. 2018). Recent literatures suggest that the total metabolic profile has potential to influence the immune system of the person, and hence immune-metabolomics is becoming an emerging field of study (Everts 2018; Leone and Powell 2020). Biochemical detection methods are both cost-effective as well as less invasive (New Non-Invasive Test Can Detect 8 Different Cancers in Their Earliest Stages (sciencealert.com)). As analysis of peripheral blood samples gives the information of the condition of distant organs such as stomach, liver, etc., it does not require any surgical interventions. The understanding of the metabolic alterations is so important that it has the potential to dictate the differential fate or level of metastasis of two individuals of similar stage of cancer progression. Hence the following section of this book chapter is very important to gain the insight into the various methods to detect the metabolites level in human body.

Glucose Sensing

Indirect Glucose Uptake measurements:

3-O-methylglucose (3-MG)

At the very early onset of quantitative metabolic studies, 3-O-methylglucose (3-MG) was radiolabeled and its uptake by the adipocytes was measured. Since this analog cannot be phosphorylated, direct measurement of the impermeable 3-MG can be achieved by scintillation counters (Foley et al. 1978). Although the test was rapid and very short incubation times were required, due to technical limitations, this is no longer considered a standard method of detection. This compound is nonmetabolizable, hence it shows a lower affinity towards the GLUT receptors. As a result, the measurement of glucose uptake is compromised significantly. Further, the method suffers from an overestimation of passive uptake of glucose in the cell.

2-deoxy-D-glucose (2-DG)

2-deoxy-D-glucose (2-DG) acts as a substrate for hexokinase, but the resultant product, 2-deoxyglucose-6-phosphate, has high stability and is not converted to fructose-6-phosphate. This accumulated derivative can then be quantified to study GLUT receptor activity. A modification of this method dispenses the need for radioisotopes and instead utilizes the inherent fluorescence of NADPH for analysis (Manchester et al. 1990). 2-DG6P is converted by glucose-6-phosphate dehydrogenase to 6-phospho-2-deoxygluconate, coupled with the reduction of $NADP^+$ to NADPH. To enhance signal intensity and precision, fluorescent dyes, which are themselves reduced by NADPH, are employed. One such example is resazurin, whose reduction to the strongly fluorescent resorufin is catalyzed by the enzyme diaphorase in the presence of NADPH (Yamamoto et al. 2006).

2-(N-(7-Nitrobenz-2-oxa-1,3-diazol-4-yl)Amino)-2-Deoxyglucose (2-NBDG)

2-NBDG, a fluorescent analog of 2-DG, allows for direct measurement of glucose accumulation. Although its fluorescence activity is relatively weak, coupling with photomultiplier tubes allows for rapid and accurate detection today (Yamamoto et al. 2015).

Direct Glucose Uptake Measurements

3-Bromopyruvate

Another novel method uses canonical glucose for measurement instead of structurally modified substrates. It utilizes the hexokinase inhibitor, 3-bromopyruvate, disallowing the entry of glucose into various metabolic pathways, and a coupled glucose oxidase-Horseradish peroxidase (HRP) enzymatic system (Yan et al. 1831). The glucose oxidase enzyme catalyzes the conversion of glucose to gluconolactone and hydrogen peroxide, while HRP, along with the hydrogen peroxide, converts a colorless substrate to the fluorescent resorufin (BioVision Cat no. K924).

Oxidative Stress

ROS Assays

Direct Mode of ROS Measurements

1. To determine ROS production, a very commonly used assay employs dichlorodihydrofluorescein diacetate (DCFH-DA), a reduced form of fluorescein. The non-fluorescent substrate (DCHF-DA) is converted to a green fluorescent product (DCF) through an intermediate (DCFH) upon removal of its acetate groups via intracellular oxidation, requiring both ROS and cellular esterases.

DCFH-DA —esterase—> DCFH —H2O2—> DCF

non fluorescent ... green fluorescent

2. Amplex Red is another substrate that produces a fluorescent product upon oxidation by Hydrogen peroxide (H_2O_2). In the presence of H_2O_2, the substrate is oxidized through horseradish peroxidase to the red resorufin (Griendling et al. 2016). This is one of the simplest, most quantitatively accurate ROS assays employed today.

Amplex Red —H2O2 + HRP—> Resorufin

non fluorescent ... red fluorescent

Indirect Mode of ROS Measurements

1. An indirect way of quantifying oxidative stress is to study the intracellular oxidized and reduced glutathione (GSSG/GSH) ratio. Generally, cells maintain a high level of reduced glutathione, and an increase in the aforementioned ratio is an indicator of cellular stress (Wu and Yotnda 2011). Detection can either be colorimetric or fluorimetric. 5,5'-dithiobis-(2-nitrobenzoic acid) (DTNB) or Ellman's reagent is reduced by glutathione-S-transferase and GSH to produce an easily detectable, yellow-colored compound 2-nitro-5-thiobenzoic acid (TNB) with the oxidized GSSG again being recycled to GSH by glutathione reductase.

$$\text{DTNB} + \text{GSH} \xrightarrow{\text{GST}} \underset{\text{yellow}}{\text{TNB}} + \text{GSSG}$$

2. Alternatively, the luciferase reaction can be coupled to glutathione reduction for a chemiluminescent assay. In two parallel reactions (i) the total glutathione pool is present, and (ii) the pre-existing GSH is blocked so that only the GSSG produced in the reaction can be scored. This provides an accurate GSSG/GSH ratio, with the GSH in both cases being used along with GST enzyme to convert luciferin-NT (a GSH probe) to luciferin, which is subsequently oxidized by luciferase to produce luminescence in proportion to the GSH being used up (Promega TM cat no. V6611) measured by a luminometer.

$$\text{GSH} + \text{Luc-NT} \longrightarrow \text{GSH-NT} + \text{Luciferin} \xrightarrow{\text{Luciferase + ATP}} \text{Luminescence}$$

RNS Assays

Detection of reactive nitrogen species (RNS) is generally done using Griess reagent, a mixture of naphthyl ethylene diamine dihydrochloride (NED) and sulphanilamide in phosphoric acid. These components react with nitrites, resulting in the formation of a purple-colored azo product (Giustarini et al. 2008).

$$\text{Sulphanilamide} + \text{NO2-} \longrightarrow \text{Diazonium salt}$$

$$\text{Diazonium salt} + \text{naphthylethylenediamine dihydrochloride} \longrightarrow \underset{\text{Purple}}{\text{Azo dye}}$$

Another fluorimetric assay uses diaminofluoresceins such as diaminofluorescein-2 (DAF2) and 4-Amino-5-Methylamino-2',7'-Difluorofluorescein (DAF-FM), which react with cellular nitric oxide to form a highly fluorescent benzotriazole product (Lacza et al. 2006).

Glycolysis/TCA Cycle

Lactate Production Assay

1. Lactate production assays provide a good estimate of glycolytic rates. The normal lactate/pyruvate concentration is altered during periods of high cellular activity and in various cell types, including different tumors, differentiating T-cells, Red Blood Corpuscles (RBCs), and neurons. This lactate is excreted and can be measured by luminescence, fluorescence, or colorimetry. Conversion of lactate to pyruvate requires lactate dehydrogenase and the reduction of NAD^+ to NADH whose intrinsic fluorescence can be harnessed for detection. Hydrazine captures the produced pyruvate to form pyruvate hydrazone. Pyruvate hydrazone, and hence the amount of lactic acid produced quantitatively, is measured at 340 nm (Grist et al. 2018).

$$\text{Lactate} + \text{NAD+} \xrightarrow{\text{LDH}} \text{Pyruvate} + \text{NADH} + \text{H+}$$

$$\text{Pyruvate} + \text{Hydrazine} \longrightarrow \text{Pyruvate hydrazone}$$

2. As described for the GSSG/GSH assay, this reaction can also be coupled to the luciferase reaction, with the NADH produced being used to reduce a pro-luciferin substrate to luciferin (Promega TM).

$$\text{Lactate} + \text{NAD+} \xrightarrow{\text{LDH}} \text{Pyruvate} + \text{NADH} + \text{H+}$$

$$\text{Pro-luciferin} + \text{NADH} \longrightarrow \text{NAD+} + \text{Luciferin} \xrightarrow{\text{Luciferase + ATP}} \text{Luminescence}$$

Extracellular Acidification Rate Measurements

Another way of measuring extracellular lactate production is to measure extracellular acidification (Box 1). Lactate that diffuses out of the cell to maintain cellular pH gets deprotonated to produce H^+ ions.

Lactic acid ——> Lactate- + H+

This acidification can be measured by Seahorse analyzers using glucose, oligomycin, and 2-DG at different time points. Oligomycin blocks the F0 subunit of mitochondrial ATP synthase, thereby inhibiting oxidative phosphorylation. Hence, the cell is driven to use glycolysis as the primary pathway for energy production, producing lactate at the end. Finally, 2-DG blocks glucose breakdown, and any residual extracellular acidification (derived from CO_2 breakdown to HCO_3^- and mitochondrial proton leak) can then be subtracted to obtain total glycolytic H^+ production and hence, the maximal cellular glycolytic capacity (Teslaa and Teitell 2014).

Gluconeogenesis Assays

Glucose production assays provide an estimate of hepatocyte conversion of gluconeogenic substrates such as lactate and pyruvate to glucose when it becomes limiting. One method employs labeled substrates (acetate, alanine, lactate), where the rate of gluconeogenesis is determined by calculating the ratio of labeled glucose to the labeled substrate. This usually does not provide a full estimate of gluconeogenesis as the labeled carbon may be exchanged with intermediates in other pathways (Chung et al. 2015).

Recently developed kit-based assays make use of colorimetric/fluorescent substrates linked to glucose oxidation. Cells are provided with sodium lactate and pyruvate in glucose-free media and allowed to anabolize glucose. A buffer-containing glucose oxidase, HRP, and a colorimetric substrate, such as 4-aminoantipyrine, or a fluorescent substrate, such as Amplex Red, is then added to the dish. The glucose released by the cells in the media is then measured using a microplate reader (Matsumoto and Sakai 2012; Sakai et al. 2012).

Box 1.

Details of the Components used in Extracellular Flux Analysis

	Name	Function
ECAR	Glucose	Used as a substrate for glycolysis.
	Oligomycin	Blocks the F0 subunit of mitochondrial ATP synthase, thereby, inhibiting oxidative phosphorylation. Hence, the cell is driven to use glycolysis as the primary pathway for energy production, producing lactate at the end.
	2-deoxyglucose	Blocks glucose breakdown and any residual extracellular acidification (derived from CO2 breakdown to HCO3- and mitochondrial proton leak) can then be subtracted to obtain total glycolytic H+ production and hence, the maximal cellular glycolytic capacity.
OCR	Oligomycin	Oligomycin addition inhibits ATP production and, hence, provides an estimate of respiration linked to ATP by subtracting from the basal respiratory rate.
	FCCP	FCCP, a proton translocator and an uncoupling agent, allows the rate of electron transport to reach its maximum and, hence, determines the maximal respiratory rate.
	Rotenone/Antimycin A	Rotenone, an inhibitor of complex I, and Antimycin A, an inhibitor of complex III, disrupt the electron transport chain, thereby abolishing mitochondrial respiration. This is used to determine non-mitochondrial respiration from which all other rates are subtracted.

Details of the Parameters measured through Extracellular Flux Analysis

	Parameters	Description
ECAR	Glycolysis	The process of converting glucose to pyruvate.
	Non-glycolytic Acidification	These measures other sources of extracellular acidification that are not attributed to Glycolysis.
	Glycolytic Capacity	This measurement is the maximum ECAR rate reached by a cell following the addition of Oligomycin, effectively shutting down oxidative phosphorylation and driving the cell to use glycolysis to its maximum capacity.
	Glycolytic Reserve	This measure indicates the capability of a cell to respond to an energetic demand as well as how close the glycolytic function is to the cell's theoretical maximum.
OCR	Basal respiratory rate	This is the amount of oxygen consumption by mitochondria in the absence of inhibitors.
	Proton leak	This measures the Oxygen consumption in the absence of proton flux through complex V and abolished ATP production.
	ATP-synthesis linked respiration	Rotenone, an inhibitor of complex I, and antimycin A, an inhibitor of The reduction in OCR in the absence of ATP production.
	Maximal respiration	The maximum capacity of mitochondria for oxygen consumption when electron-flow and proton gradient are unlinked.
	Spare respiratory capacity	This measures other sources of extracellular acidification that are not attributed to glycolysis.
	Non-mitochondrial respiration	Residual OCR in the absence of mitochondrial oxidation due to certain oxygen consuming enzymes that may be present in the sample.

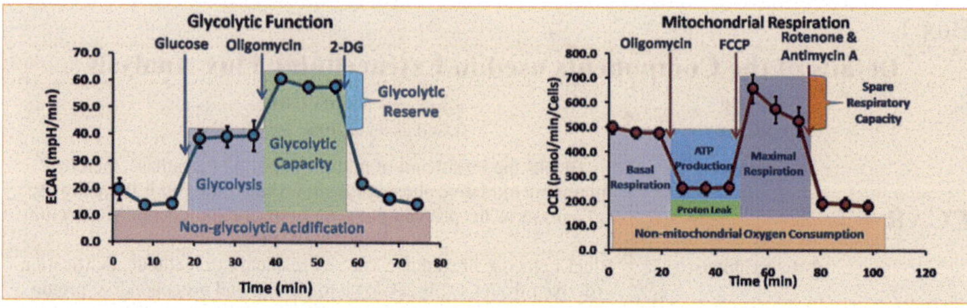

Box 1. Typical graphical representation of ECAR (Glycolytic Function) and OCR (Mitochondrial Respiration) and the detail of the components used and parameters measured through the Extracellular Flux Analysis.

Glucose —Glucose oxidase—> Gluconolactone + H2O2

H2O2 + Reduced substrate —HRP—> Oxidised product + H2O

Colourless Coloured

Oxidative Phosphorylation

Mitochondrial Membrane Potential Measurements

Mitochondrial health is vital for the proper flow of electrons and subsequent ATP production and hence, measuring the membrane potential provides an indicator for this. Healthy mitochondria generally maintain a potential of -180mV and fluorescent, lipophilic or cationic dyes such as tetraethylbenzimidazolylcarbocyanine iodide (JC-1) and Tetramethylrhodamine ethyl ester, perchlorate (TMRE) are used to stain the organelles. Depolarised mitochondria are, therefore, stained less strongly than healthy ones (Membrane and Assay 2016). Using these dyes, mitochondrial membrane potential and in turn, the active OXPHOS can be qualitatively (through confocal microscopy) or quantitatively (using FACS, colorimetry/fluorometry) measured.

Oxygen Consumption Rate Measurements

Seahorse analyzers are also used to determine the rate of oxygen consumption which allows the study of mitochondrial function and energy production from glucose (Box 1). Different drugs targeting the various inner mitochondrial membrane complexes are added at different time points to analyze overall mitochondrial function (Rose et al. 2014), including basal respiration, ATP production, proton leak, oxygen consumption, etc.

ATP Production Assay

ATP produced in mitochondria is generally measured using the luciferase reaction. Luciferase converts luciferin to oxyluciferin using ATP and oxygen.

Luciferin + ATP ---------> Luciferyl-adenylate + PPi

Luciferyl-adenylate + O2 ---------> Oxyluciferin + AMP + Luminescence

The light produced from cellular ATP is measured against a standard ATP curve to determine the content (Drew and Leeuwenburgh 2003).

Metabolomics-Based Mass Spectrometry Analysis

Analysis of metabolic flux through mass spectrometry (MS) is rapid, accurate, and highly reproducible. Generally, tandem mass spectrometry (MS/MS) is combined with powerful separation techniques such as gas chromatography (GC) or liquid chromatography (LC) to obtain precise outcomes. A stable isotope of glucose such as 13C-glucose is traced through its entry and utilization in different metabolic pathways. U-13C6-glucose, 6-13C-glucose, and 1,2-13C2-glucose are some commonly used members of this tracer family. 1,2-13C2-glucose, especially, is a fantastic substrate for distinguishing glucose entering glycolysis from that involved in the pentose phosphate pathway (Teslaa and Teitell 2014). To study the TCA cycle, intermediates of the pathway—citrate, fumarate, succinate, malate, or alpha-ketoglutarate—carrying the heavy isotope label are followed and analyzed (Koubaa et al. 2013). Rates of gluconeogenesis can also be measured using 13C-labeled precursors such as lactate, alanine, or glycerol (Chung et al. 2015).

While LC-MS samples can be processed directly, GC-MS requires an additional derivatization step to create more volatile versions of metabolites. These modifications include methoxime-trimethylsilylation, aldonitrile-pentaacetylation, and trifluoroacetylation (Project et al. 2015). Many methods are followed to create the charged particle for mass spectrometric analysis. Electrospray ionization is one of the most famous ionization techniques (Wang et al. 2018b). Other ways of ionization in gradual improvements of the technique include G (MALDI), etc. This metabolomics-based approach provides considerable insight into the various pathways that glucose and its related metabolites can enter and their Spatiotemporal fates in the cell.

Analysis of Glucose Metabolic Enzymes

Enzyme Activity

Measuring the activity of the three rate-limiting enzymes of glycolysis can provide a good estimate of cellular glycolytic rates. This is usually determined by spectrometric analysis of NADH or NADPH concentrations, both of which are strongly fluorescent at 340 nm (Teslaa and Teitell 2014). Hexokinase activity can be measured by coupling the reaction with glucose-6-phosphate dehydrogenase, a key enzyme of the pentose phosphate pathway (Crabtree and Newsholme 1972).

$$\text{Glucose} \xrightarrow{\text{Hexokinase}} \text{Glucose-6-phosphate}$$

$$\text{Glucose-6-phosphate} + \text{NADP+} \xrightarrow{\text{G6PD}} \text{6-phosphogluconolactone} + \underset{\text{Abs at 340nm}}{\text{NADPH}}$$

Phosphofructokinase, which commits glucose to the glycolytic pathway, is assayed by measuring the concentration of NADH produced in downstream steps (Opie et al. 1967). The enzymatic steps of aldolase, triosephosphate isomerase, and glyceraldehyde-3-phosphate dehydrogenase are coupled to this reaction.

$$\text{Fructose-6-phosphate} \xrightarrow{\text{PFK}} \text{Fructose-1,6-bisphosphate}$$

$$\text{Fructose-1,6-bisphosphate} \xrightarrow{\text{Aldolase}} \text{Glyceraldehyde-3-phosphate} \xleftrightarrow{\text{TIM}} \text{Dihydroxyacetone phosphate}$$

$$\text{Glyceraldehyde-3-phosphate} + \text{NAD+} + \text{Pi} \xrightarrow{\text{GAPDH}} \text{1,3-bisphosphoglycerate} + \underset{\text{Abs at 340nm}}{\text{NADH}}$$

Pyruvate kinase activity can be ascertained by coupling pyruvate production to lactate generation and measuring the subsequent loss of NADH fluorescence. Fructose-1,6-bisphosphate is also used in the reaction as it acts as an allosteric activator of Pyruvate kinase (PK) (Taber et al. 1998).

$$\text{Phosphoenol pyruvate} + \text{ADP} \xrightarrow{\text{PK}} \text{Pyruvate} + \text{ATP}$$

$$\text{'yruvate} + \underset{\text{Abs at 340nm}}{\text{NADH}} \xrightarrow{\text{LDH}} \text{Lactate} + \text{NAD+}$$

Gene Expression

Since changes in expression do not always translate into a change in function, studying gene expression may not provide the exact status of glucose in the cell, but it remains indispensable for tracing glucose regulatory networks. mRNA expression data of the genes of various enzymes involved in glucose metabolic pathways sheds light on differential expression under a plethora of conditions (Clasen et al. 2013). RT-qPCR and, on a global scale, RNA microarrays and RNA-sequencing are truly powerful techniques to understand metabolic gene expression profiles (Kolodziejczyk and Lönnberg 2018). These Next-Generation Sequencing (NGS) studies give a broad platform to understand the alteration of gene expression involved in the regulation of metabolic function in the cell (Meera Krishna et al. 2019; Thornhill 2018).

Another useful method is the luciferase reporter assay which can be used to study promoter activity and changes in gene expression (Clement et al. 2015). The luciferase reaction is extremely efficient and, hence, is highly sensitive. Single and dual reporter systems exist, with classical single reporter systems consisting of a single chimeric plasmid carrying the promoter or gene of interest fused to firefly or *Renilla* luciferase gene. Dual reporter systems utilize two plasmids, one carrying firefly luciferase along with the sequence of interest, the other harboring *Renilla* luciferase placed under a constitutive promoter which acts as the control. This confers the advantage of normalization to internal control in a single reaction which enables high-throughput analysis (McNabb et al. 2005). The gene of interest cloned in the luciferase cassette is the regulatory region of the important metabolic genes whose transcription program is needed to be studied. Introducing specific mutations in the gene sequence of the promoter region gives additional information about the sequence specificity for targeting the transcription factors to its cognate gene loci.

Altogether this section summarizes the detailed methodology to detect metabolic aberration that has tremendous implication in the context of drug designing and therapeutic regimen.

Role of Other Metabolic Pathways on Epigenetic Control of Glucose Metabolism

Apart from the glucose metabolic pathway, there are other metabolic pathways as well that are interconnected and play important role in determining the epigenetic regulation through metabolic alterations. These alternative metabolic pathways and their connection to epigenetics are summarized next.

Amino Acid Metabolism

Amino acids significantly contribute to the maintenance of chromatin signatures via histone-modifying enzymes. The various pathways involving these amphipathic molecules generate a whole host of intermediates that either act as modifying groups or as cofactors for modifiers. S-adenosylmethionine (SAM), a major donor of the methyl group, is derived from methionine and, upon losing its methyl group, forms S-adenosylhomocysteine (SAH) which acts as an inhibitor of DNMTs and HMTs (Lieu et al. 2020). Additionally, threonine dehydrogenase reduces threonine to glycine and acetyl CoA, the former being a transmethylation substrate in the SAM cycle which maintains a pool of SAM/SAH in the cell (Shyh-Chang et al. 2013). Branched-chain amino acids leucine and isoleucine are oxidized to form acetyl CoA, which is required for histone acetylation (Adeva-Andany et al. 2017). This pathway of acetyl CoA generation is especially important for the proliferation of tumor cells and activation of key genes involved in the Warburg effect.

NAD^+, apart from being involved in glycolysis, TCA cycle, and OXPHOS, is also required as a cofactor for the Sirtuin class of HDACs. The various SIRT types, especially SIRT1 and SIRT6, have been shown to have prominent roles in the regulation of glucose metabolism (Ye et al. 2017). SIRT6 acts as a tumor suppressor, competing with HIF1a to bind to the promoters of several genes involved in glucose metabolism, such as *GLUT1*, *PFK1*, *LDHA*, and *PDK1*. It subsequently reduces H3K9ac levels, leading to reduced glucose uptake and aerobic glycolysis (Zhong et al. 2010). De novo generation of NAD^+ requires tryptophan via the kynurenine pathway (Yang and Sauve 2016). Alpha-ketoglutarate is another cofactor involved in the enzymatic activity of Jumonji C domain-containing histone demethylases (Etchegaray and Mostoslavsky 2016). Several of these enzymes are known to be involved in glucose metabolic gene regulation. Jhdm1/KDM2A, for instance, has been found to demethylate H3K36me2 and reduce hepatic gluconeogenesis (Pan et al. 2012). Glutamine and glutamate are principal sources of α-KG synthesis via the enzymes glutaminase (GLS) and glutamate dehydrogenase (GDH). In addition, arginine and histidine can also be used for α-KG production, with glutamate being an intermediate in the pathway (Xiao et al. 2016).

Fatty Acid Metabolism

Mitochondrial beta-oxidation is a highly energetic process that gets activated in times of low glucose concentration. It also provides several intermediates which are directly involved in histone modification. Like branched-chain amino acids, fatty acids are also oxidized to acetyl CoA units, which subsequently act as donors of acetyl groups for modification via HATs (Etchegaray and Mostoslavsky 2016). Acetyl CoA, under limiting glucose conditions such as fasting, also results in the formation of ketone bodies, the most common of them being β-hydroxybutyrate (βOHB). βOHB is an HDAC inhibitor, similar to the fatty acid butyrate (Shimazu et al. 2013). These molecules effectively reduce glucose metabolism and cancer cell survival. For instance, HDAC1 and HDAC2 inhibition by sodium butyrate has been shown to restore H3K27ac levels, leading to increased expression of FBP1, the rate-limiting enzyme of gluconeogenesis, and reduced cellular growth in hepatocellular carcinoma (Yang et al. 2017).

Nucleotide Metabolism

Nucleotide metabolism involves the one-carbon metabolic pathway which is utilized in various biosynthetic reactions and post-translational modifications. The carriers of these groups are cofactors such as SAM and tetrahydrofolate. The folate branch of the 1C cycle provides the substrates (methylated THF derivatives) for both purine and pyrimidine synthesis and is closely linked to the methionine branch of the cycle. The cyclic transfer and distribution of 1C groups between the molecules of the two branches ensures a constant output of nucleotides and SAM, among other products, and helps maintain the epigenetic status of the cell (Shuvalov et al. 2017).

As discussed above, it is very important to understand how all the cellular metabolic pathways are intricately interconnected and has immense role to play in the context of epigenetic regulation through metabolic reprogramming.

Conclusion and Future Perspective

Maintenance of metabolic homeostasis is an important phenomenon of normal cells which goes awry in cancer. Cancer cells reprogram the metabolic pathways leading to increased amounts of glucose consumption or uptake, higher rates of both anabolic and catabolic pathways, such as glycolysis, fatty acid synthesis, and amino acid break down. It produces essential metabolic building blocks as well as a greater amount of energy for the better survival and growth of tumors. Hence metabolic milieu of a cancer cell can have the potential not only to form a tumor but also to determine its metastatic potential.

As the tumor cells uptake and process a higher amount of glucose, they produce several metabolic intermediates, also to a greater extent that favors cancer progression and metastasis. Production of ATP, H^+ ions, lactate, etc., from the metabolism of glucose helps in this process. So all these intermediates and their involvement in different steps of metabolic pathways are being studied for a better understanding of their role in promoting cancer. Despite all these pieces of information available previously, it was difficult to treat cancer cells without affecting the normal healthy cells as these metabolic processes are common for all types of cells. However, recent studies have discovered specific metabolites exclusively found in cancer cells, termed as oncometabolites, which can be a better target for the treatment of cancer. Tumor cells also aberrantly synthesize tumor-promoting genes and their gene products, which are already under extensive study for the development of targeted therapy. There are numerous examples, such as targeting hypoxia that induces tumor cell formation, by introducing hypoxic cytotoxins or HIF-1 inhibitors. This in turn enhances the potency of radiation therapy for the treatment.

As described in this book chapter, deregulation in glucose metabolism can arise from different reasons such as insulin resistance, mutations in GLUT receptors, and glucose metabolic pathway enzymes. These findings are translated into new approaches in clinical trials that are expected to lead to a better mode of cancer therapy and longer survival of tumor patients. This book chapter highlights the diverse group of epigenetic readers that have important implications in the regulation of cancer as well as metabolic pathways. The reader proteins have the potential to crosstalk with important transcription factors to alter the expression of tumor-promoting genes through which it can be beneficial for the treatment of cancer. Hence the area of epigenetic regulations in the metabolic context is now being extensively studied and can be breakthroughs in the future for the development of therapeutic targets for the treatment regimen. Recent findings are suggestive of the fact that metabolic enzymes and metabolites can dictate an immune cell how to function in the human body, which in turn results in the immune response of any individual. Different mass spectrometric or nuclear magnetic resonance (NMR)-based approaches are being useful in understanding and quantifying the difference in the metabolites obtained from patient samples to get insights into the novel functions of the immune system and to identify biomarkers for immune-related metabolic diseases. Hence, an emerging field of today's research world is immunometabolomics. Recent literature suggests that difference in the pool of metabolic intermediates among individual may have far-reaching consequences towards the predisposition of them for developing cancer.

Altogether this book chapter summarizes the importance of glucose metabolism in cancer prognosis, how metabolic intermediates influence this process and are indispensable for many enzymes to exert their functions, how the diverse family of reader proteins translates the modification in the context of regulating these metabolic pathways, the detailed methodology for the detection of the metabolic alterations and identification of probable therapeutic targets.

Acknowledgment This work has been supported in part by research grants received from S. Ramachandran National Bioscience Award for Career Development 2019 (BT/HRD-NBA-NWB/38/2019-20) and (BT/PR28920/MED/122/176/2018) from the Department of Biotechnology; SwarnaJayanti Fellowship (DST/SJF/LSA-02/2017-18) and Science & Engineering Research Board (SERB) (CRG/2018/000985) from the Department of Science & Technology; Basic and Applied Research in Biophysics and Material Science (RSI 4002) from the Department of Atomic Energy, Govt. of India to CD. AS thanks the Department of Biotechnology, Govt. of India for personal fellowship support.

References

Abeti R, Duchen MR (2012) Activation of PARP by oxidative stress induced by β-amyloid: implications for Alzheimer's disease. Neurochem Res 37(11):2589–2596. https://doi.org/10.1007/s11064-012-0895-x

Adeva-Andany MM, López-Maside L, Donapetry-García C, Fernández-Fernández C, Sixto-Leal C (2017) Enzymes involved in branched-chain amino acid metabolism in humans. Amino Acids 49(6):1005–1028. https://doi.org/10.1007/s00726-017-2412-7

Adhikary S, Sanyal S, Basu M, Sengupta I, Sen S, Srivastava DK, Roy S, Das C (2016) Selective recognition of H3.1K36 dimethylation/H4K16 acetylation facilitates the regulation of all-trans-

retinoic acid (ATRA)-responsive genes by putative chromatin reader ZMYND8. J Biol Chem. https://doi.org/10.1074/jbc.M115.679985
Alrob OA, Lopaschuk GD (2014) Role of CoA and acetyl-CoA in regulating cardiac fatty acid and glucose oxidation. Biochem Soc Trans 42(4):1043–1051. https://doi.org/10.1042/BST20140094
Anastas JN, Shi Y (2019) Histone serotonylation: can the brain have "happy" chromatin? Mol Cell 74(3):418–420. https://doi.org/10.1016/j.molcel.2019.04.017
Anderson DD, Woeller CF, Chiang EP, Shane B, Stover PJ (2012) Serine hydroxymethyltransferase anchors de Novo thymidylate synthesis pathway to nuclear lamina for DNA synthesis. J Biol Chem 287(10):7051–7062. https://doi.org/10.1074/jbc.M111.333120
Andrews FH, Shinsky SA, Shanle EK, Bridgers JB, Gest A, Tsun IK, Krajewski K, Shi X, Strahl BD, Kutateladze TG (2016) The Taf14 YEATS domain is a reader of histone crotonylation. Nat Chem Biol 12(6):396–398. https://doi.org/10.1038/NCHEMBIO.2065
Appikonda S, Thakkar KN, Barton MC (2016) Regulation of gene expression in human cancers by TRIM24. Drug Discov Today Technol 19(xx):57–63. https://doi.org/10.1016/j.ddtec.2016.05.001
Ballaré C, Lange M, Lapinaite A, Martin GM, Morey L, Pascual G, Liefke R, Simon B, Shi Y, Gozani O, Carlomagno T, Benitah SA, Di Croce L (2012) Phf19 links methylated Lys36 of histone H3 to regulation of Polycomb activity. Nat Struct Mol Biol 19(12):1257–1265. https://doi.org/10.1038/nsmb.2434
Bannister AJ, Kouzarides T (2011) Regulation of chromatin by histone modifications. Cell Res 21:381–395. https://doi.org/10.1038/cr.2011.22
Bannister AJ, Zegerman P, Partridge JF, Miska EA, Thomas JO, Allshire RC, Kouzarides T (2001) Selective recognition of methylated lysine 9 on histone H3 by the HP1 chromo domain. Nature 410(6824):120–124. https://doi.org/10.1038/35065138
Bárcena-Varela M, Caruso S, Llerena S, Álvarez-Sola G, Uriarte I, Latasa MU, Urtasun R, Rebouissou S, Alvarez L, Jimenez M, Santamaría E, Rodriguez-Ortigosa C, Mazza G, Rombouts K, San José-Eneriz E, Rabal O, Agirre X, Iraburu M, Santos-Laso A et al (2019) Dual targeting of histone methyltransferase G9a and DNA-methyltransferase 1 for the treatment of experimental hepatocellular carcinoma. Hepatology 69(2). https://doi.org/10.1002/hep.30168
Basu M, Khan MW, Chakrabarti P, Das C (2017a) Chromatin reader ZMYND8 is a key target of all trans retinoic acid-mediated inhibition of cancer cell proliferation. Biochim Biophys Acta Gene Regul Mech 1860(4):450–459. https://doi.org/10.1016/j.bbagrm.2017.02.004
Basu M, Sengupta I, Khan MW, Srivastava DK, Chakrabarti P, Roy S, Das C (2017b) Dual histone reader ZMYND8 inhibits cancer cell invasion by positively regulating epithelial genes. Biochem J 474(11):1919–1934. https://doi.org/10.1042/BCJ20170223
Becker J, Bonnet C, Clifford G, Groth A, Wilson M, Chapman JR (2020) BARD1 links histone H2A Lysine-15 ubiquitination to initiation of BRCA1-dependent homologous recombination. BioRxiv 2020(06):01.127951. https://doi.org/10.1101/2020.06.01.127951
Bian C, Xu C, Ruan J, Lee KK, Burke TL, Tempel W, Barsyte D, Li J, Wu M, Zhou BO, Fleharty BE, Paulson A, Allali-Hassani A, Zhou JQ, Mer G, Grant PA, Workman JL, Zang J, Min J (2011) Sgf29 binds histone H3K4me2/3 and is required for SAGA complex recruitment and histone H3 acetylation. EMBO J 30(14):2829–2842. https://doi.org/10.1038/emboj.2011.193
Boroughs LK, Deberardinis RJ (2015) Metabolic pathways promoting cancer cell survival and growth. Nat Cell Biol 17(4):351–359. https://doi.org/10.1038/ncb3124
Bowman GD, Poirier MG (2015) Post-translational modificationsof histones that influence nucleosome dynamics. Chem Rev 115(6):2274. https://doi.org/10.1021/CR500350X
Brent MM, Marmorstein R (2008) Ankyrin for methylated lysines. Nat Struct Mol Biol 15(3):221–222. https://doi.org/10.1038/nsmb0308-221
Brien GL, Gambero G, O'Connell DJ, Jerman E, Turner SA, Egan CM, Dunne EJ, Jurgens MC, Wynne K, Piao L, Lohan AJ, Ferguson N, Shi X, Sinha KM, Loftus BJ, Cagney G, Bracken AP (2012) Polycomb PHF19 binds H3K36me3 and recruits PRC2 and demethylase NO66 to

embryonic stem cell genes during differentiation. Nat Struct Mol Biol 19(12):1273–1281. https://doi.org/10.1038/nsmb.2449
Chen J-J, Silver D, Cantor S, Livingston DM, Scully R (1999) BRCA1, BRCA2, and Rad51 operate in a common DNA damage response pathway. Cancer Res 59(7 Suppl):1752
Chen Y, Sprung R, Tang Y, Ball H, Sangras B, Kim SC, Falck JR, Peng J, Gu W, Zhao Y (2007) Lysine propionylation and butyrylation are novel post-translational modifications in histones. Mol Cell Proteomics 6(5):812–819. https://doi.org/10.1074/mcp.M700021-MCP200
Chen JQ, Brown TR, Russo J (2009) Regulation of energy metabolism pathways by estrogens and estrogenic chemicals and potential implications in obesity associated with increased exposure to endocrine disruptors. Biochim Biophys Acta Mol Cell Res 1793(7):1128–1143. https://doi.org/10.1016/j.bbamcr.2009.03.009
Chen Q, Chen Y, Bian C, Fujiki R, Yu X (2013) TET2 promotes histone O-GlcNAcylation during gene transcription HHS public access. Nature 493(7433):561–564. https://doi.org/10.1038/nature11742
Chen Q, Meng F, Wang L, Mao Y, Zhou H, Hua D, Zhang H, Wang W (2017) A polymorphism in ABCC4 is related to efficacy of 5-FU/capecitabine-based chemotherapy in colorectal cancer patients. Sci Rep 7(1):1–7. https://doi.org/10.1038/s41598-017-07491-3
Chen R, Zhang M, Liu W, Chen H, Cai T, Xiong H, Sheng X, Liu S, Peng J, Wang F, Chen H, Lin W, Xu X, Zheng W, Jiang Q (2018a) Estrogen affects the negative feedback loop of PTENP1-miR200c to inhibit PTEN expression in the development of endometrioid endometrial carcinoma. Cell Death Dis. 10(1):4. https://doi.org/10.1038/s41419-018-1207-4
Chen Y, Zhang B, Bao L, Jin L, Yang M, Peng Y, Kumar A, Wang JE, Wang C, Zou X, Xing C, Wang Y, Luo W (2018b) ZMYND8 acetylation mediates HIF-dependent breast cancer progression and metastasis. J Clin Investig 128(5):1937–1955. https://doi.org/10.1172/JCI95089
Cheung YH, Watkinson J, Anastassiou D (2011) Conditional meta-analysis stratifying on detailed HLA genotypes identifies a novel type 1 diabetes locus around TCF19 in the MHC. Hum Genet 129(2):161–176. https://doi.org/10.1007/s00439-010-0908-2
Choudhry H, Harris AL (2018) Advances in hypoxia-inducible factor biology. Cell Metab 27(2): 281–298. https://doi.org/10.1016/j.cmet.2017.10.005
Chung ST, Chacko SK, Sunehag AL, Haymond MW (2015) Measurements of gluconeogenesis and glycogenolysis: a methodological review. Diabetes 64(12):3996–4010. https://doi.org/10.2337/db15-0640
Cieniewicz AM, Moreland L, Ringel AE, Mackintosh SG, Raman A, Gilbert TM, Wolberger C, Tackett AJ, Taverna SD (2014) The bromodomain of gcn5 regulates site specificity of lysine acetylation on histone H3. Mol Cell Proteomics 13(11):2896–2910. https://doi.org/10.1074/mcp.M114.038174
Clasen BFF, Krusenstjerna-Hafstrøm T, Vendelbo MH, Thorsen K, Escande C, Møller N, Pedersen SB, Jørgensen JOL, Jessen N (2013) Gene expression in skeletal muscle after an acute intravenous GH bolus in human subjects: identification of a mechanism regulating ANGPTL4. J Lipid Res 54(7):1988. https://doi.org/10.1194/JLR.P034520
Clement T, Salone V, Rederstorff M (2015) Dual luciferase gene reporter assays to study miRNA function. Methods Mol Biol 1296:187–198. https://doi.org/10.1007/978-1-4939-2547-6_17
Clemente-Ruiz M, González-Prieto R, Prado F (2011) Histone H3K56 acetylation, CAF1, and Rtt106 coordinate nucleosome assembly and stability of advancing replication forks. PLoS Genet 7(11):1002376. https://doi.org/10.1371/journal.pgen.1002376
Cluntun AA, Huang H, Dai L, Liu X, Zhao Y, Locasale JW (2015) The rate of glycolysis quantitatively mediates specific histone acetylation sites. Cancer Metab 3:10. https://doi.org/10.1186/s40170-015-0135-3
Collins RE, Northrop JP, Horton JR, Lee DY, Zhang X, Stallcup MR, Cheng X (2008) The ankyrin repeats of G9a and GLP histone methyltransferases are mono- and dimethyllysine binding modules. Nat Struct Mol Biol 15(3):245–250. https://doi.org/10.1038/nsmb.1384

Collins RRJ, Patel K, Putnam WC, Kapur P, Rakheja D (2017) Oncometabolites: a new paradigm for oncology, metabolism, and the clinical laboratory. Clin Chem 63(12):1812–1820. https://doi.org/10.1373/clinchem.2016.267666

Coster G, Goldberg M (2010) The cellular response to DNA damage: a focus on MDC1 and its interacting proteins. Nucleus 1(2):166–178. https://doi.org/10.4161/nucl.1.2.11176

Couture JF, Collazo E, Trievel RC (2006) Molecular recognition of histone H3 by the WD40 protein WDR5. Nat Struct Mol Biol 13(8):698–703. https://doi.org/10.1038/nsmb1116

Crabtree B, Newsholme EA (1972) The activities of phosphorylase, hexokinase, phosphofructokinase, lactate dehydrogenase and the glycerol 3-phosphate dehydrogenases in muscles from vertebrates and invertebrates. Biochem J 126(1):49–58. https://doi.org/10.1042/bj1260049

Cuesta R, Gritsenko MA, Petyuk VA, Shukla AK, Tsai C-F, Liu T, Mcdermott JE, Holz MK (2019) Phosphoproteome analysis reveals estrogen-ER pathway as a modulator of mTOR activity Via DEPTOR correspondence. Mol Cell Proteomics 18:1607–1618. https://doi.org/10.1074/mcp.RA119.001506

Dai L, Peng C, Montellier E, Lu Z, Chen Y, Ishii H, Debernardi A, Buchou T, Rousseaux S, Jin F, Sabari BR, Deng Z, Allis CD, Ren B, Khochbin S, Zhao Y (2014) Lysine 2-hydroxyisobutyrylation is a widely distributed active histone mark. Nat Chem Biol 10(5): 365–370. https://doi.org/10.1038/nchembio.1497

Dai Z, Ramesh V, Locasale JW (2020) The evolving metabolic landscape of chromatin biology and epigenetics. Nat Rev Genet 21(12):737–753. https://doi.org/10.1038/s41576-020-0270-8

De Berardinis RJ, Chandel NS (2016) Fundamentals of cancer metabolism. Sci Adv 2(5):e1600200. https://doi.org/10.1126/sciadv.1600200

Densham RM, Garvin AJ, Stone HR, Strachan J, Baldock RA, Daza-Martin M, Fletcher A, Blair-Reid S, Beesley J, Johal B, Pearl LH, Neely R, Keep NH, Watts FZ, Morris JR (2016) Human BRCA1-BARD1 ubiquitin ligase activity counteracts chromatin barriers to DNA resection. Nat Struct Mol Biol 23(7):647–655. https://doi.org/10.1038/nsmb.3236

Dhalluin C, Carlson JE, Zeng L, He C, Aggarwal AK, Zhou MM (1999) Structure and ligand of a histone acetyltransferase bromodomain. Nature 399(6735):491–496. https://doi.org/10.1038/20974

Dialynas GK, Vitalini MW, Wallrath LL (2008) Linking heterochromatin protein 1 (HP1) to cancer progression. Mutat Res Fundam Mol Mech Mutagen 647(1–2):13–20. https://doi.org/10.1016/j.mrfmmm.2008.09.007

Ding Q, He K, Luo T, Deng Y, Wang H, Liu H, Zhang J, Chen K, Xiao J, Duan X, Huang R, Xia Z, Zhou W, He J, Yu H, Jiao X, Xiang G (2016) SSRP1 contributes to the malignancy of hepatocellular carcinoma and is negatively regulated by miR-497. Mol Therapy 24(5): 903–914. https://doi.org/10.1038/MT.2016.9

Dong Y, Wang M (2017) Knockdown of TKTL1 additively complements cisplatin-induced cytotoxicity in nasopharyngeal carcinoma cells by regulating the levels of NADPH and ribose-5-phosphate. Biomed Pharmacother 85:672–678. https://doi.org/10.1016/j.biopha.2016.11.078

Dong C, Nakagawa R, Oyama K, Yamamoto Y, Zhang W, Dong A, Li Y, Yoshimura Y, Kamiya H, Nakayama JI, Ueda J, Min J (2020) Structural basis for histone variant h3tk27me3 recognition by phf1 and phf19. eLife 9:1–21. https://doi.org/10.7554/ELIFE.58675

Dover J, Schneider J, Tawiah-Boateng MA, Wood A, Dean K, Johnston M, Shilatifard A (2002) Methylation of histone H3 by COMPASS requires ubiquitination of histone H2B by Rad6. J Biol Chem 277(32):28368–28371. https://doi.org/10.1074/jbc.C200348200

Drew B, Leeuwenburgh C (2003) Method for measuring ATP production in isolated mitochondria: ATP production in brain and liver mitochondria of Fischer-344 rats with age and caloric restriction. Am J Physiol Regul Integr Comp Physiol 285(5):1259–1267. https://doi.org/10.1152/ajpregu.00264.2003

Du J, Kelly AE, Funabiki H, Patel DJ (2012) Structural basis for recognition of H3T3ph and Smac/DIABLO N-terminal peptides by human survivin. Structure 20(1):185–195. https://doi.org/10.1016/j.str.2011.12.001

Elrick H, Hlad CJ, Arai Y, Smith A (1956) The interaction of glucagon and insulin on blood glucose. J Clin Invest 35(7):757–762. https://doi.org/10.1172/JCI103327

Etchegaray JP, Mostoslavsky R (2016) Interplay between metabolism and epigenetics: a nuclear adaptation to environmental changes. Mol Cell 62(5):695–711. https://doi.org/10.1016/j.molcel.2016.05.029

Everts B (2018) Metabolomics in immunology research. Methods Mol Biol 1730:29–42. https://doi.org/10.1007/978-1-4939-7592-1_2

Fan J, Krautkramer KA, Feldman JL, Denu JM (2015) Metabolic regulation of histone post-translational modifications. ACS Chem Biol 10(1):95–108. https://doi.org/10.1021/CB500846U

Fantin VR, St-Pierre J, Leder P (2006) Attenuation of LDH-A expression uncovers a link between glycolysis, mitochondrial physiology, and tumor maintenance. Cancer Cell 9(6):425–434. https://doi.org/10.1016/J.CCR.2006.04.023

Fanucchi S, Domínguez-Andrés J, Joosten LAB, Netea MG, Mhlanga MM (2021) The intersection of epigenetics and metabolism in trained immunity. Immunity 54(1):32–43. https://doi.org/10.1016/J.IMMUNI.2020.10.011

Farrelly LA, Thompson RE, Zhao S, Lepack AE, Lyu Y, Bhanu NV, Zhang B, Loh YHE, Ramakrishnan A, Vadodaria KC, Heard KJ, Erikson G, Nakadai T, Bastle RM, Lukasak BJ, Zebroski H, Alenina N, Bader M, Berton O et al (2019) Histone serotonylation is a permissive modification that enhances TFIID binding to H3K4me3. Nature 567(7749):535–539. https://doi.org/10.1038/s41586-019-1024-7

Feldman JL, Dittenhafer-Reed KE, Denu JM (2012) Sirtuin catalysis and regulation. J Biol Chem 287(51):42419–42427. https://doi.org/10.1074/JBC.R112.378877

Fellows R, Denizot J, Stellato C, Cuomo A, Jain P, Stoyanova E, Balázsi S, Hajnády Z, Liebert A, Kazakevych J, Blackburn H, Corrêa RO, Fachi JL, Sato FT, Ribeiro WR, Ferreira CM, Perée H, Spagnuolo M, Mattiuz R et al (2018) Microbiota derived short chain fatty acids promote histone crotonylation in the colon through histone deacetylases. Nat Commun 9(1):1–15. https://doi.org/10.1038/s41467-017-02651-5

Field MS, Kamynina E, Agunloye OC, Liebenthal RP, Lamarre SG, Brosnan ME, Brosnan JT, Stover PJ (2014) Nuclear enrichment of folate cofactors and methylenetetrahydrofolate dehydrogenase 1 (MTHFD1) protect de novo thymidylate biosynthesis during folate deficiency. J Biol Chem 289(43):29642–29650. https://doi.org/10.1074/jbc.M114.599589

Filippakopoulos P, Knapp S (2014) Targeting bromodomains: epigenetic readers of lysine acetylation. Nat Rev Drug Discov 13(5):337–356. https://doi.org/10.1038/nrd4286

Filippakopoulos P, Picaud S, Mangos M, Keates T, Lambert JP, Barsyte-Lovejoy D, Felletar I, Volkmer R, Müller S, Pawson T, Gingras AC, Arrowsmith CH, Knapp S (2012) Histone recognition and large-scale structural analysis of the human bromodomain family. Cell 149(1):214–231. https://doi.org/10.1016/j.cell.2012.02.013

Fischle W, Wang Y, Jacobs SA, Kim Y, Allis CD, Khorasanizadeh S (2003) Molecular basis for the discrimination of repressive methyl-lysine marks in histone H3 by polycomb and HP1 chromodomains. Genes Dev 17(15):1870–1881. https://doi.org/10.1101/gad.1110503

Flanagan JF, Mi LZ, Chruszcz M, Cymborowski M, Clines KL, Kim Y, Minor W, Rastinejad F, Khorasanizadeh S (2005) Double chromodomains cooperate to recognize the methylated histone H3 tail. Nature 438(7071):1181–1185. https://doi.org/10.1038/nature04290

Foley JE, Cushman SW, Salans LB (1978) Glucose transport in isolated rat adipocytes with measurements of L arabinose uptake. Am J Physiol Endocrinol Metab 3(2):12–19. https://doi.org/10.1152/ajpendo.1978.234.2.e112

Fong K, Zhao JC, Song B, Zheng B, Yu J (2018) TRIM28 protects TRIM24 from SPOP-mediated degradation and promotes prostate cancer progression. Nat Commun 9(1):1–15. https://doi.org/10.1038/s41467-018-07475-5

Frezza C (2020) Metabolism and cancer: the future is now. Br J Cancer 122(2):133–135. https://doi.org/10.1038/s41416-019-0667-3

Fukuda H, Sano N, Muto S, Horikoshi M (2006) Simple histone acetylation plays a complex role in the regulation of gene expression. Brief Funct Genomics Proteomics 5(3):190–208. https://doi.org/10.1093/bfgp/ell032

Galligan JJ, Wepy JA, Streeter MD, Kingsley PJ, Mitchener MM, Wauchope OR, Beavers WN, Rose KL, Wang T, Spiegel DA, Marnett LJ, Hancock AB (2018) Methylglyoxal-derived posttranslational arginine modifications are abundant histone marks. PNAS 115(37): 9228–9233. https://doi.org/10.1073/pnas.1802901115

Gao Z, Xu CW (2011) Glucose metabolism induces mono-ubiquitination of histone H2B in mammalian cells. Biochem Biophys Res Commun 404(1):428–433. https://doi.org/10.1016/J.BBRC.2010.11.138

Geiss-Friedlander R, Melchior F (2007) Concepts in sumoylation: a decade on. Nat Rev Mol Cell Biol 8(12):947–956. https://doi.org/10.1038/nrm2293

Giancaspero TA, Busco G, Panebianco C, Carmone C, Miccolis A, Liuzzi GM, Colella M, Barile M (2013) FAD synthesis and degradation in the nucleus create a local flavin cofactor pool. J Biol Chem 288(40):29069–29080. https://doi.org/10.1074/jbc.M113.500066

Giulidori P, Galli Kienle M, Catto E, Stramentinoli G (1984) Transmethylation, transsulfuration, and aminopropylation reactions of S-adenosyl-L-methionine in vivo. J Biol Chem 259(7): 4205–4211. https://doi.org/10.1016/s0021-9258(17)43031-6

Giustarini D, Rossi R, Milzani A, Dalle-Donne I (2008) Nitrite and nitrate measurement by Griess reagent in human plasma: evaluation of interferences and standardization. Methods Enzymol 440(07):361–380. https://doi.org/10.1016/S0076-6879(07)00823-3

Goetzman ES, Prochownik EV (2018) The role for myc in coordinating glycolysis, oxidative phosphorylation, glutaminolysis, and fatty acid metabolism in normal and neoplastic tissues. Front Endocrinol 9:1. https://doi.org/10.3389/fendo.2018.00129

Goudarzi A, Zhang D, Huang H, Barral S, Kwon OK, Qi S, Tang Z, Buchou T, Vitte AL, He T, Cheng Z, Montellier E, Gaucher J, Curtet S, Debernardi A, Charbonnier G, Puthier D, Petosa C, Panne D et al (2016) Dynamic competing histone H4 K5K8 acetylation and butyrylation are hallmarks of highly active gene promoters. Mol Cell 62(2):169–180. https://doi.org/10.1016/j.molcel.2016.03.014

Gowans GJ, Bridgers JB, Zhang J, Dronamraju R, Burnetti A, King DA, Thiengmany AV, Shinsky SA, Bhanu NV, Garcia BA, Buchler NE, Strahl BD, Morrison AJ (2019) Recognition of histone crotonylation by Taf14 links metabolic state to gene expression. Mol Cell 76(6):909–921.e3. https://doi.org/10.1016/J.MOLCEL.2019.09.029

Griendling KK, Touyz RM, Zweier JL, Dikalov S, Chilian W, Chen YR, Harrison DG, Bhatnagar A (2016) Measurement of reactive oxygen species, reactive nitrogen species, and redox-dependent signaling in the cardiovascular system: a scientific statement from the American Heart Association. Circ Res 119(5):e39–e75. https://doi.org/10.1161/RES.0000000000000110

Griguer CE, Oliva CR, Gillespie GY (2005) Glucose metabolism heterogeneity in human and mouse malignant glioma cell lines. J Neuro-Oncol 74(2):123–133. https://doi.org/10.1007/S11060-004-6404-6

Grimm C, de Alonso AG, Rybin V, Steuerwald U, Ly-Hartig N, Fischle W, Müller J, Müller CW (2007) Structural and functional analyses of methyl-lysine binding by the malignant brain tumour repeat protein sex comb on midleg. EMBO Rep 8(11):1031–1037. https://doi.org/10.1038/sj.embor.7401085

Grist JT, Jarvis LB, Georgieva Z, Thompson S, Kaur Sandhu H, Burling K, Clarke A, Jackson S, Wills M, Gallagher FA, Jones JL (2018) Extracellular lactate: a novel measure of T cell proliferation. J Immunol 200(3):1220–1226. https://doi.org/10.4049/jimmunol.1700886

Guo Y, Nady N, Qi C, Allali-Hassani A, Zhu H, Pan P, Adams-Cioaba MA, Amaya MF, Dong A, Vedadi M, Schapira M, Read RJ, Arrowsmith CH, Min J (2009) Methylation-state-specific recognition of histones by the MBT repeat protein L3MBTL2. Nucleic Acids Res 37(7): 2204–2210. https://doi.org/10.1093/nar/gkp086

Haigis MC, Mostoslavsky R, Haigis KM, Fahie K, Christodoulou DC, Murphy AJJ, Valenzuela DM, Yancopoulos GD, Karow M, Blander G, Wolberger C, Prolla TA, Weindruch R, Alt FW,

Guarente L (2006) SIRT4 inhibits glutamate dehydrogenase and opposes the effects of calorie restriction in pancreatic β cells. Cell 126(5):941–954. https://doi.org/10.1016/J.CELL.2006.06.057

Han Z, Guo L, Wang H, Shen Y, Deng XW, Chai J (2006) Structural basis for the specific recognition of methylated histone H3 lysine 4 by the WD-40 protein WDR5. Mol Cell 22(1):137–144. https://doi.org/10.1016/j.molcel.2006.03.018

Han J, Zhang L, Guo H, Wysham WZ, Roque DR, Willson AK, Sheng X, Zhou C, Bae-Jump VL (2015) Glucose promotes cell proliferation, glucose uptake and invasion in endometrial cancer cells via AMPK/mTOR/S6 and MAPK signaling. Gynecol Oncol 138(3):668–675. https://doi.org/10.1016/j.ygyno.2015.06.036

Han H, Zhang T, Jin Z, Guo H, Wei X, Liu Y, Chen Q, He J (2017) Blood glucose concentration and risk of liver cancer: systematic review and meta-analysis of prospective studies. Oncotarget 8(30):50164–50173. https://doi.org/10.18632/oncotarget.16816

Haws SA, Leech CM, Denu JM (2020) Metabolism and the epigenome: a dynamic relationship. Trends Biochem Sci 45(9):731–747. https://doi.org/10.1016/J.TIBS.2020.04.002

Hay N (2016) Reprogramming glucose metabolism in cancer: can it be exploited for cancer therapy? Nat Rev Cancer. 16:635. https://doi.org/10.1038/nrc.2016.77

He F, Umehara T, Saito K, Harada T, Watanabe S, Yabuki T, Kigawa T, Takahashi M, Kuwasako K, Tsuda K, Matsuda T, Aoki M, Seki E, Kobayashi N, Güntert P, Yokoyama S, Muto Y (2010) Structural insight into the zinc finger CW domain as a histone modification reader. Structure 18(9):1127–1139. https://doi.org/10.1016/j.str.2010.06.012

He Y, Gao M, Tang H, Cao Y, Liu S, Tao Y (2019) Metabolic intermediates in tumorigenesis and progression. Int J Biol Sci 15(6):1187–1199. https://doi.org/10.7150/ijbs.33496

Hino S, Sakamoto A, Nagaoka K, Anan K, Wang Y, Mimasu S, Umehara T, Yokoyama S, Kosai KI, Nakao M (2012) FAD-dependent lysine-specific demethylase-1 regulates cellular energy expenditure. Nat Commun 3(1):1–12. https://doi.org/10.1038/ncomms1755

Hirata H, Sugimachi K, Komatsu H, Ueda M, Masuda T, Uchi R, Sakimura S, Nambara S, Saito T, Shinden Y, Iguchi T, Eguchi H, Ito S, Terashima K, Sakamoto K, Hirakawa M, Honda H, Mimori K (2016) Decreased expression of fructose-1,6-bisphosphatase associates with glucose metabolism and tumor progression in hepatocellular carcinoma. Cancer Res 76(11):3265–3276. https://doi.org/10.1158/0008-5472.CAN-15-2601

Hongs L, Schrothp GP, Matthew HR, Yaus P, Bradburysliii EM (1993) Studies of the DNA binding properties of histone H4 amino terminus. Thermal denaturation studies reveal that acetylation markedly reduces the binding constant of the H4 "tail" to DNA. J Biol Chem 268(1):305–314. https://doi.org/10.1016/S0021-9258(18)54150-8

Hou H, Yu H (2010) Structural insights into histone lysine demethylation. Curr Opin Struct Biol. 20(6):739–748. https://doi.org/10.1016/j.sbi.2010.09.006

Hou Y, Liu W, Yi X, Yang Y, Su D, Huang W, Yu H, Teng X, Yang Y, Feng W, Zhang T, Gao J, Zhang K, Qiu R, Wang Y (2020) PHF20L1 as a H3K27me2 reader coordinates with transcriptional repressors to promote breast tumorigenesis. Sci Adv 6(16). https://doi.org/10.1126/sciadv.aaz0356

Hoyal CR, Kammerer S, Roth RB, Reneland R, Marnellos G, Kiechle M, Schwarz-Boeger U, Griffiths LR, Ebner F, Rehbock J, Nelson MR, Braun A (2005) Genetic polymorphisms in DPF3 associated with risk of breast cancer and lymph node metastases. J Carcinogenesis 4:13–13. https://doi.org/10.1186/1477-3163-4-13

Hsu PP, Sabatini DM (2008) Cancer cell metabolism: Warburg and beyond. Cell 134(5):703–707. https://doi.org/10.1016/J.CELL.2008.08.021

Huang S, Yang Z, Ma Y, Yang Y, Wang S (2017) MiR-101 enhances cisplatin-induced DNA damage through decreasing nicotinamide adenine dinucleotide phosphate levels by directly repressing Tp53-induced glycolysis and apoptosis regulator expression in prostate cancer cells. DNA Cell Biol 36(4):303–310. https://doi.org/10.1089/dna.2016.3612

Huang H, Zhang D, Wang Y, Perez-Neut M, Han Z, Zheng YG, Hao Q, Zhao Y (2018) Lysine benzoylation is a histone mark regulated by SIRT2. Nat Commun 9(1): 1–11. https://doi.org/10.1038/s41467-018-05567-w

Imai S, Guarente L (2014) NAD+ and sirtuins in aging and disease. Trends Cell Biol 24(8): 464–471. https://doi.org/10.1016/j.tcb.2014.04.002

Jacobs SA, Khorasanizadeh S (2002) Structure of HP1 chromodomain bound to a lysine 9-methylated histone H3 tail. Science 295(5562):2080–2083. https://doi.org/10.1126/science.1069473

Karve TM, Cheema AK (2011) Small changes huge impact: the role of protein posttranslational modifications in cellular homeostasis and disease. J Amino Acids. https://doi.org/10.4061/2011/207691

Keating ST, El-Osta A (2015) Epigenetics and metabolism. Circ Res 116(4):715–736. https://doi.org/10.1161/CIRCRESAHA.116.303936

Kerscher O (2007) SUMO junction - what's your function? New insights through SUMO-interacting motifs. EMBO Rep. 8(6):550–555. https://doi.org/10.1038/sj.embor.7400980

Khatami F, Aghamir SMK, Tavangar SM (2019) Oncometabolites: a new insight for oncology. Mol Genet Genomic Med 7(9):e873. https://doi.org/10.1002/mgg3.873

Kikuchi M, Okumura F, Tsukiyama T, Watanabe M, Miyajima N, Tanaka J, Imamura M, Hatakeyama S (2009) TRIM24 mediates ligand-dependent activation of androgen receptor and is repressed by a bromodomain-containing protein, BRD7, in prostate cancer cells. Biochim Biophys Acta 1793(12):1828–1836. https://doi.org/10.1016/J.BBAMCR.2009.11.001

Kold-Christensen R, Johannsen M (2020) Methylglyoxal metabolism and aging-related disease: moving from correlation toward causation. Trends Endocrinol Metab 31(2):81–92. https://doi.org/10.1016/j.tem.2019.10.003

Kolodziejczyk AA, Lönnberg T (2018) Global and targeted approaches to single-cell transcriptome characterization. Brief Funct Genomics 17(4):209–219. https://doi.org/10.1093/BFGP/ELX025

Koppenol WH, Bounds PL, Dang CV (2011) Otto Warburg's contributions to current concepts of cancer metabolism. Nat Rev Cancer 11(5):325–337. https://doi.org/10.1038/nrc3038

Koubaa M, Cocuron JC, Thomasset B, Alonso AP (2013) Highlighting the tricarboxylic acid cycle: liquid and gas chromatography-mass spectrometry analyses of 13C-labeled organic acids. Anal Biochem 436(2):151–159. https://doi.org/10.1016/j.ab.2013.01.027

Krautkramer KA, Linnemann AK, Fontaine DA, Whillock AL, Harris TW, Schleis GJ, Truchan NA, Marty-Santos L, Lavine JA, Cleaver O, Kimple ME, Davis DB (2013) Tcf19 is a novel islet factor necessary for proliferation and survival in the INS-1 β-cell line. Am J Physiol Endocrinol Metab 305(5):E600–E610. https://doi.org/10.1152/ajpendo.00147.2013

Lacza Z, Kozlov AV, Pankotai E, Csordás A, Wolf G, Redl H, Kollai M, Szabó C, Busija DW, Horn TFW (2006) Mitochondria produce reactive nitrogen species via an arginine-independent pathway. Free Radic Res 40(4):369–378. https://doi.org/10.1080/10715760500539139

Ladendor NE, Wu S, Lipsick JS (2001) BS69, an adenovirus E1A-associated protein, inhibits the transcriptional activity of c-Myb. Oncogene. 20(1):125–132

Landgraf BJ, McCarthy EL, Booker SJ (2016) Radical S-Adenosylmethionine enzymes in human health and disease. Annu Rev Biochem 85:485–514. https://doi.org/10.1146/annurev-biochem-060713-035504

Lange M, Kaynak B, Forster UB, Tönjes M, Fischer JJ, Grimm C, Schlesinger J, Just S, Dunkel I, Krueger T, Mebus S, Lehrach H, Lurz R, Gobom J, Rottbauer W, Abdelilah-Seyfried S, Sperling S (2008) Regulation of muscle development by DPF3, a novel histone acetylation and methylation reader of the BAF chromatin remodeling complex. Genes Dev 22(17): 2370–2384. https://doi.org/10.1101/gad.471408

Lee J, Thompson JR, Botuyan MV, Mer G (2008) Distinct binding modes specify the recognition of methylated histones H3K4 and H4K20 by JMJD2A-tudor. Nat Struct Mol Biol 15(1):109–111. https://doi.org/10.1038/nsmb1326

Leone RD, Powell JD (2020) Metabolism of immune cells in cancer. Nat Rev Cancer 20(9): 516–531. https://doi.org/10.1038/S41568-020-0273-Y

Lepack AE, Werner CT, Stewart AF, Fulton SL, Zhong P, Farrelly LA, Smith ACW, Ramakrishnan A, Lyu Y, Bastle RM, Martin JA, Mitra S, O'Connor RM, Wang Z-J, Molina H, Turecki G, Shen L, Yan Z, Calipari ES, Dietz DM, Kenny PJ, Maze I (2020) Dopaminylation of histone H3 in ventral tegmental area regulates cocaine seeking. Science (New York, NY) 368(6487):197–201. https://doi.org/10.1126/SCIENCE.AAW8806

Lepack AE, Werner CT, Stewart AF, Fulton SL, Zhong P, Farrelly LA, Smith ACW, Ramakrishnan A, Lyu Y, Bastle RM, Martin JA, Mitra S, O'connor RM, Wang ZJ, Molina H, Turecki G, Shen L, Yan Z, Calipari ES et al. (n.d.). Dopaminylation of histone H3 in ventral tegmental area regulates cocaine seeking. http://science.sciencemag.org/. Retrieved 7 June 2021

Li H, Ilin S, Wang W, Duncan EM, Wysocka J, Allis CD, Patel DJ (2006) Molecular basis for site-specific read-out of histone H3K4me3 by the BPTF PHD finger of NURF. Nature 442(7098): 91–95. https://doi.org/10.1038/nature04802

Li H, Fischle W, Wang W, Duncan EM, Liang L, Murakami-Ishibe S, Allis CD, Patel DJ (2007) Structural basis for lower lysine methylation state-specific readout by MBT repeats of L3MBTL1 and an engineered PHD finger. Mol Cell 28(4):677–691. https://doi.org/10.1016/j.molcel.2007.10.023

Li X, Kikuchi K, Takano Y (2011) ING genes work as tumor suppressor genes in the carcinogenesis of head and neck squamous cell carcinoma. J Oncol 2011:11. https://doi.org/10.1155/2011/963614

Li Y, Wen H, Xi Y, Tanaka K, Wang H, Peng D, Ren Y, Jin Q, Dent SYR, Li W, Li H, Shi X (2014) AF9 YEATS domain links histone acetylation to DOT1L-mediated H3K79 methylation. Cell 159(3):558. https://doi.org/10.1016/J.CELL.2014.09.049

Li W, Kou J, Qin J, Li L, Zhang Z, Pan Y, Xue Y, Du W (2021) NADPH levels affect cellular epigenetic state by inhibiting HDAC3–Ncor complex. Nat Metab 3(1):75–89. https://doi.org/10.1038/s42255-020-00330-2

Liao J, Tao X, Ding Q, Liu J, Yang X, Yuan F-E, Yang J-A, Liu B, Xiang G-A, Chen Q (2017) SSRP1 silencing inhibits the proliferation and malignancy of human glioma cells via the MAPK signaling pathway. Oncol Rep 38(5):2667. https://doi.org/10.3892/OR.2017.5982

Liberti MV, Locasale JW (2016) The Warburg Effect: how does it benefit cancer cells? Trends Biochem Sci 41(3):211–218. https://doi.org/10.1016/j.tibs.2015.12.001

Lieu EL, Nguyen T, Rhyne S, Kim J (2020) Amino acids in cancer. Exp Mol Med 52(1):15–30. https://doi.org/10.1038/s12276-020-0375-3

Liszt G, Ford E, Kurtev M, Guarente L (2005) Mouse Sir2 homolog SIRT6 is a nuclear ADP-ribosyltransferase. J Biol Chem 280(22):21313–21320. https://doi.org/10.1074/JBC.M413296200

Liu K, Guo Y, Liu H, Bian C, Lam R, Liu Y, Mackenzie F, Rojas LA, Reinberg D, Bedford MT, Xu RM, Min J (2012) Crystal structure of TDRD3 and methyl-arginine binding characterization of TDRD3, SMN and SPF30. PLoS One 7(2). https://doi.org/10.1371/journal.pone.0030375

Luger K, Richmond TJ (1998) The histone tails of the nucleosome. Curr Opin Genet Dev 8(2): 140–146. https://doi.org/10.1016/S0959-437X(98)80134-2

Macdonald N, Welburn JPI, Noble MEM, Nguyen A, Yaffe MB, Clynes D, Moggs JG, Orphanides G, Thomson S, Edmunds JW, Clayton AL, Endicott JA, Mahadevan LC (2005) Molecular basis for the recognition of phosphorylated and phosphoacetylated histone H3 by 14-3-3. Mol Cell 20(2):199–211. https://doi.org/10.1016/j.molcel.2005.08.032

Mallette FA, Mattiroli F, Cui G, Young LC, Hendzel MJ, Mer G, Sixma TK, Richard S (2012) RNF8- and RNF168-dependent degradation of KDM4A/JMJD2A triggers 53BP1 recruitment to DNA damage sites. EMBO J 31(8):1865–1878. https://doi.org/10.1038/emboj.2012.47

Manchester JK, Chi M-Y, Carter JG, Ellen Pusateri M, McDougal DB, Lowry OH, Chi MMY, Carter JG, Pusateri ME, McDougal DB, Lowry OH (1990) Measurement of 2-deoxyglucose and 2-deoxyglucose 6-phosphate in tissues. Anal Biochem 185(1):118–124. https://doi.org/10.1016/0003-2697(90)90265-B

Manke IA, Lowery DM, Nguyen A, Yaffe MB (2003) BRCT repeats as phosphopeptide-binding modules involved in protein targeting. Science 302(5645):636–639. https://doi.org/10.1126/science.1088877

Mao X, Gluck N, Li D, Maine GN, Li H, Zaidi IW, Repaka A, Mayo MW, Burstein E (2009) GCN5 is a required cofactor for a ubiquitin ligase that targets NF-κB/RelA. Genes Dev 23(7):849–861. https://doi.org/10.1101/gad.1748409

Mao Z, Hine C, Tian X, Van Meter M, Au M, Vaidya A, Seluanov A, Gorbunova V (2011) SIRT6 promotes DNA repair under stress by activating PARP1. Science 332(6036):1443–1446. https://doi.org/10.1126/SCIENCE.1202723

Margueron R, Justin N, Ohno K, Sharpe ML, Son J, Drury WJ, Voigt P, Martin SR, Taylor WR, De Marco V, Pirrotta V, Reinberg D, Gamblin SJ (2009) Role of the polycomb protein EED in the propagation of repressive histone marks. Nature 461(7265):762–767. https://doi.org/10.1038/nature08398

Masson IL, Yu DY, Jensen K, Chevalier A, Courbeyrette R, Boulard Y, Smith MM, Mann C (2003) Yaf9, a novel NuA4 histone acetyltransferase subunit, is required for the cellular response to spindle stress in yeast. Mol Cell Biol 23(17):6086–6102. https://doi.org/10.1128/MCB.23.17.6086-6102.2003

Mato JM, Martínez-Chantar ML, Lu SC (2013) S-adenosylmethionine metabolism and liver disease. Ann Hepatol 12(2):183–189. https://doi.org/10.1016/s1665-2681(19)31355-9

Matsumoto M, Sakai M (2012) Glucose production assay in primary mouse hepatocytes. Bio-protocol 2(21):e284

Mattiroli F, Penengo L (2021) Histone ubiquitination: an integrative signaling platform in genome stability. Trends Genet. https://doi.org/10.1016/j.tig.2020.12.005

McNabb DS, Reed R, Marciniak RA (2005) Dual luciferase assay system for rapid assessment of gene expression in Saccharomyces cerevisiae. Eukaryot Cell 4(9):1539–1549. https://doi.org/10.1128/EC.4.9.1539-1549.2005

Meera Krishna B, Khan MA, Khan ST (2019) Next-generation sequencing (NGS) platforms: an exciting era of genome sequence analysis. Microbial Genomics Sustain Agroecosyst 2:89–109. https://doi.org/10.1007/978-981-32-9860-6_6

Membrane M, Assay P (2016) High-throughput screening assays in toxicology. Methods Mol Biol 1473(5):17–22. https://doi.org/10.1007/978-1-4939-6346-1

Messner S, Hottiger MO (2011) Histone ADP-ribosylation in DNA repair, replication and transcription. Trends Cell Biol 21(9):534–554. https://doi.org/10.1016/j.tcb.2011.06.001

Messner S, Altmeyer M, Zhao H, Pozivil A, Roschitzki B, Gehrig P, Rutishauser D, Huang D, Caflisch A, Hottiger MO (2010) PARP1 ADP-ribosylates lysine residues of the core histone tails. Nucleic Acids Res 38(19):6350–6362. https://doi.org/10.1093/nar/gkq463

Migliori V, Müller J, Phalke S, Low D, Bezzi M, Mok WC, Kumar Sahu S, Gunaratne J, Capasso P, Bassi C, Cecatiello V, De Marco A, Blackstock W, Kuznetsov V, Amati B, Mapelli M, Guccione E (2012) Symmetric dimethylation of H3R2 is a newly identified histone mark that supports euchromatin maintenance. Nat Struct Mol Biol. https://doi.org/10.1038/nsmb.2209

Mills AA (2017) The chromodomain helicase DNA-binding chromatin remodelers: family traits that protect from and promote cancer. Cold Spring Harbor Perspect Med 7(4):a026450. https://doi.org/10.1101/cshperspect.a026450

Min J, Zhang Y, Xu RM (2003) Structural basis for specific binding of polycomb chromodomain to histone H3 methylated at Lys 27. Genes Dev 17(15):1823–1828. https://doi.org/10.1101/gad.269603

Min J, Allali-Hassani A, Nady N, Qi C, Ouyang H, Liu Y, MacKenzie F, Vedadi M, Arrowsmith CH (2007) L3MBTL1 recognition of mono- and dimethylated histones. Nat Struct Mol Biol 14(12):1229–1230. https://doi.org/10.1038/nsmb1340

Mishanina TV, Kohen A (2015) Synthesis and application of isotopically labeled flavin nucleotides. J Label Compd Radiopharm 58(9):370–375. https://doi.org/10.1002/jlcr.3313

Molano-Arevalo JC, Hernandez DR, Gonzalez WG, Miksovska J, Ridgeway ME, Park MA, Fernandez-Lima F (2014) Flavin adenine dinucleotide structural motifs: from solution to gas phase. Anal Chem 86(20):10223–10230. https://doi.org/10.1021/ac5023666

Mondal P, Sen S, Klein BJ, Tiwary N, Gadad SS, Kutateladze TG, Roy S, Das C (2020) TCF19 promotes cell proliferation through binding to the histone H3K4me3 mark. Biochemistry 59(4): 389–399. https://doi.org/10.1021/acs.biochem.9b00771

Mondal P, Gadad SS, Adhikari S, Ramos EI, Sen S, Prasad P, Das C (2021) TCF19 and p53 regulate transcription of TIGAR and SCO2 in HCC for mitochondrial energy metabolism and stress adaptation. FASEB J 35(9):e21814. https://doi.org/10.1096/FJ.202002486RR

Mujtaba S, Zeng L, Zhou MM (2007) Structure and acetyl-lysine recognition of the bromodomain. Oncogene 26(37):5521–5527. https://doi.org/10.1038/sj.onc.1210618

Mukherjee S, Adhikary S, Gadad SS, Mondal P, Sen S, Choudhari R, Singh V, Adhikari S, Mandal P, Chaudhuri S, Sengupta A, Lakshmanaswamy R, Chakrabarti P, Roy S, Das C (2020) Suppression of poised oncogenes by ZMYND8 promotes chemo-sensitization. Cell Death Dis 11(12). https://doi.org/10.1038/s41419-020-03129-x

Musselman CA, Kutateladze TG (2011) Handpicking epigenetic marks with PHD fingers. Nucl Acids Res. 39(21):9061–9071. https://doi.org/10.1093/nar/gkr613

Musselman CA, Lalonde ME, Côté J, Kutateladze TG (2012) Perceiving the epigenetic landscape through histone readers. Nature Struct Mol Biol. https://doi.org/10.1038/nsmb.2436

Mutlu B, Puigserver P (2021) GCN5 acetyltransferase in cellular energetic and metabolic processes. Biochim Biophys Acta Gene Regul Mech 1864(2). https://doi.org/10.1016/j.bbagrm.2020.194626

Nady N, Lemak A, Walker JR, Avvakumov GV, Kareta MS, Achour M, Xue S, Duan S, Allali-Hassani A, Zuo X, Wang YX, Bronner C, Chédin F, Arrowsmith CH, Dhe-Paganon S (2011) Recognition of multivalent histone states associated with heterochromatin by UHRF1 protein. J Biol Chem 286(27):24300–24311. https://doi.org/10.1074/jbc.M111.234104

Nakamura K, Saredi G, Becker JR, Foster BM, Nguyen NV, Beyer TE, Cesa LC, Faull PA, Lukauskas S, Frimurer T, Chapman JR, Bartke T, Groth A (2019) H4K20me0 recognition by BRCA1–BARD1 directs homologous recombination to sister chromatids. Nat Cell Biol. 21(3): 311–318. https://doi.org/10.1038/s41556-019-0282-9

Narayanan N, Wang Z, Li L, Yang Y (2017) Arginine methylation of USP9X promotes its interaction with TDRD3 and its anti-apoptotic activities in breast cancer cells. Cell Discov 3. https://doi.org/10.1038/celldisc.2016.48

Ndiaye FK, Ortalli A, Canouil M, Huyvaert M, Salazar-Cardozo C, Lecoeur C, Verbanck M, Pawlowski V, Boutry R, Durand E, Rabearivelo I, Sand O, Marselli L, Kerr-Conte J, Chandra V, Scharfmann R, Poulain-Godefroy O, Marchetti P, Pattou F et al (2017) Expression and functional assessment of candidate type 2 diabetes susceptibility genes identify four new genes contributing to human insulin secretion. Mol Metab 6(6):459–470. https://doi.org/10.1016/j.molmet.2017.03.011

Neradil J, Pavlasová G, Veselska R (2012) New mechanisms for an old drug: DHFR- and non-DHFR-mediated effects of methotrexate in cancer cells. Klin Onkol 25:2S87–2S92

Ngo DC, Ververis K, Tortorella SM, Karagiannis TC (2015) Introduction to the molecular basis of cancer metabolism and the Warburg effect. Mol Biol Rep 42(4):819–823. https://doi.org/10.1007/s11033-015-3857-y

Nielsen PR, Nietlispach D, Mott HR, Callaghan J, Bannister A, Kouzarides T, Murzin AG, Murzina NV, Laue ED (2002) Structure of the HP1 chromodomain bound to histone H3 methylated at lysine 9. Nature 416(6876):103–107. https://doi.org/10.1038/nature722

Oishi Y, Manabe I, Tobe K, Ohsugi M, Kubota T, Fujiu K, Maemura K, Kubota N, Kadowaki T, Nagai R (2008) SUMOylation of Krüppel-like transcription factor 5 acts as a molecular switch in transcriptional programs of lipid metabolism involving PPAR-δ. Nat Med 14(6):656–666. https://doi.org/10.1038/nm1756

Oladimeji P, Skerl R, Rusch C, Diakonova M (2016) Synergistic activation of ERa by estrogen and prolactin in breast cancer cells requires tyrosyl phosphorylation of PAK1. Cancer Res 76(9): 2600–2611. https://doi.org/10.1158/0008-5472.CAN-15-1758

Opie LH, Newsholme EA, Newsholmet EA (1967) The activities of fructose 1,6-diphosphatase, phosphofructokinase and phosphoenolpyruvate carboxykinase in white muscle and red muscle. Biochem J 103(2):391–399. https://doi.org/10.1042/bj1030391

Owen DJ, Ornaghi P, Yang JC, Lowe N, Evans PR, Ballario P, Neuhaus D, Filetici P, Travers AA (2000) The structural basis for the recognition of acetylated histone H4 by the bromodomain of histone acetyltransferase Gcn5p. EMBO J 19(22):6141–6149. https://doi.org/10.1093/emboj/19.22.6141

Oya E, Nakagawa R, Yoshimura Y, Tanaka M, Nishibuchi G, Machida S, Shirai A, Ekwall K, Kurumizaka H, Tagami H, Nakayama J (2019) H3K14 ubiquitylation promotes H3K9 methylation for heterochromatin assembly. EMBO Rep 20(10):e48111

Pan D, Mao C, Zou T, Yao AY, Cooper MP, Boyartchuk V, Wang YX (2012) The histone demethylase Jhdm1a regulates hepatic gluconeogenesis. PLoS Genet 8(6). https://doi.org/10.1371/journal.pgen.1002761

Pathiraja TN, Thakkar KN, Jiang S, Stratton S, Liu Z, Gagea M, Shi X, Shah PK, Phan L, Lee MH, Andersen J, Stampfer M, Barton MC (2015) TRIM24 links glucose metabolism with transformation of human mammary epithelial cells. Oncogene 34(22):2836–2845. https://doi.org/10.1038/onc.2014.220

Pavlova NN, Thompson CB (2016) The emerging hallmarks of cancer metabolism. Cell Metab 23(1):27–47. https://doi.org/10.1016/j.cmet.2015.12.006

Peña PV, Davrazou F, Shi X, Walter KL, Verkhusha VV, Gozani O, Zhao R, Kutateladze TG (2006) Molecular mechanism of histone H3K4me3 recognition by plant homeodomain of ING2. Nature 442(7098):100–103. https://doi.org/10.1038/nature04814

Pietrocola F, Galluzzi L, Bravo-San Pedro JM, Madeo F, Kroemer G (2015) Acetyl coenzyme A: a central metabolite and second messenger. Cell Metabol. 21(6):805–821. https://doi.org/10.1016/j.cmet.2015.05.014

Project LS, Nakagawa K, Miyagawa H, Kawana S (2015) Quantitative analysis of stable isotopes of glucose in blood plasma using quadrupole GC-MS. 2:8

Qin B, Yu J, Nowsheen S, Wang M, Tu X, Liu T, Li H, Wang L, Lou Z (2019) UFL1 promotes histone H4 ufmylation and ATM activation. Nat Commun 10(1):1–13. https://doi.org/10.1038/s41467-019-09175-0

Qin B, Qin B, Yu J, Nowsheen S, Nowsheen S, Zhao F, Wang L, Lou Z (2020) STK38 promotes ATM activation by acting as a reader of histone H4 ufmylation. Sci Adv 6(23):eaax8214. https://doi.org/10.1126/sciadv.aax8214

Rivadeneira DB, Caino MC, Seo JH, Angelin A, Wallace DC, Languino LR, Altieri DC (2015) Survivin promotes oxidative phosphorylation, subcellular mitochondrial repositioning, and tumor cell invasion. Sci Signal 8(389). https://doi.org/10.1126/scisignal.aab1624

Rose S, Frye RE, Slattery J, Wynne R, Tippett M, Pavliv O, Melnyk S, James SJ (2014) Oxidative stress induces mitochondrial dysfunction in a subset of autism lymphoblastoid cell lines in a well-matched case control cohort. PLoS One 9(1). https://doi.org/10.1371/journal.pone.0085436

Rush GF, Gorski JR, Ripple MG, Sowinski J, Bugelski P, Hewitt WR (1985) Organic hydroperoxide-induced lipid peroxidation and cell death in isolated hepatocytes. Toxicol Appl Pharmacol 78(3):473–483. https://doi.org/10.1016/0041-008X(85)90255-8

Ruthenburg AJ, Wang W, Graybosch DM, Li H, Allis CD, Patel DJ, Verdine GL (2006) Histone H3 recognition and presentation by the WDR5 module of the MLL1 complex. Nat Struct Mol Biol 13(8):704–712. https://doi.org/10.1038/nsmb1119

Sabari BR, Tang Z, Huang H, Yong-Gonzalez V, Molina H, Kong HE, Dai L, Shimada M, Cross JR, Zhao Y, Roeder RG, Allis CD (2015) Intracellular crotonyl-CoA stimulates transcription through p300-catalyzed histone crotonylation. Mol Cell 58(2):203–215. https://doi.org/10.1016/J.MOLCEL.2015.02.029

Sabari BR, Zhang D, David Allis C, Zhao Y (2017) Metabolic regulation of gene expression through histone acylations. Nat Rev Mol Cell Biol. https://doi.org/10.1038/nrm.2016.140

Sakai M, Matsumoto M, Tujimura T, Yongheng C, Noguchi T, Inagaki K, Inoue H, Hosooka T, Takazawa K, Kido Y, Yasuda K, Hiramatsu R, Matsuki Y, Kasuga M (2012) CITED2 links hormonal signaling to PGC-1α acetylation in the regulation of gluconeogenesis. Nat Med 18(4): 612–617. https://doi.org/10.1038/nm.2691

Sapir A, Tsur A, Koorman T, Ching K, Mishra P, Bardenheier A, Podolsky L, Bening-Abu-Shach U, Boxem M, Chou T-F, Broday L, Sternberg PW (2014) Controlled sumoylation of the mevalonate pathway enzyme HMGS-1 regulates metabolism during aging. Proc Natl Acad Sci 111(37):E3880–E3889. https://doi.org/10.1073/PNAS.1414748111

Schnekenburger M, Diederich M (2015) Nutritional epigenetic regulators in the field of cancer: new avenues for chemopreventive approaches. Epigenetic Cancer Therapy 2015:393–425. https://doi.org/10.1016/B978-0-12-800206-3.00018-5

Schuetz A, Allali-Hassani A, Martín F, Loppnau P, Vedadi M, Bochkarev A, Plotnikov AN, Arrowsmith CH, Min J (2006) Structural basis for molecular recognition and presentation of histone H3 By WDR5. EMBO J 25(18):4245–4252. https://doi.org/10.1038/sj.emboj.7601316

Schug ZT, Peck B, Jones DT, Zhang Q, Grosskurth S, Alam IS, Goodwin LM, Smethurst E, Mason S, Blyth K, McGarry L, James D, Shanks E, Kalna G, Saunders RE, Jiang M, Howell M, Lassailly F, Thin MZ et al (2015) Acetyl-CoA synthetase 2 promotes acetate utilization and maintains cancer cell growth under metabolic stress. Cancer Cell 27(1):57–71. https://doi.org/10.1016/j.ccell.2014.12.002

Schulze JMSM, Wang AYWY, Kobor MSKS (2009) YEATS domain proteins: a diverse family with many links to chromatin modification and transcription. Biochem Cell Biol. 87(1):65–75. https://doi.org/10.1139/O08-111

Sciacovelli M, Frezza C (2016) Oncometabolites: unconventional triggers of oncogenic signalling cascades. In: Free Radic Biol Med, vol 100, pp 175–181. https://doi.org/10.1016/j.freeradbiomed.2016.04.025

Sen S, Sanyal S, Srivastava DK, Dasgupta D, Roy S, Das C (2017) Transcription factor 19 interacts with histone 3 lysine 4 trimethylation and controls gluconeogenesis via the nucleosome-remodeling-deacetylase complex. J Biol Chem 292(50):20362–20378. https://doi.org/10.1074/jbc.M117.786863

Shema-Yaacoby E, Nikolov M, Haj-Yahya M, Siman P, Allemand E, Yamaguchi Y, Muchardt C, Urlaub H, Brik A, Oren M, Fischle W (2013) Systematic identification of proteins binding to chromatin-embedded ubiquitylated H2B reveals recruitment of SWI/SNF to regulate transcription. Cell Rep 4(3):601–608. https://doi.org/10.1016/j.celrep.2013.07.014

Shi X, Hong T, Walter KL, Ewalt M, Michishita E, Hung T, Carney D, Peña P, Lan F, Kaadige MR, Lacoste N, Cayrou C, Davrazou F, Saha A, Cairns BR, Ayer DE, Kutateladze TG, Shi Y, Côté J et al (2006) ING2 PHD domain links histone H3 lysine 4 methylation to active gene repression. Nature 442(7098):96–99. https://doi.org/10.1038/nature04835

Shih HM, Chang CC, Kuo HY, Lin DY (2007) Daxx mediates SUMO-dependent transcriptional control and subnuclear compartmentalization. Biochem Soc Trans 35(6):1397–1400. https://doi.org/10.1042/BST0351397

Shimazu T, Hirschey MD, Newman J, He W, Shirakawa K, Le Moan N, Grueter CA, Lim H, Saunders LR, Stevens RD, Newgard CB, Farese RV, De Cabo R, Ulrich S, Akassoglou K, Verdin E (2013) Suppression of oxidative stress by β-hydroxybutyrate, an endogenous histone deacetylase inhibitor. Science 339(6116):211–214. https://doi.org/10.1126/science.1227166

Shuvalov O, Petukhov A, Daks A, Fedorova O, Vasileva E, Barlev NA (2017) One-carbon metabolism and nucleotide biosynthesis as attractive targets for anticancer therapy. Oncotarget 8(14):23955–23977. https://doi.org/10.18632/oncotarget.15053

Shyh-Chang N, Locasale JW, Lyssiotis CA, Zheng Y, Teo RY, Ratanasirintrawoot S, Zhang J, Onder T, Unternaehrer JJ, Zhu H, Asara JM, Daley GQ, Cantley LC (2013) Influence of threonine metabolism on S-adenosylmethionine and histone methylation. Science 339(6116): 222–226. https://doi.org/10.1126/science.1226603

Smith RL, Soeters MR, Wüst RCI, Houtkooper RH (2018) Metabolic flexibility as an adaptation to energy resources and requirements in health and disease. Endocr Rev 39(4):489. https://doi.org/10.1210/ER.2017-00211

So CC, Ramachandran S, Martin A (2019) E3 ubiquitin ligases RNF20 and RNF40 are required for double-stranded break (DSB) repair: evidence for monoubiquitination of histone H2B lysine 120 as a novel axis of DSB signaling and repair. Mol Cell Biol 39(8). https://doi.org/10.1128/mcb.00488-18

Stucki M, Clapperton JA, Mohammad D, Yaffe MB, Smerdon SJ, Jackson SP (2005) MDC1 directly binds phosphorylated histone H2AX to regulate cellular responses to DNA double-strand breaks. Cell 123(7):1213–1226. https://doi.org/10.1016/j.cell.2005.09.038

Su D, Hu Q, Li Q, Thompson JR, Cui G, Fazly A, Davies BA, Botuyan MV, Zhang Z, Mer G (2012) Structural basis for recognition of H3K56-acetylated histone H3-H4 by the chaperone Rtt106. Nature 483(7387):104–109. https://doi.org/10.1038/nature10861

Suganuma T, Pattenden SG, Workman JL (2008) Diverse functions of WD40 repeat proteins in histone recognition. Genes Dev. 22(10):1265–1268. https://doi.org/10.1101/gad.1676208

Sun ZW, Allis CD (2002) Ubiquitination of histone H2B regulates H3 methylation and gene silencing in yeast. Nature 418(6893):104–108. https://doi.org/10.1038/nature00883

Sun Z, Zhang Y, Jia J, Fang Y, Tang Y, Wu H, Fang D (2020) H3K36me3, message from chromatin to DNA damage repair. Cell Biosci. 10(1). https://doi.org/10.1186/s13578-020-0374-z

Sun L, Zhang H, Gao P (2021a) Metabolic reprogramming and epigenetic modifications on the path to cancer. Protein Cell. https://doi.org/10.1007/s13238-021-00846-7

Sun L, Zhang H, Gao P (2021b) Metabolic reprogramming and epigenetic modifications on the path to cancer. Protein Cell. https://doi.org/10.1007/S13238-021-00846-7

Taber RL, Campbell A, Spencer S (1998) A simple experiment demonstrating the allosteric regulation of yeast pyruvate kinase. Biochem Educ 26(1):73–76. https://doi.org/10.1016/S0307-4412(97)00117-9

Tan M, Peng C, Anderson KA, Chhoy P, Xie Z, Dai L, Park J, Chen Y, Huang H, Zhang Y, Ro J, Wagner GR, Green MF, Madsen AS, Schmiesing J, Peterson BS, Xu G, Ilkayeva OR, Muehlbauer MJ et al (2014) Lysine glutarylation is a protein posttranslational modification regulated by SIRT5. Cell Metab 19(4):605–617. https://doi.org/10.1016/j.cmet.2014.03.014

Teslaa T, Teitell MA (2014) Techniques to monitor glycolysis. Methods Enzymol 542:91–114. https://doi.org/10.1016/B978-0-12-416618-9.00005-4

Thakur C, Chen F (2019) Connections between metabolism and epigenetics in cancers. Semin Cancer Biol 57:52–58. https://doi.org/10.1016/j.semcancer.2019.06.006

Theillet FO-X, Liokatis S, Jost JO, Bekei B, Rose HM, Binolfi A, Schwarzer D, Selenko P (2012) Site-specific mapping and time-resolved monitoring of lysine methylation by high-resolution NMR spectroscopy. J Am Chem Soc. https://doi.org/10.1021/ja301895f

Theodorou M, Speletas M, Mamara A, Papachristopoulou G, Lazou V, Scorilas A, Katsantoni E (2013) Identification of a STAT5 target gene, Dpf3, provides novel insights in chronic lymphocytic leukemia. PLoS One 8(10):e76155. https://doi.org/10.1371/JOURNAL.PONE.0076155

Thorens B (2015) GLUT2, glucose sensing and glucose homeostasis. Diabetologia 58(2):221–232. https://doi.org/10.1007/s00125-014-3451-1

Thornhill AR (2018) Genetic analysis of the embryo: preimplantation genetic diagnosis. Encycl Reprod. https://doi.org/10.1016/B978-0-12-801238-3.64861-4

Tibbetts AS, Appling DR (2010) Compartmentalization of mammalian folate-mediated one-carbon metabolism. Annu Rev Nutr. 30:57–81. https://doi.org/10.1146/annurev.nutr.012809.104810

Treuter E, Venteclef N (2011) Transcriptional control of metabolic and inflammatory pathways by nuclear receptor SUMOylation. Biochim Biophys Acta. 1812(8):909–918. https://doi.org/10.1016/J.BBADIS.2010.12.008

Tripsianes K, Madl T, MacHyna M, Fessas D, Englbrecht C, Fischer U, Neugebauer KM, Sattler M (2011) Structural basis for dimethylarginine recognition by the Tudor domains of human SMN and SPF30 proteins. Nat Struct Mol Biol 18(12):1414–1420. https://doi.org/10.1038/nsmb.2185

Trojer P, Li G, Sims RJ, Vaquero A, Kalakonda N, Boccuni P, Lee D, Erdjument-Bromage H, Tempst P, Nimer SD, Wang YH, Reinberg D (2007) L3MBTL1, a histone-methylation-dependent chromatin lock. Cell 129(5):915–928. https://doi.org/10.1016/j.cell.2007.03.048

Uckelmann M, Densham RM, Baas R, Winterwerp HHK, Fish A, Sixma TK, Morris JR (2018) USP48 restrains resection by site-specific cleavage of the BRCA1 ubiquitin mark from H2A. Nat Commun 9(1):1–16. https://doi.org/10.1038/s41467-017-02653-3

Upadhyay M, Samal J, Kandpal M, Singh OV, Vivekanandan P (2013) The Warburg effect: insights from the past decade. Pharmacol Therap 137(3):318–330. https://doi.org/10.1016/j.pharmthera.2012.11.003

van der Knaap JA, Verrijzer CP (2016) Undercover: gene control by metabolites and metabolic enzymes. Genes Dev. 30(21):2345–2369. https://doi.org/10.1101/gad.289140.116

Vezzoli A, Bonadies N, Allen MD, Freund SMV, Santiveri CM, Kvinlaug BT, Huntly BJP, Göttgens B, Bycroft M (2010) Molecular basis of histone H3K36me3 recognition by the PWWP domain of Brpf1. Nat Struct Mol Biol 17(5):617–619. https://doi.org/10.1038/nsmb.1797

Wang AY, Schulze JM, Skordalakes E, Gin JW, Berger JM, Rine J, Kobor MS (2009a) Asf1-like structure of the conserved Yaf9 YEATS domain and role in H2A.Z deposition and acetylation. Proc Natl Acad Sci 106(51):21573–21578. https://doi.org/10.1073/PNAS.0906539106

Wang GG, Song J, Wang Z, Dormann HL, Casadio F, Li H, Luo JL, Patel DJ, Allis CD (2009b) Haematopoietic malignancies caused by dysregulation of a chromatin-binding PHD finger. Nature 459(7248):847–851. https://doi.org/10.1038/nature08036

Wang T, Chuan Pan C, Rui Yu J, Long Y, Hong Cai X, De Yin X, Qiong Hao L, Li Luo L (2013) Association between TYMS expression and efficacy of pemetrexed-based chemotherapy in advanced non-small cell lung cancer: a meta-analysis. PLoS One 8(9). https://doi.org/10.1371/journal.pone.0074284

Wang Y, Guo YR, Liu K, Yin Z, Liu R, Xia Y, Tan L, Yang P, Lee JH, Li XJ, Hawke D, Zheng Y, Qian X, Lyu J, He J, Xing D, Tao YJ, Lu Z (2017) KAT2A coupled with the α-KGDH complex acts as a histone H3 succinyltransferase. Nature 552(7684):273–277. https://doi.org/10.1038/nature25003

Wang H, Xue W, Jiang X (2018a) Overexpression of TRIM24 stimulates proliferation and glucose metabolism of head and neck squamous cell carcinoma. Biomed Res Int 2018. https://doi.org/10.1155/2018/6142843

Wang Y, Sun J, Qiao J, Ouyang J, Na N (2018b) A "soft" and "hard" ionization method for comprehensive studies of molecules. Anal Chem 90:49. https://doi.org/10.1021/acs.analchem.8b04437

Wellen KE, Hatzivassiliou G, Sachdeva UM, Bui TV, Cross JR, Thompson CB (2009) ATP-citrate lyase links cellular metabolism to histone acetylation. Science. https://doi.org/10.1126/science.1164097

Wen H, Li Y, Xi Y, Jiang S, Stratton S, Peng D, Tanaka K, Ren Y, Xia Z, Wu J, Li B, Barton MC, Li W, Li H, Shi X (2014) ZMYND11 links histone H3.3 K36 trimethylation to transcription elongation and tumor suppression. Nature 508(7495):263–268. https://doi.org/10.1038/nature13045.ZMYND11

Wheaton WW, Chandel NS (2011) Hypoxia. 2. Hypoxia regulates cellular metabolism. Am J Physiol. 300(3):C385. https://doi.org/10.1152/ajpcell.00485.2010

Winter S, Simboeck E, Fischle W, Zupkovitz G, Dohnal I, Mechtler K, Ammerer G, Seiser C (2008) 14-3-3 proteins recognize a histone code at histone H3 and are required for transcriptional activation. EMBO J 27(1):88–99. https://doi.org/10.1038/sj.emboj.7601954

Wong CC, Qian Y, Yu J (2017) Interplay between epigenetics and metabolism in oncogenesis: mechanisms and therapeutic approaches. Oncogene 36(24):3359–3374. https://doi.org/10.1038/onc.2016.485

Worden EJ, Wolberger C (2019) Activation and regulation of H2B-Ubiquitin-dependent histone methyltransferases. Curr Opin Struct Biol 59:98–106. https://doi.org/10.1016/j.sbi.2019.05.009

Worden EJ, Hoffmann NA, Hicks CW, Wolberger C (2019) Mechanism of cross-talk between H2B ubiquitination and H3 methylation by Dot1L. Cell 176(6):1490–1501.e12. https://doi.org/10.1016/j.cell.2019.02.002

Wu D, Yotnda P (2011) Production and detection of reactive oxygen species (ROS) in cancers. J Vis Exp 57:2–5. https://doi.org/10.3791/3357

Wu H, Zeng H, Lam R, Tempel W, Amaya MF, Xu C, Dombrovski L, Qiu W, Wang Y, Min J (2011) Structural and histone binding ability characterizations of human PWWP domains. PLoS One 6(6). https://doi.org/10.1371/journal.pone.0018919

Wu H, Yao S, Zhang S, Wang JR, Da Guo P, Li XM, Gan WJ, Mei L, Gao TM, Li JM (2017) Elevated expression of Erbin destabilizes ERα protein and promotes tumorigenesis in hepatocellular carcinoma. J Hepatol 66(6):1193–1204. https://doi.org/10.1016/j.jhep.2017.01.030

Wu W, He K, Guo Q, Chen J, Zhang M, Huang K, Yang D, Wu L, Deng Y, Luo X, Yu H, Ding Q, Xiang G (2019) SSRP1 promotes colorectal cancer progression and is negatively regulated by miR-28-5p. J Cell Mol Med 23(5):3118–3129. https://doi.org/10.1111/JCMM.14134

Wysocka J, Swigut T, Xiao H, Milne TA, Kwon SY, Landry J, Kauer M, Tackett AJ, Chait BT, Badenhorst P, Wu C, Allis CD (2006) A PHD finger of NURF couples histone H3 lysine 4 trimethylation with chromatin remodelling. Nature 442(7098):86–90. https://doi.org/10.1038/nature04815

Xiao M, Yang H, Xu W, Ma S, Lin H, Zhu H, Liu L, Liu Y, Yang C, Xu Y, Zhao S, Ye D, Xiong Y, Guan KL (2012) Inhibition of α-KG-dependent histone and DNA demethylases by fumarate and succinate that are accumulated in mutations of FH and SDH tumor suppressors. Genes Dev 26(12):1326–1338. https://doi.org/10.1101/gad.191056.112

Xiao D, Zeng L, Yao K, Kong X, Wu G, Yin Y (2016) The glutamine-alpha-ketoglutarate (AKG) metabolism and its nutritional implications. Amino Acids 48(9):2067–2080. https://doi.org/10.1007/s00726-016-2254-8

Xie Z, Dai J, Dai L, Tan M, Cheng Z, Wu Y, Boeke JD, Zhao Y (2012) Lysine succinylation and lysine malonylation in histones. Mol Cell Proteomics 11(5):100–107. https://doi.org/10.1074/mcp.M111.015875

Xie Z, Zhang D, Chung D, Tang Z, Huang H, Dai L, Qi S, Li J, Colak G, Chen Y, Xia C, Peng C, Ruan H, Kirkey M, Wang D, Jensen LM, Kwon OK, Lee S, Pletcher SD, Tan M, Lombard DB, White KP, Zhao H, Li J, Roeder RG, Yang X, Zhao Y (2016) Metabolic regulation of gene expression by histone lysine β-hydroxybutyrylation. Mol Cell 62(2):194–206. https://doi.org/10.1016/J.MOLCEL.2016.03.036

Xiong Y, Guan KL (2012) Mechanistic insights into the regulation of metabolic enzymes by acetylation. J Cell Biol 198(2):155–164. https://doi.org/10.1083/jcb.201202056

Xu C, Bian C, Yang W, Galka M, Ouyang H, Chen C, Qiu W, Liu H, Jones AE, MacKenzie F, Pana P, Li SSC, Wang H, Min J (2010) Binding of different histone marks differentially regulates the activity and specificity of polycomb repressive complex 2 (PRC2). Proc Natl Acad Sci U S A 107(45):19266–19271. https://doi.org/10.1073/pnas.1008937107

Xue B, Zhao J, Feng P, Xing J, Wu H, Li Y (2019) Epigenetic mechanism and target therapy of uhrf1 protein complex in malignancies. OncoTargets Therapy 12:549–559. https://doi.org/10.2147/OTT.S192234

Yamamoto N, Sato T, Kawasaki K, Murosaki S, Yamamoto Y (2006) A nonradioisotope, enzymatic assay for 2-deoxyglucose uptake in L6 skeletal muscle cells cultured in a 96-well microplate. Anal Biochem 351(1):139–145. https://doi.org/10.1016/j.ab.2005.12.011

Yamamoto N, Ueda-Wakagi M, Sato T, Kawasaki K, Sawada K, Kawabata K, Akagawa M, Ashida H (2015) Measurement of glucose uptake in cultured cells. Curr Protoc Pharmacol 71(1): 12.14.1–12.14.26. https://doi.org/10.1002/0471141755.ph1214s71

Yan K, Li S, Lin Y, Saddar S, Tchaga G, Yan G (1831) Novel method for the detection of glucose uptake : direct measurement of glucose levels in cultured cells. FASEB J 30:627

Yang XM, MacKenzie RE (1993) NAD-dependent methylenetetrahydrofolate dehydrogenase-methenyltetrahydrofolate cyclohydrolase is the mammalian homolog of the mitochondrial

enzyme encoded by the yeast MIS1 gene. Biochemistry 32(41):11118–11123. https://doi.org/10.1021/bi00092a022
Yang Y, Sauve AA (2016) NAD+ metabolism: bioenergetics, signaling and manipulation for therapy. Biochim Biophys Acta 1864(12):1787–1800. https://doi.org/10.1016/j.bbapap.2016.06.014
Yang Y, Lu Y, Espejo A, Wu J, Xu W, Liang S, Bedford MT (2010) TDRD3 is an effector molecule for arginine-methylated histone marks. Mol Cell 40(6):1016–1023. https://doi.org/10.1016/j.molcel.2010.11.024
Yang J, Jin X, Yan Y, Shao Y, Pan Y, Roberts LR, Zhang J, Huang H, Jiang J (2017) Inhibiting histone deacetylases suppresses glucose metabolism and hepatocellular carcinoma growth by restoring FBP1 expression. Sci Rep 7(February):1–13. https://doi.org/10.1038/srep43864
Yang H, Yu N, Xu J, Ding X, Deng W, Wu G, Li X, Hou Y, Liu Z, Zhao Y, Xue M, Yu S, Wang B, Li X, Niu G, Wang H, Zhu J, Zhuang T (2018) SMURF1 facilitates estrogen receptor a signaling in breast cancer cells. J Exp Clin Cancer Res 37(1):1–12. https://doi.org/10.1186/s13046-018-0672-z
Ye X, Li M, Hou T, Gao T, Zhu WG, Yang Y (2017) Sirtuins in glucose and lipid metabolism. Oncotarget 8(1):1845–1859. https://doi.org/10.18632/oncotarget.12157
Yu X, Ma R, Wu Y, Zha Y, Li S (2018) Reciprocal regulation of metabolic reprogramming and epigenetic modifications in cancer. Front Genet 9(SEP):1–14. https://doi.org/10.3389/fgene.2018.00394
Zangle TA, Chun J, Zhang J, Reed J, Teitell MA (2013) Quantification of biomass and cell motion in human pluripotent stem cell colonies. Biophys J 105(3):593–601. https://doi.org/10.1016/j.bpj.2013.06.041
Zeng L, Zhang Q, Li S, Plotnikov AN, Walsh MJ, Zhou MM (2010) Mechanism and regulation of acetylated histone binding by the tandem PHD finger of DPF3b. Nature 466(7303):258–262. https://doi.org/10.1038/nature09139
Zhang P, Du J, Sun B, Dong X, Xu G, Zhou J, Huang Q, Liu Q, Hao Q, Ding J (2006) Structure of human MRG15 chromo domain and its binding to Lys36-methylated histone H3. Nucleic Acids Res 34(22):6621–6628. https://doi.org/10.1093/nar/gkl989
Zhang S, Roche K, Nasheuer H-P, Lowndes NF (2011) Modification of histones by sugar-N-acetylglucosamine (GlcNAc) occurs on multiple residues, including histone H3 serine 10, and is cell cycle-regulated. J Biol Chem. https://doi.org/10.1074/jbc.M111.284885
Zhang X, Li B, Rezaeian AH, Xu X, Chou P-C, Jin G, Han F, Pan B-S, Wang C-Y, Long J, Zhang A, Huang C-Y, Tsai F-J, Tsai C-H, Logothetis C, Lin H-K (2017) H3 ubiquitination by NEDD4 regulates H3 acetylation and tumorigenesis. Nat Commun. https://doi.org/10.1038/ncomms14799
Zhang Q, Bai B, Mei X, Wan C, Cao H, Li D, Wang S, Zhang M, Wang Z, Wu J, Wang H, Huo J, Ding G, Zhao J, Xie Q, Wang L, Qiu Z, Zhao S, Zhang T (2018) Elevated H3K79 homocysteinylation causes abnormal gene expression during neural development and subsequent neural tube defects. Nat Commun 9(1). https://doi.org/10.1038/s41467-018-05451-7
Zhang D, Tang Z, Huang H, Zhou G, Cui C, Weng Y, Liu W, Kim S, Lee S, Perez-Neut M, Ding J, Czyz D, Hu R, Ye Z, He M, Zheng YG, Shuman HA, Dai L, Ren B et al (2019a) Metabolic regulation of gene expression by histone lactylation. Nature 574(7779):575–580. https://doi.org/10.1038/s41586-019-1678-1
Zhang X, Cao R, Niu J, Yang S, Ma H, Zhao S, Li H (2019b) Molecular basis for hierarchical histone de-β-hydroxybutyrylation by SIRT3. Cell Discov 5(1):1–15. https://doi.org/10.1038/s41421-019-0103-0
Zheng J (2012) Energy metabolism of cancer: glycolysis versus oxidative phosphorylation (review). Oncol Lett. 4(6):1151–1157. https://doi.org/10.3892/ol.2012.928
Zhong L, D'Urso A, Toiber D, Sebastian C, Henry RE, Vadysirisack DD, Guimaraes A, Marinelli B, Wikstrom JD, Nir T, Clish CB, Vaitheesvaran B, Iliopoulos O, Kurland I, Dor Y, Weissleder R, Shirihai OS, Ellisen LW, Espinosa JM, Mostoslavsky R (2010) the histone deacetylase Sirt6 regulates glucose homeostasis via Hif1α. Cell 140(2):280–293. https://doi.org/10.1016/j.cell.2009.12.041

Chapter 10
Sirtuin 6 Is a Critical Epigenetic Regulator of Cancer

Sukanya Raghu, Arathi Bangalore Prabhashankar, Bhoomika Shivanaiah, Ekta Tripathi, and Nagalingam Ravi Sundaresan

Abstract Sirtuin 6 (SIRT6) is a member of the mammalian sirtuin family with deacetylase, deacylase, and mono-ADP-ribosyl-transferase activities. It is a multi-tasking chromatin-associated protein regulating different cellular and physiological functions in cells. Specifically, SIRT6 dysfunction is implicated in several aging-related human diseases, including cancer. Studies indicate that SIRT6 has a tumor-specific role, and it is considered a tumor suppressor as well as a tumor growth inducer, depending on the type of cancer. In this chapter, we review the role of SIRT6 in metabolism, genomic stability, and cancer. Further, we provide an insight into the interplay of the tumor-suppressing and oncogenic roles of SIRT6 in cancer. Additionally, we discuss the use of small-molecule SIRT6 modulators as potential therapeutics.

Keywords SIRT6 · Metabolism · Genome instability · Aging · Cancer

---

Sukanya Raghu and Arathi Bangalore Prabhashankar contributed equally with all other contributors.

S. Raghu · A. B. Prabhashankar · B. Shivanaiah · N. R. Sundaresan (✉)
Department of Microbiology and Cell Biology, Division of Biological Sciences, Indian Institute of Science (IISc), Bengaluru, Karnataka, India
e-mail: rsundaresan@iisc.ac.in

E. Tripathi (✉)
Department of Biotechnology, Faculty of Life and Allied Health Sciences, Ramaiah University of Applied Sciences, Bengaluru, India

Department of Microbiology and Cell Biology, Division of Biological Sciences, Indian Institute of Science (IISc), Bengaluru, Karnataka, India
e-mail: ektatripathi.bt.ls@msruas.ac.in

T. K. Kundu, C. Das (eds.), *Metabolism and Epigenetic Regulation: Implications in Cancer*, Subcellular Biochemistry 100,
https://doi.org/10.1007/978-3-031-07634-3_10

The Sirtuin Family

The silent information regulator two protein (SIRT) family is comprised of class III histone deacetylases (HDACs). They are protein deacetylases, deacylases, and ADP-ribosyltransferases, well conserved from yeast to mammals (Rine et al. 1979). The sirtuin family regulates various physiological functions, from energy metabolism to stress responses (Hall et al. 2013; Singh et al. 2018; Merksamer et al. 2013). There are seven different human sirtuin isoforms, SIRT1-7, that occupy different subcellular compartments, including the cytoplasm (SIRT1 and 2), nucleus (SIRT1, 2, 3, 6, and 7), and mitochondria (SIRT3, 4, and 5) (Michishita et al. 2005; Huang et al. 2010; North and Verdin 2007; Tanno et al. 2007). The sirtuins catalyze the deacetylation reaction, which uses NAD+ as a cofactor, yielding O-acetyl-ADP-ribose, deacetylated substrate, and nicotinamide (Tong and Denu 2010).

Sirtuins have an essential role in caloric restriction, which mediates life-span extension in both lower organisms and higher mammals (Kaeberlein et al. 1999; Rogina and Helfand 2004). Most of the sirtuins are considered therapeutic targets for aging-related diseases, such as type 2 diabetes, inflammatory diseases, neurodegenerative disorders, and cancer (Milne et al. 2007; Chen et al. 2005; Albani et al. 2010; Schumacker 2010). Interestingly, sirtuins are known to play a dual role in cancer, as reported previously (Michishita et al. 2005; Wang et al. 2008; Mostoslavsky et al. 2006). They preserve genomic stability by protecting deoxyribonucleic acid (DNA) from damage and oxidative stress, thereby guarding cells against cancer. Conversely, sirtuins are also involved in cell survival during stress conditions, leading to uncontrolled cell division and tumorigenesis (Bosch-Presegue and Vaquero 2011). The expression of sirtuins is altered in different types of cancer. SIRT1, 4, and 7 are upregulated in certain types of cancers (Stunkel et al. 2007; Bradbury et al. 2005; Hida et al. 2007; Huffman et al. 2007; Ashraf et al. 2006). On the other hand, SIRT2 is downregulated in melanomas, gliomas, and gastric carcinoma (Hiratsuka et al. 2003; Inoue et al. 2007). Further, SIRT2 plays a vital role in glucose homeostasis by deacetylating and activating glucose-6-phosphatase dehydrogenase, a key enzyme in the pentose phosphate pathway, as well as the glycolytic enzyme phosphoglycerate mutase (Wang et al. 2014; Xu et al. 2014). In the case of SIRT3, studies have shown upregulation or downregulation in different types of breast cancer (Kim et al. 2010a; Ashraf et al. 2006). In addition, SIRT3 is known to regulate mitochondrial functions through the deacetylation and activation of electron transport chain complexes I and II and acetyl-CoA synthetase to protect against oxidative stress (Hallows et al. 2006). Further, during calorie restriction, SIRT3 activates superoxide dismutase 2, a critical mitochondrial antioxidant enzyme (Qiu et al. 2010). Similar to the function of SIRT3, mitochondrial SIRT4 is also known to play an important role in regulating tumorigenesis and metabolism (Haigis et al. 2006; Jeong et al. 2014; Miyo et al. 2015). Another mitochondrial sirtuin, SIRT5 preferentially targets nonacetyl lysine modifications, including malonylation, succinylation, and glutarylation (Peng et al. 2011; Park et al. 2013). Evidence suggests that SIRT5 is overexpressed in human nonsmall cell lung cancer

(Lu et al. 2014). Moreover, studies have demonstrated that SIRT5, through the demalonylation of glycolytic enzymes, can positively regulate glycolysis (Nishida et al. 2015). The nuclear sirtuin isoform SIRT7 is reported to suppress the expression of tumor suppressor genes by deacteylating acetylated histone 3 lysine 18 (H3K18Ac) (Barber et al. 2012). Moreover, several studies have demonstrated that SIRT7 is upregulated in several cancers, including breast, prostate, bladder, thyroid, liver, and colorectal cancers (Paredes et al. 2014; Geng et al. 2015; Yu et al. 2014; Kim et al. 2013). In line with the key roles of sirtuin isoforms, the nuclear SIRT6 modulates gene expression by deacetylating H3K9 and H3K56 to regulate inflammation, glucose metabolism, and genomic stability (Chang et al. 2020). Additionally, during cell division or under stress, SIRT6 is known to localize in the cytoplasm (Jedrusik-Bode et al. 2013; Ardestani and Liang 2012). These diverse roles of SIRT6 are associated with both tumor suppression and promotion (Fiorentino et al. 2021). Studies indicate that SIRT6 is a critical regulator of metabolism and aging. Interestingly, Sirt6 knockout mice exhibit signs of progeria and die prematurely from severe hypoglycemia (Peshti et al. 2017; Mostoslavsky et al. 2006), whereas the overexpression of SIRT6 in male mice extends their life span, possibly by suppressing insulin-like growth factor 1 signaling (Kanfi et al. 2012). The diverse role of SIRT6 in almost every physiological and pathological condition is being demonstrated by several recent studies. Despite this, the role of SIRT6 in tumor origin, growth, and metastasis is still a topic of debate. In the present chapter, we will discuss the cellular and molecular functions of SIRT6 in histone modifications, metabolism, and genome stability in relation to cancer.

Sirtuin 6

Among the sirtuin family of proteins, SIRT6 is one of the nuclear-localized sirtuins known primarily for its chromatin-modifying functions at basal levels. The human SIRT6 is an NAD^+-dependent deacetylase with 355 amino acids. This protein possesses a structurally conserved NAD^+-binding domain called the Rossmann fold domain and has a less homologous zinc-binding domain within its catalytic core. Notably, SIRT6 has been reported to bind to NAD^+ in the absence of acetylated substrate (Pan et al. 2011). Interestingly, calorie restriction (CR) activates SIRT6 and thereby slows cell senescence (Zhang et al. 2016). Recently, SIRT6 has been identified as an essential player in cellular protein synthesis (Ravi et al. 2019). The primary functions of SIRT6 in histone modifications, metabolism, genomic stability, aging, and cancer have been discussed in Figs. 10.1 and 10.2.

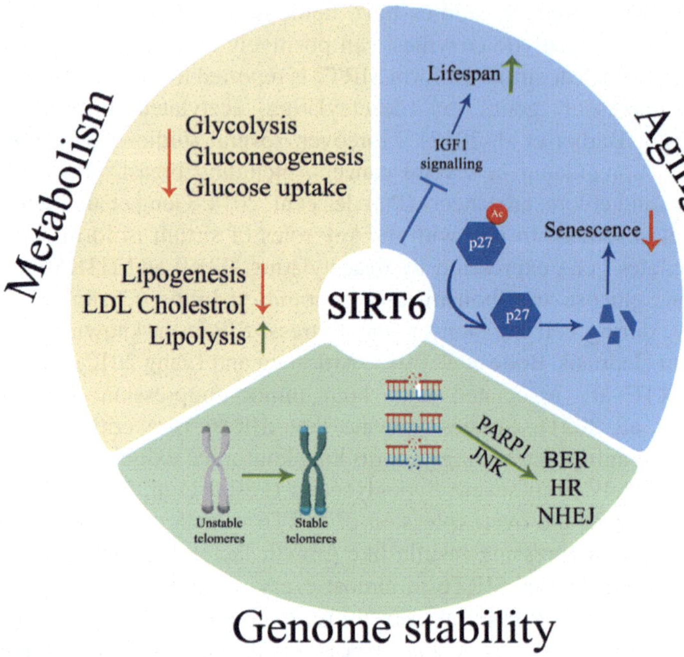

Fig. 10.1 Multifaceted role of SIRT6 in metabolism, aging, and genome stability

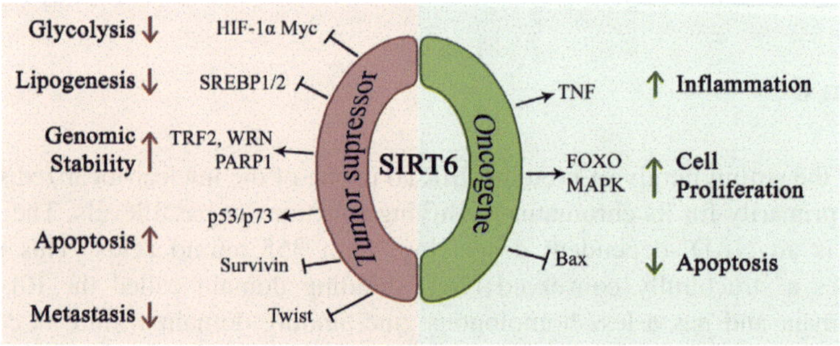

Fig. 10.2 SIRT6 can act as both tumor suppressor and tumor promoter in a tissue- and context-specific manner

Enzymatic Activity

SIRT6 was reported to have a mono-ADP-ribosyltransferase activity, where it was shown to use NAD^+ as a substrate to add ADP-ribose moiety onto itself (Liszt et al. 2005). However, the physiological relevance of this activity is still unknown. Studies have revealed ribosylation substrates of SIRT6, such as PARP1 and KAP1, the former known to regulate DNA repair in the cells (Mao et al. 2011; Van Meter et al.

2011). Moreover, the most common enzymatic activity of SIRT6 is its histone deacetylase activity. It has been shown that SIRT6 is recruited to target promoters to repress the transcription of the gene by deacetylating histone 3 lysine 9 (H3K9) and histone 3 lysine 56 (H3K56) (Kugel and Mostoslavsky 2014; Michishita et al. 2008, 2009; Yang et al. 2009). The deacetylation of H3K9 regulates telomeric chromatin and gene expression (Michishita et al. 2008), whereas H3K56 deacetylation represses gene expression and regulates DNA repair pathways through PARP1 (Mao et al. 2011). Additionally, SIRT6 has been shown to deacetylate H3K18 in pericentromeric loci (Tasselli et al. 2016). Importantly, SIRT6 is known to deacetylate nonhistone proteins that regulate several important processes in the cell, although it has weak deacetylase activity (Pan et al. 2011). Interestingly, SIRT6 has strong deacylase activity (Jiang et al. 2013), and it modulates the activity of TNFα and R-Ras2 by regulating protein secretion and localization to the plasma membrane, respectively (Jiang et al. 2013; Zhang et al. 2017; Michishita et al. 2008, 2009). Besides, it has been reported that long-chain fatty acids could regulate the deacetylase activity of SIRT6 (Feldman et al. 2013).

Regulation of SIRT6

Transcriptional Regulation

SIRT6 is known to be regulated at different levels in the cell. Under starvation, nuclear respiratory factor 1 (NRF1) binds to the NRF1-binding site on the SIRT6 promoter and increases its transcription by forming a complex with SIRT1 and Forkhead box O3a (FOXO3a) (Kim et al. 2010b). Additionally, c-FOS acts as another positive regulator of SIRT6 expression by binding to the AP-1 site on the SIRT6 promoter (Min et al. 2012). Alternatively, PARP1 and E2F1 are reported to be negative regulators of SIRT6 (Wencel et al. 2018; Wu et al. 2015), where the latter binds directly to the SIRT6 promoter and suppresses its expression. In addition to this, oxidative stress caused by diabetes also negatively regulates SIRT6 expression (Yu et al. 2016).

Regulation by miRNAs

Posttranscriptionally, SIRT6 is known to be regulated by microRNAs (miRNAs). A reduction in miR-34a levels is correlated with increased SIRT6 levels in squamous cell carcinomas (Dotto and Karine 2014). Moreover, miR-122 binds to three sites on SIRT6 3′ UTR and downregulates its expression (Elhanati et al. 2016). The downregulation of miR-122 has been previously linked to cancer progression, indicating a role for SIRT6 in tumor suppression (Ha et al. 2019; Fong et al. 2015; Dai et al. 2019). In addition, miR-33a, miR-33b, and miR-766 were shown to bind to the 3′-untranslated regions (UTR) of the SIRT6 mRNA and downregulate its expression (Chang et al. 2020). These miRNAs are also previously reported to have tumor-suppressing roles in different types of cancer, suggesting an oncogenic role for SIRT6 in these cancer types (Gao et al. 2020; Lin et al. 2015).

Posttranslational Regulation

The phosphorylation of SIRT6 at serine 10 by c-Jun N-terminal kinase (JNK) is essential for its recruitment to double-stranded breaks (DSB) on the chromatin. The S10 phosphorylation of SIRT6 activates the mono-ADP ribosylation of PARP1 to localize to DNA breaks (Van Meter et al. 2016). Additionally, a small ubiquitin-like modifier (SUMO) is known to SUMOylate SIRT6 and leads to the selective regulation of H3K56 but not H3K9 (Cai et al. 2016). Interestingly, SIRT6 requires the presence of free fatty acids (FFAs) for its efficient histone deacetylase activity. The affinity of SIRT6 toward acetylated substrates rises in the presence of FFAs, such as myristic acid, oleic acid, and linoleic acid (Feldman et al. 2013). Furthermore, the levels of SIRT6 are regulated by the phosphorylation of Ser338 by AKT1, targeting it for proteasomal degradation due to ubiquitination by MDM2 (Thirumurthi et al. 2014). The increase in SIRT6 levels during nutrient starvation has been attributed to protection against proteasomal degradation by ubiquitin ligase CHIP and ubiquitin-specific peptidase USP10 (Ronnebaum et al. 2013; Lin et al. 2013).

Cellular and Molecular Functions

Histone Modifications

As discussed earlier, one of the first established activities of SIRT6 was its capability to deacetylate H3K9. Although the deacetylase activity of SIRT6 is weaker than that of SIRT1, it also has H3K56 deacetylase activity. Notably, H3 is deacetylated in its nucleosome structure but not as a free histone (Gil et al. 2013). SIRT6 suppresses gene expression through its H3K9 deacetylase activity. It can be recruited to target promoters such as MYC, FOXO3, HIF-1α, and NF-κB, represses their transcription (Zwaans and Lombard 2014), and regulates metabolism, growth, cardiac hypertrophy, cancer, etc. Notably, the deacetylation of H3K56 by SIRT6 at the sites of DNA damage helps recruit repair factors to the site (Toiber et al. 2013).

Metabolism

Despite its involvement in multiple biological processes, one of the primary functions of SIRT6 is to maintain metabolic homeostasis. Several studies have independently shown the role of SIRT6 in regulating glycolysis, triglyceride synthesis, and lipid metabolism by modulating histone H3K9 and various other proteins (Chang et al. 2020).

Glucose Homeostasis

Various studies have shown the importance of SIRT6 in glucose metabolism. SIRT6 inhibits the expression of glycolytic genes such as GLUT1, LDHA, and PDK1, which are targets of HIF-1α (Sebastian et al. 2012; Zhong et al. 2010). Besides, SIRT6 regulates glucose homeostasis by suppressing gluconeogenesis. PGC1α activation by GCN5, which is activated by being deacetylated by SIRT6, reduces gluconeogenesis (Dominy et al. 2012). SIRT6 has also been shown to directly bind to Forkhead box protein O1 (FoxO1) and downregulates genes involved in gluconeogenesis by preventing the interaction of FoxO1 with glucose-6-phosphatase (G6p) and phosphoenolpyruvate carboxykinase (Pepck) (Zhang et al. 2014). Similarly, FoxO1 is known to be deacetylated by SIRT6, eventually suppressing gluconeogenesis through the suppression of PDK4 (Khan et al. 2018). SIRT6 is also involved in the downregulation of GLUT1 and GLUT4 to reduce glucose uptake. Similarly, SIRT6 inhibits insulin receptors IRS1 and IRS2, negatively impacting insulin signaling (Xiao et al. 2010). Additionally, pancreatic β-cell function and viability are regulated by SIRT6 (Kugel et al. 2016). In a mouse model of type 2 diabetes mellitus, administering a small molecule inhibitor of SIRT6 improved glucose tolerance, increased the expression of glucose transporters, and upregulated glycolytic activity (Sociali et al. 2017). Additionally, the loss of SIRT6 in pancreatic β-cells has been shown to induce a defect in insulin secretion (Xiong et al. 2016; Qin et al. 2018).

Lipid Homeostasis

SIRT6 plays a variety of roles in maintaining lipid homeostasis. Liver-specific SIRT6 knockout mice have increased triglycerides and reduced β-oxidation and fatty liver formation (Kim et al. 2010b). Conversely, SIRT6 overexpressing mice have lowered visceral fat and the level of low-density lipoprotein (LDL) cholesterol when challenged with a high-fat diet (Kanfi et al. 2010). SIRT6 ablation has been associated with the inflammation of adipocytes and an increase in mice body weight, leading to the induction of insulin resistance and glucose intolerance (Chen et al. 2017; Xiong et al. 2017; Yao et al. 2017). Besides, SIRT6 negatively regulates lipogenesis by inhibiting transcription factors like SREBP1 and SREBP2 (Elhanati et al. 2013). SIRT6 deacetylates the promoter of the Srebp2 gene and suppresses its expression and, therefore, its downstream targets and inhibits the cleavage of SREBP1 and SREBP2, leading to their inactivation (Tao et al. 2013). Our recent work suggested a critical role for SIRT6 in regulating lipid uptake and the development of lipotoxicity in the heart. SIRT6 transcriptionally represses key fatty acid transporter genes via the transcription factor PPARγ (Khan et al. 2021). SIRT6 is also involved in the regulation of fatty acid and cholesterol metabolism genes by recruiting SREBP-1 to the chromatin (Masri et al. 2014). Other transcription factors that regulate fat storage etc., such as PPARγ, are also controlled by SIRT6 (Kanfi et al. 2010). Not only does SIRT6 play a role in lipogenesis, but also its absence impairs lipolysis in HFD-induced obesity in mice (Kuang et al. 2017). Proprotein convertase subtilisin/kexin type 9 controls the degradation of LDL receptors in the lysosome. SIRT6 overexpression causes the deacetylation of H3K9 and H3K56 in the promoters of the Pcsk9 gene, hence lowering LDL cholesterol levels (Tao et al. 2013).

Genome Stability

DNA Repair

SIRT6 is implicated in the maintenance of genomic stability due to its chromatin-binding nature. Cells devoid of SIRT6 have chromatin aberrations, such as nicks, gaps, and DNA breaks (single stranded and double stranded) (Ciccia and Elledge 2010). SIRT6 deficiency causes impairment in the recruitment of DNA damage response factors, thereby hampering DNA repair (Antonia et al. 2018). SIRT6 is known to be phosphorylated by c-Jun NH2-terminal kinase and hence recruited to regions of DNA damage upon oxidative stress in a PARP-1 dependent manner (Van Meter et al. 2016). SIRT6 deficient/depleted cells have compromised nonhomologous end joining (NHEJ) and homologous recombination (HR) (Mao et al. 2011). Independent studies have shown that the enzymatic activity of SIRT6 is essential for its capacity to repair DSBs. SIRT6 deacetylates H3K56 and, by doing so, recruits repair factors such as 53BP1 and NBS1 to the site of DNA damage. Also, the overexpression of SIRT6 increases HR through PARP-1 activation (Mao et al. 2011). In addition to NHEJ and HR, SIRT6 absence reduces the efficiency of the base excision repair (BER) of the DNA damage (Xu et al. 2015).

Telomere Maintenance

Telomeres are long repeat sequences present at the ends of chromosomes, whose dysfunction can result in genomic instability. SIRT6 binds to the telomeric chromatin and can deacetylate H3K9 and H3K56, allowing the association of Werner syndrome protein (WRN) for efficient DNA replication and telomere metabolism (Yang et al. 2009; Michishita et al. 2008). Upon replication stress, SIRT6 interacts and deacetylates TRF2 (telomere repeat binding factor 2), targeting it for ubiquitin-dependent degradation (Rizzo et al. 2017), further maintaining normal telomere integrity. In the absence of SIRT6, cells exhibit genomic instability with fragmented chromosomes and translocations (Mostoslavsky et al. 2006).

Stress Response and Aging

SIRT6, being a multifunctional protein, is known to play a role in handling stress in the cell. It can regulate oxidative stress by binding to nuclear factor erythroid 2-related factor 2 and upregulating antioxidant genes (Pan et al. 2016). SIRT6 increases the expression of aldose reductase during hyperosmotic stress (Timucin and Basaga 2016). Additionally, SIRT6 can localize to cytoplasmic stress granules, interact with its components, and help in their disassembly after the stress is resolved (Jedrusik-Bode et al. 2013). Several studies have demonstrated the importance of SIRT6 in aging and cellular senescence. SIRT6 knockout mice show a premature aging phenotype and die by four weeks of age (Mostoslavsky et al. 2006), and when overexpressed, it can increase the life span of mice. The overexpression of SIRT6 alters the phosphorylation of the components of the IGF1 signaling pathway, playing a role in increasing longevity (Kanfi et al. 2012). In addition to this, SIRT6 deacetylates p27, promoting its degradation and thereby rescuing cells from senescence (Zhao et al. 2016). These functions of SIRT6 in metabolism, DNA repair, telomere maintenance, and aging have a major significance in several metabolic disorders, age-related diseases, and cancer. The pathways regulated by SIRT6 can be exploited for designing potential pharmacological molecules to combat several metabolic and age-related diseases.

Sirtuin 6 in Cancer

SIRT6 has a multitasking role in regulating various cellular processes, including metabolism, genome stability, cellular proliferation, and apoptosis, impact several cancer-related pathways linking SIRT6 to cancer. Over the past decades, SIRT6 has been studied for its role in cancer and other age-related diseases. In the context of cancer, SIRT6 plays a dual role as a tumor suppressor and promoter in a cell- and context-specific manner (Fig. 10.2) (Chang et al. 2020).

SIRT6 as a Tumor Suppressor

Several different groups investigated the expression of SIRT6 across various human cancers and found that SIRT6 is repressed in several cancers. Analysis of data from The Cancer Genome Atlas (TCGA) and CCLE (Cancer Cell Line Encyclopedia) revealed that SIRT6 is deleted in 20% and 35% of human cancers and cell lines, respectively, with more frequent deletions among colorectal and pancreatic cancer cell lines (Lerrer et al. 2016). In line with these findings, H3K56 acetylation is increased in multiple cancer types (Das et al. 2009). In addition, various point mutations that inactivate or destabilize SIRT6 were found in human cancers

(Kugel et al. 2015). Studies indicate that the downregulation of SIRT6 transcription is mediated by a c-Jun/c-Fos pathway in liver cancer (Min et al. 2012), runt-related transcription factor 2 in breast cancer (Choe et al. 2015), and E2F-1 in bladder and prostate cancer (Wu et al. 2015). Furthermore, the SIRT6 expression is downregulated in the early stages of colorectal cancer and maintained at low levels during cancer progression. Interestingly, SIRT6 levels correlate with poor prognosis and disease survival (Sebastian et al. 2012). In a study with breast cancer patients, the expression levels of SIRT6 were positively associated with survival, suggesting a protective role for SIRT6 against tumorigenesis (Thirumurthi et al. 2014). These studies highlight SIRT6 as a critical player in cancer development.

Cancer cells preferentially undergo aerobic glycolysis (the Warburg effect) to generate biosynthetic precursors required for rapid cell division. Increased HIF-1 and Myc expression has been observed in many different cancer types (Zwaans and Lombard 2014). HIF-1 induces the expression of several glycolytic genes and favors aerobic glycolysis, and Myc drives oncogenesis, mainly by regulating ribosomal biogenesis and glucose metabolism. SIRT6 deacetylates H3K9 and/or H3K56 at the promoters of several HIF-1 and Myc target genes, reducing their expression, thus inhibiting glycolysis and suppressing tumorigenesis (Zhong et al. 2010). In addition, SIRT6 represses the activities of the lipogenic transcription factors SREBP1 and SREBP2 (Elhanati et al. 2013; Tao et al. 2013), Jun (Xiao et al. 2012), and insulin-IGF1-like signaling (Sundaresan et al. 2012) essential for the proliferation of cancer cells. In hepatocellular carcinoma, SIRT6 deacetylates pyruvate kinase M2 (PKM2), resulting in the nuclear export of PKM2 and the suppression of cell proliferation and tumorigenesis, validating the tumor suppressor role of SIRT6 (Bhardwaj and Das 2016). Furthermore, in ovarian cancer, a decrease in SIRT6 activity leads to Notch 3 upregulation and cancer cell proliferation (Zhang et al. 2015). SIRT6-deficient cells are tumorigenic, and the resultant tumors exhibit increased aerobic glycolysis, suggesting that SIRT6 plays a role in both the initiation and maintenance of cancer. Furthermore, intestinal-specific deletion of SIRT6 led to colorectal carcinoma in APC_{min}mouse due to the upregulated expression of glycolytic genes and ribosomal genes (Sebastian et al. 2012). All these in vitro and in vivo studies reveal the undoubted role of SIRT6 as a tumor suppressor by modulating cancer metabolism.

Although the regulation of cancer metabolism is a significant contributor to its tumor-suppressive functions, SIRT6 can act as a tumor suppressor through various other mechanisms. The maintenance of genome integrity is linked to aging and cancer prevention. SIRT6 prevents genomic instability by regulating several DNA repair pathways, including base excision repair and double-strand breaks. Studies have shown the presence of dysfunctional telomeres, including chromosomal breaks and fusions, in the absence of SIRT6. Also, SIRT6-deficient mice exhibit a significant level of genomic instability and are highly prone to DNA damage (Michishita et al. 2008). Interestingly, a study showed that SIRT6 is a suppressor of LINE1 (long interspersed nuclear elements) retrotransposon activity, which has been implicated in a variety of age-related disorders, including cancer (Van Meter et al. 2014).

Apoptosis is an important process that prevents cancer cells from growing in an uncontrolled manner. Cancer cells deregulate various apoptotic mechanisms to survive longer and accumulate mutations required for tumor progression. SIRT6 is shown to inhibit cancer progression by inducing apoptosis in multiple cancer cell lines, but not in normal cells, through p53 or p73 signaling. For this process, SIRT6 uses its mono-ADP ribosyltransferase activity rather than deacetylase activity (Van Meter et al. 2011). SIRT6 suppresses an antiapoptotic protein, survivin, through histone H3K9 deacetylation and inhibits NF-kB activation in a liver cancer mouse model (Min et al. 2012). In human HepG2 hepatocarcinoma cells, the overexpression or knockdown of SIRT6 induced apoptosis or promoted the growth of these cells, respectively, demonstrating the role of SIRT6 as a tumor suppressor (Wang et al. 2017).

Another important process implicated in carcinogenesis is epithelial-mesenchymal transition (EMT), which confers metastatic properties to cancer cells. Twist1, a key EMT transcription factor, is negatively regulated by SIRT6 (Wushou et al. 2014). The loss of SIRT6 in pancreatic ductal adenocarcinoma (PDAC) resulted in the hyperacetylation and upregulation of oncofetal protein Lin28b and the consequent induction of downstream let-7 target genes, which are responsible for increased PDAC progression and metastasis (Kugel et al. 2016). In the mouse model of breast cancer with PI3K activation, the overexpression of SIRT6 reduces the grade and progression of the tumors. The transcriptomic and metabolomics analyses of tumors reveal that SIRT6 suppresses PI3K signaling and stem-like traits in this cancer model (Ioris et al. 2017). Besides, SIRT6 inhibits cell proliferation and induces cell cycle arrest in colorectal cancer stem cells by suppressing cell division cycle 25A (cdc25A) through H3K9 deacetylation (Liu et al. 2018). Taken together, these studies strongly reinforce the idea of SIRT6 being a tumor suppressor via modulating multiple pathways, among which the regulation of cancer metabolism is the main contributor.

SIRT6 as a Tumor Promoter

In contrast to the studies discussed earlier, several experimental evidence reported the overexpression of SIRT6 in multiple cancers, including breast cancer, skin squamous cell carcinoma (SCC), ovarian cancer, and leukemia, unveiling the potential oncogenic role of SIRT6 (Lee et al. 2016; Ming et al. 2014). This role has been attributed to its functions in DNA repair, inflammatory responses, and cell proliferation, making SIRT6 a tumor promoter in some instances. Notably, SIRT6 is amplified in multiple myeloma (MM) cells, which confers DNA damage resistance and promotes cell proliferation by regulating mitogen-activated protein kinase (MAPK) pathway genes (Cea et al. 2016). In SCC, the upregulation of SIRT6 promotes tumor progression by activating COX-2 and suppressing the AMPK (AMP-activated protein kinase) signaling (Ming et al. 2014). Also, SIRT6 upregulates the expression of mesenchymal markers like N-cadherin and vimentin by deacetylating FOXO3a in hepatocellular carcinoma (HCC), supporting cancer progression (Han et al. 2019). Interestingly, SIRT6 depletion enhances sensitivity to chemotherapy treatment in MM cells (Cea et al. 2016). Similarly, the overexpression of SIRT6 also increases resistance to chemotherapeutic drugs, like paclitaxel and epirubicin in breast cancer cell lines by regulating the FOXO expression (Khongkow et al. 2013). In other studies, SIRT6 promoted tumor cell migration and invasion through its regulation of pro-inflammatory cytokines via ERK1/2/MMP9 signaling in nonsmall cell lung cancer (NSCLC) tumors (Bai et al. 2016). Apart from inducing apoptosis, SIRT6 can also negatively regulate apoptosis in some cancer cell lines, including prostate cancer cells and HCC, by deregulating BCL-2 family members (Liu et al. 2013). Altogether, the above studies indicate that the dual behavior of SIRT6 in cancer is complex and more likely tissue and context dependent. Understanding these features is necessary for designing future therapies to modulate enzyme expression or activity for cancer therapeutics.

SIRT6 Modulators in Therapeutics

The therapeutic potential of sirtuins has been explored in several age-related diseases, including type 2 diabetes, cardiovascular disease, neurodegeneration, inflammation, and cancer (Yamamoto et al. 2007). The identification of small-molecule modulators of the sirtuin family of protein deacetylases has prompted the exploration of their therapeutic potential. Initial drug development efforts focused on SIRT1 and SIRT2 have yielded several potent inhibitors and activators (Dai et al. 2018). Progress in understanding the mechanisms of sirtuin modulation by such compounds provides a rational basis for further drug development. SIRT6 is a key regulator of the onset and progression of age-related diseases, including cancer (Chang et al. 2020). As we discussed earlier, the role of SIRT6 in tumorigenesis has tissue-specific functions. It plays a tumor suppressor role in pancreatic, breast, and colon cancers and hepatocellular carcinoma. On the other hand, SIRT6 acts as a tumor promoter in other cancer types, such as lung, prostate, and melanoma (Carafa et al. 2019). Therefore, both SIRT6 activators and inhibitors may be useful as therapeutics for cancer in a context-specific manner. Interestingly, several modulators of SIRT6, including both inhibitors and activators, have been identified.

SIRT6 Activators

A variety of activators and inhibitors modulating the activity of SIRT6 have been identified and listed in Tables 10.1 and 10.2.

Table 10.1 A variety of activators modulating the activity of SIRT6 have been identified and listed

SIRT6 activators	Biological role	References
Long-chain free fatty acids	Increases the deacetylating activity of Sirt6	Feldman et al. (2013)
Lamin A	Facilitates the chromatin localization of SIRT6 upon DNA damage	Ghosh et al. (2015)
N-acylethanolamines (oleoylethanolamide, myristoylethanolamide)	These are endogenous fatty acids that activate SIRT6	Rahnasto-Rilla et al. (2016)
Fucoidan	A polysaccharide that stimulates SIRT6 deacetylation activity	Rahnasto-Rilla et al. (2017)
Pyrrolo (1,2-α) quinoxaline	Synthetic SIRT6 activator	Hassanieh and Mostoslavsky (2018)
Cyanidin	Potent SIRT6 activator	Rahnasto-Rilla et al. (2018)
UBCS039	Activation of autophagy-dependent cell death by SIRT6 deacetylating activity	Iachettini et al. (2018)
MDL-800	Increased the deacetylase activity of SIRT6 and inhibited the proliferation of HCC cells through SIRT6-driven cell-cycle arrest	Huang et al. (2018)
Fluvastatin	SIRT6 activator that regulates cholesterol homeostasis	Kim et al. (2018)
2-(3-chloro-4-(2,4,6-trichloro-N-(2,4,6trichlorobenzoyl)benzamido) phenyl)-1,3-dioxoisoindoline-5-carboxylic acid	SIRT6 activator capable of enhancing the enzymatic activity	Klein et al. (2020)
MDL-811	Activation of SIRT6 deacetylase activity on histone 3	He et al. (2020)
4H-chromen	SIRT6 activating compound with anticancer activities on various breast cancer cells	Tenhunen et al. (2021)

SIRT6 Inhibitors

Table 10.2 A variety of inhibitors modulating the activity of SIRT6 have been identified and listed

SIRT6 inhibitors	Biological role	References
Thiomyristoyl peptides	Cell-permeable inhibitor for Sirt6	He et al. (2014)
2,4-dioxo-N-(4-(pyridin-3-yloxy) phenyl)-1,2,3,4-tetrahydroquinazoline-6-sulfonamide	Sirt6 inhibition displayed improved glucose tolerance in type 2 diabetes mellitus mouse model	Sociali et al. (2015)
SDK(thioAc)TM21, HKK (thioAc)LM21, AKK (thioAc)LM21	Inhibit SIRT6 activity	Kokkonen et al. (2012), Rahnasto-Rilla et al. (2018)
Trichostatin A	Reduces SIRT6-catalyzed deacetylation of lysine 382 of p53, regulates apoptosis and stress resistance	Parenti et al. (2014), Wood et al. (2018)
Quinazolinedione	Increases histone H3 lysine 9 acetylation, reduces TNF-α production, and increases glucose uptake in cultured cells	Rahnasto-Rilla et al. (2018)
Quercetin, vitexin, rutin, di quercetin and 2-chloro-1,4-naphtoquinone-quercetin, EX-527	Inhibitors of SIRT6	Singh et al. (2014), Heger et al. (2019)

Conclusion

Cancer, in a multifaceted fashion, has caused havoc in healthy lifestyles across age groups. Among several properties that initiate and sustain cancer status in an individual, epigenetics and metabolism peak the list. Sirtuins being the popular epigenetic regulators of longevity present themselves as therapeutic targets. Likewise, our chapter focusing on the role of SIRT6 in modulating cancer and its related medical or cellular properties shows its importance in the research field. Overall, the studies point out that SIRT6 modulators could be promising future therapeutics, and more research is required in understanding their potential uses against cancer. Notably, the field of sirtuin modulators is an exciting path in drug development, which holds the promise of treating age-related diseases, including cancer.

References

Albani D, Polito L, Forloni G (2010) Sirtuins as novel targets for Alzheimer's disease and other neurodegenerative disorders: experimental and genetic evidence. J Alzheimers Dis 19(1):11–26. https://doi.org/10.3233/JAD-2010-1215

Antonia C, Debora S, Stefania O, Giovanna T, Paola M, Fabio G, Veronica R, Nicoletta C, Enrico C, Marino C, Maurizio M, Micaela B, Aimable N, Michel D, Katia T, Antonino N, Mario P, Santina B, Alessio N, Francesco B, Marco G, Roberto ML, Michele C (2018) Depletion of SIRT6 enzymatic activity increases acute myeloid leukemia cells' vulnerability to DNA-damaging agents. Haematologica 103(1):80–90. https://doi.org/10.3324/haematol.2017.176248

Ardestani PM, Liang F (2012) Sub-cellular localization, expression and functions of Sirt6 during the cell cycle in HeLa cells. Nucleus 3(5):442–451. https://doi.org/10.4161/nucl.21134

Ashraf N, Zino S, Macintyre A, Kingsmore D, Payne AP, George WD, Shiels PG (2006) Altered sirtuin expression is associated with node-positive breast cancer. Br J Cancer 95(8):1056–1061. https://doi.org/10.1038/sj.bjc.6603384

Bai L, Lin G, Sun L, Liu Y, Huang X, Cao C, Guo Y, Xie C (2016) Upregulation of SIRT6 predicts poor prognosis and promotes metastasis of non-small cell lung cancer via the ERK1/2/MMP9 pathway. Oncotarget 7(26):40377–40386. https://doi.org/10.18632/oncotarget.9750

Barber MF, Michishita-Kioi E, Xi Y, Tasselli L, Kioi M, Moqtaderi Z, Tennen RI, Paredes S, Young NL, Chen K, Struhl K, Garcia BA, Gozani O, Li W, Chua KF (2012) SIRT7 links H3K18 deacetylation to maintenance of oncogenic transformation. Nature 487(7405):114–118. https://doi.org/10.1038/nature11043

Bhardwaj A, Das S (2016) SIRT6 deacetylates PKM2 to suppress its nuclear localization and oncogenic functions. Proc Natl Acad Sci U S A 113(5):E538–E547. https://doi.org/10.1073/pnas.1520045113

Bosch-Presegue L, Vaquero A (2011) The dual role of sirtuins in cancer. Genes Cancer 2(6):648–662. https://doi.org/10.1177/1947601911417862

Bradbury CA, Khanim FL, Hayden R, Bunce CM, White DA, Drayson MT, Craddock C, Turner BM (2005) Histone deacetylases in acute myeloid leukaemia show a distinctive pattern of expression that changes selectively in response to deacetylase inhibitors. Leukemia 19(10):1751–1759. https://doi.org/10.1038/sj.leu.2403910

Cai J, Zuo Y, Wang T, Cao Y, Cai R, Chen FL, Cheng J, Mu J (2016) A crucial role of SUMOylation in modulating Sirt6 deacetylation of H3 at lysine 56 and its tumor suppressive activity. Oncogene 35(37):4949–4956. https://doi.org/10.1038/onc.2016.24

Carafa V, Altucci L, Nebbioso A (2019) Dual tumor suppressor and tumor promoter action of sirtuins in determining malignant phenotype. Front Pharmacol 10:38. https://doi.org/10.3389/fphar.2019.00038

Cea M, Cagnetta A, Adamia S, Acharya C, Tai YT, Fulciniti M, Ohguchi H, Munshi A, Acharya P, Bhasin MK, Zhong L, Carrasco R, Monacelli F, Ballestrero A, Richardson P, Gobbi M, Lemoli RM, Munshi N, Hideshima T, Nencioni A, Chauhan D, Anderson KC (2016) Evidence for a role of the histone deacetylase SIRT6 in DNA damage response of multiple myeloma cells. Blood 127(9):1138–1150. https://doi.org/10.1182/blood-2015-06-649970

Chang AR, Ferrer CM, Mostoslavsky R (2020) SIRT6, a mammalian deacylase with multitasking abilities. Physiol Rev 100(1):145–169. https://doi.org/10.1152/physrev.00030.2018

Chen J, Zhou Y, Mueller-Steiner S, Chen LF, Kwon H, Yi S, Mucke L, Gan L (2005) SIRT1 protects against microglia-dependent amyloid-beta toxicity through inhibiting NF-kappaB signaling. J Biol Chem 280(48):40364–40374. https://doi.org/10.1074/jbc.M509329200

Chen Q, Hao W, Xiao C, Wang R, Xu X, Lu H, Chen W, Deng C-X (2017) SIRT6 is essential for adipocyte differentiation by regulating mitotic clonal expansion. Cell Rep 18(13):3155–3166. https://doi.org/10.1016/j.celrep.2017.03.006

Choe M, Brusgard JL, Chumsri S, Bhandary L, Zhao XF, Lu S, Goloubeva OG, Polster BM, Fiskum GM, Girnun GD, Kim MS, Passaniti A (2015) The RUNX2 transcription factor negatively regulates SIRT6 expression to alter glucose metabolism in breast cancer cells. J Cell Biochem 116(10):2210–2226. https://doi.org/10.1002/jcb.25171

Ciccia A, Elledge SJ (2010) The DNA damage response: making it safe to play with knives. Mol Cell 40(2):179–204. https://doi.org/10.1016/j.molcel.2010.09.019

Dai H, Sinclair DA, Ellis JL, Steegborn C (2018) Sirtuin activators and inhibitors: promises, achievements, and challenges. Pharmacol Ther 188:140–154. https://doi.org/10.1016/j.pharmthera.2018.03.004

Dai M, Li L, Qin X (2019) Clinical value of miRNA-122 in the diagnosis and prognosis of various types of cancer. Oncol Lett 17(4):3919–3929. https://doi.org/10.3892/ol.2019.10024

Das C, Lucia MS, Hansen KC, Tyler JK (2009) CBP/p300-mediated acetylation of histone H3 on lysine 56. Nature 459(7243):113–117. https://doi.org/10.1038/nature07861

Dominy JE Jr, Lee Y, Jedrychowski MP, Chim H, Jurczak MJ, Camporez JP, Ruan HB, Feldman J, Pierce K, Mostoslavsky R, Denu JM, Clish CB, Yang X, Shulman GI, Gygi SP, Puigserver P (2012) The deacetylase Sirt6 activates the acetyltransferase GCN5 and suppresses hepatic gluconeogenesis. Mol Cell 48(6):900–913. https://doi.org/10.1016/j.molcel.2012.09.030

Dotto GP, Karine L (2014) miR-34a/SIRT6 in squamous differentiation and cancer. Cell Cycle 13(7):1055–1056. https://doi.org/10.4161/cc.28378

Elhanati S, Kanfi Y, Varvak A, Roichman A, Carmel-Gross I, Barth S, Gibor G, Cohen HY (2013) Multiple regulatory layers of SREBP1/2 by SIRT6. Cell Rep 4(5):905–912. https://doi.org/10.1016/j.celrep.2013.08.006

Elhanati S, Ben-Hamo R, Kanfi Y, Varvak A, Glazz R, Lerrer B, Efroni S, Cohen HY (2016) Reciprocal regulation between SIRT6 and miR-122 controls liver metabolism and predicts hepatocarcinoma prognosis. Cell Rep 14(2):234–242. https://doi.org/10.1016/j.celrep.2015.12.023

Feldman JL, Baeza J, Denu JM (2013) Activation of the protein deacetylase SIRT6 by long-chain fatty acids and widespread deacylation by mammalian sirtuins. J Biol Chem 288(43):31350–31356. https://doi.org/10.1074/jbc.C113.511261

Fiorentino F, Carafa V, Favale G, Altucci L, Mai A, Rotili D (2021) The two-faced role of SIRT6 in cancer. Cancers (Basel) 13(5):1156. https://doi.org/10.3390/cancers13051156

Fong MY, Zhou W, Liu L, Alontaga AY, Chandra M, Ashby J, Chow A, O'Connor STF, Li S, Chin AR, Somlo G, Palomares M, Li Z, Tremblay JR, Tsuyada A, Sun G, Reid MA, Wu X, Swiderski P, Ren X, Shi Y, Kong M, Zhong W, Chen Y, Wang SE (2015) Breast-cancer-secreted miR-122 reprograms glucose metabolism in premetastatic niche to promote metastasis. Nature Cell Biol 17(2):183–194. https://doi.org/10.1038/ncb3094

Gao C, Wei J, Tang T, Huang Z (2020) Role of microRNA-33a in malignant cells (Review). Oncol Lett 20(3):2537–2556. https://doi.org/10.3892/ol.2020.11835

Geng Q, Peng H, Chen F, Luo R, Li R (2015) High expression of Sirt7 served as a predictor of adverse outcome in breast cancer. Int J Clin Exp Pathol 8(2):1938–1945

Ghosh S, Liu B, Wang Y, Hao Q, Zhou Z (2015) Lamin A is an endogenous SIRT6 activator and promotes SIRT6-mediated DNA repair. Cell Rep 13(7):1396–1406. https://doi.org/10.1016/j.celrep.2015.10.006

Gil R, Barth S, Kanfi Y, Cohen HY (2013) SIRT6 exhibits nucleosome-dependent deacetylase activity. Nucleic Acids Res 41(18):8537–8545. https://doi.org/10.1093/nar/gkt642

Ha SY, Yu JI, Choi C, Kang SY, Joh J-W, Paik SW, Kim S, Kim M, Park HC, Park C-K (2019) Prognostic significance of miR-122 expression after curative resection in patients with hepatocellular carcinoma. Sci Rep 9(1):14738. https://doi.org/10.1038/s41598-019-50594-2

Haigis MC, Mostoslavsky R, Haigis KM, Fahie K, Christodoulou DC, Murphy AJ, Valenzuela DM, Yancopoulos GD, Karow M, Blander G, Wolberger C, Prolla TA, Weindruch R, Alt FW, Guarente L (2006) SIRT4 inhibits glutamate dehydrogenase and opposes the effects of calorie restriction in pancreatic beta cells. Cell 126(5):941–954. https://doi.org/10.1016/j.cell.2006.06.057

Hall JA, Dominy JE, Lee Y, Puigserver P (2013) The sirtuin family's role in aging and age-associated pathologies. J Clin Invest 123(3):973–979. https://doi.org/10.1172/JCI64094

Hallows WC, Lee S, Denu JM (2006) Sirtuins deacetylate and activate mammalian acetyl-CoA synthetases. Proc Natl Acad Sci U S A 103(27):10230–10235. https://doi.org/10.1073/pnas.0604392103

Han LL, Jia L, Wu F, Huang C (2019) Sirtuin6 (SIRT6) promotes the EMT of hepatocellular carcinoma by stimulating autophagic degradation of e-cadherin. Mol Cancer Res 17(11):2267–2280. https://doi.org/10.1158/1541-7786.MCR-19-0321

Hassanieh S, Mostoslavsky R (2018) Chapter 9 - multitasking roles of the mammalian deacetylase SIRT6. In: Guarente L, Mostoslavsky R, Kazantsev A (eds) Introductory review on sirtuins in biology, aging, and disease. Academic Press, Cambridge, MA, pp 117–130. https://doi.org/10.1016/B978-0-12-813499-3.00009-5

He B, Hu J, Zhang X, Lin H (2014) Thiomyristoyl peptides as cell-permeable Sirt6 inhibitors. Org Biomol Chem 12(38):7498–7502. https://doi.org/10.1039/c4ob00860j

He T, Shang J, Gao C, Guan X, Chen Y, Zhu L, Zhang L, Zhang C, Zhang J, Pang T (2020) A novel SIRT6 activator ameliorates neuroinflammation and ischemic brain injury via EZH2/FOXC1 axis. Acta Pharm Sin B. 11(3):708–726. https://doi.org/10.1016/j.apsb.2020.11.002

Heger V, Tyni J, Hunyadi A, Horáková L, Lahtela-Kakkonen M, Rahnasto-Rilla M (2019) Quercetin based derivatives as sirtuin inhibitors. Biomed Pharmacother 111:1326–1333. https://doi.org/10.1016/j.biopha.2019.01.035

Hida Y, Kubo Y, Murao K, Arase S (2007) Strong expression of a longevity-related protein, SIRT1, in Bowen's disease. Arch Dermatol Res 299(2):103–106. https://doi.org/10.1007/s00403-006-0725-6

Hiratsuka M, Inoue T, Toda T, Kimura N, Shirayoshi Y, Kamitani H, Watanabe T, Ohama E, Tahimic CG, Kurimasa A, Oshimura M (2003) Proteomics-based identification of differentially expressed genes in human gliomas: down-regulation of SIRT2 gene. Biochem Biophys Res Commun 309(3):558–566. https://doi.org/10.1016/j.bbrc.2003.08.029

Huang JY, Hirschey MD, Shimazu T, Ho L, Verdin E (2010) Mitochondrial sirtuins. Biochim Biophys Acta 1804(8):1645–1651. https://doi.org/10.1016/j.bbapap.2009.12.021

Huang Z, Zhao J, Deng W, Chen Y, Shang J, Song K, Zhang L, Wang C, Lu S, Yang X, He B, Min J, Hu H, Tan M, Xu J, Zhang Q, Zhong J, Sun X, Mao Z, Lin H, Xiao M, Chin YE, Jiang H, Xu Y, Chen G, Zhang J (2018) Identification of a cellularly active SIRT6 allosteric activator. Nat Chem Biol 14(12):1118–1126. https://doi.org/10.1038/s41589-018-0150-0

Huffman DM, Grizzle WE, Bamman MM, Kim JS, Eltoum IA, Elgavish A, Nagy TR (2007) SIRT1 is significantly elevated in mouse and human prostate cancer. Cancer Res 67(14):6612–6618. https://doi.org/10.1158/0008-5472.CAN-07-0085

Iachettini S, Trisciuoglio D, Rotili D, Lucidi A, Salvati E, Zizza P, Di Leo L, Del Bufalo D, Ciriolo MR, Leonetti C, Steegborn C, Mai A, Rizzo A, Biroccio A (2018) Pharmacological activation of SIRT6 triggers lethal autophagy in human cancer cells. Cell Death & Disease 9(10):996. https://doi.org/10.1038/s41419-018-1065-0

Inoue T, Hiratsuka M, Osaki M, Yamada H, Kishimoto I, Yamaguchi S, Nakano S, Katoh M, Ito H, Oshimura M (2007) SIRT2, a tubulin deacetylase, acts to block the entry to chromosome condensation in response to mitotic stress. Oncogene 26(7):945–957. https://doi.org/10.1038/sj.onc.1209857

Ioris RM, Galie M, Ramadori G, Anderson JG, Charollais A, Konstantinidou G, Brenachot X, Aras E, Goga A, Ceglia N, Sebastian C, Martinvalet D, Mostoslavsky R, Baldi P, Coppari R (2017) SIRT6 suppresses cancer stem-like capacity in tumors with PI3K activation independently of its deacetylase activity. Cell Rep 18(8):1858–1868. https://doi.org/10.1016/j.celrep.2017.01.065

Jedrusik-Bode M, Studencka M, Smolka C, Baumann T, Schmidt H, Kampf J, Paap F, Martin S, Tazi J, Muller KM, Kruger M, Braun T, Bober E (2013) The sirtuin SIRT6 regulates stress granule formation in C. elegans and mammals. J Cell Sci 126(Pt 22):5166–5177. https://doi.org/10.1242/jcs.130708

Jeong SM, Lee A, Lee J, Haigis MC (2014) SIRT4 protein suppresses tumor formation in genetic models of Myc-induced B cell lymphoma. J Biol Chem 289(7):4135–4144. https://doi.org/10.1074/jbc.M113.525949

Jiang H, Khan S, Wang Y, Charron G, He B, Sebastian C, Du J, Kim R, Ge E, Mostoslavsky R, Hang HC, Hao Q, Lin H (2013) SIRT6 regulates TNF-alpha secretion through hydrolysis of long-chain fatty acyl lysine. Nature 496(7443):110–113. https://doi.org/10.1038/nature12038

Kaeberlein M, McVey M, Guarente L (1999) The SIR2/3/4 complex and SIR2 alone promote longevity in Saccharomyces cerevisiae by two different mechanisms. Genes Dev 13(19):2570–2580. https://doi.org/10.1101/gad.13.19.2570

Kanfi Y, Peshti V, Gil R, Naiman S, Nahum L, Levin E, Kronfeld-Schor N, Cohen HY (2010) SIRT6 protects against pathological damage caused by diet-induced obesity. Aging Cell 9(2): 162–173. https://doi.org/10.1111/j.1474-9726.2009.00544.x

Kanfi Y, Naiman S, Amir G, Peshti V, Zinman G, Nahum L, Bar-Joseph Z, Cohen HY (2012) The sirtuin SIRT6 regulates lifespan in male mice. Nature 483(7388):218–221. https://doi.org/10.1038/nature10815

Khan D, Sarikhani M, Dasgupta S, Maniyadath B, Pandit AS, Mishra S, Ahamed F, Dubey A, Fathma N, Atreya HS, Kolthur-Seetharam U, Sundaresan NR (2018) SIRT6 deacetylase transcriptionally regulates glucose metabolism in heart. J Cell Physiol 233(7):5478–5489. https://doi.org/10.1002/jcp.26434

Khan D, Ara T, Ravi V, Rajagopal R, Tandon H, Parvathy J, Gonzalez EA, Asirvatham-Jeyaraj N, Krishna S, Mishra S, Raghu S, Bhati AS, Tamta AK, Dasgupta S, Kolthur-Seetharam U, Etchegaray JP, Mostoslavsky R, Rao PSM, Srinivasan N, Sundaresan NR (2021) SIRT6 transcriptionally regulates fatty acid transport by suppressing PPARγ. Cell Rep 35(9):109190. https://doi.org/10.1016/j.celrep.2021.109190

Khongkow M, Olmos Y, Gong C, Gomes AR, Monteiro LJ, Yague E, Cavaco TB, Khongkow P, Man EP, Laohasinnarong S, Koo CY, Harada-Shoji N, Tsang JW, Coombes RC, Schwer B, Khoo US, Lam EW (2013) SIRT6 modulates paclitaxel and epirubicin resistance and survival in breast cancer. Carcinogenesis 34(7):1476–1486. https://doi.org/10.1093/carcin/bgt098

Kim HS, Patel K, Muldoon-Jacobs K, Bisht KS, Aykin-Burns N, Pennington JD, van der Meer R, Nguyen P, Savage J, Owens KM, Vassilopoulos A, Ozden O, Park SH, Singh KK, Abdulkadir SA, Spitz DR, Deng CX, Gius D (2010a) SIRT3 is a mitochondria-localized tumor suppressor required for maintenance of mitochondrial integrity and metabolism during stress. Cancer Cell 17(1):41–52. https://doi.org/10.1016/j.ccr.2009.11.023

Kim HS, Xiao C, Wang RH, Lahusen T, Xu X, Vassilopoulos A, Vazquez-Ortiz G, Jeong WI, Park O, Ki SH, Gao B, Deng CX (2010b) Hepatic-specific disruption of SIRT6 in mice results in fatty liver formation due to enhanced glycolysis and triglyceride synthesis. Cell Metab 12(3): 224–236. https://doi.org/10.1016/j.cmet.2010.06.009

Kim JK, Noh JH, Jung KH, Eun JW, Bae HJ, Kim MG, Chang YG, Shen Q, Park WS, Lee JY, Borlak J, Nam SW (2013) Sirtuin7 oncogenic potential in human hepatocellular carcinoma and its regulation by the tumor suppressors MiR-125a-5p and MiR-125b. Hepatology 57(3): 1055–1067. https://doi.org/10.1002/hep.26101

Kim JH, Lee JM, Kim JH, Kim KR (2018) Fluvastatin activates sirtuin 6 to regulate sterol regulatory element-binding proteins and AMP-activated protein kinase in HepG2 cells. Biochem Biophys Res Commun 503(3):1415–1421. https://doi.org/10.1016/j.bbrc.2018.07.057

Klein MA, Liu C, Kuznetsov VI, Feltenberger JB, Tang W, Denu JM (2020) Mechanism of activation for the sirtuin 6 protein deacylase. J Biol Chem 295(5):1385–1399. https://doi.org/10.1074/jbc.RA119.011285

Kokkonen P, Rahnasto-Rilla M, Kiviranta PH, Huhtiniemi T, Laitinen T, Poso A, Jarho E, Lahtela-Kakkonen M (2012) Peptides and pseudopeptides as SIRT6 deacetylation inhibitors. ACS Med Chem Lett 3(12):969–974. https://doi.org/10.1021/ml300139n

Kuang J, Zhang Y, Liu Q, Shen J, Pu S, Cheng S, Chen L, Li H, Wu T, Li R, Li Y, Zou M, Zhang Z, Jiang W, Xu G, Qu A, Xie W, He J (2017) Fat-specific Sirt6 ablation sensitizes mice to high-fat diet-induced obesity and insulin resistance by inhibiting lipolysis. Diabetes 66(5):1159–1171. https://doi.org/10.2337/db16-1225

Kugel S, Mostoslavsky R (2014) Chromatin and beyond: the multitasking roles for SIRT6. Trends Biochem Sci 39(2):72–81. https://doi.org/10.1016/j.tibs.2013.12.002

Kugel S, Feldman JL, Klein MA, Silberman DM, Sebastian C, Mermel C, Dobersch S, Clark AR, Getz G, Denu JM, Mostoslavsky R (2015) Identification of and molecular basis for SIRT6 loss-of-function point mutations in cancer. Cell Rep 13(3):479–488. https://doi.org/10.1016/j.celrep.2015.09.022

Kugel S, Sebastian C, Fitamant J, Ross KN, Saha SK, Jain E, Gladden A, Arora KS, Kato Y, Rivera MN, Ramaswamy S, Sadreyev RI, Goren A, Deshpande V, Bardeesy N, Mostoslavsky R (2016)

SIRT6 suppresses pancreatic cancer through control of Lin28b. Cell 165(6):1401–1415. https://doi.org/10.1016/j.cell.2016.04.033

Lee N, Ryu HG, Kwon JH, Kim DK, Kim SR, Wang HJ, Kim KT, Choi KY (2016) SIRT6 depletion suppresses tumor growth by promoting cellular senescence induced by DNA damage in HCC. PLoS One 11(11):e0165835. https://doi.org/10.1371/journal.pone.0165835

Lerrer B, Gertler AA, Cohen HY (2016) The complex role of SIRT6 in carcinogenesis. Carcinogenesis 37(2):108–118. https://doi.org/10.1093/carcin/bgv167

Lin Z, Yang H, Tan C, Li J, Liu Z, Quan Q, Kong S, Ye J, Gao B, Fang D (2013) USP10 antagonizes c-Myc transcriptional activation through SIRT6 stabilization to suppress tumor formation. Cell Rep 5(6):1639–1649. https://doi.org/10.1016/j.celrep.2013.11.029

Lin Y, Liu AY, Fan C, Zheng H, Li Y, Zhang C, Wu S, Yu D, Huang Z, Liu F, Luo Q, Yang CJ, Ouyang G (2015) MicroRNA-33b Inhibits Breast Cancer Metastasis by Targeting HMGA2, SALL4 and Twist1. Sci Rep 5:9995. https://doi.org/10.1038/srep09995

Liszt G, Ford E, Kurtev M, Guarente L (2005) Mouse Sir2 homolog SIRT6 is a nuclear ADP-ribosyltransferase. J Biol Chem 280(22):21313–21320. https://doi.org/10.1074/jbc.M413296200

Liu Y, Xie QR, Wang B, Shao J, Zhang T, Liu T, Huang G, Xia W (2013) Inhibition of SIRT6 in prostate cancer reduces cell viability and increases sensitivity to chemotherapeutics. Protein Cell 4(9):702–710. https://doi.org/10.1007/s13238-013-3054-5

Liu W, Wu M, Du H, Shi X, Zhang T, Li J (2018) SIRT6 inhibits colorectal cancer stem cell proliferation by targeting CDC25A. Oncol Lett 15(4):5368–5374. https://doi.org/10.3892/ol.2018.7989

Lu W, Zuo Y, Feng Y, Zhang M (2014) SIRT5 facilitates cancer cell growth and drug resistance in non-small cell lung cancer. Tumour Biol 35(11):10699–10705. https://doi.org/10.1007/s13277-014-2372-4

Mao Z, Hine C, Tian X, Van Meter M, Au M, Vaidya A, Seluanov A, Gorbunova V (2011) SIRT6 promotes DNA repair under stress by activating PARP1. Science 332(6036):1443–1446. https://doi.org/10.1126/science.1202723

Masri S, Rigor P, Cervantes M, Ceglia N, Sebastian C, Xiao C, Roqueta-Rivera M, Deng C, Osborne TF, Mostoslavsky R, Baldi P, Sassone-Corsi P (2014) Partitioning circadian transcription by SIRT6 leads to segregated control of cellular metabolism. Cell 158(3):659–672. https://doi.org/10.1016/j.cell.2014.06.050

Merksamer PI, Liu Y, He W, Hirschey MD, Chen D, Verdin E (2013) The sirtuins, oxidative stress and aging: an emerging link. Aging (Albany NY) 5(3):144–150. https://doi.org/10.18632/aging.100544

Michishita E, Park JY, Burneskis JM, Barrett JC, Horikawa I (2005) Evolutionarily conserved and nonconserved cellular localizations and functions of human SIRT proteins. Mol Biol Cell 16(10):4623–4635. https://doi.org/10.1091/mbc.e05-01-0033

Michishita E, McCord RA, Berber E, Kioi M, Padilla-Nash H, Damian M, Cheung P, Kusumoto R, Kawahara TL, Barrett JC, Chang HY, Bohr VA, Ried T, Gozani O, Chua KF (2008) SIRT6 is a histone H3 lysine 9 deacetylase that modulates telomeric chromatin. Nature 452(7186):492–496. https://doi.org/10.1038/nature06736

Michishita E, McCord RA, Boxer LD, Barber MF, Hong T, Gozani O, Chua KF (2009) Cell cycle-dependent deacetylation of telomeric histone H3 lysine K56 by human SIRT6. Cell Cycle 8(16):2664–2666. https://doi.org/10.4161/cc.8.16.9367

Milne JC, Lambert PD, Schenk S, Carney DP, Smith JJ, Gagne DJ, Jin L, Boss O, Perni RB, Vu CB, Bemis JE, Xie R, Disch JS, Ng PY, Nunes JJ, Lynch AV, Yang H, Galonek H, Israelian K, Choy W, Iffland A, Lavu S, Medvedik O, Sinclair DA, Olefsky JM, Jirousek MR, Elliott PJ, Westphal CH (2007) Small molecule activators of SIRT1 as therapeutics for the treatment of type 2 diabetes. Nature 450(7170):712–716. https://doi.org/10.1038/nature06261

Min L, Ji Y, Bakiri L, Qiu Z, Cen J, Chen X, Chen L, Scheuch H, Zheng H, Qin L, Zatloukal K, Hui L, Wagner EF (2012) Liver cancer initiation is controlled by AP-1 through SIRT6-

dependent inhibition of survivin. Nat Cell Biol 14(11):1203–1211. https://doi.org/10.1038/ncb2590

Ming M, Han W, Zhao B, Sundaresan NR, Deng CX, Gupta MP, He YY (2014) SIRT6 promotes COX-2 expression and acts as an oncogene in skin cancer. Cancer Res 74(20):5925–5933. https://doi.org/10.1158/0008-5472.CAN-14-1308

Miyo M, Yamamoto H, Konno M, Colvin H, Nishida N, Koseki J, Kawamoto K, Ogawa H, Hamabe A, Uemura M, Nishimura J, Hata T, Takemasa I, Mizushima T, Doki Y, Mori M, Ishii H (2015) Tumour-suppressive function of SIRT4 in human colorectal cancer. Br J Cancer 113(3):492–499. https://doi.org/10.1038/bjc.2015.226

Mostoslavsky R, Chua KF, Lombard DB, Pang WW, Fischer MR, Gellon L, Liu P, Mostoslavsky G, Franco S, Murphy MM, Mills KD, Patel P, Hsu JT, Hong AL, Ford E, Cheng HL, Kennedy C, Nunez N, Bronson R, Frendewey D, Auerbach W, Valenzuela D, Karow M, Hottiger MO, Hursting S, Barrett JC, Guarente L, Mulligan R, Demple B, Yancopoulos GD, Alt FW (2006) Genomic instability and aging-like phenotype in the absence of mammalian SIRT6. Cell 124(2):315–329. https://doi.org/10.1016/j.cell.2005.11.044

Nishida Y, Rardin MJ, Carrico C, He W, Sahu AK, Gut P, Najjar R, Fitch M, Hellerstein M, Gibson BW, Verdin E (2015) SIRT5 regulates both cytosolic and mitochondrial protein malonylation with glycolysis as a major target. Mol Cell 59(2):321–332. https://doi.org/10.1016/j.molcel.2015.05.022

North BJ, Verdin E (2007) Interphase nucleo-cytoplasmic shuttling and localization of SIRT2 during mitosis. PLoS One 2(8):e784. https://doi.org/10.1371/journal.pone.0000784

Pan PW, Feldman JL, Devries MK, Dong A, Edwards AM, Denu JM (2011) Structure and biochemical functions of SIRT6. J Biol Chem 286(16):14575–14587. https://doi.org/10.1074/jbc.M111.218990

Pan H, Guan D, Liu X, Li J, Wang L, Wu J, Zhou J, Zhang W, Ren R, Zhang W, Li Y, Yang J, Hao Y, Yuan T, Yuan G, Wang H, Ju Z, Mao Z, Li J, Qu J, Tang F, Liu GH (2016) SIRT6 safeguards human mesenchymal stem cells from oxidative stress by coactivating NRF2. Cell Res 26(2):190–205. https://doi.org/10.1038/cr.2016.4

Paredes S, Villanova L, Chua KF (2014) Molecular pathways: emerging roles of mammalian Sirtuin SIRT7 in cancer. Clin Cancer Res 20(7):1741–1746. https://doi.org/10.1158/1078-0432.CCR-13-1547

Parenti MD, Grozio A, Bauer I, Galeno L, Damonte P, Millo E, Sociali G, Franceschi C, Ballestrero A, Bruzzone S, Del Rio A, Nencioni A (2014) Discovery of novel and selective SIRT6 inhibitors. J Med Chem 57(11):4796–4804. https://doi.org/10.1021/jm500487d

Park J, Chen Y, Tishkoff DX, Peng C, Tan M, Dai L, Xie Z, Zhang Y, Zwaans BM, Skinner ME, Lombard DB, Zhao Y (2013) SIRT5-mediated lysine desuccinylation impacts diverse metabolic pathways. Mol Cell 50(6):919–930. https://doi.org/10.1016/j.molcel.2013.06.001

Peng C, Lu Z, Xie Z, Cheng Z, Chen Y, Tan M, Luo H, Zhang Y, He W, Yang K, Zwaans BM, Tishkoff D, Ho L, Lombard D, He TC, Dai J, Verdin E, Ye Y, Zhao Y (2011) The first identification of lysine malonylation substrates and its regulatory enzyme. Mol Cell Proteomics 10(12):M111.012658. https://doi.org/10.1074/mcp.M111.012658

Peshti V, Obolensky A, Nahum L, Kanfi Y, Rathaus M, Avraham M, Tinman S, Alt FW, Banin E, Cohen HY (2017) Characterization of physiological defects in adult SIRT6-/- mice. PLoS One 12(4):e0176371. https://doi.org/10.1371/journal.pone.0176371

Qin K, Zhang N, Zhang Z, Nipper M, Zhu Z, Leighton J, Xu K, Musi N, Wang P (2018) SIRT6-mediated transcriptional suppression of Txnip is critical for pancreatic beta cell function and survival in mice. Diabetologia 61(4):906–918. https://doi.org/10.1007/s00125-017-4542-6

Qiu X, Brown K, Hirschey MD, Verdin E, Chen D (2010) Calorie restriction reduces oxidative stress by SIRT3-mediated SOD2 activation. Cell Metab 12(6):662–667. https://doi.org/10.1016/j.cmet.2010.11.015

Rahnasto-Rilla M, Kokkola T, Jarho E, Lahtela-Kakkonen M, Moaddel R (2016) N-Acylethanolamines bind to SIRT6. Chembiochem 17(1):77–81. https://doi.org/10.1002/cbic.201500482

Rahnasto-Rilla MK, McLoughlin P, Kulikowicz T, Doyle M, Bohr VA, Lahtela-Kakkonen M, Ferrucci L, Hayes M, Moaddel R (2017) The identification of a SIRT6 activator from brown algae fucus distichus. Mar Drugs 15(6):190. https://doi.org/10.3390/md15060190

Rahnasto-Rilla M, Tyni J, Huovinen M, Jarho E, Kulikowicz T, Ravichandran S, Vilhelm AB, Ferrucci L, Lahtela-Kakkonen M, Moaddel R (2018) Natural polyphenols as sirtuin 6 modulators. Sci Rep 8(1):4163. https://doi.org/10.1038/s41598-018-22388-5

Ravi V, Jain A, Khan D, Ahamed F, Mishra S, Giri M, Inbaraj M, Krishna S, Sarikhani M, Maity S, Kumar S, Shah RA, Dave P, Pandit AS, Rajendran R, Desingu PA, Varshney U, Das S, Kolthur-Seetharam U, Rajakumari S, Singh M, Sundaresan NR (2019) SIRT6 transcriptionally regulates global protein synthesis through transcription factor Sp1 independent of its deacetylase activity. Nucleic Acids Res 47(17):9115–9131. https://doi.org/10.1093/nar/gkz648

Rine J, Strathern JN, Hicks JB, Herskowitz I (1979) A suppressor of mating-type locus mutations in Saccharomyces cerevisiae: evidence for and identification of cryptic mating-type loci. Genetics 93(4):877–901

Rizzo A, Iachettini S, Salvati E, Zizza P, Maresca C, D'Angelo C, Benarroch-Popivker D, Capolupo A, Del Gaudio F, Cosconati S, Di Maro S, Merlino F, Novellino E, Amoreo CA, Mottolese M, Sperduti I, Gilson E, Biroccio A (2017) SIRT6 interacts with TRF2 and promotes its degradation in response to DNA damage. Nucleic Acids Res 45(4):1820–1834. https://doi.org/10.1093/nar/gkw1202

Rogina B, Helfand SL (2004) Sir2 mediates longevity in the fly through a pathway related to calorie restriction. Proc Natl Acad Sci U S A 101(45):15998–16003. https://doi.org/10.1073/pnas.0404184101

Ronnebaum SM, Wu Y, McDonough H, Patterson C (2013) The ubiquitin ligase CHIP prevents SirT6 degradation through noncanonical ubiquitination. Mol Cell Biol 33(22):4461–4472. https://doi.org/10.1128/MCB.00480-13

Schumacker PT (2010) A tumor suppressor SIRTainty. Cancer Cell 17(1):5–6. https://doi.org/10.1016/j.ccr.2009.12.032

Sebastian C, Zwaans BM, Silberman DM, Gymrek M, Goren A, Zhong L, Ram O, Truelove J, Guimaraes AR, Toiber D, Cosentino C, Greenson JK, MacDonald AI, McGlynn L, Maxwell F, Edwards J, Giacosa S, Guccione E, Weissleder R, Bernstein BE, Regev A, Shiels PG, Lombard DB, Mostoslavsky R (2012) The histone deacetylase SIRT6 is a tumor suppressor that controls cancer metabolism. Cell 151(6):1185–1199. https://doi.org/10.1016/j.cell.2012.10.047

Singh N, Ravichandran S, Spelman K, Fugmann SD, Moaddel R (2014) The identification of a novel SIRT6 modulator from Trigonella foenum-graecum using ligand fishing with protein coated magnetic beads. J Chromatogr B Analyt Technol Biomed Life Sci 968:105–111. https://doi.org/10.1016/j.jchromb.2014.03.016

Singh CK, Chhabra G, Ndiaye MA, Garcia-Peterson LM, Mack NJ, Ahmad N (2018) The role of sirtuins in antioxidant and redox signaling. Antioxid Redox Signal 28(8):643–661. https://doi.org/10.1089/ars.2017.7290

Sociali G, Galeno L, Parenti MD, Grozio A, Bauer I, Passalacqua M, Boero S, Donadini A, Millo E, Bellotti M, Sturla L, Damonte P, Puddu A, Ferroni C, Varchi G, Franceschi C, Ballestrero A, Poggi A, Bruzzone S, Nencioni A, Del Rio A (2015) Quinazolinedione SIRT6 inhibitors sensitize cancer cells to chemotherapeutics. Eur J Med Chem 102:530–539. https://doi.org/10.1016/j.ejmech.2015.08.024

Sociali G, Magnone M, Ravera S, Damonte P, Vigliarolo T, Von Holtey M, Vellone VG, Millo E, Caffa I, Cea M, Parenti MD, Del Rio A, Murone M, Mostoslavsky R, Grozio A, Nencioni A, Bruzzone S (2017) Pharmacological Sirt6 inhibition improves glucose tolerance in a type 2 diabetes mouse model. FASEB J 31(7):3138–3149. https://doi.org/10.1096/fj.201601294R

Stunkel W, Peh BK, Tan YC, Nayagam VM, Wang X, Salto-Tellez M, Ni B, Entzeroth M, Wood J (2007) Function of the SIRT1 protein deacetylase in cancer. Biotechnol J 2(11):1360–1368. https://doi.org/10.1002/biot.200700087

Sundaresan NR, Vasudevan P, Zhong L, Kim G, Samant S, Parekh V, Pillai VB, Ravindra PV, Gupta M, Jeevanandam V, Cunningham JM, Deng CX, Lombard DB, Mostoslavsky R, Gupta

MP (2012) The sirtuin SIRT6 blocks IGF-Akt signaling and development of cardiac hypertrophy by targeting c-Jun. Nat Med 18(11):1643–1650. https://doi.org/10.1038/nm.2961

Tanno M, Sakamoto J, Miura T, Shimamoto K, Horio Y (2007) Nucleocytoplasmic shuttling of the NAD+-dependent histone deacetylase SIRT1. J Biol Chem 282(9):6823–6832. https://doi.org/10.1074/jbc.M609554200

Tao R, Xiong X, DePinho RA, Deng CX, Dong XC (2013) Hepatic SREBP-2 and cholesterol biosynthesis are regulated by FoxO3 and Sirt6. J Lipid Res 54(10):2745–2753. https://doi.org/10.1194/jlr.M039339

Tasselli L, Xi Y, Zheng W, Tennen RI, Odrowaz Z, Simeoni F, Li W, Chua KF (2016) SIRT6 deacetylates H3K18ac at pericentric chromatin to prevent mitotic errors and cellular senescence. Nature Struct Mol Biol 23(5):434–440. https://doi.org/10.1038/nsmb.3202

Tenhunen J, Kucera T, Huovinen M, Kublbeck J, Bisenieks E, Vigante B, Ogle Z, Duburs G, Dolezal M, Moaddel R, Lahtela-Kakkonen M, Rahnasto-Rilla M (2021) Screening of SIRT6 inhibitors and activators: A novel activator has an impact on breast cancer cells. Biomed Pharmacother 138:111452. https://doi.org/10.1016/j.biopha.2021.111452

Thirumurthi U, Shen J, Xia W, LaBaff AM, Wei Y, Li CW, Chang WC, Chen CH, Lin HK, Yu D, Hung MC (2014) MDM2-mediated degradation of SIRT6 phosphorylated by AKT1 promotes tumorigenesis and trastuzumab resistance in breast cancer. Sci Signal 7(336):ra71. https://doi.org/10.1126/scisignal.2005076

Timucin AC, Basaga H (2016) SIRT6 is a positive regulator of aldose reductase expression in U937 and HeLa cells under osmotic stress: in vitro and in silico insights. PLoS One 11(8):e0161494. https://doi.org/10.1371/journal.pone.0161494

Toiber D, Erdel F, Bouazoune K, Silberman DM, Zhong L, Mulligan P, Sebastian C, Cosentino C, Martinez-Pastor B, Giacosa S, D'Urso A, Naar AM, Kingston R, Rippe K, Mostoslavsky R (2013) SIRT6 recruits SNF2H to DNA break sites, preventing genomic instability through chromatin remodeling. Mol Cell 51(4):454–468. https://doi.org/10.1016/j.molcel.2013.06.018

Tong L, Denu JM (2010) Function and metabolism of sirtuin metabolite O-acetyl-ADP-ribose. Biochim Biophys Acta 1804(8):1617–1625. https://doi.org/10.1016/j.bbapap.2010.02.007

Van Meter M, Mao Z, Gorbunova V, Seluanov A (2011) SIRT6 overexpression induces massive apoptosis in cancer cells but not in normal cells. Cell Cycle 10(18):3153–3158. https://doi.org/10.4161/cc.10.18.17435

Van Meter M, Kashyap M, Rezazadeh S, Geneva AJ, Morello TD, Seluanov A, Gorbunova V (2014) SIRT6 represses LINE1 retrotransposons by ribosylating KAP1 but this repression fails with stress and age. Nat Commun 5:5011. https://doi.org/10.1038/ncomms6011

Van Meter M, Simon M, Tombline G, May A, Morello TD, Hubbard BP, Bredbenner K, Park R, Sinclair DA, Bohr VA, Gorbunova V, Seluanov A (2016) JNK Phosphorylates SIRT6 to stimulate DNA double-strand break repair in response to oxidative stress by recruiting PARP1 to DNA breaks. Cell Rep 16(10):2641–2650. https://doi.org/10.1016/j.celrep.2016.08.006

Wang RH, Sengupta K, Li C, Kim HS, Cao L, Xiao C, Kim S, Xu X, Zheng Y, Chilton B, Jia R, Zheng ZM, Appella E, Wang XW, Ried T, Deng CX (2008) Impaired DNA damage response, genome instability, and tumorigenesis in SIRT1 mutant mice. Cancer Cell 14(4):312–323. https://doi.org/10.1016/j.ccr.2008.09.001

Wang YP, Zhou LS, Zhao YZ, Wang SW, Chen LL, Liu LX, Ling ZQ, Hu FJ, Sun YP, Zhang JY, Yang C, Yang Y, Xiong Y, Guan KL, Ye D (2014) Regulation of G6PD acetylation by SIRT2 and KAT9 modulates NADPH homeostasis and cell survival during oxidative stress. EMBO J 33(12):1304–1320. https://doi.org/10.1002/embj.201387224

Wang Y, Pan T, Wang H, Li L, Li J, Zhang D, Yang H (2017) Overexpression of SIRT6 attenuates the tumorigenicity of hepatocellular carcinoma cells. Oncotarget 8(44):76223–76230. https://doi.org/10.18632/oncotarget.19297

Wencel PL, Lukiw WJ, Strosznajder JB, Strosznajder RP (2018) Inhibition of Poly(ADP-ribose) Polymerase-1 Enhances Gene Expression of Selected Sirtuins and APP Cleaving Enzymes in

Amyloid Beta Cytotoxicity. Mol Neurobiol 55(6):4612–4623. https://doi.org/10.1007/s12035-017-0646-8

Wood M, Rymarchyk S, Zheng S, Cen Y (2018) Trichostatin A inhibits deacetylation of histone H3 and p53 by SIRT6. Arch Biochem Biophys 638:8–17. https://doi.org/10.1016/j.abb.2017.12.009

Wu M, Seto E, Zhang J (2015) E2F1 enhances glycolysis through suppressing Sirt6 transcription in cancer cells. Oncotarget 6(13):11252–11263. https://doi.org/10.18632/oncotarget.3594

Wushou A, Hou J, Zhao YJ, Shao ZM (2014) Twist-1 up-regulation in carcinoma correlates to poor survival. Int J Mol Sci 15(12):21621–21630. https://doi.org/10.3390/ijms151221621

Xiao C, Kim HS, Lahusen T, Wang RH, Xu X, Gavrilova O, Jou W, Gius D, Deng CX (2010) SIRT6 deficiency results in severe hypoglycemia by enhancing both basal and insulin-stimulated glucose uptake in mice. J Biol Chem 285(47):36776–36784. https://doi.org/10.1074/jbc.M110.168039

Xiao C, Wang RH, Lahusen TJ, Park O, Bertola A, Maruyama T, Reynolds D, Chen Q, Xu X, Young HA, Chen WJ, Gao B, Deng CX (2012) Progression of chronic liver inflammation and fibrosis driven by activation of c-JUN signaling in Sirt6 mutant mice. J Biol Chem 287(50):41903–41913. https://doi.org/10.1074/jbc.M112.415182

Xiong X, Sun X, Wang Q, Qian X, Zhang Y, Pan X, Dong XC (2016) SIRT6 protects against palmitate-induced pancreatic β-cell dysfunction and apoptosis. J Endocrinol 231(2):159–165. https://doi.org/10.1530/joe-16-0317

Xiong X, Zhang C, Zhang Y, Fan R, Qian X, Dong XC (2017) Fabp4-Cre-mediated Sirt6 deletion impairs adipose tissue function and metabolic homeostasis in mice. J Endocrinol 233(3):307–314. https://doi.org/10.1530/JOE-17-0033

Xu Y, Li F, Lv L, Li T, Zhou X, Deng CX, Guan KL, Lei QY, Xiong Y (2014) Oxidative stress activates SIRT2 to deacetylate and stimulate phosphoglycerate mutase. Cancer Res 74(13):3630–3642. https://doi.org/10.1158/0008-5472.CAN-13-3615

Xu Z, Zhang L, Zhang W, Meng D, Zhang H, Jiang Y, Xu X, Van Meter M, Seluanov A, Gorbunova V, Mao Z (2015) SIRT6 rescues the age related decline in base excision repair in a PARP1-dependent manner. Cell Cycle 14(2):269–276. https://doi.org/10.4161/15384101.2014.980641

Yamamoto H, Schoonjans K, Auwerx J (2007) Sirtuin functions in health and disease. Mol Endocrinol 21(8):1745–1755. https://doi.org/10.1210/me.2007-0079

Yang B, Zwaans BM, Eckersdorff M, Lombard DB (2009) The sirtuin SIRT6 deacetylates H3 K56Ac in vivo to promote genomic stability. Cell Cycle 8(16):2662–2663. https://doi.org/10.4161/cc.8.16.9329

Yao L, Cui X, Chen Q, Yang X, Fang F, Zhang J, Liu G, Jin W, Chang Y (2017) Cold-inducible SIRT6 regulates thermogenesis of brown and beige fat. Cell Rep 20(3):641–654. https://doi.org/10.1016/j.celrep.2017.06.069

Yu H, Ye W, Wu J, Meng X, Liu RY, Ying X, Zhou Y, Wang H, Pan C, Huang W (2014) Overexpression of sirt7 exhibits oncogenic property and serves as a prognostic factor in colorectal cancer. Clin Cancer Res 20(13):3434–3445. https://doi.org/10.1158/1078-0432.CCR-13-2952

Yu J, Wu Y, Yang P (2016) High glucose-induced oxidative stress represses sirtuin deacetylase expression and increases histone acetylation leading to neural tube defects. J Neurochem 137(3):371–383. https://doi.org/10.1111/jnc.13587

Zhang P, Tu B, Wang H, Cao Z, Tang M, Zhang C, Gu B, Li Z, Wang L, Yang Y, Zhao Y, Wang H, Luo J, Deng C-X, Gao B, Roeder RG, Zhu W-G (2014) Tumor suppressor p53 cooperates with SIRT6 to regulate gluconeogenesis by promoting FoxO1 nuclear exclusion. Proc Nat Acad Sci 111(29):10684. https://doi.org/10.1073/pnas.1411026111

Zhang J, Yin XJ, Xu CJ, Ning YX, Chen M, Zhang H, Chen SF, Yao LQ (2015) The histone deacetylase SIRT6 inhibits ovarian cancer cell proliferation via down-regulation of Notch 3 expression. Eur Rev Med Pharmacol Sci 19(5):818–824

Zhang N, Li Z, Mu W, Li L, Liang Y, Lu M, Wang Z, Qiu Y, Wang Z (2016) Calorie restriction-induced SIRT6 activation delays aging by suppressing NF-kappaB signaling. Cell Cycle 15(7): 1009–1018. https://doi.org/10.1080/15384101.2016.1152427

Zhang X, Spiegelman NA, Nelson OD, Jing H, Lin H (2017) SIRT6 regulates Ras-related protein R-Ras2 by lysine defatty-acylation. Elife 6:25158. https://doi.org/10.7554/eLife.25158

Zhao G, Wang H, Xu C, Wang P, Chen J, Wang P, Sun Z, Su Y, Wang Z, Han L, Tong T (2016) SIRT6 delays cellular senescence by promoting p27Kip1 ubiquitin-proteasome degradation. Aging (Albany NY) 8(10):2308–2323. https://doi.org/10.18632/aging.101038

Zhong L, D'Urso A, Toiber D, Sebastian C, Henry RE, Vadysirisack DD, Guimaraes A, Marinelli B, Wikstrom JD, Nir T, Clish CB, Vaitheesvaran B, Iliopoulos O, Kurland I, Dor Y, Weissleder R, Shirihai OS, Ellisen LW, Espinosa JM, Mostoslavsky R (2010) The histone deacetylase Sirt6 regulates glucose homeostasis via Hif1alpha. Cell 140(2):280–293. https://doi.org/10.1016/j.cell.2009.12.041

Zwaans BMM, Lombard DB (2014) Interplay between sirtuins, MYC and hypoxia-inducible factor in cancer-associated metabolic reprogramming. Dis Model Mech 7(9):1023–1032. https://doi.org/10.1242/dmm.016287

Chapter 11
Epigenetic Regulation During Hypoxia and Its Implications in Cancer

Deepak Pant, Srinivas Abhishek Mutnuru, and Sanjeev Shukla

Abstract Hypoxia is defined as a cellular stress condition caused by a decrease in oxygen below physiologically normal levels. Cells in the core of a rapidly growing solid tumor are faced with the challenge of inadequate supply of oxygen through the blood, owing to improper vasculature inside the tumor. This hypoxic microenvironment inside the tumor initiates a gene expression program that alters numerous signaling pathways, allowing the cancer cell to eventually evade adverse conditions and attain a more aggressive phenotype. A multitude of studies covering diverse aspects of gene regulation has tried to uncover the mechanisms involved in hypoxia-induced tumorigenesis. The role of epigenetics in executing widespread and dynamic changes in gene expression under hypoxia has been gaining an increasing amount of support in recent years. This chapter discusses, in detail, various epigenetic mechanisms driving the cellular response to hypoxia in cancer.

Keywords Hypoxia · Epigenetics · Cancer · DNA methylation · Histone modifications · Chromatin modifications · Noncoding RNA · RNA methylation

Hypoxia

Hypoxia can be defined as a low-oxygen condition that is often experienced by certain mammalian tissues during many pathological conditions, like stroke, ischemia, inflammation, and solid tumors, and also during normal mammalian developmental morphogenesis. O_2, in general, is present in ambient air at 21% (150 mm Hg); about 2–9% (40 mm Hg) is found in mammalian tissue, and about $\leq 2\%$ is usually defined as hypoxic (Bertout et al. 2008). The cell can tolerate a range of hypoxic oxygen tension when compared to ambient atmospheric conditions. Oxygen availability of $<1\%$ can be termed "pathological hypoxia," which is a condition where the

D. Pant · S. A. Mutnuru · S. Shukla (✉)
Epigenetics and RNA Processing Lab (ERPL), Indian Institute of Science Education and Research Bhopal, Bhopal, India
e-mail: sanjeevs@iiserb.ac.in

T. K. Kundu, C. Das (eds.), *Metabolism and Epigenetic Regulation: Implications in Cancer*, Subcellular Biochemistry 100,
https://doi.org/10.1007/978-3-031-07634-3_11

cell's metabolic capabilities are compromised (Okazaki and Maltepe 2006). Often, various cells that are present in different tissues experience hypoxia in cases of developmental morphogenesis, cardiovascular disease, cancer, high-intensity exercise, and high altitude. These events trigger the cells to adapt to a low-oxygen condition by regulating various cellular events to survive the stress and by inducing certain physiological changes, such as erythropoiesis and angiogenesis, to counter the effects of a lack of oxygen (Chen et al. 2020).

Hypoxia in Cancer

A cell can enter a phase of stress due to various reasons, including high temperature, deoxyribonucleic acid (DNA) damage (genotoxic stress), ROS (reactive oxygen species), and the lack of nutrients or oxygen (Abramowicz et al. 2019). Hypoxia is one such stress faced by cells during the course of development and many other pathogenic outcomes, including cancer. In the growing tumor, due to a disturbed and poorly organized vasculature, a subset of cells experiences a state of hypoxia, where oxygen availability drastically decreases, which triggers the cells to alter a number of signaling pathways to maintain cellular homeostasis (Folkman et al. 2000).

The responses to hypoxia are majorly mediated by a group of master regulators named hypoxia-inducible factors: HIF-1α, HIF-2α, and HIF-3α, which are transcription factors and key modulators of the cellular hypoxic response (Semenza 2014). HIF proteins are heterodimers consisting of the HIF-α and HIF-β subunits. The HIF-α subunit functions as an oxygen sensor in the cell, and more specifically, HIF-1α is rapidly degraded under normoxic conditions by Von-Hippel Lindau (VHL)-dependent ubiquitination, followed by proteasomal degradation (Semenza 2012). The degradation of HIF-1α is induced by the oxygen-dependent hydroxylation of its two proline residues (PRO402, PRO564), present in the oxygen-dependent degradation domain (ODDD), by prolyl hydroxylation domain proteins (PHDs) (Nguyen et al. 2013). The hydroxylated proline residues are recognized by VHL, which then ubiquitinates the protein and marks it for subsequent degradation by the proteasome complex. Unlike HIF-α, HIF-β is constantly produced inside the cell but not degraded. During hypoxia, the PHDs are inactive owing to the unavailability of oxygen molecules, leading to the stabilization of HIF-1α. Stable HIF-1α dimerizes with HIF-1β and translocates to the nucleus and binds to gene promoters containing hypoxia-responsive elements (HREs) that have the consensus sequence 5′-[A/G]CGTG-3′ (Choudhry and Harris 2018a) (Fig. 11.1). This causes the regulation of genes which are essential to adapting to low oxygen availability and maintaining cellular homeostasis. HIFs help in maintaining oxygen homeostasis by regulating genes that would either increase O_2 delivery (VEGF, EPO) or decrease O_2 consumption (LDHA, PDK1). Vascular endothelial growth factor (VEGF) helps in increased oxygen delivery by inducing the growth of new blood vessels (angiogenesis), whereas erythropoietin (EPO) promotes red blood cell (RBC) production (erythropoiesis), thereby enhancing the oxygen-carrying capacity of the blood

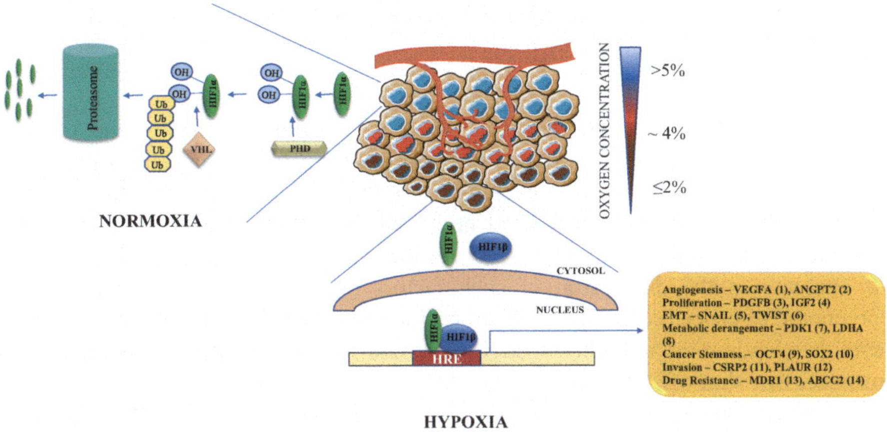

Fig. 11.1 The HIF transcription factors have two fates, depending on oxygen availability. Under normoxic conditions, HIF-1α is hydroxylated by PHD, followed by subsequent ubiquitination by VHL targeting it for proteasomal degradation. Under hypoxia, HIF-1α is stabilized, and both HIF-1α and HIF-1β translocate to the nucleus, where they dimerize and bind to the HRE elements of target genes and induce the transcription of various cancer hallmark genes. The list indicates a few examples of genes induced by the HIF pathway (1) (Büchler et al. 2003) (2) (Simon et al. 2008) (3) (Schito et al. 2012) (4) (Feldser et al. 1999) (5) (Yang et al. 2017) (6) (He et al. 2017) (7) (Kim et al. 2006) (8) (Cui et al. 2017) (9)–(10) (Mathieu et al. 2011) (11) (Hoffmann et al. 2018) (12) (Büchler et al. 2009) (13) (Rehman et al. 2019) (14) (He et al. 2016)

(Hashimoto and Shibasaki 2015). An increase in lactate dehydrogenase (LDHA) and pyruvate dehydrogenase kinase 1 (PDK1) ensures that cellular metabolism is shifted from oxidative phosphorylation, which produces energy in the presence of O_2, to glycolysis, which can take place even in low-oxygen conditions (Nagao et al. 2019). In addition to rewiring cellular energetics, hypoxic signaling involving HIF transcription factors is known to promote stemness and epithelial-mesenchymal transition (EMT) in cancer cells, thus promoting the metastatic and invasive abilities of cancer cells (Tam et al. 2020).

Epigenetic alteration in tumor hypoxia is an advancing field in cancer biology and has a lot of clinical importance as emerging therapeutic targets revolve around eliminating these particular groups of cells residing in the hypoxic niche of the tumor because of their increased resistance to currently available therapeutic options. Hypoxia-mediated signaling not only allows cells to withstand stressful conditions but also confers on them vicious life-threatening potency to stay dormant and gain the ability to migrate and cause metastasis (Tsai and Wu 2014).

Epigenetic Pathways Involved in Driving Hypoxia-Induced Cellular Responses in Cancer

Over the past two decades, there has been substantial support for the role of epigenetics in mediating the hypoxia-driven response of tumor cells. Epigenetic mechanisms have been found to be a part of both HIF-dependent and -independent cellular pathways. There are several epigenetic modifiers that are known to be upregulated in response to hypoxia through HIF-mediated transcriptional activation. An increase in the expression of such factors results in global changes in the epigenome of cancer cells. The escalation of a repressive histone mark at the global level in response to a higher load of hypoxia-responsive histone modifier, for instance, can shut off the expression of tumor suppressor genes and lead to tumor progression. Some of the classes of epigenetic regulators induced by HIF are histone methyltransferases, histone demethylases, histone deacetylases, DNA demethylases (ten-eleven translocases), and so on (Batie et al. 2018; Camuzi et al. 2019; Kim and Park 2020).

Apart from being induced transcriptionally, the function of certain epigenetic enzymes can be perturbed by the lack of oxygen in the tumor microenvironment itself. The 2-oxoglutarate (2-OG)- and Fe(II)-dependent family of dioxygenases, which includes PHD and factor-inhibiting HIF-1 (FIH), requires molecular oxygen to perform its function. Jumonji-C-domain-containing histone demethylases comprise the largest group within the 2-OG dioxygenase family. With the help of Fe (II) and 2-oxoglutarate acting as cofactors, these enzymes catalyze oxygen-dependent hydroxylation and the subsequent removal of the methyl group from a histone lysine residue (Kim and Park 2020). Another group of epigenetic enzymes, called ten-eleven translocases (TETs), falls into this category. TETs carry out the oxygenation of the 5-methyl group present on cytosine into the 5-hydroxymethyl group. Increasing evidence suggests that the activity of these epigenetic modifiers is decreased during tumor hypoxia, leading to a large-scale change in the histone and DNA methylation landscape (Thienpont et al. 2016a). Apart from HIF-1α, PHD and FIH can also mediate the degradation of other substrates under normoxic conditions. Recently published studies have described the stabilization of certain repressive histone methyltransferases under hypoxia, which leads to reduced expression of antimetastasis genes (Chopra et al. 2020).

In this chapter, the different classes of epigenetic modifiers involved in driving hypoxia-induced changes in cancer cells will be discussed. The epigenetic phenomena involved in hypoxia-driven cellular responses can be broadly divided into chromatin remodeling, histone modifications, DNA methylation and hydroxymethylation, ribonucleic acid (RNA) methylation, and, lastly, posttranscriptional gene regulation by noncoding RNAs (ncRNAs) (Fig. 11.2).

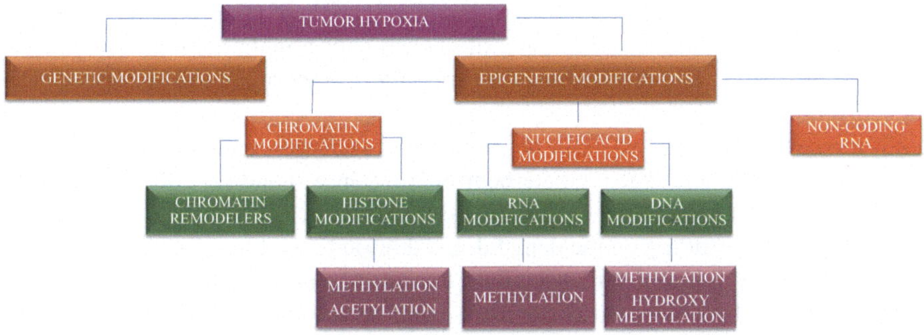

Fig. 11.2 Cellular response to hypoxic stress in the tumor microenvironment can lead to changes in both the genetic and epigenetic landscape. Epigenetic modifications can be broadly classified into chromatin modifications, nucleic acid modifications, and the production of noncoding RNAs. Chromatin remodelers physically alter the architecture of the chromatin itself, whereas histones, which are the core proteins of the chromatin, get chemically modified to allow the existence of chromatin in an active or repressed state. DNA and RNA also get chemically modified, allowing or inhibiting repressors and activators to bind, respectively

Chromatin Remodelers in Tumor Hypoxia

The tight wrapping of DNA around histone proteins renders a few regulatory regions of the DNA inaccessible to transcription factors and other DNA-binding proteins. Chromatin remodelers are precisely those proteins whose function is to resolve the higher-order structures of chromatin and make the regulatory regions of the DNA accessible to other DNA-binding proteins (Clapier et al. 2017). The plasticity of the chromatin is majorly regulated by two distinct mechanisms. Firstly, the core histone proteins can undergo a variety of posttranslational modifications allowing the local unwinding of the DNA, which exposes the regulatory regions in order for the proteins to bind (Längst and Manelyte 2015). Secondly, certain chromatin remodelers can physically slide, eject, or restructure the nucleosome core, making the DNA available for regulation (Clapier and Cairns 2009).

Chromatin remodeling complexes are a conglomeration of multiple proteins, which consist of a conserved ATPase subunit and different unique epigenetic reader subunits, allowing them to identify different signals on the chromatin and perform different functions in accordance with the signal. Based on the different flanking subunits present along with the conserved ATPase subunit, these remodelers are classified into four different families. They are SWI/SNF (SWItch/sucrose nonfermentable), CHD (chromodomain-helicase-DNA-binding protein), ISWI (imitation SWI), and the INO80 family (Clapier and Cairns 2009; Längst and Manelyte 2015; Tyagi et al. 2016). All chromatin remodelers function by hydrolyzing adenosine triphosphate (ATP) to either translocate DNA or remodel nucleosomes, allowing different regulatory proteins to access DNA and execute various actions, such as transcription and DNA repair. The epigenetic reader subunits help the complex in identifying the exact regions on the chromatin where restructuring is required (Saha et al. 2006). Thus, chromatin remodeling is an intricate interplay between the identification of a particular chemical signal, such as acetylation, methylation, etc., on the chromatin and the physical mobilization of the nucleosome unit.

In the cell, chromatin exists in two functional subtypes, called transcriptionally active euchromatin and inactive heterochromatin (Grewal and Jia 2007). In euchromatin, the dynamic packaging of DNA is essential for the cell to activate and repress different genes depending on the demand. In addition to the transcription factors and other DNA binding proteins that regulate transcriptional activities, the structure of chromatin must be continually altered to cater to the different needs of the cell (Strahl and Allis 2000). In this regard, the chromatin remodelers play a crucial role in regulating the complex framework of the chromatin to facilitate DNA access. The technological advancement in the field of nucleic acid sequencing has enabled our understanding of the importance of chromatin-regulating proteins and their role in tumorigenesis (Nair and Kumar 2012). In this part of the chapter, we will look at how different chromatin remodelers regulate gene expression under hypoxic stress faced by tumor cells.

A very interesting observation was made in Hep3B cells, wherein they found that Brahma (BRM) and Brahma-related gene 1 (BRG1), which are ATPase-harboring SWI/SNF chromatin remodeling complexes, promoted an increased expression of a subset of HIF-1α and HIF-2α target genes, such as CA9 and EPO, under hypoxic conditions. While the BRG1 complex regulates *HIF2A* gene expression in Hep3B and SW13 cells, the BRM complex regulates *HIF2A* gene expression in SW13 cells. In addition, they also promoted the induction of these HIF target genes in SW13 cells by modifying the nucleosome in an ATPase-dependent manner after being recruited by HIF transcription factors. However, the expression of some of the major metabolic HIF target genes, such as *GLUT1*, *LDHA*, *PDK1*, and *PGK1*, remained unaltered irrespective of the presence or absence of BRM or BRG1. Also, BRG1 was reported to have opposite functions in normoxic and hypoxic conditions. In the RCC4T renal cell carcinoma cell line, it was observed that the BRG1 complex inhibits cell proliferation and promotes cell migration under normoxia, whereas it increases cell proliferation and decreases cell migration in hypoxia (Sena et al. 2013). In U2OS osteosarcoma and HeLa cervical cancer cells, ISW1 (imitation switch) regulates cellular hypoxic response by affecting some of the HIF target genes. It was found that ISW1 depletion resulted in an increased expression of *CA9*, *PHD2*, and *BNIP3* and a reduced expression of *GLUT3*. Mechanistically, it was observed that ISW1 regulated HIF expression by altering RNA polymerase II loading onto FIH (factor-inhibiting hypoxia) mRNA. FIH regulates HIF activity by affecting its association with CBP/p300, thereby indirectly tinkering the expression of a subset of HIF target genes. Also, the loss of ISW1 expression resulted in decreased autophagy and increased apoptosis, indicating that ISW1 might be an important factor in cell survival during hypoxic stress (Melvin et al. 2011). Pontin, which is a chromatin remodeler with dual functions of DNA helicase and ATPase activity, is methylated by G9a/GLP (G9a-like protein) methyltransferases under hypoxic conditions, leading to the enhanced recruitment of p300 by HIF-1α on the promoter region of a few selected HIF-regulated genes. Ets1 is a promoter of EMT in cancer and is one of the HIF-1α target genes regulated by pontin methylation under hypoxia, highlighting the role of chromatin remodelers in tumor progression (Lee et al. 2011). On a contrasting note, reptin, which is also an ATPase and DNA helicase activity containing a chromatin remodeler methylated by G9a/GLP methyltransferases under hypoxia, repressed some HIF target genes, such as VEGF, in breast cancer cells and mouse xenograft models. Reptin methylation also inhibited the invasion, motility, and in vivo growth of breast cancer cells, indicating that reptin ensures the maintenance of cellular homeostasis under hypoxia by functioning as a repressor (Lee et al. 2010). These two contrasting routes which arise out of the same stimulus i.e., hypoxia, set an oxymoronic scheme with regards to how a cell adapts to this stress. While one protein (Pontin) is able to activate tumor-promoting genes, the other (Reptin) is making sure that the tumor promoting-genes are repressed via the same solitary factor HIF. How do cells in hypoxia tip in favor of one route? And what determines the tipping point would be an exhilarating study because one promotes tumor progression while the other represses it. TP53, which is the most commonly mutated protein in numerous cancers, forms a complex with hypoxia-induced HIF-1α in NSCLC (nonsmall cell lung carcinoma) and propagates a protumorigenic signaling cascade. It was observed that under hypoxia, a mutant p53/HIF-1α complex was formed that was capable of binding to SWI/SNF chromatin remodeler complex, enhancing the expression of a subset of HIF target genes, mainly involving those that encode components of the ECM (extracellular matrix). The depletion of this mutant p53 resulted in the upregulation of type VIIa1 collagen and laminin-γ2, enhancing the invading potential of NSCLC cells in vitro and in vivo. This observation was consistent with a previous study that reported the involvement of the formation of a complex between SWI/SNF chromatin remodeler and mutant p53 in regulating VEGFR2 expression in multiple breast cancer cell lines (Amelio et al. 2018; Pfister et al. 2015).

In the hypoxic pockets of the growing tumor, chromatin remodelers play a distinct role in partnering with HIFs and other proteins to facilitate protumorigenic as well as antitumorigenic properties. They mainly exert their function by helping HIFs access their target regions by modifying the nucleosome/chromatin architecture to promote the hallmark properties of cancer, such as EMT and metastasis, which make these remodelers lucrative therapeutic targets in combination with HIF-1α.

Histone Modifications in Tumor Hypoxia

The state of tumor cells undergoing hypoxia is governed by a multitude of factors that synergistically work together in achieving increased invasive properties. There is a complex interplay between hypoxia-specific transcriptional regulation, epigenetics, and metabolism, which induces aggressive tumor phenotype in cells experiencing low oxygen tension. The regulation of gene expression by posttranslational histone modification is one of the major epigenetic phenomena involved in deciding the transcriptional fate of numerous genes involved in various cellular pathways under hypoxia. Modifications like acetylation, methylation, and others on N-terminal tails of histones can alter gene expression under hypoxia in both a global and context-specific manner. These modifications are brought about by either a change in the expression or the modulation of the activity of histone modifiers (Batie et al. 2018; Camuzi et al. 2019; Li et al. 2020a). The event that triggers changes in the expression or activity of these factors may or may not be a direct result of the canonical HIF pathway. Furthermore, the activity of some histone-modifying enzymes is altered in response to hypoxia as a direct result of low O_2 availability or due to changes in the levels of certain metabolites (Kim and Park 2020).

The stabilization of the α-subunit of HIF-1 is the event that initiates cellular response toward hypoxia. Under normoxia, HIF-1α is degraded via proteasomal degradation in a PHD- and VHL-dependent manner (Petrova et al. 2018). However, these two are not the only factors that negatively regulate the activity of HIF-1α. Even after rescue from degradation under hypoxia, HIF-1α is inhibited from interacting with p300 by NAD^+-dependent histone deacetylases, called sirtuins. HIF-1α is a nonhistone substrate of SIRT1 (sirtuin 1), which deacetylates it at Lys674. Apart from SIRT1, there are several other histone modifiers that posttranslationally modify HIF-1α, like SIRT2 (sirtuin2), SETD7 (SET domain containing 7, histone lysine methyltransferase), and LSD1 (lysine-specific histone demethylase 1). Acetylation at Lys674 of HIF-1α is essential for the recruitment of p300 at the promoter of its target genes. Since metabolic reprogramming under hypoxia causes a decrease in NAD^+ levels as a result of low oxidative phosphorylation, the activity of SIRT1 is abrogated, leading to sustained acetylation and subsequent recruitment of p300 by HIF-1α (Camuzi et al. 2019). This is one of the ways in which the metabolic state of the cell can regulate response toward hypoxia.

The genes that are transcriptionally upregulated under hypoxia form the primary response of the cell toward low O_2 availability. Several histone modifiers belonging to more than one subtype are induced at the mRNA level under hypoxia (Table 11.1). Many members of the Jumonji-C-domain-containing family of histone demethylases, like JMJD2C (Jumonji-domain-containing protein 2C), are upregulated by HIF-1α-induced transcription (Kim and Park 2020). In breast cancer, JMJD2C acts as a coactivator for HIF-1α by decreasing H3K9me3 marks on HIF-1 target genes, like LDHA, PDK1, LOXL2 (lysyl oxidase-like 2), and L1CAM (L1 cell adhesion molecule), thus promoting metabolic reprogramming and metastasis (Luo et al. 2012). Histone deacetylase 3 (HDAC3), a histone deacetylase, is another HIF-1 target and is essential for turning on the expression of mesenchymal genes during epithelial-to-mesenchymal transition (Wu et al. 2011). Induction under hypoxia can also be mediated by decreased proteasomal degradation, leading to higher protein stability. Histone methyltransferases G9a and GLP are induced by increased protein stabilization or enzymatic activity under hypoxic conditions. G9a/GLP stabilization has been shown to increase dimethyl and trimethyl marks at H3K9 on specific genes involved in metastasis (Chopra et al. 2020) (Fig. 11.3).

Table 11.1 List of all histone modifiers that are upregulated in tumor hypoxia (*IGFBP3*: insulin-like growth-factor-binding protein 3, *PSA*: prostate-specific antigen, *NKX3.1*: NK3 homeobox 1)

Name	Targets on histone	Effect on target gene expression	Cancer type	References
EZH2	H3K27me1/2/3	Repression of *RAD51*	Breast cancer	Chang et al. (2011)
KDM2A	H3K4me3, H3K36me1/2	Repression	Various cancer cell lines	Batie et al. (2017)
KDM2B	H3K4me3, H3K36me1/2	Repression	Various cancer cell lines	Batie et al. (2017)
KDM3A	H3K9me1/2	Activation of *ADM* and *GDF15*	Renal cell carcinoma	Krieg et al. (2010)
KDM3B	H3K9me1/2	Activation	Hepatocellular carcinoma/glioblastoma	Xia et al. (2009)
KDM3C	H3K9me1/2	Activation	Hepatocellular carcinoma/glioblastoma	Xia et al. (2009)
KDM4B	H3K9me2/3	Activation	Cervical/prostate cancer	Beyer et al. (2008)
KDM4C	H3K9me2/3	Activation	Cervical/prostate cancer	Beyer et al. (2008)
KDM5B	H3K4me2/3	Repression	Prostate cancer	Lee et al. (2013)
KDM5C	H3K4me2/3	Repression of *IGFBP3*	Clear cell renal cell carcinoma	Niu et al. (2012)
KDM5D	H3K4me2/3	Repression	Hepatocellular carcinoma/glioblastoma	Xia et al. (2009)
KDM6A	H3K27me2/3	Activation	Various cancer cell lines	Chakraborty et al. (2019)
KDM6B	H3K27me2/3	Activation	Hepatocellular carcinoma/glioblastoma	Xia et al. (2009)
KDM7A	H3K9me1/2, H3K27me1/2	Activation	Hepatocellular carcinoma/glioblastoma	Xia et al. (2009)
KDM7B	H3K9me1/2, H3K27me1/2	Activation of *PSA* and *NKX3.1*	Prostate cancer	Tong et al. (2016)
KDM8	H3K36me2/3	Activation of *GLUT1* and *LDHA*	Breast cancer	Wang et al. (2014)
HDAC1	H3K18Ac	Repression of *TP53* and *VHL*	Various cancer cell lines	Kim et al. (2001)
HDAC3	H3K4Ac	Activation of N-cadherin and vimentin via WDR5	Various cancer cell lines	Wu et al. (2011)

Histone Methylation

The regulation of gene expression under hypoxia is accomplished by a number of epigenetic mechanisms, of which histone methylation is the most explored phenomenon. Histone methylation is a unique epigenetic mark as it can lead to both activation and repression of gene expression. Methylation at H3K4 and H3K36

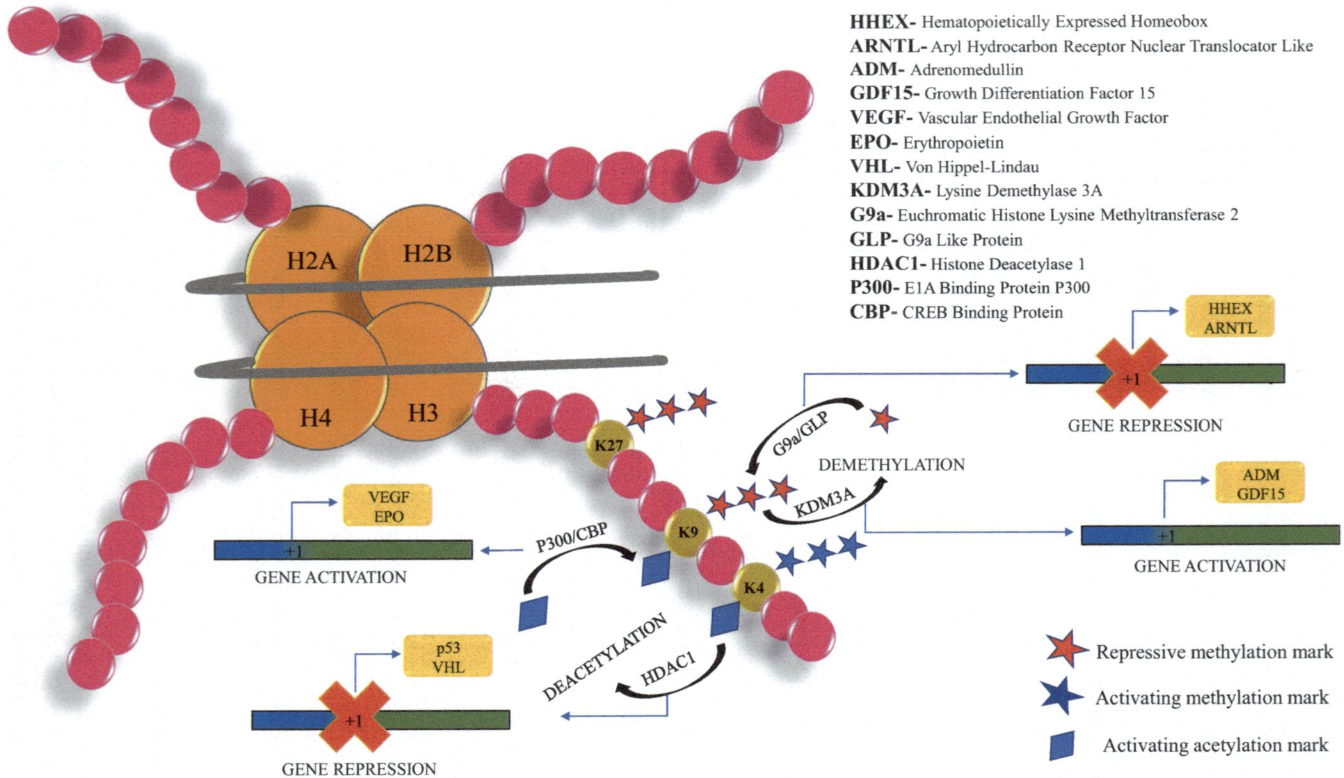

Fig. 11.3 Histone modifications facilitate the activation or repression of oncogenes and tumor suppressor genes, respectively. The trimethylation mark on K4 is an active transcription mark, whereas K9 and K27 are repressive marks. The figure depicts a histone methyltransferase (G9a/GLP) and a histone demethylase (KDM3A), both of which are upregulated in response to hypoxia. Histone acetylation is an active transcription mark facilitated by histone acetyltransferases (HATs), and the reverse process is carried out by histone deacetylases (HDACs), which repress transcription. Under hypoxia, p300 and CBP bind to HIF-1α and acetylate histones at the promoters of HIF target genes. HDAC1 is one of the HDACs that remove the acetylation mark at the promoters of TSG, e.g., p53 under hypoxia. *HHEX* hematopoietically expressed homeobox, *ARNTL* aryl hydrocarbon receptor nuclear translocator like, *ADM* adrenomedullin, *GDF15* growth differentiation factor 15, *VEGF* vascular endothelial growth factor, *EPO* erythropoietin, *VHL* Von Hippel-Lindau, *KDM3A* lysine demethylase 3A, *G9a* euchromatic histone lysine methyltransferase 2, *GLP* G9a-like protein, *HDAC1* histone deacetylase1, *P300* E1A-binding protein P300, *CBP* CREB-binding protein

leads to gene activation, whereas methylated H3K9 and H3K27 residues lead to the inhibition of gene transcription. Several extensive, and at times, conflicting studies have explored the trends in histone methylation change under hypoxia and its significance in cancer progression in the past decade. In order to study the role of histone methylation in regulating gene expression under hypoxia, we first need to be cognizant of the processes that lead to the perturbations in the histone methylation status of cancer cells experiencing a state of low oxygen tension. In summary, the changes in histone methylation induced by hypoxia can be attributed to the transcriptional upregulation of histone methyltransferase or demethylase via the HIF-1 pathway, increase in histone methyltransferase activity by enhanced protein stability or enzymatic activity, and, lastly, decreased histone demethylase activity of O_2-dependent dioxygenases. In general, there is an increase in the levels of histone methylation in response to hypoxia. Histone methylation at H3K4, H3K9, H3K27, H3K36, H3K79, and H4R3 has been reported to be induced under hypoxia (Kim and Park 2020).

To date, only one histone methyltransferase has been shown to be transcriptionally upregulated under hypoxia. The enhancer of zeste homolog 2 (EZH2), which catalyzes methylation at the H3K27 residue, has been shown to increase in hypoxia via a HIF response element in its promoter. EZH2 induces genomic aberration in a hypoxia-dependent manner by the silencing of the *RAD51* expression in breast-tumor-initiating cells (Chang et al. 2011). On the other hand, several histone demethylases have been reported to be induced under hypoxia. *KDM3A* (lysine demethylase 3A), *KDM4B*, and *KDM4C* are some of the demethylases that are upregulated during hypoxia (Pollard et al. 2008). KDM3A targets demethylation at H3K9me1/2, whereas KDM4B and KDM4C target H3K9me2/3 marks. As methylation at H3K9 is a repressive mark, the HIF-1-dependent upregulation of these demethylases under hypoxia leads to gene activation at specific loci, which favors cancer progression. For instance, Krieg et al. have shown that the upregulation of KDM3A in HCT116 colon cancer cells undergoing hypoxia causes an increase in the transcription of tumor-promoting genes, like *ADM* (adrenomedullin), and *GDF15* (growth differentiation factor 15) (Krieg et al. 2010). KDM5B and KDM5C, which target methylation at H3K4 residues leading to gene repression, are also reported to increase under hypoxia. KDM5B has been shown to negatively affect the expression of tumor suppressor genes, such as *CAV1* (caveolin-1), *HOXA5* (homeobox A cluster 5), and *BRCA1* (breast cancer type 1 susceptibility protein) (Yamane et al. 2007). KDM8, an H3K36me2 demethylase, has been shown by Wang et al. to increase the expression of hypoxia-responsive genes by enhancing the noncanonical coactivator role of the glycolytic enzyme PKM2 (pyruvate kinase M2) in breast cancer. KDM8 physically interacts with PKM2 and facilitates its nuclear translocation, thereby increasing the enrichment of PKM2 on the promoters of HIF-1-induced genes, like *LDHA* and *GLUT1* (glucose transporter 1), leading to their enhanced transcription (Wang et al. 2014).

Another mechanism by which changes in histone methylation can occur in cancer cells experiencing hypoxia is through increased protein stability or enhanced enzymatic activity of certain histone methyltransferases. G9a and GLP (G9a-like protein) fall under this category. G9a and GLP are two of the most extensively studied lysine methyltransferases with both histone and nonhistone substrates. Both the enzymes catalyze the addition of mono- or dimethyl groups at H3K9 residues and exist in the cell as a heterodimer. Although G9a and GLP are capable of functioning independently of each other, the global levels of H3K9me1 and H3K9me2 are decreased when either of the factors is knocked out (Shinkai and Tachibana 2011). There is a global increase in H3K9 methylation levels under hypoxia, which can be partly attributed to the posttranslational regulation of G9a and GLP. Similar to HIF-1α, under normoxia, G9a is hydroxylated at P676 and P1207 residues, leading to VHL-dependent ubiquitination and proteasomal degradation by PHD1. Furthermore, both G9a and GLP are targeted by FIH for hydroxylation at asparagine residues that fall within the conserved methyl-lysine-binding ankyrin repeat domain (ARD), leading to a decrease in their methyltransferase activity. When there is a decrease in oxygen levels, PHD1 and FIH are unable to function, thus leading to increased stability or activity of G9a/GLP (Chopra et al. 2020; Kim and Park 2020). This increase in G9a/GLP activity in hypoxia was shown to be essential for the survival of breast cancer cells undergoing hypoxia. In a study by Casciello et al., it was shown that G9a/GLP causes the downregulation of a number of genes via increased H3K9 methylation at their promoters, including tumor suppressors *HHEX* and *ARNTL*, and this decrease in expression was reversible when a chemical inhibitor of G9a was used (Casciello et al. 2017) (Fig. 11.3).

The 2-oxoglutarate-dependent dioxygenases are a large family of proteins consisting of about 60 proteins. These proteins catalyze the addition of the hydroxyl group to their substrate in a reaction that is dependent on Fe(II) and 2-oxoglutarate as cofactors. Molecular oxygen is utilized in this hydroxylation reaction, with CO_2 and water being the by-products. PHDs and FIH are the most well-known members of this family that can be considered oxygen sensors of the cell. Jumonji-C-domain-containing demethylases are the largest group of 2-OG-dependent dioxygenases. Apart from HIF-mediated upregulation, these demethylases are also subjected to decreased enzymatic activity because of the lack of oxygen during hypoxic conditions in cancer cells (Kim and Park 2020). However, not all JmjC demethylases are known to be sensitive to a decrease in the levels of oxygen. KDM5A and KDM6A have been recently shown to be directly affected by decreased oxygen levels in cancer cells. KDM5A, an H3K4 demethylase, is known to be upregulated at the transcriptional level in various kinds of cancers and is also associated with chemotherapeutic resistance by the shutting of the expression of tumor suppressor genes. The decrease in the activity of KDM5A under hypoxia and its role as a direct oxygen sensor have been studied by Batie and colleagues. It has been argued that as histone methylation changes induced by low levels of oxygen occur prior to the HIF-mediated upregulation of demethylases, it is essential to look at the changes in the activity of demethylases due to their dependence on oxygen (Batie et al. 2019). In another study, the decreased activity of KDM6A under hypoxia was found to be responsible for a global increase in H3K27me3 levels under hypoxia (Chakraborty et al. 2019). Despite growing evidence, further investigation into the change in activity of histone demethylases under hypoxia is needed to cement their role as novel epigenetic oxygen sensors.

Histone Acetylation

The acetylation of histone proteins is carried out by histone acetyltransferases (HATs), where an acetyl group is transferred from acetyl-coenzyme A to the NH_3+ group of lysine residues located on the tails of the histones. The removal of the acetyl group is performed by a group of enzymes known as histone deacetylases (HDACs). Acetylation leads to a change in the charge of histone tails from positive to neutral, thereby disrupting the compact association between histones and DNA, leading to the loosening of chromatin. This makes the DNA more accessible to transcription factors and thus results in enhanced transcription of genes with acetylated histone marks on their promoters. The removal of the acetyl group by HDACs reverses this process, making the DNA tightly packed and eventually leading to gene silencing (Bannister and Kouzarides 2011).

The acetylation of histones is one of the major mechanisms by which HIF-mediated transcriptional activation takes place in cells undergoing hypoxia. The cAMP response element-binding protein binding protein (CREBBP or CBP) and p300 (also known as EP300) are the coactivators that function alongside HIF. Both CBP and p300 are known to have histone acetyltransferase activity and activate transcription in a HIF-dependent manner (Fauquier et al. 2018) (Fig. 11.3). It has been shown that promoters of genes suppressed under hypoxia have low H3K9ac marks, whereas those that are activated bear high H3K9ac levels, like VEGF (vascular endothelial growth factor) and EGR1 (early growth response 1) (Johnson et al. 2008). Also, an increase in acetylated histone marks was found to be responsible for the upregulation of oncogenic long noncoding RNA (lncRNA) *HOTAIR* (HOX antisense intergenic RNA) via an MLL1-p300-HIF-1α-dependent mechanism in hypoxic colon cancer cells. Hypoxia may cause an increase in acetylation at target promoters indirectly, as shown by Hait et al. High nuclear levels of sphingosine-1-phosphate, which is an inhibitor of HDACs (histone deacetylases), causes an increase in H3 acetylation at the promoters of HIF targets in triple-negative breast cancer cells (Hait et al. 2020).

Several histone deacetylases are also known to play an important role in gene regulation under hypoxia. For example, HDAC1 is upregulated under hypoxia and is essential for stimulating angiogenesis. The inhibition of HDAC1 using trichostatin A led to an increase in tumor suppressors *TP53* and *VHL*, thereby decreasing HIF-1α and VEGF levels (Kim et al. 2001). HDAC3 also gets upregulated in hypoxia and recruits WDR5, a component of H3K4 methyltransferase complexes, to the promoters of genes associated with mesenchymal phenotypes. This causes an increase in the transcription of EMT-related genes, like N-cadherin and vimentin (Wu et al. 2011). A global decrease in hypoxia-mediated H3K9ac and H3K27ac has also been reported. PDK1, which is one of the genes upregulated in a HIF-dependent manner under hypoxia, functions by phosphorylating the pyruvate dehydrogenase complex, leading to its inactivation and, subsequently, to low levels of acetyl-CoA. Since acetyl-CoA is the substrate required for histone acetylation, the decrease of this metabolite under hypoxia causes a decrease in global H3K9ac and H3K27ac levels in neuroblastoma cells, blocking their differentiation. Furthermore, supplementing cells with acetate or glycerol triacetate rescued the expression of differentiation-related genes (Li et al. 2020b).

Increase in the expression of oncogenic MSMP (microseminoprotein, prostate-associated) protein through histone-acetylation-dependent mechanism was found to be the cause of resistance to anti-VEGF therapy in ovarian cancer. Under hypoxia, a decrease in histone acetylation inhibits the binding of transcriptional repressor CTCF (CCCTC-binding factor) to the enhancer of *MSMP*, which in turn leads to its increased expression and enhanced MAPK (mitogen-activated protein kinase) signaling, along with increased tube formation capability (Mitamura et al. 2018). Response to hypoxia via HDACs can also be mediated by the acetylation of nonhistone substrates. It was shown that for HIF-1α stabilization and nuclear accumulation to happen, deacetylation of Hsp70 was necessary for cells undergoing hypoxia. AMPK (AMP-activated protein kinase) activation in response to hypoxia led to the export of HDAC5 to the cytoplasm, where it deacetylates Hsp70, inhibiting its degradation and thereby leading to its interaction with Hsp90 (a positive regulator of HIF-1α folding) and its subsequent folding and stabilization (Chen et al. 2015).

Other Histone Modifications

Apart from acetylation and methylation, histone modifications like phosphorylation and ubiquitination have also been reported in hypoxic tumor cells. Both of these modifications have been found to be associated with the activation of DNA damage and repair pathways (DDRs) under hypoxia. One of the most prominent signs of early double-stranded breaks in DNA is the phosphorylation of the histone H2A variant H2AX at the Ser139 residue (also known as γH2AX). The phosphorylation of H2AX initiates DNA repair through homologous recombination by repositioning nucleosomes around the damage site and recruiting cohesins to keep the damaged ends in close proximity during repair (Mah et al. 2010).

Although there is no substantial evidence of hypoxia-mediated DNA damage, it may induce genomic instability indirectly through the downregulation of genes involved in DDR pathways, like *RAD51* (Bindra et al. 2004). Hammond and coworkers first demonstrated the presence of γH2AX under severe hypoxia in the HCT116 cell line, which was abrogated on the downregulation of ATR (ATM and RAD3 related), one of the kinases responsible for this phosphorylation event (Hammond et al. 2003). Furthermore, a highly proliferating subpopulation of hypoxic endothelial cells was found to be positive for γH2AX in an ATR-dependent pathway (Economopoulou et al. 2009). Another study by Storci et al. demonstrated that the activation of γH2AX in a hypoxic microenvironment was dependent on the transcriptional repressor Slug. This led to NF-κB-mediated cytokine upregulation and the maintenance of the pro-inflammatory phenotype in breast cancer stem cells (Storci et al. 2013).

An important study published in 2017 tried to explore the interplay of H2AX phosphorylation and ubiquitination in sustaining the effects of HIF-1α for a prolonged period of time. Rezaeian et al. have shown that HIF-inducible TRAF6 (tumor necrosis factor receptor-associated factor 6) mediates the monoubiquitination of H2AX at Lys119/120. It further leads to the recruitment of ATM (ataxia telangiectasia mutated) kinase on mUb-H2AX and the subsequent phosphorylation of H2AX at Ser139. γH2AX then binds to HIF-1α at its target promoters and maintains the transcriptional activation of hypoxia-inducible genes. γH2AX also protects HIF-1α from VHL-independent proteasomal degradation and maintains a stable pool of HIF-1α inside the nucleus by binding to it (Rezaeian et al. 2017).

DNA Methylation and Hydroxymethylation in Tumor Hypoxia

DNA methylation is one of the most well-studied epigenetic modifications situated on the fifth carbon of the cytosine present in the CpG dinucleotide of the DNA (Zhu et al. 2016). The addition of methyl groups to the cytosine base is catalyzed by a family of methyltransferases called DNA methyltransferases (DNMTs), which transfer a methyl group from the universal methyl group donor S-adenosyl methionine (SAM) (Mahmood and Rabbani 2019). DNA methylation marks are considered to be essential checkpoints governing gene regulation, making it possible for different cells to perform different activities by keeping in check what genes should be actively transcribed and how and when they are going to be transcribed (Moore et al. 2012). So far, the importance of DNA methylation has been very well studied during embryonic development, genomic imprinting, and X-chromosome inactivation and, more recently, even in tumorigenesis and progression. The two-way function of DNA methylation marks ensures that gene regulation takes place by either blocking the binding of transcription factors essential for transcriptional regulation or allowing methylation-binding proteins (MBPs) to dock at the methyl sites, enabling them to recruit other repressors or activator proteins.

In cancers, DNA methylation has been reported to play a decisive role in the repression of important gene loci, which protect cells from adversities. Global hypomethylation and hypermethylation at promoter regions of specific genes have been an important feature of cancer cells in comparison to their normal counterparts. The effect of such contrasting conditions, in principle, can lead to two possible outcomes, they being the active transcription of previously silent protooncogenes due to global hypomethylation and the repression of tumor suppressor genes caused by the local hypermethylation of promoters (Ross 2019). Hence, the cancer cell gains the ability to continuously proliferate with no spatiotemporal regulation. In addition to being able to divide unchecked, aberrant methylation patterns can promote transcriptional relief to previously quiescent genes, which can now aid the growing tumor and meet the rigorous demands of the continuously proliferating cancer cell.

The creation of hypoxic niches inside the tumor mass forces the cells to adapt to low oxygen availability by initiating angiogenesis and a myriad of other signaling pathways to sustain the stress (Krock et al. 2011). Changes in DNA methylation and hydroxymethylation patterns in tumor hypoxia are being pursued extensively and are emerging hallmarks of cancers, and how these epigenetic alterations enable the tumor cells to adapt to the vicious condition of hypoxia is a major area of interest to many researchers around the globe (Rasmussen and Helin 2016). In the coming section, we will look at how various enzymes that are involved in methylation and hydroxymethylation processes are governed in tumor hypoxia and its subsequent effect on how changes in the pattern of these particular epigenetic alterations affect the adaptation of cancer cells.

In colorectal cancer cells, it has been reported that under hypoxic and hypoglycemic conditions, the mRNA transcript levels and activity of DNMT1, DNMT3a, and DNMT3b enzymes were downregulated, leading to a reduction in the global methylation status (Skowronski et al. 2010). On the contrary, a study performed in hepatoma cells showed that under hypoxia, DNMT1 and DNMT3a, and not DNMT3b activity, were upregulated. (Liu et al. 2011). These observations highlight the fact that intertumoral heterogeneity is an important trademark feature of cancer and tumor cells of varied origins have unique responses and mechanisms to cope with hypoxic stress. Also, these studies highlight the effect of the dysregulated epigenetic landscape, which promotes tumor progression by disrupting the regulated expression of oncogenes or tumor suppressor genes. In a study performed on renal cell carcinoma, it was observed that DNMT3a served as a gatekeeper tumor suppressor by preventing the cells from adapting to hypoxia. The loss of DNMT3a resulted in the ability of renal cell carcinoma (RCC) cells to access the HIF-2α signaling cascade to acclimatize to the hypoxic condition. This study portrays the dynamic nature of DNMTs, wherein they not only methylate DNA but also play an important role as tumor suppressors, keeping tumor progression in check. Sprouty (SPRY2), a tumor suppressor gene that shunts the receptor tyrosine kinase pathway, is often downregulated in multiple cancers, including liver, lung, prostate, and breast. In Hep3B hepatoma cells, HIF-1α/HIF-2α and DNMT1 bind to the promoter and first intron of the SPRY2 gene and increase the methylation of those regulatory regions, resulting in reduced levels of SPRY2 mRNA/protein (Gao et al. 2017). In another study performed on a panel of 31 breast cancer cell lines, researchers found that epidermal growth factor receptor (EGFR) gene intron 18, which houses a putative/consensus HRE-binding site (5’ACGTG 3′), is often methylated in some cancer cell lines and normal cells. Hypoxic induction resulted in increased expression of the EGFR (epidermal growth factor receptor) gene, leading to elevated cell proliferation and migration. In addition to supposedly aiding tumor progression, increased EGFR expression also increased the sensitivity of cancer cells in hypoxia to EGFR inhibitors. The inhibition of DNMTs using DNMT inhibitors reversed the methylation status generally observed at the HRE consensus sequence located at intron 18 of the EGFR gene. These observations imply that the current EGFR inhibitors available in the clinic could be potential therapeutic options for patients with hypoxic breast tumors (Mamo et al. 2020).

So far, in the various studies discussed above, it can be inferred that DNA methylation is a very dynamic process, and under hypoxia, the methylation status varies in different cancers. DNMTs are the major proteins regulating the methylation status of the genome, and it is imperative that any misregulation of these proteins could lead to aberrant methylation patterns, which are often seen in multiple cancers. Also, the dynamic nature of DNMTs is intriguing since one of the family members (DNMT1) aids in the downregulation of a tumor suppressor promoting tumorigenesis, while the other (DNMT3a) functions as a guardian and inhibits the cells from adapting to hypoxic stress. It would be interesting to see if there is any correlation involved between the distinctive activities that these DNMTs perform and their contribution in the event of carcinogenesis since one of them is a maintenance methyltransferase (DNMT1) while the other is a de novo methyltransferase (DNMT3a). HIF-1α is the master regulator of hypoxic response and, in combination with DNMTs, leads to distinctive methylation patterns, which can prove to be pivotal in giving cancer cells an undue advantage.

In addition to aberrant DNA methylation, recent studies have explored the role of 5-hydroxymethylation as well in promoting oncogenic signaling. The involvement of the TET (ten-eleven translocation) family of proteins in the process of demethylation by oxidizing 5-methylcytosine to 5-hydroxymethylcytosine opened up new doors for researchers to look deeper into elusive demethylation mechanisms and their role in cancer. A decrease in the hydroxymethylation and an increase in the 5-mc status of tumor suppressor genes have been reported to be prevalent in hypoxic niches of multiple solid tumors (Thienpont et al. 2016b). TET proteins are centrally involved in regulating 5-hmc levels, and numerous studies have been undertaken to assess their role in regulating the hypoxic response. TET proteins are dioxygenases that require α-ketoglutarate and Fe(II) to carry out their functions (Shen et al. 2014). Studies performed in neuroblastoma have reported an increase in global 5-hmc levels in hypoxia via TET1 activation mediated by HIF-1, and a specific increase was observed at the promoters, enhancers, and untranslated regions (UTRs) of putative hypoxia response genes (Mariani et al. 2014). In a similar study performed on HepG2 cells of hepatoblastoma origin, upregulation of TET1 was observed upon hypoxic induction via HIF-1α (Lin et al. 2017). In another study performed on breast cancer cells, TET-1 and TET-3 were upregulated in hypoxia via HIF-1α, leading to a global increase in 5hmc levels. Also, a TNFα-p38-MAPK signaling axis was implicated in this study, which was found to be responsible for the retention of breast-tumor-initiating cell (BTIC) properties that have an adverse effect on prognostic and therapeutic programs employed for treating breast cancer (Wu et al. 2015b). These studies indicate that hypoxia induces epigenetic alterations by modifying the 5-hmc status majorly mediated by HIF-1α protein on a global scale as well as specifically in the HIF-1α regulatory regions. It is interesting to note that HIF-1α mediates the upregulation of TET proteins, but how these dioxygenases execute their function in an oxygen-deficient environment is a subject of an ironical enigma. In glioblastoma, the IDH^{R132H} mutations are quite frequent, which results in the accumulation of 2-hydroxyglutarate, a competitive inhibitor of α-ketoglutarate. Hence, due to a lack of either oxygen or a cofactor, TET activity is compromised.

Moreover, in glioblastoma, an overlap was observed between genes hypermethylated in hypoxia and IDH^{R132H} mutant subgroups, indicating that the loss of TET activity is the major cause of the loss of 5-hmc in hypoxia. In an experiment involving recombinant TET1 and TET2 enzymes, to assess the conversion of 5-mc to 5-hmc on double-stranded genomic DNA, it was observed that decreasing oxygen concentration impacted the oxidative capacity of TET proteins, reducing their activity by 45%±7–52%±8. This study further strengthens the argument about whether the upregulation of TET is a real indication of its involvement in 5-hmc regulation in hypoxia. On an opposing note, the knockdown of HIF-1α increased TET2 gene and protein expression in both WM9 human metastatic melanoma cells and T98G glioblastoma cells. Further, ascorbic acid supplementation of these two cell lines resulted in increased 5-hmc levels via TET-2 in the absence of HIF-1α (Fischer and Miles 2017). In addition to being a dioxygenase enzyme catalyzing the conversion of 5-methylcytosine (5-mc) to 5-hydroxymethyl cytosine (5-hmc), TET1 acts as a transcriptional coactivator of HIF-1α and HIF-2α in hypoxia-promoting EMT (epithelial to mesenchymal transition) (Tsai et al. 2014).

As observed in the case of DNA methylation events, DNA hydroxymethylation also involves high levels of heterogeneity. TET enzymes, just like DNMTs, are very dynamic in their function, from catalyzing the conversion of 5-mc to 5-hmc to acting as transcriptional coactivators. Further, HIF-1α and TET enzymes can be attractive therapeutic targets in the treatment of various cancers because of their combined involvement in promoting stemness and EMT phenotypes, which are roadblocks to an effective treatment strategy.

RNA Methylation in Tumor Hypoxia

RNA methylation is one of the most common chemical modifications present on mRNA which alters its stability, translocation, translation, and degradation patterns. In the recent past, N^6-methyladenosine (m^6A) modification has gained much attention after the advent of advanced sequencing technology (Zheng et al. 2020). It is majorly found at the beginning of the 3′-UTR region within the consensus sequence 5′-RRACU-3′, where R can be A or G and accounts for about 0.1%–0.4% of all adenosine residues present in mRNA (Zhang et al. 2016). Similar to DNA methylation, RNA m^6A modification also consists of "readers," "writers," and "eraser" proteins. METTL3 (methyltransferase-like 3), METTL14 (methyltransferase-like 14), and WTAP (Wilm's-tumor-associated protein) together form a methyltransferase complex that acts as a writer, whereas FTO (fat mass and obesity-associated protein) and ALKBH5 (alkylation repair homolog protein 5) are the demethylases that act as erasers. m^6A marks on RNA are read by a class of YT521-B homology (YTH)-domain-containing proteins, which also include YTHDF1–3, YTHDC1, and YTHDC2 (Pan et al. 2018). Several studies have reported the important role that the m^6A modification of mRNA play in regulating various cellular events in health and disease, including tumorigenesis and progression. The induction of the stemness program in the cancer cell milieu is an important feature of tumor sustenance and progression. A subpopulation of tumor cells, termed cancer stem cells (CSCs), is capable of self-renewal and differentiation similar to normal stem cells, which contributes to the growth and sustenance of the tumor population (Yu et al. 2012). The origin of cancer stem cells (CSCs) is still elusive, and there have been a number of probable theories that have been projected as possible contributors to the induction of stemness in a subpopulation of cancer cells (Sell 2010; Beck and Blanpain 2013; Ayob and Ramasamy 2018). These cancer stem cells display resistance to therapeutic treatment and have been tipped as one of the most coveted targets in the field of cancer therapeutics for the achievement of complete ablation of tumor growth (Ayob and Ramasamy 2018). Tumor hypoxia is one of the major inducers of stemness in cancer cells, which promotes stress survival, drug resistance, and metastasis (Zhang et al. 2016). A study conducted in breast cancer cells showed that HIF-1α induced the expression of ALKBH5 (AlkB homolog 5, RNA demethylase) demethylase, which causes the demethylation of *NANOG* mRNA, improving its stability. This leads to the increased expression of NANOG protein, which is a pluripotency factor required for the maintenance of breast cancer stem cell (BCSC) properties (Zhang et al. 2016). In a similar study performed in breast cancer cells to understand the role of m^6A-modified RNA in the maintenance of pluripotency factors in hypoxic cancer cells, it was observed that HIF-1α induction under hypoxia upregulated ZNF217 (zinc finger protein 217), which inhibits METTL3 methyltransferase, reducing the methylation marks on the mRNAs of NANOG and KLF4 (Kruppel-like factor 4), leading to their stabilization and increased protein production. These two results indicate the importance of the epigenetic regulation of mRNA similar to DNA, wherein m^6A modification, which is crucial for the maintenance of pluripotency in embryonic stem cells (ESCs), is hijacked by cancer cells under hypoxic stress to promote tumor progression (Fig. 11.4).

Noncoding RNAs

It is a well-established fact that a majority of the human genome does not code for protein coding transcripts. At more than 95%, the noncoding part of the human transcriptome was considered earlier to be "transcriptional noise." However, gigantic growth in studies dedicated to uncovering the function of ncRNAs in the past two decades has now cemented their role as one of the major players in gene regulation. We will cover two of the most important classes of ncRNAs in the subsequent sections: microRNAs and long noncoding RNAs.

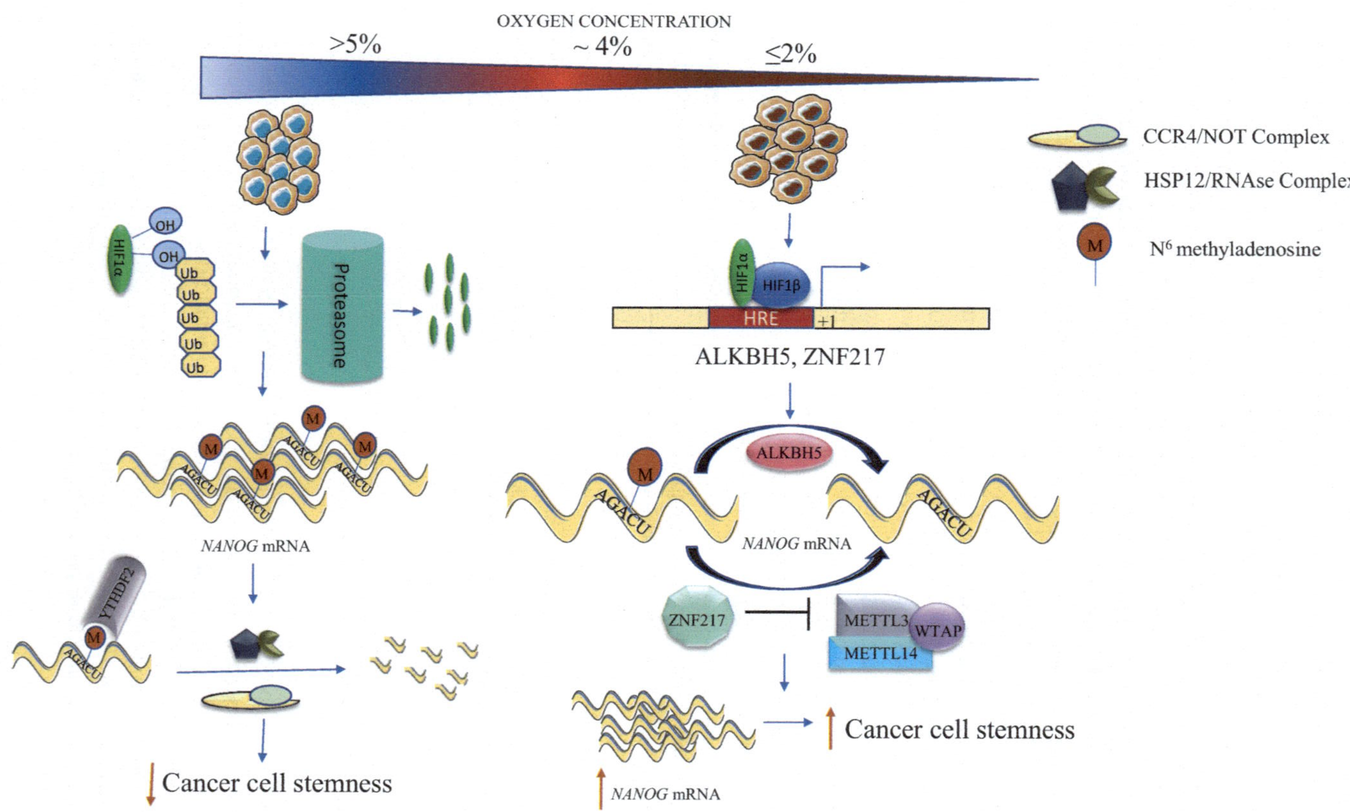

Fig. 11.4 The abolishment or inhibition of the RNA N^6 methyladenosine mark increases cancer stemness. Under normal conditions, m^6A mRNA is targeted for degradation by YTHFD2 m^6A reader protein via the HSP-12/RNAse complex or CCR4/NOT complex, among a few others. Under hypoxia, the N^6 methyladenosine mark present on the mRNA of the *NANOG* gene is abolished through a combined activity of ALKBH5, which is a methyladenosine eraser, and the inhibition of the METTL3 methyladenosine writer by ZNF217. This leads to the stabilization of *NANOG* mRNA, promoting stemness in cancer cells

Micro-RNAs and Tumor Hypoxia

MicroRNAs or miRNAs are small ncRNAs that are 20–24 nucleotides in length. They are transcribed as longer primary miRNAs and are further processed into precursor and eventually mature miRNAs. They regulate gene expression by binding to the 3′-UTR of various genes in cells. If the sequence (seed sequence) required to bind to target mRNAs shows perfect complementarity with mRNA, mRNA is degraded through cleavage mediated by a microprocessor complex. On the other hand, if the seed sequence does not show perfect complementarity, the translation of that transcript will be inhibited (Peng et al. 2020).

A large number of miRNAs show a change in expression upon hypoxic induction. They can be induced by HIF-mediated transactivation or through other factors that are induced under hypoxia. On the other hand, miRNAs are also known to affect the stability of HIF-1α. This function can be mediated through several other factors that are directly regulated by certain miRNAs. One report suggests that an increase in the expression of miR-543 decreases the destabilization of HIF-1α by its target PRMT9 (protein arginine methyltransferases) in osteosarcoma cells (Zhang et al. 2017). Another report has explored the role of miR-592, which is downregulated in hepatocellular carcinoma, due to which its target *WSB1* (WD repeat and SOCS box containing 1) gets stabilized. WSB1 protein, in turn, increases the stability of HIF-1α protein (Jia et al. 2016). miR-18a is another miRNA that negatively regulates HIF-1α by directly binding to its 3′-UTR under normoxia in various cancer types (Wu et al. 2015a). miR-183 has been known to inhibit IDH2 in glioma, which causes a decrease in α-ketoglutarate (α-KG) levels. Since α-KG is required by PHDs to hydroxylate HIF-1α, they are unable to function, thus leading to HIF-1α stabilization (Tanaka et al. 2013).

miR-210 is one of the most widely studied miRNAs that is upregulated under hypoxia through HIF-1-mediated transcriptional activation. Its targets are involved in several key cellular processes. One of its targets is mitochondrial iron-sulfur scaffold protein (ISCU). ISCU is a component of iron-sulfur clusters that act as a cofactor for enzymes involved in the TCA (tri-carboxylic acid) cycle and oxidative phosphorylation. The downregulation of ISCU leads to decreased oxidative phosphorylation and increased glycolysis (the Warburg effect) in hypoxic cancer cell lines (Chan et al. 2009; Favaro et al. 2010). miR-210 targets *EFNA3* and *PTP1B*, which are inhibitors of VEGF, thus enhancing angiogenesis (Fasanaro et al. 2008; Kim et al. 2013). It also regulates apoptosis by downregulating several proapoptotic proteins, such as AIFM3, CASP8AP2, and SIN3A (Dang and Myers 2015). Furthermore, miR-210 promotes metastasis by downregulating tumor suppressor vacuole-membrane protein 1 (VMP1) in many cancer types (Ying et al. 2011).

miRNAs can also be upregulated in a HIF-independent mechanism under hypoxia. For instance, the activation of Akt signaling in hypoxic cells leads to the upregulation of miR-21 via NF-κB (nuclear factor kappa B subunit 1) and CREB (Polytarchou et al. 2011). The HIF-regulated prometastatic transcription factor Twist leads to an increase in oncogenic miR-10b in breast cancer cells (Haque et al. 2011).

Some recent publications have looked into the regulation of the miRNA biogenesis pathway in tumor hypoxia on a global scale. The miRNA biogenesis pathway is, in general, downregulated under hypoxia (Choudhry and Harris 2018b). Shen and colleagues have discovered that an increase in EGFR signaling under hypoxia leads to the phosphorylation of AGO2 protein at Tyr 393. This disrupts the interaction of DICER1 with AGO2, thereby downregulating the miRNA maturation process. Moreover, the hypoxia-mediated phosphorylation of AGO2 at Tyr 393 is associated with poor clinical outcomes in breast cancer patients (Shen et al. 2013). The factors involved in miRNA biogenesis, like DICER, Drosha, and DGCR8 (DiGeorge syndrome critical region gene 8), are transcriptionally downregulated under hypoxia (Bandara et al. 2014). miR-630 targets the 3′-UTR of *DICER* under hypoxia and has also shown promise as a novel therapeutic target in combination with anti-VEGF therapy in ovarian cancer (Rupaimoole et al. 2016). One study has explored the decrease in DICER in hypoxic breast cancer cells via epigenetic mechanisms. The decrease in the activity of oxygen-dependent H3K27me3 demethylases KDM6A/B leads to the promoter of *DICER* getting hypermethylated, therefore causing a decrease in its expression (Van Den Beucken et al. 2014).

Long Noncoding RNAs and Tumor Hypoxia

Long noncoding RNAs (lncRNAs) are another class of ncRNAs that are known to play a regulatory role in tumor hypoxia. Belonging to a highly heterogeneous group, lncRNAs are more than 200 base pairs in length and are known to affect various processes inside the cell, including transcription, mRNA splicing, RNA interference, genomic imprinting, and translational control. The expression of lncRNAs in cancer cells undergoing hypoxia is often dictated by the presence of HRE sequence. The tumor suppressor *lncRNA-LET* (lncRNA-low expression in tumor) is one of the lncRNAs that are known to get repressed under hypoxia in hepatocellular carcinoma. The low expression of *lncRNA-LET* attributed to HDAC3 induction was shown to rescue the degradation of nuclear factor 9 protein (NF9), which enhances hypoxic response by stabilizing *HIF1A* mRNA (Yang et al. 2013).

The wide variety of mechanisms employed by lncRNAs to regulate cellular responses is well supported by a wealth of literature. Similarly, lncRNAs are known to function via several mechanisms in hypoxia also. lncRNAs function in association with transcription factors to enhance gene expression. *lncHIFCAR* (long noncoding HIF-1α coactivating RNA) interacts directly with HIF-1α and acts as a coactivator to enhance HIF-1α and p300 binding at HIF targets (Shih et al. 2017). Several lncRNAs have been reported to affect the stability of the *HIF1A* transcript. It was demonstrated that during prolonged hypoxic treatment, HIF-1α levels decreased, whereas HIF-2α remained upregulated because of a negative feedback regulatory mechanism involving antisense lncRNA *HIF1A-AS2*. The interaction of *HIF1A-AS2* with *HIF1A* exposes AU-rich elements in its 3′-UTR, which are absent in *HIF2A* mRNA, decreasing the stability of mRNA, leading to faster degradation (Uchida et al. 2004).

Another well-characterized mechanism by which lncRNAs can function is by acting as "RNA sponges" or competing endogenous RNAs (ceRNAs). ceRNAs competitively bind miRNAs and sequester them, leading to the attenuated repression of their mRNA targets. Let-7 miRNA, which targets *HIF1A* mRNA, has shown to be sequestered by hypoxia-inducible *H19* lncRNA in breast cancer cells (Peng et al. 2018). Similarly, HIF-1α target *GAPLINC* (gastric-adenocarcinoma-associated, positive CD44 regulator, long intergenic noncoding RNA) absorbs miR-211–3p, rescuing the expression of its target *CD44* and thus contributing to increased tumorigenic potential in gastric cancer (Hu et al. 2014). lncRNAs can also regulate protein stability and function in multiple and complicated processes. One such way is by blocking the interaction between two proteins. Yang et al. have demonstrated that *lincRNA-p21* manages HIF-1α protein stabilization by blocking its interaction with VHL, thereby preventing its ubiquitination and degradation (Yang et al. 2014).

Concluding Remarks

Epigenetics plays an essential and indispensable role in achieving enhanced tumorigenic properties in cancer cells that undergo hypoxia. Due to their broad specificity, changes in the expression of epigenetic regulators can effectively bring about global changes in gene expression in response to hypoxia. On the other hand, these factors can function in a context-dependent manner, as seen in the case of many histone demethylases. The importance of epigenetic factors in mediating hypoxia-driven cellular response is emphasized by the fact that some of them are directly dependent on intracellular oxygen levels for their function. The activity of enzymes like TET proteins and JmjC histone demethylases is inhibited under hypoxia. The consequences of this decrease in activity are rapid and can potentially be observed even before the stabilization of HIF-1α, which constitutes a primary response to hypoxia.

Epigenetic factors mediating cellular response to hypoxia are considered great targets for the development of novel anticancer therapeutics. There have also been reports that talk about existing anticancer drugs targeting new epigenetic factors. HDACs are potential targets for therapy because of their role in increasing HIF-1α stability. Nakazawa et al. have shown that treating soft tissue sarcomas with HDAC inhibitor Vorinostat causes a decrease in HIF-1α levels, leading to suppressed tumor growth in vivo (Nakazawa et al. 2016). Another HDAC inhibitor, NVP-LAQ824, when given in combination with an anti-VEGF drug, was shown to decrease angiogenesis and increase the expression of tumor suppressor *CDKN1A* (which codes for p21) gene (Qian et al. 2004). Further, histone acetyltransferase inhibitors have also been explored as targets for therapeutics. CTK7A, an inhibitor of p300, was shown to induce apoptosis in hypoxic gastric cancer cells through a mechanism that involves the proapoptotic protein Noxa (Rath et al. 2017). Another p300 inhibitor, FM19G11, was shown to decrease hypoxia-associated stemness-related gene-like *Oct4* and *Sox2* in human embryonic stem cells (Moreno-Manzano et al. 2010).

Several chemical compounds that target miRNAs have also been discovered. A study by Rustum et al. has found that the treatment of clear cell renal cell carcinoma with selenium-based compounds like methylselenocysteine can downregulate miR-155 and miR-210, along with VEGF (Rustum et al. 2018). Also, a synthetic derivative of curcumin showed antitumor properties in pancreatic cancer by decreasing miR-210 and miR-21 expression (Bao et al. 2012). All of these examples are proof of the enormous potential of epigenetic modifiers as therapeutic targets. Such inhibitors often show greater efficacy when used in combination with antiangiogenic or anti-HIF drugs. Further mechanistic insights into the functioning of hypoxia-induced epigenetic factors will pave the way for improved and safer therapeutic options.

Acknowledgment We thank DBT/Wellcome Trust India Alliance Fellowship Grant IA/I/16/2/502719 for the financial assistance. DP is supported by the University Grants Commission (UGC), Government of India fellowship. SAM is supported by an institutional fellowship provided by IISER-Bhopal.

References

Abramowicz A, Widłak P, Pietrowska M (2019) Different types of cellular stress affect the proteome composition of small extracellular vesicles: a mini review. Proteomes 7(2):23. https://doi.org/10.3390/proteomes7020023

Amelio I, Mancini M, Petrova V et al (2018) p53 mutants cooperate with HIF-1 in transcriptional regulation of extracellular matrix components to promote tumor progression. Proc Natl Acad Sci U S A 115:10869–10878. https://doi.org/10.1073/pnas.1808314115

Ayob AZ, Ramasamy TS (2018) Cancer stem cells as key drivers of tumour progression. J Biomed Sci 25:20. https://doi.org/10.1186/s12929-018-0426-4

Bandara V, Michael MZ, Gleadle JM (2014) Hypoxia represses microRNA biogenesis proteins in breast cancer cells. BMC Cancer 14:1–18

Bannister AJ, Kouzarides T (2011) Regulation of chromatin by histone modifications. Cell Res 21: 381–395. https://doi.org/10.1038/cr.2011.22

Bao B, Ali S, Ahmad A et al (2012) Hypoxia-induced aggressiveness of pancreatic cancer cells is due to increased expression of VEGF, IL-6 and miR-21, which can be attenuated by CDF treatment. PLoS One 7:e50165

Batie M, Druker J, D'Ignazio L, Rocha S (2017) KDM2 family members are regulated by HIF-1 in hypoxia. Cell 6(1):8

Batie M, Del Peso L, Rocha S (2018) Hypoxia and chromatin: a focus on transcriptional repression mechanisms. Biomedicine 6:47

Batie M, Frost J, Frost M et al (2019) Hypoxia induces rapid changes to histone methylation and reprograms chromatin. Science (80-) 363:1222–1226

Beck B, Blanpain C (2013) Unravelling cancer stem cell potential. Nat Rev Cancer 13:727–738

Bertout JA, Patel SA, Simon MC (2008) The impact of O2 availability on human cancer. Nat Rev Cancer 8:967–975. https://doi.org/10.1038/nrc2540

Beyer S, Kristensen MM, Jensen KS et al (2008) The histone demethylases JMJD1A and JMJD2B are transcriptional targets of hypoxia-inducible factor HIF*. J Biol Chem 283:36542–36552. https://doi.org/10.1074/jbc.M804578200

Bindra RS, Schaffer PJ, Meng A et al (2004) Down-regulation of Rad51 and decreased homologous recombination in hypoxic cancer cells. Mol Cell Biol 24:8504LP–8518LP. https://doi.org/10.1128/MCB.24.19.8504-8518.2004

Büchler P, Reber HA, Büchler M et al (2003) Hypoxia-inducible factor 1 regulates vascular endothelial growth factor expression in human pancreatic cancer. Pancreas 26:56–64. https://doi.org/10.1097/00006676-200301000-00010

Büchler P, Reber HA, Tomlinson JS et al (2009) Transcriptional regulation of urokinase-type plasminogen activator receptor by hypoxia-inducible factor 1 is crucial for invasion of pancreatic and liver cancer. Neoplasia 11:196–206. https://doi.org/10.1593/neo.08734

Camuzi D, de Amorim ÍSS, Ribeiro Pinto LF et al (2019) Regulation is in the air: the relationship between hypoxia and epigenetics in cancer. Cell 8:300

Casciello F, Al-Ejeh F, Kelly G et al (2017) G9a drives hypoxia-mediated gene repression for breast cancer cell survival and tumorigenesis. Proc Natl Acad Sci 114:7077LP–7082LP. https://doi.org/10.1073/pnas.1618706114

Chakraborty AA, Laukka T, Myllykoski M et al (2019) Histone demethylase KDM6A directly senses oxygen to control chromatin and cell fate. Science (80-) 363:1217–1222

Chan SY, Zhang Y-Y, Hemann C et al (2009) MicroRNA-210 controls mitochondrial metabolism during hypoxia by repressing the iron-sulfur cluster assembly proteins ISCU1/2. Cell Metab 10: 273–284

Chang C-J, Yang J-Y, Xia W et al (2011) EZH2 promotes expansion of breast tumor initiating cells through activation of RAF1-β-catenin signaling. Cancer Cell 19:86–100. https://doi.org/10.1016/j.ccr.2010.10.035

Chen S, Yin C, Lao T et al (2015) AMPK-HDAC5 pathway facilitates nuclear accumulation of HIF-1α and functional activation of HIF-1 by deacetylating Hsp70 in the cytosol. Cell Cycle 14: 2520–2536. https://doi.org/10.1080/15384101.2015.1055426

Chen PS, Chiu WT, Hsu PL et al (2020) Pathophysiological implications of hypoxia in human diseases. J Biomed Sci 27:1–19. https://doi.org/10.1186/s12929-020-00658-7

Chopra A, Cho WC, Willmore WG, Biggar KK (2020) Hypoxia-inducible lysine methyltransferases: G9a and GLP hypoxic regulation, non-histone substrate modification, and pathological relevance. Front Genet 11:579636

Choudhry H, Harris AL (2018a) Review advances in hypoxia-inducible factor biology. Cell Metab 27:281–298. https://doi.org/10.1016/j.cmet.2017.10.005

Choudhry H, Harris AL (2018b) Advances in hypoxia-inducible factor biology. Cell Metab 27: 281–298. https://doi.org/10.1016/j.cmet.2017.10.005

Clapier CR, Cairns BR (2009) The biology of chromatin remodeling complexes. Annu Rev Biochem 78:273–304. https://doi.org/10.1146/annurev.biochem.77.062706.153223

Clapier CR, Iwasa J, Cairns BR, Peterson CL (2017) Mechanisms of action and regulation of ATP-dependent chromatin-remodelling complexes. Nat Publ Gr 18:407–422. https://doi.org/10.1038/nrm.2017.26

Cui X-G, Han Z-T, He S-H et al (2017) HIF1/2α mediates hypoxia-induced LDHA expression in human pancreatic cancer cells. Oncotarget 8:24840–24852. https://doi.org/10.18632/oncotarget.15266

Dang K, Myers KA (2015) The role of hypoxia-induced miR-210 in cancer progression. Int J Mol Sci 16:6353–6372

Economopoulou M, Langer HF, Celeste A et al (2009) Histone H2AX is integral to hypoxia-driven neovascularization. Nat Med 15:553–558. https://doi.org/10.1038/nm.1947

Fasanaro P, D'Alessandra Y, Di Stefano V et al (2008) MicroRNA-210 modulates endothelial cell response to hypoxia and inhibits the receptor tyrosine kinase ligand Ephrin-A3. J Biol Chem 283:15878–15883

Fauquier L, Azzag K, Parra MAM et al (2018) CBP and P300 regulate distinct gene networks required for human primary myoblast differentiation and muscle integrity. Sci Rep 8:12629. https://doi.org/10.1038/s41598-018-31102-4

Favaro E, Ramachandran A, McCormick R et al (2010) MicroRNA-210 regulates mitochondrial free radical response to hypoxia and Krebs cycle in cancer cells by targeting iron sulfur cluster protein ISCU. PLoS One 5:e10345

Feldser D, Agani F, Iyer NV et al (1999) Reciprocal positive regulation of hypoxia-inducible factor 1α and insulin-like growth factor 2. Cancer Res 59:3915LP–3918LP

Fischer AP, Miles SL (2017) Biochemical and biophysical research communications silencing HIF-1 a induces TET2 expression and augments ascorbic acid induced 5-hydroxymethylation of DNA in human metastatic melanoma cells. Biochem Biophys Res Commun 490:176–181. https://doi.org/10.1016/j.bbrc.2017.06.017

Folkman J, Hahnfeldt P, Hlatky L (2000) Cancer: looking outside the genome. Nat Rev Mol Cell Biol 1:76–79. https://doi.org/10.1038/35036100

Gao X, Hicks KC, Neumann P, Patel TB (2017) Hypoxia inducible factors regulate the transcription of the sprouty2 gene and expression of the sprouty2 protein. 1–22. PLoS One 12(2):e0171616. https://doi.org/10.1371/journal.pone.0171616

Grewal SIS, Jia S (2007) Heterochromatin revisited. Nat Rev Genet 8:35–46. https://doi.org/10.1038/nrg2008

Hait NC, Maiti A, Xu P et al (2020) Regulation of hypoxia-inducible factor functions in the nucleus by sphingosine-1-phosphate. FASEB J 34:4293–4310

Hammond EM, Dorie MJ, Giaccia AJ (2003) ATR/ATM targets are phosphorylated by ATR in response to hypoxia and ATM in response to Reoxygenation*. J Biol Chem 278:12207–12213. https://doi.org/10.1074/jbc.M212360200

Haque I, Banerjee S, Mehta S et al (2011) Cysteine-rich 61-connective tissue growth factor-nephroblastoma-overexpressed 5 (CCN5)/Wnt-1-induced signaling protein-2 (WISP-2) regulates microRNA-10b via hypoxia-inducible factor-1α-TWIST signaling networks in human breast cancer cells. J Biol Chem 286:43475–43485

Hashimoto T, Shibasaki F (2015) Hypoxia-inducible factor as an Angiogenic master switch. Front Pediatr 3:1–15. https://doi.org/10.3389/fped.2015.00033

He X, Wang J, Wei W et al (2016) Hypoxia regulates ABCG2 activity through the activation of ERK1/2/HIF-1α and contributes to chemoresistance in pancreatic cancer cells. Cancer Biol Ther 17:188–198. https://doi.org/10.1080/15384047.2016.1139228

He X, Wang JJ, Wei W et al (2017) TWIST activation by hypoxia inducible factor-1 (HIF-1): implications in metastasis and development. Oncotarget 8:2090–2096. https://doi.org/10.1016/j.cmet.2006.02.002

Hoffmann C, Mao X, Brown-Clay J et al (2018) Hypoxia promotes breast cancer cell invasion through HIF-1α-mediated up-regulation of the invadopodial actin bundling protein CSRP2. Sci Rep 8:10191. https://doi.org/10.1038/s41598-018-28637-x

Hu Y, Wang J, Qian J et al (2014) Long noncoding RNA GAPLINC regulates CD44-dependent cell invasiveness and associates with poor prognosis of gastric cancer. Cancer Res 74:6890–6902

Jia Y-Y, Zhao J-Y, Li B-L et al (2016) miR-592/WSB1/HIF-1α axis inhibits glycolytic metabolism to decrease hepatocellular carcinoma growth. Oncotarget 7:35257

Johnson AB, Denko N, Barton MC (2008) Hypoxia induces a novel signature of chromatin modifications and global repression of transcription. Mutat Res 640:174–179. https://doi.org/10.1016/j.mrfmmm.2008.01.001

Kim I, Park JW (2020) Hypoxia-driven epigenetic regulation in cancer progression: a focus on histone methylation and its modifying enzymes. Cancer Lett 489:41–49

Kim MS, Kwon HJ, Lee YM et al (2001) Histone deacetylases induce angiogenesis by negative regulation of tumor suppressor genes. Nat Med 7:437–443. https://doi.org/10.1038/86507

Kim J, Tchernyshyov I, Semenza GL, Dang CV (2006) HIF-1-mediated expression of pyruvate dehydrogenase kinase: a metabolic switch required for cellular adaptation to hypoxia. Cell Metab 3:177–185. https://doi.org/10.1016/j.cmet.2006.02.002

Kim JH, Park SG, Song SY et al (2013) Reactive oxygen species-responsive miR-210 regulates proliferation and migration of adipose-derived stem cells via PTPN2. Cell Death Dis 4:e588–e588

Krieg AJ, Rankin EB, Chan D et al (2010) Regulation of the histone demethylase JMJD1A by hypoxia-inducible factor 1α enhances hypoxic gene expression and tumor growth. Mol Cell Biol 30:344LP–353LP. https://doi.org/10.1128/MCB.00444-09

Krock BL, Skuli N, Simon MC (2011) Hypoxia-induced angiogenesis: good and evil. Genes Cancer 2(12):1117–1133. https://doi.org/10.1177/1947601911423654

Längst G, Manelyte L (2015) Chromatin remodelers: from function to dysfunction. Genes (Basel) 6(2):299–324. https://doi.org/10.3390/genes6020299

Lee JS, Kim Y, Kim IS et al (2010) Negative regulation of hypoxic responses via induced Reptin methylation. Mol Cell 39:71–85. https://doi.org/10.1016/j.molcel.2010.06.008

Lee JS, Kim Y, Bhin J et al (2011) Hypoxia-induced methylation of a pontin chromatin remodeling factor. Proc Natl Acad Sci U S A 108:13510–13515. https://doi.org/10.1073/pnas.1106106108

Lee H-Y, Yang EG, Park H (2013) Hypoxia enhances the expression of prostate-specific antigen by modifying the quantity and catalytic activity of Jumonji C domain-containing histone demethylases. Carcinogenesis 34:2706–2715. https://doi.org/10.1093/carcin/bgt256

Li T, Mao C, Wang X et al (2020a) Epigenetic crosstalk between hypoxia and tumor driven by HIF regulation. J Exp Clin Cancer Res 39:224. https://doi.org/10.1186/s13046-020-01733-5

Li Y, Gruber JJ, Litzenburger UM et al (2020b) Acetate supplementation restores chromatin accessibility and promotes tumor cell differentiation under hypoxia. Cell Death Dis 11:102. https://doi.org/10.1038/s41419-020-2303-9

Lin G, Sun W, Yang Z et al (2017) Hypoxia induces the expression of TET enzymes in HepG2 cells. Oncol Lett 14:6457–6462. https://doi.org/10.3892/ol.2017.7063

Liu Q, Liu L, Zhao Y et al (2011) Hypoxia induces genomic DNA demethylation through the activation of HIF-1 a and transcriptional upregulation of MAT2A in hepatoma cells. Mol Cancer Ther 10(6):1113–1124. https://doi.org/10.1158/1535-7163.MCT-10-1010

Luo W, Chang R, Zhong J et al (2012) Histone demethylase JMJD2C is a coactivator for hypoxia-inducible factor 1 that is required for breast cancer progression. Proc Natl Acad Sci U S A 109(49):E3367–E3376. https://doi.org/10.1073/pnas.1217394109

Mah L-J, El-Osta A, Karagiannis TC (2010) γH2AX: a sensitive molecular marker of DNA damage and repair. Leukemia 24:679–686. https://doi.org/10.1038/leu.2010.6

Mahmood N, Rabbani SA (2019) DNA methylation readers and cancer: mechanistic and therapeutic applications. Front Oncol 9:489. https://doi.org/10.3389/fonc.2019.00489

Mamo M, Ye IC, Digiacomo JW et al (2020) Hypoxia alters the response to anti-EGFR therapy by regulating EGFR expression and downstream signaling in a DNA methylation-specific and HIF-dependent manner. Cancer Res 80(22):4998–5010. https://doi.org/10.1158/0008-5472.CAN-20-1232

Mariani CJ, Vasanthakumar A, Madzo J et al (2014) Report TET1-mediated Hydroxymethylation facilitates hypoxic gene induction in neuroblastoma. Cell Rep 7(5):1343–1352. https://doi.org/10.1016/j.celrep.2014.04.040

Mathieu J, Zhang Z, Zhou W et al (2011) HIF induces human embryonic stem cell markers in cancer cells. Cancer Res 71:4640–4652. https://doi.org/10.1158/0008-5472.CAN-10-3320

Melvin A, Mudie S, Rocha S, Tansey WP (2011) The chromatin remodeler ISWI regulates the cellular response to hypoxia: role of FIH. Mol Biol Cell 22:4171–4181. https://doi.org/10.1091/mbc.E11-02-0163

Mitamura T, Pradeep S, McGuire M et al (2018) Induction of anti-VEGF therapy resistance by upregulated expression of microseminoprotein (MSMP). Oncogene 37:722–731. https://doi.org/10.1038/onc.2017.348

Moore LD, Le T, Fan G (2012) DNA methylation and its basic function. Neuropsychopharmacology 38:23–38. https://doi.org/10.1038/npp.2012.112

Moreno-Manzano V, Rodríguez-Jiménez FJ, Aceña-Bonilla JL et al (2010) FM19G11, a new hypoxia-inducible factor (HIF) modulator, affects stem cell differentiation status. J Biol Chem 285:1333–1342. https://doi.org/10.1074/jbc.M109.008326

Nagao A, Kobayashi M, Koyasu S, Chow CCT (2019) HIF-1-dependent reprogramming of glucose metabolic pathway of cancer cells and its therapeutic significance. Int J Mol Sci 20(2):238. https://doi.org/10.3390/ijms20020238

Nair SS, Kumar R (2012) Chromatin remodeling in cancer: a gateway to regulate gene transcription. Mol Oncol 6(6):611–619

Nakazawa MS, Eisinger-Mathason TSK, Sadri N et al (2016) Epigenetic re-expression of HIF-2α suppresses soft tissue sarcoma growth. Nat Commun 7:10539. https://doi.org/10.1038/ncomms10539

Nguyen MP, Lee S, Lee YM (2013) Epigenetic regulation of hypoxia inducible factor in diseases and therapeutics. Arch Pharm Res 36:252–263. https://doi.org/10.1007/s12272-013-0058-x

Niu X, Zhang T, Liao L et al (2012) The von Hippel–Lindau tumor suppressor protein regulates gene expression and tumor growth through histone demethylase JARID1C. Oncogene 31:776–786

Okazaki K, Maltepe E (2006) Oxygen, epigenetics and stem cell fate. Regen Med 1:71–83. https://doi.org/10.2217/17460751.1.1.71

Pan Y, Ma P, Liu Y et al (2018) Multiple functions of m 6 A RNA methylation in cancer. J Hematol Oncol 11(1):48

Peng F, Wang J-H, Fan W-J et al (2018) Glycolysis gatekeeper PDK1 reprograms breast cancer stem cells under hypoxia. Oncogene 37:1062–1074

Peng X, Gao H, Xu R et al (2020) The interplay between HIF-1α and noncoding RNAs in cancer. J Exp Clin Cancer Res 39:27. https://doi.org/10.1186/s13046-020-1535-y

Petrova V, Annicchiarico-Petruzzelli M, Melino G, Amelio I (2018) The hypoxic tumour microenvironment. Oncogenesis 7:10. https://doi.org/10.1038/s41389-017-0011-9

Pfister NT, Fomin V, Regunath K et al (2015) Mutant p53 cooperates with the SWI/SNF chromatin remodeling complex to regulate VEGFR2 in breast cancer cells. Genes Dev 29(12):1298–1315. https://doi.org/10.1101/gad.263202.115

Pollard PJ, Loenarz C, Mole DR et al (2008) Regulation of Jumonji-domain-containing histone demethylases by hypoxia-inducible factor (HIF)-1α. Biochem J 416:387–394

Polytarchou C, Iliopoulos D, Hatziapostolou M et al (2011) Akt2 regulates all Akt isoforms and promotes resistance to hypoxia through induction of miR-21 upon oxygen deprivation. Cancer Res 71:4720–4731

Qian DZ, Wang X, Kachhap SK et al (2004) The histone deacetylase inhibitor NVP-LAQ824 inhibits angiogenesis and has a greater antitumor effect in combination with the vascular endothelial growth factor receptor tyrosine kinase inhibitor PTK787/ZK222584. Cancer Res 64:6626LP–6634LP. https://doi.org/10.1158/0008-5472.CAN-04-0540

Rasmussen KD, Helin K (2016) Role of TET enzymes in DNA methylation, development, and cancer. Genes Dev 30:733–750. https://doi.org/10.1101/gad.276568.115

Rath S, Das L, Kokate SB et al (2017) Inhibition of histone/lysine acetyltransferase activity kills CoCl2-treated and hypoxia-exposed gastric cancer cells and reduces their invasiveness. Int J Biochem Cell Biol 82:28–40. https://doi.org/10.1016/j.biocel.2016.11.014

Rehman Z, Fahim A, Bhatti A et al (2019) Co-expression of HIF-1α, MDR1 and LAPTM4B in peripheral blood of solid tumors. PeerJ 7:e6309–e6309. https://doi.org/10.7717/peerj.6309

Rezaeian A-H, Li C-F, Wu C-Y et al (2017) A hypoxia-responsive TRAF6–ATM–H2AX signalling axis promotes HIF1α activation, tumorigenesis and metastasis. Nat Cell Biol 19:38–51. https://doi.org/10.1038/ncb3445

Ross JP (2019) DNA methylation cancer biomarkers: translation to the clinic. Front Genet 10:1150. https://doi.org/10.3389/fgene.2019.01150

Rupaimoole R, Ivan C, Yang D et al (2016) Hypoxia-upregulated microRNA-630 targets Dicer, leading to increased tumor progression. Oncogene 35:4312–4320

Rustum YM, Chintala S, Durrani FA, Bhattacharya A (2018) Non-coding micro RNAs and hypoxia-inducible factors are selenium targets for development of a mechanism-based combination strategy in clear-cell renal cell carcinoma—bench-to-bedside therapy. Int J Mol Sci 19:3378

Saha A, Wittmeyer J, Cairns BR (2006) Chromatin remodelling: the industrial revolution of DNA around histones. Nat Rev Mol Cell Biol 7:437–447. https://doi.org/10.1038/nrm1945
Schito L, Rey S, Tafani M et al (2012) Hypoxia-inducible factor 1-dependent expression of platelet-derived growth factor B promotes lymphatic metastasis of hypoxic breast cancer cells. Proc Natl Acad Sci U S A 109:E2707–E2716. https://doi.org/10.1073/pnas.1214019109
Sell S (2010) On the stem cell origin of cancer. Am J Pathol 176:2494–2584. https://doi.org/10.2353/ajpath.2010.091064
Semenza GL (2012) Review hypoxia-inducible factors in physiology and medicine. Cell 148:399–408. https://doi.org/10.1016/j.cell.2012.01.021
Semenza GL (2014) Oxygen sensing, hypoxia-inducible factors, and disease Pathophysiology. Annu Rev Pathol 9:47–71. https://doi.org/10.1146/annurev-pathol-012513-104720
Sena JA, Wang L, Hu C (2013) BRG1 and BRM chromatin-remodeling complexes regulate the hypoxia response by acting as coactivators for a subset of hypoxia-inducible transcription factor target genes. Mol Cell Biol 33:3849–3863. https://doi.org/10.1128/MCB.00731-13
Shen J, Xia W, Khotskaya YB et al (2013) EGFR modulates microRNA maturation in response to hypoxia through phosphorylation of AGO2. Nature 497:383–387
Shen L, Song CX, He C, Zhang Y (2014) Mechanism and function of oxidative reversal of DNA and RNA methylation. Annu Rev Biochem 83:585–614. https://doi.org/10.1146/annurev-biochem-060713-035513
Shih J-W, Chiang W-F, Wu ATH et al (2017) Long noncoding RNA LncHIFCAR/MIR31HG is a HIF-1α co-activator driving oral cancer progression. Nat Commun 8:1–16
Shinkai Y, Tachibana M (2011) H3K9 methyltransferase G9a and the related molecule GLP. Genes Dev 25:781–788
Simon M-P, Tournaire R, Pouyssegur J (2008) The angiopoietin-2 gene of endothelial cells is up-regulated in hypoxia by a HIF binding site located in its first intron and by the central factors GATA-2 and Ets-1. J Cell Physiol 217:809–818. https://doi.org/10.1002/jcp.21558
Skowronski K, Dubey S, Rodenhiser DI et al (2010) Ischemia dysregulates DNA methyltransferases in human colorectal cancer cells. Epigenetics 5(6):547–556. https://doi.org/10.4161/epi.5.6.12400
Storci G, Bertoni S, De Carolis S et al (2013) Slug/β-catenin–dependent Proinflammatory phenotype in hypoxic breast cancer stem cells. Am J Pathol 183:1688–1697. https://doi.org/10.1016/j.ajpath.2013.07.020
Strahl BD, Allis CD (2000) The language of covalent histone modifications. Nature 403:41–45. https://doi.org/10.1038/47412
Tam SY, Wu VWC, Law HKW (2020) Hypoxia-induced epithelial-mesenchymal transition in cancers: HIF-1α and beyond. Front Oncol 10:1–11. https://doi.org/10.3389/fonc.2020.00486
Tanaka H, Sasayama T, Tanaka K et al (2013) MicroRNA-183 upregulates HIF-1α by targeting isocitrate dehydrogenase 2 (IDH2) in glioma cells. J Neuro-Oncol 111:273–283
Thienpont B, Steinbacher J, Zhao H et al (2016a) Tumour hypoxia causes DNA hypermethylation by reducing TET activity. Nature 537:63–68. https://doi.org/10.1038/nature19081
Thienpont B, Steinbacher J, Zhao H et al (2016b) Tumour hypoxia causes DNA hyper - methylation by reducing TET activity. Nature 537(7618):63–68. https://doi.org/10.1038/nature19081
Tong D, Liu Q, Liu G et al (2016) The HIF/PHF8/AR axis promotes prostate cancer progression. Oncogenesis 5:e283–e283. https://doi.org/10.1038/oncsis.2016.74
Tsai YP, Wu KJ (2014) Epigenetic regulation of hypoxia-responsive gene expression: focusing on chromatin and DNA modifications. Int J Cancer 134:249–256. https://doi.org/10.1002/ijc.28190
Tsai Y, Chen H, Chen S et al (2014) TET1 regulates hypoxia-induced epithelial-mesenchymal transition by acting as a coactivator. Genome Biol 15:513. https://doi.org/10.1186/s13059-014-0513-0
Tyagi M, Imam N, Verma K, Patel AK (2016) Chromatin remodelers: we are the drivers !! Nucleus 7:388–404. https://doi.org/10.1080/19491034.2016.1211217

Uchida T, Rossignol F, Matthay MA et al (2004) Prolonged hypoxia differentially regulates hypoxia-inducible factor (HIF)-1α and HIF-2α expression in lung epithelial cells: IMPLICATION OF NATURAL ANTISENSE HIF-1α*. J Biol Chem 279:14871–14878. https://doi.org/10.1074/jbc.M400461200

Van Den Beucken T, Koch E, Chu K et al (2014) Hypoxia promotes stem cell phenotypes and poor prognosis through epigenetic regulation of DICER. Nat Commun 5:1–13

Wang H-J, Hsieh Y-J, Cheng W-C et al (2014) JMJD5 regulates PKM2 nuclear translocation and reprograms HIF-1α–mediated glucose metabolism. Proc Natl Acad Sci 111:279LP–284LP. https://doi.org/10.1073/pnas.1311249111

Wu M-Z, Tsai Y-P, Yang M-H et al (2011) Interplay between HDAC3 and WDR5 is essential for hypoxia-induced epithelial-mesenchymal transition. Mol Cell 43:811–822. https://doi.org/10.1016/j.molcel.2011.07.012

Wu F, Huang W, Wang X (2015a) microRNA-18a regulates gastric carcinoma cell apoptosis and invasion by suppressing hypoxia-inducible factor-1α expression. Exp Ther Med 10:717–722

Wu MZ, Chen SF, Nieh S et al (2015b) Hypoxia drives breast tumor malignancy through a TET-TNFα-p38-MAPK signaling axis. Cancer Res 75:3912–3924

Xia X, Lemieux ME, Li W et al (2009) Integrative analysis of HIF binding and transactivation reveals its role in maintaining histone methylation homeostasis. Proc Natl Acad Sci U S A 106(11):4260–4265

Yamane K, Tateishi K, Klose RJ et al (2007) PLU-1 is an H3K4 demethylase involved in transcriptional repression and breast cancer cell proliferation. Mol Cell 25:801–812

Yang F, Huo X, Yuan S et al (2013) Repression of the long noncoding RNA-LET by histone deacetylase 3 contributes to hypoxia-mediated metastasis. Mol Cell 49:1083–1096. https://doi.org/10.1016/j.molcel.2013.01.010

Yang F, Zhang H, Mei Y, Wu M (2014) Reciprocal regulation of HIF-1α and lincRNA-p21 modulates the Warburg effect. Mol Cell 53:88–100

Yang S-W, Zhang Z-G, Hao Y-X et al (2017) HIF-1α induces the epithelial-mesenchymal transition in gastric cancer stem cells through the snail pathway. Oncotarget 8:9535–9545. https://doi.org/10.18632/oncotarget.14484

Ying Q, Liang L, Guo W et al (2011) Hypoxia-inducible microRNA-210 augments the metastatic potential of tumor cells by targeting vacuole membrane protein 1 in hepatocellular carcinoma. Hepatology 54:2064–2075

Yu Z, Pestell TG, Lisanti MP, Pestell RG (2012) Cancer stem cells. Int J Biochem Cell Biol 44:2144–2151. https://doi.org/10.1016/j.biocel.2012.08.022

Zhang C, Samanta D, Lu H et al (2016) Hypoxia induces the breast cancer stem cell phenotype by HIF-dependent and ALKBH5-mediated m6A-demethylation of NANOG mRNA. Proc Natl Acad Sci U S A 113:E2047–E2056. https://doi.org/10.1073/pnas.1602883113

Zhang H, Guo X, Feng X et al (2017) MiRNA-543 promotes osteosarcoma cell proliferation and glycolysis by partially suppressing PRMT9 and stabilizing HIF-1α protein. Oncotarget 8:2342

Zheng Y, Ling Y, Zhang DY et al (2020) Regulating tumor N6-Methyladenosine methylation landscape using hypoxia-modulating OsSx nanoparticles. Small 2005086:1–12. https://doi.org/10.1002/smll.202005086

Zhu H, Wang G, Qian J (2016) Transcription factors as readers and effectors of DNA methylation. Nat Rev Genet 17(9):551–565. https://doi.org/10.1038/nrg.2016.83

Part III
Epigenetic Regulation in Cancer

Chapter 12
Metabolic Regulation of Lysine Acetylation: Implications in Cancer

Siddharth Singh, Parijat Senapati, and Tapas K. Kundu

Abstract Lysine acetylation is the second most well-studied post-translational modification after phosphorylation. While phosphorylation regulates signaling cascades, one of the most significant roles of acetylation is regulation of chromatin structure. Acetyl-coenzyme A (acetyl-CoA) serves as the acetyl group donor for acetylation reactions mediated by lysine acetyltransferases (KATs). On the other hand, NAD+ serves as the cofactor for lysine deacetylases (KDACs). Both acetyl-CoA and NAD+ are metabolites integral to energy metabolism, and therefore, their metabolic flux can regulate the activity of KATs and KDACs impacting the epigenome. In this chapter, we review our current understanding of how metabolic pathways regulate lysine acetylation in normal and cancer cells.

Keywords Acetylation · Acetyl-CoA · NAD+ · Metabolism · Epigenome · Metabolic reprogramming · Oral cancer

---

Siddharth Singh and Parijat Senapati contributed equally to this article.

S. Singh
Transcription and Disease Laboratory, Molecular Biology and Genetics Unit, Jawaharlal Nehru Centre for Advanced Scientific Research, Bangalore, Karnataka, India

P. Senapati
Transcription and Disease Laboratory, Molecular Biology and Genetics Unit, Jawaharlal Nehru Centre for Advanced Scientific Research, Bangalore, Karnataka, India

Present Address: Beckman Research Institute, City of Hope, Duarte, CA, USA

T. K. Kundu (✉)
Transcription and Disease Laboratory, Molecular Biology and Genetics Unit, Jawaharlal Nehru Centre for Advanced Scientific Research, Bangalore, Karnataka, India

Division of Cancer Biology, CSIR-Central Drug Research Institute, Lucknow, Uttar Pradesh, India
e-mail: tapas@jncasr.ac.in

T. K. Kundu, C. Das (eds.), *Metabolism and Epigenetic Regulation: Implications in Cancer*, Subcellular Biochemistry 100,
https://doi.org/10.1007/978-3-031-07634-3_12

Introduction

In eukaryotic cells, access to the genetic information contained within DNA is regulated by nucleosomes, the smallest unit of chromatin. A nucleosome consists of DNA wrapped around an octamer of histones which are positively charged proteins with globular domains and flexible N-terminal tails. The N-terminal tails protrude from the nucleosome structure and are subject to multiple post-translational modifications, additions of chemical moieties that alter the structure, and compactness of chromatin. Post-translational modifications can either modify the chemical properties of the amino acids or be recognized by binding proteins that recruit additional effectors, thereby altering chromatin structure. Acetylation is one such post-translational modification that is added to the lysine residues in histone tails. Although other PTMs such as methylation are known to help in the loosening as well as compaction of chromatin, acetylation has been unequivocally shown to help in opening up of chromatin and, as such, exists only in euchromatin where genes are actively transcribed. Apart from histones, many nonhistone proteins are also known to be acetylated; several of them are chromatin- and transcription-associated proteins, and acetylation is known to regulate their function. Thus, histone and nonhistone acetylation together contribute to the epigenetic state of the cell. The acetyl group for lysine acetylation is added by enzymes known as lysine acetyltransferases and removed by lysine deacetylases. Both these enzymes require central metabolites, namely acetyl-coenzyme A (acetyl-CoA) and NAD+, as cofactors for their activity. Thus, metabolic pathways that produce or utilize these central metabolites can impact or regulate the epigenome by altering the levels of these metabolites. This allows the cell to coordinate gene expression programs with extrinsic factors such as nutrient availability. Moreover, the discovery that acetyl-CoA-producing enzymes can moonlight to the nucleus to regulate specific genes by increasing the local acetyl-CoA levels adds an additional layer of complexity to the relationship between metabolism and epigenome. In this chapter, we review our understanding of how metabolic pathways regulate protein acetylation. We focus on acetylation of histones and nonhistone proteins with transcriptional consequences, which thereby regulate the epigenetic state of the cell. We also discuss physiological conditions that lead to the altered metabolism of cofactors and their impact on the epigenome, with special emphasis on cancer. We further provide a summary of our current understanding of metabolic and epigenetic alterations in oral cancer.

Introduction to Lysine Acetylation: A Key Epigenetic Modification to Regulate Chromatin Structure and Gene Regulation

Reversible and Irreversible Lysine Acetylation

Acetylation is one of the most well-studied post-translational modifications, second only to phosphorylation. Acetylation is a process in which the acetyl group is transferred from the acetyl donor acetyl-coenzyme A (Ac-CoA) to a specific site on a polypeptide chain. Acetylation of proteins is carried out in the cell in two distinct ways. Most proteins get acetylated at the N^{α}-terminus of the nascent polypeptide chain co-translationally or post-translationally. In contrast, the other form of protein acetylation occurs post-translationally on the ε-amino group of a lysine residue. N^{α}-acetylation reactions are catalyzed by Nt-acetyltransferases (NATs) and are considered to be irreversible. On the contrary, lysine acetylation reactions are reversible and well-regulated. About 85% of human proteins are N^{α}-acetylated, and this is important for their localization, stability, and function (Kinnaird et al. 2016). N^{ε}-acetylation can lead to altered interaction with other proteins, subcellular localization, stability, and protein function such as enzyme activity.

Writers, Erasers, and Readers of Lysine Acetylation

Vincent Allfrey and group first discovered lysine acetylation on histones in 1964; the enzymes catalyzing lysine acetylation were therefore termed histone acetyltransferases (HATs). Several nonhistone substrates have been discovered since then, and the enzymes have been renamed as lysine (K) acetyltransferases (KATs). These enzymes are termed "writers," and about 37 different KATs have been identified in mammals (Sheikh and Akhtar 2019) that are broadly classified as nuclear A-type KATs and cytoplasmic B-type KATs. A-type KATs are grouped into three prominent families: GNAT (GCN5-related N-acetyltransferases), MYST (MOZ, YBF2, SAS2, and TIP60), and p300/CBP (CREB-binding protein) that share a conserved acetyl-CoA binding site (Berndsen and Denu 2008). N^{ε}-acetylation can also occur nonenzymatically in the cell, usually in alkaline compartments such as the mitochondrial matrix. Lysine deacetylases (KDACs) work as "erasers" and counteract the function of KATs. A total of 18 KDACs have been identified in humans and mice that are grouped into two families based on their catalytic mechanism: Zn^{2+}-dependent histone deacetylases (HDAC 1–11) and NAD^{+}-dependent sirtuin deacetylases (SIRT 1–7) (Choudhary et al. 2014). There is a class of bromodomain-containing proteins, many of which have been shown to bind to acetylated lysine residues on proteins selectively. They function as "readers" of acetyl lysine and play a role in substrate recruitment for acetyltransferases or assisting in complex formation at the site of active transcription. Acetyl lysine is recognized by bromodomain, YEATS domain, and double PHD finger domain-containing proteins and are therefore designated as "readers" of acetylation marks. These proteins can recruit or interact with other effector proteins such as transcription factors (TFs), ATP-dependent chromatin remodelers, and super-elongation complexes that can support transcriptional activation at specific loci.

Impact of Lysine Acetylation on the Structure of Chromatin and Regulation of Gene Expression

Histone acetylation has a detrimental impact on the compactness of chromatin. In general, acetylation is associated with increased transcriptional activity. Upon acetylation, the lysine residue loses its positive charge, thereby decreasing the ability of histone to interact with the negatively charged phosphate backbone leading to destabilized DNA–histone interaction. This results in an open, decondensed chromatin structure (euchromatin) that leads to enhanced gene transcription (Kouzarides 2007). In addition to chromatin decompaction, KATs themselves act as transcriptional coactivators. They are involved in the acetylation of several transcription-regulating proteins such as TFs, transcriptional coactivators, and nuclear receptors. Through the specific acetylation of a plethora of nonhistone targets, lysine acetylation, impacts diverse cellular processes such as DNA damage repair, cell cycle, cellular signaling, protein folding and aggregation, and autophagy.

Metabolic Regulation of Acetylation

Acetylation of histone tails impacts chromatin structure through neutralization of positive charge of histones and by enabling the binding of TFs, chromatin remodeling proteins, etc. These events, in turn, regulate gene expression. Acetylation of nonhistone proteins also regulates the function of several chromatin-associated proteins and proteins involved in energy metabolism. As metabolites such as acetyl-CoA and NAD+ are used as substrates or cofactors for KATs/KDACs, their enzyme activities depend on the local concentration of such metabolites. Thus, cellular metabolism can influence histone modifications, and therefore, metabolism and epigenetics are intimately connected.

Metabolic Regulation of Acetylation through Acetyl-CoA

KATs are sensitive to acetyl-CoA levels as the cellular concentration of acetyl-CoA is comparable to the Km values of KAT enzymes (Locasale and Cantley 2011). Km or the Michaelis–Menten constant is the substrate concentration at which half the maximum velocity of an enzyme is achieved. Local changes in acetyl-CoA therefore can impact activity of KAT enzymes. In cultured human cells, the acetyl-CoA concentration is 2–13 μM (Lee et al. 2014), which is greater than the Km of Gcn5 and P/CAF but near the Km of p300 (Table 12.1). This indicates that the p300 enzyme activity is regulated by cellular acetyl-CoA concentrations, and metabolic pathways that impact acetyl-CoA levels could regulate p300 mediated acetylation reactions. Other enzymes such as kinases and E3 ligases are not limited by substrate concentration as the cellular ATP levels (in mM range) are much higher than their Km values which are in the μM range (Reid et al. 2017). KATs are also sensitive to the concentrations of coenzyme A as it is a competitive inhibitor for many KATs, and its Ki values are comparable to the Km of acetyl-CoA (Table 12.1).

Acetyl-CoA, the substrate used by KATs for histone and nonhistone acetylation, is a central metabolite connecting catabolism, anabolism, and energy generation. Acetyl-CoA is produced by oxidative decarboxylation of pyruvate by the pyruvate dehydrogenase complex (PDC) in mitochondria (Patel and Roche 1990). The other sources of acetyl-CoA are β-oxidation of fatty acids or degradation of amino acids. Acetyl-CoA can also be synthesized from acetate by mitochondrial acetyl-CoA synthetases (ACSS1/ACSS3) (Haws et al. 2020) (Fig. 12.1). Most proteins are acetylated co-translationally in the cytoplasm; however, acetylation of proteins including histones also occurs in the nucleus (Drazic et al. 2016). Acetyl-CoA is

Table 12.1 Kinetic and thermodynamic parameters of KAT and KDAC enzymes

		Km [μM]	Kd [μM]	Ki [μM]
Gcn5 (yeast)	Acetyl-CoA	2.5 (Langer et al. 2002)	8.5 (Langer et al. 2002)	6.7 (Tanner et al. 2000b)
Gcn5 (human)		0.62 (Langer et al. 2002), 0.98 (Tanner et al. 2000a)	0.56 (Langer et al. 2002), 0.64 (Tanner et al. 2000a)	–
P/CAF (human)		0.64 (Lau et al. 2000)	–	0.57, 0.44 (Tanner et al. 2000a)
Esa1 (yeast)		0.9 (Berndsen et al. 2007)	–	1.8 (Berndsen et al. 2007)
p300 (human)		6.7 (Liu et al. 2008)	–	21 (Liu et al. 2008)
Sir2	NAD+	~100 (Imai et al. 2000)	–	–
SIRT1		~150–170 (Smith et al. 2009)	–	–
SIRT2		100	–	–
SIRT3		280 (Jin et al. 2009)	–	–
SIRT6		~100–300	27 (Pan et al. 2011)	–

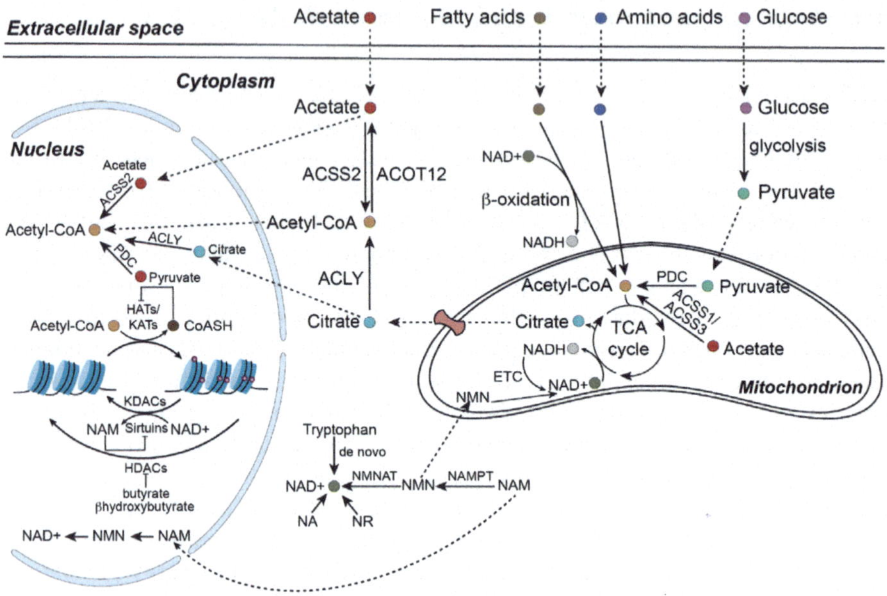

Fig. 12.1 Metabolic pathways that regulate acetyl-CoA and NAD+ levels: Acetyl-CoA can be produced from multiple nutrients directly taken up by cells, for example, glucose, amino acids, fatty acids, and acetate. Most of the acetyl-CoA is produced in the mitochondria. In the cytoplasm, acetyl-CoA is produced from citrate that is transported across the mitochondrial membrane. ACSS2 can also produce acetyl-CoA from acetate in the cytoplasm. Nuclear acetyl-CoA pool is derived mainly through diffusion of the cytoplasmic pool. Local acetyl Co-A is also produced through nuclear localization of acetyl-CoA-producing enzymes, PDC, ACLY, and ACSS2. NAD+, a cofactor for deacetylase reactions, can be produced de novo from dietary tryptophan or from nicotinic acid (NA) (Preiss–Handler pathway). However, most commonly NAD+ is produced using the salvage pathway by conversion of precursors such as nicotinamide (NAM) or nicotinamide riboside (NR)

required for these cytoplasmic and nuclear acetylation reactions, but till date there is no transporter known that can transport the mitochondrial acetyl-CoA to the cytoplasm. Instead, acetyl-CoA is converted into citrate in the mitochondria by citrate synthase and transported using the citrate shuttle to the cytoplasm. In the cytoplasm, it is converted back to acetyl-CoA and oxaloacetate by ATP citrate lyase (ACLY). Acetyl-CoA can also be produced in the cytoplasm from acetate by cytosolic acetyl-CoA synthetase (ACSS2). Acetate can also be generated from pyruvate, providing another pathway of acetyl-CoA production from glucose (Fig. 12.1). Moreover, ethanol metabolism in the liver can generate acetate as well as acetyl-CoA that supports histone acetylation in the brain. The nuclear pool of acetyl-CoA required for histone acetylation is generated by passive diffusion of the cytosolic pool through the nuclear pores as well as nuclear shuttling of acetyl-CoA generating enzymes ACLY, ACSS2, and PDC (Fig. 12.1). The redundant pathways to generate acetyl-CoA ensure that the cell can respond to any nutritional changes or deficiencies accordingly. The relative contribution of each pathway to generate acetyl-CoA depends on the environment and the cell type. For example, cancer cells grown in a high glucose culture medium produce acetyl-CoA through glycolysis-generated pyruvate. The acetyl-CoA is shuttled as citrate and converted back to acetyl-CoA in the cytoplasm by ACLY (Wellen et al. 2009). However, yeast cells grown on acetate mainly produce acetyl-CoA from acetate (De Virgilio et al. 1992).

ACLY, PDC, and ACSS2 can maintain global histone acetylation levels; however, ACLY and ACSS2 can also regulate locus-specific histone acetylation. For example, ACLY drives the expression of key glycolytic genes during the differentiation of preadipocytes to mature adipocytes (Wellen et al. 2009). ACSS2 is a critical regulator of memory-regulated neuronal genes in the murine hippocampus (Mews et al. 2017).

The rate and pathway of acetyl-CoA production are dependent on many external and internal factors, including hormone stimulation, e.g., insulin, oxygen tension, and the availability of carbon sources. Internal factors include genetic perturbations that influence the activity of acetyl-CoA metabolic pathways. Oncogenes Myc, Akt, and KRas enhance acetyl-CoA production as well as histone acetylation through increased acetyl-CoA (Lee et al. 2014; Morrish et al. 2010) (discussed in further detail in section "Altered signaling"). On the other hand, quiescent cells with lower acetyl-CoA production rates have a lower rate of glucose-derived histone acetylation (Evertts et al. 2013).

Metabolic Regulation of Acetylation through NAD+

The lysine deacetylases (KDACs) balance the activities of KATs by removing acetyl groups from acetylated lysines. The sirtuin family of KDACs (SIRT1-SIRT7) utilizes NAD+ (nicotinamide adenine dinucleotide) as a substrate for the deacetylation process, and a growing body of literature suggests that NAD+ levels can regulate sirtuin activity (Borra et al. 2004). SIRT3–5 are localized in the mitochondria and function as protein deacetylases, whereas SIRT1, SIRT2, SIRT6, and SIRT7 can deacetylate histones in the nucleus (Haigis and Sinclair 2010). NAD+ is an important redox cofactor required by many enzymes in catabolic or oxidative pathways such as glycolysis, the TCA cycle, and β-oxidation of fatty acids. Thus, the NAD+ pool available to nuclear sirtuins that alter histone acetylation could be limited based on the activity of other pathways. Intracellular NAD+ levels are controlled by (a) NAD+ turnover balanced by rates of production (via de novo biosynthesis and salvage pathways) and consumption by enzymes (such as sirtuins and PARP) and (b) redox conversion between NAD+ and NADH during oxidative phosphorylation by the electron transport chain. In most physiological conditions, the NAD+ flux is controlled by the NAD+ and NADH redox conversion that exceeds the rate of structural NAD+ turnover. There are three pathways for NAD+ biosynthesis (a) the de novo pathway, (b) the salvage pathway, and (c) the canonical Preiss–Handler pathway (Verdin 2015). The de novo synthesis pathway uses tryptophan to produce NAD+ in the cytoplasm. NAD+ can also be synthesized from nicotinic acid (NA) through the Preiss–Handler pathway. However, the primary cellular source of NAD+ is through the salvage pathway where precursor molecules such as nicotinamide (NAM) or nicotinamide riboside (NR) are converted to NAD+ (Newman and Verdin 2014; Revollo et al. 2004) (Fig. 12.1).

Nicotinamide phosphoribosyltransferase (NAMPT), an enzyme that converts NAM into NAM mononucleotide (NMN), is the rate-limiting enzyme for the NAD+ salvage pathway (Revollo et al. 2004). NMN is then converted into NAD+ by NMN adenylyltransferase (NMNAT). In mammalian cells, three NMNAT enzymes with various subcellular localizations have been identified. NMNAT1 is nuclear, NMNAT2 resides in the cytosol and Golgi apparatus whereas NMNAT3 is present in both cytosol and mitochondria (Berger et al. 2005; Zhang et al. 2003). These enzymes likely regulate the NAD+ levels in the various subcellular compartments. Although direct evidence showing the exchange of NAD+ between the cytosol and nucleus is lacking, these two compartments have similar NAD+ concentrations and therefore considered interchangeable. However, neither NAD+ nor its reduced form NADH can diffuse through membranes and no NAD+ transporter has been identified. Therefore, shuttling of NAD+ precursors such as NAM and NMN likely maintain the local NAD+ pool through NMNAT activity. Deletion of *NMNAT1* is embryonic lethal showing that nuclear NAD+ pools are important during embryonic development and are not compensated by cytosolic or mitochondrial NAD+ synthesis by NMNAT2 or NMNAT3. The mitochondrial NAD+ pool is also thought to be separate as depletion of cytosolic NAD+ does not lead to changes to the mitochondrial NAD+ pool (Yang et al. 2007; Pittelli et al. 2010).

Most of the cellular NAD+ is protein-bound, making it challenging to determine the subcellular concentrations of NAD+ accurately. Nuclear NAD+ levels are estimated to be 70 μM using 2-photon microscopy (Fjeld et al. 2003). Mitochondrial NAD+ was determined to be 256 μM using a mass spectrometry approach (Yang et al. 2007). The total NAD+ pool in HEK293 and erythrocytes cells was measured to be ~365 μM and 368 μM respectively (Yang et al. 2007). The mammalian sirtuin SIRT1 has a Km value of 150–170 μM for NAD+ (Smith et al. 2009) similar to the yeast homolog Sir2, which is 100 μM (Imai et al. 2000) (Table 12.1). SIRT6 has a higher affinity for NAD+ (Kd 27 μM) and can bind without an acetylated substrate suggesting it exists in NAD+-bound poised state. Mitochondrial sirtuins such as SIRT3 have Km values that range from 100 to 300 μM. As the mitochondrial NAD+ levels are higher, they show smaller changes in activity in response to changes in NAD+ levels than non-mitochondrial sirtuins. Changes in nuclear NAD+ levels might lead to dramatic increases in SIRT1 deacetylation activity in the nucleus.

Metabolic Regulation of Acetylation through Other Metabolites

Class I and Class IIa family of KDACs are NAD+-independent; however, they are inhibited by β-hydroxybutyrate (βOHB), an endogenous metabolite and ketone body. βOHB has been shown to induce histone acetylation in cultured cells and mice to a similar extent observed with fasting and caloric restriction. Fasting and caloric restriction increase ketogenesis that elevate levels of βOHB (Shimazu et al. 2013). Physiological concentrations of βOHB in human serum are in the low μM range; however, fasting, caloric restriction, strenuous exercise, and ketogenic diet can increase βOHB levels to 1–2 mM. Since the IC50 of βOHB for Class I and IIa KDACs is 2-5 mM, it is possible that fluctuations in the concentration of this metabolite can lead to an increase in histone acetylation (Newman and Verdin 2014).

Butyrate, another metabolite, gained interest in the 1970s due to its potential role in inducing cell cycle arrest and in gene regulation (Davie 2003). It was only later shown that sodium butyrate increased histone acetylation in HeLa cells by inhibiting KDACs (Riggs et al. 1977). Butyrate was found to be a broad-spectrum inhibitor of most KDACs except Class III KDACs 6 and 10. It is a non-competitive inhibitor of KDACs with Ki of ~60 μM and IC50 of 10–100 μM (Cousens et al. 1979; Sekhavat et al. 2007; Waldecker et al. 2008). Butyrate is produced in the gut by the bacterial fermentation of plant polysaccharides. In colon tumors, the butyrate concentrations are estimated to be in the range of 100 μM (Cummings et al. 1987; Donohoe et al. 2014). In the colon, normal colonocytes utilize butyrate as the primary carbon source and convert it to acetyl-CoA, which is then used as a substrate for the TCA cycle and KATs (Bultman 2014). In colon cancer cells, due to the Warburg effect, there is a metabolic shift toward increased glycolysis, leading to accumulation of butyrate, which functions as a KDAC inhibitor and thereby increases histone acetylation at pro-apoptotic genes (Donohoe et al. 2012). Butyrate can also serve as a precursor of acetyl-CoA through the β-oxidation pathway, which might lead to an increase in acetyl-CoA levels and thereby acetylation (Donohoe et al. 2014). Propionate and polyphenol metabolites produced by gut bacteria have also been found to inhibit KDAC activity (Waldecker et al. 2008).

In addition to direct substrate-level regulation of enzyme activity of writers and erasers of acetylation, other small-molecule metabolites have been shown to regulate activity by regulating the PTM of the enzyme itself. For example, oleic acid induces the cAMP/PKA signaling pathway that leads to phosphorylation of the Ser-434 residue of SIRT1, leading to increased deacetylase activity (Lim et al. 2013). Furthermore, SIRT6 has been reported to exhibit increased deacetylation activity toward an H3K9ac substrate in the presence of a physiological concentration of several free long-chain fatty acids that bind to SIRT6 (Feldman et al. 2013).

Metabolic Rewiring by Acetylation

In addition to the role of acetylation on chromatin structure and gene expression, acetylation of metabolic enzymes can, in turn, rewire or alter metabolic pathways. For example, acetylation promotes mitochondrial suppression and leads to the upregulation of glycolysis through multiple mechanisms. Acetylation of pyruvate dehydrogenase alpha subunit and pyruvate dehydrogenase phosphatase directly inhibits PDC activity thereby promoting glycolysis (Fan et al. 2014). Hyperacetylation of mitochondrial matrix proteins such as long-chain acyl-CoA dehydrogenase can also inhibit β-oxidation (Hirschey et al. 2010; Bharathi et al. 2013). Acetylation also inhibits the activity of Krebs cycle enzymes IDH2 (Yu et al. 2012) and SDH (Finley et al. 2011; Cimen et al. 2010) and suppresses Complex I of the electron transport chain (Ahn et al. 2008). Furthermore, it inhibits mitochondrial superoxide dismutase (MnSOD or SOD2) (Tao et al. 2010) and increases the nuclear localization of HIF1α-ARNT complex (Lim et al. 2010), which suppresses mitochondria by upregulating pyruvate dehydrogenase kinase (PDK) that inhibits PDC (Kim et al. 2006). Similarly, acetylation of TFs that control metabolic enzyme genes is one more mechanism by which acetylation can rewire metabolism. For example, acetylation of PGC1-α leads to its inhibition (Lerin et al. 2006), whereas NRF1 and NRF2 are activated by acetylation (Izumi et al. 2003). HIF1 and HIF2, the hypoxia-induced TFs, are also activated and inhibited by acetylation, respectively (Keith et al. 2011). Acetylation also promotes the translocation of several metabolic enzymes and TFs to the nucleus, where they have been suggested to play roles as transcriptional regulators. For example, metabolic enzymes GAPDH (Ventura et al. 2010) and PKM2 (Lv et al. 2013), translocate to the nucleus and TFs MYC (Vervoorts et al. 2003) and STAT3 (Yuan et al. 2005) show increased transcriptional activity upon acetylation. RICTOR, a member of the mTOR complex 2, is also known to be acetylated in response to metabolic conditions, promoting resistance to cancer therapies (Masui et al. 2015). Furthermore, acetylation of 6-phosphogluconate (PGD), an enzyme in the pentose phosphate pathway, leads to increased activity, inducing the production of nucleosides (Shan et al. 2014). ACLY can itself be acetylated, leading to increased stability and activity, which leads to an increase in citrate-derived acetyl-CoA levels that can be utilized to produce lipids (Lin et al. 2013). On the other hand, ACSS2 is activated when deacetylated by SIRT1, indicating that this might be a mechanism to sense low nutrient conditions and promote a compensatory pathway of acetyl-CoA production from acetate (Hallows et al. 2006).

Cellular and Organismal Variable Regulating Acetylation

Aging

Changes in the metabolic activity during aging orchestrates local or global changes in the histone acetylation pattern. These changes lead to age-associated alterations in transcriptome thereby affecting the lifespan of an individual and several age-related disorders. Several age-associated alterations in histone acetylation have been reported. For instance, the levels of H3K14ac, H4K8ac and H4K12ac were gradually increased during postovulatory aging in mouse oocytes (Huang et al. 2007). Similarly, aging-induced infertility of mouse oocytes was associated with the occurrence of higher levels of H4K12ac in the metaphase II stage oocytes (Manosalva and

Gonzalez 2009). Reduced acetylation of histone at various sites has been implicated in age-dependent memory impairment in mice. Low levels of acetylated H4K12 in aged mice resulted in failure to initiate a hippocampal gene expression program associated with memory consolidation during learning (Peleg et al. 2010). Low levels of acetylated H3K9 and H3K14 at the promoter regions of schizophrenia-related genes were observed in early age and throughout aging in human subjects with schizophrenia (Tang et al. 2011).

Acetyl-CoA metabolism is directly linked with the age-associated deregulation in histone acetylation. In Drosophila, brain-specific knockdown of acetyl-CoA synthetase led to enhanced autophagy and prolonged lifespan (Eisenberg et al. 2014). Similarly, reducing the activity of ATP citrate lyase (low acetyl-CoA levels) and decreasing the level of H4K12-specific acetyltransferase Chameau resulted in extended lifespan (Peleg et al. 2016). Decrease in ACLY and ACSS2 expression and reduced levels of cytoplasmic acetyl-CoA due to fasting or calorie restriction decreases p300 activity thereby inducing pro-longevity autophagy. On the contrary, nuclear acetyl-CoA has longevity-promoting role, wherein nuclear ACSS2 associates with acetyltransferases such as CBP and PCAF and acetylate H3K9 and H3K27 on the promoter of neuroprotective genes resulting in increased lysosomal biogenesis and autophagy (Bradshaw 2021).

Dietary Alterations

Dietary habits have profound impact over metabolism to which several histone modifications are responsive to. Consumption of high-fat diet led to the reduction in acetyl-CoA levels in white adipose tissue, liver and pancreas of mice which can be attributed to decrease in the levels of ATP citrate lyase in HFD-fed mice. Consequently, reduced levels of histone acetylation (acetylated H3K23 in particular) were observed in white adipose tissue and pancreas but not in liver (Carrer et al. 2017). In another study, consumption of high-fat western diet (WD) resulted in reduced acetylation of histones in the liver of nonalcoholic fatty liver disease mice model. It was shown that WD caused increased flux of histones without any effect on acetyl-CoA turnover (Arias-Alvarado et al. 2021). Ketogenic diet (KD) has been used for the treatment of epilepsy for many years. Recently, it is observed that KD resulted in the increase of neuregulin1 (NRG1) expression in the mice hippocampus through elevated levels of histone acetylation at its promoter (Wang et al. 2021).

Hormones

Several incidences of hormone-induced changes in histone acetylation have been reported. A very striking example includes global changes in the histone acetylation levels during remodeling of endometrium during menstrual cycle. The levels of acetylated H2AK5, H3K9, and H4K8 are elevated during early proliferative phase but decline during ovulation and late secretory phase, and acetylated H4K8 levels increases during early till mid-secretory phase (Gujral et al. 2020). Estradiol treatment of MCF7 cells induced hyperacetylation at the promoter regions of estrogen receptor target genes through acetyltransferase activity of p300/CBP (Chen et al. 1999). In response to different physiological and pathological states such as fasting or liver cancer, glucagon induced acetylation of several of the energy-sensing factors to control hepatic metabolism and it involved several of the acetyltransferases (Zhang et al. 2019).

Bacterial Infection

Bacterial pathogens have developed several complex strategies to evade the host immune response. One such efficient strategy involves alterations in the host cellular epigenetic machineries particularly through modulation of histone acetylation in host cells. Bacterial components such as metabolites and virulence factors have a direct and indirect effects over alterations in the histone acetylation marks and modulation of the HAT and HDAC expression (Grabiec and Potempa 2018).

Mycobacterium tuberculosis and its products inhibit IFN-γ-dependent HLA-DR gene expression in human macrophages through a repressor pathway that involves histone hypoacetylation at HLA-DRα promoter. Though the infection had no effect on HDAC expression, it led to the upregulation of mammalian Sin3A, a corepressor protein, which promoted HDAC-containing complex formation at HLA-DRα promoter, resulting in histone deacetylation and gene silencing (Wang et al. 2005). Infection of *Helicobacter pylori* in gastric epithelial cells led to global deacetylation of histone H3K23. This effect was observed to be dependent on virulence factor *cag*PAI (Ding et al. 2010) (Fig. 12.2). *Listeria monocytogenes*, upon infection of HeLa cells, induces a global deacetylation of histone H4 (along with reduced H3S10 phosphorylation) during early phases of infection. This effect is mediated through major listerial toxin listeriolysin-O (LLO). These changes in histone modifications correlated with reduced expression of a subset of host genes, including key regulators of immune response. Two such genes are CXCL2 (chemokine) and DUSP4 (phosphatase) (Hamon et al. 2007). *L. monocytogenes* also induces H3K18 deacetylation which is dependent host deacetylase Sirt2 (Fig. 12.2). Upon infection, the cell surface receptor c-Met gets activated by the bacterial protein InlB, thereby triggering a signaling cascade involving (PI3K)/AKT signaling that results in nuclear translocation and recruitment of Sirt2 to the promoters of subset of genes involved in immune responses (Eskandarian et al. 2013). Sirt2-mediated deacetylation of H3K18 supports intracellular survival of *L. monocytogenes* by maintaining genome integrity (by counteracting infection induced DNA damage) and host cell viability (Eldridge and Hamon 2021).

Anaplasma phagocytophilum infection led to increase in HDAC1 expression and activity that caused reduction of H3 acetylation and therefore silencing of host

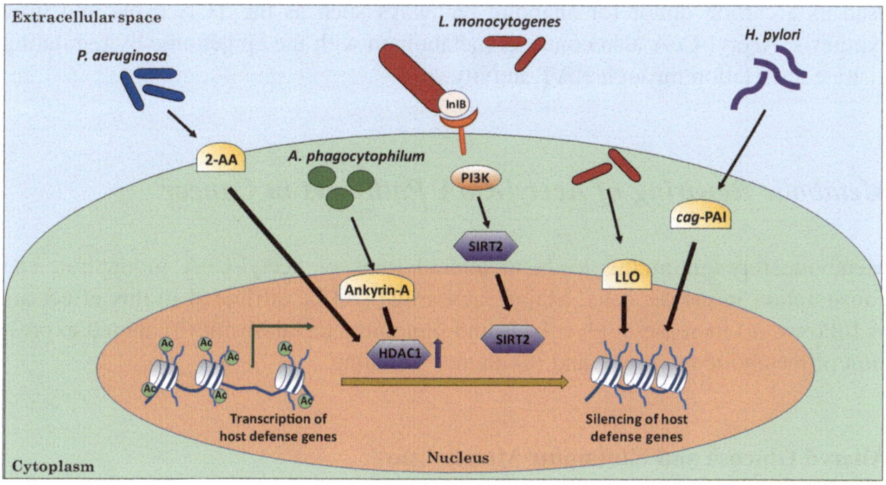

Fig. 12.2 Bacterial infection induces modulation of histone acetylation in host cells. Figure shows the suppression of host defense genes due to reduction in histone acetylation levels in the host cells induced by several effector molecules during bacterial infection. Hypoacetylation of histones in host cells is either global or promoter-specific which leads to condensed chromatin and inhibition of genes involved in host defense against infection (*Figure modified from Grabiec, A.M., Potempa, J., 2018. Crit. Rev. Microbiol. 44 (3), 336 -350. Taylor and Francis*)

defense gene expression (Garcia-Garcia et al. 2009). Effector molecule ankyrin A promoted recruitment of HDAC1 to modify host defense gene expression (Rennoll-Bankert et al. 2015) (Fig. 12.2). A quorum-sensing signal molecule 2-aminoacetophenone (2-AA), released by *Pseudomonas aeruginosa*, modulates host metabolism and impairs host cell immune response (Fig. 12.2). 2-AA promoted expression of HDAC1 in human THP-1 monocytes, resulting in histone H3K18 hypoacetylation at pro-inflammatory cytokine loci. This epigenetic reprogramming in host cell dampened host response to subsequent exposure to 2-AA or other pathogen-associated molecules, thereby allowing pathogen persistence in host tissues (Bandyopadhaya et al. 2016).

Acetylation and Cancer: A Metabolism View

Cancer cells undergo extensive metabolic and epigenetic reprogramming to support the higher metabolic needs associated with a high proliferation rate. It is now well-established that energy metabolism in cancer cells is rewired to facilitate increased uptake and utilization of nutrients to produce macromolecules needed for making more cells (Ward and Thompson 2012). Acetyl-CoA takes center stage in cancer cell metabolism as it connects catabolism to anabolic pathways needed for biomass production. Acetyl-CoA is produced from pyruvate, a product of glycolysis and is used as a carbon donor for anabolic pathways such as the TCA cycle and lipid synthesis. Acetyl-CoA also connects metabolism with the epigenome by regulating histone acetylation through KAT activity.

Metabolic Rewiring of Acetyl-CoA Pathways in Cancer

Metabolic reprogramming leads to altered rates of acetyl-CoA production and consumption in cancer cells. Major causes that can be attributed to this effect are as follows: (a) increase in glycolysis and glutamine metabolism; (b) altered expression of metabolic enzymes, and (c) altered signaling.

Altered Glucose and Glutamine Metabolism

Most cancer cells display enhanced glucose uptake as compared to normal cells which is metabolized through aerobic glycolysis, a phenomenon commonly known as "the Warburg effect." The glycolytic end product pyruvate enters the mitochondria where it is converted into acetyl-CoA by the PDC. Any excess pyruvate is converted into lactate and secreted out by the cell. The pyruvate-derived acetyl-CoA then enters the TCA cycle where it is utilized to make metabolic intermediates that are precursors for macromolecules such as nucleotides, amino acids, and fatty acids. Citrate is the first product of the TCA cycle generated from combining oxaloacetate and acetyl-CoA. The enhanced glycolytic flux leads to an excess of citrate in the mitochondria which is exported to the cytoplasm by the citrate shuttle to support fatty acid synthesis (Vander Heiden et al. 2009).

Fatty acids are the building blocks for all lipids and thus important for building membrane structure. Normal cells take up most fatty acids from the circulation using specific receptors (Zaidi et al. 2013). However, cancer cells mainly derive all their intracellular fatty acids from de novo fatty acid synthesis (Zaidi et al. 2013). In addition to the increased glycolytic flux that supports fueling of the TCA cycle, cancer cells exhibit increased fatty acid synthesis activity. Citrate derived from the TCA cycle is converted back to acetyl-CoA and oxaloacetate by ACLY in the cytoplasm. This cytoplasmic pool of acetyl-CoA is the primary substrate for fatty acid synthesis of different chain lengths. Acetyl-CoA is utilized by acetyl-CoA carboxylase (ACC) to produce malonyl-CoA which is used by fatty acid synthase (FASN) to carry out step-wise chain elongation to generate fatty acids (Rohrig and Schulze 2016). Malonyl-CoA is also utilized for the production of cholesterol.

To compensate for the diversion of TCA cycle intermediates such as citrate to other biosynthetic pathways, cancer cells use a process called as anaplerosis (DeBerardinis et al. 2007). Anaplerosis involves supplementing the TCA cycle with α-ketoglutarate (αKG) produced from glutamine. Cancer cells exhibit enhanced glutamine uptake which is converted to glutamate by glutaminase enzyme. Glutamate is further converted into αKG which enters the TCA cycle to be eventually converted into oxaloacetate. Acetyl-CoA combining with oxaloacetate to produce citrate refuels the TCA cycle (DeBerardinis et al. 2007). In idealized cell culture conditions where cancer cells are grown in excess glucose and glutamine levels, nucleocytoplasmic acetyl-CoA are derived mainly from citrate through the activity of ACLY (Sivanand et al. 2018). However, in conditions of hypoxia or limited glucose availability such as the case in tumors, nuclear acetyl-CoA and cytoplasmic acetyl-CoA are derived from acetate through the action of ACSS2 (Bulusu et al. 2017; Comerford et al. 2014; Schug et al. 2015). ACSS2 also localizes to the nucleus where it can recycle acetate derived from histone deacetylation reactions back to acetyl-CoA (Li et al. 2017; Bulusu et al. 2017). Moreover, in hypoxic conditions, acetyl-CoA can also be derived from reductive carboxylation of glutamine-derived αKG which produces citrate (Metallo et al. 2011; Wise et al. 2011; Mullen et al. 2011).

Altered Expression of Metabolic Enzymes

ATP citrate lyase (ACLY) that produces acetyl-CoA from citrate is upregulated or activated in several cancer types (Hatzivassiliou et al. 2005; Carrer et al. 2019). Under high glucose conditions, ACLY gets acetylated by PCAF, which increases its stability, promoting lipid synthesis and tumor growth (Fig. 12.3). Enhanced acetylation of ACLY was found in lung cancers (Lin et al. 2013). Glutamine metabolism also contributes to the production of acetyl Co-A and α-KG. It was observed that inhibition of glutamate oxidation resulted in the downregulation of several critical cancer-related genes through significant downregulation of epigenetic marks such as H4K16ac and H3K4me3 (Simpson et al. 2012). Certain cancer cells such as glioblastoma and hepatocellular carcinoma cells utilize acetate as an alternative carbon source for sustaining acetyl Co-A production (Comerford et al. 2014; Mashimo et al. 2014). In cancer cells during hypoxic conditions, when the contribution of glucose drops for the production of acetyl-CoA, acetate emerges as a primary alternative carbon source that, despite at low concentration, contributes to acetyl Co-A production so that a constant supply of fatty acids is maintained for tumor growth and proliferation (Fig. 12.3) (Kamphorst et al. 2014). In addition to being a metabolic precursor for fatty acid synthesis, acetate has been shown to induce a hyperacetylated state of histone H3 in hypoxic cells that activated the expression of lipogenic genes ACACA and FASN by increasing H3K9ac, H3K27ac, and H3K56ac levels at their promoter regions (Fig. 12.3). Acetyl-CoA synthetases (ACSS1, ACSS2) are the enzymes involved in these acetate-mediated epigenetic alterations and were found to be highly expressed in hepatocellular carcinoma (Gao et al. 2016).

Fatty acids are another very prominent carbon source that can alter the epigenetic landscape. Fatty acid oxidation leads to an increase in global H3K9ac. Interestingly, even in the presence of excess glucose, the majority of the acetylation was derived from fatty acid carbon, and this lipid-derived acetyl-CoA promoted lipid-specific gene expression (McDonnell et al. 2016). A high-fat diet was shown to reduce

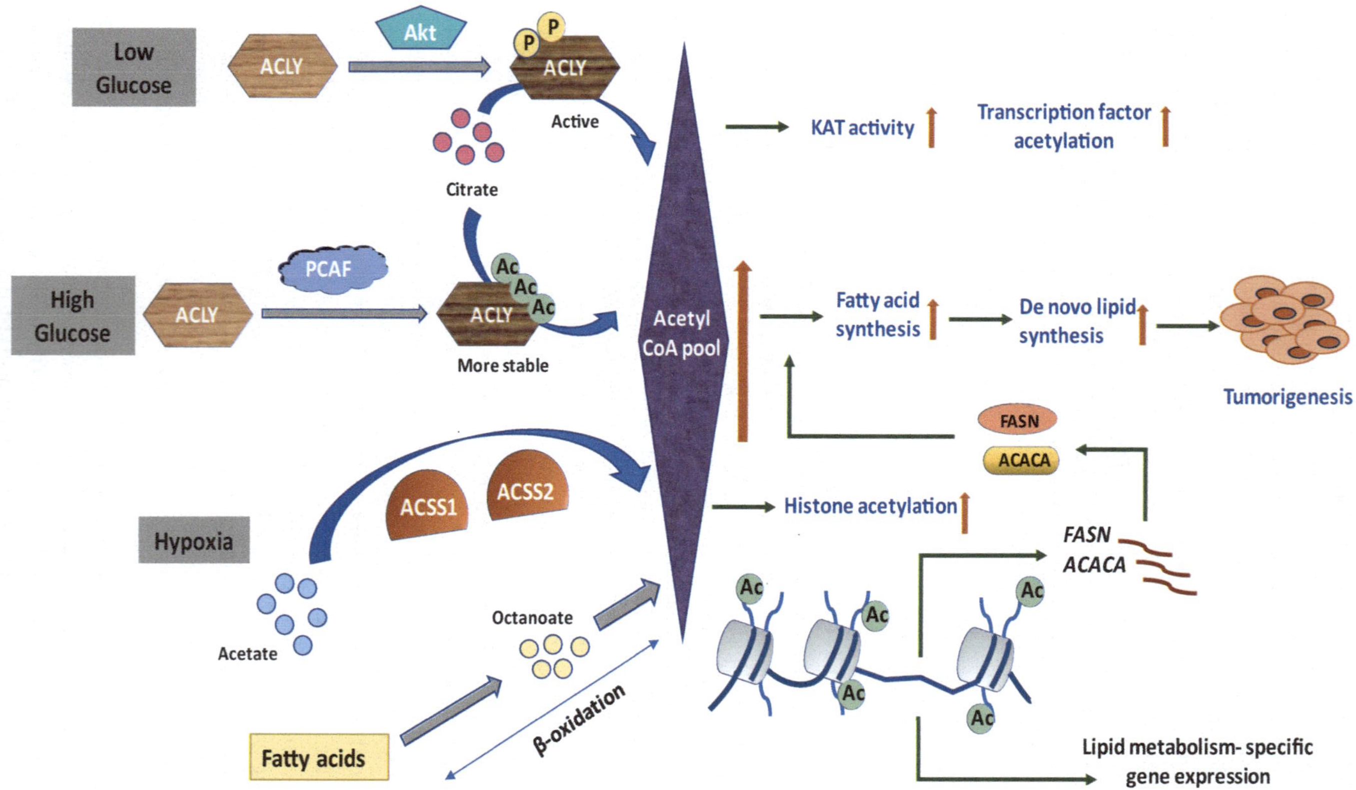

Fig. 12.3 Role of metabolites and associated rewiring of acetyl-CoA metabolism in cancer promotion: Figure shows the enrichment of cellular pool of acetyl-CoA in cancer cells under specific conditions and through different metabolites. This increase in acetyl-CoA production is utilized for several of the downstream processes in different types of cancer for cancer cell survival, sustained proliferation, and growth leading to tumorigenesis

acetyl-CoA levels and decreased H3K23ac in white adipose tissue but not the liver. Since obesity predisposes to the development of multiple cancers, lipids can significantly impact cancer risk through epigenetic regulation (Carrer et al. 2017).

Altered Signaling

ACLY and ACSS2, two key enzymes responsible for Acetyl-CoA production from citrate and acetate, respectively, are directly regulated by signaling pathways activated upon growth factor stimulation or nutrient availability. ACLY is phosphorylated by several kinases at serine 455 (Ser 455), leading to an increase in its activity (Potapova et al. 2000). Kinases such as AKT, PKA, and BCKDK are known to phosphorylate this site of ACLY (Guy et al. 1980; Berwick et al. 2002; White et al. 2018). The PI3K-AKT signaling pathway commonly activated in several types of cancers regulates cell survival, proliferation, and growth (Manning and Toker 2017). PI3K-AKT signaling thus regulates histone acetylation through increased phosphorylation and activity of ACLY. Oncogenic KRAS signaling also leads to increased AKT and ACLY phosphorylation and elevates histone acetylation prior to pancreatic carcinogenesis. Deletion of Acly suppresses pancreatic carcinogenesis (Carrer et al. 2019). Insulin signaling that activates the PI3K-AKT pathway also increases global histone acetylation levels (Senapati et al. 2019).

AMPK is a cellular stress sensor and is rapidly activated upon cellular stress or nutrient starvation. It helps shut down anabolic processes and stimulates nutrient uptake and catabolic processes (Gonzalez et al. 2020). AMPK directly phosphorylates acetyl-CoA carboxylase 1 (ACC1) and ACSS2. Phosphorylation of ACC1 by AMPK inhibits its activity which stops utilization of acetyl-CoA to make malonyl-CoA, thereby increasing acetyl-CoA pools and global histone acetylation levels (Galdieri and Vancura 2012). Activating AMPK with metformin, a commonly used drug for diabetes, also leads to an increase in global histone acetylation (Galdieri et al. 2016). Phosphorylation of ACSS2 by AMPK leads to its increased nuclear localization and interaction with TF TFEB that promotes histone acetylation at autophagy and lysosome biogenesis genes (Li et al. 2017). In addition to Akt, Myc is also known to promote acetyl-CoA production through ATP citrate lyase (ACLY). Myc regulates acetyl-CoA production for its use in lipid synthesis and histone acetylation, whereas Akt maintains histone acetylation even if glucose availability is limited (Fig. 12.3) (Morrish et al. 2010; Lee et al. 2014; Dang 2013).

Role of Histone Acetylation in Cancer Progression

Certain histone marks get globally altered during the cancer progression and are closely related to the cancer phenotype. Several among them are biomarkers for specific cancers and have a prognostic value too. One such mark is H4K16ac, whose overall loss was shown in a variety of cancers. Additionally, it was found to be associated with the hypomethylation of DNA repetitive sequences. Thus, loss of acetylation at H4K16 can be considered an epigenetic marker for malignant transformation that can aid in disease prognosis (Table 12.2) (Fraga et al. 2005). Reduced H4K16ac levels were also observed in medulloblastomas and breast carcinomas due to hMOF downregulation (Pfister et al. 2008). In prostate tumors, a higher level of H3K18ac along with H3K4me2 was shown to be correlated with better prognosis and disease-free survival. A classic study demonstrated that a combination of histone modifications could be utilized as a prognostic biomarker to predict tumor recurrence (Seligson et al. 2005). Low levels of H3K18ac were observed to be necessary for maintaining the malignancy and drive tumorigenesis by promoting anchorage-dependent growth and escaping contact inhibition (Barber et al. 2012). Low H3K18ac levels were a significant and independent predictor of renal cell carcinoma (RCC) progression (Mosashvilli et al. 2010) and pancreatic adenocarcinoma patients (Table 12.2) (Manuyakorn et al. 2010). In non-small cell lung cancer, patients with tumors with low levels of H2AK5ac had worse survival than patients with a high level of H2AK5ac expression in their tumors (Barlesi et al. 2007). In an extensive study conducted in human breast carcinomas, H4K16ac was observed to be either low or absent in the majority of the breast cancer cases, and low levels of H3K9ac and H4K16ac were significantly associated with large tumor size. Additionally, moderate to low levels of several histone acetylation marks such as H3K9ac, H3K18ac, and H4K12ac was associated with poorer prognosis in basal carcinomas and HER-2 positive tumors (Table 12.2) (Elsheikh et al. 2009).

In contrast to the hypoacetylation of several histone marks being associated with poor prognosis, there are numerous contradictory reports for certain histone marks such as H3K9ac and H3K18ac. For instance, an increase in H3K9ac was reported in hepatocellular carcinoma, and low levels of H3K9ac were observed to be associated with better prognosis in non-small cell lung adenocarcinoma and astrocytoma (Barlesi et al. 2007; Fullgrabe et al. 2011). Similarly, low levels of H3K18ac were associated with a better prognosis for patients with esophageal squamous cell carcinoma and glioblastoma (Table 12.2) (Fullgrabe et al. 2011). Therefore, alterations in each of the histone modifications are specific to tissue types and can predict opposite prognoses in different cancer types. H3K56 mark has a unique location on the nucleosome, both at its entry and exit points. Acetylated H3K56, in addition to transcription, is involved in chromatin assembly during DNA replication and repair. In the context of cancer, elevated levels of H3K56ac were observed in multiple types of cancers and correlated with tumorigenicity (Table 12.2). Upon DNA damage, H3K56 gets acetylated and incorporated into chromatin at the sites of DNA repair. Histone chaperone ASF1A was found to be required for the increase in H3K56 acetylation in vivo following DNA damage, whereas histone chaperone CAF1 is required for assembly of acetylated H3K56 onto DNA. This increase in the acetylated H3K56 levels was found to be proportional to tumor grade in many cancer tissues. In line with these observations, non-tumorigenic breast cancer cells (MCF10A) showed lower levels of H3K56ac than tumorigenic breast cancer cells (MCF7). Additionally, H3K56ac levels positively correlated with dedifferentiated cancer cells and undifferentiated cells (Das et al. 2009).

Table 12.2 Changes in histone acetylation in cancer and their associated prognosis

Histone	Modification	Levels in cancer	Cancer type	Disease prognosis/biomarker	Refs.
H2A	H2AK5ac	Hypoacetylation	Non-small cell lung cancer	Poor survival	Barlesi et al. (2007)
		Hyperacetylation	Oral squamous cell carcinoma	–	Arif et al. (2010)
H3	H3K4ac	Hypoacetylation	Oral squamous cell carcinoma	Correlated with lower disease-free survival	Chen et al. (2013)
	H3K9ac	Hypoacetylation	Non-small cell lung cancer	Better survival	Barlesi et al. (2007)
		Hypoacetylation	Astrocytoma	Better prognosis	Fullgrabe et al. (2011)
		Hypoacetylation	Breast cancer (subtype basal carcinoma and HER-2 positive)	Associated with a poor prognostic subtype of breast cancer and adverse patient outcome	Elsheikh et al. (2009)
		Hypoacetylation	Oral squamous cell carcinoma	A biomarker for poor prognosis and lower survival rate	Webber et al. (2017)
		Hyperacetylation	Oral squamous cell carcinoma	–	Arif et al. (2010)
	H3K14ac	Hyperacetylation	Oral squamous cell carcinoma	–	Arif et al. (2010)
	H3K18ac	Hyperacetylation	Prostate cancer	Correlated with high grade of tumors but better prognosis and disease-free survival	Seligson et al. (2005)
		Hypoacetylation	Renal cell carcinoma	Poor prognosis and shorter disease-free survival	Mosashvilli et al. (2010)
		Hypoacetylation	Pancreatic adenocarcinoma	Worse overall and disease-free survival	16
		Hypoacetylation	Breast cancer (subtype basal carcinoma and HER-2 positive)	Associated with a poor prognostic subtype of breast cancer and adverse patient outcome	Elsheikh et al. (2009)
		Hypoacetylation	Glioblastoma	Better prognosis	Fullgrabe et al. (2011)

(continued)

Table 12.2 (continued)

Histone	Modification	Levels in cancer	Cancer type	Disease prognosis/biomarker	Refs.
		Hypoacetylation	Esophageal squamous cell carcinoma	Better prognosis	Fullgrabe et al. (2011)
	H3K56ac	Hyperacetylation	Multiple cancer types	Correlated with dedifferentiated cancer cells and undifferentiated cells	Das et al. (2009)
		Hyperacetylated	Oral squamous cell carcinoma	–	Arif et al. (2010)
H4	H4K12ac	Hypoacetylation	Breast cancer (subtype basal carcinoma and HER-2 positive)	Associated with a poor prognostic subtype of breast cancer and adverse patient outcome	Elsheikh et al. (2009)
	H4K16ac	Hypoacetylation	Leukemia cell lines, primary lymphomas	Universal epigenetic mark for malignant transformation	Fraga et al. (2005)
		Hypoacetylation	Medulloblastoma, breast carcinoma	Associated with poor prognosis	Pfister et al. (2008)
		Hypoacetylation	Breast cancer	Correlated with poor prognostic characteristics such as large tumor size, vascular invasion	Elsheikh et al. (2009)
		Hypoacetylation	Oral squamous cell carcinoma	–	Arif et al. (2010)

Metabolic Alterations of Acetyl-CoA Pathways and Its Implications in Oral Cancer

Oral cancers include malignancies of the oral cavity and form a subset of head and neck cancers. Squamous cell carcinoma accounts for 90% of head and neck cancer and therefore is referred to as head and neck squamous cell carcinoma (HNSCC). It is the sixth most common cancer worldwide; however, India has a relatively high incidence especially of oral squamous cell carcinoma (OSCC) (Bose et al. 2013). Although the risk factors and mechanisms of HNSCC carcinogenesis have been studied to some extent, the metabolic alterations are understudied especially in OSCC as compared to HNSCC and other types of cancer. Here, we summarize the current knowledge about metabolic alterations in HNSSC/OSCC.

Metabolic Alterations of Acetyl-CoA Pathways in Oral Cancer

Increased glucose uptake and glycolysis are characteristic features of head and neck cancer and are routinely used for clinical diagnosis. For example, ^{18}F-deoxyglucose positron emission tomography (FDG-PET) that utilizes a glucose analog FDG for imaging tumors is regularly used for evaluation of HNSCC and monitoring the response to chemotherapy and radiotherapy. It is based on the principle that tumor tissues exhibit high glucose uptake and thus can be easily distinguished from other tissues due to the high FDG uptake. Indeed glucose receptors GLUT1 and GLUT3 are upregulated in OSCC and correlate with poor outcomes (Ayala et al. 2010; Grimm et al. 2014a, b; Li et al. 2016). A comparison of GLUT1 expression in oral dysplastic lesions and oral invasive tumors with normal mucosal tissues revealed that GLUT1 mRNA and protein expression increased with dysplasia and was highest in the invasive tumors (Nakazato et al. 2019). Interestingly, in the oral dysplasia samples, GLUT1 expression was restricted to the dysplastic tissue and depended on the extent of dysplasia. While GLUT1 expression did not show any correlation with tumor grade for the invasive tumors, high GLUT1 expression did correlate significantly with increased lymph node metastasis and poor disease-free survival (Nakazato et al. 2019). Other glycolytic enzymes such as PKM2 (Wang et al. 2015), LDHA (Grimm et al. 2013), HK-2, and PFK-1 have also been reported to be upregulated in OSCC and associated with clinical aggressiveness and worse prognosis (Grimm et al. 2014a). Together, these reports suggest that oral tumors indeed exhibit increased rates of glycolysis. Apart from glycolysis, fatty acid synthesis also has been reported to be upregulated in OSCC/HNSCC (Sant'Anna-Silva et al. 2018; Krontiras et al. 1999; Silva et al. 2004) and inhibition of FASN leads to decreased proliferation and apoptosis of OSCC cell lines (Agostini et al. 2004; Zhang et al. 2005). In orthotopic tumor models, FASN inhibition leads to decreased tumor volume and number of lymph node metastasis (Agostini et al. 2014). Together these studies show that oral tumors show increased rates of glycolysis and fatty acid synthesis indicating metabolic reprogramming and altered rates of acetyl-CoA production.

Acetylated Histone Marks Associated with Oral Cancer and Its Poor Prognosis

Though histone acetylation is generally involved in chromatin relaxation and provides more accessibility to DNA for transcription, some histone marks are observed to be hypoacetylated in oral cancer. One such histone modification is H3K4ac, low levels of which were found to have a positive correlation with T (primary tumor) status, N (nodal invasion) status, advanced tumor stage, and perineural invasion (PNI), though H3K18ac positively correlated with T status and tumor stage (Chen et al. 2013). Another histone mark observed to be frequently hypoacetylated in oral cancer is H3K9. In oral rinse samples from OSCC patients and several OSCC cell lines, H3K9 was observed to be hypoacetylated (Kusumoto et al. 2012; Giudice et al. 2013).

On the other side of the story, we have H3, which is observed to be hyperacetylated at several residues, namely K9, K14, and K56 in an oral cancer cell line as well as in the tumor tissues of grade II OSCC patients. In addition to enhanced H3K9 acetylation, hyperacetylation of H2AK5 was also observed, whereas the H4K16ac level was significantly low and the level of H4K8ac was minimally altered in the same set of tumor tissues (Table 12.2) (Arif et al. 2010). However, an increase in H3K9ac and H4K16ac was observed to be associated with the absence of lymph node metastasis, initial tumors, and well-differentiated carcinomas of the head and neck (Noguchi et al. 2013).

Acetylated Histone Marks as Biomarkers for Oral Cancer

Hypoacetylation of H3K9 was observed to be associated with poor prognosis in OSCC and, therefore, may be used as a novel prognostic marker for OSCC. H3K9 was found to be hypoacetylated in OSCC compared to oral leukoplakia (OL), an early stage of carcinogenesis. A decrease in H3K9ac was concomitant with the increase in proliferation and EMT and was associated with the lower survival rates in patients (Webber et al. 2017). Furthermore, the patient group with a low score for H3K4ac exhibited poor survival with less percentage of cancer-specific survival/disease-free survival (CCS/DFS) as compared to the high score group (Chen et al. 2013).

Involvement of KATs in Imparting the Acetylation Signature Associated with Oral Cancer

Mutations or altered expression of lysine acetyltransferases have been observed in several tumors that contribute to cancer development. Mutations in EP300 and CBP in many HNSCC cell lines might contribute to HNSCC initiation and progression (Martin et al. 2014). In one of the studies, p300 was identified to be overexpressed in tumor tissues of OSCC patients, though no significant change was observed for CBP and PCAF. Not only the expression but the activity of p300 itself was found to be drastically elevated since the level of autoacetylated p300 (antibody against p300 acetylated at K1499) was more in tumor tissues than adjacent normal. This enhanced activity of p300 was attributed to an increase in protein levels of NPM1 and GAPDH (inducers of p300 autoacetylation) in tumor tissues which was nitric oxide (NO) signal-dependent (Arif et al. 2010). NPM1 and GAPDH, among several others, are the inducers of p300 autoacetylation whereby the basal catalytic activity of p300 is stimulated by trans-autoacetylation of several of its key lysine residues within activation loop motif (Thompson et al. 2004). Molecular chaperone activity of NPM1 through its reversible binding to p300 and oligomerization of NPM1 was shown to be essential for induction of p300 autoacetylation (Kaypee et al. 2018b). NPM1 itself gets acetylated by p300 that further enhances the transcriptional activation potential of NPM1. Levels of both NPM1 and acetylated form of NPM1 (AcNPM1) have been observed to be elevated in oral squamous cell carcinoma. A recent study has directly implicated the role of AcNPM1 in promoting oral tumorigenesis, whereby AcNPM1 was shown to function as coactivator during RNA Pol II-driven transcription of plethora of genes that includes several key genes involved in promoting oral carcinogenesis (Senapati et al. 2022). There are several other inducers of p300 autoacetylation identified so far. One such bona fide inducer is tumor suppressor p53. p53-mediated induction of p300 autoacetylation resulted in redistribution of ac-p300 through preferred enrichment of ac-p300 on the proximal promoter regions as against p300. This differential occupancy of ac-p300 upon p53 expression might assist in p53-mediated transcriptional initiation. Not only wild-type p53 but several of the gain-of-function mutant p53 were found to induce p300 autoacetylation through which they can modulate p300 functions in order to execute their oncogenic gain-of-function activities (Kaypee et al. 2018a).

Placenta-specific protein 2 (PLAC2), a long non-coding RNA, was identified to be upregulated in OSCC whose overexpression promoted proliferation, migration, and invasion of OSCC cells. H3K27ac was shown to be enriched at the promoter of the PLAC2 gene, which got reduced upon CBP depletion leading to reduced PLAC2 levels in OSCC cells (Chen et al. 2019). TIP60, a MYST family KAT, was found to be haploinsufficient in primary HNSCC since its reduced mRNA level was identified in tumors of HNSCC as compared to matched normal tissue (Gorrini et al. 2007). Another MYST family member, hMOF, was observed to be upregulated in human oral tongue squamous cell carcinoma (OTSCC), which predicted poor overall survival (OS) and disease-free survival (DFS). hMOF, which acetylates H4K16, regulated OTSCC growth by modulating promoter activity of enhancer of zeste homolog 2 (EZH2), a polycomb group protein (Li et al. 2015).

Concluding Remarks and Future Perspectives

Considerable progress has been made in the recent years toward understanding the links between metabolism and the epigenome. In this review, we mainly focused on metabolic control of acetylation through acetyl-CoA and NAD+ levels. Altered cofactor availability provides a direct path for modulation of the epigenome in response to altered environmental conditions such as nutrient availability, stress, and circadian cycles. This allows the cell to change its transcriptional output accordingly to adapt to the environmental stimuli. However, how locus-specific regulation of some genomic loci occurs in response to metabolic fluctuation of cofactor availability is not well understood. It can be partially explained by localized acetyl-CoA production by ACSS2, ACLY, and PDC. However, the mechanisms of locus-specific recruitment of these metabolic enzymes are not clear. In addition, there are other epigenetic modifications such as histone and DNA methylation that are also dependent on metabolic cofactors such as SAM controlled by the TCA cycle enzymes. The combinatorial effect of histone acetylation and methylation may determine the transcriptional outcome for some genes. Further studies are needed to understand the complex mechanisms by which metabolism regulates the epigenome.

We also discussed the metabolic alterations observed with age and in cancer cells that have consequences for transcriptional regulation and cell proliferation. Especially, in cancer cells, rewired metabolism ensures a continuous supply of acetyl-CoA through enhanced glycolysis that supports increased fatty acid synthesis for cell proliferation. Altered growth factor signaling in cancer cells also ensures increased acetyl-CoA production through increased ACLY activity. Due to the rewired metabolic program, cancer cells are highly dependent on glucose and glutamine for their survival and proliferation. Leveraging the link between metabolism and epigenetics for designing therapeutic targets and identifying distinct metabolic vulnerabilities would be key future areas of development. In addition, dietary changes can also modify the therapeutic response to some cancers by impacting the levels of metabolites such as acetate and butyrate. Diet can modify the risk for some types of cancer; however, the mechanisms are not well understood. Caloric restriction, for example, has been shown to promote longevity through altered NAD+ levels. Further studies would be needed to identify specific vulnerabilities in various tumor types that could be used to design strategies that include nutritional interventions in addition to therapeutics.

References

Agostini M, Silva SD, Zecchin KG, Coletta RD, Jorge J, Loda M, Graner E (2004) Fatty acid synthase is required for the proliferation of human oral squamous carcinoma cells. Oral Oncol 40(7):728–735. https://doi.org/10.1016/j.oraloncology.2004.01.011

Agostini M, Almeida LY, Bastos DC, Ortega RM, Moreira FS, Seguin F, Zecchin KG, Raposo HF, Oliveira HC, Amoedo ND, Salo T, Coletta RD, Graner E (2014) The fatty acid synthase inhibitor orlistat reduces the growth and metastasis of orthotopic tongue oral squamous cell carcinomas. Mol Cancer Ther 13(3):585–595. https://doi.org/10.1158/1535-7163.MCT-12-1136

Ahn BH, Kim HS, Song S, Lee IH, Liu J, Vassilopoulos A, Deng CX, Finkel T (2008) A role for the mitochondrial deacetylase Sirt3 in regulating energy homeostasis. Proc Natl Acad Sci U S A 105(38):14447–14452. https://doi.org/10.1073/pnas.0803790105

Arias-Alvarado A, Aghayev M, Ilchenko S, Rachdaoui N, Lepp J, Tsai TH, Zhang GF, Previs S, Kasumov T (2021) Measuring acetyl-CoA and acetylated histone turnover in vivo: effect of a high fat diet. Anal Biochem 615:114067. https://doi.org/10.1016/j.ab.2020.114067

Arif M, Vedamurthy BM, Choudhari R, Ostwal YB, Mantelingu K, Kodaganur GS, Kundu TK (2010) Nitric oxide-mediated histone hyperacetylation in oral cancer: target for a water-soluble HAT inhibitor, CTK7A. Chem Biol 17(8):903–913. https://doi.org/10.1016/j.chembiol.2010.06.014

Ayala FR, Rocha RM, Carvalho KC, Carvalho AL, da Cunha IW, Lourenco SV, Soares FA (2010) GLUT1 and GLUT3 as potential prognostic markers for Oral squamous cell carcinoma. Molecules 15(4):2374–2387. https://doi.org/10.3390/molecules15042374

Bandyopadhaya A, Tsurumi A, Maura D, Jeffrey KL, Rahme LG (2016) A quorum-sensing signal promotes host tolerance training through HDAC1-mediated epigenetic reprogramming. Nat Microbiol 1:16174. https://doi.org/10.1038/nmicrobiol.2016.174

Barber MF, Michishita-Kioi E, Xi Y, Tasselli L, Kioi M, Moqtaderi Z, Tennen RI, Paredes S, Young NL, Chen K, Struhl K, Garcia BA, Gozani O, Li W, Chua KF (2012) SIRT7 links H3K18 deacetylation to maintenance of oncogenic transformation. Nature 487(7405):114–118. https://doi.org/10.1038/nature11043

Barlesi F, Giaccone G, Gallegos-Ruiz MI, Loundou A, Span SW, Lefesvre P, Kruyt FA, Rodriguez JA (2007) Global histone modifications predict prognosis of resected non small-cell lung cancer. J Clin Oncol 25(28):4358–4364. https://doi.org/10.1200/JCO.2007.11.2599

Berger F, Lau C, Dahlmann M, Ziegler M (2005) Subcellular compartmentation and differential catalytic properties of the three human nicotinamide mononucleotide adenylyltransferase isoforms. J Biol Chem 280(43):36334–36341. https://doi.org/10.1074/jbc.M508660200

Berndsen CE, Denu JM (2008) Catalysis and substrate selection by histone/protein lysine acetyltransferases. Curr Opin Struct Biol 18(6):682–689. https://doi.org/10.1016/j.sbi.2008.11.004

Berndsen CE, Albaugh BN, Tan S, Denu JM (2007) Catalytic mechanism of a MYST family histone acetyltransferase. Biochemistry 46(3):623–629. https://doi.org/10.1021/bi602513x

Berwick DC, Hers I, Heesom KJ, Moule SK, Tavare JM (2002) The identification of ATP-citrate lyase as a protein kinase B (Akt) substrate in primary adipocytes. J Biol Chem 277(37):33895–33900. https://doi.org/10.1074/jbc.M204681200

Bharathi SS, Zhang Y, Mohsen AW, Uppala R, Balasubramani M, Schreiber E, Uechi G, Beck ME, Rardin MJ, Vockley J, Verdin E, Gibson BW, Hirschey MD, Goetzman ES (2013) Sirtuin 3 (SIRT3) protein regulates long-chain acyl-CoA dehydrogenase by deacetylating conserved lysines near the active site. J Biol Chem 288(47):33837–33847. https://doi.org/10.1074/jbc.M113.510354

Borra MT, Langer MR, Slama JT, Denu JM (2004) Substrate specificity and kinetic mechanism of the Sir2 family of NAD+-dependent histone/protein deacetylases. Biochemistry 43(30):9877–9887. https://doi.org/10.1021/bi049592e

Bose P, Brockton NT, Dort JC (2013) Head and neck cancer: from anatomy to biology. Int J Cancer 133(9):2013–2023. https://doi.org/10.1002/ijc.28112

Bradshaw PC (2021) Acetyl-CoA metabolism and histone acetylation in the regulation of aging and lifespan. Antioxidants (Basel) 10(4):doi:10.3390/antiox10040572

Bultman SJ (2014) Molecular pathways: gene-environment interactions regulating dietary fiber induction of proliferation and apoptosis via butyrate for cancer prevention. Clin Cancer Res 20(4):799–803. https://doi.org/10.1158/1078-0432.CCR-13-2483

Bulusu V, Tumanov S, Michalopoulou E, van den Broek NJ, MacKay G, Nixon C, Dhayade S, Schug ZT, Vande Voorde J, Blyth K, Gottlieb E, Vazquez A, Kamphorst JJ (2017) Acetate recapturing by nuclear acetyl-CoA Synthetase 2 prevents loss of histone acetylation during oxygen and serum limitation. Cell Rep 18(3):647–658. https://doi.org/10.1016/j.celrep.2016.12.055

Carrer A, Parris JLD, Trefely S, Henry RA, Montgomery DC, Torres A, Viola JM, Kuo YM, Blair IA, Meier JL, Andrews AJ, Snyder NW, Wellen KE (2017) Impact of a high-fat diet on tissue acyl-CoA and histone acetylation levels. J Biol Chem 292(8):3312–3322. https://doi.org/10.1074/jbc.M116.750620

Carrer A, Trefely S, Zhao S, Campbell SL, Norgard RJ, Schultz KC, Sidoli S, Parris JLD, Affronti HC, Sivanand S, Egolf S, Sela Y, Trizzino M, Gardini A, Garcia BA, Snyder NW, Stanger BZ, Wellen KE (2019) Acetyl-CoA metabolism supports multistep pancreatic tumorigenesis. Cancer Discov 9(3):416–435. https://doi.org/10.1158/2159-8290.CD-18-0567

Chen H, Lin RJ, Xie W, Wilpitz D, Evans RM (1999) Regulation of hormone-induced histone hyperacetylation and gene activation via acetylation of an acetylase. Cell 98(5):675–686. https://doi.org/10.1016/s0092-8674(00)80054-9

Chen YW, Kao SY, Wang HJ, Yang MH (2013) Histone modification patterns correlate with patient outcome in oral squamous cell carcinoma. Cancer 119(24):4259–4267. https://doi.org/10.1002/cncr.28356

Chen F, Qi S, Zhang X, Wu J, Yang X, Wang R (2019) lncRNA PLAC2 activated by H3K27 acetylation promotes cell proliferation and invasion via the activation of Wnt/betacatenin pathway in oral squamous cell carcinoma. Int J Oncol 54(4):1183–1194. https://doi.org/10.3892/ijo.2019.4707

Choudhary C, Weinert BT, Nishida Y, Verdin E, Mann M (2014) The growing landscape of lysine acetylation links metabolism and cell signalling. Nat Rev Mol Cell Biol 15(8):536–550. https://doi.org/10.1038/nrm3841

Cimen H, Han MJ, Yang Y, Tong Q, Koc H, Koc EC (2010) Regulation of succinate dehydrogenase activity by SIRT3 in mammalian mitochondria. Biochemistry 49(2):304–311. https://doi.org/10.1021/bi901627u

Comerford SA, Huang Z, Du X, Wang Y, Cai L, Witkiewicz AK, Walters H, Tantawy MN, Fu A, Manning HC, Horton JD, Hammer RE, McKnight SL, Tu BP (2014) Acetate dependence of tumors. Cell 159(7):1591–1602. https://doi.org/10.1016/j.cell.2014.11.020

Cousens LS, Gallwitz D, Alberts BM (1979) Different accessibilities in chromatin to histone acetylase. J Biol Chem 254(5):1716–1723

Cummings JH, Pomare EW, Branch WJ, Naylor CP, Macfarlane GT (1987) Short chain fatty acids in human large intestine, portal, hepatic and venous blood. Gut 28(10):1221–1227. https://doi.org/10.1136/gut.28.10.1221

Dang CV (2013) MYC, metabolism, cell growth, and tumorigenesis. Cold Spring Harb Perspect Med 3(8):a014217. https://doi.org/10.1101/cshperspect.a014217

Das C, Lucia MS, Hansen KC, Tyler JK (2009) CBP/p300-mediated acetylation of histone H3 on lysine 56. Nature 459(7243):113–117. https://doi.org/10.1038/nature07861

Davie JR (2003) Inhibition of histone deacetylase activity by butyrate. J Nutr 133(7 Suppl):2485S–2493S. https://doi.org/10.1093/jn/133.7.2485S

De Virgilio C, Burckert N, Barth G, Neuhaus JM, Boller T, Wiemken A (1992) Cloning and disruption of a gene required for growth on acetate but not on ethanol: the acetyl-coenzyme a synthetase gene of Saccharomyces cerevisiae. Yeast 8(12):1043–1051. https://doi.org/10.1002/yea.320081207

DeBerardinis RJ, Mancuso A, Daikhin E, Nissim I, Yudkoff M, Wehrli S, Thompson CB (2007) Beyond aerobic glycolysis: transformed cells can engage in glutamine metabolism that exceeds the requirement for protein and nucleotide synthesis. Proc Natl Acad Sci U S A 104(49): 19345–19350. https://doi.org/10.1073/pnas.0709747104

Ding SZ, Fischer W, Kaparakis-Liaskos M, Liechti G, Merrell DS, Grant PA, Ferrero RL, Crowe SE, Haas R, Hatakeyama M, Goldberg JB (2010) Helicobacter pylori-induced histone modification, associated gene expression in gastric epithelial cells, and its implication in pathogenesis. PLoS One 5(4):e9875. https://doi.org/10.1371/journal.pone.0009875

Donohoe DR, Collins LB, Wali A, Bigler R, Sun W, Bultman SJ (2012) The Warburg effect dictates the mechanism of butyrate-mediated histone acetylation and cell proliferation. Mol Cell 48(4):612–626. https://doi.org/10.1016/j.molcel.2012.08.033

Donohoe DR, Holley D, Collins LB, Montgomery SA, Whitmore AC, Hillhouse A, Curry KP, Renner SW, Greenwalt A, Ryan EP, Godfrey V, Heise MT, Threadgill DS, Han A, Swenberg JA, Threadgill DW, Bultman SJ (2014) A gnotobiotic mouse model demonstrates that dietary fiber protects against colorectal tumorigenesis in a microbiota- and butyrate-dependent manner. Cancer Discov 4(12):1387–1397. https://doi.org/10.1158/2159-8290.CD-14-0501

Drazic A, Myklebust LM, Ree R, Arnesen T (2016) The world of protein acetylation. Biochim Biophys Acta 1864(10):1372–1401. https://doi.org/10.1016/j.bbapap.2016.06.007

Eisenberg T, Schroeder S, Andryushkova A, Pendl T, Kuttner V, Bhukel A, Marino G, Pietrocola F, Harger A, Zimmermann A, Moustafa T, Sprenger A, Jany E, Buttner S, Carmona-Gutierrez D, Ruckenstuhl C, Ring J, Reichelt W, Schimmel K, Leeb T, Moser C, Schatz S, Kamolz LP, Magnes C, Sinner F, Sedej S, Frohlich KU, Juhasz G, Pieber TR, Dengjel J, Sigrist SJ, Kroemer G, Madeo F (2014) Nucleocytosolic depletion of the energy metabolite acetyl-coenzyme a stimulates autophagy and prolongs lifespan. Cell Metab 19(3):431–444. https://doi.org/10.1016/j.cmet.2014.02.010

Eldridge MJG, Hamon MA (2021) Histone H3 deacetylation promotes host cell viability for efficient infection by listeria monocytogenes. PLoS Pathog 17(12):e1010173. https://doi.org/10.1371/journal.ppat.1010173

Elsheikh SE, Green AR, Rakha EA, Powe DG, Ahmed RA, Collins HM, Soria D, Garibaldi JM, Paish CE, Ammar AA, Grainge MJ, Ball GR, Abdelghany MK, Martinez-Pomares L, Heery DM, Ellis IO (2009) Global histone modifications in breast cancer correlate with tumor phenotypes, prognostic factors, and patient outcome. Cancer Res 69(9):3802–3809. https://doi.org/10.1158/0008-5472.CAN-08-3907

Eskandarian HA, Impens F, Nahori MA, Soubigou G, Coppee JY, Cossart P, Hamon MA (2013) A role for SIRT2-dependent histone H3K18 deacetylation in bacterial infection. Science 341(6145):1238858. https://doi.org/10.1126/science.1238858

Evertts AG, Zee BM, Dimaggio PA, Gonzales-Cope M, Coller HA, Garcia BA (2013) Quantitative dynamics of the link between cellular metabolism and histone acetylation. J Biol Chem 288(17): 12142–12151. https://doi.org/10.1074/jbc.M112.428318

Fan J, Shan C, Kang HB, Elf S, Xie J, Tucker M, Gu TL, Aguiar M, Lonning S, Chen H, Mohammadi M, Britton LM, Garcia BA, Aleckovic M, Kang Y, Kaluz S, Devi N, Van Meir EG, Hitosugi T, Seo JH, Lonial S, Gaddh M, Arellano M, Khoury HJ, Khuri FR, Boggon TJ, Kang S, Chen J (2014) Tyr phosphorylation of PDP1 toggles recruitment between ACAT1 and SIRT3 to regulate the pyruvate dehydrogenase complex. Mol Cell 53(4):534–548. https://doi.org/10.1016/j.molcel.2013.12.026

Feldman JL, Baeza J, Denu JM (2013) Activation of the protein deacetylase SIRT6 by long-chain fatty acids and widespread deacylation by mammalian sirtuins. J Biol Chem 288(43): 31350–31356. https://doi.org/10.1074/jbc.C113.511261

Finley LW, Haas W, Desquiret-Dumas V, Wallace DC, Procaccio V, Gygi SP, Haigis MC (2011) Succinate dehydrogenase is a direct target of sirtuin 3 deacetylase activity. PLoS One 6(8): e23295. https://doi.org/10.1371/journal.pone.0023295

Fjeld CC, Birdsong WT, Goodman RH (2003) Differential binding of NAD+ and NADH allows the transcriptional corepressor carboxyl-terminal binding protein to serve as a metabolic sensor. Proc Natl Acad Sci U S A 100(16):9202–9207. https://doi.org/10.1073/pnas.1633591100

Fraga MF, Ballestar E, Villar-Garea A, Boix-Chornet M, Espada J, Schotta G, Bonaldi T, Haydon C, Ropero S, Petrie K, Iyer NG, Perez-Rosado A, Calvo E, Lopez JA, Cano A, Calasanz MJ, Colomer D, Piris MA, Ahn N, Imhof A, Caldas C, Jenuwein T, Esteller M (2005) Loss of acetylation at Lys16 and trimethylation at Lys20 of histone H4 is a common hallmark of human cancer. Nat Genet 37(4):391–400. https://doi.org/10.1038/ng1531

Fullgrabe J, Kavanagh E, Joseph B (2011) Histone onco-modifications. Oncogene 30(31): 3391–3403. https://doi.org/10.1038/onc.2011.121

Galdieri L, Vancura A (2012) Acetyl-CoA carboxylase regulates global histone acetylation. J Biol Chem 287(28):23865–23876. https://doi.org/10.1074/jbc.M112.380519

Galdieri L, Gatla H, Vancurova I, Vancura A (2016) Activation of AMP-activated protein kinase by metformin induces protein acetylation in prostate and ovarian cancer cells. J Biol Chem 291(48): 25154–25166. https://doi.org/10.1074/jbc.M116.742247

Gao X, Lin SH, Ren F, Li JT, Chen JJ, Yao CB, Yang HB, Jiang SX, Yan GQ, Wang D, Wang Y, Liu Y, Cai Z, Xu YY, Chen J, Yu W, Yang PY, Lei QY (2016) Acetate functions as an epigenetic metabolite to promote lipid synthesis under hypoxia. Nat Commun 7:11960. https://doi.org/10.1038/ncomms11960

Garcia-Garcia JC, Barat NC, Trembley SJ, Dumler JS (2009) Epigenetic silencing of host cell defense genes enhances intracellular survival of the rickettsial pathogen Anaplasma phagocytophilum. PLoS Pathog 5(6):e1000488. https://doi.org/10.1371/journal.ppat.1000488

Giudice FS, Pinto DS Jr, Nor JE, Squarize CH, Castilho RM (2013) Inhibition of histone deacetylase impacts cancer stem cells and induces epithelial-mesenchyme transition of head and neck cancer. PLoS One 8(3):e58672. https://doi.org/10.1371/journal.pone.0058672

Gonzalez A, Hall MN, Lin SC, Hardie DG (2020) AMPK and TOR: the yin and Yang of cellular nutrient sensing and growth control. Cell Metab 31(3):472–492. https://doi.org/10.1016/j.cmet.2020.01.015

Gorrini C, Squatrito M, Luise C, Syed N, Perna D, Wark L, Martinato F, Sardella D, Verrecchia A, Bennett S, Confalonieri S, Cesaroni M, Marchesi F, Gasco M, Scanziani E, Capra M, Mai S, Nuciforo P, Crook T, Lough J, Amati B (2007) Tip60 is a haplo-insufficient tumour suppressor required for an oncogene-induced DNA damage response. Nature 448(7157):1063–1067. https://doi.org/10.1038/nature06055

Grabiec AM, Potempa J (2018) Epigenetic regulation in bacterial infections: targeting histone deacetylases. Crit Rev Microbiol 44(3):336–350. https://doi.org/10.1080/1040841X.2017.1373063

Grimm M, Alexander D, Munz A, Hoffmann J, Reinert S (2013) Increased LDH5 expression is associated with lymph node metastasis and outcome in oral squamous cell carcinoma. Clin Exp Metastasis 30(4):529–540. https://doi.org/10.1007/s10585-012-9557-2

Grimm M, Cetindis M, Lehmann M, Biegner T, Munz A, Teriete P, Kraut W, Reinert S (2014a) Association of cancer metabolism-related proteins with oral carcinogenesis - indications for chemoprevention and metabolic sensitizing of oral squamous cell carcinoma? J Transl Med 12: 208. https://doi.org/10.1186/1479-5876-12-208

Grimm M, Munz A, Teriete P, Nadtotschi T, Reinert S (2014b) GLUT-1(+)/TKTL1(+) coexpression predicts poor outcome in oral squamous cell carcinoma. Oral Surg Oral Med Oral Pathol Oral Radiol 117(6):743–753. https://doi.org/10.1016/j.oooo.2014.02.007

Gujral P, Mahajan V, Lissaman AC, Ponnampalam AP (2020) Histone acetylation and the role of histone deacetylases in normal cyclic endometrium. Reprod Biol Endocrinol 18(1):84. https://doi.org/10.1186/s12958-020-00637-5

Guy PS, Cohen P, Hardie DG (1980) Rat mammary gland ATP-citrate lyase is phosphorylated by cyclic amp-dependent protein kinase. FEBS Lett 109(2):205–208. https://doi.org/10.1016/0014-5793(80)81087-8

Haigis MC, Sinclair DA (2010) Mammalian sirtuins: biological insights and disease relevance. Annu Rev Pathol 5:253–295. https://doi.org/10.1146/annurev.pathol.4.110807.092250

Hallows WC, Lee S, Denu JM (2006) Sirtuins deacetylate and activate mammalian acetyl-CoA synthetases. Proc Natl Acad Sci U S A 103(27):10230–10235. https://doi.org/10.1073/pnas.0604392103

Hamon MA, Batsche E, Regnault B, Tham TN, Seveau S, Muchardt C, Cossart P (2007) Histone modifications induced by a family of bacterial toxins. Proc Natl Acad Sci U S A 104(33): 13467–13472. https://doi.org/10.1073/pnas.0702729104

Hatzivassiliou G, Zhao F, Bauer DE, Andreadis C, Shaw AN, Dhanak D, Hingorani SR, Tuveson DA, Thompson CB (2005) ATP citrate lyase inhibition can suppress tumor cell growth. Cancer Cell 8(4):311–321. https://doi.org/10.1016/j.ccr.2005.09.008

Haws SA, Leech CM, Denu JM (2020) Metabolism and the epigenome: a dynamic relationship. Trends Biochem Sci 45(9):731–747. https://doi.org/10.1016/j.tibs.2020.04.002

Hirschey MD, Shimazu T, Goetzman E, Jing E, Schwer B, Lombard DB, Grueter CA, Harris C, Biddinger S, Ilkayeva OR, Stevens RD, Li Y, Saha AK, Ruderman NB, Bain JR, Newgard CB, Farese RV Jr, Alt FW, Kahn CR, Verdin E (2010) SIRT3 regulates mitochondrial fatty-acid oxidation by reversible enzyme deacetylation. Nature 464(7285):121–125. https://doi.org/10.1038/nature08778

Huang JC, Yan LY, Lei ZL, Miao YL, Shi LH, Yang JW, Wang Q, Ouyang YC, Sun QY, Chen DY (2007) Changes in histone acetylation during postovulatory aging of mouse oocyte. Biol Reprod 77(4):666–670. https://doi.org/10.1095/biolreprod.107.062703

Imai S, Armstrong CM, Kaeberlein M, Guarente L (2000) Transcriptional silencing and longevity protein Sir2 is an NAD-dependent histone deacetylase. Nature 403(6771):795–800. https://doi.org/10.1038/35001622

Izumi H, Ohta R, Nagatani G, Ise T, Nakayama Y, Nomoto M, Kohno K (2003) p300/CBP-associated factor (P/CAF) interacts with nuclear respiratory factor-1 to regulate the UDP-N-acetyl-alpha-d-galactosamine: polypeptide N-acetylgalactosaminyltransferase-3 gene. Biochem J 373(Pt 3):713–722. https://doi.org/10.1042/BJ20021902

Jin L, Galonek H, Israelian K, Choy W, Morrison M, Xia Y, Wang X, Xu Y, Yang Y, Smith JJ, Hoffmann E, Carney DP, Perni RB, Jirousek MR, Bemis JE, Milne JC, Sinclair DA, Westphal CH (2009) Biochemical characterization, localization, and tissue distribution of the longer form of mouse SIRT3. Protein Sci 18(3):514–525. https://doi.org/10.1002/pro.50

Kamphorst JJ, Chung MK, Fan J, Rabinowitz JD (2014) Quantitative analysis of acetyl-CoA production in hypoxic cancer cells reveals substantial contribution from acetate. Cancer Metab 2:23. https://doi.org/10.1186/2049-3002-2-23

Kaypee S, Sahadevan SA, Patil S, Ghosh P, Roy NS, Roy S, Kundu TK (2018a) Mutant and wild-type tumor suppressor p53 induces p300 autoacetylation. iScience 4:260–272. https://doi.org/10.1016/j.isci.2018.06.002

Kaypee S, Sahadevan SA, Sudarshan D, Halder Sinha S, Patil S, Senapati P, Kodaganur GS, Mohiyuddin A, Dasgupta D, Kundu TK (2018b) Oligomers of human histone chaperone NPM1 alter p300/KAT3B folding to induce autoacetylation. Biochim Biophys Acta Gen Subj 1862(8): 1729–1741. https://doi.org/10.1016/j.bbagen.2018.05.003

Keith B, Johnson RS, Simon MC (2011) HIF1alpha and HIF2alpha: sibling rivalry in hypoxic tumour growth and progression. Nat Rev Cancer 12(1):9–22. https://doi.org/10.1038/nrc3183

Kim JW, Tchernyshyov I, Semenza GL, Dang CV (2006) HIF-1-mediated expression of pyruvate dehydrogenase kinase: a metabolic switch required for cellular adaptation to hypoxia. Cell Metab 3(3):177–185. https://doi.org/10.1016/j.cmet.2006.02.002

Kinnaird A, Zhao S, Wellen KE, Michelakis ED (2016) Metabolic control of epigenetics in cancer. Nat Rev Cancer 16(11):694–707. https://doi.org/10.1038/nrc.2016.82

Kouzarides T (2007) Chromatin modifications and their function. Cell 128(4):693–705. https://doi.org/10.1016/j.cell.2007.02.005

Krontiras H, Roye GD, Beenken SE, Myers RB, Mayo MS, Peters GE, Grizzle WE (1999) Fatty acid synthase expression is increased in neoplastic lesions of the oral tongue. Head Neck 21(4): 325–329. https://doi.org/10.1002/(sici)1097-0347(199907)21:4<325::aid-hed6>3.0.co;2-p

Kusumoto T, Hamada T, Yamada N, Nagata S, Kanmura Y, Houjou I, Kamikawa Y, Yonezawa S, Sugihara K (2012) Comprehensive epigenetic analysis using oral rinse samples: a pilot study. J Oral Maxillofac Surg 70(6):1486–1494. https://doi.org/10.1016/j.joms.2011.04.021

Langer MR, Fry CJ, Peterson CL, Denu JM (2002) Modulating acetyl-CoA binding in the GCN5 family of histone acetyltransferases. J Biol Chem 277(30):27337–27344. https://doi.org/10.1074/jbc.M203251200

Lau OD, Courtney AD, Vassilev A, Marzilli LA, Cotter RJ, Nakatani Y, Cole PA (2000) p300/CBP-associated factor histone acetyltransferase processing of a peptide substrate. Kinetic analysis of the catalytic mechanism. J Biol Chem 275(29):21953–21959. https://doi.org/10.1074/jbc.M003219200

Lee JV, Carrer A, Shah S, Snyder NW, Wei S, Venneti S, Worth AJ, Yuan ZF, Lim HW, Liu S, Jackson E, Aiello NM, Haas NB, Rebbeck TR, Judkins A, Won KJ, Chodosh LA, Garcia BA, Stanger BZ, Feldman MD, Blair IA, Wellen KE (2014) Akt-dependent metabolic reprogramming regulates tumor cell histone acetylation. Cell Metab 20(2):306–319. https://doi.org/10.1016/j.cmet.2014.06.004

Lerin C, Rodgers JT, Kalume DE, Kim SH, Pandey A, Puigserver P (2006) GCN5 acetyltransferase complex controls glucose metabolism through transcriptional repression of PGC-1alpha. Cell Metab 3(6):429–438. https://doi.org/10.1016/j.cmet.2006.04.013

Li Q, Sun H, Shu Y, Zou X, Zhao Y, Ge C (2015) hMOF (human males absent on the first), an oncogenic protein of human oral tongue squamous cell carcinoma, targeting EZH2 (enhancer of zeste homolog 2). Cell Prolif 48(4):436–442. https://doi.org/10.1111/cpr.12177

Li CX, Sun JL, Gong ZC, Lin ZQ, Liu H (2016) Prognostic value of GLUT-1 expression in oral squamous cell carcinoma: a prisma-compliant meta-analysis. Medicine (Baltimore) 95(45): e5324. https://doi.org/10.1097/MD.0000000000005324

Li X, Yu W, Qian X, Xia Y, Zheng Y, Lee JH, Li W, Lyu J, Rao G, Zhang X, Qian CN, Rozen SG, Jiang T, Lu Z (2017) Nucleus-translocated ACSS2 promotes gene transcription for lysosomal biogenesis and autophagy. Mol Cell 66 (5):684-697:e689. https://doi.org/10.1016/j.molcel.2017.04.026

Lim JH, Lee YM, Chun YS, Chen J, Kim JE, Park JW (2010) Sirtuin 1 modulates cellular responses to hypoxia by deacetylating hypoxia-inducible factor 1alpha. Mol Cell 38(6):864–878. https://doi.org/10.1016/j.molcel.2010.05.023

Lim JH, Gerhart-Hines Z, Dominy JE, Lee Y, Kim S, Tabata M, Xiang YK, Puigserver P (2013) Oleic acid stimulates complete oxidation of fatty acids through protein kinase A-dependent activation of SIRT1-PGC1alpha complex. J Biol Chem 288(10):7117–7126. https://doi.org/10.1074/jbc.M112.415729

Lin R, Tao R, Gao X, Li T, Zhou X, Guan KL, Xiong Y, Lei QY (2013) Acetylation stabilizes ATP-citrate lyase to promote lipid biosynthesis and tumor growth. Mol Cell 51(4):506–518. https://doi.org/10.1016/j.molcel.2013.07.002

Liu X, Wang L, Zhao K, Thompson PR, Hwang Y, Marmorstein R, Cole PA (2008) The structural basis of protein acetylation by the p300/CBP transcriptional coactivator. Nature 451(7180): 846–850. https://doi.org/10.1038/nature06546

Locasale JW, Cantley LC (2011) Metabolic flux and the regulation of mammalian cell growth. Cell Metab 14(4):443–451. https://doi.org/10.1016/j.cmet.2011.07.014

Lv L, Xu YP, Zhao D, Li FL, Wang W, Sasaki N, Jiang Y, Zhou X, Li TT, Guan KL, Lei QY, Xiong Y (2013) Mitogenic and oncogenic stimulation of K433 acetylation promotes PKM2 protein kinase activity and nuclear localization. Mol Cell 52(3):340–352. https://doi.org/10.1016/j.molcel.2013.09.004

Manning BD, Toker A (2017) AKT/PKB signaling: navigating the network. Cell 169(3):381–405. https://doi.org/10.1016/j.cell.2017.04.001

Manosalva I, Gonzalez A (2009) Aging alters histone H4 acetylation and CDC2A in mouse germinal vesicle stage oocytes. Biol Reprod 81(6):1164–1171. https://doi.org/10.1095/biolreprod.109.078386

Manuyakorn A, Paulus R, Farrell J, Dawson NA, Tze S, Cheung-Lau G, Hines OJ, Reber H, Seligson DB, Horvath S, Kurdistani SK, Guha C, Dawson DW (2010) Cellular histone modification patterns predict prognosis and treatment response in resectable pancreatic

adenocarcinoma: results from RTOG 9704. J Clin Oncol 28(8):1358–1365. https://doi.org/10.1200/JCO.2009.24.5639
Martin D, Abba MC, Molinolo AA, Vitale-Cross L, Wang Z, Zaida M, Delic NC, Samuels Y, Lyons JG, Gutkind JS (2014) The head and neck cancer cell oncogenome: a platform for the development of precision molecular therapies. Oncotarget 5(19):8906–8923. https://doi.org/10.18632/oncotarget.2417
Mashimo T, Pichumani K, Vemireddy V, Hatanpaa KJ, Singh DK, Sirasanagandla S, Nannepaga S, Piccirillo SG, Kovacs Z, Foong C, Huang Z, Barnett S, Mickey BE, DeBerardinis RJ, Tu BP, Maher EA, Bachoo RM (2014) Acetate is a bioenergetic substrate for human glioblastoma and brain metastases. Cell 159(7):1603–1614. https://doi.org/10.1016/j.cell.2014.11.025
Masui K, Tanaka K, Ikegami S, Villa GR, Yang H, Yong WH, Cloughesy TF, Yamagata K, Arai N, Cavenee WK, Mischel PS (2015) Glucose-dependent acetylation of Rictor promotes targeted cancer therapy resistance. Proc Natl Acad Sci U S A 112(30):9406–9411. https://doi.org/10.1073/pnas.1511759112
McDonnell E, Crown SB, Fox DB, Kitir B, Ilkayeva OR, Olsen CA, Grimsrud PA, Hirschey MD (2016) Lipids reprogram metabolism to become a major carbon source for histone acetylation. Cell Rep 17(6):1463–1472. https://doi.org/10.1016/j.celrep.2016.10.012
Metallo CM, Gameiro PA, Bell EL, Mattaini KR, Yang J, Hiller K, Jewell CM, Johnson ZR, Irvine DJ, Guarente L, Kelleher JK, Vander Heiden MG, Iliopoulos O, Stephanopoulos G (2011) Reductive glutamine metabolism by IDH1 mediates lipogenesis under hypoxia. Nature 481(7381):380–384. https://doi.org/10.1038/nature10602
Mews P, Donahue G, Drake AM, Luczak V, Abel T, Berger SL (2017) Acetyl-CoA synthetase regulates histone acetylation and hippocampal memory. Nature 546(7658):381–386. https://doi.org/10.1038/nature22405
Morrish F, Noonan J, Perez-Olsen C, Gafken PR, Fitzgibbon M, Kelleher J, VanGilst M, Hockenbery D (2010) Myc-dependent mitochondrial generation of acetyl-CoA contributes to fatty acid biosynthesis and histone acetylation during cell cycle entry. J Biol Chem 285(47):36267–36274. https://doi.org/10.1074/jbc.M110.141606
Mosashvilli D, Kahl P, Mertens C, Holzapfel S, Rogenhofer S, Hauser S, Buttner R, Von Ruecker A, Muller SC, Ellinger J (2010) Global histone acetylation levels: prognostic relevance in patients with renal cell carcinoma. Cancer Sci 101(12):2664–2669. https://doi.org/10.1111/j.1349-7006.2010.01717.x
Mullen AR, Wheaton WW, Jin ES, Chen PH, Sullivan LB, Cheng T, Yang Y, Linehan WM, Chandel NS, DeBerardinis RJ (2011) Reductive carboxylation supports growth in tumour cells with defective mitochondria. Nature 481(7381):385–388. https://doi.org/10.1038/nature10642
Nakazato K, Mogushi K, Kayamori K, Tsuchiya M, Takahashi KI, Sumino J, Michi Y, Yoda T, Uzawa N (2019) Glucose metabolism changes during the development and progression of oral tongue squamous cell carcinomas. Oncol Lett 18(2):1372–1380. https://doi.org/10.3892/ol.2019.10420
Newman JC, Verdin E (2014) Beta-hydroxybutyrate: much more than a metabolite. Diabetes Res Clin Pract 106(2):173–181. https://doi.org/10.1016/j.diabres.2014.08.009
Noguchi A, Li X, Kubota A, Kikuchi K, Kameda Y, Zheng H, Miyagi Y, Aoki I, Takano Y (2013) SIRT1 expression is associated with good prognosis for head and neck squamous cell carcinoma patients. Oral Surg Oral Med Oral Pathol Oral Radiol 115(3):385–392. https://doi.org/10.1016/j.oooo.2012.12.013
Pan PW, Feldman JL, Devries MK, Dong A, Edwards AM, Denu JM (2011) Structure and biochemical functions of SIRT6. J Biol Chem 286(16):14575–14587. https://doi.org/10.1074/jbc.M111.218990
Patel MS, Roche TE (1990) Molecular biology and biochemistry of pyruvate dehydrogenase complexes. FASEB J 4(14):3224–3233. https://doi.org/10.1096/fasebj.4.14.2227213
Peleg S, Sananbenesi F, Zovoilis A, Burkhardt S, Bahari-Javan S, Agis-Balboa RC, Cota P, Wittnam JL, Gogol-Doering A, Opitz L, Salinas-Riester G, Dettenhofer M, Kang H, Farinelli L, Chen W, Fischer A (2010) Altered histone acetylation is associated with age-dependent memory impairment in mice. Science 328(5979):753–756. https://doi.org/10.1126/science.1186088

Peleg S, Feller C, Forne I, Schiller E, Sevin DC, Schauer T, Regnard C, Straub T, Prestel M, Klima C, Schmitt Nogueira M, Becker L, Klopstock T, Sauer U, Becker PB, Imhof A, Ladurner AG (2016) Life span extension by targeting a link between metabolism and histone acetylation in drosophila. EMBO Rep 17(3):455–469. https://doi.org/10.15252/embr.201541132

Pfister S, Rea S, Taipale M, Mendrzyk F, Straub B, Ittrich C, Thuerigen O, Sinn HP, Akhtar A, Lichter P (2008) The histone acetyltransferase hMOF is frequently downregulated in primary breast carcinoma and medulloblastoma and constitutes a biomarker for clinical outcome in medulloblastoma. Int J Cancer 122(6):1207–1213. https://doi.org/10.1002/ijc.23283

Pittelli M, Formentini L, Faraco G, Lapucci A, Rapizzi E, Cialdai F, Romano G, Moneti G, Moroni F, Chiarugi A (2010) Inhibition of nicotinamide phosphoribosyltransferase: cellular bioenergetics reveals a mitochondrial insensitive NAD pool. J Biol Chem 285(44): 34106–34114. https://doi.org/10.1074/jbc.M110.136739

Potapova IA, El-Maghrabi MR, Doronin SV, Benjamin WB (2000) Phosphorylation of recombinant human ATP:citrate lyase by cAMP-dependent protein kinase abolishes homotropic allosteric regulation of the enzyme by citrate and increases the enzyme activity. Allosteric activation of ATP:citrate lyase by phosphorylated sugars. Biochemistry 39(5):1169–1179. https://doi.org/10.1021/bi992159y

Reid MA, Dai Z, Locasale JW (2017) The impact of cellular metabolism on chromatin dynamics and epigenetics. Nat Cell Biol 19(11):1298–1306. https://doi.org/10.1038/ncb3629

Rennoll-Bankert KE, Garcia-Garcia JC, Sinclair SH, Dumler JS (2015) Chromatin-bound bacterial effector ankyrin a recruits histone deacetylase 1 and modifies host gene expression. Cell Microbiol 17(11):1640–1652. https://doi.org/10.1111/cmi.12461

Revollo JR, Grimm AA, Imai S (2004) The NAD biosynthesis pathway mediated by nicotinamide phosphoribosyltransferase regulates Sir2 activity in mammalian cells. J Biol Chem 279(49): 50754–50763. https://doi.org/10.1074/jbc.M408388200

Riggs MG, Whittaker RG, Neumann JR, Ingram VM (1977) N-butyrate causes histone modification in HeLa and friend erythroleukaemia cells. Nature 268(5619):462–464. https://doi.org/10.1038/268462a0

Rohrig F, Schulze A (2016) The multifaceted roles of fatty acid synthesis in cancer. Nat Rev Cancer 16(11):732–749. https://doi.org/10.1038/nrc.2016.89

Sant'Anna-Silva ACB, Santos GC, Campos SPC, Oliveira Gomes AM, Perez-Valencia JA, Rumjanek FD (2018) Metabolic profile of Oral squamous carcinoma cell lines relies on a higher demand of lipid metabolism in metastatic cells. Front Oncol 8:13. https://doi.org/10.3389/fonc.2018.00013

Schug ZT, Peck B, Jones DT, Zhang Q, Grosskurth S, Alam IS, Goodwin LM, Smethurst E, Mason S, Blyth K, McGarry L, James D, Shanks E, Kalna G, Saunders RE, Jiang M, Howell M, Lassailly F, Thin MZ, Spencer-Dene B, Stamp G, van den Broek NJ, Mackay G, Bulusu V, Kamphorst JJ, Tardito S, Strachan D, Harris AL, Aboagye EO, Critchlow SE, Wakelam MJ, Schulze A, Gottlieb E (2015) Acetyl-CoA synthetase 2 promotes acetate utilization and maintains cancer cell growth under metabolic stress. Cancer Cell 27(1):57–71. https://doi.org/10.1016/j.ccell.2014.12.002

Sekhavat A, Sun JM, Davie JR (2007) Competitive inhibition of histone deacetylase activity by trichostatin a and butyrate. Biochem Cell Biol 85(6):751–758. https://doi.org/10.1139/o07-145

Seligson DB, Horvath S, Shi T, Yu H, Tze S, Grunstein M, Kurdistani SK (2005) Global histone modification patterns predict risk of prostate cancer recurrence. Nature 435(7046):1262–1266. https://doi.org/10.1038/nature03672

Senapati P, Kato H, Lee M, Leung A, Thai C, Sanchez A, Gallagher EJ, LeRoith D, Seewaldt VL, Ann DK, Schones DE (2019) Hyperinsulinemia promotes aberrant histone acetylation in triple-negative breast cancer. Epigenetics Chromatin 12(1):44. https://doi.org/10.1186/s13072-019-0290-9

Senapati P, Bhattacharya A, Das S, Dey S, Sudarshan D, Shyla G, Vishwakarma J, Sudevan S, Ramachandran R, Maliekal TT, Kundu TK (2022) Histone chaperone Nucleophosmin regulates transcription of key genes involved in Oral tumorigenesis. Mol Cell Biol 42(2):e0066920. https://doi.org/10.1128/MCB.00669-20

Shan C, Elf S, Ji Q, Kang HB, Zhou L, Hitosugi T, Jin L, Lin R, Zhang L, Seo JH, Xie J, Tucker M, Gu TL, Sudderth J, Jiang L, DeBerardinis RJ, Wu S, Li Y, Mao H, Chen PR, Wang D, Chen GZ, Lonial S, Arellano ML, Khoury HJ, Khuri FR, Lee BH, Brat DJ, Ye K, Boggon TJ, He C, Kang S, Fan J, Chen J (2014) Lysine acetylation activates 6-phosphogluconate dehydrogenase to promote tumor growth. Mol Cell 55(4):552–565. https://doi.org/10.1016/j.molcel.2014.06.020

Sheikh BN, Akhtar A (2019) The many lives of KATs - detectors, integrators and modulators of the cellular environment. Nat Rev Genet 20(1):7–23. https://doi.org/10.1038/s41576-018-0072-4

Shimazu T, Hirschey MD, Newman J, He W, Shirakawa K, Le Moan N, Grueter CA, Lim H, Saunders LR, Stevens RD, Newgard CB, Farese RV Jr, de Cabo R, Ulrich S, Akassoglou K, Verdin E (2013) Suppression of oxidative stress by beta-hydroxybutyrate, an endogenous histone deacetylase inhibitor. Science 339(6116):211–214. https://doi.org/10.1126/science.1227166

Silva SD, Agostini M, Nishimoto IN, Coletta RD, Alves FA, Lopes MA, Kowalski LP, Graner E (2004) Expression of fatty acid synthase, ErbB2 and Ki-67 in head and neck squamous cell carcinoma. A clinicopathological study. Oral Oncol 40(7):688–696. https://doi.org/10.1016/j.oraloncology.2004.01.004

Simpson NE, Tryndyak VP, Pogribna M, Beland FA, Pogribny IP (2012) Modifying metabolically sensitive histone marks by inhibiting glutamine metabolism affects gene expression and alters cancer cell phenotype. Epigenetics 7(12):1413–1420. https://doi.org/10.4161/epi.22713

Sivanand S, Viney I, Wellen KE (2018) Spatiotemporal control of acetyl-CoA metabolism in chromatin regulation. Trends Biochem Sci 43(1):61–74. https://doi.org/10.1016/j.tibs.2017.11.004

Smith BC, Hallows WC, Denu JM (2009) A continuous microplate assay for sirtuins and nicotinamide-producing enzymes. Anal Biochem 394(1):101–109. https://doi.org/10.1016/j.ab.2009.07.019

Tang B, Dean B, Thomas EA (2011) Disease- and age-related changes in histone acetylation at gene promoters in psychiatric disorders. Transl Psychiatry 1:e64. https://doi.org/10.1038/tp.2011.61

Tanner KG, Langer MR, Denu JM (2000a) Kinetic mechanism of human histone acetyltransferase P/CAF. Biochemistry 39(39):11961–11969. https://doi.org/10.1021/bi001272h

Tanner KG, Langer MR, Kim Y, Denu JM (2000b) Kinetic mechanism of the histone acetyltransferase GCN5 from yeast. J Biol Chem 275(29):22048–22055. https://doi.org/10.1074/jbc.M002893200

Tao R, Coleman MC, Pennington JD, Ozden O, Park SH, Jiang H, Kim HS, Flynn CR, Hill S, Hayes McDonald W, Olivier AK, Spitz DR, Gius D (2010) Sirt3-mediated deacetylation of evolutionarily conserved lysine 122 regulates MnSOD activity in response to stress. Mol Cell 40(6):893–904. https://doi.org/10.1016/j.molcel.2010.12.013

Thompson PR, Wang D, Wang L, Fulco M, Pediconi N, Zhang D, An W, Ge Q, Roeder RG, Wong J, Levrero M, Sartorelli V, Cotter RJ, Cole PA (2004) Regulation of the p300 HAT domain via a novel activation loop. Nat Struct Mol Biol 11(4):308–315. https://doi.org/10.1038/nsmb740

Vander Heiden MG, Cantley LC, Thompson CB (2009) Understanding the Warburg effect: the metabolic requirements of cell proliferation. Science 324(5930):1029–1033. https://doi.org/10.1126/science.1160809

Ventura M, Mateo F, Serratosa J, Salaet I, Carujo S, Bachs O, Pujol MJ (2010) Nuclear translocation of glyceraldehyde-3-phosphate dehydrogenase is regulated by acetylation. Int J Biochem Cell Biol 42(10):1672–1680. https://doi.org/10.1016/j.biocel.2010.06.014

Verdin E (2015) NAD(+) in aging, metabolism, and neurodegeneration. Science 350(6265):1208–1213. https://doi.org/10.1126/science.aac4854

Vervoorts J, Luscher-Firzlaff JM, Rottmann S, Lilischkis R, Walsemann G, Dohmann K, Austen M, Luscher B (2003) Stimulation of c-MYC transcriptional activity and acetylation by recruitment of the cofactor CBP. EMBO Rep 4(5):484–490. https://doi.org/10.1038/sj.embor.embor821

Waldecker M, Kautenburger T, Daumann H, Busch C, Schrenk D (2008) Inhibition of histone-deacetylase activity by short-chain fatty acids and some polyphenol metabolites formed in the colon. J Nutr Biochem 19(9):587–593. https://doi.org/10.1016/j.jnutbio.2007.08.002

Wang Y, Curry HM, Zwilling BS, Lafuse WP (2005) Mycobacteria inhibition of IFN-gamma induced HLA-DR gene expression by up-regulating histone deacetylation at the promoter region in human THP-1 monocytic cells. J Immunol 174(9):5687–5694. https://doi.org/10.4049/jimmunol.174.9.5687

Wang Y, Zhang X, Zhang Y, Zhu Y, Yuan C, Qi B, Zhang W, Wang D, Ding X, Wu H, Cheng J (2015) Overexpression of pyruvate kinase M2 associates with aggressive clinicopathological features and unfavorable prognosis in oral squamous cell carcinoma. Cancer Biol Ther 16(6):839–845. https://doi.org/10.1080/15384047.2015.1030551

Wang J, Huang J, Yao S, Wu JH, Li HB, Gao F, Wang Y, Huang GB, You QL, Li J, Chen X, Sun XD (2021) The ketogenic diet increases Neuregulin 1 expression via elevating histone acetylation and its anti-seizure effect requires ErbB4 kinase activity. Cell Biosci 11(1):93. https://doi.org/10.1186/s13578-021-00611-7

Ward PS, Thompson CB (2012) Metabolic reprogramming: a cancer hallmark even Warburg did not anticipate. Cancer Cell 21(3):297–308. https://doi.org/10.1016/j.ccr.2012.02.014

Webber LP, Wagner VP, Curra M, Vargas PA, Meurer L, Carrard VC, Squarize CH, Castilho RM, Martins MD (2017) Hypoacetylation of acetyl-histone H3 (H3K9ac) as marker of poor prognosis in oral cancer. Histopathology 71(2):278–286. https://doi.org/10.1111/his.13218

Wellen KE, Hatzivassiliou G, Sachdeva UM, Bui TV, Cross JR, Thompson CB (2009) ATP-citrate lyase links cellular metabolism to histone acetylation. Science 324(5930):1076–1080. https://doi.org/10.1126/science.1164097

White PJ, McGarrah RW, Grimsrud PA, Tso SC, Yang WH, Haldeman JM, Grenier-Larouche T, An J, Lapworth AL, Astapova I, Hannou SA, George T, Arlotto M, Olson LB, Lai M, Zhang GF, Ilkayeva O, Herman MA, Wynn RM, Chuang DT, Newgard CB (2018) The BCKDH kinase and phosphatase integrate BCAA and lipid metabolism via regulation of ATP-citrate Lyase. Cell Metab 27(6):1281–1293.e7. https://doi.org/10.1016/j.cmet.2018.04.015

Wise DR, Ward PS, Shay JE, Cross JR, Gruber JJ, Sachdeva UM, Platt JM, DeMatteo RG, Simon MC, Thompson CB (2011) Hypoxia promotes isocitrate dehydrogenase-dependent carboxylation of alpha-ketoglutarate to citrate to support cell growth and viability. Proc Natl Acad Sci U S A 108(49):19611–19616. https://doi.org/10.1073/pnas.1117773108

Yang H, Yang T, Baur JA, Perez E, Matsui T, Carmona JJ, Lamming DW, Souza-Pinto NC, Bohr VA, Rosenzweig A, de Cabo R, Sauve AA, Sinclair DA (2007) Nutrient-sensitive mitochondrial NAD+ levels dictate cell survival. Cell 130(6):1095–1107. https://doi.org/10.1016/j.cell.2007.07.035

Yu W, Dittenhafer-Reed KE, Denu JM (2012) SIRT3 protein deacetylates isocitrate dehydrogenase 2 (IDH2) and regulates mitochondrial redox status. J Biol Chem 287(17):14078–14086. https://doi.org/10.1074/jbc.M112.355206

Yuan ZL, Guan YJ, Chatterjee D, Chin YE (2005) Stat3 dimerization regulated by reversible acetylation of a single lysine residue. Science 307(5707):269–273. https://doi.org/10.1126/science.1105166

Zaidi N, Lupien L, Kuemmerle NB, Kinlaw WB, Swinnen JV, Smans K (2013) Lipogenesis and lipolysis: the pathways exploited by the cancer cells to acquire fatty acids. Prog Lipid Res 52(4):585–589. https://doi.org/10.1016/j.plipres.2013.08.005

Zhang X, Kurnasov OV, Karthikeyan S, Grishin NV, Osterman AL, Zhang H (2003) Structural characterization of a human cytosolic NMN/NaMN adenylyltransferase and implication in human NAD biosynthesis. J Biol Chem 278(15):13503–13511. https://doi.org/10.1074/jbc.M300073200

Zhang Y, Guo C, Yu G (2005) A pilot study of fatty acid metabolism in oral squamous cell carcinoma. Int J Oral Maxillofac Surg 34(1):78–81. https://doi.org/10.1016/j.ijom.2004.03.001

Zhang L, Yao W, Xia J, Wang T, Huang F (2019) Glucagon-induced acetylation of energy-sensing factors in control of hepatic metabolism. Int J Mol Sci 20(8):doi:10.3390/ijms20081885

Chapter 13
The Cross-Talk between Epigenetic Gene Regulation and Signaling Pathways Regulates Cancer Pathogenesis

Snigdha Samanta, Rumpa Mahata, and Manas Kumar Santra

Abstract Cancer begins due to uncontrolled cell division. Cancer cells are insensitive to the signals that control normal cell proliferation. This uncontrolled cell division is due to the accumulation of abnormalities in different factors associated with the cell division, including different cyclins, cell cycle checkpoint inhibitors, and cellular signaling. Cellular signaling pathways are aberrantly activated in cancer mainly due to epigenetic regulation and post-translational regulation. In this chapter, the role of epigenetic regulation in aberrant activation of PI3K/AKT, Ras, Wnt, Hedgehog, Notch, JAK/STAT, and mTOR signaling pathways in cancer progression is discussed. The role of epigenetic regulators in controlling the upstream regulatory proteins and downstream effector proteins responsible for abnormal cellular signaling-mediated cancer progression is covered in this chapter. Similarly, the role of signaling pathways in controlling epigenetic gene regulation-mediated cancer progression is also discussed. We have tried to ascertain the current status of potential epigenetic drugs targeting several epigenetic regulators to prevent different cancers.

Keywords Promoter hypermethylation · Histone modification · Wnt/β-catenin · Hedgehog · Notch · PI3K/AKT · Ras · MAPK · JAK/STAT · mTOR · Epigenetic drugs · Epigenetic Biomarkers

---

Snigdha Samanta and Rumpa Mahata contributed equally with all other contributors.

S. Samanta · R. Mahata
Molecular Oncology Laboratory, National Centre for Cell Science, NCCS Complex, S. P. Pune University Campus, Ganeshkhind Road, Pune, Maharashtra, India

Department of Biotechnology, Savitribai Phule Pune University, Pune, Maharashtra, India

M. K. Santra (✉)
Molecular Oncology Laboratory, National Centre for Cell Science, NCCS Complex, S. P. Pune University Campus, Ganeshkhind Road, Pune, Maharashtra, India
e-mail: manas@nccs.res.in

T. K. Kundu, C. Das (eds.), *Metabolism and Epigenetic Regulation: Implications in Cancer*, Subcellular Biochemistry 100,
https://doi.org/10.1007/978-3-031-07634-3_13

Introduction

The term 'epigenetic' is the study of all mitotically and meiotically heritable changes in gene expression without alteration in nucleotide sequences (Holliday 1994). Basically, it is due to either tagging of the methyl group to the cytosine in the CpG island or post-translational modification in histone tails or both types of modification. These modifications alter DNA accessibility and chromatin structure without any variation of DNA sequence (Zuo et al. 2011; Bird 2002; Strahl and Allis 2000). In normal cells, epigenetic mechanisms play a central role in the maintenance of tissue-specific gene expression. Epigenetic alteration by different chromatin modifiers has been considered as the most important mechanism to promote oncogenic growth and malignancy.

Transcription alteration could be regulated by epigenetic modification, including DNA methylation and histone post-translational modifications. DNA methylation is heritable either in progeny cells or in the organism. Enzyme DNA methyltransferase 1 (DNMT1) is predominantly responsible for CpG methylation during cell division (Jones and Liang 2009). Two other DNMT family members, DNMT3a and DNMT3b, also play essential for DNA de novo methylation (Okano et al. 1999). The genome of cancer cells is hypomethylated as compared to their normal counterpart, but several CpG islands in the promoter regions of tumor suppressor genes are aberrantly methylated.

In addition to DNA methylation, post-translation modifications of histone determine the chromatin structure and gene function. The major post-translational histone modifications responsible for chromatin structure alteration are methylation, acetylation, and ubiquitination. The trimethylation of H3 at lysine 4 (H3K4me3), trimethylation of H3 at lysine 36 (H3K36me3), acetylation of H3 at lysine 39 (H3K9Ac), and acetylation of H3 at lysine 27 (H3K27Ac) on the promoter are the marker of genes to be functional (Wu et al. 2011). On the other hand, H3K27me3, H3K9me2/3, H4K20me3, and H4K12Ac) are the marker for genes to be non-functional (Grunstein 1997). Interestingly, epigenetic modifications through post-translation modification of histone may repress or activate the gene function, whereas DNA methylation always represses the gene function.

Several cellular signaling pathways, including Wnt, Hedgehog, PI3K/AKT, and Ras, Notch, JAK/STAT, were deregulated in cancer to promote the malignancy. Many genetic and epigenetic factors are associated with aberrant activation of these signaling pathways in cancer. On the other hand, these activated signaling pathways, directly or indirectly, modulate the function of epigenetic modifiers to activate oncogenes and inactivate tumor suppressors in multiple cancers. Dysregulation of several downstream effectors of these signaling pathways, which are crucial for maintaining cellular homeostasis, leads to induction of cancer by activating the expression of oncogenes or by inactivating tumor suppressor genes.

The evolutionarily conserved wingless-Int-1 (Wnt) pathway plays a crucial role in the developmental processes, including cell proliferation, cell migration, cell fate decisions in development, body axis patterning, development of bone, heart, muscles, and apoptosis (Miller 2002; Cadigan and Nusse 1997). There are 19 Wnt genes present in mammals and all of them encode cysteine-rich secreted glycoproteins. Due to their structural similarities, many Wnt proteins share a common signaling pathway, in which cytoplasmic β-catenin acts as a key signaling intermediate. This signaling pathway is known as the Wnt/β-catenin signaling pathway or 'canonical' Wnt signaling pathway. Besides this, there is another Wnt signaling pathway, known as the 'non-canonical' pathway in which β-catenin does not involve. Canonical Wnt cascade is considered as a critical regulator of stem cells. Wnt signaling allows malignant proliferation by mediating the self-renewal of cancerous stem cells (Reya and Clevers 2005).

Hedgehog (Hh) signaling pathway also plays critical roles in fundamental processes including embryogenesis, cell proliferation, tissue homeostasis maintenance, and tissue regeneration (Katoh and Katoh 2005). Hh proteins are secreted glycoproteins and they are categorized into three subgroups, including Sonic Hedgehog (SHh), Indian Hedgehog (IHh), and Desert Hedgehog (DHh), of which SHh is the best studied. The main difference among these three groups of Hh proteins is their tissue location. SHh has a crucial role in developed and adult tissues, whereas DHh and IHh have a key effect on developing tissues (Rimkus et al. 2016). Both Wnt signaling and Hedgehog (Hh) signaling pathway remain dormant in adult tissues but get activated under specific condition. It was reported that anomalous signaling by WNT pathways is linked to many aggressive human cancers, including colon carcinomas and melanomas (Moon et al. 2004). Epigenetic changes lead to the progression of several types of cancers, including cervical cancer, breast cancer, endometrial cancer, ovarian cancer, colorectal cancer, renal cancer, hepatoblastoma, hepatoma, hepatocellular carcinoma, and pancreatic cell lines (Surana et al. 2014; Tada et al. 2008).

Ras family proteins are GTPase, which act as a molecular switch and are involved in cellular signal transduction. The Ras subfamily consists of the well-known classical p21 Ras proteins, H (Harvey)-Ras, N (Neuroblastoma)-Ras, and the two splice-variants of K (Kirsten)-Ras. Classical Ras proteins are found to be mutated in 30% of all human cancers. Among the oncogenic Ras oncogenes (K-Ras, H-Ras, and N-Ras), K-Ras mutations are prevalent in nearly all pancreatic adenocarcinomas, in lung cancer and colorectal cancer. On the other hand, N-Ras mutations are frequently found leukemia and H-Ras mutations are only found in some types of bladder and kidney cancer (Giehl 2005). A study in polypoid-type and flat-type early-stage CCRs 9 colorectal carcinoma) suggested that BRaf mutations are most frequent in de novo colorectal carcinogenesis, K-Ras mutations in the adenoma-carcinoma of colorectal carcinogenesis (Noda et al. 2006). In addition to this, Ras oncogenes are hypomethylated and overexpressed in primary human cancers.

Phosphatidylinositol-3-kinase (PI3K) signaling is stimulated by diverse oncogenes and growth factor receptors and contributes to tumorigenesis (Fruman et al. 2017; Thorpe et al. 2015). Hyperactivated PI3K signaling by epigenetic modifications causes abnormal cell cycle progression, altered cell motility and adhesion, apoptosis inhibition, and induction of angiogenesis. Therefore, hyperactivation of this signaling is considered a hallmark of various types of human cancer. Activated

PI3K further directly or indirectly modulates the critical epigenetic modifiers, including DNA methyltransferases (DNMTs), cetyl-CoA synthetases to establish oncogenicity.

The Janus kinase (JAK) and signal transducer and activator of transcription (STAT) play a central role in many cellular processes, including development. There are four JAK family members (*JAK1, JAK2, JAK3,* and *TYK2)* and seven STAT family members (*STAT 1-4, STAT5A, STAT5B,* and *STAT6)* found in humans. It is already documented that the JAK/STAT pathway is highly deregulated due to either gain-of-function mutation of JAKs/STATs or other regulatory components in hematological malignancies, head and neck squamous cell carcinoma, breast cancer, colorectal cancer, high-grade ovarian epithelial tumors, granular lymphocytic leukemia, prostate cancer, and glioblastoma (Thomas et al. 2015).

The Notch signaling pathway is an evolutionarily conserved pathway involved in proliferation, apoptosis, cell fate determination, and development of a diverse group of organisms (Hori et al. 2013). The canonical Notch cascade is initiated when the Notch receptor on one cell physically interacts with the membrane-bound ligand present on the neighbor cell. Then, the intracellular part of the Notch receptor is released and translocated into the nucleus to activate the transcription of its target genes (Andersson et al. 2011). Four Notch receptors (NOTCH1, NOTCH2, NOTCH3, and NOTCH4) and five ligands of the Delta–Serrate–Lag (DSL) family (jagged 1 (JAG1), JAG2, delta-like 1 (DLL1), DLL3, and DLL4) are reported to be present in mammal (Nowell and Radtke 2017). Interestingly, the Notch signaling pathway may promote or suppress tumor progression depending on the cellular context and the cross-talk with other signal-transduction pathways. The tumor-suppressive role of the Notch pathway is widely documented in various solid tumors, including squamous cell carcinoma in several epithelial tissues, subtypes of brain cancer, liver cancer, and small-cell lung cancer (Nowell and Radtke 2017). On the other hand, Notch signaling plays an oncogenic role in human T-cell leukemia, T-cell neoplasia, and different epithelial tumors, including mammary and salivary gland adenocarcinomas, breast cancer progression (Radtke and Raj 2003).

Epigenetic alteration of Notch pathway components also has been extensively studied in various types of cancers, including clear cell renal carcinoma (CCRC) (Bhagat et al. 2017), gastrointestinal cancer (Piazzi et al. 2012), breast cancer, and human B-cell acute lymphoblastic leukemia (Kuang et al. 2013). Similarly, there are few studies suggesting that the epigenetic alterations generated by abnormal hyperactivated Notch signaling is also playing a significant role in carcinogenesis.

Here, we are revisiting the mechanisms of how different signaling pathways modulating epigenetic reprogramming responsible for aberrant activation of signaling pathways as well as how these signaling pathways utilize epigenetic processes to deregulate the effector proteins to promote cancer pathogenesis. Overall, we attempt to establish a complex connection between various signaling cascades and the epigenome during the development of different malignancies.

Epigenetic Modifications Responsible for the Activation of Wnt Signaling Pathway

The key molecules associated with the canonical Wnt signaling pathway are glycogen synthase kinase (GSK-3β), adenomatous polyposis coli (APC), β-catenin, and Axin (Widelitz 2005). β-Catenin functions as a transcriptional cofactor encoded by CTNNB1 gene (Kraus et al. 1994). In the absence of the Wnt ligand, expression of cytoplasmic β-catenin is maintained at the low level via proteasomal degradation by the actions of a 'destruction complex' that contains APC and AXIN proteins (Nelson and Nusse 2004). The destruction complex facilitates phosphorylation of β-catenin by GSK3β and casein-kinase 1α (CK1α) (Moon et al. 2004). The phosphorylated β-catenin is then targeted by ubiquitin ligase $SCF^{\beta\text{-}TRCP}$ for ubiquitination-mediated proteasomal degradation. In the presence of the Wnt ligand, it binds to its receptor frizzled (Fz) and co-receptor low-density lipoprotein receptor-related protein (LRP) and activates downstream adaptor protein Disheveled (Dvl), which promotes the phosphorylation of LRP. The phosphorylated LRP then recruits Axin to the plasma membrane from the destruction complex and inhibits the activity of GSK-3β (Widelitz 2005). Non-phosphorylated β-catenin is then translocated into the nucleus, binds with the T-cell factor (Tcf)/lymphoid enhancing factor (Lef) family of transcription factors, and stimulates the transcription of its target genes (Staal and Clevers 2005).

It is well reported that most human cancers have elevated expression levels of β-catenin. Therefore, the Wnt/β-catenin pathway is abnormally activated in most cancers. Constitutive activation of this pathway has become a hallmark of a specific type of solid tumor (Serman et al. 2014). Aberrant activation of Wnt signaling occurs due to acquiring a mutation in the component of the Wnt/β-catenin pathway to drive the signaling pathway in a ligand-independent manner. For example, approximately 85% of colorectal cancer cases show the loss-of-function mutations in tumor suppressor gene APC (Polakis 2000). In addition, overexpression of oncogenic components including Fzd receptors (Fzd7 and Fzd10), Wnt ligands (WNT1, WNT2, WNT2B2, WNT3A), Dvl family members is also responsible for aberrant activation of Wnt/β-catenin pathway (Uematsu et al. 2003; Terasaki et al. 2002).

Additionally, epigenetic silencing of the Wnt/β-catenin pathway inhibitory proteins is also responsible for the anomalous activation of Wnt/β-catenin signaling (Moskalev et al. 2012). Genome-wide analysis of promoter DNA hypermethylation in several human cancers showed that Wnt pathway inhibitory factors such as secreted frizzled-related protein (SFRP), Dickkopf-related protein (DKK3), WNT inhibitory factor-1 (WIF1) and HDPR1 are common targets for epigenetic silencing to activate Wnt/β-catenin signaling (Batra et al. 2006; Liu et al. 2006; Kawano and Kypta 2003). The promoter hypermethylation of these genes is closely associated with the malignancy of several human cancers. Figure 13.1 and Table 13.1 demonstrate the epigenetic regulation of this signaling pathway in cancer.

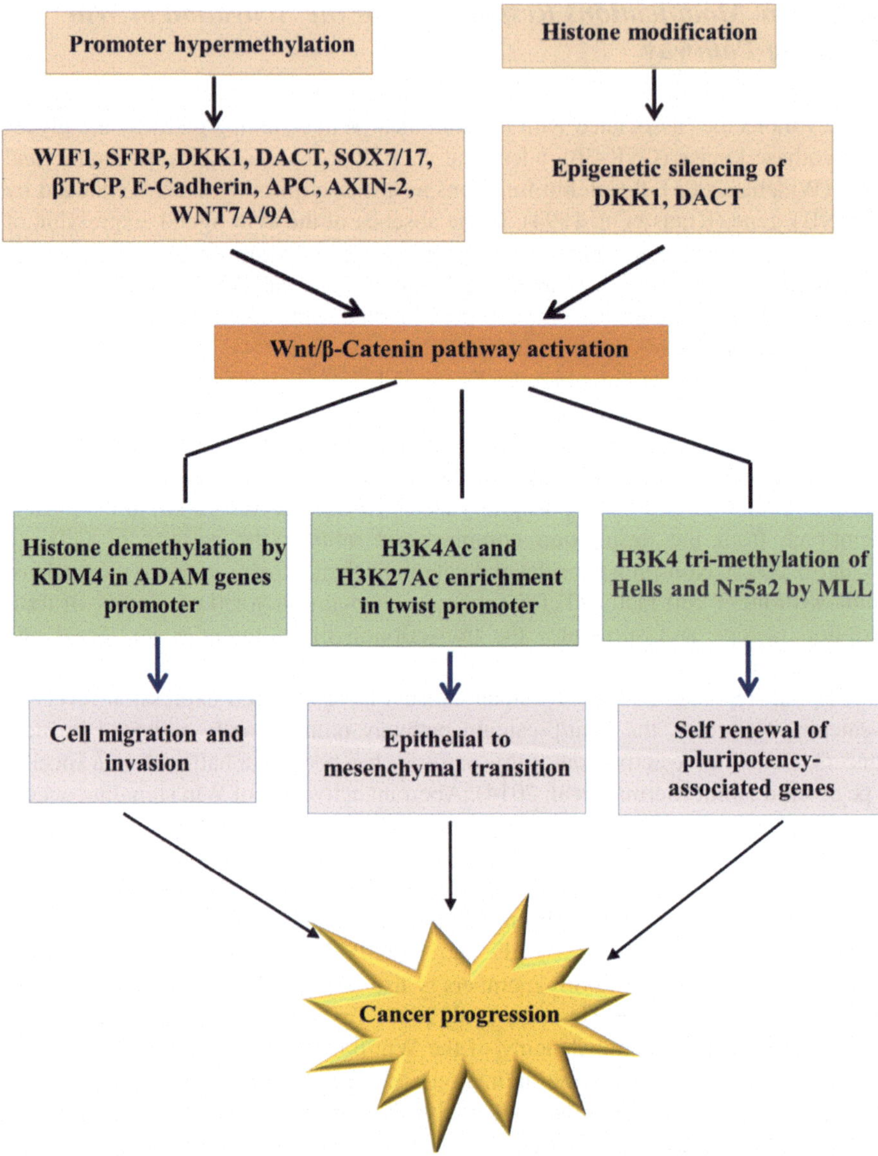

Fig. 13.1 Model demonstrates the epigenetic modifications (promoter hypermethylation and histone modification) of different genes are responsible for aberrant activation of the Wnt pathway. On the other hand, the activated Wnt pathway induces histone acetylation, methylation, and demethylation of different gene promoters, which results in increased pluripotency, EMT (epithelial-to-mesenchymal transition) process, migration, and invasion in different cancers

Table 13.1 Epigenetic modifications of Wnt signaling pathway in cancer

Classification	Factors	Epigenetic modifications	Cancer type
Wnt/β-catenin pathway antagonists	Secreted frizzled-related proteins (SFRP)	Promoter hypermethylation	Colon cancer (Caldwell et al. 2004), lung cancer (Fukui et al. 2005), breast cancer (Veeck et al. 2008), colorectal cancer (Suzuki et al. 2004)
	Dickkopf (DKK) family	Promoter hypermethylation	Colon cancer, breast cancer (Kawano and Kypta 2003)
	DACT gene family (DACT3)	Histone modification	Colon cancer (Jiang et al. 2008)
	AXIN2	Promoter hypermethylation	Colon cancer (Koinuma et al. 2006)
Nuclear antagonists	SOX7 and SOX17	Promoter hypermethylation	Colon cancer (Fu et al. 2010), breast cancer (Zhang et al. 2008), lung cancer (Zhang et al. 2009)
WNT non-transforming ligands	WNT5A, WNT7A, and WNT9A	Promoter hypermethylation	Glioblastoma (Okamoto et al. 2014), (pancreatic tumor (Sato et al. 2003), colorectal carcinoma, acute lymphoid leukemia (Shu et al. 2006)
Membrane-associated molecules	E-cadherin (CDH1)	Promoter hypermethylation	Breast cancer (Prasad et al. 2008)

Epigenetic Silencing of Extracellular Wnt/β-Catenin Pathway Inhibitors SFRP Family

Secreted frizzled-related proteins (SFRP) are a family of Wnt antagonists. Five members of this family protein have a cysteine-rich domain, homologous to the frizzled receptors (Uren et al. 2000). SFRPs can bind with Wnt proteins in the extracellular compartment, thereby preventing Wnt ligands from interacting with their receptor frizzled (Bafico et al. 1999). Thus, SFRP plays a vital role in restricting the activation of the Wnt/β-catenin signaling pathway in normal cells. Previous studies showed that Wnt/β-catenin pathway is constitutively activated due to hypermethylation of promoter of SFRP gene in colon cancer (Caldwell et al. 2004), lung cancer (Fukui et al. 2005), breast cancer (Veeck et al. 2008), and colorectal cancer (Suzuki et al. 2004).

Dickkopf (DKK)

The evolutionarily conserved Dickkopf (Dkk) genes family comprises of four members (DKK1-4) and a unique DKK3-related gene, DKKl1 (soggy). The extracellular protein products of these genes bind to the co-receptor LRP5/6 of Wnt and prevent the formation of the Wnt-FZD-LRP5/6 complex (Kawano and Kypta 2003). Additionally, Dkks can also bind with the transmembrane proteins Kremen1 and

Kremen2 with high affinity and can modulate Wnt signaling. DKK1 binds to LRP5/6 and Kremen proteins and inducing the endocytosis of LRP and prevents the activation of Wnt signaling (Niehrs 2006; Semënov et al. 2001).

However, DKK genes are epigenetically silenced in cancer to activate Wnt/β-catenin signaling. It is reported that the DKK1 is epigenetically silent in colon and breast cancer due to hypermethylation of CpG island present in its promoter and prevents Dkk1-mediated blocking of LRP5/6 (Kawano and Kypta 2003). Additionally, DKK1 is epigenetically regulated by polycomb protein chromo box 7 (CBX7). CBX7 inhibits the Wnt/β-catenin pathway by increasing the expression of DKK1. In particular, CBX7 increases DKK1 transcription in association with p300 acetyltransferase and subsequently enhances the histone acetylation of the DKK1 promoter (Kim et al. 2015).

Furthermore, prostate tumor overexpressed 1 (PTOV1) also epigenetically silence DKK1 by recruiting HDAC1 and HDAC2 on DKK1 promoter and reduces DKK1 promoter histone acetylation (Cui et al. 2016). In addition to acetylation and deacetylation, DKK1 is also inactivated through demethylation of $H3K4Me^3$ by lysine demethylase 1A (KDM1A) (Kim et al. 2013). Thus, downregulation of DKK1 by either DNA methylation or histone modifications leads to constitutive activation of canonical Wnt/β-catenin signaling in various tumors (Suzuki et al. 2004).

Epigenetic Silencing of Cytosolic Wnt/β-Catenin Pathway Inhibitors APC, AXIN2, and DACT Gene Family

The well-reported cytosolic Wnt antagonists are the members of the DACT gene family (which antagonize Dvl), including APC and AXIN2. DACT3, a member of the DACT gene family, interacts with Dvl2 (a Dvl member), facilitating the degradation of Dvl2 and thereby decreasing the nuclear translocation of β-catenin and thus acts as a negative modulator of Wnt/β-catenin signaling (Wang et al. 2015). In several cancers, DACT3 is epigenetically silenced through histone modifications. Epigenetic repression of DACT3 is associated with 'bivalent' histone modifications. In normal cells, the transcription start site region of DACT3 is enriched with only the $H3K4me^3$ histone mark. In contrast, in addition to $H3K4Me^3$ enrichment, the transcription start site region of DACT3 is enriched with repressive $H3K27me^3$ histone mark in cancer colon (Jiang et al. 2008). Thus, erasing of $H3K4Me^3$ histone mark by demethylase is not essential for epigenetic silencing of DACT3. In colon cancer, it was reported that the epigenetic silencing of APC and AXIN2 by promoter hypermethylation results in constitutive activation of β-catenin signaling (Wang et al. 2012a; Koinuma et al. 2006). Figure 13.1 and Table 13.1 demonstrate the epigenetic regulation of this signaling pathway in cancer.

Epigenetic Silencing of Nuclear Factors SOX7 and SOX17

Nuclear proteins SOX7 and SOX17 (member of SRY-box containing gene) are members of the high mobility group (HMG) transcription factor superfamily. SOX7 interacts with β-catenin and inhibits Wnt/beta-catenin-stimulated transcription. In contrast, SOX17 transcriptionally silences β-catenin by directly binding to the specific region of the β-catenin promoter. Thus, they inhibit the transcription of Wnt signaling target genes. It is reported that epigenetic silencing of SOX7 and SOX17 by promoter hypermethylation activates Wnt signaling pathway in colon, breast, and lung cancer (Fu et al. 2010; Zhang et al. 2008, 2009).

Epigenetic Silencing of WNT Non-Transforming Ligands WNT5A, WNT7A, and WNT9A

It is reported that WNT families such as WNT5A and WNT7A antagonize WNT signaling through GSK3-independent β-catenin degradation (Topol et al. 2003). Therefore, WNT5A and WNT7A are inactivated in cancer by the epigenetic gene silencing pathway. However, it is also reported that WNT5A is epigenetically activated in Glioblastoma (Okamoto et al. 2014). Therefore, WNT5A functions as both a tumor suppressor and oncogene in a context-dependent manner. Epigenetic gene inactivation of WNT7A and WNT9A occurs through promoter DNA hyper-methylation in several cancers. WNT7A is epigenetically silenced in pancreatic tumors, whereas WNT9A is epigenetically silenced in colorectal carcinoma and acute lymphoid leukemia (Sato et al. 2003; Shu et al. 2006).

Epigenetic Silencing of Epithelial Adhesion Molecules

The epithelial adhesion molecule E-cadherin (CDH1) also acts as a negative regulator of the Wnt signaling pathway. CDH1 binds to the β-catenin and sequesters it from the nucleus (Nelson and Nusse 2004). Promoter DNA hyper-methylation of CDH1 causes epigenetic silencing of this gene, leading to elevated expression of Wnt/β-catenin signaling in breast cancer (Prasad et al. 2008).

Wnt Signaling Induces Epigenetic Alteration to Promote Cancer

The canonical Wnt signaling pathway maintains the tissue homeostasis but deregulated Wnt signaling facilitates human malignancies through transcriptional regulation of several genes. Gene transcription and expression are regulated by Wnt/β-catenin signaling by β-catenin/LEF/TCF complex. This complex recognizes

Wnt response elements (WRE) in the promoter and modulates the gene expression through chromatin remodeling (Mosimann et al. 2009). It is reported that once TCF and β-catenin complex is recruited onto the promoter, it helps in the recruitments of other coactivators, including CREB-binding protein (CBP) and its close relative p300. Both of them have histone acetyltransferase (HAT) activity and enhance the transcription (Parker et al. 2008).

It was reported that β-catenin C-terminal activation domain and adjacent ARM 11/12 repeats (CTARM) are involved in H3K4 trimethylation at the *c-Myc* gene in association with nuclear chromatin remodeling subunits of TRRAP/TIP 60 histone acetyltransferase (HAT), ISW1, and mixed-lineage-leukemia (MLL1/MLL2) SET1-type histone methyltransferase (HMT) complexes in order to promote H3K4 trimethylation (H3K4Me3)-mediated gene transcription (Sierra et al. 2006).

A study by Peter Wend group demonstrated that activation of Wnt/β-catenin and attenuation of BMP (bone morphogenetic protein) signaling result in high-grade salivary gland SCC (squamous cell carcinomas) and head and neck SCC in mice (Wend et al. 2013). In this case, β-catenin activation leads to induction and stabilization of histone methyltransferase MLL. β-Catenin and CBP are associated with MLL on the promoter of self-renewal genes like Hells and Nr5a2 in tumor-propagating cells and subsequently increase H3K4 tri-methylation (Fig. 13.1). The inactivation of Wnt/β-catenin signaling restores differentiation of CD24+CD29+ tumor-propagating cells. Wnt signaling and PI3K/AKT signaling promote epithelial-to-mesenchymal transition (EMT) in gastric cancer (Song et al. 2017). EMT is a closely associated characteristic in gastric cancer where epithelial cells are converted to migratory and invasive mesenchymal cells. PI3K/AKT and Wnt/β-catenin signaling increased the levels of H3K4Ac and H3K27Ac in the Twist promoter, thereby activating its expression in gastric cancer at the epigenetic level (Fig. 13.1). Because of the increased expression of mesenchymal marker Twist, EMT is promoted in gastric cancer.

A disintegrin and metalloproteinases (ADAMs) are a group of transmembrane proteolytic enzymes. Overexpression of ADAM10/17/19 is closely correlated with the metastasis of CRC. They promote cell migration and invasion of CRC. A recent study showed that β-catenin augments the recruitment of the chromatin remodeling protein BRG1 on the promoter of ADAM genes (Fig. 13.1) (Sun et al. 2020). BRG1 then recruits histone demethylase KDM4 to demethylate $H3K9Me^3$ to activate ADAM genes at the epigenetic level transcriptionally.

Epigenetic Modifications Cause Activation of the Hedgehog (Hh) Signaling Pathway

The initiation of the Hh signaling pathway occurs via binding of one of the three Hh-secreted ligands to the extracellular domain of a conserved 12-pass transmembrane protein receptor, called 'patched' (PTCH). The binding of Hh ligands to PTCH

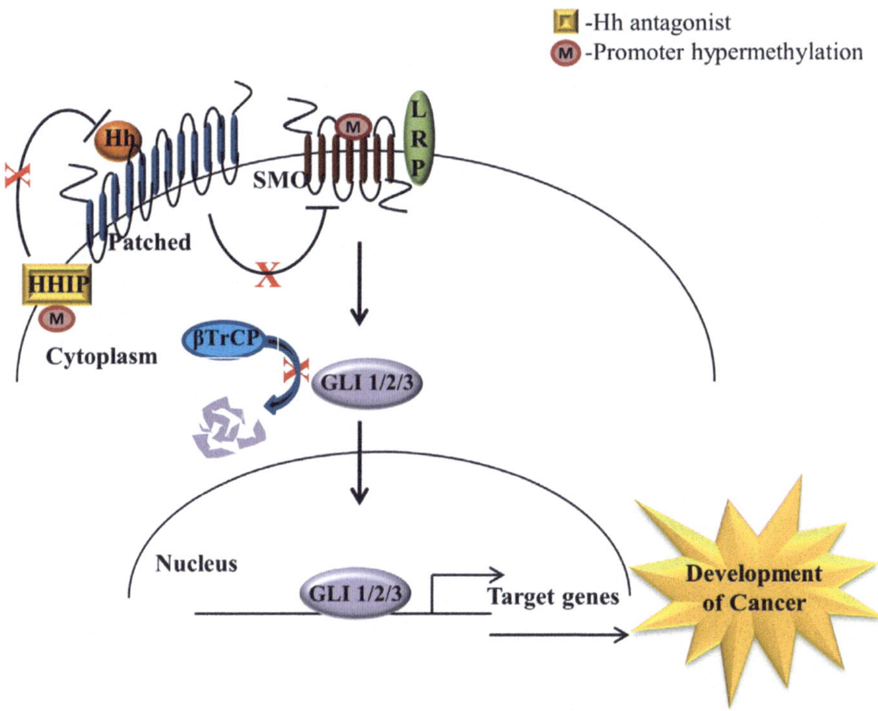

Fig. 13.2 Role of epigenetic process in Hedgehog signaling. Promoter hypermethylation of Hedgehog pathway antagonists (HHIP) and Hh pathway members (PATCH, SMO) inhibit GLI1/2/3 from βTrCP-mediated proteasomal degradation. Epigenetic gene silencing increases the expression level of Hh targeted genes to promote cancer development

causes the release of smoothened (SMO). Released SMO then activates glioma-associated oncoproteins (GLI) transcription factors (GLI 1, 2, 3) by separating GLIs from proteolytic complex (Hui et al. 2013). Deshpande et al. demonstrated that, upon Hh stimulation, cholesterol modification was required for releasing SMO from PTCH inhibition (Deshpande et al. 2019). The activated GLI transcription factor then enters into the nucleus and controls the transcription of Hedgehog target genes (Taipale and Beachy 2001). Among them, some are oncogenic genes like Myc, VEGFA, and Bmi1 to promote cancer cell survival, invasion, and angiogenesis (Mill et al. 2005; Wang et al. 2012b; Chen et al. 2011).

Like the Wnt signaling pathway, the Hedgehog (Hh) signaling pathway also remains dormant in adult tissues but gets activated under specific conditions. Aberrant activation of the Hh signaling pathway by genetic or epigenetic changes leads to the progression of several cancer types (ten Haaf et al. 2011). Here, epigenetic changes that alter Hh signaling pathway are discussed (Fig. 13.2 and Table 13.2).

Table 13.2 Epigenetic modifications of Hedgehog signaling pathway in cancer

Classification	Factors	Epigenetic modifications	Cancer type
Ligands	Sonic Hedgehog (SHh)	Promoter hypermethylation, histone modification	Gastric carcinoma (Wang et al. 2006), breast cancer (Duan et al. 2015)
Hh pathway members	Patched-1 (PTCH1)	Promoter hypermethylation	Basal cell carcinoma (BCC) (Heitzer et al. 2010), breast cancer (Wolf et al. 2007), rhabdomyosarcoma, medulloblastomas (Ecke et al. 2009), gastric cancer (Zuo and Song 2013)
	Smoothened (SMO)	Promoter hypermethylation	Glioblastoma, medulloblastoma (Shahi et al. 2008)
	Hh-interacting protein (HHIP)	Promoter hypermethylation	Hepatoblastoma, hepatoma, hepatocellular carcinoma (Tada et al. 2008)

Epigenetic Modifications of Ligands

In gastric carcinoma and breast cancer, it is reported that promoter hypomethylation of SHh (Sonic Hedgehog) genes increases expression of SHh at the transcriptional level by enhancing recruitment of transcription factor NF-κB (Wang et al. 2006; Duan et al. 2015). Though hypermethylation was observed in exon 2 and exon 3 of SHh, activation of the upstream SHh promoter supersedes the exon methylation in breast cancer cell lines (ten Haaf et al. 2011). Additionally, histone modifications also play a vital role in transcription activation of the SHh gene (ten Haaf et al. 2011).

Epigenetic Modifications of Hh Pathway Members PTCH1

The tumor suppressor patched-1 (PTCH1) is the primary receptor for Hh signaling pathway (Marigo et al. 1996). PTCH-1, PTCH-1A, PTCH-1B, and PTCH-1C are the four different isoforms of PTCH1. The loss of function of PTCH1 by epigenetic silencing causes aberrant ligand-independent activation of the Hh pathway. Heitzer et al. reported that PTCH promoter is hypermethylated in basal cell carcinoma (BCC) (Heitzer et al. 2010). The promoter hypermethylation of PTCH1 is also reported in breast cancer (Wolf et al. 2007), rhabdomyosarcoma, medulloblastomas (Ecke et al. 2009), and in gastric cancer (Zuo and Song 2013). Thus, epigenetic silencing plays a critical role in the hyperactivation of Hh signaling pathway in cancer.

SMO

The G-protein-coupled receptor smoothened (SMO) plays a critical role in Hh signaling. It activates glioma-associated oncoprotein transcription factors GLI1, GLI2, and GLI3. The SMO promoter is hypermethylated in several glioblastomas and medulloblastoma cell lines to overproduce GLI through the non-canonical pathway (Shahi et al. 2008). Overproduction of GLI facilitates aberrant Hh signaling. These observations indicate that Hh signaling is altered in cancer through epigenetic silencing of SMO.

HHIP

Hh-interacting protein (HHIP) is also a negative regulator of the Hh signaling pathway. Membrane-bound HHIP can regulate the availability of ligand and attenuates signaling via binding with Hh proteins with an affinity similar to that of PTCH1. Therefore, HHIP promoter is hypermethylated in many cancers, including hepatoblastoma, hepatoma, hepatocellular carcinoma, pancreatic cell lines, and pancreatic tissue sample (Tada et al. 2008) (Table 13.2).

Epigenetic Modifications Cause Activation of PI3K/AKT Signaling Pathway

The oncogenic AKT kinase pathway is one of the significant downstream effectors in the PI3K pathway and is frequently dysregulated in cancer (Thorpe et al. 2015). Its activation is correlated with its membrane localization through interaction with PIP3 (phosphatidylinositol-3, 4, 5-triphosphate) and subsequently phosphorylation by PDK1 and mTORC2. In addition to the phosphorylation, several other post-translational modifications are associated with the activation of AKT-mediated cancer progression. In 2019, Wang et al. reported that AKT could be methylated by SETDB1 at the K64 position and subsequently causes Thr308 phosphorylation and activation. They further revealed that JMJD2A, a histone demethylase, can recognize the methylated form of AKT at Lys64 and subsequently recruits E3 ligase SCF^{SKP2} and TRAF6 to promote the Lys63-linked polyubiquitination of AKT to facilitate its membrane recruitment-mediated activation (Wang et al. 2019). Interestingly, the demethylase activity of JMJD2A is dispensable. In the same year, Guo et al. also reported that SETDB1 plays a vital role in the kinase activity of AKT. They found that SETDB1 preferentially recognizes the PIP3 (phosphatidylinositol (3,4,5)-trisphosphate) bound form of AKT and then methylates AKT to promote its activation (Guo et al. 2019). In non-small-cell lung carcinoma (NSCLC), SETDB1-mediated AKT methylation at Lys64 is upregulated and is closely associated with AKT hyperactivation (Wang et al. 2019).

In addition to post-translational modifications, epigenetic regulation, including DNA methylation of upstream and downstream players of the AKT signaling pathway, also plays an essential role in controlling the activation of the AKT kinase pathway (Fig. 13.3 and Table 13.3).

Epigenetic Silencing of Phosphatase and Tensin Homolog (PTEN)

The tumor suppressor gene PTEN encodes the protein, Phosphatase and Tensin homolog (PTEN), which negatively regulates the PI3K/AKT pathway by removing the phosphate group at the D3 position of PIP3. Epigenetic modifications, including histone modifications and DNA methylation of PTEN promoter, lead to aberrant PI3K/AKT pathway activation. Aberrantly activated AKT is responsible for the progression and malignancy of the majority of cancer. Therefore, promoter hypermethylation-mediated inactivation of PTEN promotes the progression of many cancers like gastric carcinoma (Kang et al. 2002), melanoma (Mirmohammadsadegh et al. 2006), breast cancer (García et al. 2004), and sporadic colorectal cancers (CRCs) (Goel et al. 2004).

Emerging evidence suggests that other than PTEN, many other tumor suppressor genes (TSGs) including CHRDL1, ROR2, FBP2, Sema3E, REG1A are epigenetically silenced to facilitate the aberrant activation of AKT in gastric carcinoma (Fattahi et al. 2020). These TSGs act as the antagonist to inhibit the PI3K/AKT pathway by abrogating the phosphorylation of essential proteins. Therefore, promoter hypermethylation of these TSGs is closely associated with aberrant PI3K pathway activation in gastric carcinoma.

Epigenetic Silencing of Rab Protein

Rab proteins are the largest subgroup of the Ras-like small GTPase family and they control several processes including membrane trafficking (Olkkonen and Stenmark 1997), intracellular signal transduction (Bucci and Chiariello 2006), apoptosis (Guo et al. 2010), autophagy (Munafó and Colombo 2002). A previous study showed that Rab39a, a member of Rab family protein, is epigenetically silenced through DNA hypermethylation in cervical cancer, which activates the AKT signaling pathway. It mainly prevents the phosphorylation of AKT at Ser473 to inhibit the cancer cell EMT, migration, and invasion. However, a detailed mechanism is yet to be discovered how Rab39a alters the phosphorylation status of AKT (Zou et al. 2019).

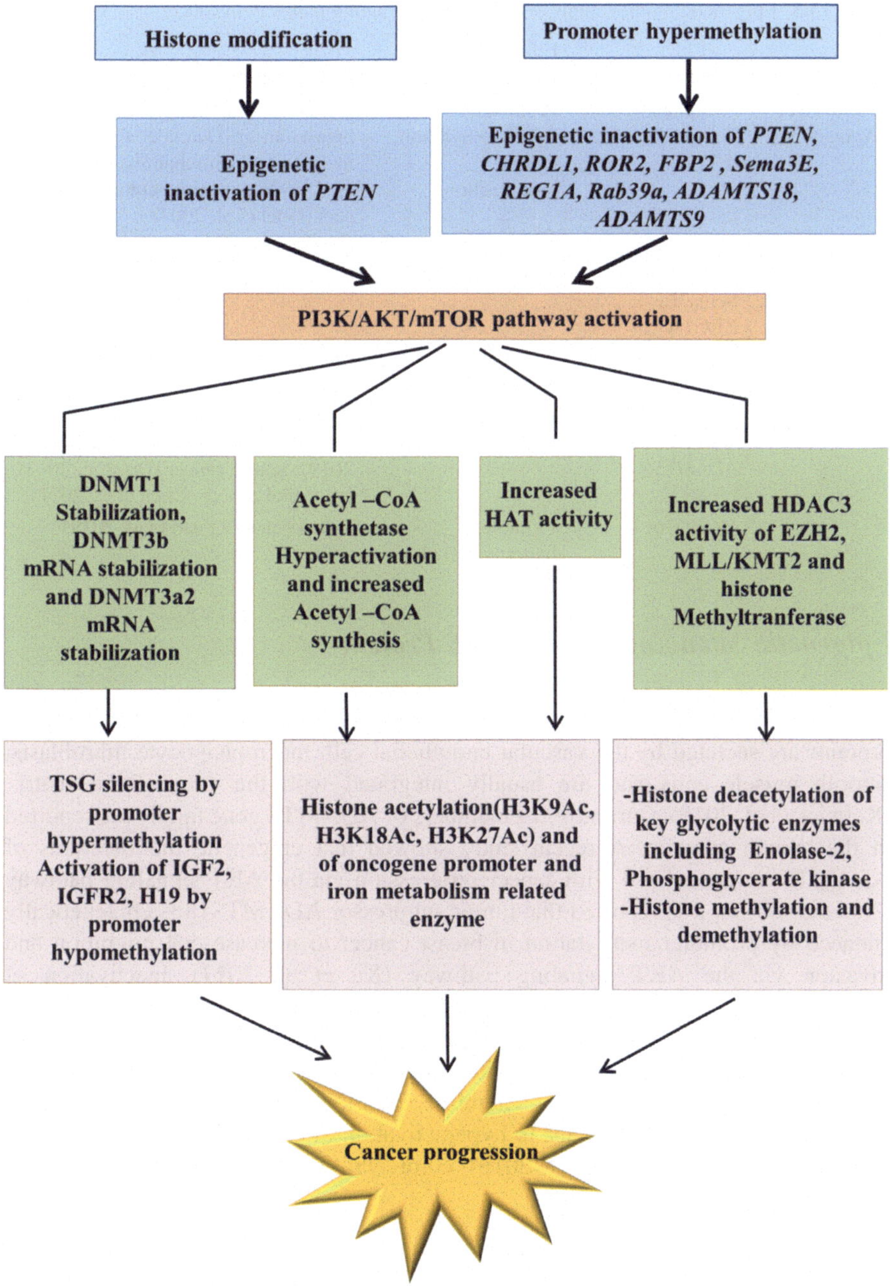

Fig. 13.3 Role of epigenetic regulation on activation of PI3K/AKT signaling is elaborated. DNA methylation and histone modification-mediated inactivation of negative regulatory proteins, leading to the hyperactivation of PI3K/AKT pathway. Hyperactivated PI3K/AKT/mTOR signaling pathway promotes the stabilization of DNA-methylating enzymes. These enzymes silence tumor suppressor genes (TGS) and activate oncogenes to initiate cancer. Activity of Histone acetyltransferase (HAT), Histone methylftransferase, histone deacetylase enzymes and availability of acetyl–CoA in the cell are increased, resulting the enhanced expression of some iron metabolism related enzyme, glycolytic enzymes to support cancer development

Table 13.3 Epigenetic modifications of PI3K/AKT signaling pathway in cancer

Classification	Factors	Epigenetic modifications	Cancer type
PI3K/AKT Antagonists	Phosphatase and Tensin homolog (PTEN)	Promoter hypermethylation, histone modification	Gastric cancer (Kang et al. 2002), breast cancer (García et al. 2004), melanoma (Mirmohammadsadegh et al. 2006), sporadic colorectal cancers (Goel et al. 2004)
	CHRDL1, ROR2, FBP2, Sema3E, REG1A	Promoter hypermethylation	Gastric cancer (Fattahi et al. 2020)
	Rab family: Rab39a	Promoter hypermethylation	Cervical cancer (Zou et al. 2019)
	ADAMT family: *ADAMTS18*, *ADAMTS9*	Promoter hypermethylation	Breast cancer (Xu et al. 2017; Shao et al. 2018), lung cancer (Zhang et al. 2019), gastric cancer (Du et al. 2013), colorectal cancer (Chen et al. 2017)
	HOXD10	Promoter hypermethylation	Colon cancer (Yuan et al. 2019)

Epigenetic Silencing of ADAMTS Protein

A disintegrin and metalloproteinase with thrombospondin motifs (ADAMTS) family proteins are secreted by the vascular endothelial cell, macrophagocyte, fibroblasts, smooth muscle cells and are usually integrated with the extracellular matrix (Kelwick et al. 2015). Currently, 19 members of ADAMTS gene family are reported in the human genome. A recent study showed that epigenetic modifications of ADAMTS are associated with cancer progression via the AKT signaling pathway (Xu et al. 2017). It is reported that tumor suppressor ADAMTS18 is epigenetically silenced by promoter methylation in breast cancer to increase cell migration and invasion via the AKT signaling pathway (Xu et al. 2017). Inactivation of ADAMTS18 by aberrant promoter hypermethylation also causes lung cancer progression via EGFR/AKT signaling pathway (Zhang et al. 2019). ADAMTS9 also acts as a tumor suppressor by promoting the inactivation of the oncogenic AKT/mTOR signaling pathway. Therefore, promoter hypermethylation-mediated inactivation of ADAMTS9 leads to hyperactivation of AKT/mTOR signaling pathway in gastric cancer (Du et al. 2013), colorectal cancer (Chen et al. 2017), and breast cancer (Shao et al. 2018).

Epigenetic Silencing of HOXD10 Protein

Homeobox D10 (HOXD10) is a member of homeobox family and functions as a tumor suppressor in many cancers, including breast, cholangio-cellular carcinoma, and colorectal cancer, which acts as an important transcription factor (Yang et al. 2015). HOXD10 facilities the dephosphorylation of AKT and MAPK to inactivate these kinase pathways. It is reported that HOXD10 is epigenetically silenced in colon cancer by promoter hypermethylation (Yuan et al. 2019) (Table 13.3).

PI3K/AKT Signaling Induces Epigenetic Alterations to Promote Cancer

The activation of the PI3K/AKT kinase signaling pathway is initiated by the activation of membranous receptors like receptor tyrosine kinases (RTKs), epidermal growth factor receptor (EGFR), fibroblast growth factor receptor (FGFR), and insulin-like growth factor I receptor (IGF-IR). In response to growth factors or cytokine stimulation, RTKs or G-protein-coupled receptors (GPCRs) are activated. This results in the recruitment of PI3Ks on the cell membrane and subsequent phosphorylation of PIP2 to generate PIP3 (Liu et al. 2009). PIP3 acts as a second messenger to activate AKT. Previous studies showed that the activated PI3K/AKT kinase signaling pathway modulates the function of crucial chromatin modifiers by promoting post-translational modification. Chromatin modifier plays a vital role in activating/repressing the gene function (Fig. 13.3).

Ten eleven translocation (TET) family functions as DNA methylcytosine dioxygenases. These enzymes catalyze the DNA demethylation of 5-methyl-cytosine to cytosine (Xu et al. 2011). The precise balance between DNA methylation and demethylation pattern is crucial for the transcription of any gene. Constitutively activated signaling pathways including AKT and MAPK are majorly involved in the deregulation of epigenetic alteration in cancer. It is reported that 180 gene loci were epigenetically repressed due to the DNA methylation and histone H3 lysine 27 (H3K27) methylation in cancer upon aberrant activation of AKT signaling (Zuo et al. 2011)

AKT Directs Stabilization and Transcriptional Regulation of DNA-Methylating Enzymes

PI3K/AKT kinase signaling pathway regulates DNA methylation by either stabilizing DNMT1 proteins or transcription upregulation of DNMT3a/DNMT3b.

A previous study reported that methylation at lysine-142 of DNMT1 is marked for its proteasomal degradation (Estève et al. 2009; Estève et al. 2011). Methylation of DNMT1 is catalyzed by the SET domain containing 7, a histone lysine methyltransferase. Interestingly, AKT abrogates the lysine-142 methylation by directly phosphorylating Ser-143 of DNMT1 to prevent its proteasomal degradation.

Inhibition of methylation prevents proteasomal degradation of DNMT1 (Hervouet et al. 2010).

GSK3β is a serine/threonine kinase and it phosphorylates an N-terminal serine/threonine residue, located four residues prior to a pre-phosphorylated serine/threonine S/T-X-X-X-S/T(P). GSK3β kinase functions as tumor suppression by promoting proteasomal degradation of many oncoproteins. It is reported that GSK3β phosphorylates DNMT1 at Ser-410 and Ser-414 to promote the proteasomal degradation of DNMT1 by F-box protein β-TrCP(β-transducin repeat-containing protein) through SCF (SKP1, Cullin 1, and F-box protein) complex (Lin et al. 2010). However, tumor suppressor activity of GSK3β is countered by AKT. AKT phosphorylates GSK3β at Ser9, resulting in its inactivation. Thus, AKT increases the stability of DNMT1 in hepatocellular carcinoma (Fang et al. 2015).

IL-6 cross-talk with multiple signaling cascades to reply to the inflammatory signal. It is also reported that IL-6 plays a crucial role in the nuclear translocation of DNMT1 through the AKT signaling pathway. In the presence of IL-6, AKT directly phosphorylates Ser-396 present in the nuclear localization signal (NLS) sequence of DNMT1. Phosphorylation at Ser-396 results in importin-mediated nuclear translocation of DNMT1 to facilitate targeted gene DNA methylation (Hodge et al. 2007).

In addition to DNMT1, the other two DNA-methylating enzymes DNMT3a and DNMT3b, are transcriptionally regulated by PI3K/AKT pathway through different mediators (Fig. 13.3). PM2.5 (particulate matter 2.5) induces the production of intracellular reactive oxygen species (ROS) production, resulting in the induction of AKT activity and upregulation of DNMT3b. DNMT3B then promotes the methylation of the *TP53* promoter region in human bronchial epithelial cells (BEAS-2B). Chronic epigenetic changes are often associated with inhibition of *TP53*, a common finding in lung cancer (Zhou et al. 2016).

In gastric tumors, NF-κB activates transcription of RNA-binding factor HuR via phosphatidylinositol 3-kinase/AKT signaling (Kang et al. 2008). HuR binds DNMT3B 3′UTR on HuR-recognition motif to increase the stability of DNMT3b mRNA and increase the expression of DNMT3B (López de Silanes et al. 2009).

Overexpression of insulin-like growth factor 2 (IGF2), long non-coding RNA, H19, and insulin-like growth factor II receptor (IGF2R) is associated with the development of several carcinogeneses such as colorectal cancer (CRC), breast carcinoma, and oral squamous cell carcinoma (Kasprzak and Adamek 2019). H19 is associated with an increased risk of breast carcinogenesis (Wang et al. 2019; Zavras et al. 2003; Kasprzak and Adamek 2019).

Epigenetic Regulation of Gene by AKT through Histone Modification

Post-translational modifications of histone on the promoter site determine the functionality of the gene. PI3K/AKT pathway also regulates the different types of post-translational modifications of histone to promote tumorigenesis.

Histone Acetylation and Deacetylation

Histone is acetylated by histone acetyltransferase (HAT) and deacetylated by histone deacetylase (HDAC) (Gräff and Tsai 2013). Detection of histone acetylation on the promoter of the gene is used as a marker for an active gene. Recent studies showed that tumor suppressor genes are epigenetically silenced in cancer through histone deacetylation, mediated by histone deacetylase.

In premalignant and cancer cells, AKT enhances the glucose uptake and phosphorylates ATP-citrate lyase (ACLY) to promote the production of acetyl-CoA in gliomas and prostate tumor (Lee et al. 2014). Acetyl-CoA is essential for the acetylation of histone by HAT (Das et al. 2017). In parallel, AKT kinase pathway also plays a crucial role in the hyperactivation of acetyl-CoA synthetases (ACSS1) to promote the production of acetyl-CoA (Gopal and Pizzo 2017). Overall, PI3K/AKT regulates histone acetylation by activating acetyl-CoA synthetase ACLY and ACSS1 (Fig. 13.3), which is reflected in cancer like prostate. In prostate cancer, there is a strong positive correlation between AKT activity and histone acetylation levels (Lee et al. 2014).

In addition to acetyl-CoA production, activated AKT also increases the HAT activity of p300/CBP by phosphorylating p300 and CBP at Ser-1834 and Thr-1871, respectively (Liu et al. 2020). Phosphorylated p300/CBP promotes acetylation of histones at the promoter region of heat-shock protein HSP70 to augment its mRNA levels. HSP70 then protects AKT from its proteasomal degradation. Therefore, the inactivation of p300/CBP (HAT) results in inactivation of AKT activity via transcription repression of HSP70 (Liu et al. 2020).

Intercellular adhesion molecule 1 (ICAM-1) promotes cancer metastasis. A study has suggested that AKT increases the expression levels of ICAM-1 through epigenetic modification of ICAM-1 promoter (Huang and Chen 2005). In response to TNF-α, nuclear-localized AKT promotes histone acetylation on the promoter of ICAM-1 by p300, resulting in the enhanced recruitment of basal transcription complex to increase the expression levels of ICAM-1.

AKT Regulates Histone Methylation

Histone methylation marks for both epigenetic gene silencing as well as gene activation based on the position of the lysine residue in histone. Methylation of histone is frequently disrupted in cancer. EZH2 functions as histone methyltransferase in the polycomb repressive complex 2 (PRC2) and forms a complex with EED, SUZ12, JARID2, AEBP2, RbAp46/48, and PCL (Margueron and Reinberg 2011). EZH2 is responsible for transcriptional gene silencing by stimulating promoter-associated histone H3 Lys27 trimethylation (H3K27me3). Therefore, overexpression of EHZ2 is closely correlated with cancer malignancy (Visser et al. 2001; Kleer et al. 2003; Collett et al. 2006; Gonzalez et al. 2011). A previous study showed AKT plays an important role in the stability of EZH2 in cancer (Riquelme et al. 2016). Conversely, EZH2 also promotes the activation of AKT in cancer. Thus, AKT and EZH2 positively regulate each other to promote cancer pathogenesis by epigenetic gene silencing (Fig. 13.3).

Wolf–Hirschhorn syndrome candidate 1 (WHSC1) is a histone dimethyl transferase. It catalyzes the demethylation of histone H3 at lys36 (H3K36me2). H3K36me^2 is a marker associated with transcriptional activation. It is reported that the accumulation of WHSC1 is closely correlated with cancer progression and metastasis (Wu et al. 2019; Hudlebusch et al. 2011). Activated AKT phosphorylates WHSC1 at Ser-172 to prevent its proteasomal degradation by CRL4Cdt2 E3 ligase in prostate cancer (Li et al. 2017). In a feed-forward pathway, stabilized WHSC1 enhances the AKT activity by upregulating RICTOR.

H3K4 Trimethylation

Promoter-associated H3K4me3 marks for transcriptional activation of euchromatin. PI3K-pathway-mediated H3K4me3 has been reported to be elevated in breast and colorectal cancers (Benard et al. 2014; Mungamuri et al. 2013).

H3K4me3 is regulated by MLL/KMT2 histone methyltransferases (HMTs) and the KDM5 family of histone demethylases (KDMs). In breast cancer, PI3K activation inhibits nuclear localization of KDM5A to limit the exposure of KDM5A to its substrate H3K4me2/3. Therefore, high occupancy of H3K4me2/3 is observed in AKT1/3 activated breast cancer (Spangle et al. 2016). Another study suggested that activated PI3K/AKT cascade phosphorylates and stabilizes Pygo2 protein, which promotes H3K4 methylation by interacting with MLL2 histone methyltransferase (HMT) complex (Li et al. 2015). Thus, AKT plays an important role in the methylation of H3 at lysine 4.

H2A Ubiquitination by AKT Signaling

Bmi1 is a member of the PRC1 core complex and it directs ubiquitination of H2A at lysine-119 (Cao et al. 2005). Ubiquitination of H2A results in repression of gene expression. A previous study showed that Bmi1 plays a vital role in the inactivation of INK4A/ARF gene locus encoding p16^{INK4A} and p19^{ARF} tumor suppressors by recruiting the ubiquitinated H2A on the promoter (Jacobs et al. 1999). Interestingly, AKT phosphorylates Bmi1 at Ser-316 to inhibit its ubiquitin ligase activity and prevent H2A ubiquitination on the INK4A/ARF gene locus. Thus, in many cancer cells, AKT prevents epigenetic gene silencing of p16^{INK4A} and p19^{ARF} (Liu et al. 2012).

AKT-mTOR Signaling-Mediated Histone Acetylation in Cancer Progression

Recently, it is demonstrated that the mechanistic target of rapamycin complex 2 (mTORC2) functions as a strong acetylation driver in cancer. mTORC2-mediated acetylation contributes to metabolic reprogramming by controlling glycolytic metabolism and regulating the cellular level of c-Myc in cancer. mTORC2 facilitates the acetyl-CoA production and nuclear translocation of histone-modifying enzymes like pyruvate dehydrogenase (PDH). Thus, mTORC2 promotes histone acetylation (H3K9ac, H3K18ac, H3K27ac) to increase the transcriptional activation of iron metabolism-related enzymes, including ferritin light chain (FTL), ferritin heavy chain (FTH1), transferrin receptor (TFR), and divalent metal transporter 1 (DMT1) in order to drive cell survival and tumor growth in glioblastoma (Masui et al. 2019; 2020).

mTORC2 also downregulates global histone H3 lysine 56 acetylation (H3K56Ac) on the promoter of glycolytic genes to control their expression via regulation of Sirtuin 6 (SIRT6) in glioma cells (Fig. 13.3). SIRT6 plays a crucial role in the maintenance of telomere integrity and in the regulation of metabolic homeostasis and DNA repair (Vadla and Haldar 2018). It is also reported that metabolic syndrome like hyperinsulinemic enhances risk factors in triple-negative breast cancer through PI3K/AKT/mTOR pathway-mediated histone acetylation of (H3K9ac) on promoters of insulin-induced genes (Senapati et al. 2019).

Ras Signaling Pathway Induces Epigenetic Alterations to Promote Cancer

Stimulation of the Ras signal pathway plays an essential role in the epigenetic inactivation of several genes in human cancers. In the H-Ras-transformed rat fibroblast FE-8 cells, MEK/ERK target genes such as clusterin, matrix metalloproteinase 2 (Mmp2), peptidylprolyl isomerase C-associated protein, syndecan-4, Timp2, and Thbs1 are repressed in a methylation-dependent manner (Lund et al. 2006). It was reported that K-Ras and N-Ras are activated in acute myeloid leukemia without activating Ras mutation (Ehmann et al. 2006). Ras, along with AP-1, promotes the transcriptional level of DNMT1 in cancer cells (MacLeod et al. 1995).

A previous study reported that H-Ras epigenetically silence pro-apoptotic death receptor Fas expression in order to promote tumor progression by inhibiting Fas-mediated apoptosis in mouse NIH3T3 cells (Peli et al. 1999). In this case, Ras triggers PI3K pathway-mediated suppression of Fas expression (Fenton et al. 1998). Although Ras-mediated epigenetically silencing of Fas is well documented, factors required for Ras-mediated silencing of Fas were largely unknown for a long time. A genome-wide RNA interference (RNAi) screening was performed by Michael

Green's group in K-Ras-transformed NIH 3T3 cells, in which they discovered 28 genes to be required for Ras-mediated epigenetic silencing of Fas gene (Gazin et al. 2007). These RESEs (Ras epigenetic silencing effectors) include DNA methyltransferase (DNMT1) and several polycomb group proteins directly associated with specific regions of the Fas promoter. This study identified three regions of Fas promoter, which are hypermethylated in K-Ras NIH 3T3 cells. An RNAi-based epistasis experiment suggested that these RAS-directed silencing of tumor suppressor genes (TSGs) occurs through a highly ordered pathway that is initiated by RAS signaling through the MAPK and PI3K/AKT pathways. They identified ZFP354B as an important initiator of epigenetic gene silencing in the Ras signaling pathway. ZFP354B is a zinc finger protein containing a KRAB transcriptional repressor domain. The binding of ZFP354B on the promoter initiates the recruitment of additional cofactors DNA-binding proteins, multisubunit complexes, Polycomb repressive complex (EZH2 and EED), *etc.* Finally, this ordered RAS-directed transcriptional silencing pathway culminates in the recruitment of the DNA methyltransferase DNMT1 leading to methylation of the promoter and silencing of TSG expression (Wajapeyee et al. 2013). In addition to Ras, BRAF(V600E) mediates widespread epigenetic silencing in melanoma and colorectal cancer (CRC) (Fang et al. 2016). In both CRC and melanoma, BRAF(V600E) promotes epigenetic silencing through upregulation of a transcriptional repressor v-maf avian musculoaponeurotic fibrosarcoma oncogene homolog G (MAFG). Increased BRAF/MEK/ERK signaling results in ERK-directed phosphorylation of MAFG at Ser-124. Phosphorylation at Ser-124 prevents its polyubiquitination and subsequent proteasomal-mediated degradation. MAFG then binds to DNA and facilitates the recruitment of other corepressor complexes, including BACH1, CHD8, and DNMT3b followed by promoter hypermethylation and transcriptional silencing of CIMP (CpG island methylator phenotype) and HIM (hypermethylated in melanoma) genes in BRAF-positive CRCs and melanomas (Fang et al. 2016). The apoptosis-promoting protein Par-4 is epigenetically silenced by promoter hypermethylation in Ras-transformed NIH3T3 fibroblasts through the Raf/MEK/ERK MAPK pathway in association with upregulation of DNMT1 and DNMT3 (Pruitt et al. 2005).

Six proteins belonging from the RASSF family, containing conserved Ras Association domains, serve as Ras effectors. Frequent epigenetic inactivation of RASSF genes in human tumors disconnects Ras from pro-death signaling systems and enhances Ras-driven transformation and metastasis. One of the best-characterized members of the RASSF family is RASSF1A, a tumor suppressor gene, and is epigenetically silenced in a variety of cancers. Some sporadic malignant melanomas and melanoma cell lines exhibit reduced RASSF1A transcript with the hypermethylated promoter. RASSF1A promoter methylation is associated with BRaf or N-Ras mutations in most tumors (Reifenberger et al. 2004). A study, utilizing a genome-wide human shRNA screen, has shown that homeobox protein HOXB3 promotes epigenetic silencing of RASSF1A in human A549 NSCLC (non-small-cell lung cancers) and NCI-H1437 cell (Palakurthy et al. 2009). HOXB3 enhances upregulation and the recruitment of DNA methyltransferase DNMT3b to the RASSF1A promoter. DNMT3B physically associates with EZH2, a subunit of polycomb repressive complex 2 (PRC2) forming Polycomb repressive complex 2 and DNMT3B complex (PRC2/DNMT3B. An oncoprotein c-Myc first binds near the RASSF1A transcription start site and then promotes the recruitment of PRC2/DNMT3B complex to the RASSF1A promoter. Therefore, hypermethylation and epigenetic silencing of RASSF1A frequently occurs in lung adenocarcinomas and other cancer types. Similarly, another RASSF family protein, RASSF2 is methylated and inactivated by K-Ras/BRaf mutations in primary colorectal cancer (CRC) and in colorectal adenomas (Akino et al. 2005). There, RASSF2 enhances K-*Ras*-induced oncogenic transformation. In malignant astrocytoma, activated Ras signaling frequently harbors DNA-methylations that inactivate p53 (Akino et al. 2005).

Tumor suppressor OPCML (opioid-binding protein/cell adhesion molecule like gene) is frequently inactivated by methylation of CpG island present in the promoter (Mei et al. 2006). They showed that RAS^{V12}-mediated oncogenic transformation was accompanied by a concomitant loss of OPCML expression by promoter hypermethylation in human epithelial ovarian cells (T29H). Thus, the identification of new components that act downstream of Ras and are required for epigenetic silencing and transformation could be a potential anti-cancer target.

Epigenetic Modification of JAK/STAT Signaling in Cancer Progression

Upon prognosis in several tumor types, inappropriate activation of JAK/STAT signaling has already been studied to a large extent. STAT3 and STAT5A/B are the critical members of this pathway, often constitutively activated in numerous cancers. Several studies have also suggested the role of the JAK/STAT pathway in cancer progression through epigenetic modification.

SOCS5 (suppressor of cytokine signaling 5) silences the JAK/STAT signaling pathway. Sharma et al. have demonstrated that epigenetic silencing of *SOCS5* (suppressor of cytokine signaling 5) potentiates the aberrant JAK/STAT signal to initiate T-cell lineage acute lymphoblastic leukemia (T-ALL) (Sharma et al. 2019). In T-ALL, SOCS5 plays a tumor-suppressive role by suppressing MYC expression. However, SOCS5 is epigenetically silenced by *DNMT3A*-mediated DNA methylation and HDAC-mediated (HDAC1 and HDAC2) histone deacetylation. This SOCS downregulation reinforces the abnormal JAK/STAT signaling in response to IL-7 stimulation.

In addition to SOCS5 silencing, JAK/STAT pathway is also involved in the epigenetic silencing of p53 (Tan et al. 2014). In this instance, the natural variant of prolactin receptor Delta-S2 PRLRs lacking its S2 extracellular domain is constitutively activated in breast cancer. This delta S2 PRLR induces the expression of histone methyltransferase EZH2, which increases the trimethylation of histone 3 on lysine 27 (H3K27Me3) on the promoter of p53 to silence it (Tan et al. 2014).

Another study by Wu et al. demonstrated that STAT3 plays a major role in neurofibroma progression. STAT3 promotes aberrant expression of β-catenin by facilitating epigenetic silencing of *Arid1b* through histone modification (Wu et al. 2016a).

Investigations have established that hyperactivation of JAK/STAT signaling is responsible for the progression of gastric cancer (GC) (Wu et al. 2016b; Wei et al. 2021; Liu et al. 2011). Under normal circumstances, two members of trefoil factor family, TFF1 and TFF2 (TFF1/2), are secreted by gastric epithelial cells and function as a tumor suppressor. They play an important role in maintaining the mucosal integrity of the gastrointestinal tract and protect the epithelia against mucosal damage. Their expression levels are transcriptionally regulated by GATA6. However, in hyperactivated JAK/STAT signaling, STAT3 binds to the upstream of the GATA6 promoter and enhances the promoter methylation, thereby silencing of GATA6 expression as well as downregulation of TFF1/2 in gastric cancer AGS cell line. In addition to TEF, NR4A3 (nuclear receptor subfamily 4 group A member 3) functions as a tumor suppressor in GC. It is a nuclear receptor and transcription factor. GC cell line MKN28 has constitutively activated STAT mutant, S16, which induces the promoter hypermethylation and subsequent epigenetic silencing of NR4A3. There is a good correlation of epigenetic silencing of NR4A3 and activated STAT3 in gastric cancer patients with poor survival outcomes (Yeh et al. 2016). Recent study has also reported the constitutive activation of JAK/STAT signaling in AGS gastric cancer cells, contributing to epigenetic silencing of tumor suppressor microRNA, miR-193a. This transcriptional repression of miR-193a may result in overexpression of (YWHAZ) tyrosine 3-monooxygenase/tryptophan 5-monooxygenase activation protein zeta, which was previously known to be involved in cancer metastasis (Wei et al. 2021). JAK2 signaling majorly controls myeloproliferative neoplasms (MPNs), the disease of blood cells, and clonal proliferation of hematopoietic cell lineages in the bone marrow. Activated JAK2 inactivates arginine methyltransferase, PRMT5, by phosphorylation in order to prevent histone methylation-mediated chromatin remodeling in MPNs (Liu et al. 2011).

Epigenetic Modification of NOTCH Signaling Pathway in Cancer

Dysregulated Notch signaling has contributed to many types of cancer progression via both genetic and epigenetic alterations. Both hypermethylation and hypomethylation of Notch genes have been reported in several human cancers. It is reported that the promoter of *NOTCH1* is hypermethylated in gastric cancer while the promoter of *NOTCH4* gene is hypomethylation in hepatocellular carcinoma. Hypomethylation of the promoter of notch pathway genes such as *DLL1*, *HEY1*, *DTX1*, *HDAC1*, *NOTCH2,* and *JAG1* in hepatoblastoma and promoter of *JAG2* in multiple myeloma patients and cell lines has been documented (Aithal and Rajeswari

2013). Zampieri et al. reported that H3K4Me3 is enriched on the promoter of NOTCH3 (Zampieri et al. 2014). They discovered that the chromatin regulator protein BORIS/CTCFL plays an important role in regulating NOTCH3 gene expression. Their data suggested that BORIS binds to NOTCH3 promoter in cancer cells and induce as well as maintains an active/permissive chromatin conformation. Moreover, NOTCH1 is also highly expressed in glioma cell lines through epigenetic gene regulation (Liu et al. 2014). The expression of Delta-like 3 (DLL3), a ligand of NOTCH receptor, is silenced in HuH2 hepatocellular carcinoma (HCC) cells by DNA methylation (Mizuno et al. 2018).

Similar to the epigenetic modification-mediated abnormal activation of NOTCH signaling in cancer, the aberrantly activated NOTCH pathway also promotes cancer progression by controlling the genes at the epigenetic levels. It is reported that the activated delta-NOTCH signaling pathway promotes metastatic tumor in Drosophila through the regulation of two Polycomb group epigenetic silencers, lola and pipsqueak (psq). These genes induce the activity of histone methyltransferase E (z)–Esc complex and switch the target genes from the active H3K4me3 state to H3K9 and H3K27 methylation silent state. The unusual histone modification inhibits the transcription termination of the Rbf gene, the fly homolog of the tumor suppressor RB/Rb. Thus, hyperactivation of the NOTCH pathway results in these tumors commences by repressing the Rbf gene (Ferres-Marco et al. 2006).

The role of Ephrin type-B receptor 2 precursor (EPHB2) and Ephrin type-B receptor 4 precursor (EPHB4) in colorectal cancer (CRC) has been studied (Lian et al. 2018). Their study demonstrated that EPHB2 functions as a tumor suppressor, whereas EPHB4 functions as an oncogene. Higher expression of EPHB4 in CRC patients has been shown to be correlated with higher tumor stage and grade. Further, they showed that Notch signaling plays a crucial role in epigenetic regulation of EPHB4 to promote CRC.

Epigenetics and Cancer Therapeutics

From the above discussion, it is obvious that promoter hypermethylation and histone modifications are heavily altered in several cancer cells (Sharma et al. 2010). Further, the uncontrolled epigenetics-mediated altered gene expression is responsible for the deregulation of cellular signaling in cancer, indicating that epigenetic drugs could be important for cancer therapeutics (Jones et al. 2016).

Indeed, two classes (broad reprogrammers and precision medicines) of epigenetic drugs are in the clinical trial. Precision medicines are developed to treat specific patient subsets. Broad reprogrammers are developed to cause large-scale changes in gene expression. Histone deacetylase inhibitors (HDACi), DNMT inhibitors (DNMTi), extra-terminal motif proteins (iBETs), and inhibitors of the bromodomain are under this category (Bhadury et al. 2014; Yang et al. 2014). On the other hand, mIDH inhibitors, lysine-specific demethylase 1 (LSD1), SRC family kinase (SFK) inhibitors are under the category of precision medicines. Some of the potential drugs and their mode of action in different cancer types are listed below in Table 13.4.

Application of Epigenetic Biomarkers for the Detection of Cancer at the Early Stage

Epigenetic reprogramming plays a predominant role in the initiation of malignancy. Therefore, specific abnormal epigenetic changes are prevalent in a specific tumor type and occur at an initial stage of tumorigenesis.

Numerous inexpensive, non-invasive, reliable, and highly specific biomarker detection methods have been developed. Methylation-specific PCR(MCP)-based approach is the conventional sensitive method that uses PCR primers that comprise CpG dinucleotides. MethyLight and SMART-MSP are more powerful quantitative MSP-based approaches for molecular detection. ChIP-Seq and MeDIP-seq are two chromatin immunoprecipitation-based techniques. Other methods like affinity capture techniques, multiplexed probe-based detection, bisulfite sequencing (BS-Seq), liquid chromatography-electrospray ionization tandem mass spectrometry with multiple reaction monitoring (LC–ESI–MS/MS–MRM), and immunoquantification tools (ELISA) are also widely used in methylation analysis for their ease of adaptability in the clinic (Pan et al. 2018). Moreover, many liquid biopsy tests are currently registered in the marketplace, such as GynTect for cervical cancer, Cologuard for colorectal cancer, Therascreen® MGMT Pyro for glioblastoma, Epi proLung for lung cancer, ConfirmMDx for prostate cancer, and HCCBloodTest for liver cancer (Locke et al. 2019).

It is reported that methylation-based clinically relevant biomarkers including *BRCA1* in breast cancer (Liu et al. 2015), *MGMT* in glioblastoma multiform (GBM) (Hegi et al. 2008), and *MLH1* in colon cancer (Kuismanen et al. 2000) are under clinical investigation. Also, epigenetic modification of immune-associated genes including CD96, HHLA2, CCR5, CXCR5, and CCL5 in CD8+ T cell and CTLA-4 methylation, PD-L1 methylation can lead to a promising and accurate predictor of patient outcome. Thus, the combined treatments of epi-drugs can modulate the immunotherapy responses, maximize the treatment success, and minimize unwanted effects (Villanueva et al. 2020).

Epigenetic modifications can be detected as cancer biomarkers in circulating cell-free DNA (cfDNA) present in the blood, plasma, stool, urine, bronchial aspirates, saliva, and other body fluids in cancer patients. Numerous studies over the years have explored circulating non-invasive biomarkers for early diagnosis and prognosis of breast cancer, endometrial cancer, cervical cancer, colorectal cancer, gastric cancer, prostate cancer, pancreatic cancer, lung cancer with several translated examples already (Duruisseaux and Esteller 2018; Zamani et al. 2018; Haldrup et al. 2018; Pan et al. 2018; Duruisseaux and Esteller 2018; Dumitrescu 2018). So far, a

Table 13.4 Epigenetic drugs and their mode of action in different cancer

Inhibitors	Name of the drugs	Mode of action	Cancer type
DNMTi	Cytarabine	Targets DNMTs and inhibits DNA methylation	Acute myeloid leukemia (Wei et al. 2020)
	5-aza-2′-deoxycytidine (5-aza-CdR)	It can inhibit cell growth significantly, induce significant apoptosis, down-regulate DNMT1, and upregulate estrogen receptor alpha (ERα) and estrogen receptor beta (ERβ) gene expression at different periods	Colon cancer (Sanaei and Kavoosi 2020)
	Zebularine	It alters DNA methylation status and induces suppression of the Wnt signaling pathway, which leads to apoptotic cell death in CCA	Cholangiocarcinoma (CCA) (Nakamura et al. 2015)
HDACi	Romidepsin	Targets HDACs and Inhibits histone deacetylation	Cutaneous T-cell lymphoma (Frye et al. 2012)
	Aza-PBHA	It increased Aryl hydrocarbon receptor (AHR) levels to suppress HDAC activity and inhibit cell migration by activating PKCα activation	Gastric cancer (Tsai et al. 2020)
	Suberoylanilide hydroxamic acid (SAHA)	It restored acetylation of histone 4 lysine 12 (H4K12Ac) and histone 3 lysine 9 (H3K9Ac) with induction of G1 or G2 cell cycle arrest and apoptosis	Esophageal squamous cell cancer (ESCC) (Tzao et al. 2014)
	Trichostatin A	It induces apoptosis and autophagy by enhancing beclin 1 protein expression and LC3 II/I ratio. It reduces p62, PRMT5, and TRPV6 levels and enhances STC1 and p-JNK levels	Cervical cancer (Liu et al. 2021)
Multi Kinase Inhibitor	Sorafenib	It induced apoptosis and decreased cellular NAD concentrations in HCC cell lines. It also can decrease the activity of mTOR	Hepatocellular carcinomas (HCCs) (Garten et al. 2019)
	Midostaurin	It inhibits protein kinase C	Acute myeloid leukemia (Stone et al. 2017)
Dual inhibitor of HDACs and PI3K/AKT	CUDC-907	It inhibits myofibroblasts/CAF cell proliferation, migration, and apoptosis in a dose-dependent manner and induces cell cycle arrest at G1-S phase. It also inhibits the expression of myofibroblast markers and significantly inhibits the phosphorylation level of AKT, mTOR, Smad 2/3. Besides these, it promotes the acetylation of histones	Lung cancer (Zhang et al. 2020)

(continued)

Table 13.4 (continued)

Inhibitors	Name of the drugs	Mode of action	Cancer type
mIDH Inhibitors	Vorasidenib	It is a brain penetrant and dual inhibitor of the mutant isocitrate dehydrogenase 1 (IDH1) and mutant IDH2 enzymes. It also affects erasers of DNA methylation (TET enzymes) and histone methylation	Lower-grade gliomas (LGGs) (Mellinghoff et al. 2021)
JAK-STAT signaling Inhibitors	Ruxolitinib	It inhibits JAK1/2	Chronic neutrophilic leukemia (CNL), atypical chronic myeloid leukemia (a CML) (Dao et al. 2020)
	Momelotinib	It inhibits paclitaxel-induced JAK2/STAT3 pathway activation and cancer stem cell-like development	Ovarian cancer (Chan et al. 2018)
Wnt signaling inhibitors	CWP232291	It induces ER stress and upregulates the expression of proapoptotic transcription factor CHOP. ER stress is associated with WNT pathway modulation. It also suppresses the expression of β-catenin and AR in castration-resistant prostate cancer (CRPC)	Prostate cancer (Pak et al. 2019)
	PRI-724(C-82)	It is a selective inhibitor of Wnt/β-catenin/CBP signaling	Hepatocellular carcinoma (Gabata et al. 2020)
Lysine-specific demethylase 1 (LSD1) Inhibitors	GSK2879552	It targets the FAD domain of LSD1	Triple-negative breast cancer (Tan et al. 2019)
	CC90011	It is a reversible inhibitor of LSD1	Acute myeloid leukemia (AML) and small cell lung cancer (SCLC) cell lines. (Kanouni et al. 2020)
BET inhibitors	JQ1	JQ1 increased DNA damage in pancreatic ductal adenocarcinoma cells	Pancreatic ductal adenocarcinoma (Miller et al. 2019)
PI3K/Akt/mTOR inhibitors	Pictilisib (GDC-0941)	It is an oral inhibitor of class I PI3K that prevents the formation of phosphatidylinositol (3,4,5)-trisphosphate by binding to the adenosine triphosphate-binding pocket of PI3K	Metastatic breast cancer (Schöffski et al. 2018)
	Rapamycin	In combination with melatonin (a potent free radical scavenger), Rapamycin blocked the negative feedback loop from the specific downstream effector of mTOR activation S6K1 to Akt signaling, which decreased cell proliferation and cell viability	Head and neck squamous cell carcinoma (HNSCC) (Shen et al. 2018)

Notch signaling inhibitors	MK-0752	It is a gamma-secretase inhibitor (GSI). In combination with tocilizumab (IL6R antagonist), it decreases breast cancer stem cell (BCSC) and inhibits tumor growth	Notch 3-expressing breast cancer (Wang et al. 2018)
	Quinomycin A	It affects pancreatosphere formation. It decreased the expression of cancer stem cell marker proteins CD24, DCLK1, CD44, and EPCAM. It also reduced the levels of Notch 1-4 receptors, their ligands Jagged1, Jagged2, DLL1, DLL3, DLL4, and the down-stream target protein Hes-1	Pancreatic cancer (Ponnurangam et al. 2016)
	Gliotoxin	It inhibits nuclear Notch 2. It induced apoptosis by inhibiting the *NOTCH2/FCER2* (CD23) axis together with concomitant upregulation of the *NOTCH3/NR4A1* axis	Chronic lymphocytic leukemia (CLL) (Hubmann et al. 2020)
Hh signaling inhibitors	Sonidegib	It is a potent inhibitor of Smoothened (SMO). It induces apoptosis in the NKTCL cells	Natural killer/T-cell lymphoma (NKTCL) (Liu et al. 2019)
	Ciclesonide	It reduces the protein levels of GL1, GL2, and Smoothened (SMO). It also reduces the expression of SOX2 and inhibits cell proliferation, as well as CSC formation	Lung cancer (Choi et al. 2020)
Ras/Raf/MAPK signaling inhibitors	AMG 510	It is the first K-RAS (G12C) inhibitor in clinical development. It reduces K-RAS^{G12C} tumors and improves the anti-tumor efficacy of chemotherapy and targeted agents.	K-RAS^{G12C} mutant cancer (Canon et al. 2019)

blood-based methylated biomarker, *SEPT9* has been approved by the Food and Drug Administration (FDA) to diagnose colon cancer (Tham et al. 2014).

According to recent studies, DNA methylation, combined with other epigenetic markers, can improve diagnostic accuracy compared to evaluating biomarkers individually. In conclusion, epigenetic biomarkers have tremendous potential due to their accuracy, specificity, and ease of collection. Further research is still necessary for the standardization of detection methods to establish the widespread use of epigenetic biomarkers in the clinic to diagnose different malignancies.

Summary

Cancer is a heterogeneous disease accompanied by multiple epigenetic changes in DNA and histone within the cell. DNA methylation regulates gene expression by recruiting non-transcribing protein complexes or by preventing the binding of transcription activating factors at a specific site on DNA. As a consequence, oncogenes are activated and tumor suppressor genes are turned off. Different post-translational modifications in histone N-terminal tails are responsible for the activation or repression of gene expression. In this chapter, we have demonstrated how epigenetic regulations are responsible for the activation of cellular signaling pathways, including AKT, Ras, Wnt and Hedgehog, Notch, JAK-STAT in different cancers. Similarly, numerous studies have provided the evidence of abnormal activation of different signaling pathways involved in the induction of the epigenetic signature in various types of cancers (Fig. 13.4).

It is known that epigenetic modifications are stable, heritable, and reversible changes. Relying on these phenomena, studies over the years have developed lots of potential drug candidates which are targeting the epigenetic regulators. Epigenetic drugs like azacytidine, decitabine (targeting DNMT1), veronistat, and panobistat (targeting histone deacetylase) have already been approved by the U.S. FDA to treat hematological malignancies. However, to date, epigenetic drugs (epi-drugs) for the treatment of solid tumors need to be developed. Currently, it has been observed that epigenetic modulation by epi-drugs remains unsuccessful in solid tumors due to its instability, toxicity, and off-target side effects. To overcome these problems, nanocarriers-based epi-drugs delivery systems are in a developmental stage. Thus, the combined treatment of nano-based delivery systems with new-generation epi-drugs will contribute to the personalized-targeted therapy. The efficacy of nano-based delivery systems should be further tested in different cancer types. In combination with chemotherapy (cisplatin, doxorubicin) or radiotherapy or immunotherapy drugs, these epigenetic drugs show more efficacy in various types of cancers, including lung, breast, bladder, glioma, prostate, colon, esophageal and head and neck carcinoma, and gastrointestinal tract carcinoma. Thus, disease prediction and cancer prevention have turned out to be promising areas of epigenetic study in the future.

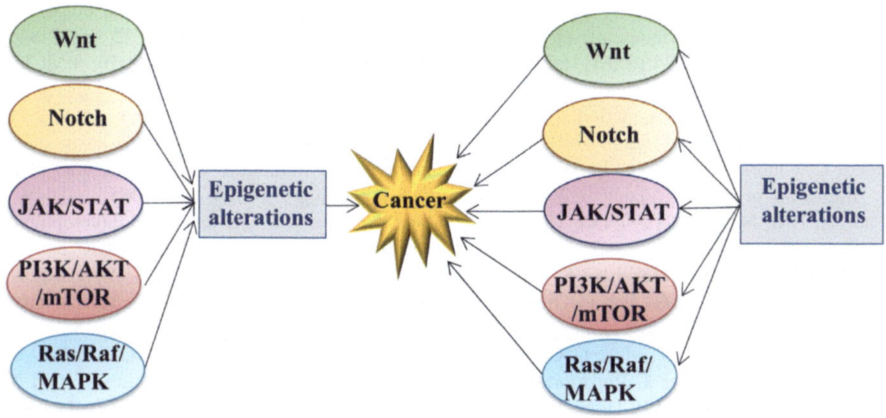

Fig. 13.4 Model demonstrates the epigenetic regulation of cellular signaling pathways responsible for cancer pathogenesis

References

Aithal MG, Rajeswari N (2013 Dec) Role of Notch signalling pathway in cancer and its association with DNA methylation. J Genet 92(3):667-675. https://doi.org/10.1007/s12041-013-0284-5. PMID: 24371188

Akino K, Toyota M, Suzuki H, Mita H, Sasaki Y, Ohe-Toyota M, Issa JP, Hinoda Y, Imai K, Tokino T (2005 Jul) The Ras effector RASSF2 is a novel tumor-suppressor gene in human colorectal cancer. Gastroenterology 129(1):156-169. https://doi.org/10.1053/j.gastro.2005.03.051. PMID: 16012945

Andersson ER, Sandberg R, Lendahl U (2011 Sep) Notch signaling: simplicity in design, versatility in function. Development 138(17):3593-3612. https://doi.org/10.1242/dev.063610. PMID: 21828089

Bafico A, Gazit A, Pramila T, Finch PW, Yaniv A, Aaronson SA (1999 Jun 4) Interaction of frizzled-related protein (FRP) with Wnt ligands and the frizzled receptor suggests alternative mechanisms for FRP inhibition of Wnt signaling. J Biol Chem 274(23):16180-16187. https://doi.org/10.1074/jbc.274.23.16180. PMID: 10347172

Batra S, Shi Y, Kuchenbecker KM, He B, Reguart N, Mikami I, You L, Xu Z, Lin YC, Clément G, Jablons DM (2006 Apr 21) Wnt inhibitory factor-1, a Wnt antagonist, is silenced by promoter hypermethylation in malignant pleural mesothelioma. Biochem Biophys Res Commun 342(4):1228-1232. https://doi.org/10.1016/j.bbrc.2006.02.084. Epub 2006 Feb 24. PMID: 16516163

Benard A, Goossens-Beumer IJ, van Hoesel AQ, de Graaf W, Horati H, Putter H, Zeestraten EC, van de Velde CJ, Kuppen PJ (2014 Jul 22) Histone trimethylation at H3K4, H3K9 and H4K20 correlates with patient survival and tumor recurrence in early-stage colon cancer. BMC Cancer 14:531. https://doi.org/10.1186/1471-2407-14-531. PMID: 25047223; PMCID: PMC4223547

Bhadury J, Nilsson LM, Muralidharan SV, Green LC, Li Z, Gesner EM, Hansen HC, Keller UB, McLure KG, Nilsson JA (2014 Jul 1) BET and HDAC inhibitors induce similar genes and biological effects and synergize to kill in Myc-induced murine lymphoma. Proc Natl Acad Sci U S A 111(26):E2721-30. https://doi.org/10.1073/pnas.1406722111. Epub 2014 Jun 16. PMID: 24979794; PMCID: PMC4084424

Bhagat TD, Zou Y, Huang S, Park J, Palmer MB, Hu C, Li W, Shenoy N, Giricz O, Choudhary G, Yu Y, Ko YA, Izquierdo MC, Park AS, Vallumsetla N, Laurence R, Lopez R, Suzuki M, Pullman J, Kaner J, Gartrell B, Hakimi AA, Greally JM, Patel B, Benhadji K, Pradhan K,

Verma A, Susztak K (2017 Jan 20) Notch pathway is activated via genetic and epigenetic alterations and is a therapeutic target in clear cell renal cancer. J Biol Chem. 292(3):837-846. https://doi.org/10.1074/jbc.M116.745208. Epub 2016 Dec 1. PMID: 27909050; PMCID: PMC5247657

Bird A (2002 Jan 1) DNA methylation patterns and epigenetic memory. Genes Dev. 16(1):6-21. https://doi.org/10.1101/gad.947102. PMID: 11782440

Bucci C, Chiariello M (2006 Jan) Signal transduction gRABs attention. Cell Signal. 18(1):1-8. https://doi.org/10.1016/j.cellsig.2005.07.001. Epub 2005 Aug 9. PMID: 16084065

Cadigan KM, Nusse R (1997 Dec 15) Wnt signaling: a common theme in animal development. Genes Dev. 11(24):3286-3305. https://doi.org/10.1101/gad.11.24.3286. PMID: 9407023

Caldwell GM, Jones C, Gensberg K, Jan S, Hardy RG, Byrd P, Chughtai S, Wallis Y, Matthews GM, Morton DG (2004 Feb 1) The Wnt antagonist sFRP1 in colorectal tumorigenesis. Cancer Res 64(3):883-888. https://doi.org/10.1158/0008-5472.can-03-1346. PMID: 14871816

Canon J, Rex K, Saiki AY, Mohr C, Cooke K, Bagal D, Gaida K, Holt T, Knutson CG, Koppada N, Lanman BA, Werner J, Rapaport AS, San Miguel T, Ortiz R, Osgood T, Sun JR, Zhu X, McCarter JD, Volak LP, Houk BE, Fakih MG, O'Neil BH, Price TJ, Falchook GS, Desai J, Kuo J, Govindan R, Hong DS, Ouyang W, Henary H, Arvedson T, Cee VJ, Lipford JR (2019 Nov) The clinical KRAS(G12C) inhibitor AMG 510 drives anti-tumour immunity. Nature 575 (7781):217-223. https://doi.org/10.1038/s41586-019-1694-1. Epub 2019 Oct 30. PMID: 31666701

Cao R, Tsukada Y, Zhang Y (2005 Dec 22) Role of Bmi-1 and Ring1A in H2A ubiquitylation and Hox gene silencing. Mol Cell 20(6):845-854. https://doi.org/10.1016/j.molcel.2005.12.002. PMID: 16359901

Chan E, Luwor R, Burns C, Kannourakis G, Findlay JK, Ahmed N (2018 Mar 30) Momelotinib decreased cancer stem cell associated tumor burden and prolonged disease-free remission period in a mouse model of human ovarian cancer. Oncotarget 9(24):16599-16618. https://doi.org/10.18632/oncotarget.24615. PMID: 29682172; PMCID: PMC5908273

Chen W, Tang T, Eastham-Anderson J, Dunlap D, Alicke B, Nannini M, Gould S, Yauch R, Modrusan Z, DuPree KJ, Darbonne WC, Plowman G, de Sauvage FJ, Callahan CA (2011 Jun 7) Canonical hedgehog signaling augments tumor angiogenesis by induction of VEGF-A in stromal perivascular cells. Proc Natl Acad Sci U S A. 108(23):9589-94. https://doi.org/10.1073/pnas.1017945108. Epub 2011 May 19. PMID: 21597001; PMCID: PMC3111273

Chen L, Tang J, Feng Y, Li S, Xiang Q, He X, Ren G, Peng W, Xiang T (2017) ADAMTS9 is silenced by epigenetic disruption in colorectal cancer and inhibits cell growth and metastasis by regulating Akt/p53 signaling. Cell Physiol Biochem. 44(4):1370-1380. https://doi.org/10.1159/000485534. Epub 2017 Nov 30. PMID: 29186710

Choi HS, Kim SL, Kim JH, Lee DS (2020 Feb 4). The FDA-approved anti-asthma medicine ciclesonide inhibits lung cancer stem cells through hedgehog signaling-mediated SOX2 regulation. Int J Mol Sci. 21(3):1014. https://doi.org/10.3390/ijms21031014. PMID: 32033067; PMCID: PMC7038186

Collett K, Eide GE, Arnes J, Stefansson IM, Eide J, Braaten A, Aas T, Otte AP, Akslen LA (2006 Feb 15) Expression of enhancer of zeste homologue 2 is significantly associated with increased tumor cell proliferation and is a marker of aggressive breast cancer. Clin Cancer Res 12(4): 1168-1174. https://doi.org/10.1158/1078-0432.CCR-05-1533. PMID: 16489070

Cui Y, Ma W, Lei F, Li Q, Su Y, Lin X, Lin C, Zhang X, Ye L, Wu S, Li J, Yuan Z, Song L (2016 Jul) Prostate tumour overexpressed-1 promotes tumourigenicity in human breast cancer via activation of Wnt/β-catenin signalling. J Pathol 239(3):297-308. https://doi.org/10.1002/path.4725. Epub 2016 May 30. PMID: 27060981

Dao KT, Gotlib J, Deininger MMN, Oh ST, Cortes JE, Collins RH Jr, Winton EF, Parker DR, Lee H, Reister A, Schultz, Savage S, Stevens, Brockett C, Subbiah N, Press RD, Raess PW, Cascio M, Dunlap J, Chen Y, Degnin C, Maxson JE, Tognon CE, Macey T, Druker BJ, Tyner JW (2020 Apr 1) Efficacy of ruxolitinib in patients with chronic neutrophilic leukemia and atypical chronic myeloid leukemia. J Clin Oncol 38(10):1006-1018. https://doi.org/10.1200/JCO.19.00895. Epub 2019 Dec 27. PMID: 31880950; PMCID: PMC7106977

Das S, Morvan F, Morozzi G, Jourde B, Minetti GC, Kahle P, Rivet H, Brebbia P, Toussaint G, Glass DJ, Fornaro M. ATP citrate lyase regulates myofiber differentiation and increases regeneration by altering histone acetylation. Cell Rep. 2017 Dec 12;21(11):3003-3011. https://doi.org/10.1016/j.celrep.2017.11.038. PMID: 29241530

Deshpande I, Liang J, Hedeen D, Roberts KJ, Zhang Y, Ha B, Latorraca NR, Faust B, Dror RO, Beachy PA, Myers BR, Manglik A (2019 Jul) Smoothened stimulation by membrane sterols drives Hedgehog pathway activity. Nature 571(7764):284-288. https://doi.org/10.1038/s41586-019-1355-4. Epub 2019 Jul 1. PMID: 31263273; PMCID: PMC6709672

Du W, Wang S, Zhou Q, Li X, Chu J, Chang Z, Tao Q, Ng EK, Fang J, Sung JJ, Yu J (2013 Jul 11) ADAMTS9 is a functional tumor suppressor through inhibiting AKT/mTOR pathway and associated with poor survival in gastric cancer. Oncogene. 32(28):3319-28. https://doi.org/10.1038/onc.2012.359. Epub 2012 Aug 20. PMID: 22907434

Duan ZH, Wang HC, Zhao DM, Ji XX, Song M, Yang XJ, Cui W (2015 Aug) Cooperatively transcriptional and epigenetic regulation of sonic hedgehog overexpression drives malignant potential of breast cancer. Cancer Sci 106(8):1084–1091. https://doi.org/10.1111/cas.12697. Epub 2015 Jul 25. PMCID: PMC4556399

Dumitrescu RG (2018) Early epigenetic markers for precision medicine. Methods Mol Biol 1856: 3-17. https://doi.org/10.1007/978-1-4939-8751-1_1. PMID: 30178243

Duruisseaux M, Esteller M (2018 Aug) Lung cancer epigenetics: from knowledge to applications. Semin Cancer Biol 51:116-128. https://doi.org/10.1016/j.semcancer.2017.09.005. Epub 2017 Sep 14. PMID: 28919484

Ecke I, Petry F, Rosenberger A, Tauber S, Mönkemeyer S, Hess I, Dullin C, Kimmina S, Pirngruber J, Johnsen SA, Uhmann A, Nitzki F, Wojnowski L, Schulz-Schaeffer W, Witt O, Hahn H (2009 Feb 1) Antitumor effects of a combined 5-aza-2′deoxycytidine and valproic acid treatment on rhabdomyosarcoma and medulloblastoma in Ptch mutant mice. Cancer Res 69(3): 887-895. https://doi.org/10.1158/0008-5472.CAN-08-0946. Epub 2009 Jan 20. PMID: 19155313

Ehmann F, Horn S, Garcia-Palma L, Wegner W, Fiedler W, Giehl K, Mayr GW, Jücker M (2006 Jul) Detection of N-RAS and K-RAS in their active GTP-bound form in acute myeloid leukemia without activating RAS mutations. Leuk Lymphoma 47(7):1387-1391. https://doi.org/10.1080/10428190600565925. PMID: 16923573

Estève PO, Chin HG, Benner J, Feehery GR, Samaranayake M, Horwitz GA, Jacobsen SE, Pradhan S (2009 Mar 31) Regulation of DNMT1 stability through SET7-mediated lysine methylation in mammalian cells. Proc Natl Acad Sci U S A 106(13):5076-81. https://doi.org/10.1073/pnas.0810362106. Epub 2009 Mar 12. PMID: 19282482; PMCID: PMC2654809

Estève PO, Chang Y, Samaranayake M, Upadhyay AK, Horton JR, Feehery GR, Cheng X, Pradhan S (2011 Jan) A methylation and phosphorylation switch between an adjacent lysine and serine determines human DNMT1 stability. Nat Struct Mol Biol 18(1):42-8. https://doi.org/10.1038/nsmb.1939. Epub 2010 Dec 12. PMID: 21151116; PMCID: PMC3048033

Fang QL, Yin YR, Xie CR, Zhang S, Zhao WX, Pan C, Wang XM, Yin ZY (2015 Feb) Mechanistic and biological significance of DNA methyltransferase 1 upregulated by growth factors in human hepatocellular carcinoma. Int J Oncol 46(2):782-790. doi: https://doi.org/10.3892/ijo.2014.2776. Epub 2014 Nov 24. PMID: 25420499

Fang M, Hutchinson L, Deng A, Green MR (2016 Feb 2) Common BRAF(V600E)-directed pathway mediates widespread epigenetic silencing in colorectal cancer and melanoma. Proc Natl Acad Sci U S A 113(5):1250-5. https://doi.org/10.1073/pnas.1525619113. Epub 2016 Jan 19. PMID: 26787892; PMCID: PMC4747705

Fattahi S, Amjadi-Moheb F, Tabaripour R, Ashrafi GH, Akhavan-Niaki H (2020 Dec 1) PI3K/AKT/mTOR signaling in gastric cancer: Epigenetics and beyond. Life Sci 262:118513. https://doi.org/10.1016/j.lfs.2020.118513. Epub 2020 Oct 1. PMID: 33011222

Fenton RG, Hixon JA, Wright PW, Brooks AD, Sayers TJ (1998 Aug 1) Inhibition of Fas (CD95) expression and Fas-mediated apoptosis by oncogenic Ras. Cancer Res 58(15): 3391-3400. PMID: 9699671

Ferres-Marco D, Gutierrez-Garcia I, Vallejo DM, Bolivar J, Gutierrez-Aviño FJ, Dominguez M (2006 Jan 26) Epigenetic silencers and Notch collaborate to promote malignant tumours by Rb silencing. Nature 439(7075):430-436. https://doi.org/10.1038/nature04376. PMID: 16437107

Fruman DA, Chiu H, Hopkins BD, Bagrodia S, Cantley LC, Abraham RT (2017 Aug 10) The PI3K pathway in human disease. Cell 170(4):605-635. https://doi.org/10.1016/j.cell.2017.07.029. PMID: 28802037; PMCID: PMC5726441

Frye R, Myers M, Axelrod KC, Ness EA, Piekarz RL, Bates SE, Booher S (2012 Apr) Romidepsin: a new drug for the treatment of cutaneous T-cell lymphoma. Clin J Oncol Nurs 16(2):195-204. https://doi.org/10.1188/12.CJON.195-204. PMID: 22459529; PMCID: PMC7243910

Fu DY, Wang ZM, Li-Chen, Wang BL, Shen ZZ, Huang W, Shao ZM (2010 Feb) Sox17, the canonical Wnt antagonist, is epigenetically inactivated by promoter methylation in human breast cancer. Breast Cancer Res Treat 119(3):601-612. https://doi.org/10.1007/s10549-009-0339-8. Epub 2009 Mar 20. PMID: 19301122

Fukui T, Kondo M, Ito G, Maeda O, Sato N, Yoshioka H, Yokoi K, Ueda Y, Shimokata K, Sekido Y (2005 Sep 15) Transcriptional silencing of secreted frizzled related protein 1 (SFRP 1) by promoter hypermethylation in non-small-cell lung cancer. Oncogene 24(41):6323-6327. https://doi.org/10.1038/sj.onc.1208777. PMID: 16007200

Gabata R, Harada K, Mizutani Y, Ouchi H, Yoshimura K, Sato Y, Kitao A, Kimura K, Kouji H, Miyashita T, Tajima H, Ohta T (2020 Sep) Anti-tumor activity of the small molecule inhibitor PRI-724 against β-Catenin-activated Hepatocellular Carcinoma. Anticancer Res. 40(9): 5211-5219. https://doi.org/10.21873/anticanres.14524. PMID: 32878809

García JM, Silva J, Peña C, Garcia V, Rodríguez R, Cruz MA, Cantos B, Provencio M, España P, Bonilla F (2004 Oct) Promoter methylation of the PTEN gene is a common molecular change in breast cancer. Genes Chromosomes Cancer 41(2):117-124. https://doi.org/10.1002/gcc.20062. PMID: 15287024

Garten A, Grohmann T, Kluckova K, Lavery GG, Kiess W, Penke M (2019 Aug 19) Sorafenib-induced apoptosis in hepatocellular carcinoma is reversed by SIRT1. Int J Mol Sci 20(16):4048. https://doi.org/10.3390/ijms20164048. PMID: 31430957; PMCID: PMC6719220

Gazin C, Wajapeyee N, Gobeil S, Virbasius CM, Green MR (2007 Oct 25) An elaborate pathway required for Ras-mediated epigenetic silencing. Nature 449(7165):1073-7. https://doi.org/10.1038/nature06251. PMID: 17960246; PMCID: PMC2147719

Giehl K (2005 Mar) Oncogenic Ras in tumour progression and metastasis. Biol Chem 386(3): 193-205. https://doi.org/10.1515/BC.2005.025. PMID: 15843165

Goel A, Arnold CN, Niedzwiecki D, Carethers JM, Dowell JM, Wasserman L, Compton C, Mayer RJ, Bertagnolli MM, Boland CR (2004 May 1) Frequent inactivation of PTEN by promoter hypermethylation in microsatellite instability-high sporadic colorectal cancers. Cancer Res 64 (9):3014-3021. https://doi.org/10.1158/0008-5472.can-2401-2. PMID: 15126336

Gonzalez ME, DuPrie ML, Krueger H, Merajver SD, Ventura AC, Toy KA, Kleer CG (2011 Mar 15) Histone methyltransferase EZH2 induces Akt-dependent genomic instability and BRCA1 inhibition in breast cancer. Cancer Res 71(6):2360-70. https://doi.org/10.1158/0008-5472.CAN-10-1933. PMID: 21406404; PMCID: PMC3071296

Gopal U, Pizzo SV (2017 Nov 14) Cell surface GRP78 promotes tumor cell histone acetylation through metabolic reprogramming: a mechanism which modulates the Warburg effect. Oncotarget 8(64):107947-107963. https://doi.org/10.18632/oncotarget.22431. PMID: 29296215; PMCID: PMC5746117

Gräff J, Tsai LH (2013 Feb) Histone acetylation: molecular mnemonics on the chromatin. Nat Rev Neurosci 14(2):97-111. https://doi.org/10.1038/nrn3427. Epub 2013 Jan 17. PMID: 23324667

Grunstein M (1997 Sep 25) Histone acetylation in chromatin structure and transcription. Nature 389 (6649):349-352. https://doi.org/10.1038/38664. PMID: 9311776

Guo P, Hu T, Zhang J, Jiang S, Wang X (2010 Oct 19) Sequential action of Caenorhabditis elegans Rab GTPases regulates phagolysosome formation during apoptotic cell degradation. Proc Natl Acad Sci U S A 107(42):18016-21. https://doi.org/10.1073/pnas.1008946107. Epub 2010 Oct 4. PMID: 20921409; PMCID: PMC2964220

Guo J, Dai X, Laurent B, Zheng N, Gan W, Zhang J, Guo A, Yuan M, Liu P, Asara JM, Toker A, Shi Y, Pandolfi PP, Wei W (2019 Feb) AKT methylation by SETDB1 promotes AKT kinase activity and oncogenic functions. Nat Cell Biol 21(2):226-237. https://doi.org/10.1038/s41556-018-0261-6. Epub 2019 Jan 28. PMID: 30692625; PMCID: PMC6377565

Haldrup C, Pedersen AL, Øgaard N, Strand SH, Høyer S, Borre M, Ørntoft TF, Sørensen KD (2018 Apr) Biomarker potential of ST6GALNAC3 and ZNF660 promoter hypermethylation in prostate cancer tissue and liquid biopsies. Mol Oncol 12(4):545-560. https://doi.org/10.1002/1878-0261.12183. Epub 2018 Mar 13. PMID: 29465788; PMCID: PMC5891052

Hegi ME, Liu L, Herman JG, Stupp R, Wick W, Weller M, Mehta MP, Gilbert MR (2008 Sep 1) Correlation of O6-methylguanine methyltransferase (MGMT) promoter methylation with clinical outcomes in glioblastoma and clinical strategies to modulate MGMT activity. J Clin Oncol 26(25):4189-4199. https://doi.org/10.1200/JCO.2007.11.5964. PMID: 18757334

Heitzer E, Bambach I, Dandachi N, Horn M, Wolf P (2010 Oct) PTCH promoter methylation at low level in sporadic basal cell carcinoma analysed by three different approaches. Exp Dermatol 19 (10):926-8. https://doi.org/10.1111/j.1600-0625.2010.01120.x. Epub 200 Aug 31. PMID: 20849535

Hervouet E, Lalier L, Debien E, Cheray M, Geairon A, Rogniaux H, Loussouarn D, Martin SA, Vallette FM, Cartron PF (2010 Jun 29) Disruption of Dnmt1/PCNA/UHRF1 interactions promotes tumorigenesis from human and mice glial cells. PLoS One 5(6):e11333. https://doi.org/10.1371/journal.pone.0011333. PMID: 20613874; PMCID: PMC2894052

Hodge DR, Cho E, Copeland TD, Guszczynski T, Yang E, Seth AK, Farrar WL (2007 Nov-Dec) IL-6 enhances the nuclear translocation of DNA cytosine-5-methyltransferase 1 (DNMT1) via phosphorylation of the nuclear localization sequence by the AKT kinase. Cancer Genomics Proteomics 4(6):387-398. PMID: 18204201

Holliday R (1994) Epigenetics: an overview. Dev Genet 15(6):453-457. https://doi.org/10.1002/dvg.1020150602. PMID: 7834903

Hori K, Sen A, Artavanis-Tsakonas S (2013 May 15). Notch signaling at a glance. J Cell Sci. 126 (Pt 10):2135-40. https://doi.org/10.1242/jcs.127308. Epub 2013 May 31. PMID: 23729744; PMCID: PMC3672934

Huang WC, Chen CC (2005 Aug) Akt phosphorylation of p300 at Ser-1834 is essential for its histone acetyltransferase and transcriptional activity. Mol Cell Biol 25(15):6592-602. https://doi.org/10.1128/MCB.25.15.6592-6602.2005. PMID: 16024795; PMCID: PMC1190347

Hubmann R, Schnabl S, Araghi M, Schmidl C, Rendeiro AF, Hilgarth M, Demirtas D, Ali F, Staber PB, Valent P, Zielinski C, Jäger U, Shehata M (2020 Jun 18) Targeting nuclear NOTCH2 by gliotoxin recovers a tumor-suppressor NOTCH3 activity in CLL. Cells 9(6):1484. https://doi.org/10.3390/cells9061484. PMID: 32570839; PMCID: PMC7348714

Hudlebusch HR, Santoni-Rugiu E, Simon R, Ralfkiær E, Rossing HH, Johansen JV, Jørgensen M, Sauter G, Helin K (2011 May 1) The histone methyltransferase and putative oncoprotein MMSET is overexpressed in a large variety of human tumors. Clin Cancer Res 17(9):2919-2933. https://doi.org/10.1158/1078-0432.CCR-10-1302. Epub 2011 Mar 8.

Hui M, Cazet A, Nair R, Watkins DN, O'Toole SA, Swarbrick A (2013 Mar 28) The Hedgehog signalling pathway in breast development, carcinogenesis and cancer therapy. Breast Cancer Res 15(2):203. https://doi.org/10.1186/bcr3401. PMID: 23547970; PMCID: PMC3672663

Jacobs JJ, Kieboom K, Marino S, DePinho RA, van Lohuizen M (1999 Jan 14) The oncogene and Polycomb-group gene bmi-1 regulates cell proliferation and senescence through the ink4a locus. Nature 397(6715):164-168. https://doi.org/10.1038/16476. PMID: 9923679

Jiang X, Tan J, Li J, Kivimäe S, Yang X, Zhuang L, Lee PL, Chan MT, Stanton LW, Liu ET, Cheyette BN, Yu Q (2008 Jun) DACT3 is an epigenetic regulator of Wnt/beta-catenin signaling in colorectal cancer and is a therapeutic target of histone modifications. Cancer Cell 13(6):529-41. https://doi.org/10.1016/j.ccr.2008.04.019. PMID: 18538736; PMCID: PMC2577847

Jones PA, Liang G (2009 Nov) Rethinking how DNA methylation patterns are maintained. Nat Rev Genet 10(11):805-11. https://doi.org/10.1038/nrg2651. Epub 2009 Sep 30. PMID: 19789556; PMCID: PMC2848124.

Jones PA, Issa JP, Baylin S (2016 Sep 15) Targeting the cancer epigenome for therapy. Nat Rev Genet 17(10):630-641. https://doi.org/10.1038/nrg.2016.93. PMID: 27629931

Kang YH, Lee HS, Kim WH (2002 Mar) Promoter methylation and silencing of PTEN in gastric carcinoma. Lab Invest 82(3):285-291. https://doi.org/10.1038/labinvest.3780422. PMID: 11896207

Kang MJ, Ryu BK, Lee MG, Han J, Lee JH, Ha TK, Byun DS, Chae KS, Lee BH, Chun HS, Lee KY, Kim HJ, Chi SG (2008 Dec) NF-kappaB activates transcription of the RNA-binding factor HuR, via PI3K-AKT signaling, to promote gastric tumorigenesis. Gastroenterology 135(6): 2030-42, 2042.e1-3. https://doi.org/10.1053/j.gastro.2008.08.009. Epub 2008 Aug 20. PMID: 18824170

Kanouni T, Severin C, Cho RW, Yuen NY, Xu J, Shi L, Lai C, Del Rosario JR, Stansfield RK, Lawton LN, Hosfield D, O'Connell S, Kreilein MM, Tavares-Greco P, Nie Z, Kaldor SW, Veal JM, Stafford JA, Chen YK (2020 Dec 10) Discovery of CC-90011: a potent and selective reversible inhibitor of lysine specific demethylase 1 (LSD1). J Med Chem 63(23):14522-14529. https://doi.org/10.1021/acs.jmedchem.0c00978. Epub 2020 Oct 9. PMID: 33034194

Kasprzak A, Adamek A (2019 Oct 3) Insulin-like growth factor 2 (IGF2) signaling in colorectal cancer-from basic research to potential clinical applications. Int J Mol Sci 20(19):4915. https://doi.org/10.3390/ijms20194915. PMID: 31623387; PMCID: PMC6801528

Katoh Y, Katoh M. Comparative genomics on Sonic hedgehog orthologs. Oncol Rep. 2005 Oct;14 (4):1087-1090. PMID: 16142377

Kawano Y, Kypta R. Secreted antagonists of the Wnt signalling pathway. J Cell Sci. 2003 Jul 1;116 (Pt 13):2627-2634. https://doi.org/10.1242/jcs.00623. PMID: 12775774

Kelwick R, Desanlis I, Wheeler GN, Edwards DR (2015 May 30) The ADAMTS (A disintegrin and metalloproteinase with thrombospondin motifs) family. Genome Biol 16(1):113. https://doi.org/10.1186/s13059-015-0676-3. PMID: 26025392; PMCID: PMC4448532

Kim JT, Li J, Jang ER, Gulhati P, Rychahou PG, Napier DL, Wang C, Weiss HL, Lee EY, Anthony L, Townsend CM Jr, Liu C, Evers BM (2013 May) Deregulation of Wnt/β-catenin signaling through genetic or epigenetic alterations in human neuroendocrine tumors. Carcinogenesis 34(5):953-61. https://doi.org/10.1093/carcin/bgt018. Epub 2013 Jan 25. PMID: 23354304; PMCID: PMC3643417

Kim BK, Yoo HI, Kim I, Park J, Kim Yoon S (2015 Jun) FZD6 expression is negatively regulated by miR-199a-5p in human colorectal cancer. BMB Rep 48(6):360-6. https://doi.org/10.5483/bmbrep.2015.48.6.031. PMID: 25772759; PMCID: PMC4578624

Kleer CG, Cao Q, Varambally S, Shen R, Ota I, Tomlins SA, Ghosh D, Sewalt RG, Otte AP, Hayes DF, Sabel MS, Livant D, Weiss SJ, Rubin MA, Chinnaiyan AM (2003 Sep 30) EZH2 is a marker of aggressive breast cancer and promotes neoplastic transformation of breast epithelial cells. Proc Natl Acad Sci U S A 100(20):11606-11. https://doi.org/10.1073/pnas.1933744100. Epub 2003 Sep 19. PMID: 14500907; PMCID: PMC208805

Koinuma K, Yamashita Y, Liu W, Hatanaka H, Kurashina K, Wada T, Takada S, Kaneda R, Choi YL, Fujiwara SI, Miyakura Y, Nagai H, Mano H (2006 Jan 5) Epigenetic silencing of AXIN2 in colorectal carcinoma with microsatellite instability. Oncogene. 25(1):139-146. https://doi.org/10.1038/sj.onc.1209009. PMID: 16247484

Kraus C, Liehr T, Hülsken J, Behrens J, Birchmeier W, Grzeschik KH, Ballhausen WG (1994 Sep 1) Localization of the human beta-catenin gene (CTNNB1) to 3p21: a region implicated in tumor development. Genomics 23(1):272-274. https://doi.org/10.1006/geno.1994.1493. PMID: 7829088

Kuang SQ, Fang Z, Zweidler-McKay PA, Yang H, Wei Y, Gonzalez-Cervantes EA, Boumber Y, Garcia-Manero G (2013 Apr 26) Epigenetic inactivation of Notch-Hes pathway in human B-cell acute lymphoblastic leukemia. PLoS One 8(4):e61807. https://doi.org/10.1371/journal.pone.0061807. PMID: 23637910; PMCID: PMC3637323

Kuismanen SA, Holmberg MT, Salovaara R, de la Chapelle A, Peltomäki P (2000 May) Genetic and epigenetic modification of MLH1 accounts for a major share of microsatellite-unstable colorectal cancers. Am J Pathol 156(5):1773-9. https://doi.org/10.1016/S0002-9440(10)65048-1. PMID: 10793088; PMCID: PMC1876911

Lee JV, Carrer A, Shah S, Snyder NW, Wei S, Venneti S, Worth AJ, Yuan ZF, Lim HW, Liu S, Jackson E, Aiello NM, Haas NB, Rebbeck TR, Judkins A, Won KJ, Chodosh LA, Garcia BA, Stanger BZ, Feldman MD, Blair IA, Wellen KE (2014 Aug 5) Akt-dependent metabolic reprogramming regulates tumor cell histone acetylation. Cell Metab 20(2):306-319. https://doi.org/10.1016/j.cmet.2014.06.004. Epub 2014 Jul 3. PMID: 24998913; PMCID: PMC4151270

Li Q, Li Y, Gu B, Fang L, Zhou P, Bao S, Huang L, Dai X (2015 Aug 28) Akt phosphorylates Wnt coactivator and chromatin effector Pygo2 at serine 48 to antagonize its ubiquitin/proteasome-mediated degradation. J Biol Chem 290(35):21553-67. https://doi.org/10.1074/jbc.M115.639419. Epub 2015 Jul 13. PMID: 26170450; PMCID: PMC4571880

Li N, Xue W, Yuan H, Dong B, Ding Y, Liu Y, Jiang M, Kan S, Sun T, Ren J, Pan Q, Li X, Zhang P, Hu G, Wang Y, Wang X, Li Q, Qin J (2017 Apr 3) AKT-mediated stabilization of histone methyltransferase WHSC1 promotes prostate cancer metastasis. J Clin Invest 127(4): 1284-1302. https://doi.org/10.1172/JCI91144. Epub 2017 Mar 20. PMID: 28319045; PMCID: PMC5373871

Lian H, Jia X, Shi N, Xie S, Wang J, Wang W, Ma F, Liu H, Wang A, Cheng X, Liu C (2018 Nov 22) Notch signaling promotes serrated neoplasia pathway in colorectal cancer through epigenetic modification of EPHB2 and EPHB4. Cancer Manag Res 10:6129-6141. https://doi.org/10.2147/CMAR.S178126. PMID: 30538561; PMCID: PMC6257864

Lin RK, Hsieh YS, Lin P, Hsu HS, Chen CY, Tang YA, Lee CF, Wang YC (2010 Feb) The tobacco-specific carcinogen NNK induces DNA methyltransferase 1 accumulation and tumor suppressor gene hypermethylation in mice and lung cancer patients. J Clin Invest 120(2): 521-32. https://doi.org/10.1172/JCI40706. Epub 2010 Jan 19. PMID: 20093774; PMCID: PMC2810088

Liu TH, Raval A, Chen SS, Matkovic JJ, Byrd JC, Plass C (2006 Jan 15) CpG island methylation and expression of the secreted frizzled-related protein gene family in chronic lymphocytic leukemia. Cancer Res 66(2):653-658. https://doi.org/10.1158/0008-5472.CAN-05-3712. PMID: 16423993

Liu P, Cheng H, Roberts TM, Zhao JJ (2009 Aug) Targeting the phosphoinositide 3-kinase pathway in cancer. Nat Rev Drug Discov. 8(8):627-44. https://doi.org/10.1038/nrd2926. PMID: 19644473; PMCID: PMC3142564

Liu F, Zhao X, Perna F, Wang L, Koppikar P, Abdel-Wahab O, Harr MW, Levine RL, Xu H, Tefferi A, Deblasio A, Hatlen M, Menendez S, Nimer SD (2011 Feb 15) JAK2V617F-mediated phosphorylation of PRMT5 downregulates its methyltransferase activity and promotes myeloproliferation. Cancer Cell 19(2):283-94. https://doi.org/10.1016/j.ccr.2010.12.020. PMID: 21316606; PMCID: PMC4687747

Liu Y, Liu F, Yu H, Zhao X, Sashida G, Deblasio A, Harr M, She QB, Chen Z, Lin HK, Di Giandomenico S, Elf SE, Yang Y, Miyata Y, Huang G, Menendez S, Mellinghoff IK, Rosen N, Pandolfi PP, Hedvat CV, Nimer SD (2012 Oct 23) Akt phosphorylates the transcriptional repressor bmi1 to block its effects on the tumor-suppressing ink4a-arf locus. Sci Signal 5(247):ra77. https://doi.org/10.1126/scisignal.2003199. PMID: 23092893

Liu Q, Zheng JM, Chen JK, Yan XL, Chen HM, Nong WX, Huang HQ (2014 Oct) Histone deacetylase 5 promotes the proliferation of glioma cells by upregulation of Notch 1. Mol Med Rep 10(4):2045-2050. https://doi.org/10.3892/mmr.2014.2395. Epub 2014 Jul 18. PMID: 25050565

Liu L, Sun L, Li C, Li X, Zhang Y, Yu Y, Xia W (2015) Quantitative detection of methylation of FHIT and BRCA1 promoters in the serum of ductal breast cancer patients. Biomed Mater Eng 26 Suppl 1:S2217-S2222. https://doi.org/10.3233/BME-151527. PMID: 26406001

Liu S, Li C, Xin P, Zheng Y, Peng Q, Xu Y, Luo Y, Wu Y, Zhu X (2019 Nov 14) Sonidegib, a smoothened inhibitor, promotes apoptosis and suppresses proliferation of natural Killer/T-cell lymphoma. Med Sci Monit. 25:8579-8586. https://doi.org/10.12659/MSM.918812. PMID: 31724562; PMCID: PMC6873646

Liu Y, Yang EJ, Shi C, Mou PK, Zhang B, Wu C, Lyu J, Shim JS (2020 Mar 25) Histone Acetyltransferase (HAT) P300/CBP inhibitors induce synthetic lethality in PTEN-deficient colorectal cancer cells through destabilizing AKT. Int J Biol Sci 16(11):1774-1784. https://doi.org/10.7150/ijbs.42197. PMID: 32398948; PMCID: PMC7211175.19

Liu JH, Cao YM, Rong ZP, Ding J, Pan X (2021) Trichostatin A induces autophagy in cervical cancer cells by regulating the PRMT5-STC1-TRPV6-JNK pathway. Pharmacology 106(1-2): 60-69. https://doi.org/10.1159/000507937. Epub 2020 Nov 3. PMID: 33142290

Locke WJ, Guanzon D, Ma C, Liew YJ, Duesing KR, Fung KYC, Ross JP (2019 Nov 14) DNA methylation cancer biomarkers: translation to the clinic. Front Genet 10:1150. https://doi.org/10.3389/fgene.2019.01150. PMID: 31803237; PMCID: PMC6870840

López de Silanes I, Gorospe M, Taniguchi H, Abdelmohsen K, Srikantan S, Alaminos M, Berdasco M, Urdinguio RG, Fraga MF, Jacinto FV, Esteller M (2009 May) The RNA-binding protein HuR regulates DNA methylation through stabilization of DNMT3b mRNA. Nucleic Acids Res 37(8):2658-71. https://doi.org/10.1093/nar/gkp123. Epub 2009 Mar 6. PMID: 19270063; PMCID: PMC2677878

Lund P, Weisshaupt K, Mikeska T, Jammas D, Chen X, Kuban RJ, Ungethüm U, Krapfenbauer U, Herzel HP, Schäfer R, Walter J, Sers C (2006 Aug 10) Oncogenic HRAS suppresses clusterin expression through promoter hypermethylation. Oncogene 25(35):4890-4903. https://doi.org/10.1038/sj.onc.1209502. Epub 2006 Mar 27. PMID: 16568090

MacLeod AR, Rouleau J, Szyf M (1995 May 12) Regulation of DNA methylation by the Ras signaling pathway. J Biol Chem 270(19):11327-11337. https://doi.org/10.1074/jbc.270.19.11327. PMID: 7744770

Margueron R, Reinberg D (2011 Jan 20) The Polycomb complex PRC2 and its mark in life. Nature 469(7330):343-9. https://doi.org/10.1038/nature09784. PMID: 21248841; PMCID: PMC3760771

Marigo V, Davey RA, Zuo Y, Cunningham JM, Tabin CJ (1996 Nov 14) Biochemical evidence that patched is the Hedgehog receptor. Nature 384(6605):176-179. https://doi.org/10.1038/384176a0. PMID: 8906794

Masui K, Harachi M, Ikegami S, Yang H, Onizuka H, Yong WH, Cloughesy TF, Muragaki Y, Kawamata T, Arai N, Komori T, Cavenee WK, Mischel PS, Shibata N (2019 Dec 20) mTORC2 links growth factor signaling with epigenetic regulation of iron metabolism in glioblastoma. J Biol Chem 294(51):19740-19751. https://doi.org/10.1074/jbc.RA119.011519. Epub 2019 Nov 11. PMID: 31712311; PMCID: PMC6926466

Masui K, Harachi M, Cavenee WK, Mischel PS, Shibata N (2020 May 28) mTOR complex 2 is an integrator of cancer metabolism and epigenetics. Cancer Lett 478:1-7. https://doi.org/10.1016/j.canlet.2020.03.001. Epub 2020 Mar 5. PMID: 32145344

Mei FC, Young TW, Liu J, Cheng X (2006 Mar) RAS-mediated epigenetic inactivation of OPCML in oncogenic transformation of human ovarian surface epithelial cells. FASEB J 20(3):497-499. https://doi.org/10.1096/fj.05-4586fje. Epub 2005 Dec 29. PMID: 16384911

Mellinghoff IK, Penas-Prado M, Peters KB, Burris HA 3rd, Maher EA, Janku F, Cote GM, de la Fuente MI, Clarke JL, Ellingson BM, Chun S, Young RJ, Liu H, Choe S, Lu M, Le K, Hassan I, Steelman L, Pandya SS, Cloughesy TF, Wen PY (2021 Jun 2) Vorasidenib, a dual inhibitor of mutant IDH1/2, in recurrent or progressive glioma; results of a first-in-human phase I trial. Clin Cancer Res https://doi.org/10.1158/1078-0432.CCR-21-0611. Epub ahead of print. PMID: 34078652

Mill P, Mo R, Hu MC, Dagnino L, Rosenblum ND, Hui CC (2005 Aug) Shh controls epithelial proliferation via independent pathways that converge on N-Myc. Dev Cell. 9(2):293-303. https://doi.org/10.1016/j.devcel.2005.05.009. PMID: 16054035

Miller JR (2002) The Wnts. Genome Biol 3(1):REVIEWS3001. https://doi.org/10.1186/gb-2001-3-1-reviews3001. Epub 2001 Dec 28. PMID: 11806834; PMCID: PMC150458

Miller AL, Fehling SC, Garcia PL, Gamblin TL, Council LN, van Waardenburg RCAM, Yang ES, Bradner JE, Yoon KJ (2019 Jun) The BET inhibitor JQ1 attenuates double-strand break repair and sensitizes models of pancreatic ductal adenocarcinoma to PARP inhibitors. EBioMedicine

44:419-430. https://doi.org/10.1016/j.ebiom.2019.05.035. Epub 2019 May 22. PMID: 31126889; PMCID: PMC6604668

Mirmohammadsadegh A, Marini A, Nambiar S, Hassan M, Tannapfel A, Ruzicka T, Hengge UR (2006 Jul 1) Epigenetic silencing of the PTEN gene in melanoma. Cancer Res. 66(13): 6546-6552. https://doi.org/10.1158/0008-5472.CAN-06-0384. PMID: 16818626

Mizuno Y, Maemura K, Tanaka Y, Hirata A, Futaki S, Hamamoto H, Taniguchi K, Hayashi M, Uchiyama K, Shibata MA, Otsuki Y, Kondo Y (2018 May) Expression of delta-like 3 is downregulated by aberrant DNA methylation and histone modification in hepatocellular carcinoma. Oncol Rep 39(5):2209-2216. https://doi.org/10.3892/or.2018.6293. Epub 2018 Mar 5. PMID: 29512761

Moon RT, Kohn AD, De Ferrari GV, Kaykas A (2004 Sep) WNT and beta-catenin signalling: diseases and therapies. Nat Rev Genet 5(9):691-701. https://doi.org/10.1038/nrg1427. PMID: 15372092

Mosimann C, Hausmann G, Basler K (2009 Apr) Beta-catenin hits chromatin: regulation of Wnt target gene activation. Nat Rev Mol Cell Biol 10(4):276–286. https://doi.org/10.1038/nrm2654

Moskalev EA, Luckert K, Vorobjev IA, Mastitsky SE, Gladkikh AA, Stephan A, Schrenk M, Kaplanov KD, Kalashnikova OB, Pötz O, Joos TO, Hoheisel JD (2012 Jun 6) Concurrent epigenetic silencing of wnt/β-catenin pathway inhibitor genes in B cell chronic lymphocytic leukaemia. BMC Cancer 12:213. https://doi.org/10.1186/1471-2407-12-213. PMID: 22672427; PMCID: PMC3489542

Munafó DB, Colombo MI (2002 Jul) Induction of autophagy causes dramatic changes in the subcellular distribution of GFP-Rab24. Traffic 3(7):472-482. https://doi.org/10.1034/j.1600-0854.2002.30704.x. PMID: 12047555

Mungamuri SK, Murk W, Grumolato L, Bernstein E, Aaronson SA (2013 Oct 31) Chromatin modifications sequentially enhance ErbB2 expression in ErbB2-positive breast cancers. Cell Rep 5(2):302-13. https://doi.org/10.1016/j.celrep.2013.09.009. Epub 2013 Oct 10. PMID: 24120871; PMCID: PMC3905738

Nakamura K, Nakabayashi K, Htet Aung K, Aizawa K, Hori N, Yamauchi J, Hata K, Tanoue A (2015 Mar 23) DNA methyltransferase inhibitor zebularine induces human cholangiocarcinoma cell death through alteration of DNA methylation status. PLoS One 10(3):e0120545. https://doi.org/10.1371/journal.pone.0120545. PMID: 25799509; PMCID: PMC4370694

Nelson WJ, Nusse R (2004 Mar 5) Convergence of Wnt, beta-catenin, and cadherin pathways. Science 303(5663):1483-7. https://doi.org/10.1126/science.1094291. PMID: 15001769; PMCID: PMC3372896.

Niehrs C (2006 Dec 4) Function and biological roles of the Dickkopf family of Wnt modulators. Oncogene 25(57):7469-7481. https://doi.org/10.1038/sj.onc.1210054. PMID: 17143291

Noda H, Kato Y, Yoshikawa H, Arai M, Togashi K, Nagai H, Konishi F, Miki Y (2006 Jun) Frequent involvement of ras-signalling pathways in both polypoid-type and flat-type early-stage colorectal cancers. J Exp Clin Cancer Res 25(2):235-242. PMID: 16918136

Nowell CS, Radtke F (2017 Mar) Notch as a tumour suppressor. Nat Rev Cancer 17(3):145-159. https://doi.org/10.1038/nrc.2016.145. Epub 2017 Feb 3. PMID: 28154375

Okamoto M, Udagawa N, Uehara S, Maeda K, Yamashita T, Nakamichi Y, Kato H, Saito N, Minami Y, Takahashi N, Kobayashi Y (2014 Mar 27) Noncanonical Wnt5a enhances Wnt/β-catenin signaling during osteoblastogenesis. Sci Rep 4:4493. https://doi.org/10.1038/srep04493. PMID: 24670389; PMCID: PMC3967152

Okano M, Bell DW, Haber DA, Li E (1999 Oct 29) DNA methyltransferases Dnmt3a and Dnmt3b are essential for de novo methylation and mammalian development. Cell 99(3):247-257. https://doi.org/10.1016/s0092-8674(00)81656-6. PMID: 10555141

Olkkonen VM, Stenmark H (1997) Role of RabGTPases in membrane traffic. Int Rev Cytol. 176: 1-85. https://doi.org/10.1016/s0074-7696(08)61608-3. PMID: 9394917

Pak S, Park S, Kim Y, Park JH, Park CH, Lee KJ, Kim CS, Ahn H (2019 Aug 6) The small molecule WNT/β-catenin inhibitor CWP232291 blocks the growth of castration-resistant prostate cancer by activating the endoplasmic reticulum stress pathway. J Exp Clin Cancer Res 38(1):342.

https://doi.org/10.1186/s13046-019-1342-5. Erratum in: J Exp Clin Cancer Res. 2019 Oct 31;38 (1):440. PMID: 31387608; PMCID: PMC6685284

Parker DS, Ni YY, Chang JL, Li J, Cadigan KM (2008 Mar) Wingless signaling induces widespread chromatin remodeling of target loci. Mol Cell Biol 28(5):1815–1828. https://doi.org/10.1128/MCB.01230-07. Epub 2007 Dec 26; PMCID: PMC2258762

Palakurthy RK, Wajapeyee N, Santra MK, Gazin C, Lin L, Gobeil S, Green MR (2009 Oct 23) Epigenetic silencing of the RASSF1A tumor suppressor gene through HOXB3-mediated induction of DNMT3B expression. Mol Cell. 36(2):219–30. https://doi.org/10.1016/j.molcel.2009.10.009. PMID: 19854132; PMCID: PMC2776651

Pan Y, Liu G, Zhou F, Su B, Li Y (2018 Feb) DNA methylation profiles in cancer diagnosis and therapeutics. Clin Exp Med 18(1):1-14. https://doi.org/10.1007/s10238-017-0467-0. Epub 2017 Jul 27. PMID: 28752221.

Peli J, Schröter M, Rudaz C, Hahne M, Meyer C, Reichmann E, Tschopp J (1999 Apr 1) Oncogenic Ras inhibits Fas ligand-mediated apoptosis by downregulating the expression of Fas. EMBO J 18(7):1824-31. https://doi.org/10.1093/emboj/18.7.1824. PMID: 10202146; PMCID: PMC1171268.

Piazzi G, Bazzoli F, Ricciardiello L (2012 Dec 1) Epigenetic silencing of Notch signaling in gastrointestinal cancers. Cell Cycle 11(23):4323-7. https://doi.org/10.4161/cc.22388. Epub 2012 Oct 19. PMID: 23085543; PMCID: PMC3552914

Polakis P (2000 Aug 1) Wnt signaling and cancer. Genes Dev 14(15):1837-1851. PMID: 10921899

Ponnurangam S, Dandawate PR, Dhar A, Tawfik OW, Parab RR, Mishra PD, Ranadive P, Sharma R, Mahajan G, Umar S, Weir SJ, Sugumar A, Jensen RA, Padhye SB, Balakrishnan A, Anant S, Subramaniam D (2016 Jan 19) Quinomycin A targets Notch signaling pathway in pancreatic cancer stem cells. Oncotarget 7(3):3217–32. https://doi.org/10.18632/oncotarget.6560. PMID: 26673007; PMCID: PMC4823101

Prasad CP, Mirza S, Sharma G, Prashad R, DattaGupta S, Rath G, Ralhan R (2008 Aug 29) Epigenetic alterations of CDH1 and APC genes: relationship with activation of Wnt/beta-catenin pathway in invasive ductal carcinoma of breast. Life Sci 83(9–10):318-325. https://doi.org/10.1016/j.lfs.2008.06.019. Epub 2008 Jul 1. PMID: 18662704

Pruitt K, Ulkü AS, Frantz K, Rojas RJ, Muniz-Medina VM, Rangnekar VM, Der CJ, Shields JM (2005 Jun 17) Ras-mediated loss of the pro-apoptotic response protein Par-4 is mediated by DNA hypermethylation through Raf-independent and Raf-dependent signaling cascades in epithelial cells. J Biol Chem 280(24):23363-23370. https://doi.org/10.1074/jbc.M503083200. Epub 2005 Apr 14. PMID: 15831492.

Radtke F, Raj K (2003 Oct) The role of Notch in tumorigenesis: oncogene or tumour suppressor? Nat Rev Cancer 3(10):756-767. https://doi.org/10.1038/nrc1186. PMID: 14570040

Reifenberger J, Knobbe CB, Sterzinger AA, Blaschke B, Schulte KW, Ruzicka T, Reifenberger G (2004 Apr 10) Frequent alterations of Ras signaling pathway genes in sporadic malignant melanomas. Int J Cancer 109(3):377-384. https://doi.org/10.1002/ijc.11722. PMID: 14961576

Reya T, Clevers H (2005 Apr 14) Wnt signalling in stem cells and cancer. Nature 434(7035): 843-850. https://doi.org/10.1038/nature03319. PMID: 15829953

Rimkus TK, Carpenter RL, Qasem S, Chan M, Lo HW (2016 Feb 15) Targeting the sonic hedgehog signaling pathway: review of smoothened and GLI inhibitors. Cancers (Basel) 8(2):22. https://doi.org/10.3390/cancers8020022. PMID: 26891329; PMCID: PMC4773745

Riquelme E, Behrens C, Lin HY, Simon G, Papadimitrakopoulou V, Izzo J, Moran C, Kalhor N, Lee JJ, Minna JD, Wistuba II (2016 Feb 1) Modulation of EZH2 expression by MEK-ERK or PI3K-AKT signaling in lung cancer is dictated by different KRAS oncogene mutations. Cancer Res 76(3):675-85. https://doi.org/10.1158/0008-5472.CAN-15-1141. Epub 2015 Dec 16. PMID: 26676756; PMCID: PMC4738155

Sanaei M, Kavoosi F (2020 Sep 10) Effect of 5-aza-2′-deoxycytidine on estrogen receptor alpha/beta and DNA methyltransferase 1 genes expression, apoptosis induction, and cell growth prevention of the colon cancer HT 29 Cell Line. Int J Prev Med 11:147. https://doi.org/10.4103/ijpvm.IJPVM_140_19. PMID: 33209217; PMCID: PMC7643574

Sato N, Fukushima N, Maitra A, Matsubayashi H, Yeo CJ, Cameron JL, Hruban RH, Goggins M (2003 Jul 1) Discovery of novel targets for aberrant methylation in pancreatic carcinoma using high-throughput microarrays. Cancer Res 63(13):3735-3742. PMID: 12839967

Schöffski P, Cresta S, Mayer IA, Wildiers H, Damian S, Gendreau S, Rooney I, Morrissey KM, Spoerke JM, Ng VW, Singel SM, Winer E (2018 Sep 5) A phase Ib study of pictilisib (GDC-0941) in combination with paclitaxel, with and without bevacizumab or trastuzumab, and with letrozole in advanced breast cancer. Breast Cancer Res 20(1):109. https://doi.org/10.1186/s13058-018-1015-x. PMID: 30185228; PMCID: PMC6125885

Semënov MV, Tamai K, Brott BK, Kühl M, Sokol S, He X (2001 Jun 26) Head inducer Dickkopf-1 is a ligand for Wnt coreceptor LRP6. Curr Biol 11(12):951-961. https://doi.org/10.1016/s0960-9822(01)00290-1. PMID: 11448771

Senapati P, Kato H, Lee M, Leung A, Thai C, Sanchez A, Gallagher EJ, LeRoith D, Seewaldt VL, Ann DK, Schones DE (2019 Jul 17) Hyperinsulinemia promotes aberrant histone acetylation in triple-negative breast cancer. Epigenetics Chromatin 12(1):44. https://doi.org/10.1186/s13072-019-0290-9. PMID: 31315653; PMCID: PMC6636093

Serman L, Nikuseva Martic T, Serman A, Vranic S (2014 Nov 12) Epigenetic alterations of the Wnt signaling pathway in cancer: a mini review. Bosn J Basic Med Sci 14(4):191-4. https://doi.org/10.17305/bjbms.2014.4.205. PMID: 25428669; PMCID: PMC4333971

Shahi MH, Lorente A, Castresana JS (2008 Mar) Hedgehog signalling in medulloblastoma, glioblastoma and neuroblastoma. Oncol Rep 19(3):681-688. PMID: 18288402

Shao B, Feng Y, Zhang H, Yu F, Li Q, Tan C, Xu H, Ying J, Li L, Yang D, Peng W, Tang J, Li S, Ren G, Tao Q, Xiang T (2018 Feb) The 3p14.2 tumour suppressor ADAMTS9 is inactivated by promoter CpG methylation and inhibits tumour cell growth in breast cancer. J Cell Mol Med 22 (2):1257-1271. https://doi.org/10.1111/jcmm.13404. Epub 2017 Nov 29. PMID: 29193730; PMCID: PMC5783842

Sharma S, Kelly TK, Jones PA (2010 Jan) Epigenetics in cancer. Carcinogenesis 31(1):27-36. https://doi.org/10.1093/carcin/bgp220. Epub 2009 Sep 13. PMID: 19752007; PMCID: PMC2802667

Sharma ND, Nickl CK, Kang H, Ornatowski W, Brown R, Ness SA, Loh ML, Mullighan CG, Winter SS, Hunger SP, Cannon JL, Matlawska-Wasowska K (2019 Jun) Epigenetic silencing of SOCS5 potentiates JAK-STAT signaling and progression of T-cell acute lymphoblastic leukemia. Cancer Sci 110(6):1931-1946. https://doi.org/10.1111/cas.14021. Epub 2019 May 3. PMID: 30974024; PMCID: PMC6549933

Shen YQ, Guerra-Librero A, Fernandez-Gil BI, Florido J, García-López S, Martinez-Ruiz L, Mendivil-Perez M, Soto-Mercado V, Acuña-Castroviejo D, Ortega-Arellano H, Carriel V, Diaz-Casado ME, Reiter RJ, Rusanova I, Nieto A, López LC, Escames G (2018 Apr) Combination of melatonin and rapamycin for head and neck cancer therapy: suppression of AKT/mTOR pathway activation, and activation of mitophagy and apoptosis via mitochondrial function regulation. J Pineal Res 64(3). https://doi.org/10.1111/jpi.12461. Epub 2018 Jan 9. PMID: 29247557.

Shu J, Jelinek J, Chang H, Shen L, Qin T, Chung W, Oki Y, Issa JP (2006 May 15) Silencing of bidirectional promoters by DNA methylation in tumorigenesis. Cancer Res 66(10):5077-5084. https://doi.org/10.1158/0008-5472.CAN-05-2629. PMID: 16707430

Sierra J, Yoshida T, Joazeiro CA, Jones KA (2006 Mar 1) The APC tumor suppressor counteracts beta-catenin activation and H3K4 methylation at Wnt target genes. Genes Dev 20(5):586–600. https://doi.org/10.1101/gad.1385806. PMCID: PMC1410807

Song Y, Li ZX, Liu X, Wang R, Li LW, Zhang Q (2017 Jul) The Wnt/β-catenin and PI3K/Akt signaling pathways promote EMT in gastric cancer by epigenetic regulation via H3 lysine 27 acetylation. Tumour Biol 39(7):1010428317712617. https://doi.org/10.1177/1010428317712617

Spangle JM, Dreijerink KM, Groner AC, Cheng H, Ohlson CE, Reyes J, Lin CY, Bradner J, Zhao JJ, Roberts TM, Brown M (2016 Jun 21) PI3K/AKT signaling regulates H3K4 methylation in breast cancer. Cell Rep 15(12):2692-704. https://doi.org/10.1016/j.celrep.2016.05.046. Epub 2016 Jun 9. PMID: 27292631; PMCID: PMC5094353

Staal FJ, Clevers HC (2005 Jan) WNT signalling and haematopoiesis: a WNT-WNT situation. Nat Rev Immunol 5(1):21-30. https://doi.org/10.1038/nri1529. PMID: 15630426

Stone RM, Mandrekar SJ, Sanford BL, Laumann K, Geyer S, Bloomfield CD, Thiede C, Prior TW, Döhner K, Marcucci G, Lo-Coco F, Klisovic RB, Wei A, Sierra J, Sanz MA, Brandwein JM, de Witte T, Niederwieser D, Appelbaum FR, Medeiros BC, Tallman MS, Krauter J, Schlenk RF, Ganser A, Serve H, Ehninger G, Amadori S, Larson RA, Döhner H (2017 Aug 3) Midostaurin plus chemotherapy for acute myeloid leukemia with a FLT3 mutation. N Engl J Med 377(5): 454-464. https://doi.org/10.1056/NEJMoa1614359. Epub 2017 Jun 23. PMID: 28644114; PMCID: PMC5754190

Strahl BD, Allis CD (2000 Jan 6) The language of covalent histone modifications. Nature 403 (6765):41-45. https://doi.org/10.1038/47412. PMID: 10638745

Sun L, Chen B, Wu J, Jiang C, Fan Z, Feng Y, Xu Y (2020 Sep 11) Epigenetic regulation of a disintegrin and metalloproteinase (ADAM) transcription in colorectal cancer cells: involvement of β-catenin, BRG1, and KDM4. Front Cell Dev Biol 8:581692. https://doi.org/10.3389/fcell.2020.581692. PMCID: PMC7517301

Surana R, Sikka S, Cai W, Shin EM, Warrier SR, Tan HJ, Arfuso F, Fox SA, Dharmarajan AM, Kumar AP (2014 Jan) Secreted frizzled related proteins: implications in cancers. Biochim Biophys Acta 1845(1):53-65. https://doi.org/10.1016/j.bbcan.2013.11.004. Epub 2013 Dec 5. PMID: 24316024

Suzuki H, Watkins DN, Jair KW, Schuebel KE, Markowitz SD, Chen WD, Pretlow TP, Yang B, Akiyama Y, Van Engeland M, Toyota M, Tokino T, Hinoda Y, Imai K, Herman JG, Baylin SB (2004 Apr) Epigenetic inactivation of SFRP genes allows constitutive WNT signaling in colorectal cancer. Nat Genet 36(4):417-422. https://doi.org/10.1038/ng1330. Epub 2004 Mar 14. PMID: 15034581

Tada M, Kanai F, Tanaka Y, Tateishi K, Ohta M, Asaoka Y, Seto M, Muroyama R, Fukai K, Imazeki F, Kawabe T, Yokosuka O, Omata M (2008 Jun 15) Down-regulation of hedgehog-interacting protein through genetic and epigenetic alterations in human hepatocellular carcinoma. Clin Cancer Res 14(12):3768-3776. https://doi.org/10.1158/1078-0432.CCR-07-1181. PMID: 18559595

Taipale J, Beachy PA (2001 May 17) The Hedgehog and Wnt signalling pathways in cancer. Nature 411(6835):349-354. https://doi.org/10.1038/35077219. PMID: 11357142

Tan D, Tang P, Huang J, Zhang J, Zhou W, Walker AM. Expression of a constitutively active prolactin receptor causes histone trimethylation of the p53 gene in breast cancer. Chin Med J (Engl). 2014;127(6):1077-1083. PMID: 24622438

Tan AHY, Tu W, McCuaig R, Hardy K, Donovan T, Tsimbalyuk S, Forwood JK, Rao S (2019 Jun 12) Lysine-specific histone demethylase 1A regulates macrophage polarization and checkpoint molecules in the tumor microenvironment of triple-negative breast cancer. Front Immunol 10: 1351. https://doi.org/10.3389/fimmu.2019.01351. PMID: 31249575; PMCID: PMC6582666

ten Haaf A, Franken L, Heymann C, von Serenyi S, Cornelissen C, de Hoon JP, Veeck J, Lüscher B, Knüchel R, Dahl E (2011 Apr) Paradox of sonic hedgehog (SHH) transcriptional regulation: alternative transcription initiation overrides the effect of downstream promoter DNA methylation. Epigenetics 6(4):465-477. https://doi.org/10.4161/epi.6.4.14952. Epub 2011 Apr 1. PMID: 21372638.

Terasaki H, Saitoh T, Shiokawa K, Katoh M (2002 Feb) Frizzled-10, up-regulated in primary colorectal cancer, is a positive regulator of the WNT - beta-catenin - TCF signaling pathway. Int J Mol Med 9(2):107-112. PMID: 11786918

Tham C, Chew M, Soong R, Lim J, Ang M, Tang C, Zhao Y, Ong SY, Liu Y (2014 Oct 15) Postoperative serum methylation levels of TAC1 and SEPT9 are independent predictors of recurrence and survival of patients with colorectal cancer. Cancer 120(20):3131-3141. https://doi.org/10.1002/cncr.28802. Epub 2014 Jun 12 PMID: 24925595

Thomas SJ, Snowden JA, Zeidler MP, Danson SJ (2015 Jul 28) The role of JAK/STAT signalling in the pathogenesis, prognosis and treatment of solid tumours. Br J Cancer 113(3):365-71. https://doi.org/10.1038/bjc.2015.233. Epub 2015 Jul 7. PMID: 26151455; PMCID: PMC4522639

Thorpe LM, Yuzugullu H, Zhao JJ (2015 Jan) PI3K in cancer: divergent roles of isoforms, modes of activation and therapeutic targeting. Nat Rev Cancer 15(1):7-24. https://doi.org/10.1038/nrc3860. PMID: 25533673; PMCID: PMC4384662

Topol L, Jiang X, Choi H, Garrett-Beal L, Carolan PJ, Yang Y (2003 Sep 1) Wnt-5a inhibits the canonical Wnt pathway by promoting GSK-3-independent beta-catenin degradation. J Cell Biol 162(5):899-908. https://doi.org/10.1083/jcb.200303158. PMID: 12952940; PMCID: PMC2172823

Tsai CH, Li CH, Liao PL, Chang YW, Cheng YW, Kang JJ (2020 Feb) Aza-PBHA, a potent histone deacetylase inhibitor, inhibits human gastric-cancer cell migration via PKCα-mediated AHR-HDAC interactions. Biochim Biophys Acta Mol Cell Res 1867(2):118564. https://doi.org/10.1016/j.bbamcr.2019.118564. Epub 2019 Oct 28. PMID: 31672612

Tzao C, Jin JS, Chen BH, Chung HY, Chang CC, Hsu TY, Sun GH (2014 Sep-Oct) Anticancer effects of suberoylanilide hydroxamic acid in esophageal squamous cancer cells in vitro and in vivo. Dis Esophagus 27(7):693-702. https://doi.org/10.1111/dote.12127. Epub 2013 Sep 4. PMID: 24033428

Uematsu K, Kanazawa S, You L, He B, Xu Z, Li K, Peterlin BM, McCormick F, Jablons DM (2003 Aug 1) Wnt pathway activation in mesothelioma: evidence of Dishevelled overexpression and transcriptional activity of beta-catenin. Cancer Res 63(15):4547-4551. PMID: 12907630

Uren A, Reichsman F, Anest V, Taylor WG, Muraiso K, Bottaro DP, Cumberledge S, Rubin JS (2000 Feb 11) Secreted frizzled-related protein-1 binds directly to Wingless and is a biphasic modulator of Wnt signaling. J Biol Chem 275(6):4374-4382. https://doi.org/10.1074/jbc.275.6.4374. PMID: 10660608

Vadla R, Haldar D (2018) Mammalian target of rapamycin complex 2 (mTORC2) controls glycolytic gene expression by regulating Histone H3 Lysine 56 acetylation. Cell Cycle 17(1):110-123. https://doi.org/10.1080/15384101.2017.1404207. Epub 2018 Jan 8. PMID: 29143563; PMCID: PMC5815439

Veeck J, Noetzel E, Bektas N, Jost E, Hartmann A, Knüchel R, Dahl E (2008 Nov 6) Promoter hypermethylation of the SFRP2 gene is a high-frequent alteration and tumor-specific epigenetic marker in human breast cancer. Mol Cancer 7:83. https://doi.org/10.1186/1476-4598-7-83. PMID: 18990230; PMCID: PMC2613402

Villanueva L, Álvarez-Errico D, Esteller M (2020 Aug) The contribution of epigenetics to cancer immunotherapy. Trends Immunol 41(8):676-691. https://doi.org/10.1016/j.it.2020.06.002. Epub 2020 Jul 2. PMID: 32622854

Visser HP, Gunster MJ, Kluin-Nelemans HC, Manders EM, Raaphorst FM, Meijer CJ, Willemze R, Otte AP (2001 Mar) The Polycomb group protein EZH2 is upregulated in proliferating, cultured human mantle cell lymphoma. Br J Haematol 112(4):950-958. https://doi.org/10.1046/j.1365-2141.2001.02641.x. PMID: 11298590

Wajapeyee N, Malonia SK, Palakurthy RK, Green MR (2013 Oct 15) Oncogenic RAS directs silencing of tumor suppressor genes through ordered recruitment of transcriptional repressors. Genes Dev 27(20):2221-6. https://doi.org/10.1101/gad.227413.113. Epub 2013 Oct 8. PMID: 24105743; PMCID: PMC3814642.

Wang LH, Choi YL, Hua XY, Shin YK, Song YJ, Youn SJ, Yun HY, Park SM, Kim WJ, Kim HJ, Choi JS, Kim SH (2006 May) Increased expression of sonic hedgehog and altered methylation of its promoter region in gastric cancer and its related lesions. Mod Pathol 19(5):675–683. https://doi.org/10.1038/modpathol.3800573

Wang X, Venugopal C, Manoranjan B, McFarlane N, O'Farrell E, Nolte S, Gunnarsson T, Hollenberg R, Kwiecien J, Northcott P, Taylor MD, Hawkins C, Singh SK (2012a Jan 12) Sonic hedgehog regulates Bmi1 in human medulloblastoma brain tumor-initiating cells. Oncogene 31(2):187-199. doi: https://doi.org/10.1038/onc.2011.232. Epub 2011 Jun 20. PMID: 21685941

Wang ZK, Liu J, Liu C, Wang FY, Chen CY, Zhang XH (2012b Oct) Hypermethylation of adenomatous polyposis coli gene promoter is associated with novel Wnt signaling pathway in gastric adenomas. J Gastroenterol Hepatol 27(10):1629-1634. https://doi.org/10.1111/j.1440-1746.2012.07219.x. PMID: 22741528

Wang S, Dong Y, Zhang Y, Wang X, Xu L, Yang S, Li X, Dong H, Xu L, Su L, Ng SS, Chang Z, Sung JJ, Zhang X, Yu J (2015 May 14) DACT2 is a functional tumor suppressor through inhibiting Wnt/β-catenin pathway and associated with poor survival in colon cancer. Oncogene 34(20):2575-85. https://doi.org/10.1038/onc.2014.201. Epub 2014 Jul 14. PMID: 25023701; PMCID: PMC4761644.

Wang D, Xu J, Liu B, He X, Zhou L, Hu X, Qiao F, Zhang A, Xu X, Zhang H, Wicha MS, Zhang L, Shao ZM, Liu S (2018 Feb) IL6 blockade potentiates the anti-tumor effects of γ-secretase inhibitors in Notch3-expressing breast cancer. Cell Death Differ 25(2):330-339. https://doi.org/10.1038/cdd.2017.162. Epub 2017 Oct 13. PMID: 29027990; PMCID: PMC5762847.

Wang G, Long J, Gao Y, Zhang W, Han F, Xu C, Sun L, Yang SC, Lan J, Hou Z, Cai Z, Jin G, Hsu CC, Wang YH, Hu J, Chen TY, Li H, Lee MG, Lin HK (2019 Feb) SETDB1-mediated methylation of Akt promotes its K63-linked ubiquitination and activation leading to tumorigenesis. Nat Cell Biol 21(2):214-225. https://doi.org/10.1038/s41556-018-0266-1. Epub 2019 Jan 28. PMID: 30692626; PMCID: PMC6414065.

Wei H, Wang Y, Gale RP, Lin D, Zhou C, Liu B, Qiu S, Gu R, Li Y, Zhao X, Wei S, Gong B, Liu K, Gong X, Liu Y, Zhang G, Song Z, Wang Y, Li W, Mi Y, Wang J (2020 Jul 1) Randomized trial of intermediate-dose cytarabine in induction and consolidation therapy in adults with acute myeloid leukemia. Clin Cancer Res 26(13):3154-3161. https://doi.org/10.1158/1078-0432.CCR-19-3433. Epub 2020 Feb 6. PMID: 32029439

Wei KL, Chou JL, Chen YC, Low JT, Lin GL, Liu JL, Chang TS, Chen WM, Hsieh YY, Yan PS, Chuang YM, Lin JMJ, Wu SF, Chiang MK, Li C, Wu CS, Chan MWY (2021 Feb 25) Epigenetic silencing of STAT3-targeted miR-193a, by constitutive activation of JAK/STAT signaling, leads to tumor progression through overexpression of YWHAZ in gastric cancer. Front Oncol 11:575667. https://doi.org/10.3389/fonc.2021.575667. PMID: 33718136; PMCID: PMC7951088.

Wend P, Fang L, Zhu Q, Schipper JH, Loddenkemper C, Kosel F, Brinkmann V, Eckert K, Hindersin S, Holland JD, Lehr S, Kahn M, Ziebold U, Birchmeier W (2013 Jul 17) Wnt/β-catenin signalling induces MLL to create epigenetic changes in salivary gland tumours. EMBO J 32(14):1977–1989. https://doi.org/10.1038/emboj.2013.127. Epub 2013 Jun 4. PMCID: PMC3715856

Widelitz R (2005 Jun) Wnt signaling through canonical and non-canonical pathways: recent progress. Growth Factors 23(2):111-116. https://doi.org/10.1080/08977190500125746. PMID: 16019432

Wolf I, Bose S, Desmond JC, Lin BT, Williamson EA, Karlan BY, Koeffler HP (2007 Oct) Unmasking of epigenetically silenced genes reveals DNA promoter methylation and reduced expression of PTCH in breast cancer. Breast Cancer Res Treat 105(2):139-155. https://doi.org/10.1007/s10549-006-9440-4. Epub 2007 Feb 13. PMID: 17295047

Wu Y, Kikuchi S, Yan H, Zhang W, Rosenbaum H, Iniguez AL, Jiang J (2011 Nov) Euchromatic subdomains in rice centromeres are associated with genes and transcription. Plant Cell 23(11): 4054-64. https://doi.org/10.1105/tpc.111.090043. Epub 2011 Nov 11. PMID: 22080597; PMCID: PMC3246336.

Wu CS, Wei KL, Chou JL, Lu CK, Hsieh CC, Lin JM, Deng YF, Hsu WT, Wang HM, Leung CH, Ma DL, Li C, Chan MW (2016a Sep 2) Aberrant JAK/STAT signaling suppresses TFF1 and TFF2 through epigenetic silencing of GATA6 in gastric cancer. Int J Mol Sci 17(9):1467. https://doi.org/10.3390/ijms17091467. PMID: 27598141; PMCID: PMC5037745

Wu J, Keng VW, Patmore DM, Kendall JJ, Patel AV, Jousma E, Jessen WJ, Choi K, Tschida BR, Silverstein KA, Fan D, Schwartz EB, Fuchs JR, Zou Y, Kim MO, Dombi E, Levy DE, Huang G, Cancelas JA, Stemmer-Rachamimov AO, Spinner RJ, Largaespada DA, Ratner N (2016b Mar 1) Insertional mutagenesis identifies a STAT3/Arid1b/β-catenin pathway driving neurofibroma initiation. Cell Rep 14(8):1979-90. https://doi.org/10.1016/j.celrep.2016.01.074. Epub 2016 Feb 18. PMID: 26904939; PMCID: PMC4782770

Wu Z, Yang Z, Dai Y, Zhu Q, Chen LA (2019 Jul 1) Update on liquid biopsy in clinical management of non-small cell lung cancer. Onco Targets Ther 12:5097-5109. https://doi.org/10.2147/OTT.S203070. PMID: 31303765; PMCID: PMC6611714.

Xu Y, Wu F, Tan L, Kong L, Xiong L, Deng J, Barbera AJ, Zheng L, Zhang H, Huang S, Min J, Nicholson T, Chen T, Xu G, Shi Y, Zhang K, Shi YG (2011 May 20) Genome-wide regulation of 5hmC, 5mC, and gene expression by Tet1 hydroxylase in mouse embryonic stem cells. Mol Cell 42(4):451-64. https://doi.org/10.1016/j.molcel.2011.04.005. Epub 2011 Apr 21. PMID: 21514197; PMCID: PMC3099128.

Xu H, Xiao Q, Fan Y, Xiang T, Li C, Li C, Li S, Hui T, Zhang L, Li H, Li L, Ren G (2017 Jun) Epigenetic silencing of ADAMTS18 promotes cell migration and invasion of breast cancer through AKT and NF-κB signaling. Cancer Med 6(6):1399-1408. https://doi.org/10.1002/cam4.1076. Epub 2017 May 15. PMID: 28503860; PMCID: PMC5463072.

Yang X, Han H, De Carvalho DD, Lay FD, Jones PA, Liang G (2014 Oct 13) Gene body methylation can alter gene expression and is a therapeutic target in cancer. Cancer Cell 26(4):577-90. https://doi.org/10.1016/j.ccr.2014.07.028. Epub 2014 Sep 25. PMID: 25263941; PMCID: PMC4224113.

Yang H, Zhou J, Mi J, Ma K, Fan Y, Ning J, Wang C, Wei X, Zhao H, Li E (2015 Oct) HOXD10 acts as a tumor-suppressive factor via inhibition of the RHOC/AKT/MAPK pathway in human cholangiocellular carcinoma. Oncol Rep 34(4):1681-91. https://doi.org/10.3892/or.2015.4194. Epub 2015 Aug 10. PMID: 26260613; PMCID: PMC4564083.

Yeh CM, Chang LY, Lin SH, Chou JL, Hsieh HY, Zeng LH, Chuang SY, Wang HW, Dittner C, Lin CY, Lin JM, Huang YT, Ng EK, Cheng AS, Wu SF, Lin J, Yeh KT, Chan MW (2016 Aug 16) Epigenetic silencing of the NR4A3 tumor suppressor, by aberrant JAK/STAT signaling, predicts prognosis in gastric cancer. Sci Rep 6:31690. https://doi.org/10.1038/srep31690. PMID: 27528092; PMCID: PMC4985659.

Yuan YH, Wang HY, Lai Y, Zhong W, Liang WL, Yan FD, Yu Z, Chen JK, Lin Y (2019 Jan 25) Epigenetic inactivation of HOXD10 is associated with human colon cancer via inhibiting the RHOC/AKT/MAPK signaling pathway. Cell Commun Signal 17(1):9. https://doi.org/10.1186/s12964-018-0316-0. PMID: 30683109; PMCID: PMC6347846.

Zamani M, Hosseini SV, Mokarram P (2018 Mar) Epigenetic biomarkers in colorectal cancer: premises and prospects. Biomarkers 23(2):105-114. https://doi.org/10.1080/1354750X.2016.1252961. Epub 2016 Nov 10. PMID: 27788596

Zampieri M, Ciccarone F, Palermo R, Cialfi S, Passananti C, Chiaretti S, Nocchia D, Talora C, Screpanti I, Caiafa P (2014 Sep). The epigenetic factor BORIS/CTCFL regulates the NOTCH3 gene expression in cancer cells. Biochim Biophys Acta 1839(9):813-825. https://doi.org/10.1016/j.bbagrm.2014.06.017. Epub 2014 Jun 28. PMID: 24984200

Zavras AI, Pitiphat W, Wu T, Cartsos V, Lam A, Douglass CW, Diehl SR (2003 Jan 15) Insulin-like growth factor II receptor gene-167 genotype increases the risk of oral squamous cell carcinoma in humans. Cancer Res 63(2):296-297. PMID: 12543777

Zhang W, Glöckner SC, Guo M, Machida EO, Wang DH, Easwaran H, Van Neste L, Herman JG, Schuebel KE, Watkins DN, Ahuja N, Baylin SB (2008 Apr 15) Epigenetic inactivation of the canonical Wnt antagonist SRY-box containing gene 17 in colorectal cancer. Cancer Res 68(8):2764-72. https://doi.org/10.1158/0008-5472.CAN-07-6349. Erratum in: Cancer Res. 2008 Jul 15;68(14):6030. PMID: 18413743; PMCID: PMC2823123.

Zhang Y, Huang S, Dong W, Li L, Feng Y, Pan L, Han Z, Wang X, Ren G, Su D, Huang B, Lu J (2009 May 8) SOX7, down-regulated in colorectal cancer, induces apoptosis and inhibits proliferation of colorectal cancer cells. Cancer Lett 277(1):29–37

Zhang Y, Xu H, Mu J, Guo S, Ye L, Li D, Peng W, He X, Xiang T (2019 May) Inactivation of ADAMTS18 by aberrant promoter hypermethylation contribute to lung cancer progression. J Cell Physiol 234(5):6965-6975. https://doi.org/10.1002/jcp.27439. Epub 2018 Nov 11. PMID: 30417422

Zhang W, Zhang Y, Tu T, Schmull S, Han Y, Wang W, Li H (2020 Sep 17) Dual inhibition of HDAC and tyrosine kinase signaling pathways with CUDC-907 attenuates TGFβ1 induced lung and tumor fibrosis. Cell Death Dis 11(9):765. https://doi.org/10.1038/s41419-020-02916-w. PMID: 32943605; PMCID: PMC7499263.

Zhou W, Tian D, He J, Wang Y, Zhang L, Cui L, Jia L, Zhang L, Li L, Shu Y, Yu S, Zhao J, Yuan X, Peng S (2016 Apr 12) Repeated PM2.5 exposure inhibits BEAS-2B cell P53 expression through ROS-Akt-DNMT3B pathway-mediated promoter hypermethylation. Oncotarget 7 (15):20691-703. https://doi.org/10.18632/oncotarget.7842. PMID: 26942697; PMCID: PMC4991485.

Zou C, Fan J, He M, Xu Y, Wang K, Cai Y, Li M (2019 May 15) Epigenetic silencing of Rab39a promotes epithelial to mesenchymal transition of cervical cancer through AKT signaling. Exp Cell Res 378(2):139-148. doi: https://doi.org/10.1016/j.yexcr.2019.02.025. Epub 2019 Feb 28. PMID: 30826396

Zuo Y, Song Y (2013 Dec) Detection and analysis of the methylation status of PTCH1 gene involved in the hedgehog signaling pathway in a human gastric cancer cell line. Exp Ther Med 6 (6):1365-1368. https://doi.org/10.3892/etm.2013.1334. Epub 2013 Oct 9. PMID: 24255663; PMCID: PMC3829737.

Zuo T, Liu TM, Lan X, Weng YI, Shen R, Gu F, Huang YW, Liyanarachchi S, Deatherage DE, Hsu PY, Taslim C, Ramaswamy B, Shapiro CL, Lin HJ, Cheng AS, Jin VX, Huang TH (2011 Mar 1) Epigenetic silencing mediated through activated PI3K/AKT signaling in breast cancer. Cancer Res 71(5):1752-62. https://doi.org/10.1158/0008-5472.CAN-10-3573. Epub 2011 Jan 7. PMID: 21216892; PMCID: PMC3048165.

Chapter 14
Epigenetic Regulation Towards Acquired Drug Resistance in Cancer

Madhura Ketkar and Shilpee Dutt

Abstract Therapy resistance remains the most challenging obstacle in cancer treatment. Substantial efforts and evidences have accumulated over decades suggesting not only genetic but non-genomic mechanisms underlying this adaptation of tumor cells. Alterations in epigenome can have a fundamental effect on cellular functions and response to stresses like anticancer therapy. This chapter discusses the principal mechanisms by which epigenetic modifications in the genome and transcriptome aid tumor cells toward acquisition of resistance to chemotherapy.

Keywords Chemoresistance · Epigenetic modification · Cancer · DNA methylation · Histone methylation · HDAC · Histone ubiquitination · Chromatin remodeling · lncRNA · miRNA

Introduction

Functional heterogeneity in cells is attributed to the numerous combinations of genetic and epigenetic alterations in genome. This heterogeneity is nowhere more pertinent than neoplasms and anticancer therapy. In spite of tremendous efforts being put into cancer therapeutics, in almost all types of cancers, patients come back with refractory disease. As the cancer treatments developed over the years, so did the

---

M. Ketkar
Shilpee Dutt Laboratory, Advanced Centre for Treatment, Research and Education in Cancer (ACTREC), Tata Memorial Centre, Kharghar, Navi Mumbai, India

Homi Bhabha National Institute, Mumbai, India

S. Dutt (✉)
Shilpee Dutt Laboratory, Advanced Centre for Treatment, Research and Education in Cancer (ACTREC), Tata Memorial Centre, Kharghar, Navi Mumbai, India

Homi Bhabha National Institute, Mumbai, India

ACTREC, Tata Memorial Centre, Kharghar, Navi Mumbai, Maharashtra, India
e-mail: sdutt@actrec.gov.in

T. K. Kundu, C. Das (eds.), *Metabolism and Epigenetic Regulation: Implications in Cancer*, Subcellular Biochemistry 100,
https://doi.org/10.1007/978-3-031-07634-3_14

tumor's resistance to them. Though genetic mutations are the basis for this resistance in many cases, their inadequacy in explaining the rapid appearance of drug tolerance after the drug exposure and more fascinatingly its reversibility with no particular DNA sequence alterations suggested an epigenetic basis (Wilting and Dannenberg 2012; Banelli et al. 2015). Epigenetic modifications are heritable changes that occur without altering the principal DNA sequence. Epigenetic instability is well-associated with carcinogenesis, progression of tumor, and metastasis since last five decades (Tsanev and Sendov 1971; Feinberg et al. 2016). More recently, link between cancer treatment response, non-genomic changes, and players behind them is also established. Hundreds of studies have evidenced the epigenetic tweaking of genome during the course of treatment turning residual disease into a full relapse in multiple cancers. This chapter will discuss the four central players in epigenetic regulation of genome, namely DNA methylation, histone modifications, chromatin remodeling complexes, and non-coding RNAs in the context of acquired drug resistance in cancer.

DNA Methylation

DNA methylation is an epigenetic heritable mark involving covalent transfer of a methyl group to adenine and the C5 position on cytosine residues in CpG dinucleotide sequences of DNA, by DNA methyltransferases (DNMTs) (Jin et al. 2011; Wilting and Dannenberg 2012; Robertson 2005). Historically, methylated cytosines were discovered in 1948 by Rollin Hotchkiss who observed that a derivative of cytosine separated from cytosine on paper chromatography in a way that thymine (also known as 5-methyluracil) separated from uracil (Hotchkiss 1948). Methylated cytosines can serve as a docking site for methyl-binding proteins that can recruit transcription factors or co-repressors, thereby modulating gene expression (Wilting and Dannenberg 2012).

Hypomethylation and hypermethylation of DNA are part of the epigenetic hallmarks of cancers (Suzuki and Bird 2008). Global hypomethylation predominantly occurs at repetitive elements like retrotransposons and satellites leading to genomic instability, while hypermethylation of CpG islands occurs at promoters of tumor suppressor genes leading to silencing of their expression stimulating tumorigenicity (Jin et al. 2011). The evidence that exposure of tumor cells to antineoplastic agents induces changes in DNA methylation was reported by Jonathan Nyce in 1997. He observed hypermethylation in DNA following treatment with 5-fluorouracil, methotrexate, 2′,3′-dideoxy-3′-azidothymidine (AZT), and cisplatin, leading to inactivation of genes required to be expressed for the action of drugs (Nyce 1997).

Most extensively studied mechanism of multidrug resistance in tumors like acute myeloid leukemia, chronic lymphocytic leukemia, breast carcinoma, and ovarian cancers is overexpression of P-glycoprotein (PgP), an ATP-binding cassette sub-family B member-1 (ABCB1) encoded by *MDR1* gene (Kantharidis et al. 1997). It is a plasma membrane glycoprotein transporter facilitating increased efflux of chemotherapeutic drugs resulting in lower intracellular drug accumulation (Kantharidis et al. 1997; Chekhun et al. 2006). With no significant mutations found in the *MDR1* gene, reason for its overexpression was unknown until in an in vitro drug-resistant model of CLL cell lines, PgP overexpression was found to be proportional to drug insensitivity and inversely correlated with the extent of DNA methylation at *MDR1* promoter (Kantharidis et al. 1997). Similar observations were made in sensitive and resistant breast cancer cell line MCF7 where significant hypomethylation in the *MDR1* gene promoter occurred along with acquisition of resistance (Chekhun et al. 2006). In bladder cancer patients, 3.5–5.7-fold increase in *MDR1*mRNA levels was found after chemotherapeutic treatment correlating with decreased methylation at the gene promoter (Tada et al. 2000). Hypomethylation of *MDR1* gene promoter as the cause of overexpression of PgP and drug insensitivity was confirmed by treatment with demethylating agent 5-azacytidine in PgP non-expressing cells which turned these PgP negative cells into PgP positive cells accompanied by acquisition of resistance to doxorubicin in CLL (Kantharidis et al. 1997).

How chemotherapy induces hypomethylation of *MDR1* promoter is not well-known, but some mechanisms have been postulated. For example, chemotherapeutic drugs may induce demethylation by inducing the DNA demethylases like TET enzymes or by exerting selection pressure on cells harboring hypomethylated gene sequences. Although the latter is less supported, primary tumors have been shown to have hypermethylation of *MDR1* gene promoter which significantly changes along with acquisition of drug resistance (Nakayama et al. 1998; Tada et al. 2000).

Another mechanism that leads to multidrug resistance is modulation of apoptotic machinery of cancer cells. Accordingly, aberrant epigenetic changes involving hypo- and hypermethylation in promoter sequences of apoptotic genes have been reported as a mechanism of insensitivity toward chemotherapeutic drugs (Dive 1997). In a comparative patient study of chemotherapy responsive and nonresponsive group of cervical squamous cell carcinoma, methylation frequencies of 6 apoptosis-related genes (*APAF1, CASP8, DAPK, FAS, survivin,* and *TRAILR1*) were identified by methylation-specific PCR. Of the 6 genes, *DAPK, FAS*, and *TRAILR1* were highly methylated in resistant and advanced cases of cervical cancer (Chaopatchayakul et al. 2010). Similarly, following cytotoxic drug treatment in leukemic blasts loss of *APAF1* gene expression due to hypermethylation at its promoter was found leading to drug resistance (Fu et al. 2003). Accordingly, in a study by Nakatsuka et al., the treatment of resistant leukemic cells with demethylating agent decitabine (5′-aza-2′-deoxycytidine) resulted in re-expression of *DAPK* and *APAF1* genes leading to resensitization of leukemic cells to the chemotherapy (Nakatsuka et al. 2003). In chemotherapy-treated pancreatic cancer, expression of *BNIP3* (another pro-apoptotic protein) is seen to be inversely proportional to methylation at its promoter CpG islands acting as key regulator behind the resistance to hypoxia-mediated apoptosis. Expectedly enough, treatment with 5aza2dC restores the *BNIP3* expression and drug sensitivity in these cells (Abe et al. 2005). Moreover, in 80 gastric cancer patients following fluoropyrimidine-based therapy, hypermethylation at *DAPK* and *BNIP3* promoter sequences has been correlated with lower drug response rate, reduced progression free, and overall survival (Sugita et al. 2011).

While trying to identify a link between DNA hypermethylation and loss of gene expression, Conway and co-workers overexpressed DNA methyltransferase DNMT1 in human fibroblasts and breast cancer cells and found *TMS1*, a novel pro-apoptotic CpG island-associated gene to be significantly downregulated. In 40% of primary breast tumors, *TMS1* was found hypermethylated and silenced at mRNA levels. Ectopic expression of *TMS1* in human breast cancer cell lines inhibited cell growth and induced efficient apoptosis confirming its involvement in regulating resistance via hypermethylation (Conway et al. 2000). In response to chemotherapeutic agents, gemcitabine and docetaxel, *TMS1* gene is seen to be hypermethylated in pancreatic cancer cell lines as well (Ramachandran et al. 2010). Intriguingly, in a complex multifactorial process like apoptosis, concurrent methylation of multiple genes is seen to be preferential than single-gene methylation while acquiring drug resistance.

Since cytotoxic drugs majorly act by damaging the DNA of cancer cells, modulation of DNA damage repair pathway is another prominent function seen to be dysregulated in resistant cancer cells. Correspondingly, experimental evidence from multiple sources has identified a link between altered methylation pattern in one or more of the DNA damage repair genes and chemotherapeutic drug resistance in cancer (Li et al. 2015). For example, hypermethylation at *MLH1* (DNA mismatch repair gene) promoter was acquired in about 25–35% of ovarian cancer patients following chemotherapy. 90% of in vitro selected cisplatin-resistant cell lines too expressed loss of *MLH1* attributed to hypermethylation of both the alleles in comparison to the *MLH1* expressing parental cells which had methylation at only single allele (Strathdee et al. 1999). In another study, acquisition of drug resistance following chemotherapy in ovarian cancer patients was monitored in context of *MLH1* gene promoter methylation and expression in 36 paired primary and secondary tumors. Interestingly, none of the primary tumors showed methylation of *MLH1* promoter, whereas 56.3% of secondary tumors exhibited hypermethylation that occurred more frequently in patients with four or more courses of chemotherapy (Watanabe et al. 2007). Loss of MMR proteins has been reported in 10–15% of colorectal cancer patients with hypermethylation of *MLH1* gene and resistance to 5′-fluorouracil. Treatment of these resistant cells with demethylating agent 5-′-azacytidine leads to reexpression of the MMR protein and sensitivity to cisplatin and 5′-FU (Arnold et al. 2003). Such reports demonstrate a direct link between epigenetic modulation of MMR genes and insensitivity of cancer cells toward cytotoxic drugs. Another major DNA repair gene to be modulated in cancer via DNA methylation at its promoter is O6-methylguanine-DNA methyltransferase (MGMT). MGMT removes the adducts formed in the DNA by alkylating agents like temozolomide, BCNU (1,3-bis(2-chloroethyl)-1-nitrosourea), and CCNU (1-(2-chloroethyl)-3-cyclohexyl-1-nitrosourea that are also used as chemotherapeutic agent (Chen et al. 2018). Resistance to alkylating agents is shown via two mechanisms, via hypomethylation of *MGMT* gene promoter and thus resulting in its overexpression and by selection of innately resistant subclones (Chen et al. 2018).

On the other hand, acquisition of drug resistance to fotemustine, a chloroethylating agent, has been observed after repeated treatment of the drug in malignant melanoma cell lines. The resistant cells show cross-resistance to O6-alkylating agents and an upregulation of *MGMT* gene expression (Christmann et al. 2001). In an independent study on melanoma, repeated temozolomide treatment is seen to cause resistance acquisition in vitro accompanied by upregulation of *MGMT* expression and loss of a key MMR protein *MSH6* (Chen et al. 2018) though a relation of this change to DNA methylation is not established.

Histone Modifications

Histones are the core element of nucleosomes that form an octamer consisting of four core histone proteins (H2A, H2B, H3, and H4) around which a 147 base pair segment of DNA is wrapped (Jenuwein and Allis 2001; Wilting and Dannenberg 2012; Audia and Campbell 2016). The access of DNA to the transcription factors, enhancers, or corepressors largely depends on posttranslational modifications in arginine- and lysine-rich regions of histone tails which allow small charge variations on the N-terminus so as to open or close the chromatin (Bruce 1989; Audia and Campbell 2016).

First evidence for histones being post translationally modified was reported by two groups independently (Phillips 1963) (Allfrey and Mirsky 1964). They identified an acetylation marked arginine rich and methyl group attached lysine rich histone fraction in column chromatography. These acetylation and methylation marks on histones in gene body or gene promoter sequences are now well-studied and exemplify in most cases either activation or repression of gene expression (Allfrey et al. 1964; Wilting and Dannenberg 2012) leading to cancer progression as well as treatment resistance (Audia and Campbell 2016). In this section, we will look at the most prominent of histone PTMs and their implications in growing tolerance of tumors to antineoplastic agents.

Histone Acetylation

Addition of acetyl group to histones can occur on multiple lysine residues at N-terminal tail. This acetylation alters the charge on histone, neutralizing its basic nature and allowing an open access of transcription factors to DNA (Eberharter and Becker 2002). Acetylation of histones is unequivocally the epigenetic activation mark on gene promoter. Deacetylation thus represents a repression of gene expression.

Since the earliest identification of histone acetylation in 1963 (Phillips 1963), various acetylation marks at histones H1, H3, and H4 are identified and studied in context of cancer as well as response to anticancer therapy (He et al. 2020; You et al.

2020; Lee et al. 2014b; Ye et al. 2018; Song et al. 2017; Liu et al. 2016). Hyperacetylation of H3 and H4 lysine residues including H3K14 and H3K9 is known as the activation marks on chromatin and are most commonly associated with ongoing or active transcription of genes nearby along with H3K27ac (Wang et al. 2008), whereas acetylation of H4K16 directly affects compaction of nucleosomes by altering the charge and thereby opening up the chromatin (Shogren-Knaak et al. 2006). Although few specific functions are known for these histone acetylations, they act in coherence with each other and chemotherapeutic agents are seen to induce a global acetylation change in a chromatin rather than an isolated effect. Doxorubicin, a well-known antineoplastic agent, is shown to induce hyperacetylation of histone H1 leading to acquisition of 30-fold higher resistance to the drug in colon carcinoma (Mannironi and D'Incalci 1988). Similarly, doxorubicin is attributed to hyperacetylate H3 and H4 in *ABCB4* promoter resulting in increased drug efflux and drug resistance in breast cancer (Huang et al. 2018). In esopharyngeal, epidermoid and prostate cancer cell lines increased acetylation at H3K14 and H4K16 mediated by *TIP60*. A MYST family histone acetyltransferase leads to cisplatin resistance (Miyamoto et al. 2008; Su et al. 2017). Mechanistically, this *TIP60*-mediated H4 hyperacetylation is localized to the promoters of *BRCA1* and *FANCD2* genes leading to efficient HR and Fanconi anemia pathway DNA damage repair (Su et al. 2017). In acute myeloid leukemia, our group has reported an upregulation of histone acetyltransferase GCN5/KAT2A leading to acquired resistance to doxorubicin by modulating the DDR in resistant cells (Salunkhe et al. 2018). Another HAT p300/CBP is also shown to be recruited to the DDR gene promoters in response to gemcitabine treatment in pancreatic cancer (Ono et al. 2016). On the other hand, in colorectal carcinoma, p300/CBP loss results in resistance to 5-FU treatment through HAT degradation-mediated hyperdeacetylation (Du et al. 2017). Recently, a role of p300/CBP hyperactivity-mediated increased acetylation of histone H3 at *CD274* (PD-L1) promoter has been associated with resistance to PD-L1 blockade immunotherapy in prostate cancer (Liu et al. 2020).

After DNA methylation, the second most studied epigenetic regulation of gene silencing is histone deacetylation (Lin et al. 2008). HDAC inhibitors have shown tremendous potential as an epigenetic drug to increase sensitivity of tumor cells to conventional anticancer agents. In a gastric cancer study, five of the most commonly used chemotherapeutic drugs 5-fluorouracil, paclitaxel, oxaliplatin, irinotecan, and gemcitabine showed synergy with HDACi TSA (trichostatin A) reversing drug resistance in gastric cancer cells (Zhang et al. 2006). Similar results have been observed in refractory melanoma (Monte et al. 2006), colorectal carcinoma (Ye et al. 2018), ovarian cancer (Lin et al. 2008), prostate cancer (Xu et al. 2018a), and head and neck squamous cell carcinoma (Whang et al. 2005). Furthermore, in synergy with irinotecan treatment, TSA is shown to increase H4K16 acetylation and reverse the drug resistance in colorectal carcinoma (Meisenberg et al. 2017). In retinoic acid refractory HNSCC and lung cancer, TSA treatment is shown to induce increased acetylation of H3 and H4 at the promoter of retinoic acid receptor beta (RARbeta), thus reversing the resistance to RA (Whang et al. 2005) (Suh et al. 2002). Pro-apoptotic protein Fas is frequently downregulated in ovarian cancer giving it a drug tolerance advantage through hypoacetylation at Fas promoter (Cacan 2016). HDACi treatment like HDAC6 inhibition by SAHA (suberoylanilide hydroxamic acid) along with paclitaxel, however, reverses this phenotype and increases sensitivity to chemotherapeutic agents (Angelucci et al. 2010). SAHA, vorinostat, and panobinostat act as pan-HDAC inhibitors and induce H3 acetylation, p21 expression, G1 cell cycle arrest, and sensitivity to rituximab in refractory B-cell lymphoma (Xue et al. 2016). Likewise, in breast cancer, HDAC4-mediated hyperacetylation of H3 at SMAD4 gene promoter results in acquisition of resistance to 5-FU treatment (Yu et al. 2013b), while an inhibitor to SIRT6, quinazolinedione, and a novel HDACi CG200745 increase sensitivity of pancreatic cells to gemcitabine-/erlotinib-induced apoptosis (Sociali et al. 2015; Lee et al. 2017). HDACs have also been shown to regulate expression of PD-1/PD-L1 genes and mitogen growth signaling (Yeon et al. 2020). PD-L1 overexpression in cisplatin-resistant non-small cell lung cancer and doxorubicin-resistant breast cancer cell lines is seen to be due to downregulation of HDAC3 resulting in hyperacetylation of H3 at PD-L1 gene promoter (Wang et al. 2020a).

Histone Methylation

Unlike histone acetylation, histone methylation does not alter amino acid charge but change the hydrophobicity of histones (Wilting and Dannenberg 2012). Gene activation or repression following methylation of histones depends on the residues modified and the number of methyl groups added (Brown et al. 2014). For example, monomethylated lysine 9 on histone H3 (H3K9me) represents activation mark, while tri-methylation at the same residue (H3K9me3) predicts repression of gene expression. Histone methylation is brought about by two classes of enzymes: histone methyltransferases (HMTs) and histone demethylases. HMTs that transfer methyl group onto a lysine residue are called KMTs, while arginine methyltransferases are known as PRMTs (Yang et al. 2020a; Orouji and Utikal 2018). Histone lysine demethylation occurs via KDM1 (KDM1A and KDM1B) and JmjC domain containing histone demethylases (encompassing about 20 members of KDM2, KDM3, KDM4, KDM5, and KDM6 subfamilies) (D'Oto et al. 2016).

Lysine tri-methylation at histone H3 (H3K27me3) is the chiefly studied modification in context of chemotherapeutic drug resistance. Repression of expression through H3K27me3 occurs via EZH2 (enhancer of zeste homolog 2) which deposits a methylation mark that recruits corepressors for gene inactivation (Kelly et al. 2010; Chen et al. 2015). Overexpression of H3K27me3 thus leads to repression of genes associated with metastasis, advanced clinical stage, and acquired chemotherapy resistance in multiple cancers (Cai et al. 2011; Hu et al. 2010; Hung et al. 2013; Cui et al. 2014; Qiu et al. 2019). Cisplatin is shown to induce increased expression of EZH2, thereby increasing H3K27me3 levels globally as well as locally on promoter of cell cycle inhibitor *CDKN1C* in small cell lung cancer resulting in acquisition of resistance (Qiu et al. 2019). Accordingly, EZH2 inhibitor GSK-126 is able to derepress *CDKN1C* and re-sensitize cells to cisplatin. Similar results are seen in ovarian cancer, where EZH2 knockdown leads to decreased methylation at H3K27 and resensitization of tumor xenografts to cisplatin in vivo (Hu et al. 2010). In a panel of cancer cell lines including cervical adenocarcinoma, colorectal cancer, liver cancer, and colon carcinoma, EZH2 overexpression and increased H3K27me3 have been observed in response to 5-fluorouracil treatment in a dose–dependent manner, substantiating the association of this histone modification with acquisition of drug resistance (Wu et al. 2016). On the other hand, loss of H3K27me3 mark has also been associated with chemotherapy resistance in non-small cell lung cancer (Ávila-Moreno et al. 2014). Ávila-Moreno et al. observed that patients deprived of H3K27me3 repression mark show poor overall survival with inadequate response to cisplatin/carboplatin. Hypomethylation at H3K27 leads to upregulation of transcription factors *TWIST1* and *MEOX2* and resistance to cisplatin in patients and cell lines (Ávila-Moreno et al. 2014). Similarly, loss of histone methylation mark on H3K79 (by DOT1L methyltransferase) has been reported in increased resistance to PARP inhibitor olaparib in colorectal cancer patients (Kari et al. 2019). Conversely, gain of expression of DOT1L, higher methylation at H3K79, and therefore upregulation of *MDR1* gene are associated with increased drug efflux and cisplatin resistance in ovarian cancer (Liu et al. 2018). Recently, we have reported upregulation of H3K36me2 via histone methyltransferase SETMAR, leading to efficient NHEJ DNA repair and imparting chemo-radio resistance in glioblastoma (Kaur et al. 2020).

Alteration in histone demethylases is also shown to result in aberrant methylation pattern promoting cellular resistance phenotype. Overexpression of KDM5A demethylase results in decreased H3K4 di- and tri-methylated marks, increased CDK inhibitor expression and erlotinib resistance in breast cancer (Hou et al. 2012) and temozolomide resistance in glioblastoma (Banelli et al. 2015). In prostate cancer, histone demethylase KDM4B is overexpressed in response to enzalutamide (a next-generation antiandrogen) increasing demethylation of repressive mark H3K9me3, upregulation of *MYC* gene expression, and tolerance to enzalutamide (Tang et al. 2020). KDM6A is recently reported as a novel relapse-associated gene in acute myeloid leukemia as well (Stief et al. 2020). Mechanistically, resistance to cytarabine in these refractory AML patients is due to downregulation of *ENT1* (nucleoside membrane transporter) caused by high levels of H3K27me3 at its promoter as a result of loss of KDM6A (Stief et al. 2020). Similarly, oxaliplatin induces upregulation of KDM6A/6B and decreases H3K27me3 repressive mark at *NOTCH2* promoter, thereby increasing stemness in colorectal cancer cells (Wang et al. 2020b). Accordingly, small-molecule inhibitor of KDM6A/6B, GSK-J4, increases oxaliplatin-induced cell apoptosis, while an inhibitor of EZH2 leads to decreased H3K27me3 and resistance acquisition in sensitive cells. Parallelly, in gastric cancer, cisplatin-resistant cell line is seen to acquire downregulation of H3K4 methylation through overexpression of KDM5B. Gene silencing of KDM5B or treatment of H3K4 demethylase inhibitor JIB-04 expectedly leads to resensitization of gastric cancer cells to cisplatin (Xu et al. 2018b).

Other Histone Modifications: Phosphorylation, Ubiquitination, ADP-Ribosylation, and Biotinylation

In addition to acetylation and methylation of histone tails, other posttranslational modifications of histones include phosphorylation, biotinylation, sumoylation, ADP-ribosylation, ubiquitination, and deamination (Lawrence et al. 2016; Kouzarides 2007). These PTMs can occur in the core histone as well as at N-terminal tail and influence the compaction or accessibility of the chromatin. Histone phosphorylation was evidenced first by Stevely and Stocken in 1966 when incorporation of ^{32}P into nuclear histones was observed in rat thymus (Stevely and Stocken 1966). Histone phosphorylation is a highly dynamic process involving transfer of a phosphate group from ATP to the hydroxyl group of amino acid chain of serine, threonine, and tyrosine residues on histones by histone kinases (Bannister and Kouzarides 2011; Oki et al. 2007). These phosphorylation modifications in a few cases are correlated with chemotherapy drug response and tumor progression. For instance, in hepatocellular carcinoma, phosphorylation at S10 of histone H3 mediated by histone kinases MSK1 and RSK has been associated with resistance to doxorubicin treatment (Li et al. 2013). Phosphorylation at serine 139 on C-terminus tails of H2AX mediated by a group of PIKK kinases including ATM, ATR, and DNAPK is one of the most commonly investigated histone PTMs and is an early indicator of DNA damage. In imatinib-treated CML, persistent γH2AX foci correlate with higher sensitivity to drug and genetic perturbation of H2AX expression or S139 mutation in H2AX results in acquisition of imatinib resistance (Dong et al. 2014). We have also reported persistent γH2AX foci and ATM kinase activity correlating with poor patient survival in therapy-resistant GBM cells (Kaur et al. 2020, 2015). These reports suggest a strong ability of H2AX phosphorylation in predicting pathological response to chemotherapeutic drugs in cancer patients.

Ubiquitination is the formation of an isopeptide bond between the terminal glycine and ε-amino group of lysine residue on the target protein. All core histone proteins H2A, H2B, H3, H4, and the linker histone H1 are mono- or polyubiquitinylated on multiple lysine residues and dictate influence on gene expression (Tweedie-Cullen et al. 2009). H2Bub1 is seen to be enriched downstream of transcription start sites of many cancer-associated genes including *XPC* and *POLH* involved in platinum drug resistance (Cole et al. 2020). In ER-positive breast cancer cells, proteasomal degradation induces global reduction in H2Bub1, reducing gene expression of ER targets and ER independent cell growth which is speculated to have a role in resistance to ER targeting therapy (Prenzel et al. 2011). Similarly, monoubiquitination on histone H4K91 by an E3 ligase BBAP has been shown to impart protective response to DNA damaging agents in B-cell lymphoma (Yan et al. 2009).

ADP-ribosylation ($<1\%$ of histone proteins) and biotinylation are rather rarely occurring posttranslational modifications of histones (Audia and Campbell 2016; Boulikas 1989), though in cancer, there are few reports suggesting their role in regulating drug resistance or sensitivity. Elevated H1 histone poly-ADP-ribosylation is shown to correlate with higher internucleosomal DNA fragmentation and better response to adriamycin, mitomycin C, and cisplatin in promyelocytic leukemia cell lines (Yoon et al. 1996). In human choriocarcinoma and lymphoma, H4K12-biotinylation mark is observed to reduce by 50% in response to etoposide treatment in vitro. Interestingly, biotin-deficient choriocarcinoma cells exhibit higher sensitivity to etoposide implying a possibility of higher histone biotinylation in acquisition of resistance to the drug (Kothapalli et al. 2005).

Chromatin Remodeling Complexes

Apart from the processes discussed above, ATP-dependent chromatin remodeling can also impart a distinct nongenetic code to the genome. Proteins carrying out these ATP-dependent remodeling are multifactorial units, known as chromatin remodeling complexes (CRCs). They can displace DNA wrapped around a nucleosome by utilizing ATP hydrolysis to alter histone-DNA connection and reposition the chromatin, thereby changing gene accessibility to DNA binding proteins (Lafon-Hughes et al. 2008). CRCs belong to the SWI/SNF2 family of proteins first identified in yeast (Winston and Carlson 1992; Laurent et al. 1993) and are divided into 4 complexes: SWI/SNF (switching defective/sucrose non-fermenting), ISWI (imitation SWI), CHD (chromodomain helicase DNA binding), INO80 (inositol requiring 80) (Ferraro 2016). Evidences of drug insensitivity in tumor population due to altered chromatin organizations have accumulated in large numbers in the past decade (Sharma et al. 2010; Gurard-Levin et al. 2016) including breast cancer (Magnani et al. 2013), ovarian carcinoma (Luo et al. 2018), lymphoma (Agarwal et al. 2019), melanoma (Dar et al. 2015), lung cancer (Strobeck et al. 2002), acute promyelocytic leukemia (Nichol et al. 2016), and prostate cancer (Yu et al. 2013a).

BRG1 (encoded by *SMARCA4*), a SWI/SNF complex ATPase, is one of the most studied molecules with respect to tumorigenesis and drug resistance. Loss of BRG1 re-sensitizes docetaxel-resistant breast cancer cells to chemotherapy (Gurard-Levin et al. 2016). BRG1 is also found to be upregulated in GICs (glioma initiating cells) resulting in downregulation of *TXNIP* (negative regulator of glycolysis) and resistance to temozolomide and carmustine (Ganguly et al. 2018). SWI-/SNF-mediated chromatin remodeling is required for transcriptional activity of glucocorticoid (a therapeutic agent in acute lymphoblastic leukemia) as well. Expectedly, in prednisolone- and dexamethasone-resistant ALL patients, SWI/SNF components *ARID1A*, *SMARCA4,* and *SMARCAB1* are downregulated (Pottier et al. 2008). On the other hand, in neuroblastoma, *SMARCA2*/BRG1 overexpression is associated with chemotherapy resistance (Jubierre et al. 2016). High expression of another ATPase from SWI/SNF complex, BRM (encoded by *SMARCA2*), correlates with resistance to gemcitabine via activation of *STAT3* pathway in PDAC patients (Zhang et al. 2017) as well as cisplatin resistance via metabolic reprograming in ovarian cancer (Xu et al. 2018c). While on the one hand BAF57 (encoded by *SMARCE1*) loss causes upregulation of EGFR resulting in resistance to MET and ALK inhibitors in non-small cell lung cancer (Papadakis et al. 2015), on the other hand, in ovarian cancer cell lines, in response to four chemotherapeutic agents cisplatin, paclitaxel, doxorubicin, and 5-fluorouracil, BAF57 expression was observed to be elevated promoting resistance to all the drugs (Yamaguchi and Hung 2014).

AT-rich interaction domain containing protein-1A (ARID1A) is one of the most commonly mutated genes in human cancers (Tomihara et al. 2020; Kadoch et al. 2013). ARID1A inactivation mutations are found in ER+ breast cancer causing loss of ARID1A expression and resistance to fulvestrant (Xu et al. 2020). Furthermore, in cases where resistance to hormone therapy in ER+ breast cancer cells is attributed to switching from ER-dependent luminal type to ER independent basal type, loss of ARID1A is shown to be responsible for this epigenetic modulation resulting in hormone therapy resistance (Tomihara et al. 2020). Similarly, ARID1A loss in ovarian cancer cells leads to decreased occupancy of ARID1A onto MRP2 (multidrug resistance protein 2) promoter while elevating the recruitment of ATPase BRG1 onto MRP2 promoter, thereby increasing its expression (Luo et al. 2018) demonstrating that these chromatin reorganization events occur in cooperation rather than in isolation.

NuRD (nucleosome remodeling deacetylation complex) is the only CRC clubbing the functions of reorganizing ATPases and histone deacetylases. Overexpression of NuRD components MTA1 (methyl-CpG binding domain protein 1) and CHD4 (chromodomain helicase DNA binding protein 4) has been implicated clinically in cancers and therapy resistance (Kaur et al. 2014). In breast cancer, CHD4 deficiency impairs the recruitment of HDAC1 to the promoter of cell cycle checkpoint protein p21 and sensitizes tumor cells to cisplatin and PARP inhibitors (Hou et al. 2017).

Unmistakably, alterations in chromatin accessibility due to the modifications summarized above have a distinguished influence on gene expression and therefore subsequent cellular functions culminating in a discrete phenotypic response of tumor cells to chemotherapy (Fig. 14.1).

Non-Coding RNAs

With recent advances in nucleotide sequencing techniques, a vast majority of posttranscriptional modifications of RNA molecules have been detected. These modifications decide the fate of that RNA in the cellular processes and function. Along with conventional ways including DNA methylation and histone modifications, actions of the once considered "junk" of transcriptome, non-coding RNAs, have emerged to be prominent means by which the epigenome of a cell can be altered (Yang et al. 2020b; Guttman et al. 2009). Non-coding RNAs do not code for a functional protein as the name suggests; however, they do possess regulatory functions and are divided into small (miRNA, siRNA, and piRNA) and long non-coding RNAs based on their length (Wei et al. 2017; Zaratiegui et al. 2007).

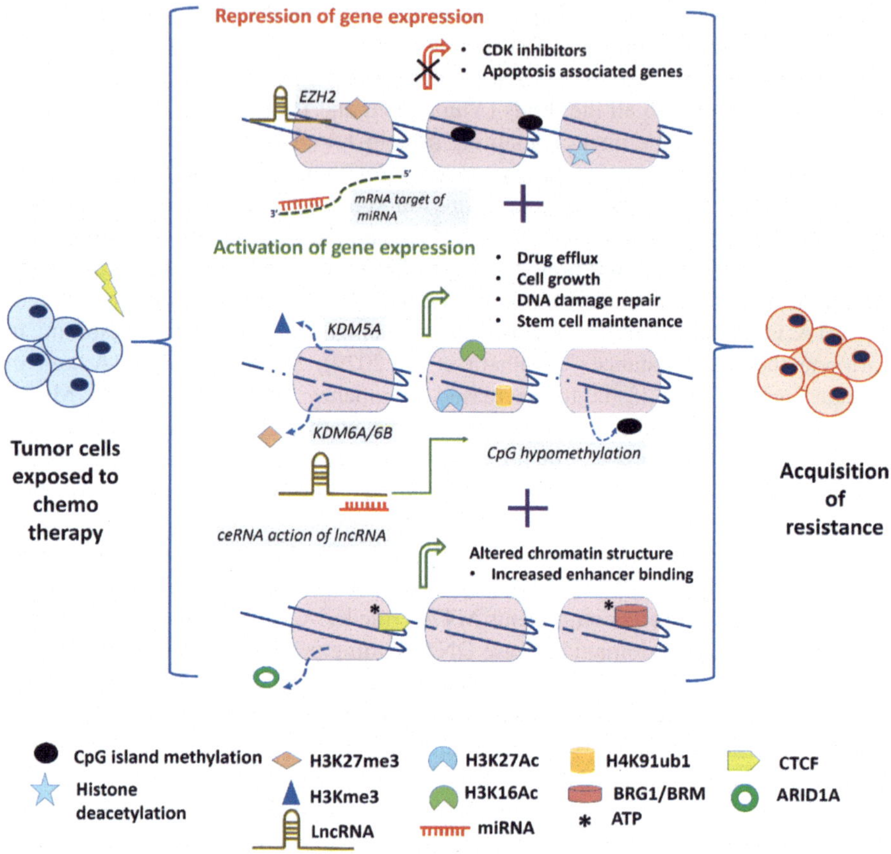

Fig. 14.1 Epigenetic regulation mediating acquisition of chemoresistance in cancer cells. CpG hypermethylation, repressive histone lysine methylation, histone deacetylation, and miRNA-mediated mRNA targeting events lead to downregulation of gene expression, while histone acetylation, ubiquitylation, activating histone methylation, competitive inhibition of miRNA by lncRNA, and chromatin reorganization via CRC ATPases lead to activation of gene expression. A combined effect of one or more of these results in an advantageous epigenome of cancer cells making them chemo-tolerant

Since the first report of *lin14* found in *C.elegans* (Chalfie et al. 1981), microRNAs are known for their silencing function by binding to 3′-UTR of a transcribed mRNA and inhibiting its translation (El-Awady et al. 2015; Lee et al. 1993). MicroRNAs are small RNAs of 20–23nt length, and their epigenetic action on the target mRNA subsequently altering gene function is widely observed in malignancies and their insensitivity to antineoplastic agents (Giovannetti et al. 2012; Kovalchuk et al. 2008; Chen and Xue 2016; He and Hannon 2004; de Souza Rocha Simonini et al. 2010). The most prominently detected mechanism of chemotherapeutic drug resistance by miRNAs is aberrantly methylated promoters leading to downregulation of the respective miRNAs that affects target gene mRNA expression which in majority of the cases are tumor-promoting genes (Braga et al. 2019). Methylation-specific PCR, DNA methylome, and miRNome studies have identified miRNAs being regulated by DNA methyltransferases where treatment with 5-Aza-2-dc reverses the resistance and resensitizes the cells to chemotherapeutic agents due to miRNA re-expression (Chen et al. 2014; Lv et al. 2014; Lu et al. 2020). The target genes for these miRNAs are expectedly found to be involved in cell proliferation, cell cycle progression, drug efflux, DNA damage repair, and mitogenic signaling (*MAPK, ErBb2, Notch, Wnt, Myc, NFkB* led pathways), imparting a survival advantage to cancer cells post-chemotherapeutic treatment (Table 14.1). Cisplatin is the most widely utilized chemotherapeutic drug with applications in multiple cancer types. It is therefore no surprise that cancer cells have evolved approaches other than the abovementioned epigenetic mechanisms to evade the cytotoxic effects of cisplatin. MicroRNAs miR-370, miR-130b, miR-199b-5p, miR-320a, and miR-302 are strongly associated with cisplatin resistance (Chen et al. 2014; Yang et al. 2012; Liu et al. 2014; Yan et al. 2018; Bourguignon et al. 2012). Their absence of expression in response to cisplatin leads to acquisition of drug tolerance which is confirmed by resensitization post-miRNA overexpression. Aside from these, several other miRNAs have been implicated in conferring chemo-insensitivity in response to various classes of chemotherapeutic drugs which are summarized in Table 14.1. The expansive reach of miRNAs in cellular functions and their diversity along with specificity has proven to be fascinatingly acting in acquiring drug resistance via epigenetically modulating translation of the proteins. Since miRNAs act on multiple pathways and genes at the same time, they prove to be an outstanding candidate for combating with therapy resistance in cancer.

Unlike microRNAs, long non-coding RNAs are poorly understood in terms of acquired chemo-drug resistance in cancer. lncRNAs are named so for their length of more than 200nt and are transcribed as a non-coding transcripts (Kung et al. 2013). The very first identification of lncRNA H19 occurred in mice and was observed to act in gene imprinting three decades ago (Brannan et al. 1990; Bartolomei et al. 1991). lncRNAs based on their location, e.g., intergenic, antisense, and intronic, get transcribed into distinct forms which are evolutionarily conserved. They can form complex secondary structures allowing them to interact with proteins, DNA and RNA, thereby deciding their function (Wapinski and Chang 2011; Chen et al. 2020). Due to this tremendous potential of forming interactions and regulating the genome epigenetically with respect to key functions like cell cycle progression, apoptosis, and differentiation, lncRNAs have transpired as an epigenome modulator of tumorigenesis and response to anticancer therapies (Fang and Fullwood 2016; Vafadar et al. 2019). However, the precise mechanisms by which these lncRNAs may be directing cellular resistance to chemotherapy are still an area of intense research. Nonetheless, few well-characterized lncRNAs like HOTAIR (HOX transcript antisense RNA) are seen overexpressed in pancreatic cancer (Yang et al. 2017) and endometrial carcinoma (Chi et al. 2019) in response to TNF receptor antagonist and progesterone therapy, respectively. In endometrial carcinoma, HOTAIR acts by recruiting KDM1A, removing the activation mark H3K4me2 and repressing the expression of *PRB*, required for progesterone dependent apoptosis. While in PDAC

Table 14.1 List of non-coding RNAs associated with acquired chemoresistance in cancer

Non-coding RNA	Cancer type	Resistance to chemotherapeutic agent	Expression	Mechanism of action	References
MicroRNA					
miR-451	Breast cancer	Doxorubicin	↓	–	Kovalchuk et al. (2008)
miR-124	Renal cell carcinoma	Doxorubicin; vinblastine	↓	Wnt pathway activation	Gebauer et al. (2013)
miR-335	Melanoma; hepato-cellular carcinoma	Taxol; vinblastine	↓	Upregulation of E3ubiquitin ligase SIAH2	Kim et al. (2015)
miR-155; miR-200c	Non-small cell lung cancer	Gefitinib	↓	Increased EMT	Narita et al. (2017)
miR-21	Breast cancer	Topotecan	↑	Decreased apoptosis	Si et al. (2007)
	Pancreatic cancer	Gemcitabine	↑	Activation of Akt signaling	Song et al. (2013)
miR-638	Breast cancer	Docetaxel	↓	Upregulation of target STARD10 lipid-binding protein	Zhao et al. (2017)
miR-34a	Lung cancer	ALK inhibitors	↓	–	Yun et al. (2018)
miR-146b	Endothelial ovarian carcinoma	Paclitaxel; cisplatin	↓	Upregulation of target FBXL10	Yan et al. (2018)
miR-302a	Colorectal cancer	Cetuximab	↓	Upregulation of targets NF1B and CD44	Sun et al. (2019)
	Non-small cell lung cancer	Cisplatin	↓	Upregulation of targets MAPK, ErbB pathway	Lu et al. (2020)
miR-193a-2p	Esophageal cancer	Paclitaxel; Docetaxel; 5-fluorouracil	↑	Downregulation of target PSEN1 led apoptosis	Meng et al. (2016)
miR-101-3p	Pancreatic cancer	gemcitabine	↓	Upregulation of target gene RRM1	Fan et al. (2016)

miR-27b	Breast cancer	Tamoxifen	↓	Upregulation of target gene HMGB3 and EMT	Li et al. (2016)
miR-135b; miR-196b	Leukemia	Etoposide; topotecan; doxorubicin	↑	Upregulation of drug efflux transporter ABCB1	Ho et al. (2016)
miR-199b 5p	Ovarian cancer	Cisplatin	↓	Upregulation of target JAG1 led NOTCH signaling	Liu et al. (2014)
miR-375	Breast cancer	Trastuzumab	↓	Upregulation of target gene IGF1R	Ye et al. (2014)
miR-302	HNSCC	Cisplatin	↑	Activation of stem cell factors Oct4-Sox2-Nanog	Bourguigno n et al. (2012)
Long non-coding RNA					
SnaR; BACE1AS	Colon cancer	5-fluorouracil	↓	–	Lee et al. (2014a)
H19	Breast cancer	Paclitaxel	↑	Recruitment of EZH2 onto pro-apoptotic gene promoters	Si et al. (2016)
HOTAIR	Pancreatic cancer	TRAIL	↑	Recruitment of EZH2 onto DR5 promoter	Yang et al. (2017)
	Endometria l carcinoma	Medroxyprogesterone acetate (MPA)	↑	Interaction with LSD1 to remove activation histone methylation from apoptotic gene promoter	Chi et al. (2019)
CLARA	Breast cancer	Epirubicin; cyclophos-phamide, docetaxel	↑	–	Li et al. (2017)
MIR100HG	Colorectal cancer	cetuximab	↑	Overexpression of two embedded miRNAs	Lu et al. (2017)
UCA1	Breast cancer	Trastuzumab	↑	Inhibition of miR-18a and upregulation of YAP1	Zhu et al. (2018)
KCNQ1OT 1	Tongue squamous cell carcinoma	Cisplatin	↑	Competing endogenous RNA to miR-211-5p	Zhang et al. (2018)
PANDAR	Ovarian cancer	Cisplatin	↑	Through SRF2-p53 feedback loop	Wang et al. (2018)
TUG1	Acute myeloid leukemia	Adriamycin	↑	Epigenetic suppression of miR-34a	Li et al. (2019)
MPRL	Tongue squamous cell carcinoma	Cisplatin	↑	Interaction with miR-483-5p	Tian et al. (2019)

(continued)

Table 14.1 (continued)

Non-coding RNA	Cancer type	Resistance to chemotherapeutic agent	Expression	Mechanism of action	References
HOXA11-AS1	Lung cancer	Cisplatin	↑	Hypermethylation at HOXA11	Zhang et al. (2019)
HOXD-AS1	Gastric cancer	Cisplatin	↑	Recruitment of EZH2 onto PDCD4 promoter	Ye et al. (2019)
GAS5	Cervical cancer	Cisplatin	↓	Competing RNA to miR-21	Fang et al. (2020)
PVT1	Cervical cancer	Paclitaxel	↑	Interaction with EZH2 and competing RNA to miR-195	Shen et al. (2017)

samples, HOTAIR acts by recruiting DNA methyltransferase EZH2 and repressing *DR5* expression via tri-methylation of H3K27 imparting resistance to apoptosis. EZH2 has been associated with multiple such lncRNAs leading to deposition of histone methylation repressive mark at the tumor suppressor gene promoters, thereby conferring chemo tolerance in cancer cells, for example, lncRNA H19 expression in paclitaxel-resistant ER+ve breast cancer (Si et al. 2016), TUG1 upregulation and miR-34a suppression in adriamycin-resistant acute myeloid leukemia (Li et al. 2019), and HXD-AS1 (antisense to HOXD mRNA) overexpression in response to cisplatin leading to suppression of *PDCD4* (programmed cell death 4). More such lnRNAs and their roles in chemoresistance are summarized in Table 14.1.

In spite of 98% of the RNAs in a cell not being translated into proteins (Schliesser et al. 2016), these above reports suggest that the diversity and flexibility of functions of these non-coding RNAs evidently can alter genome as well as epigenome of a cell in response to chemotherapy.

Future Perspective

Evidently, epigenetic modulation of genes involved in cell cycle progression, chromatin remodeling, drug efflux, apoptosis, and DNA damage repair contributes to drug resistance in several cancer types. With the advances in technology as well as knowledge about chromatin organization and functioning, scientific endeavors are not restricted to the traditional approaches and have increased the accuracy of observations. Several studies with high throughput analysis have recently identified DNA methylation changes as a significant player in carcinogenesis and increased chemo-tolerance (Costello et al. 2000). A genome-wide DNA methylation profiling of resistant clones from ovarian cancer cell lines has suggested that less than 1% hypermethylated genes in resistant ovarian cancer cells actually account for acquisition of platinum drug resistance and can be probable "drivers" of resistance (Zeller et al. 2012). Moreover, three-dimensional genome organization is one of the new additional levels of epigenetic regulation being investigated. In mammals, chromatin loops created by CRCs are stabilized by architectural proteins cohesion and CTCF (CCCTC-binding factor). Tethered chromatin conformation and high-depth sequencing approaches have led to the observation that estrogen induction reduces CTCF binding on DERE (a distant ER response element) causing tamoxifen resistance in time-dependent manner (Zhou et al. 2019). Although four major epigenetic modulations DNA methylation, histone modifications, chromatin reorganization, and non-coding RNAs have been discussed separately here, it is important to note that they work in conjunction with each other. The growing efforts in the field of cancer therapeutics are focusing on utilizing these mechanisms as targets and devising combinatory approaches to chemotherapy. With a tremendous success of HDAC inhibitors and DNMT inhibitors in clinical trials, epigenetic therapies to combat drug resistance in cancer show a promising future (Sermer et al. 2019; Zeller and Brown 2010; Kelly et al. 2010; Imai et al. 2016).

References

Abe T, Toyota M, Suzuki H, Murai M, Akino K, Ueno M, Nojima M, Yawata A, Miyakawa H, Suga T, Ito H, Endo T, Tokino T, Hinoda Y, Imai K (2005) Upregulation of BNIP3 by 5-aza-2′-deoxycytidine sensitizes pancreatic cancer cells to hypoxia-mediated cell death. Journal of gastroenterology 40(5):504–510. https://doi.org/10.1007/s00535-005-1576-1

Agarwal R, Chan Y-C, Tam CS, Hunter T, Vassiliadis D, Teh CE, Thijssen R, Yeh P, Wong SQ, Ftouni S, Lam EYN, Anderson MA, Pott C, Gilan O, Bell CC, Knezevic K, Blombery P, Rayeroux K, Zordan A, Li J, Huang DCS, Wall M, Seymour JF, Gray DHD, Roberts AW, Dawson MA, Dawson S-J (2019) Dynamic molecular monitoring reveals that SWI–SNF mutations mediate resistance to ibrutinib plus venetoclax in mantle cell lymphoma. Nature Medicine 25(1):119–129. https://doi.org/10.1038/s41591-018-0243-z

Allfrey VG, Mirsky AE (1964) Structural modifications of histones and their possible role in the regulation of RNA synthesis., J Sci 144(3618):559–559. https://doi.org/10.1126/science.144.3618.559

Allfrey VG, Faulkner R, Mirsky AE (1964) Acetylation and methylation of histones and their possible role in the regulation of RNA synthesis. Proc Nat Acad Sci U S A 51(5):786–794. https://doi.org/10.1073/pnas.51.5.786

Angelucci A, Mari M, Millimaggi D, Giusti I, Carta G, Bologna M, Dolo V (2010) Suberoylanilide hydroxamic acid partly reverses resistance to paclitaxel in human ovarian cancer cell lines. Gynecol Oncol 119(3):557–563. https://doi.org/10.1016/j.ygyno.2010.07.036

Arnold CN, Goel A, Boland CR (2003) Role of hMLH1 promoter hypermethylation in drug resistance to 5- fluorouracil in colorectal cancer cell lines. Int J Cancer 106(1):66–73. https://doi.org/10.1002/ijc.11176

Audia JE, Campbell RM (2016) Histone modifications and cancer. Cold Spring Harb Perspect Biol 8(4):a019521. https://doi.org/10.1101/cshperspect.a019521

Ávila-Moreno F, Armas-López L, Álvarez-Moran AM, López-Bujanda Z, Ortiz-Quintero B, Hidalgo-Miranda A, Urrea-Ramírez F, Rivera-Rosales RM, Vázquez-Manríquez E, Peña-Mirabal E, Morales-Gómez J, Vázquez-Minero JC, Téllez-Becerra JL, Ramírez-Mendoza R, Ávalos-Bracho A, de Alba EG, Vázquez- Santillán K, Maldonado-Lagunas V, Santillán-Doherty P, Piña-Sánchez P, Zúñiga-Ramos J (2014) Overexpression of MEOX2 and TWIST1 is associated with H3K27me3 levels and determines lung cancer chemoresistance and prognosis. PloS One 9(12):e114104. https://doi.org/10.1371/journal.pone.0114104

Banelli B, Carra E, Barbieri F, Würth R, Parodi F, Pattarozzi A, Carosio R, Forlani A, Allemanni G, Marubbi D, Florio T, Daga A, Romani M (2015) The histone demethylase KDM5A is a key factor for the resistance to temozolomide in glioblastoma. Cell Cycle (Georgetown, Tex) 14(21): 3418–3429. https://doi.org/10.1080/15384101.2015.1090063

Bannister AJ, Kouzarides T (2011) Regulation of chromatin by histone modifications. Cell Res 21(3):381–395. https://doi.org/10.1038/cr.2011.22

Bartolomei MS, Zemel S, Tilghman SM (1991) Parental imprinting of the mouse H19 gene. Nature 351(6322):153–155. https://doi.org/10.1038/351153a0

Boulikas T (1989) DNA strand breaks alter histone ADP-ribosylation. Proc Nat Acad Sci U S A 86(10):3499–3503. https://doi.org/10.1073/pnas.86.10.3499

Bourguignon LY, Wong G, Earle C, Chen L (2012) Hyaluronan-CD44v3 interaction with Oct4-Sox2-Nanog promotes miR-302 expression leading to self-renewal, clonal formation, and cisplatin resistance in cancer stem cells from head and neck squamous cell carcinoma. J Biol Chem 287(39):32800–32824. https://doi.org/10.1074/jbc.M111.308528

Braga EA, Fridman MV, Loginov VI, Dmitriev AA, Morozov SG (2019) Molecular mechanisms in clear cell renal cell carcinoma: role of miRNAs and hypermethylated miRNA genes in crucial oncogenic pathways and processes. Front Genet 10:320–320. https://doi.org/10.3389/fgene.2019.00320

Brannan CI, Dees EC, Ingram RS, Tilghman SM (1990) The product of the H19 gene may function as an RNA. Mol Cell Biol 10(1):28–36. https://doi.org/10.1128/mcb.10.1.28

Brown R, Curry E, Magnani L, Wilhelm-Benartzi CS, Borley J (2014) Poised epigenetic states and acquired drug resistance in cancer. Nature reviews Cancer 14(11):747–753. https://doi.org/10.1038/nrc3819

Bruce A (1989) Molecular biology of the cell, 2nd edn. Garland Pub. ©1989, New York

Cacan E (2016) Histone deacetylase-1-mediated suppression of FAS in Chemoresistant ovarian cancer cells. Anticancer Res 36(6):2819–2826

Cai MY, Tong ZT, Zhu W, Wen ZZ, Rao HL, Kong LL, Guan XY, Kung HF, Zeng YX, Xie D (2011) H3K27me3 protein is a promising predictive biomarker of patients' survival and chemoradioresistance in human nasopharyngeal carcinoma. Mol Med (Cambridge, Mass) 17(11-12):1137–1145. https://doi.org/10.2119/molmed.2011.00054

Chalfie M, Horvitz HR, Sulston JE (1981) Mutations that lead to reiterations in the cell lineages of C. elegans. Cell 24(1):59–69. https://doi.org/10.1016/0092-8674(81)90501-8

Chaopatchayakul P, Jearanaikoon P, Yuenyao P, Limpaiboon T (2010) Aberrant DNA methylation of apoptotic signaling genes in patients responsive and nonresponsive to therapy for cervical carcinoma. Am J Obstet Gynecol 202(3):281.e281–281.e289. https://doi.org/10.1016/j.ajog.2009.11.037

Chekhun VF, Kulik GI, Yurchenko OV, Tryndyak VP, Todor IN, Luniv LS, Tregubova NA, Pryzimirska TV, Montgomery B, Rusetskaya NV, Pogribny IP (2006) Role of DNA hypomethylation in the development of the resistance to doxorubicin in human MCF-7 breast adenocarcinoma cells. Cancer letters 231(1):87–93. https://doi.org/10.1016/j.canlet.2005.01.038

Chen J, Xue Y (2016) Emerging roles of non-coding RNAs in epigenetic regulation. Sci China Life Sci 59(3):227–235. https://doi.org/10.1007/s11427-016-5010-0

Chen XP, Chen YG, Lan JY, Shen ZJ (2014) MicroRNA-370 suppresses proliferation and promotes endometrioid ovarian cancer chemosensitivity to cDDP by negatively regulating ENG. Cancer letters 353(2):201–210. https://doi.org/10.1016/j.canlet.2014.07.026

Chen X, Xie W, Gu P, Cai Q, Wang B, Xie Y, Dong W, He W, Zhong G, Lin T, Huang J (2015) Upregulated WDR5 promotes proliferation, self-renewal and chemoresistance in bladder cancer via mediating H3K4 trimethylation. Scientific reports 5:8293. https://doi.org/10.1038/srep08293

Chen TC, Chan N, Minea RO, Hartman H, Hofman FM, Schönthal AH (2018) Rare stochastic expression of O6-methylguanine-DNA methyltransferase (MGMT) in MGMT-negative melanoma cells determines immediate emergence of drug-resistant populations upon treatment with temozolomide in vitro and in vivo. Cancers 10(10):362. https://doi.org/10.3390/cancers10100362

Chen J, Wang Y, Wang C, Hu J-F, Li W (2020) LncRNA functions as a new emerging epigenetic factor in determining the fate of. Stem Cells. 11:277. https://doi.org/10.3389/fgene.2020.00277

Chi S, Liu Y, Zhou X, Feng D, Xiao X, Li W, Zhao Y, Wang H (2019) Knockdown of long non-coding HOTAIR enhances the sensitivity to progesterone in endometrial cancer by epigenetic regulation of progesterone receptor isoform B. Cancer chemotherapy and pharmacology 83(2):277–287. https://doi.org/10.1007/s00280-018-3727-0

Christmann M, Pick M, Lage H, Schadendorf D, Kaina B (2001) Acquired resistance of melanoma cells to the antineoplastic agent fotemustine is caused by reactivation of the DNA repair gene MGMT. International journal of cancer 92(1):123–129

Cole AJ, Dickson KA, Liddle C, Stirzaker C, Shah JS, Clifton-Bligh R, Marsh DJ (2020) Ubiquitin chromatin remodelling after DNA damage is associated with the expression of key cancer genes and pathways. Cellular and molecular life sciences: CMLS. https://doi.org/10.1007/s00018-020-03552-5

Conway KE, McConnell BB, Bowring CE, Donald CD, Warren ST, Vertino PM (2000) *TMS1*, a novel proapoptotic caspase recruitment domain protein, is a target of methylation-induced gene silencing in human breast cancers. Cancer Res 60(22):6236–6242

Costello JF, Frühwald MC, Smiraglia DJ, Rush LJ, Robertson GP, Gao X, Wright FA, Feramisco JD, Peltomäki P, Lang JC, Schuller DE, Yu L, Bloomfield CD, Caligiuri MA, Yates A,

Nishikawa R, Su Huang H, Petrelli NJ, Zhang X, O'Dorisio MS, Held WA, Cavenee WK, Plass C (2000) Aberrant CpG-island methylation has non-random and tumour-type-specific patterns. Nature genetics 24(2):132–138. https://doi.org/10.1038/72785

Cui B, Yang Q, Guan H, Shi B, Hou P, Ji M (2014) PRIMA-1, a mutant p53 reactivator, restores the sensitivity of TP53 mutant-type thyroid cancer cells to the histone methylation inhibitor 3-Deazaneplanocin A. The Journal of clinical endocrinology and metabolism 99(6):E962–E970. https://doi.org/10.1210/jc.2013-3147

Dar AA, Nosrati M, Bezrookove V, de Semir D, Majid S, Thummala S, Sun V, Tong S, Leong SP, Minor D, Billings PR, Soroceanu L, Debs R, Miller JR 3rd, Sagebiel RW, Kashani-Sabet M (2015) The role of BPTF in melanoma progression and in response to BRAF-targeted therapy. Journal of the National Cancer Institute 107(5):djv034. https://doi.org/10.1093/jnci/djv034

de Souza Rocha Simonini P, Breiling A, Gupta N, Malekpour M, Youns M, Omranipour R, Malekpour F, Volinia S, Croce CM, Najmabadi H, Diederichs S, Sahin O, Mayer D, Lyko F, Hoheisel JD, Riazalhosseini Y (2010) Epigenetically deregulated microRNA-375 is involved in a positive feedback loop with estrogen receptor alpha in breast cancer cells. Cancer Res 70(22): 9175–9184. https://doi.org/10.1158/0008-5472.Can-10-1318

Dive C (1997) Avoidance of apoptosis as a mechanism of drug resistance. Journal of internal medicine Supplement 740:139–145

Dong Y, Xiong M, Duan L, Liu Z, Niu T, Luo Y, Wu X, Xu C, Lu C (2014) H2AX phosphorylation regulated by p38 is involved in Bim expression and apoptosis in chronic myelogenous leukemia cells induced by imatinib. Apoptosis: an international journal on programmed cell death 19(8): 1281–1292. https://doi.org/10.1007/s10495-014-0997-9

D'Oto A, Tian Q-W, Davidoff AM, Yang J (2016) Histone demethylases and their roles in cancer epigenetics. J Med Oncol Ther 1(2):34–40

Du C, Huang D, Peng Y, Yao Y, Zhao Y, Yang Y, Wang H, Cao L, Zhu WG, Gu J (2017) 5-Fluorouracil targets histone acetyltransferases p300/CBP in the treatment of colorectal cancer. Cancer letters 400:183–193. https://doi.org/10.1016/j.canlet.2017.04.033

Eberharter A, Becker PB (2002) Histone acetylation: a switch between repressive and permissive chromatin. Second in review series on chromatin dynamics. EMBO Rep 3(3):224–229. https://doi.org/10.1093/embo-reports/kvf053

El-Awady RA, Hersi F, Al-Tunaiji H, Saleh EM, Abdel-Wahab AH, Al Homssi A, Suhail M, El-Serafi A, Al-Tel T (2015) Epigenetics and miRNA as predictive markers and targets for lung cancer chemotherapy. Cancer biology & therapy 16(7):1056–1070. https://doi.org/10.1080/15384047.2015.1046023

Fan P, Liu L, Yin Y, Zhao Z, Zhang Y, Amponsah PS, Xiao X, Bauer N, Abukiwan A, Nwaeburu CC, Gladkich J, Gao C, Schemmer P, Gross W, Herr I (2016) MicroRNA-101-3p reverses gemcitabine resistance by inhibition of ribonucleotide reductase M1 in pancreatic cancer. Cancer letters 373(1):130–137. https://doi.org/10.1016/j.canlet.2016.01.038

Fang Y, Fullwood MJ (2016) Roles, functions, and mechanisms of long non-coding RNAs in cancer. Genomics, Proteomics & Bioinformatics 14(1):42–54. https://doi.org/10.1016/j.gpb.2015.09.006

Fang X, Zhong G, Wang Y, Lin Z, Lin R, Yao T (2020) Low GAS5 expression may predict poor survival and cisplatin resistance in cervical cancer. Cell Death & Disease 11(7):531. https://doi.org/10.1038/s41419-020-2735-2

Feinberg AP, Koldobskiy MA, Göndör A (2016) Epigenetic modulators, modifiers and mediators in cancer aetiology and progression. Nature reviews Genetics 17(5):284–299. https://doi.org/10.1038/nrg.2016.13

Ferraro A (2016) Altered primary chromatin structures and their implications in cancer development. Cellular oncology (Dordrecht) 39(3):195–210. https://doi.org/10.1007/s13402-016-0276-6

Fu WN, Bertoni F, Kelsey SM, McElwaine SM, Cotter FE, Newland AC, Jia L (2003) Role of DNA methylation in the suppression of Apaf-1 protein in human leukaemia. Oncogene 22(3): 451–455. https://doi.org/10.1038/sj.onc.1206147

Ganguly D, Sims M, Cai C, Fan M, Pfeffer LM (2018) Chromatin Remodeling Factor BRG1 Regulates Stemness and Chemosensitivity of Glioma Initiating Cells. Stem cells (Dayton, Ohio) 36(12):1804–1815. https://doi.org/10.1002/stem.2909

Gebauer K, Peters I, Dubrowinskaja N, Hennenlotter J, Abbas M, Scherer R, Tezval H, Merseburger AS, Stenzl A, Kuczyk MA, Serth J (2013) Hsa-mir-124-3 CpG island methylation is associated with advanced tumours and disease recurrence of patients with clear cell renal cell carcinoma. British journal of cancer 108(1):131–138. https://doi.org/10.1038/bjc.2012.537

Giovannetti E, Erozenci A, Smit J, Danesi R, Peters GJ (2012) Molecular mechanisms underlying the role of microRNAs (miRNAs) in anticancer drug resistance and implications for clinical practice. Critical Reviews in Oncology/Hematology 81(2):103–122. https://doi.org/10.1016/j.critrevonc.2011.03.010

Gurard-Levin ZA, Wilson LO, Pancaldi V, Postel-Vinay S, Sousa FG, Reyes C, Marangoni E, Gentien D, Valencia A, Pommier Y, Cottu P, Almouzni G (2016) Chromatin Regulators as a Guide for Cancer Treatment Choice. Molecular cancer therapeutics 15(7):1768–1777. https://doi.org/10.1158/1535-7163.Mct-15-1008

Guttman M, Amit I, Garber M, French C, Lin MF, Feldser D, Huarte M, Zuk O, Carey BW, Cassady JP, Cabili MN, Jaenisch R, Mikkelsen TS, Jacks T, Hacohen N, Bernstein BE, Kellis M, Regev A, Rinn JL, Lander ES (2009) Chromatin signature reveals over a thousand highly conserved large non-coding RNAs in mammals. Nature 458(7235):223–227. https://doi.org/10.1038/nature07672

He L, Hannon GJ (2004) MicroRNAs: small RNAs with a big role in gene regulation. Nature Reviews Genetics 5(7):522–531. https://doi.org/10.1038/nrg1379

He B, Wang Q, Liu X, Lu Z, Han J, Pan C, Carter BZ, Liu Q, Xu N, Zhou H (2020) A novel HDAC inhibitor chidamide combined with imatinib synergistically targets tyrosine kinase inhibitor resistant chronic myeloid leukemia cells. Biomedicine & pharmacotherapy = Biomedecine & pharmacotherapie 129:110390. https://doi.org/10.1016/j.biopha.2020.110390

Ho TT, He X, Mo YY, Beck WT (2016) Transient resistance to DNA damaging agents is associated with expression of microRNAs-135b and -196b in human leukemia cell lines. International journal of biochemistry and molecular biology 7(2):27–47

Hotchkiss RD (1948) The quantitative separation of purines, pyrimidines, and nucleosides by paper chromatography. The Journal of biological chemistry 175(1):315–332

Hou J, Wu J, Dombkowski A, Zhang K, Holowatyj A, Boerner JL, Yang ZQ (2012) Genomic amplification and a role in drug-resistance for the KDM5A histone demethylase in breast cancer. American journal of translational research 4(3):247–256

Hou MF, Luo CW, Chang TM, Hung WC, Chen TY, Tsai YL, Chai CY, Pan MR (2017) The NuRD complex- mediated p21 suppression facilitates chemoresistance in BRCA-proficient breast cancer. Experimental cell research 359(2):458–465. https://doi.org/10.1016/j.yexcr.2017.08.029

Hu S, Yu L, Li Z, Shen Y, Wang J, Cai J, Xiao L, Wang Z (2010) Overexpression of EZH2 contributes to acquired cisplatin resistance in ovarian cancer cells in vitro and in vivo. Cancer biology & therapy 10(8):788–795. https://doi.org/10.4161/cbt.10.8.12913

Huang JF, Wen CJ, Zhao GZ, Dai Y, Li Y, Wu LX, Zhou HH (2018) Overexpression of ABCB4 contributes to acquired doxorubicin resistance in breast cancer cells in vitro. Cancer chemotherapy and pharmacology 82(2):199–210. https://doi.org/10.1007/s00280-018-3603-y

Hung SW, Mody H, Marrache S, Bhutia YD, Davis F, Cho JH, Zastre J, Dhar S, Chu CK, Govindarajan R (2013) Pharmacological reversal of histone methylation presensitizes pancreatic cancer cells to nucleoside drugs: in vitro optimization and novel nanoparticle delivery studies. PloS one 8(8):e71196. https://doi.org/10.1371/journal.pone.0071196

Imai Y, Maru Y, Tanaka J (2016) Action mechanisms of histone deacetylase inhibitors in the treatment of hematological malignancies. Cancer science 107(11):1543–1549. https://doi.org/10.1111/cas.13062

Jenuwein T, Allis CD (2001) Translating the histone code. J Science 293(5532):1074–1080. https://doi.org/10.1126/science.1063127

Jin B, Li Y, Robertson KD (2011) DNA methylation: superior or subordinate in the epigenetic hierarchy? Genes Cancer 2(6):607–617. https://doi.org/10.1177/1947601910393957

Jubierre L, Soriano A, Planells-Ferrer L, París-Coderch L, Tenbaum SP, Romero OA, Moubarak RS, Almazán- Moga A, Molist C, Roma J, Navarro S, Noguera R, Sánchez-Céspedes M, Comella JX, Palmer HG, Sánchez de Toledo J, Gallego S, Segura MF (2016) BRG1/SMARCA4 is essential for neuroblastoma cell viability through modulation of cell death and survival pathways. Oncogene 35(39):5179–5190. https://doi.org/10.1038/onc.2016.50

Kadoch C, Hargreaves DC, Hodges C, Elias L, Ho L, Ranish J, Crabtree GR (2013) Proteomic and bioinformatic analysis of mammalian SWI/SNF complexes identifies extensive roles in human malignancy. Nature genetics 45(6):592–601. https://doi.org/10.1038/ng.2628

Kantharidis P, El-Osta A, deSilva M, Wall DM, Hu XF, Slater A, Nadalin G, Parkin JD, Zalcberg JR (1997) Altered methylation of the human MDR1 promoter is associated with acquired multidrug resistance. Clinical cancer research : an official journal of the American Association for Cancer Research 3(11):2025–2032

Kari V, Raul SK, Henck JM, Kitz J, Kramer F, Kosinsky RL, Übelmesser N, Mansour WY, Eggert J, Spitzner M, Najafova Z, Bastians H, Grade M, Gaedcke J, Wegwitz F, Johnsen SA (2019) The histone methyltransferase DOT1L is required for proper DNA damage response, DNA repair, and modulates chemotherapy responsiveness. Clinical epigenetics 11(1):4. https://doi.org/10.1186/s13148-018-0601-1

Kaur E, Gupta S, Dutt S (2014) Clinical implications of MTA proteins in human cancer. Cancer and Metastasis Reviews 33(4):1017–1024. https://doi.org/10.1007/s10555-014-9527-z

Kaur E, Rajendra J, Jadhav S, Shridhar E, Goda JS, Moiyadi A, Dutt S (2015) Radiation-induced homotypic cell fusions of innately resistant glioblastoma cells mediate their sustained survival and recurrence. Carcinogenesis 36(6):685–695. https://doi.org/10.1093/carcin/bgv050

Kaur E, Nair J, Ghorai A, Mishra SV, Achareker A, Ketkar M, Sarkar D, Salunkhe S, Rajendra J, Gardi N, Desai S, Iyer P, Thorat R, Dutt A, Moiyadi A, Dutt S (2020) Inhibition of SETMAR-H3K36me2-NHEJ repair axis in residual disease cells prevents glioblastoma recurrence. Neuro-oncology 22(12):1785–1796. https://doi.org/10.1093/neuonc/noaa128

Kelly TK, De Carvalho DD, Jones PA (2010) Epigenetic modifications as therapeutic targets. Nature biotechnology 28(10):1069–1078. https://doi.org/10.1038/nbt.1678

Kim Y, Kim H, Park D, Jeoung D (2015) miR-335 Targets SIAH2 and Confers Sensitivity to Anti-Cancer Drugs by Increasing the Expression of HDAC3. Molecules and cells 38(6):562–572. https://doi.org/10.14348/molcells.2015.0051

Kothapalli N, Sarath G, Zempleni J (2005) Biotinylation of K12 in histone H4 decreases in response to DNA double-strand breaks in human JAr choriocarcinoma cells. The Journal of nutrition 135(10):2337–2342. https://doi.org/10.1093/jn/135.10.2337

Kouzarides T (2007) Chromatin modifications and their function. Cell 128(4):693–705. https://doi.org/10.1016/j.cell.2007.02.005

Kovalchuk O, Filkowski J, Meservy J, Ilnytskyy Y, Tryndyak VP, Chekhun VF, Pogribny IP (2008) Involvement of microRNA-451 in resistance of the MCF-7 breast cancer cells to chemotherapeutic drug doxorubicin. Molecular cancer therapeutics 7(7):2152–2159. https://doi.org/10.1158/1535-7163.Mct-08-0021

Kung JTY, Colognori D, Lee JT (2013) Long noncoding RNAs: past, present, and future. Genetics 193(3):651–669. https://doi.org/10.1534/genetics.112.146704

Lafon-Hughes L, Di Tomaso MV, Méndez-Acuña L, Martínez-López W (2008) Chromatin-remodelling mechanisms in cancer. Mutation research 658(3):191–214. https://doi.org/10.1016/j.mrrev.2008.01.008

Laurent BC, Treich I, Carlson M (1993) Role of yeast SNF and SWI proteins in transcriptional activation. Cold Spring Harbor symposia on quantitative biology 58:257–263. https://doi.org/10.1101/sqb.1993.058.01.030

Lawrence M, Daujat S, Schneider R (2016) Lateral Thinking: How Histone Modifications Regulate Gene Expression. Trends in genetics : TIG 32(1):42–56. https://doi.org/10.1016/j.tig.2015.10.007

Lee RC, Feinbaum RL, Ambros V (1993) The C. elegans heterochronic gene lin-4 encodes small RNAs with antisense complementarity to lin-14. Cell 75(5):843–854. https://doi.org/10.1016/0092-8674(93)90529-Y

Lee H, Kim C, Ku JL, Kim W, Yoon SK, Kuh HJ, Lee JH, Nam SW, Lee EK (2014a) A long non-coding RNA snaR contributes to 5-fluorouracil resistance in human colon cancer cells. Molecules and cells 37(7):540–546. https://doi.org/10.14348/molcells.2014.0151

Lee J, Bartholomeusz C, Mansour O, Humphries J, Hortobagyi GN, Ordentlich P, Ueno NT (2014b) A class I histone deacetylase inhibitor, entinostat, enhances lapatinib efficacy in HER2-overexpressing breast cancer cells through FOXO3-mediated Bim1 expression. Breast cancer research and treatment 146(2):259–272. https://doi.org/10.1007/s10549-014-3014-7

Lee HS, Park SB, Kim SA, Kwon SK, Cha H, Lee DY, Ro S, Cho JM, Song SY (2017) A novel HDAC inhibitor, CG200745, inhibits pancreatic cancer cell growth and overcomes gemcitabine resistance. Scientific reports 7:41615. https://doi.org/10.1038/srep41615

Li T, Liu Z, Jiang K, Ruan Q (2013) Angiopoietin2 enhances doxorubin resistance in HepG2 cells by upregulating survivin and Ref-1 via MSK1 activation. Cancer letters 337(2):276–284. https://doi.org/10.1016/j.canlet.2013.04.028

Li Y, Yang Y, Lu Y, Herman JG, Brock MV, Zhao P, Guo M (2015) Predictive value of CHFR and MLH1 methylation in human gastric cancer. Gastric Cancer: official journal of the International Gastric Cancer Association and the Japanese Gastric Cancer Association 18(2):280–287. https://doi.org/10.1007/s10120-014-0370-2

Li X, Wu Y, Liu A, Tang X (2016) MiR-27b is epigenetically downregulated in tamoxifen resistant breast cancer cells due to promoter methylation and regulates tamoxifen sensitivity by targeting HMGB3. Biochemical and biophysical research communications 477(4):768–773. https://doi.org/10.1016/j.bbrc.2016.06.133

Li Y, Wang B, Lai H, Li S, You Q, Fang Y, Li Q, Liu Y (2017) Long non-coding RNA CRALA is associated with poor response to chemotherapy in primary breast cancer. Thoracic cancer 8(6):582–591. https://doi.org/10.1111/1759-7714.12487

Li Q, Song W, Wang J (2019) TUG1 confers Adriamycin resistance in acute myeloid leukemia by epigenetically suppressing miR-34a expression via EZH2. Biomedicine & pharmacotherapy = Biomedecine & pharmacotherapie 109:1793–1801. https://doi.org/10.1016/j.biopha.2018.11.003

Lin CT, Lai HC, Lee HY, Lin WH, Chang CC, Chu TY, Lin YW, Lee KD, Yu MH (2008) Valproic acid resensitizes cisplatin-resistant ovarian cancer cells. Cancer science 99(6):1218–1226. https://doi.org/10.1111/j.1349-7006.2008.00793.x

Liu MX, Siu MK, Liu SS, Yam JW, Ngan HY, Chan DW (2014) Epigenetic silencing of microRNA-199b-5p is associated with acquired chemoresistance via activation of JAG1-Notch1 signaling in ovarian cancer. Oncotarget 5(4):944–958. https://doi.org/10.18632/oncotarget.1458

Liu T, Guo Q, Guo H, Hou S, Li J, Wang H (2016) Quantitative analysis of histone H3 and H4 post-translational modifications in doxorubicin-resistant leukemia cells. Biomedical chromatography : BMC 30(4):638–644. https://doi.org/10.1002/bmc.3608

Liu D, Zhang XX, Li MC, Cao CH, Wan DY, Xi BX, Tan JH, Wang J, Yang ZY, Feng XX, Ye F, Chen G, Wu P, Xi L, Wang H, Zhou JF, Feng ZH, Ma D, Gao QL (2018) C/EBPβ enhances platinum resistance of ovarian cancer cells by reprogramming H3K79 methylation. Nature communications 9(1):1739. https://doi.org/10.1038/s41467-018-03590-5

Liu J, He D, Cheng L, Huang C, Zhang Y, Rao X, Kong Y, Li C, Zhang Z, Liu J, Jones K, Napier D, Lee EY, Wang C, Liu X (2020) p300/CBP inhibition enhances the efficacy of programmed death-ligand 1 blockade treatment in prostate cancer. Oncogene 39(19):3939–3951. https://doi.org/10.1038/s41388-020-1270-z

Lu Y, Zhao X, Liu Q, Li C, Graves-Deal R, Cao Z, Singh B, Franklin JL, Wang J, Hu H, Wei T, Yang M, Yeatman TJ, Lee E, Saito-Diaz K, Hinger S, Patton JG, Chung CH, Emmrich S, Klusmann JH, Fan D, Coffey RJ (2017) lncRNA MIR100HG-derived miR-100 and miR-125b

mediate cetuximab resistance via Wnt/β-catenin signaling. Nat Med 23(11):1331–1341. https://doi.org/10.1038/nm.4424
Lu M, Hu C, Wu F, Shu L, Pan Y, Liu X, Liu P, Ma F, Deng C, Huang M (2020) MiR-320a is associated with cisplatin resistance in lung adenocarcinoma and its clinical value in non-small cell lung cancer: A comprehensive analysis based on microarray data. Lung cancer (Amsterdam, Netherlands) 147:193–197. https://doi.org/10.1016/j.lungcan.2020.06.020
Luo Q, Wu X, Zhang Y, Shu T, Ding F, Chen H, Zhao P, Chang W, Zhu X, Liu Z (2018) ARID1A ablation leads to multiple drug resistance in ovarian cancer via transcriptional activation of MRP2. Cancer letters 427:9–17. https://doi.org/10.1016/j.canlet.2018.04.013
Lv L, Deng H, Li Y, Zhang C, Liu X, Liu Q, Zhang D, Wang L, Pu Y, Zhang H, He Y, Wang Y, Yu Y, Yu T, Zhu J (2014) The DNA methylation-regulated miR-193a-3p dictates the multi-chemoresistance of bladder cancer via repression of SRSF2/PLAU/HIC2 expression. Cell Death Dis 5(9):e1402. https://doi.org/10.1038/cddis.2014.367
Magnani L, Stoeck A, Zhang X, Lánczky A, Mirabella AC, Wang T-L, Gyorffy B, Lupien M (2013) Genome-wide reprogramming of the chromatin landscape underlies endocrine therapy resistance in breast cancer. Proceedings of the National Academy of Sciences of the United States of America 110(16):E1490–E1499. https://doi.org/10.1073/pnas.1219992110
Mannironi C, D'Incalci M (1988) Doxorubicin induces the acetylation of histone H1 in a human colon cancer cell line (LoVo/DX) selected for resistance to the drug, but not in the sensitive parental line (LoVo). Biochemical and biophysical research communications 155(3): 1221–1229. https://doi.org/10.1016/s0006-291x(88)81270-1
Meisenberg C, Ashour ME, El-Shafie L, Liao C, Hodgson A, Pilborough A, Khurram SA, Downs JA, Ward SE, El- Khamisy SF (2017) Epigenetic changes in histone acetylation underpin resistance to the topoisomerase I inhibitor irinotecan. Nucleic Acids Res 45(3):1159–1176. https://doi.org/10.1093/nar/gkw1026
Meng F, Qian L, Lv L, Ding B, Zhou G, Cheng X, Niu S, Liang Y (2016) miR-193a-3p regulation of chemoradiation resistance in oesophageal cancer cells via the PSEN1 gene. Gene 579(2): 139–145. https://doi.org/10.1016/j.gene.2015.12.060
Miyamoto N, Izumi H, Noguchi T, Nakajima Y, Ohmiya Y, Shiota M, Kidani A, Tawara A, Kohno K (2008) Tip60 is regulated by circadian transcription factor clock and is involved in cisplatin resistance. The Journal of biological chemistry 283(26):18218–18226. https://doi.org/10.1074/jbc.M802332200
Monte M, Simonatto M, Peche LY, Bublik DR, Gobessi S, Pierotti MA, Rodolfo M, Schneider C (2006) MAGE-A tumor antigens target p53 transactivation function through histone deacetylase recruitment and confer resistance to chemotherapeutic agents. Proceedings of the National Academy of Sciences of the United States of America 103(30):11160–11165. https://doi.org/10.1073/pnas.0510834103
Nakatsuka S, Takakuwa T, Tomita Y, Hoshida Y, Nishiu M, Yamaguchi M, Nishii K, Yang WI, Aozasa K (2003) Hypermethylation of death-associated protein (DAP) kinase CpG island is frequent not only in B-cell but also in T- and natural killer (NK)/T-cell malignancies. Cancer science 94(1):87–91. https://doi.org/10.1111/j.1349-7006.2003.tb01357.x
Nakayama M, Wada M, Harada T, Nagayama J, Kusaba H, Ohshima K, Kozuru M, Komatsu H, Ueda R, Kuwano M (1998) Hypomethylation status of CpG sites at the promoter region and overexpression of the human MDR1 gene in acute myeloid leukemias. Blood 92(11): 4296–4307
Narita M, Shimura E, Nagasawa A, Aiuchi T, Suda Y, Hamada Y, Ikegami D, Iwasawa C, Arakawa K, Igarashi K, Kuzumaki N, Yoshioka Y, Ochiya T, Takeshima H, Ushijima T, Narita M (2017) Chronic treatment of non-small-cell lung cancer cells with gefitinib leads to an epigenetic loss of epithelial properties associated with reductions in microRNA-155 and -200c. PloS one 12(2):e0172115. https://doi.org/10.1371/journal.pone.0172115
Nichol JN, Galbraith MD, Kleinman CL, Espinosa JM, Miller WH Jr (2016) NPM and BRG1 Mediate Transcriptional Resistance to Retinoic Acid in Acute Promyelocytic Leukemia. Cell reports 14(12):2938–2949. https://doi.org/10.1016/j.celrep.2016.02.074

Nyce JW (1997) Drug-induced DNA hypermethylation: a potential mediator of acquired drug resistance during cancer chemotherapy. Mutation research 386(2):153–161. https://doi.org/10.1016/s1383-5742(96)00051-8

Oki M, Aihara H, Ito T (2007) Role of histone phosphorylation in chromatin dynamics and its implications in diseases. Sub-cellular biochemistry 41:319–336

Ono H, Basson MD, Ito H (2016) P300 inhibition enhances gemcitabine-induced apoptosis of pancreatic cancer. Oncotarget 7(32):51301–51310. https://doi.org/10.18632/oncotarget.10117

Orouji E, Utikal J (2018) Tackling malignant melanoma epigenetically: histone lysine methylation. Clinical epigenetics 10(1):145–145. https://doi.org/10.1186/s13148-018-0583-z

Papadakis AI, Sun C, Knijnenburg TA, Xue Y, Grernrum W, Hölzel M, Nijkamp W, Wessels LF, Beijersbergen RL, Bernards R, Huang S (2015) SMARCE1 suppresses EGFR expression and controls responses to MET and ALK inhibitors in lung cancer. Cell Res 25(4):445–458. https://doi.org/10.1038/cr.2015.16

Phillips DM (1963) The presence of acetyl groups of histones. Biochem J 87(2):258–263. https://doi.org/10.1042/bj0870258

Pottier N, Yang W, Assem M, Panetta JC, Pei D, Paugh SW, Cheng C, Den Boer ML, Relling MV, Pieters R, Evans WE, Cheok MH (2008) The SWI/SNF chromatin-remodeling complex and glucocorticoid resistance in acute lymphoblastic leukemia. Journal of the National Cancer Institute 100(24):1792–1803. https://doi.org/10.1093/jnci/djn416

Prenzel T, Begus-Nahrmann Y, Kramer F, Hennion M, Hsu C, Gorsler T, Hintermair C, Eick D, Kremmer E, Simons M, Beissbarth T, Johnsen SA (2011) Estrogen-dependent gene transcription in human breast cancer cells relies upon proteasome-dependent monoubiquitination of histone H2B. Cancer research 71(17):5739–5753. https://doi.org/10.1158/0008-5472.Can-11-1896

Qiu Z, Zhu W, Meng H, Tong L, Li X, Luo P, Yi L, Zhang X, Guo L, Wei T, Zhang J (2019) CDYL promotes the chemoresistance of small cell lung cancer by regulating H3K27 trimethylation at the CDKN1C promoter. Theranostics 9(16):4717–4729. https://doi.org/10.7150/thno.33680

Ramachandran K, Miller H, Gordian E, Rocha-Lima C, Singal R (2010) Methylation-mediated silencing of TMS1 in pancreatic cancer and its potential contribution to chemosensitivity. Anticancer research 30(10):3919–3925

Robertson KD (2005) DNA methylation and human disease. Nature reviews Genetics 6(8): 597–610. https://doi.org/10.1038/nrg1655

Salunkhe S, Mishra SV, Nair J, Ghosh S, Choudhary N, Kaur E, Shah S, Patkar K, Anand D, Khattry N, Hasan SK, Dutt S (2018) Inhibition of novel GCN5-ATM axis restricts the onset of acquired drug resistance in leukemia. International journal of cancer 142(10):2175–2185. https://doi.org/10.1002/ijc.31242

Schliesser MG, Claus R, Hielscher T, Grimm C, Weichenhan D, Blaes J, Wiestler B, Hau P, Schramm J, Sahm F, Weiß EK, Weiler M, Baer C, Schmidt-Graf F, Schackert G, Westphal M, Hertenstein A, Roth P, Galldiks N, Hartmann C, Pietsch T, Felsberg J, Reifenberger G, Sabel MC, Winkler F, von Deimling A, Meisner C, Vajkoczy P, Platten M, Weller M, Plass C, Wick W (2016) Prognostic relevance of miRNA-155 methylation in anaplastic glioma. Oncotarget 7(50):82028–82045. https://doi.org/10.18632/oncotarget.13452

Sermer D, Pasqualucci L, Wendel HG, Melnick A, Younes A (2019) Emerging epigenetic-modulating therapies in lymphoma. Nature reviews Clinical oncology 16(8):494–507. https://doi.org/10.1038/s41571-019-0190-8

Sharma SV, Lee DY, Li B, Quinlan MP, Takahashi F, Maheswaran S, McDermott U, Azizian N, Zou L, Fischbach MA, Wong KK, Brandstetter K, Wittner B, Ramaswamy S, Classon M, Settleman J (2010) A chromatin- mediated reversible drug-tolerant state in cancer cell subpopulations. Cell 141(1):69–80. https://doi.org/10.1016/j.cell.2010.02.027

Shen CJ, Cheng YM, Wang CL (2017) LncRNA PVT1 epigenetically silences miR-195 and modulates EMT and chemoresistance in cervical cancer cells. Journal of drug targeting 25(7): 637–644. https://doi.org/10.1080/1061186x.2017.1307379

Shogren-Knaak M, Ishii H, Sun JM, Pazin MJ, Davie JR, Peterson CL (2006) Histone H4-K16 acetylation controls chromatin structure and protein interactions. Science (New York, NY) 311(5762):844–847. https://doi.org/10.1126/science.1124000

Si ML, Zhu S, Wu H, Lu Z, Wu F, Mo YY (2007) miR-21-mediated tumor growth. Oncogene 26(19):2799–2803. https://doi.org/10.1038/sj.onc.1210083

Si X, Zang R, Zhang E, Liu Y, Shi X, Zhang E, Shao L, Li A, Yang N, Han X, Pan B, Zhang Z, Sun L, Sun Y (2016) LncRNA H19 confers chemoresistance in ERα-positive breast cancer through epigenetic silencing of the pro-apoptotic gene BIK. Oncotarget 7(49):81452–81462

Sociali G, Galeno L, Parenti MD, Grozio A, Bauer I, Passalacqua M, Boero S, Donadini A, Millo E, Bellotti M, Sturla L, Damonte P, Puddu A, Ferroni C, Varchi G, Franceschi C, Ballestrero A, Poggi A, Bruzzone S, Nencioni A, Del Rio A (2015) Quinazolinedione SIRT6 inhibitors sensitize cancer cells to chemotherapeutics. European journal of medicinal chemistry 102: 530–539. https://doi.org/10.1016/j.ejmech.2015.08.024

Song WF, Wang L, Huang WY, Cai X, Cui JJ, Wang LW (2013) MiR-21 upregulation induced by promoter zone histone acetylation is associated with chemoresistance to gemcitabine and enhanced malignancy of pancreatic cancer cells. Asian Pacific journal of cancer prevention: APJCP 14(12):7529–7536. https://doi.org/10.7314/apjcp.2013.14.12.7529

Song KH, Choi CH, Lee HJ, Oh SJ, Woo SR, Hong SO, Noh KH, Cho H, Chung EJ, Kim JH, Chung JY, Hewitt SM, Baek S, Lee KM, Yee C, Son M, Mao CP, Wu TC, Kim TW (2017) HDAC1 upregulation by NANOG promotes multidrug resistance and a stem-like phenotype in immune edited tumor cells. Cancer research 77(18):5039–5053. https://doi.org/10.1158/0008-5472.Can-17-0072

Stevely WS, Stocken LA (1966) Phosphorylation of rat-thymus histone. Biochem J 100(2):20C–21C. https://doi.org/10.1042/bj1000020c

Stief SM, Hanneforth AL, Weser S, Mattes R, Carlet M, Liu WH, Bartoschek MD, Domínguez Moreno H, Oettle M, Kempf J, Vick B, Ksienzyk B, Tizazu B, Rothenberg-Thurley M, Quentmeier H, Hiddemann W, Vosberg S, Greif PA, Metzeler KH, Schotta G, Bultmann S, Jeremias I, Leonhardt H, Spiekermann K (2020) Loss of KDM6A confers drug resistance in acute myeloid leukemia. Leukemia 34(1):50–62. https://doi.org/10.1038/s41375-019-0497-6

Strathdee G, MacKean MJ, Illand M, Brown R (1999) A role for methylation of the hMLH1 promoter in loss of hMLH1 expression and drug resistance in ovarian cancer. Oncogene 18(14): 2335–2341. https://doi.org/10.1038/sj.onc.1202540

Strobeck MW, Reisman DN, Gunawardena RW, Betz BL, Angus SP, Knudsen KE, Kowalik TF, Weissman BE, Knudsen ES (2002) Compensation of BRG-1 function by Brm: insight into the role of the core SWI- SNF subunits in retinoblastoma tumor suppressor signaling. The Journal of biological chemistry 277(7):4782–4789. https://doi.org/10.1074/jbc.M109532200

Su W-P, Ho Y-C, Wu C-K, Hsu S-H, Shiu J-L, Huang J-C, Chang S-B, Chiu W-T, Hung J-J, Liu T-L, Wu W-S, Wu P-Y, Su W-C, Chang J-Y, Liaw H (2017) Chronic treatment with cisplatin induces chemoresistance through the TIP60-mediated Fanconi anemia and homologous recombination repair pathways. Scientific reports 7(1):3879–3879. https://doi.org/10.1038/s41598-017-04223-5

Sugita H, Iida S, Inokuchi M, Kato K, Ishiguro M, Ishikawa T, Takagi Y, Enjoji M, Yamada H, Uetake H, Kojima K, Sugihara K (2011) Methylation of BNIP3 and DAPK indicates lower response to chemotherapy and poor prognosis in gastric cancer. Oncology reports 25(2): 513–518. https://doi.org/10.3892/or.2010.1085

Suh YA, Lee HY, Virmani A, Wong J, Mann KK, Miller WH Jr, Gazdar A, Kurie JM (2002) Loss of retinoic acid receptor beta gene expression is linked to aberrant histone H3 acetylation in lung cancer cell lines. Cancer research 62(14):3945–3949

Sun L, Fang Y, Wang X, Han Y, Du F, Li C, Hu H, Liu H, Liu Q, Wang J, Liang J, Chen P, Yang H, Nie Y, Wu K, Fan D, Coffey RJ, Lu Y, Zhao X, Wang X (2019) miR-302a inhibits metastasis and cetuximab resistance in colorectal cancer by targeting NFIB and CD44. Theranostics 9(26): 8409–8425. https://doi.org/10.7150/thno.36605

Suzuki MM, Bird A (2008) DNA methylation landscapes: provocative insights from epigenomics. Nature reviews Genetics 9(6):465–476. https://doi.org/10.1038/nrg2341

Tada Y, Wada M, Kuroiwa K, Kinugawa N, Harada T, Nagayama J, Nakagawa M, Naito S, Kuwano M (2000) *MDR1* gene overexpression and altered degree of methylation at the promoter region in bladder cancer during chemotherapeutic treatment. Clin Cancer Res 6(12): 4618–4627

Tang DE, Dai Y, He JX, Lin LW, Leng QX, Geng XY, Fu DX, Jiang HW, Xu SH (2020) Targeting the KDM4B-AR-c- Myc axis promotes sensitivity to androgen receptor-targeted therapy in advanced prostate cancer. The Journal of pathology 252(2):101–113. https://doi.org/10.1002/path.5495

Tian T, Lv X, Pan G, Lu Y, Chen W, He W, Lei X, Zhang H, Liu M, Sun S, Ou Z, Lin X, Cai L, He L, Tu Z, Wang X, Tannous BA, Ferrone S, Li J, Fan S (2019) Long noncoding RNA MPRL promotes mitochondrial fission and cisplatin chemosensitivity via disruption of pre-miRNA processing. Clinical cancer research: an official journal of the American Association for Cancer Research 25(12):3673–3688. https://doi.org/10.1158/1078-0432.Ccr-18-2739

Tomihara H, Carbone F, Perelli L, Huang JK, Soeung M, Rose JL, Robinson FS, Lissanu Deribe Y, Feng N, Takeda M, Inoue A, Deem AK, Maitra A, Msaouel P, Tannir NM, Draetta GF, Bristow CA, Carugo A, Genovese G (2020) Loss of ARID1A promotes epithelial-mesenchymal transition and sensitizes pancreatic tumors to proteotoxic stress. Cancer research. https://doi.org/10.1158/0008-5472.Can-19-3922

Tsanev R, Sendov B (1971) An epigenetic mechanism for carcinogenesis. Zeitschrift für Krebsforschung und Klinische Onkologie 76(4):299–319. https://doi.org/10.1007/BF00304033

Tweedie-Cullen RY, Reck JM, Mansuy IM (2009) Comprehensive mapping of post-translational modifications on synaptic, nuclear, and histone proteins in the adult mouse brain. Journal of Proteome Research 8(11):4966–4982. https://doi.org/10.1021/pr9003739

Vafadar A, Shabaninejad Z, Movahedpour A, Mohammadi S, Fathullahzadeh S, Mirzaei HR, Namdar A, Savardashtaki A, Mirzaei H (2019) Long non-coding RNAs as epigenetic regulators in cancer. Current pharmaceutical design 25(33):3563–3577. https://doi.org/10.2174/1381612825666190830161528

Wang Z, Zang C, Rosenfeld JA, Schones DE, Barski A, Cuddapah S, Cui K, Roh TY, Peng W, Zhang MQ, Zhao K (2008) Combinatorial patterns of histone acetylations and methylations in the human genome. Nature genetics 40(7):897–903. https://doi.org/10.1038/ng.154

Wang H, Fang L, Jiang J, Kuang Y, Wang B, Shang X, Han P, Li Y, Liu M, Zhang Z, Li P (2018) The cisplatin- induced lncRNA PANDAR dictates the chemoresistance of ovarian cancer via regulating SFRS2- mediated p53 phosphorylation. Cell Death Dis 9(11):1103. https://doi.org/10.1038/s41419-018-1148-y

Wang H, Fu C, Du J, Wang H, He R, Yin X, Li H, Li X, Wang H, Li K, Zheng L, Liu Z, Qiu Y (2020a) Enhanced histone H3 acetylation of the PD-L1 promoter via the COP1/c-Jun/HDAC3 axis is required for PD-L1 expression in drug-resistant cancer cells. Journal of experimental & clinical cancer research: CR 39(1):29. https://doi.org/10.1186/s13046-020-1536-x

Wang Q, Chen X, Jiang Y, Liu S, Liu H, Sun X, Zhang H, Liu Z, Tao Y, Li C, Hu Y, Liu D, Ye D, Liu Y, Wang M, Zhang X (2020b) Elevating H3K27me3 level sensitizes colorectal cancer to oxaliplatin. Journal of molecular cell biology 12(2):125–137. https://doi.org/10.1093/jmcb/mjz032

Wapinski O, Chang HY (2011) Long noncoding RNAs and human disease. Trends in cell biology 21(6):354–361. https://doi.org/10.1016/j.tcb.2011.04.001

Watanabe Y, Ueda H, Etoh T, Koike E, Fujinami N, Mitsuhashi A, Hoshiai H (2007) A change in promoter methylation of hMLH1 is a cause of acquired resistance to platinum-based chemotherapy in epithelial ovarian cancer. Anticancer research 27(3b):1449–1452

Wei JW, Huang K, Yang C, Kang CS (2017) Non-coding RNAs as regulators in epigenetics (Review). Oncology reports 37(1):3–9. https://doi.org/10.3892/or.2016.5236

Whang YM, Choi EJ, Seo JH, Kim JS, Yoo YD, Kim YH (2005) Hyperacetylation enhances the growth-inhibitory effect of all-trans retinoic acid by the restoration of retinoic acid receptor beta

expression in head and neck squamous carcinoma (HNSCC) cells. Cancer chemotherapy and pharmacology 56(5):543–555. https://doi.org/10.1007/s00280-004-0970-3
Wilting RH, Dannenberg JH (2012) Epigenetic mechanisms in tumorigenesis, tumor cell heterogeneity and drug resistance. Drug resistance updates: reviews and commentaries in antimicrobial and anticancer chemotherapy 15(1-2):21–38. https://doi.org/10.1016/j.drup.2012.01.008
Winston F, Carlson M (1992) Yeast SNF/SWI transcriptional activators and the SPT/SIN chromatin connection. Trends in genetics : TIG 8(11):387–391. https://doi.org/10.1016/0168-9525(92)90300-s
Wu R, Nie Q, Tapper EE, Jerde CR, Dunlap GS, Shrestha S, Elraiyah TA, Offer SM, Diasio RB (2016) Histone H3K27 Trimethylation Modulates 5-Fluorouracil Resistance by Inhibiting PU.1 Binding to the DPYD Promoter. Cancer research 76(21):6362–6373. https://doi.org/10.1158/0008-5472.Can-16-1306
Xu Q, Liu X, Zhu S, Hu X, Niu H, Zhang X, Zhu D, Nesa EU, Tian K, Yuan H (2018a) Hyper-acetylation contributes to the sensitivity of chemo-resistant prostate cancer cells to histone deacetylase inhibitor Trichostatin A. Journal of cellular and molecular medicine 22(3): 1909–1922. https://doi.org/10.1111/jcmm.13475
Xu W, Zhou B, Zhao X, Zhu L, Xu J, Jiang Z, Chen D, Wei Q, Han M, Feng L, Wang S, Wang X, Zhou J, Jin H (2018b) KDM5B demethylates H3K4 to recruit XRCC1 and promote chemoresistance. International journal of biological sciences 14(9):1122–1132. https://doi.org/10.7150/ijbs.25881
Xu X, Zheng Z, Jia L, Suo S, Liu B, Shao T, Tu Q, Hua Y, Xu H (2018c) Overexpression of SMARCA2 or CAMK2D is associated with cisplatin resistance in human epithelial ovarian cancer. Oncology letters 16(3):3796–3804. https://doi.org/10.3892/ol.2018.9109
Xu G, Chhangawala S, Cocco E, Razavi P, Cai Y, Otto JE, Ferrando L, Selenica P, Ladewig E, Chan C, Da Cruz PA, Witkin M, Cheng Y, Park J, Serna-Tamayo C, Zhao H, Wu F, Sallaku M, Qu X, Zhao A, Collings CK, D'Avino AR, Jhaveri K, Koche R, Levine RL, Reis-Filho JS, Kadoch C, Scaltriti M, Leslie CS, Baselga J, Toska E (2020) ARID1A determines luminal identity and therapeutic response in estrogen-receptor- positive breast cancer. Nature genetics 52(2):198–207. https://doi.org/10.1038/s41588-019-0554-0
Xue K, Gu JJ, Zhang Q, Mavis C, Hernandez-Ilizaliturri FJ, Czuczman MS, Guo Y (2016) Vorinostat, a histone deacetylase (HDAC) inhibitor, promotes cell cycle arrest and re-sensitizes rituximab- and chemo- resistant lymphoma cells to chemotherapy agents. Journal of cancer research and clinical oncology 142(2):379–387. https://doi.org/10.1007/s00432-015-2026-y
Yamaguchi H, Hung MC (2014) Regulation and Role of EZH2 in Cancer. Cancer research and treatment : official journal of Korean Cancer Association 46(3):209–222. https://doi.org/10.4143/crt.2014.46.3.209
Yan Q, Dutt S, Xu R, Graves K, Juszczynski P, Manis JP, Shipp MA (2009) BBAP monoubiquitylates histone H4 at lysine 91 and selectively modulates the DNA damage response. Molecular cell 36(1):110–120. https://doi.org/10.1016/j.molcel.2009.08.019
Yan M, Yang X, Shen R, Wu C, Wang H, Ye Q, Yang P, Zhang L, Chen M, Wan B, Zhang Q, Xia S, Lu X, Shao G, Zhou X, Yu J, Shao Q (2018) miR-146b promotes cell proliferation and increases chemosensitivity, but attenuates cell migration and invasion via FBXL10 in ovarian cancer. Cell Death Dis 9(11):1123. https://doi.org/10.1038/s41419-018-1093-9
Yang C, Cai J, Wang Q, Tang H, Cao J, Wu L, Wang Z (2012) Epigenetic silencing of miR-130b in ovarian cancer promotes the development of multidrug resistance by targeting colony-stimulating factor 1. Gynecologic oncology 124(2):325–334. https://doi.org/10.1016/j.ygyno.2011.10.013
Yang S-Z, Xu F, Zhou T, Zhao X, McDonald JM, Chen Y (2017) The long non-coding RNA HOTAIR enhances pancreatic cancer resistance to TNF-related apoptosis-inducing ligand. The Journal of biological chemistry 292(25):10390–10397. https://doi.org/10.1074/jbc.M117.786830

Yang C, Zhang J, Ma Y, Wu C, Cui W, Wang L (2020a) Histone methyltransferase and drug resistance in cancers. Journal of Experimental & Clinical Cancer Research 39(1):173. https://doi.org/10.1186/s13046-020-01682-z

Yang X, Liu M, Li M, Zhang S, Hiju H, Sun J, Mao Z, Zheng M, Feng B (2020b) Epigenetic modulations of noncoding RNA: a novel dimension of Cancer biology. Molecular cancer 19(1): 64. https://doi.org/10.1186/s12943-020-01159-9

Ye XM, Zhu HY, Bai WD, Wang T, Wang L, Chen Y, Yang AG, Jia LT (2014) Epigenetic silencing of miR-375 induces trastuzumab resistance in HER2-positive breast cancer by targeting IGF1R. BMC cancer 14:134. https://doi.org/10.1186/1471-2407-14-134

Ye C, Han K, Lei J, Zeng K, Zeng S, Ju H, Yu L (2018) Inhibition of histone deacetylase 7 reverses concentrative nucleoside transporter 2 repression in colorectal cancer by up-regulating histone acetylation state. British journal of pharmacology 175(22):4209–4217. https://doi.org/10.1111/bph.14467

Ye Y, Yang S, Han Y, Sun J, Xv L, Wu L, Ming L (2019) HOXD-AS1 confers cisplatin resistance in gastric cancer through epigenetically silencing PDCD4 via recruiting EZH2. Open biology 9(9):190068. https://doi.org/10.1098/rsob.190068

Yeon M, Kim Y, Jung HS, Jeoung D (2020) Histone Deacetylase Inhibitors to Overcome Resistance to Targeted and Immuno Therapy in Metastatic Melanoma. Frontiers in cell and developmental biology 8:486. https://doi.org/10.3389/fcell.2020.00486

Yoon YS, Kim JW, Kang KW, Kim YS, Choi KH, Joe CO (1996) Poly(ADP-ribosyl)ation of histone H1 correlates with internucleosomal DNA fragmentation during apoptosis. The Journal of biological chemistry 271(15):9129–9134. https://doi.org/10.1074/jbc.271.15.9129

You D, Richardson JR, Aleksunes LM (2020) Epigenetic Regulation of Multidrug Resistance Protein 1 and Breast Cancer Resistance Protein Transporters by Histone Deacetylase Inhibition. Drug Metab Dispos: the biological fate of chemicals 48(6):459–480. https://doi.org/10.1124/dmd.119.089953

Yu L, Su YS, Zhao J, Wang H, Li W (2013a) Repression of NR4A1 by a chromatin modifier promotes docetaxel resistance in PC-3 human prostate cancer cells. FEBS letters 587(16): 2542–2551. https://doi.org/10.1016/j.febslet.2013.06.029

Yu SL, Lee DC, Son JW, Park CG, Lee HY, Kang J (2013b) Histone deacetylase 4 mediates SMAD family member 4 deacetylation and induces 5-fluorouracil resistance in breast cancer cells. Oncology reports 30(3):1293–1300. https://doi.org/10.3892/or.2013.2578

Yun MR, Lim SM, Kim SK, Choi HM, Pyo KH, Kim SK, Lee JM, Lee YW, Choi JW, Kim HR, Hong MH, Haam K, Huh N, Kim JH, Kim YS, Shim HS, Soo RA, Shih JY, Yang JC, Kim M, Cho BC (2018) Enhancer remodeling and MicroRNA alterations are associated with acquired resistance to ALK inhibitors. Cancer research 78(12):3350–3362. https://doi.org/10.1158/0008-5472.Can-17-3146

Zaratiegui M, Irvine DV, Martienssen RA (2007) Noncoding RNAs and gene silencing. Cell 128(4):763–776. https://doi.org/10.1016/j.cell.2007.02.016

Zeller C, Brown R (2010) Therapeutic modulation of epigenetic drivers of drug resistance in ovarian cancer. Therapeutic advances in medical oncology 2(5):319–329. https://doi.org/10.1177/1758834010375759

Zeller C, Dai W, Steele NL, Siddiq A, Walley AJ, Wilhelm-Benartzi CS, Rizzo S, van der Zee A, Plumb JA, Brown R (2012) Candidate DNA methylation drivers of acquired cisplatin resistance in ovarian cancer identified by methylome and expression profiling. Oncogene 31(42): 4567–4576. https://doi.org/10.1038/onc.2011.611

Zhang X, Yashiro M, Ren J, Hirakawa K (2006) Histone deacetylase inhibitor, trichostatin A, increases the chemosensitivity of anticancer drugs in gastric cancer cell lines. Oncology reports 16(3):563–568

Zhang Z, Wang F, Du C, Guo H, Ma L, Liu X, Kornmann M, Tian X, Yang Y (2017) BRM/SMARCA2 promotes the proliferation and chemoresistance of pancreatic cancer cells by targeting JAK2/STAT3 signaling. Cancer letters 402:213–224. https://doi.org/10.1016/j.canlet.2017.05.006

Zhang S, Ma H, Zhang D, Xie S, Wang W, Li Q, Lin Z, Wang Y (2018) LncRNA KCNQ1OT1 regulates proliferation and cisplatin resistance in tongue cancer via miR-211-5p mediated Ezrin/Fak/Src signaling. Cell Death Dis 9(7):742. https://doi.org/10.1038/s41419-018-0793-5

Zhang Y, Yuan Y, Li Y, Zhang P, Chen P, Sun S (2019) An inverse interaction between HOXA11 and HOXA11-AS is associated with cisplatin resistance in lung adenocarcinoma. Epigenetics 14(10):949–960. https://doi.org/10.1080/15592294.2019.1625673

Zhao G, Li Y, Wang T (2017) Potentiation of docetaxel sensitivity by miR-638 via regulation of STARD10 pathway in human breast cancer cells. Biochemical and biophysical research communications 487(2):255–261. https://doi.org/10.1016/j.bbrc.2017.04.045

Zhou Y, Gerrard DL, Wang J, Li T, Yang Y, Fritz AJ, Rajendran M, Fu X, Stein G, Schiff R, Lin S, Frietze S, Jin VX (2019) Temporal dynamic reorganization of 3D chromatin architecture in hormone-induced breast cancer and endocrine resistance. Nature communications 10(1):1522. https://doi.org/10.1038/s41467-019-09320-9

Zhu HY, Bai WD, Ye XM, Yang AG, Jia LT (2018) Long non-coding RNA UCA1 desensitizes breast cancer cells to trastuzumab by impeding miR-18a repression of Yes-associated protein 1. Biochemical and biophysical research communications 496(4):1308–1313. https://doi.org/10.1016/j.bbrc.2018.02.006

Chapter 15
Structural Basis of Targeted Imaging and Therapy in Cancer Explorations with the Epigenetic Drugs

Parimal Pandit and Varsha Brahmkhatri

This book chapter is dedicated to Prof. Hanudatta S. Atreya, NMR Research Centre, IISc Bangalore, India. He passed away on July 30, 2020.

Abstract Origin of cancer is strongly related to the unusual epigenetic regulation of gene function as indicated by recent reports. The covalent modifications to DNA or histones without affecting genomes that finally lead to phenotypical changes in cells or organisms are referred as "Epigenetics." The possibility to reprogram the epigenetics in the cancer epigenome is the most important target for cancer treatment and drug resistance. The development of epigenetic drugs holds a great potential for the current cancer therapeutic approaches. Nevertheless, targeting cancer epigenetic pathways is still exciting due to the lack of selective and effective small molecule compounds or drug molecules. Therefore, the current book chapter highlights epigenetic pathways for cancer and potential small molecule inhibitors and epidrugs targeting DNA methyltransferase, histone modification, and more new therapies with nanomaterials and imaging to improve the effectiveness of cancer treatment. The structural aspects on discovery of novel small molecules or drugs targeting epigenetic pathways in cancer exploration as promising strategies will be also discussed.

Keywords Epigenetic drugs · Cancer · DNA methylation · Targeted therapy · Imaging

---

P. Pandit · V. Brahmkhatri (✉)
Centre for Nano and Material Sciences, Jain University, Jain Global Campus, Bengaluru, Karnataka, India
e-mail: b.varsha@jainuniversity.ac.in

T. K. Kundu, C. Das (eds.), *Metabolism and Epigenetic Regulation: Implications in Cancer*, Subcellular Biochemistry 100,
https://doi.org/10.1007/978-3-031-07634-3_15

Introduction

Recent advances in significant recognition of the involvement of epigenetic factors and the identification of somatic alterations highlight epigenetic deregulation as a reason for many cancers. "Epigenetics" is defined as the genetic information that is inherited without any change in the original nucleotide sequence of DNA (Miranda Furtado et al. 2019; Mondal et al. 2020; Cheng et al. 2019; Liu et al. 2019). The genetic changes happening during the multiple cellular biological processes with the same genetic information require precise epigenetic modifications, which commonly including DNA methylation, histone, or chromatin post-translational modifications, as well as noncoding RNAs regulations. Catastrophe of heritability of epigenetic markers may result in unsuitable instigation or inhibition of gene expressions leading to pathological changes, together with cancers (Egger et al. 2004; Cao et al. 2020). Epigenetics allows us to explore the potential mechanism underlying cancer types and provides possible treatment options. The mechanisms of epigenetic pathways that control gene expression through regulation of transcription are shown in Fig. 15.1. The field of epigenetic modifications in cancer has tremendously advanced into an extent that holds great promise for unraveling important mechanisms involving cancer and the discovery of targeted therapies. Furthermore, the patterns associated with epigenetic pathways in some cancers may lead to the discovery of new

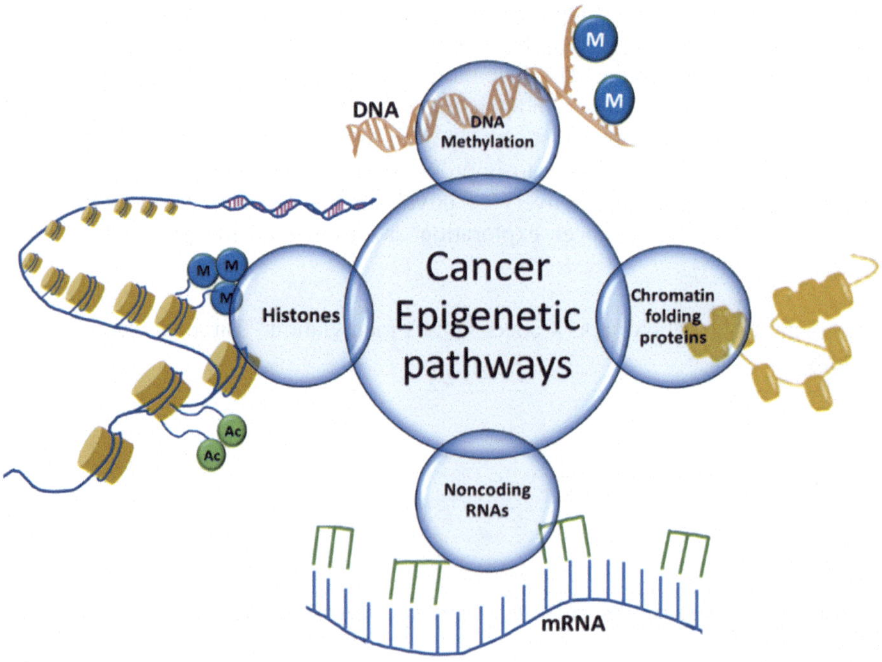

Fig. 15.1 Mechanisms of epigenetic pathways that control gene expression through regulation of transcription

biomarkers and therapeutic targets that may lead to the personalized medicine treatment options (Pan et al. 2018; Tiwari 2020; Morel et al. 2020; Fattahi et al. 2020).

In this book chapter, we have discussed some important targets of epigenetic modifications and the corresponding small molecule drugs and compounds in human cancers. The use of combination therapies such as chemotherapy and immunotherapy may improve the therapeutic strategies. These discoveries will enlighten some hints on discovery of more small molecule drugs targeting cancer epigenetic pathways as upcoming therapeutic strategies.

Cancer Exploration with Epigenetic Drugs

Cancer cells change the structure and stability of the genome due to epimutations. These mutations have been proposed as driver mutation in tumor instigation, and along with genetic lesions, they spread carcinogenesis (Cao et al. 2020). The main mechanisms and elements involved in epigenetic alterations in cancer are DNA methylation, histone modification, noncoding RNA, and chromatin accessibility as shown in Fig. 15.2.

Epigenetic drugs (epidrugs) are small molecules or chemical compounds that modify DNA and chromatin structure, interrupting transcriptional and post-transcriptional modifications, mainly by regulating the enzymes necessary for their maintenance, and re-activating epigenetically silenced tumor suppressor and DNA repair genes (Salarinia et al. 2016; Lu et al. 2020). The design and structural aspects of epidrugs are an emerging field in drug discovery that emphasize on the cancer epigenome to develop pharmacological compounds which could restore a regular

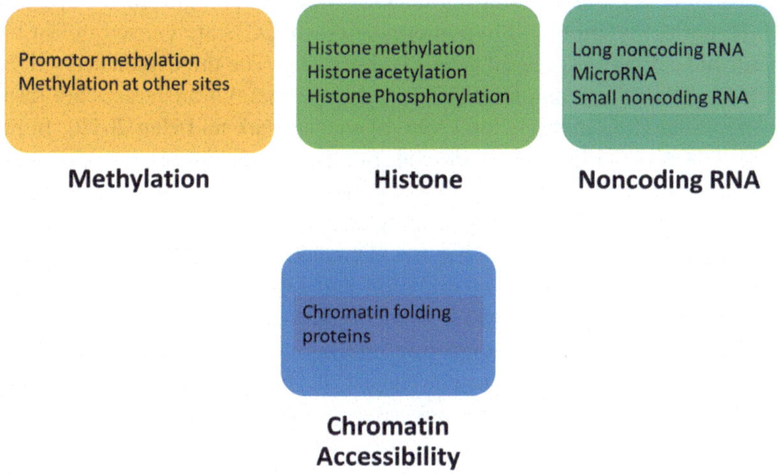

Fig. 15.2 Mechanism in Epigenetic alterations

epigenetic setup (Montalvo-Casimiro et al. 2020). Epidrugs target on the enzymes essential for the preservation and establishment of epigenetic modifications, with the main strategy being the inhibition of DNA methylation and histone modification (Gnyszka et al. 2013). Various epigenetic drugs have been approved for cancer therapy by Food and Drug Administration (FDA) (Lu et al. 2020) as shown in Table 15.1. The new therapeutic advancement in epidrugs is being evaluated for its cytotoxicity and its mechanism of action in clinical and preclinical trials for advancement of new therapies.

Classification of Epigenetic Cancer Drugs

Epigenetic drugs are classified based on their anticancer mechanism and phase of therapy which have been approved by the FDA. Thus, the main categories of epidrugs such as DNA methyltransferase inhibitors, histone deacetylase inhibitors (HDACIs), and noncoding RNA-based therapeutics are discussed below and the mechanism of action of various epidrugs as inhibitors is depicted in Fig. 15.3.

DNA Methyltransferase Inhibitors

DNA methylation was the first-ever epigenetic modification discovered in humans in early 1980(Felsenfeld 2014). Basically, DNA methylation takes place in cytosines of CpG (cytosine-phosphate-guanine) dinucleotide sequences to generate 5-methylcytosine (5mC). 5-methylcytosine (5mC) catalyzed by DNA methyltransferases (DNMTs), where S-adenyl methionine (SAM) acts as the methyl donor. Promoter regions contain higher GC which is recognized as CpG islands (CGIs). Hypermethylation and hypomethylation of CGIs are carried out in heterochromatin and actively expressed genes, respectively. The three main active DNMTs are DNMT1, DNMT3A, and DNMT3B, and these methyltransferases are responsible for the initial CpG methylation embryogenesis (Park and Han 2019). In normal cells, the DNMTs caried out a crucial function maintaining the chromosomal stability, regulation of embryonic development, and silencing of genes; on the other hand, in cancerous cells, hypermethylation of CpG causes the inactivation of several important tumor suppressor genes (Dhanak et al. 2014). DNMT1 activity is associated with the inheritance methylation and DNMT3A, DNMT3B are related with denovo methylation. DNMT1 is the major enzyme involved in the DNA methylation and the most abundant one and maintains methylation during DNA replication. DNMT1 is a large protein consist of 1616 amino acids which predominantly catalyzes DNA methylation inheritance activity. DNMT3A and DNMT3B enzymes mainly associated with the de novo DNA methylation which is essential during embryogenic development or gametogenesis (Hervouet et al. 2018). DNMT inhibitors are broadly categorized into nucleoside-derived and non-nucleoside derived inhibitors. The nucleoside-derived inhibitors contain two FDA-approved

Table 15.1 FDA-approved epigenetic cancer drugs/compounds and their mechanism of action

FDA-approved drugs/ compounds	Structure	Cancer therapy	Target	References
Azacytidine (Vidaza)	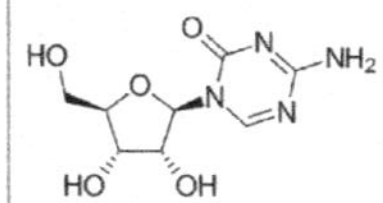	Myelodysplastic syndrome, AML	DNA methylation	Silverman et al. (2002)
Decitabine (Dacogen®)	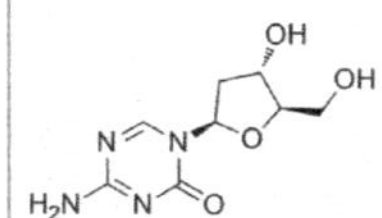	Myelodysplastic syndrome, AML	DNA methylation	Kantarjian et al. (2006)
Belinostat (Beleodaq/ PXD101)	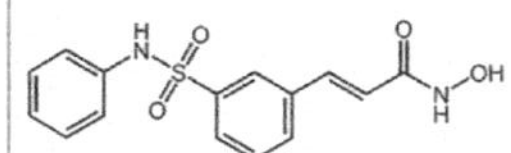	PTCL; HCC, Burkitt lymphoma, DLBCL, thymic carcinoma, MDS	Histone deacetylases	Mcdermott and Jimeno (2014)
Panobinostat (LBH-589)	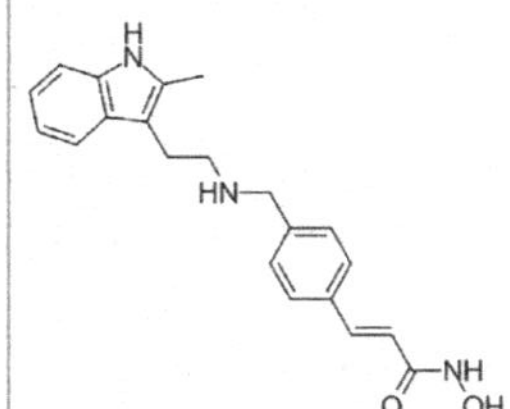	MM; thyroid carcinoma, RCC, breast cancer, AML	Histone deacetylases	Richardson et al. (2015)

(continued)

Table 15.1 (continued)

FDA-approved drugs/ compounds	Structure	Cancer therapy	Target	References
Romidepsin (Depsipeptide/ FK228)	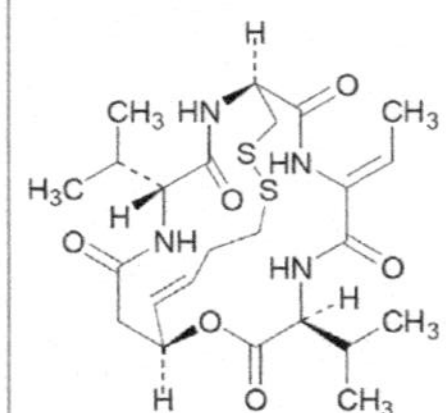	Cutaneous T-cell lymphoma	Histone deacetylases	Frye et al. (2012)
Patisiran	21-bp double-stranded siRNA; all PO, and 20-OMe; lipid nanoparticle M	Rare polyneuropathy	ncRNA	Wood (2018)

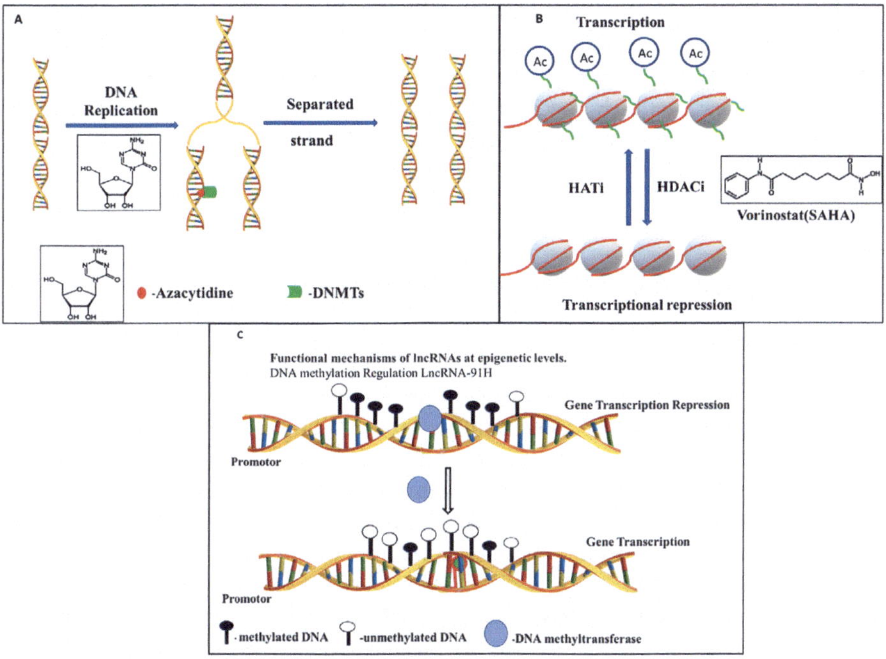

Fig. 15.3 Mechanism of action of various epidrugs as DNA methyltransferase inhibitors, histone deacetylase inhibitors (HDACIs), and noncoding RNA

drugs which are azacitidine (Vidaza) and decitabine (Dacogen) (Dhanak et al. 2014). The 5-azacytidine and 5-aza-2′-deoxycytidine are the two forms of one drug; the only difference between these two drugs is that 5-azacytidine is preferentially incorporated into RNA, whereas 5-aza-2′-deoxycytidine is incorporated into DNA. Azacitidine and decitabine (5- aza-2-deoxycytidine) are FDA-approved drugs for the treatment of myelodysplastic syndromes (MDS), chronic myelomonocytic leukemia (CMML), and acute myeloid leukemia (AML) (Derissen et al. 2013). Azacitidine (Vidaza; Pharmion) is an DNA methylation inhibitor which was approved by the US FDA in 2004 to treat the myelodysplastic syndromes (MDS). Azacitidine is a nucleoside analogue of cytidine which precisely inhibits DNA methylation by trapping DNA methyltransferases. Azacitidine acts as a DNA methyltransferase (DNMTs) and catalyzes the methylation of the 5-positions of the cytosine ring, and then after conversion to 5-aza-2′-deoxycytidine (decitabine), azacitidine (Z) is incorporated in place of cytidine into DNA, where it acts as a direct and irreversible inhibitor of DNMTs. Cells then divide in the absence of DNMTs which in turn gives progressive DNA hypomethylation and reactivation of previously silenced genes (Issa et al. 2005). Zebularine drug is a cytidine analogue containing a 2-(1H)-pyrimidinone ring that was originally developed as a cytidine deaminase inhibitor. Zebularine is highly stable, nontoxic, effective DNMT inhibitor. Upon action, Zebularine forms tight covalent complexes between DNMT protein and zebularine-substituted DNA (Billam et al. 2010). Due to the less frequent use of nucleoside analogues, non-nucleoside analogues are now in the use, which inhibit DNMT via other mechanisms rather than DNA incorporation. Non-nucleoside inhibitors are less toxic. Some non-nucleoside inhibitors are developed recently which are (−)-epigallocatechin-3- gallate EGCG, RG108, and procaine for clinical trials (Ghasemi 2020).

Histone Deacetylase Inhibitors (HDACIs)

In epigenetic regulation, acetylation is the well-studied modification, where it is controlled by two main enzymes histone acetyltransferases (HATs) and the histone deacetylases (HDACs), expressing activity on the 3-amino group of Lys residues. HDACs chemical structure is made up of three parts: zinc-chelating group; spacer group (hydrophobic); and enzyme binding group. So far, 18 HDACs were identified which are classified into four subgroups. The classification includes Class I (HDAC 1, 2, 3, 8), Class II (HDAC 4, 5, 7, 9), class III (SIRT1–7), and Class IV (HDAC 11) enzymes. Class I, class II, and class IV HDACs are Zn^{2+}-dependent enzymes due to single isoform (HDAC11). Class I HDACs are basically Zn^{+}-dependent and include HDACs 1, 2, 3, and 8, which are basically located in the nucleus and are homologous to the yeast RPD3 proteins, which are expressed ubiquitously in various human tissues. Class I HDAC plays crucial role in cell proliferation, differentiation, and cell cycle progression. Class III HDACs are structurally different from remaining three and are identified by their nicotinamide adenine dinucleotide (NAD+) dependency (Ropero and Esteller 2007; Glozak and Seto 2007). Histone deacetylase inhibitors are classified into six groups based on the short-chain fatty acids, hydroxamates, cyclic tetrapeptides, benzamides, electrophilic ketones, and miscellaneous, among these inhibitors containing hydroxamates are mostly used and are potent. There are four histone deacetylases inhibitors till date approved by the US FDA which are Belinostat (PXD101), Vorinostat (SAHA), Panobinostat (LBH589), and Romidepsin (FK228) for the treatments of hematologic malignancies (Yang et al. 2019). Epigenetic enzyme lysine methyl transferase (KMTase) attaches to the histone protein via methyl linkage so finding the inhibitors for the identification of lysine methyltransferases and demethylases leads to the discovery of new inhibitors, such as the histone methyltransferases DOT1L and EZH2, and demethylase LSD1 which has already reached to the clinical trials (Morera et al. 2016). HDAC inhibitors show high potential to be implemented in combinational therapy for the cancer. Vorinostat structurally belongs to the hydroxymate group acts as a histone deacetylase (HDAC) inhibitor, which is also named as suberoylanilide hydroxamic acid (SAHA), belonging to the inhibitor of class I and II HDACs. Vorinostat inhibits class I and II HDACs enzymes, but does not inhibit class III HDAC enzymes. Phenyl ring of Vorinostat binds to the zinc atom of catalytic site of the HDAC enzyme onto the surface of HDAC enzyme. After binding to the HDAC enzyme, acylated protein and histone get accumulated which in turn results in the multiple cellular effects.

Cellular effect can be transcriptional or non-transcriptional. Transcription effects may result in the alteration in the expression of certain genes like acetylation of Bcl-6 transcriptional activator. Non transcriptional effects of Vorinostat may result in the reduced cyclin-dependent kinase activity via mechanism of downregulation of cyclins or apoptosis (Bubna 2015).

Noncoding RNA-Based Therapeutics

Non-coding RNA are the type of RNA which are not translated after transcription. This noncoding RNA can be categorized into two subtypes as short ncRNA and long ncRNA. Short ncRNA comprises the small nucleolar RNAs (snoRNAs), PIWI interacting RNAs (piRNAs), small interfering RNAs (siRNAs), and microRNAs (miRNAs). These ncRNA show high degree of sequence conversion and are responsible for the transcriptional and post-transcriptional gene silencing. While in long ncRNA poor cross-species sequence conservation is observed, long ncRNA act as a chaperones or scaffolds for various chromatin regulators (Peschansky and Wahlestedt 2014). Small non-coding miRNA regulate gene expression where miRNAs target the 3′-untranslated region (3′-UTR) or the 5’-UTR of the target mRNA which eventually results in the mRNA degradation or inhibition of translation. miRNA is involved in various biological events of body like development, differentiation, cell cycle regulation, and metabolism, so miRNA can indirectly regulate the epigenetic process. miR-15 and miR-16 are located on chromosome 13q14, and deletion of this miR-15 and miR-16 leads to the chronic lymphocytic leukemia (CLL). Oncogenic miRNAs are often upregulated in cancer. MiR-155 and miR-21 are the oncomers that are induced in several neoplasms. There are several miRNA like miR-1, miR-124a, and miR-127 which are under epigenetic control as they are embedded in CpG island regions and epigenetically silenced. Epi-miRNAs are regulators of epigenetic effectors; epi-miRNAs regulate the expression of DNMT3a, DNMT3b, and DNMT1 (miR-29a, miR-29b, miR-29c) in lung cancer and AML, RBL2 the inhibitor of DNMT3 genes (miR-290), HDACs (miR-1, miR-140, and miR-449a) in prostate cancer (Suzuki et al. 2012; Esmaeili et al. 2010). Long intergenic noncoding RNAs (lincRNAs) can regulate dosage compensation, imprinting and developmental gene expression by chromatin domains in an allele. lincRNA HOTAIR is commonly work as a scaffold for two distinct histone modification complexes; HOTAIR reprograms chromatin state to promote cancer metastasis (Gibb et al. 2011).

Epigenetics: Cancer Targeting, Imaging, and Therapy

The first generation of drugs targeting the epigenetic pathways as discussed in previous section was established using a multipurpose approach and found to have unsatisfactory efficiency in solid tumors (Majchrzak-Celińska et al. 2021). Recently, to enhance the potential of epidrugs and to overcome drug resistance in solid tumors, combination of epidrug with chemotherapeutic drug exhibits new routes for clinical examinations. The efficacy of epidrugs in combination with various other anticancer therapies can be improved by consecutive treatment plans (Chen et al. 2020; Roberti et al. 2019; Li et al. 2020). The upcoming section of the potential of epidrugs in combination with other anticancer drugs and drug delivery agents is described briefly. The use of efficient drug delivery agents like nanoparticles helps to improve the limitations of current epigenetic therapies as shown in Fig. 15.4.

Long noncoding RNAs (lncRNAs) can be used as a target to treat triple-negative breast cancer (TNBC) due to their activity like multiple gene regulation, but for conventional approaches, systemic targeting with these undruggable macromolecules remains the challenge. To conquer this challenge, Vaidya and co-workers synthesized tumor-targeting RGD-PEG-ECO/siDANCR nanoparticles for effective TNBC therapy (Vaidya et al. 2019). RGD-PEG-ECO/siDANCR nanoparticles were synthesized by multifunctional pH-sensitive amino lipid, ECO, where ECO self-assembles with siRNA to form stable ECO/siRNA nanoparticles, and DANCR is the oncogenic lncRNA which is highly overexpressed in TNBC. Therapy with the RGD-PEG-ECO/siDANCR nanoparticles arbitrates the robust DANCR silencing which in turn effects in the TNBC, concomitant gene expression varies in cancer-driving pathways. They have demonstrated a significant overexpression of DANCR in human breast cancer (BCa) cells, tumor, and The Cancer Genome Atlas (TCGA) database and then studied the DANCR as an effective therapeutic target for TNBC therapy. Studies were conducted for two cell lines MDA-MB-231 and BT549 and mouse model by silencing expression of the therapeutic target using RGD-PEG-ECO/siDANCR nanoparticles. RGD-PEG-ECO/siDANCR nanoparticles show the effective intracellular siRNA delivery and sustained target silencing and sustained target silencing by efficient cytosolic delivery of siDANCR. RGD-PEG-ECO/siDANCR nanoparticles exhibit excellent in vitro therapeutic efficacy which can be analyzed by significant reduction in the invasion, migration, survival, tumor spheroid formation, and proliferation of the TNBC cell lines. At the molecular level, these RGD-PEG-ECO/siDANCR nanoparticles work by functional ablation of DANCR which dynamically impacted the oncogenic nexus by downregulating PRC2-mediated H3K27-trimethylation and Wnt/EMT signaling, causing alteration in the phosphorylation profiles of several kinases in the TNBC cells. RGD-PEG-ECO/siDANCR shows no side effects while used as dose of 1 mg/kg siRNA in nude mice-bearing TNBC xenografts. These study gives the nanoparticle-mediated modulation of onco-lncRNAs as a promising formulation for developing curative therapeutic for TNBC.

Epigenetic alterations like DNA methylation and histone modifications can lead to silencing of the tumor suppressor genes. To overcome this limitation, Vijayaraghavalu and his co-workers made a hypothesis where they used drug-resistant breast cancer cells (MCF-7/ADR) and two epigenetic drugs like 5-aza-2-′-deoxycytidine (decitabine, DAC), a demethylating agent, and suberoylanilide hydroxamic acid (SAHA), a histone deacetylase inhibitor in combination with doxorubicin (Vijayaraghavalu et al. 2013). The sequential treatment of resistant

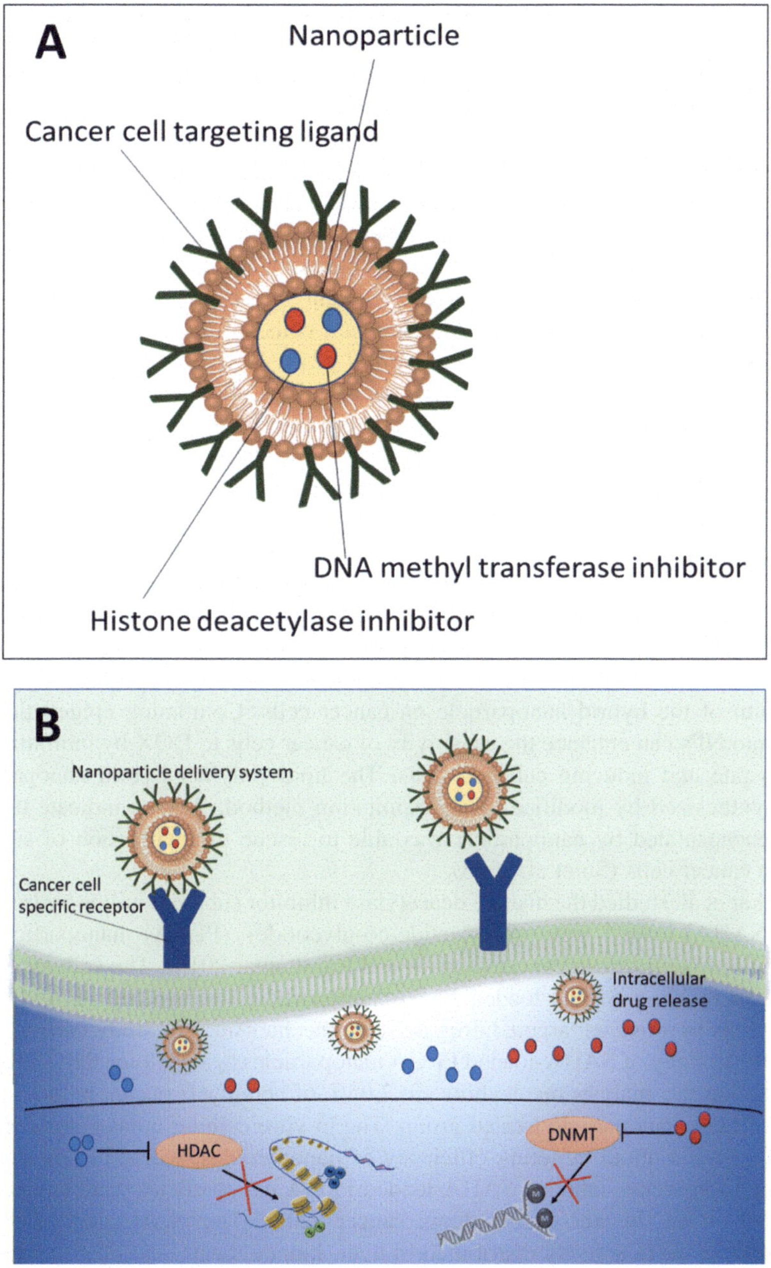

Fig. 15.4 Combining nanoparticle-based drug delivery systems with epigenetic therapy, by incorporating epigenetic inhibitors along with chemo drug, a promising approach to get high therapeutic efficacy in cancer

cells starts with epigenetic drug (DAC or SAHA) followed by doxorubicin tested on drug-sensitive breast cancer cells (MCF-7, MDA-MB 231, and BT-459). 5-aza-2-′-deoxycytidine (DAC), (a DNA methyltransferase inhibitor), and suberoylanilide hydroxamic acid (SAHA), (a histone deacetylase inhibitor) work in the manner where they reactivate the expression of transcriptionally silenced genes by inhibiting the DNMT activity and DAC reverses DNA hypermethylation and the expression of multidrug resistance 1 (mdr1) efflux protein which eventually enhances the intracellular doxorubicin levels. Sequential treatment of resistant cancer cells, first with an epigenetic drug (DAC), and then with doxorubicin, results in highly synergistic effect, which reduces the IC50 of doxorubicin by several thousand-fold. This sequential treatment causes over 90% resistant cells to undergo G2/M cell cycle arrest, which is caused due to upregulation of p21WAF1/CIP1 expression, responsible for cell cycle regulation. The synergistic effect shows that DAC was more effective than SAHA in overcoming drug resistance. Thus, combination therapy with the epigenetic drug and doxorubicin can demonstrate an effective therapeutic strategy to overcome drug resistance that can potentially minimize doxorubicin-induced cardiotoxicity.

Su et al. proposed a nanoparticle-based drug delivery system for cancer treatment, where they synthesized a lipid polymer nanoparticle which is used for the co-delivery of epigenetic drug DAC (2′-deoxy-5-azacytidine) that are responsible for the reactive expression of tumor suppressor genes (TSG) and chemotherapeutic drug doxorubicin (DOX) to cancer cells and then monitored this strategy as growth inhibition of the hybrid nanoparticle on cancer cells. Combining epigenetic drug DAC into NPs can enhance the sensitivity of cancer cells to DOX by inhibiting cell growth rate and inducing cell apoptosis. The lipid–polymer hybrid nanoparticles were synthesized by modified nanoprecipitation method. Results indicate that the DAC encapsulated by nanoparticle was able to rescue the expression of silenced TSG in cancer cells (Su et al. 2013).

Sankar et al. studied the histone deacetylase inhibitor suberoylanilide hydroxamic acid (SAHA)-loaded poly (DL-lactide-co-glycolide) (PLGA) nanoparticle for targeted drug delivery in cancer (Sankar and Ravikumar 2014). The cellular uptake of epigenetic drug SAHA-loaded PLGA nanoparticle was studied in A549 lung cancer cells to prove its targeted drug delivery mechanism. The study carried out to check the toxicity of SAHA-loaded PLGA nanoparticles against mice RBCs. Results indicate that the analysis result show cut 2.05% of hemolysis occurs in the SAHA-loaded PLGA nanoparticle-treated group. The in vitro cellular uptake study reveals the information about targeting efficiency of nanoparticles, where in this study in 2 hours fluorescence-labelled SAHA-loaded PLGA nanoparticles were antagonistically taken up by the A549 lung cancer cells. The SAHA-loaded PLGA nanoparticles were actively distributed in liver, kidney, heart, and lung tissues, and SAHA-loaded PLGA nanoparticles were greatly taken up in the lung cancer A549 cells.

Meng et al. have developed a stable, highly sensitive HDAC targeting probe, where near-infrared (NIR) fluorescence optical imaging is used for visualizing disease at the molecular level in vivo with high sensitivity. Herein, they have mentioned design, synthesis, and bioactivation evaluation of LBH589-Cy5.5 for in vivo NIR fluorescence imaging of HDACs (Meng et al. 2015). They have chosen LBH589 as the targeting ligand for the fabrication of the HDACs imaging probe for triple-native breast cancer detection as well as for the HDACi therapy evaluation. LBH589 is a potent hydroxamic acid pan-HDAC inhibitor having nanomolar range IC50 values for all HDACs, with some advantages like high affinity to target, low immunogenicity, and flexibility in chemical modification. Successfully synthesized LBH589-Cy5.5 by conjugating NHS-activated ester of Cy5.5 to the secondary amine group of LBH589 which is adjacent to the capping motif. The synthesized probe shows high binding affinity to HDACs with an IC50 value of 9.6 nM. The imaging studies in vivo and ex vivo LBH589-Cy5.5 probe were capable of imaging HDACs, while showing high specificity and sensitivity, which is due to the fast and selective accumulation in MDA-MB-231 xenograft tumors. To study the availability of LBH589-Cy5.5 probe in the HDACi therapy evaluation, the MDA-MB-231 tumor-bearing mice were injected with LBH589-Cy5.5, and later these mice were subjected to the fluorescent imaging before and after 3-day treatment with epigenetic drug SAHA. Where the results indicated the continuous administration of SAHA can be harmful and affect the tumor cells by blocking the expression HDACs this leads reduced uptake of LBH589-Cy5.5 in the tumors, causing the reduced activity of drug. So, the LBH589-Cy5.5 probe shows the potential to evaluate the response of HDACi therapies in early stages of treatment in vivo. For treating the triple-negative breast cancer, LBH589-Cy5.5 can be a sensitive probe and also as therapy response DAC inhibitors contribute to the development of HDAC inhibitors as anticancer agents.

Cisplatin (CDDP) is an anticancer drug in clinical practice, which shows limitation in targeting the tumor site, which may lead to clinical failure and side effects. To diminish these restrictions, Xu et al. synthesized a supramolecular conjugate by combining cisplatin with Vorinostat. Vorinostat (SAHA) is the FDA-approved histone deacetylase inhibitor (Xu et al. 2017). They report conjugation of the CDDP and SAHA drug together through coordination bonds, and then, supramolecular CDDP-SAHA conjugate can self-assemble into nano-micelles. This supramolecular drug shows 100 percent drug loading and simple administration with enhanced permeation and retention (EPR) effect. Studies show that after cellular uptake CDDP-SAHA nanodrug gets hydrolyzed inside cancer cells and releases CDDP and SAHA simultaneously for synergistic therapy. After releasing SAHA drug inhibits activity of HDAC, which gives advantage to CDDP to bind DNA forming strand cross-links, these interns result in the induced apoptosis of cancer cells and also work for in vitro inhibition of cell proliferation of A549/DR tumor cells. Lipid nanoemulsions (LNEs) encapsulating both DAC (decitabine) and PAN (Panobinostat) to treat breast cancer cells reported by Kim et al., where DAC and PAN show reverse abnormal methylation of DNA which altering chromatin structure, leading to the increased expression of tumor suppressor genes and decreased expression of oncogene. These studies demonstrate dual delivery of DAC and PAN, cellular uptake, and biodistribution of LNEs in TNBCs (Kim et al. 2019). The results suggest that the nanoparticle-mediated modulation of oncogenes through epigenetic inhibitors in combination with chemotherapeutic drugs and nanoparticle-based drug delivery systems is the promising formulation for emerging curative therapeutics for cancer.

Monitoring Epigenetic Changes in Cancer by Various Imaging Techniques

The detailed understanding of epigenetic modifications is still minute, but impending management of the epigenetic mechanisms for cancer will be a key component in upcoming therapies. The in-depth investigations of epigenetic pathways is mandatory for the further enhancement in the field of epigenetic research on cancer. The technological advancements have improved both genetic insights and new tactics in monitoring epigenetic mechanisms of cancer. Cancer imaging along with targeted therapies in epigenetic modulations may overcome the limitations of traditional methods. The various imaging methods along with targeted epigenetic therapies can accelerate the progress in the field that was just unimaginable a few years ago to apprehend many unanswered epigenomics queries (Abramczyk et al. 2016). To monitor cancer epigenetics meticulously fluorescence imaging, surface-enhanced Raman scattering (SERS), atomic force microscopy (AFM), and scanning near-field optical microscopy (SNOM) techniques have been in practice and some of the important findings are discussed here.

In case of advanced melanoma, attributed to the low tumor-associated antigen (TAA), PD-1/PD-L1 blockade therapy is less effective. To overcome this difficulty, Ruan and team synthesized a dual bioresponsive depot consisting of drug Zebularine (Zeb), a demethylation agent, and anti-PD1 antibody (aPD1) encapsulated into the pH-sensitive calcium carbonate nanoparticles ($CaCO_3$ NPs). In situ synthesis method is performed where aPD1 loaded into the CaCO3 nanoparticles; then, Zebularine was encapsulated into the ROS-responsive hydrogel (Zeb-aPD1-NPs-Gel) crosslinked by blending polyvinyl alcohol (PVA) and N1 -(4-boronobenzyl)-N3-(4-boronophenyl)-N1, N1, N3, N3-tetramethylpropane-1,3-diaminium (TSPBA) linker. Zeb-aPD1-NPs-Gel were synthesized for combination therapy to treat cancer cells. The retention ability of aPD1 in vivo was studied in the B16F10 melanoma-bearing mice model. aPD1 labeled with Cyanine5.5 (Cy5.5) was loaded into the gel depot. The fluorescence signal remained detectable six days after gel injection and no signal for the groups of free Cy5.5-aPD1 and Cy5.5-aPD1-NPs, which indicates the encapsulation of aPD1-NPs increases the retention time. To check the combination strategy of epigenetic modulation and immune checkpoint blockade, ICB therapy could enhance the inhibition of tumor growth so different formulations were injected including blank Gel. The in vivo tumor growth was monitored by capturing the bioluminescence images of luciferase-tagged cancer cells, where the Zeb-aPD1-NPs-Gel-treated group showed the most notable tumor inhibition effect (Ruan et al. 2019).

Yuan and team synthesized the hyaluronic acid-coated, pH/redox dual-responsive nanosystem (HPMSNs) encapsulated with doxorubicin (DOX) and GCN5 siRNA (siGCN5). This system was effective towards preventing premature leakage and releases the therapeutics intracellularly via its pH/redox dual responsiveness. DOX/siGCN5@HPMSNs enhances the drug internalization in CD44-overexpressing cancer cells due to the CD44-mediated targeting and promotes DOX retention by downregulating P-gp expression through silencing GCN5. Hyaluronic acid is a natural CD44-binding polysaccharide that can enhance drug delivery efficacy in CD44-overexpressing cancer cells. To check the targeting property of HPMSN confocal images of MCF7/ADR cells were exposed to DOX and naked siRNAFAM, DOX/siRNAFAM@HPMSNs, or DOX/siRNAFAM@HPMSNs. It shows that the co-delivery of DOX and siGCN5 using HPMSNs could enhance DOX intracellular accumulation by simultaneously increasing drug influx and suppressing drug efflux, likely due to HA-mediated targeting (Yuan et al. 2021).

Ledezma and co-workers synthesized poly (lactic-co-glycolic) acid (PLGA)-based nanoparticles encapsulated with the indocyanine green (ICG), a PTT agent, and Nexturastat A (NextA), an epigenetic drug for combined photothermal therapy (PTT) with epigenetic therapy for melanoma. The simple nanoemulsion method for the synthesis of ICG and NextA within stable and monodispersed INAPs was used and characterized the size, stability, and encapsulation efficiency of INAPs. This formulation shows concentration-dependent near-infrared (NIR) laser power-dependent photothermal heating characteristics and works as a single-use agent for PTT of melanoma cells in vitro. INAPs work by inhibiting the expression of pan-histone deacetylase (HDAC) and HDAC6-specific activity in melanoma cells in vitro. The studies were carried out to check the activity of INAPs for both PTT and epigenetic therapy in vitro. The photothermal heating profiles for INAPs were studied using INAPs nanoparticle concentration (0.5–4.0 mg/mL) and irradiated with an 808 nm near-infrared (NIR) laser for 5 min, which demonstrate that INAPs were effective single-use PTT agents (Ledezma et al. 2020).

In another study, Goswami and her team demonstrated combination drug therapy luminescent Au nanoclusters where AuNCs were formulated into spherical polyethylene glycol (PEG) NC. PEG−AuNC were again treated to encapsulate the drug sodium butyrate (NaB) resulting in composite nanoparticles (PEG−Au NC − NaB-NPs). The effect of nanocomposites on the cancer cells was studied using bioimaging that clearly explains the endocytosis pathway and combination therapy along with the plant based antimalarial drug artesunate (ART). Upon the uptake of PEG−Au NC − NaB-NPs bright red luminescence was observed in the compartment of the cells upon uptake predominantly through a caveolae-mediated pathway. The polymer-coated HDACi drug encapsulated composite NPs were studied for the first time for combination therapy with ART. Action of both NaB and ART was studied in vivo on the Swiss albino mice with Dalton's lymphoma Ascites (DLA) to endorse it for real-life applications (Goswami et al. 2018).

Structural Aspects of Epigenetic Drugs/Small Molecules

The combinatorial drug therapies have extended the cancer treatment to achieve prolonged disease control and management. However, the main difficulties ascend due to the dose and schedule of drug when employed in combinations. This eventually results in unexpected pharmacokinetics and possible drug–drug interactions may arise difficult conditions. In contradictory, a single-drug molecule with multitarget ability has the advantages of higher synergistic effect, predictable pharmacokinetics, and reduced difficulties. However, rational designing of such multitargeting drugs is highly challenging mission, explicitly for the targets which are from different protein families. For instance, multitarget antitumor drugs or small molecule inhibitors are considered as an important approach to overcome drug resistance and improve therapeutic efficacy. For example, a first report on small molecule that simultaneously inhibits NAMPT and HDAC, two important targets of cancer epigenetics, was identified recently through structure-based drug design, chemical synthesis, and biological assays, namely, N-(2-aminophenyl)-4-((4-(4-((pyridin-3-ylmethyl)carbamoyl)-phenyl)-1H-1,2,3 triazol-1 yl)methyl) benzamide. This compound demonstrated excellent activities against both NAMPT (IC50 = 31 nM) and HDAC1 (IC50 = 55 nM) (Dong et al. 2017). Hence, the search for novel multitargeting small molecular drugs and their structural and designing aspects are of key interest for researcher to develop new therapeutic agents for epigenetic modulation of cancer.

Future Perspectives

Epigenetic mechanisms and cancer are closely associated. The unique discovery of DNA methylation, histone acetylation, RNA methylation, and the uses of different RNAs for epigenetic reprogramming have been essential for the development of more cost-effective and efficient antitumor drugs and treatments. The use of epidrugs in solitary or multidrug therapy are promising avenues for cancer treatment. While various studies have been carried out using diverse combinations of traditional epigenetic inhibitors along with chemotherapeutics for cancer treatment, these studies result in fewer side effects from chemotherapy and effective cancer therapy due to synergism of traditional epidrugs and chemotherapy. On the other hand, it is still challenging to understand the mechanism of action of these epidrugs when administered in solitary or in combination drug therapies.

Acknowledgments We are thankful to Jain University, Bangalore, India, for providing facilities. V. Brahmkhatri also acknowledges TARE-SERB File NO:TAR/2018/000547.

References

Abramczyk H, Surmacki J, Kopeć M, Olejnik AK, Kaufman-Szymczyk A, Fabianowska-Majewska K (2016) Epigenetic changes in cancer by Raman imaging, fluorescence imaging, AFM and scanning near-field optical microscopy (SNOM). Acetylation in normal and human cancer breast cells MCF10A, MCF7 and MDA-MB-231. Analyst 141:5646–5658

Billam M, Sobolewski MD, Davidson NE (2010) Effects of a novel DNA methyltransferase inhibitor zebularine on human breast cancer cells. Breast Cancer Res Treat 120:581–592

Bubna AK (2015) Vorinostat—an overview. Indian J Dermatol 60:419

Cao W, Lee H, Wu W, Zaman A, Mccorkle S, Yan M, Chen J, Xing Q, Sinnott-Armstrong N, Xu H, Sailani MR, Tang W, Cui Y, Liu J, Guan H, Lv P, Sun X, Sun L, Han P, Lou Y, Chang J, Wang J, Gao Y, Guo J, Schenk G, Shain AH, Biddle FG, Collisson E, Snyder M, Bivona TG (2020) Multi-faceted epigenetic dysregulation of gene expression promotes esophageal squamous cell carcinoma. Nat Commun 11:3675

Chen F, Shi Y, Zhang J, Liu Q (2020) Nanoparticle-based drug delivery Systems for Targeted Epigenetics Cancer Therapy. Curr Drug Targets 21:1084–1098

Cheng Y, He C, Wang M, Ma X, Mo F, Yang S, Han J, Wei X (2019) Targeting epigenetic regulators for cancer therapy: mechanisms and advances in clinical trials. Signal Transduct Target Ther 4:62

Derissen EJ, Beijnen JH, Schellens JHM (2013) Concise drug review: azacitidine and decitabine. Oncologist 18:619

Dhanak D, Jackson PJB, Communications BR (2014) Development and classes of epigenetic drugs for cancer. Biochem Biophys Res Commun 455:58–69

Dong G, Chen W, Wang X, Yang X, Xu T, Wang P, Zhang W, Rao Y, Miao C, Sheng C (2017) Small molecule inhibitors simultaneously targeting cancer metabolism and epigenetics: discovery of novel nicotinamide Phosphoribosyltransferase (NAMPT) and histone deacetylase (HDAC) dual inhibitors. J Med Chem 60:7965–7983

Egger G, Liang G, Aparicio A, Jones PA (2004) Epigenetics in human disease and prospects for epigenetic therapy. Nature 429:457–463

Esmaeili F, Bamdad T, Ghasemi S (2010) Stable suppression of gene expression by short interfering RNAs targeted to promoter in a mouse embryonal carcinoma stem cell line. In Vitro Cell Dev Biol Anim 46:834–840

Fattahi S, Kosari-Monfared M, Golpour M, Emami Z, Ghasemiyan M, Nouri M, Akhavan-Niaki H (2020) LncRNAs as potential diagnostic and prognostic biomarkers in gastric cancer: a novel approach to personalized medicine. J Cell Physiol 235:3189–3206

Felsenfeld G (2014) A brief history of epigenetics. Cold Spring Harb Perspect Biol 6:a018200

Frye R, Myers M, Axelrod KC, Ness EA, Piekarz RL, Bates SE, Booher S (2012) Romidepsin: a new drug for the treatment of cutaneous T-cell lymphoma. Clin J Oncol Nurs 16:195–204

Ghasemi S (2020) Cancer's epigenetic drugs: where are they in the cancer medicines? Pharmacogenomics J 20:367–379

Gibb EA, Brown CJ, Lam WL (2011) The functional role of long non-coding RNA in human carcinomas. Mol Cancer 10:38. https://doi.org/10.1186/1476-4598-10-38

Glozak M, Seto EJO (2007) Histone deacetylases and cancer. Oncogene 26:5420–5432

Gnyszka A, Jastrzebski Z, Flis S (2013) DNA methyltransferase inhibitors and their emerging role in epigenetic therapy of cancer. Anticancer Res 33:2989–2996

Goswami U, Kandimalla R, Kalita S, Chattopadhyay A, Ghosh SS (2018) Polyethylene glycol-encapsulated histone deacetylase inhibitor drug-composite nanoparticles for combination therapy with artesunate. ACS Omega 3:11504–11516

Hervouet E, Peixoto P, Delage-Mourroux R, Boyer-Guittaut M, Cartron P-F (2018) Specific or not specific recruitment of DNMTs for DNA methylation, an epigenetic dilemma. Clin Epigenetics 10:1–18

Issa J-PJ, Kantarjian HM, Kirkpatrick P (2005) Azacitidine. Nat Rev Drug Discov 4:275

Kantarjian H, Issa JP, Rosenfeld CS, Bennett JM, Albitar M, Dipersio J, Klimek V, Slack J, De Castro C, Ravandi F, Helmer R 3rd, Shen L, Nimer SD, Leavitt R, Raza A, Saba H (2006) Decitabine improves patient outcomes in myelodysplastic syndromes: results of a phase III randomized study. Cancer 106:1794–1803

Kim B, Pena CD, Auguste DT (2019) Targeted lipid Nanoemulsions encapsulating epigenetic drugs exhibit selective cytotoxicity on CDH1–/FOXM1+ triple negative breast cancer cells. Mol Pharm 16:1813–1826

Ledezma DK, Balakrishnan PB, Cano-Mejia J, Sweeney EE, Hadley M, Bollard CM, Villagra A, Fernandes R (2020) Indocyanine green-nexturastat A-PLGA nanoparticles combine photothermal and epigenetic therapy for melanoma. Nanomaterials (Basel) 10:161

Li Y, Ding J, Xu X, Shi R, Saw PE, Wang J, Chung S, Li W, Aljaeid BM, Lee RJ, Tao W, Teng L, Farokhzad OC, Shi J (2020) Dual hypoxia-targeting RNAi nanomedicine for precision cancer therapy. Nano Lett 20:4857–4863

Liu Z, Gao Y, Li X (2019) Cancer epigenetics and the potential of epigenetic drugs for treating solid tumors. Expert Rev Anticancer Ther 19:139–149

Lu Y, Chan Y-T, Tan H-Y, Li S, Wang N, Feng Y (2020) Epigenetic regulation in human cancer: the potential role of epi-drug in cancer therapy. Mol Cancer 19:79

Majchrzak-Celińska A, Warych A, Szoszkiewicz M (2021) Novel approaches to epigenetic therapies: from drug combinations to epigenetic editing. Genes (Basel) 12:208

Mcdermott J, Jimeno A (2014) Belinostat for the treatment of peripheral T-cell lymphomas. Drugs Today (Barcelona, Spain: 1998) 50:337–345

Meng Q, Liu Z, Li F, Ma J, Wang H, Huan Y, Li Z (2015) An HDAC-targeted imaging probe LBH589–Cy5. 5 for tumor detection and therapy evaluation. Mol Pharm 12:2469–2476

Miranda Furtado CL, Dos Santos Luciano MC, Silva Santos RD, Furtado GP, Moraes MO, Pessoa C (2019) Epidrugs: targeting epigenetic marks in cancer treatment. Epigenetics 14:1164–1176

Mondal P, Natesh J, Penta D, Meeran SM (2020) Progress and promises of epigenetic drugs and epigenetic diets in cancer prevention and therapy: a clinical update. Semin Cancer Biol. https://doi.org/10.1016/j.semcancer.2020.12.006

Montalvo-Casimiro M, González-Barrios R, Meraz-Rodriguez MA, Juárez-González VT, Arriaga-Canon C, Herrera LA (2020) Epidrug repurposing: discovering new faces of old acquaintances in cancer therapy. Front Oncol 10:605386

Morel D, Jeffery D, Aspeslagh S, Almouzni G, Postel-Vinay S (2020) Combining epigenetic drugs with other therapies for solid tumours - past lessons and future promise. Nat Rev Clin Oncol 17: 91–107

Morera L, Lübbert M, Jung M (2016) Targeting histone methyltransferases and demethylases in clinical trials for cancer therapy. Clin Epigenetics 8:1–16

Pan Y, Liu G, Zhou F, Su B, Li Y (2018) DNA methylation profiles in cancer diagnosis and therapeutics. Clin Exp Med 18:1–14

Park JW, Han J-W (2019) Targeting epigenetics for cancer therapy. Arch Pharm Res 42:159–170

Peschansky VJ, Wahlestedt C (2014) Non-coding RNAs as direct and indirect modulators of epigenetic regulation. Epigenetics 9:3–12

Richardson PG, Laubach JP, Lonial S, Moreau P, Yoon SS, Hungria VT, Dimopoulos MA, Beksac M, Alsina M, San-Miguel JF (2015) Panobinostat: a novel pan-deacetylase inhibitor for the treatment of relapsed or relapsed and refractory multiple myeloma. Expert Rev Anticancer Ther 15:737–748

Roberti A, Valdes AF, Torrecillas R, Fraga MF, Fernandez AF (2019) Epigenetics in cancer therapy and nanomedicine. Clin Epigenetics 11:81

Ropero S, Esteller M (2007) The role of histone deacetylases (HDACs) in human cancer. Mol Oncol 1:19–25

Ruan H, Hu Q, Wen D, Chen Q, Chen G, Lu Y, Wang J, Cheng H, Lu W, Gu Z (2019) A dual-bioresponsive drug-delivery depot for combination of epigenetic modulation and immune checkpoint blockade. Adv Mater 31:1806957

Salarinia R, Sahebkar A, Peyvandi M, Mirzaei HR, Jaafari MR, Riahi MM, Ebrahimnejad H, Nahand JS, Hadjati J, Asrami MO, Fadaei S, Salehi R, Mirzaei H (2016) Epi-drugs and epi-miRs: moving beyond current cancer therapies. Curr Cancer Drug Targets 16:773–788

Sankar R, Ravikumar V (2014) Biocompatibility and biodistribution of suberoylanilide hydroxamic acid loaded poly (DL-lactide-co-glycolide) nanoparticles for targeted drug delivery in cancer. Biomed Pharmacother 68:865–871

Silverman LR, Demakos EP, Peterson BL, Kornblith AB, Holland JC, Odchimar-Reissig R, Stone RM, Nelson D, Powell BL, Decastro CM, Ellerton J, Larson RA, Schiffer CA, Holland JF (2002) Randomized controlled trial of azacitidine in patients with the myelodysplastic syndrome: a study of the cancer and leukemia group B. J Clin Oncol 20:2429–2440

Su X, Wang Z, Li L, Zheng M, Zheng C, Gong P, Zhao P, Ma Y, Tao Q, Cai L (2013) Lipid–polymer nanoparticles encapsulating doxorubicin and 2′-deoxy-5-azacytidine enhance the sensitivity of cancer cells to chemical therapeutics. Mol Pharm 10:1901–1909

Suzuki H, Maruyama R, Yamamoto E, Kai M (2012) DNA methylation and microRNA dysregulation in cancer. Mol Oncol 6:567–578

Tiwari PK (2020) Epigenetic biomarkers in gallbladder cancer. Trends Cancer 6:540–543

Vaidya AM, Sun Z, Ayat N, Schilb A, Liu X, Jiang H, Sun D, Scheidt J, Qian V, He S, Gilmore H, Schiemann WP, Lu Z-R (2019) Systemic delivery of tumor-targeting siRNA nanoparticles against an oncogenic LncRNA facilitates effective triple-negative breast cancer therapy. Bioconjug Chem 30:907–919

Vijayaraghavalu S, Dermawan JK, Cheriyath V, Labhasetwar V (2013) Highly synergistic effect of sequential treatment with epigenetic and anticancer drugs to overcome drug resistance in breast cancer cells is mediated via activation of p21 gene expression leading to G2/M cycle arrest. Mol Pharm 10:337–352

Wood H (2018) FDA approves patisiran to treat hereditary transthyretin amyloidosis. Nat Rev Neurol 14:570

Xu S, Zhu X, Huang W, Zhou Y, Yan D (2017) Supramolecular cisplatin-vorinostat nanodrug for overcoming drug resistance in cancer synergistic therapy. J Control Release 266:36–46

Yang F, Zhao N, Ge D, Chen Y (2019) Next-generation of selective histone deacetylase inhibitors. RSC Adv 9:19571–19583

Yuan Y, Liu J, Yu X, Liu X, Cheng Y, Zhou C, Li M, Shi L, Deng Y, Liu H et al (2021) Tumor-targeting pH/redox dual-responsive nanosystem epigenetically reverses cancer drug resistance by co-delivering doxorubicin and GCN5 siRNA. Acta Biomater 135:556–566

Chapter 16
Epigenetic Small-Molecule Modulators Targeting Metabolic Pathways in Cancer

Dipanwita Das, Nandini Karthik, and Reshma Taneja

Abstract Metabolic deregulation is a key factor in cancer progression. Epigenetic changes and metabolic rewiring are intertwined in cancer. Deregulated epigenetic modifiers cause metabolic aberrations by targeting the expression of metabolic enzymes. Conversely, metabolites and cofactors affect the expression and activity of epigenetic regulators. Small molecules are promising therapeutic approaches to target the epigenetic-metabolomic crosstalk in cancer. Here, we focus on the interplay between metabolic rewiring and epigenetic landscape in the context of tumourigenesis and highlight recent advances in the use of small-molecule drug targets for therapy.

Keywords Small molecules · Epigenetics · Metabolism · Cancer · Signalling pathways

Introduction

In the field of cancer research, traditional focus on genetic mutations in cancer cells for the purpose of developing therapy has shifted to include reprogramming of the epigenetic landscape of these cells (Baylin and Jones 2011). Rapid advancement of technologies to study epigenetic modifiers and mechanisms has led to a surge in the interest in epigenetic deregulation in cancer. The differences between the epigenetic landscape of cancer cells and their normal counterparts show that this is a critical component in malignant transformation and progression of the disease, thereby representing a promising avenue for development of therapeutics. The expression or function of several modifiers has been shown to be deregulated in various forms of cancer, with crucial downstream effects on epigenetic marks and expression of important drivers of disease. Therefore, many cancer-associated phenotypes,

D. Das · N. Karthik · R. Taneja (✉)
Department of Physiology and Healthy Longevity Translational Research Program Yong Loo Lin School of Medicine, National University of Singapore, Singapore, Singapore
e-mail: phsrt@nus.edu.sg

T. K. Kundu, C. Das (eds.), *Metabolism and Epigenetic Regulation: Implications in Cancer*, Subcellular Biochemistry 100,
https://doi.org/10.1007/978-3-031-07634-3_16

including metabolic reprogramming, uncontrolled proliferation, reversion to a stem-cell-like state, migration and metastasis, can be linked to alterations in the expression or activity of epigenetic modifiers. Thus, epigenetic modifiers represent an important class of druggable targets for cancer therapy, especially in the absence of druggable genetic alterations or transcription factors, and many small-molecule inhibitors of these modifiers have been developed for cancer therapy. The specific and reversible nature of epigenetic modifications makes these modifiers attractive therapeutic targets (Best and Carey 2010; Dawson and Kouzarides 2012; Baylin and Jones 2016). Events that are reprogrammed by cancer cells to allow them to survive, proliferate, metastasize and colonize can potentially be reversed by resetting the activity of aberrantly functioning epigenetic modifiers that regulate some of these processes. One such process is metabolic rewiring.

Metabolic Deregulation in Cancer

Metabolic pathways refer to the ways by which cells can break down nutrients for acquiring energy required for growth, proliferation and sustenance of critical cellular processes. Metabolic profiles are significantly different between tumour and normal cells. The cancer cells that adapt to the hostile tumour microenvironment and have high proliferative state compared to normal cells have an elevated metabolic demand, which is fulfilled by an altered metabolic program (Yu and Li 2017).

A phenomenon called "Warburg effect", observed by Otto Warburg in the 1920s, showed that tumour cells prefer "aerobic glycolysis" over oxidative phosphorylation to generate energy. Warburg showed that tumour cells take up significantly higher levels of glucose compared to normal tissues. Furthermore, he showed that cancer cells produced more lactate than normal tissues, even in the presence of sufficient oxygen, indicating that glucose was shuttled through a glycolytic fermentation pathway (Warburg 1956). This accelerated glycolysis in cancer cells efficiently facilitates the accumulation of macromolecule precursors (nucleic acids, proteins, lipids) to fulfil the high anabolic demands. Cancer cells continue to carry out oxidative phosphorylation in addition to aerobic glycolysis to sustain the high needs of rapidly dividing cells (Kim and Dang 2006; Vander Heiden et al. 2009).

Besides the preferred Warburg effect, tumour cells show a broad range of metabolic alterations. In addition to increased uptake of glucose, tumour cells also show a high dependence on glutamine, which is the chief source of nitrogen and is required for nucleotide and amino acid synthesis (Wise et al. 2008; Dang 2010; Jin et al. 2016). Interactions between various candidates involved in the multistep processes of glycolysis, TCA cycle, ETC, pentose phosphate pathway as well as lipid metabolism are modulated in tumour cells and have direct role in tumourigenesis (Cantor and Sabatini 2012).

Metabolic changes can be a consequence of genetic aberrations in metabolic enzymes as well as activating mutations in different growth factors and oncogenes. It can also be triggered by the loss of tumour suppressor signalling and epigenetic alterations. Metabolic modulation of tumour cells is listed as one of the ten hallmarks of cancer (Hanahan and Weinberg 2011; Pavlova and Thompson 2016).

Though glucose and glutamine metabolism have been the primary focus for metabolic alterations in cancer cells, lipid metabolism is also targeted in proliferating and cancer cells. Fatty acids are additional source of energy and are required for membrane synthesis during cell growth and division.

Lipids operate as secondary messengers and hormones in cell signalling pathways (Santos and Schulze 2012). Fatty acids are synthesized within cells de novo or obtained from environmental sources. It has been observed that de novo synthesis of fatty acids is upregulated in cancer cells (Currie et al. 2013). Fatty acid synthase (FASN) enzyme complex has been observed to remain upregulated in breast tumours. The sterol regulatory element-binding proteins (SREBPs), which are upstream regulators of lipid synthesis, are likewise overexpressed in cancer cells. They stand as promising therapeutic targets for reducing tumour growth and proliferation, and the inhibitors of the above-mentioned molecules are in preclinical trials (Röhrig and Schulze 2016).

The influence of metabolic pathways and reprogramming in tumour cell growth and proliferation has become an active area of research to develop therapeutic interventions. Though there is little effect of metabolite profiles at the genetic level, they have fundamental roles in epigenetic regulation of gene expression. Metabolic reprogramming can control the epigenome machinery in cancer cells and transmit a mitogenic gene expression profile (Gupta et al. 2014; Johnson et al. 2015). Conversely, epigenetic deregulation in cancer alters expression of genes involved in cellular metabolism.

Crosstalk Between Metabolomics and Epigenetics

Epigenetic and metabolic alterations in cancer cells are highly interrelated. Metabolic reprogramming affects the availability of cofactors and generate oncometabolites that function as agonists or antagonists of enzyme activity involved in epigenetic modifications. Reciprocally, epigenetic changes can affect cellular metabolism by altering the expression of genes involved in metabolic pathways (Fig. 16.1). Therefore, strategies for developing therapeutics must take into account this crosstalk.

Impact of Metabolic Changes on Enzymes that Catalyse Epigenetic Modifications

DNA methylation has been extensively explored in cancer. DNA methylation in the CpG sites of gene promoters represses gene expression. In cancer, DNA methylation

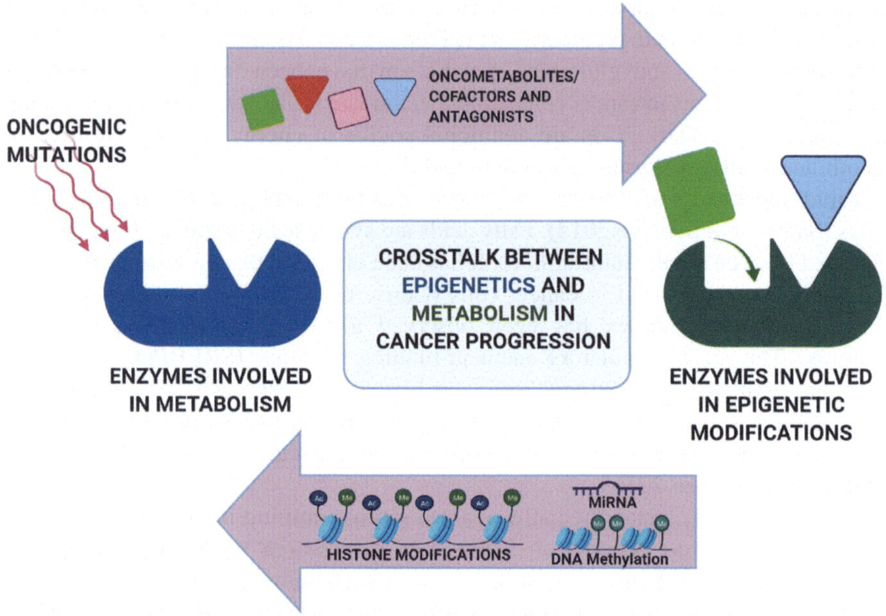

Fig. 16.1 Crosstalk between metabolism and epigenetics in the context of cancer progression. A two-way crosstalk exists between metabolism and epigenetics. Metabolic reprograming affects the availability of cofactors that are necessary for functioning of epigenetic modifying enzymes. The oncometabolites that are generated function as agonists/antagonists for the epigenetic enzymes, rewiring the epigenetic scenario. Epigenetic dysfunction in turn impacts metabolism by deregulating the metabolic enzymes and thus changing signalling pathways involved in cell metabolism

is observed in the promoter sites of tumour suppressor genes. Methylation marks on lysine residues of different histone proteins likewise have crucial roles in controlling gene transcription and chromatin structure. Histone methyl transferases (HMTs) mediating histone-lysine methylation basically consist of two enzyme families: SET-domain (SETD) and Dot1-like (DOT1L) proteins (Greer and Shi 2012). Mono-, di- or trimethylation of these lysine residues on histone H3 may be potential readers of complex histone methylation codes (Zhang and Reinberg 2001). The DNMT isoforms including DNMT1, DNMT3A and DNMT3B are overexpressed in different cancers (Robertson et al. 1999). DNMTs and HMTs use an activated methyl donor from SAM, which is a product of one-carbon metabolism and is synthesized by methionine adenosyl transferase (methionine and ATP are utilized as substrates) (Mentch et al. 2015). Nicotinamide N-methyl-transferase (NNMT) uses SAM as methyl donor to catalyse the conversion of nicotinamide to 1-methylnicotinamide (1-MNA). NNMT shows oncogenic effect and is overexpressed in different cancers including lungs, liver, kidney, bladder and colon (Wu et al. 2008; Tang et al. 2011; Thomas et al. 2013; Zhang et al. 2014). NNMT remains upregulated in human embryonic stem cells and plays an essential role in maintaining pluripotency. NNMT also serves as a sink for SAM, drastically reducing cellular SAM pool, making it unavailable for HMTs (Ulanovskaya et al. 2013). S-adenosyl-homocysteine (SAH) is produced on donation of methyl group from SAM, which in turn acts as an inhibitor of different methyltransferases. Thus the SAM/SAH ratio regulates methyltransferase activity. Dysregulation of carbon metabolism pathways can alter levels of SAM and methyl donors in cancers, which in turn can modify epigenetic alterations in tumour cells. SAH is homeostatically maintained at low levels through hydrolysis to homocysteine, which in turn is recycled to methionine via methyl group transfer from 5-methyl-tetrathydrofolate. In an alternative pathway, homocysteine can be catabolized to glutathione and inorganic sulphate. Thus, DNMT and HMT activity is crucially regulated by SAM/SAH ratio. The SAM/SAH ratio alterations largely affect aberrant histone methylations in cancers (Wong et al. 2017). On the other hand, excess levels of SAM can cause DNA hypermethylation at CpG sites resulting in undesirable gene silencing.

TCA cycle metabolites are prime targets through which metabolic pathways can affect epigenetic landscape. α-Ketoglutarate (α-KG), a TCA cycle intermediate, is an important cofactor for the activity of several histone demethylases. Inhibition of demethylases by TCA cycles intermediates can hypermethylate genes that lead to repression of tumour suppressor genes (Kooistra and Helin 2012). Acetyl CoA is involved in different aspects of cellular metabolism and is also a cofactor for enzymes that transfer acetyl groups, including histone acetyl transferases. Since histone acetylation is associated with transcriptional activation of genes, acetyl CoA availability is highly influenced by cellular metabolism pathways and broadly alters the epigenetic landscape. Oncoprotein transcription factor MYC is a key player of histone acetylation. MYC drives gene expression of metabolic enzymes linked to acetyl CoA synthesis, including glycolysis and glutaminolysis. MYC upregulates the expression of GCN5, which belongs to the family of histone acetyl transferases (HAT), and regulates protein function by acetylation and in turn induces mono-, di-, tri- and tetra-acetylation at the N-terminal of histone H4. MYC thus plays a role in gene expression of HATs and availability of acetyl CoA to promote histone acetylation in response to proliferative signals (Knoepfler et al. 2006). TCA cycle intermediate enzymes FH and succinate dehydrogenase (SDH) are driver mutations in a subset of human cancers like renal cell carcinoma, paraganglioma, etc., and they mediate epigenetic rewiring (Pollard et al. 2003). Oncogenic-driven metabolic rewiring in cancers promotes increased levels of glycolytic flux and mitochondrial production of citrate which in turn is translocated to the cytosol and nucleus.

Cancer cells harbouring mutations in metabolic genes can accumulate or synthesize metabolites like 2-hydroxylglutarate (2-HG), succinate and fumarate. Mutations in isoforms of metabolic enzymes like isocitrate dehydrogenase (IDH1/2) are seen in gliomas, AML and angioimmunoblastic T-cell lymphoma (Turcan et al. 2012; Venneti et al. 2013). Mutant IDH1/2 possess oncogenic properties, and their ectopic expression increases cancer cell proliferation and colony formation and inhibits cellular differentiation in vitro (Wardell et al. 2009; Dang et al. 2009). These mutations facilitate the conversion of α-KG to 2-HG. 2-HG can function as a competitive agonist of α-KG. 2-HG levels in IDH1/2 mutant tumours are associated with histone demethylase activity and strongly inhibit several histone demethylases (Dang et al. 2009). Human glioma cells show that 2-HG or overexpression of mutant IDH1 can increase global H3K9, H3K27, H3K79 dimethylation and H3K4 trimethylation levels (Xu et al. 2011). IDH1 knockdown in hematopoietic cells led to increased dimethylation of H3K79 and trimethylation of H3K4, H3K9, H3K27 and H3K36. Modulations of histone marks, specifically H3K9Me3, promote pluripotency and inhibit differentiation (Lu et al. 2012).

NAD^+ plays an essential role in deacetylation activity of sirtuins (SIRTs), a subgroup of HDACs. $NAD^+/NADH^+$ ratio, which is closely associated with energy status in cells, positively regulates the activity of SIRTs. Thus, by sensing the ratio, the sirtuins can bridge between energy status and gene expression (Li and Kazgan 2011). This ratio is starkly reduced during high glycolytic activity in cancers, which in turn inhibits SIRTs. Increased activity of HAT, induced by acetyl CoA along with repressed SIRTs, facilitates histone hyperacetylation and aberrant gene transcription (Wong et al. 2017).

O-linked N-acetylglucosaminylation (O-GlcNAcylation) is a nutrient and stress-driven post-translational modification by attachment of O-linked β-D-N-acetylglucosamine (O-GlcNAc) to serine and threonine residues. O-GlcNAc is synthesized by the hexosamine biosynthetic pathway (HBP), where glucose entering glycolytic pathway is initially metabolized to glucose-6-phosphate (G6P) and then fructose-6-phosphate. The HBP integrates substrates from carbohydrates, amino acid, fat and nucleotide metabolism. There is an upregulation of HBP in cancer cells and is primarily driven by increased glucose uptake and metabolism. O-GlcNAcylation is controlled by competitive action of O-GlcNAc transferase (OGT) and O-GlcNAcase (OGA) promoting addition and removal of O-GlcNAc from proteins, respectively. Modifications of all core histone proteins by O-GlcNAc generate a part of the histone code (Sakabe and Hart 2010; Myers et al. 2011; Fujiki et al. 2011; Fong et al. 2012). Ten-eleven translocation (TET) family of proteins are putative tumour suppressors that catalyse the demethylation of methylated cytosine residues (Tahiliani et al. 2009; Ito et al. 2010; Guo et al. 2011). Downregulation of TETs has important roles in a variety of human cancers (Kudo et al. 2012; Yang et al. 2013; Gambichler et al. 2013). OGT complexes with TETs to modulate O-GlcNAcylation of histone H2B for active transcription(Chen et al. 2013a; Ito et al. 2014). Conversely, its interaction with Ezh2 leads to tumour suppressor gene silencing via H3K27me3 (Chu et al. 2014). Increased OGT levels and protein hyper-O-GlcNAcylation is commonly seen in human cancers (Pinho and Reis 2015).

Role of Epigenetic Rewiring on Metabolic Gene Expression

The opposite is also true: epigenetic rewiring in cancer cells can significantly alter metabolic gene expression and disrupt metabolic pathways. Here, we summarize some of these key events.

DNA Methylation

DNA methyltransferases (DNMTs) comprising of DNMT1, DNMT2, DNMT3a, DNMT3b and DNMT3L catalyse the formation of 5-methylcytosine (5mC) through the transfer of a methyl group from the cofactor S-adenyl methionine (SAM) to the cytosine residues in DNA. This is a major repressive mark that represses transcription. DNA methylation can alter expression of metabolic genes by dysregulation of various oncogenic pathways that include PI3K/AKT/mTOR complex 1, adenosine monophosphate kinase (AMPK) and hypoxia inducible factor-1α (HIF1α). DNA hypermethylation results in GLUT1 overexpression by epigenetic repression of Derlin-3, a key gene involved in GLUT1 proteasomal degradation (Lopez-Serra et al. 2014). GLUT1 overexpression allows tumour cells to increase absorption and metabolism of glucose, thus allowing tumour growth. DNA hypermethylation also facilitates silencing of fructose-1,6-bisphosphatase 1 (FBP1) and fructose-1,6-bisphosphatase 2 (FBP2), which are rate-limiting enzymes in gluconeogenesis, via promoter methylation in breast, gastric, liver and colorectal cancers (Chen et al. 2011; Yang et al. 2017a; Dong et al. 2018; Li et al. 2020b). These enzymes antagonize glycolysis, and thus their downregulated levels promote glycolytic flux for ATP synthesis and production (Li et al. 2014). Hypermethylation also leads to the downregulation of tumour suppressors that regulate the AMPK, PI3K/Akt/mTORC and HIF1α pathway, including SKT11 or liver kinase B1 (LKB1) (Qanungo et al. 2003), PTEN, von Hippel-Lindau protein (VHL) (Herman et al. 1994) and prolyl hydroxylase 3 (PHD3) (Rawluszko et al. 2013). Indeed, PTEN loss through hypermethylation is a common aberration across many different forms of cancer (García et al. 2004). Thus, these glycolysis-promoting pathways are activated through DNA hypermethylation. In addition, promoter hypomethylation is responsible for the upregulation of PKM2, which drives glucose flux towards macromolecule biosynthesis (Desai et al. 2013). Hypomethylation also results in the upregulation of Type II hexokinase (HKII), which accelerates glycolytic flux in liver cancer (Goel et al. 2003). Deregulated DNA methylation, therefore, contributes towards a glycolytic flux in cancer cells.

Histone Modifications

Histones can acquire a variety of post-translational modifications in the nucleosomal state, particularly on arginine and lysine residues, which determines their interactions with other histone subunits in the nucleosome, as well as with the associated chromatin. The effect of a particular histone modification depends on the type of modification, such as the group/combination of groups and the number of groups added, and the residue that is modified. Among these, the most well-studied and best-understood modifications are methylation, acetylation and phosphorylation. However, recently, there has been increasing attention paid to lesser known but potentially critical modifications, such as formylation, propionylation, butyrylation, crotonylation, malonylation, succinylation, hydroxylation, ubiquitination, sumoylation, adenosine diphosphate (ADP)-ribosylation, citrullination and glycosylation (Rothbart and Strahl 2014). SIRTs are important histone modification enzymes involved in regulating glucose homeostasis and cell metabolism. The SIRT family consists of seven members (SIRT1–7), which affect glucose metabolism in different ways.

SIRT6 inhibits transcription of HIF1α and MYC by functioning as a co-repressor through histone deacetylation (Zhong et al. 2010, p. 6; Sebastián et al. 2012, p. 6). Thus, SIRT6 represses HIF1-dependent glycolytic switch and MYC-dependent ribosome biogenesis and glutaminolysis. Frequent deletion of SIRT6 has been observed in colon, pancreatic and hepatocellular cancer cell lines in line with its tumour-suppressive role (Zwaans and Lombard 2014). SIRT7 also opposes MYC-dependent gene regulation and metabolic alterations through its ability to catalyse H3K18Ac marks (Shin et al. 2013, p. 7). SIRT2, on the contrary, suppresses the expression of E3 protein ubiquitin ligase NEDD4, a negative regulator of MYC, by deacetylating histones at H4K16 (Liu et al. 2013, p. 2). Therefore, SIRT6/7 and SIRT2 have opposing roles in MYC-dependent metabolism. Other histone deacetylases involved in MYC activity and stabilization are HDAC1 and 2, which also upregulate MYC and thereby promote oncogenic signalling (Nguyen et al. 2020a). HDACs also cause the downregulation of FBP1 and HKI (Wardell et al. 2009; Yang et al. 2017b) and the upregulation of HKII and GLUT1 (Wardell et al. 2009; Nakajima et al. 2019).

Methyltransferases also play critical roles. EHMT2 is known to activate the serine-glycine biosynthetic pathway through its H3K9 monomethylation activity, thus increasing proliferation and tumourigenicity of cancer cells (Ding et al. 2013). Among the many targets of Prc2/Ezh2 is PTEN, a classic tumour suppressor lost in many forms of cancer (Gan et al. 2018; Jarome et al. 2018). The histone demethylase LSD1 can reduce expression of the gluconeogenesis-related enzymes FBP1 and glucose-6-phosphatase (G6Pase) through demethylation; removal of H3K4me2 marks reduces the expression of these genes (Pan et al. 2013).

RNA Epigenetics

Non-coding RNAs (ncRNAs) have likewise come to be recognized as crucial regulators of gene expression and cell function, often deregulated in expression under pathological circumstances. Several areas of cellular metabolism are regulated by miRs. miRs regulate glycolysis and mitochondrial TCA cycle, as well as impact signal transduction pathways via PI3K/AKT, HIF1α and MYC, which that affect metabolism in different human cancers. Genes related to glucose uptake are regulated by numerous miRs. Among these are miR-1291, which inhibits GLUT1 (Yamasaki et al. 2013, p. 12), miR-195-5p and miR-106a which regulate GLUT3 (Fei et al. 2012; Dai et al. 2013) and miR-93, miR-223, miR-106b, miR-27a and miR-30d which directly regulate GLUT4 (Lu et al. 2010; Chen et al. 2013b; Chuang et al. 2015; Zhou et al. 2016). HKII is regulated by miR-125/a (Jin et al. 2017; Sun et al. 2017b; Luo et al. 2020), miR513a-3p (Li et al. 2020a), miR185 (Liu et al. 2019), miR-143 (Gregersen et al. 2012; Fang et al. 2012; Peschiaroli et al. 2013; Sun and Zhang 2017) and miR-145 (Yoshino et al. 2013), as well miR-155, which indirectly upregulates HKII by downregulating miR143 (Jiang et al. 2012; Jin et al. 2017). miR-200 members regulate glucose-6-P isomerase (GPI)(Ahmad et al. 2011). miR-15a/16-1 and miR-122 regulate aldolase A (ALDOA) (Calin et al. 2008; Li et al. 2019). miR-326, miR148a and miR133a/b regulate PKM2 (Wu et al. 2016; Yu et al. 2018). PKM2 is also indirectly regulated by miRs; miR140-5p inhibits the transcription factor Sine oculis homeobox homolog 1 (SIX1) (Nie et al. 2019), while miR361-5p inhibits Specificity protein 1 (Sp1) (Ling et al. 2017), both of which induce PKM2 expression. Further down the glucose metabolism pathway, miR-23a/b targets glutaminolysis via glutaminase (GLS) (Gao et al. 2009; Rathore et al. 2012). In addition, miR-210 inhibits mitochondrial function by suppressing iron-sulphur cluster assembly proteins (ISCU1/2) (Sun et al. 2017a). It therefore facilitates a shift towards glycolytic phenotype and lactate production, crucial for adaptation to hypoxic tumour microenvironment. miR-29 and miR-451 inhibit AKT and AMPK, respectively (Godlewski et al. 2010; Teng et al. 2015). Therefore, deregulation of the above-mentioned miRs contributes to upregulated glycolysis and glutaminolysis in different cancers.

We have summarized above a few examples in which metabolic alterations regulate the epigenetic landscape. Thus, targeting metabolic pathways alters epigenetic control of gene expression. Reciprocally, targeting epigenetic pathways alters expression of enzymes involved in metabolic pathways. Future investigations will unfold the complex interactions between metabolism and chromatin dynamics in normal and cancer cells.

Therapeutic Strategies to Target the Epigenetic-Metabolomic Crosstalk

Small-molecule drugs targeting epigenetic modifiers have been developed and used for the purpose of targeting deregulated pathways in cancer. Drugs for treating metabolic disorders have also been repurposed to treat cancer, given the importance of metabolic reprogramming in cancer. In this section, we summarize both drugs that alter metabolic enzymes which are linked to epigenetic modifiers and drugs that alter epigenetic modifiers that control metabolic enzymes.

Glycolytic flux and glutaminolysis interfere with histone acetylation and thus contribute to modulation of acetyl CoA levels (Cluntun et al. 2015). 2-deoxyglucose (2-DG), a glucose analogue, reduces acetyl CoA levels which in turn leads to global suppression of histone acetylation and sensitization of cancer cells to DNA damaging agents (Liu et al. 2015). Similarly, glycolytic inhibitor 3-bromopyruvate can induce embryonic stem cell differentiation by reducing acetyl CoA levels (Moussaieff et al. 2015). GLS inhibitors including BPTES (Robinson et al. 2007),

CB-839 (Robinson et al. 2007) and compound 968 (Wang et al. 2010) can significantly modulate H3 and H4 acetylation, and in turn suppress expression of relevant tumour-related genes in breast cancer (Gross et al. 2014). Zaprinast, which is an IDH2 inhibitor, shows GLS inhibitory activity. The drug can modulate histone methylation instead of histone acetylation (Wong et al. 2017). To reshape modified DNA and histone methylation in cancer cells, one strategy is to target the SAM/SAH ratio which is part of SAM synthesis. DZNep (3-deazaneplanocin A), the inhibitor of s-adenosylhomocysteine hydrolase (AHCY), is the enzyme responsible for SAH hydrolysis to adenosine and homocysteine (Glazer et al. 1986). Interestingly, DZNep participates in the maintenance of methylation homeostasis, which in turn causes suppression of SAM, the essential mediator of DNMT and HMT activity. DZNep in combination with the DNMT inhibitors showed synergistic anti-cancer activity in leukaemia by reactivating genes that were silenced by histone and DNA methylation (Miranda et al. 2009). To counteract modified DNA hypermethylation facilitated by 2-HG production in IDH1/2 mutant cancers, promising inhibitors targeting mutant IDH1/2 have been approved for acute myeloid leukaemia (Golub et al. 2019). AG-120, AG-881, ML309, GSK321 and GSK864 are some of the IDH1/2 mutant inhibitors that have histone and DNA demethylation roles and show anti-tumour activity. Significantly, AG-221, an inhibitor of IDH2 mutant, is under clinical trial (phase I and II), which significantly reduces 2HG levels in the bone marrow and plasma and diminishes disease severity in patients with IDH2 mutant advanced haematologic malignancies (Popovici-Muller et al. 2018) (Table 16.1 and Fig. 16.2).

Drugs that target epigenetic modifiers involved in deregulated metabolism also have significant therapeutic potential. Many of these drugs are either approved for use by the FDA or are at various stages of clinical trials, often in combination with other standard chemotherapy drugs. These drugs provide a valuable alternate means of targeting metabolic reprogramming (the other being the direct targeting of various enzymes involved in the processes themselves) and have the additional advantage of targeting various cancer phenotypes that lie downstream of the modifiers. Often, epigenetic modifiers have a range of targets which can be involved in different phenotypes. Thus, these modifiers provide a way to regulate multiple targets and phenotypes, including metabolic deregulation. For example, the DNMT inhibitor azacitidine, in combination with other drugs such as venetoclax, has been shown to disrupt energy metabolism in leukaemia stem cells (Pollyea et al. 2018). However, it also targets a number of other phenotypes, such as proliferation and immune evasion by de-repressing various other targets of hypermethylation (Issa et al. 2005; Derissen et al. 2013). In addition, whereas the effect of azacitidine on some metabolic targets has been directly studied (see Table 16.1), others, which are repressed by hypermethylation, and presumably de-repressed by azacitidine treatment and DMNT inhibition, have not been directly studied. Finally, there are drugs that have been shown to affect epigenetic marks that are known to deregulate metabolic targets but have not been studied directly for their effect on these targets. For example, BET proteins affect the expression of MYC proteins, which play an important role in glycolytic shift (Delmore et al. 2011; Mertz et al. 2011). However, not all BET inhibitors have been directly studied for their effect on MYC expression.

Table 16.1 Drugs targeting tumour metabolome that impact epigenetic landscape

Drug	Chemical structures	Target	Method of action	Used in the treatment of	Stage of trial	References
2-Deoxyglucose (2-DG)	$C_6H_{12}O_5$ 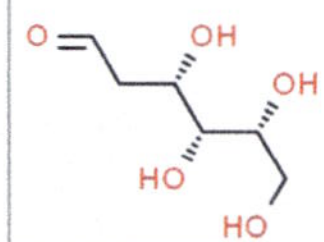	Hexokinases	2-DG inhibits hexokinase and diminishes acetyl CoA levels. Reduces the acetylation levels of histones H3, H4, H2A and H2B	Multiple cancer cell lines	Phase I/II	(Chen and Guéron (1992), Liu et al. (2015)
3-Bromopyruvate (BrPA)	$C_3H_3BrO_3$ 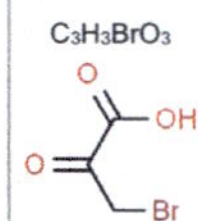	Glyceraldehyde 3-phosphate dehydrogenase	BrPA suppresses GAPDH and acetyl CoA levels. Inhibits histone acetylation. Induces differentiation of embryonic stem cells	Prostate cancer cell lines, breast cancer cell lines and hepatocellular carcinoma cell lines	NA	Moussaieff et al. (2015)
Bis-2-(5-phenylacetamido-1,2,4-thiadiazol- 2-yl) ethyl sulfide (BPTES)	$C_{24}H_{24}N_6O_2S_3$ 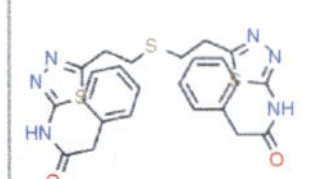	Glutaminase (GLS)	Suppresses acetyl CoA and 2-HG levels. Can alter acetylation levels of histones H3 and H4	Breast cancer	NA	Robinson et al. (2007)
CB-839 (in combination with standard chemotherapy)	$C_{26}H_{24}F_3N_7O_3S$ 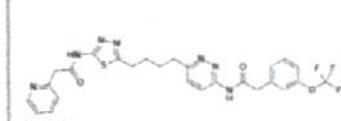	Glutaminase (GLS)	Suppresses acetyl CoA and 2-HG levels. Can alter acetylation levels of histones H3 and H4	Solid tumours Triple-negative breast cancer Non-small cell lung cancer Renal cell carcinoma Mesothelioma	Phase I/II	Calithera Biosciences, Inc. (2017, 2018), Bajor (2020), National Cancer Institute (NCI) (2021)
Compound 968		Glutaminase (GLS)	Suppresses acetyl CoA and 2-HG levels.	Breast cancer	NA	Yuan et al. (2016), Wang et al. (2016)

(continued)

Table 16.1 (continued)

Drug	Chemical structures	Target	Method of action	Used in the treatment of	Stage of trial	References
	$C_{27}H_{27}BrN_2O$ $C_{27}H_{27}BrN_2O$ 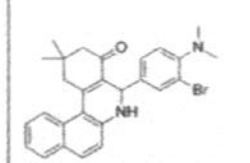		Reduces H3K4me3 levels			
Zaprinast	$C_{13}H_{13}N_5O_2$ $C_{13}H_{13}N_5O_2$ 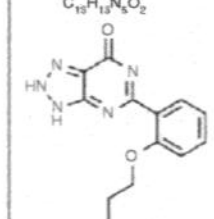	Glutaminase (GLS)	Suppresses acetyl CoA and 2-HG levels. Reduces H3K27me2/3 levels	IDH1 mutant cancer cells	NA	Elhammali et al. (2014)
AG-120	$C_{28}H_{22}ClF_3N_6O_3$ 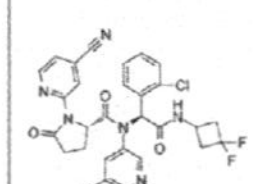 AG-120	Mutant IDH1	Suppresses 2-HG production in IDH1 mutant cells. It is a regulator of methylation	Glioma, acute myeloid leukaemia, hepatic impairment, chondrosarcoma	Phase I/II/III	Agios Pharmaceuticals, Inc. (2018a, b, 2020a)
AG-881	$C_{14}H_{13}ClF_6N_6$ $C_{14}H_{13}ClF_6N_6$ 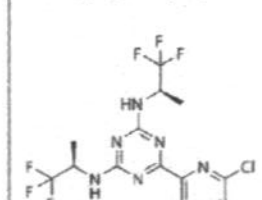	Mutant IDH1	Suppresses 2-HG production in IDH1 mutant cells	Acute myeloid leukaemia (AML), myelodysplastic syndrome, haematologic malignancies, glioma	Phase I	Agios Pharmaceuticals, Inc. (2019, 2020b)
BAY1436032		Mutant IDH1	Suppresses 2-HG production in IDH1 mutant cells. AML cell	Acute myeloid leukaemia	Phase I	Bayer (2019, 2020)

	$C_{26}H_{30}F_3N_3O_3$ 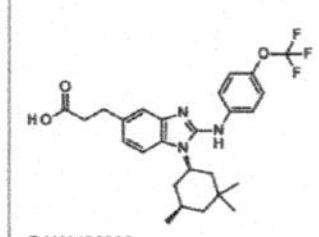 BAY1436032		lines show the efficacy of this drug in reducing histone methylation			
FT-2102	$C_{18}H_{15}ClN_4O_2$ $C_{18}H_{15}ClN_4O_2$ 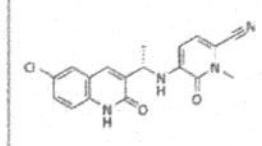	Mutant IDH1	Suppresses 2-HG production in IDH1 mutant cells. It can inhibit DNA and histone demethylation, leading to hypermethylation	Glioma Myelodysplastic syndrome Acute myeloid leukaemia	Phase I/II	forma Therapeutics, Inc. (2019, 2020)
IDH305	$C_{23}H_{22}F_4N_6O$ $C_{23}H_{22}F_4N_6O$ 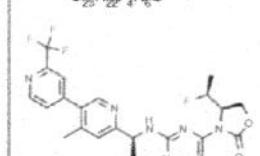	Mutant IDH1	Suppresses 2-HG production and induces H3K9me3/H3K27me3 levels in IDH1 mutant cells	Glioma	Phase I	Novartis Pharmaceutica ls (2020)
AGI-5198	$C_{27}H_{31}FN_4O_2$ $C_{27}H_{31}FN_4O_2$ 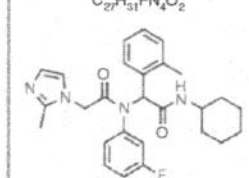	Mutant IDH1	Suppresses 2-HG production in IDH1 mutant cells and induces demethylation of H3K9me3 and H3K27me3.	IDH1 mutant chondrosarcoma cells	NA	Johannessen et al. (2016); Popovici-Muller et al. (2018)
bis-Imidazole phenol (cpd 1)	$C_{22}H_{30}N_4O$ $C_{22}H_{30}N_4O$ 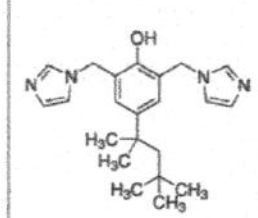	Mutant IDH1	Suppresses 2-HG production in IDH1 mutant cells		NA	Deng et al. (2015)

(continued)

Table 16.1 (continued)

Drug	Chemical structures	Target	Method of action	Used in the treatment of	Stage of trial	References
GSK321	$C_{28}H_{28}FN_5O_3$ 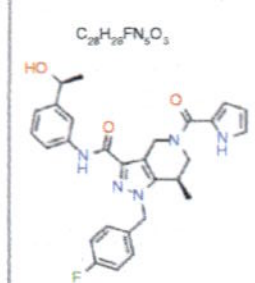	Mutant IDH1	Suppresses 2-HG production and enhances genome-wide DNA hypomethylation in IDH1 mutant cells	IDH1 mutant acute myeloid leukaemia cells	NA	Dang et al. (2009)
GSK864 (drug with improved pharmacokinetic properties of GSK321)	$C_{30}H_{31}FN_6O_4$ 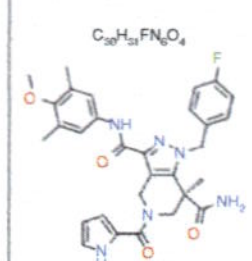	Mutant IDH1	Suppresses 2-HG production and enhances genome-wide DNA hypomethylation in IDH1 mutant cells	IDH1 mutant acute myeloid leukaemia cells	NA	Calvert et al. (2017)
1-hydroxypyridin- 2-one compounds	C_5H_5NO 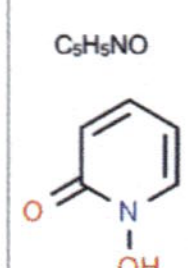	Mutant IDH1	Suppresses 2-HG production in IDH1 mutant cells	IDH1 mutant cells	NA	Zheng et al. (2013)
ML309	$C_{29}H_{29}FN_4O_2$ 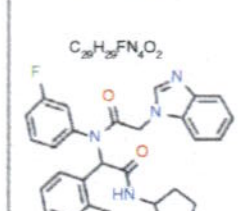	Mutant IDH1	Suppresses 2-HG production in IDH1 mutant cells. Shows histone and DNA demethylation activity	IDH1 mutant acute myeloid leukaemia cells	NA	Davis et al. (2014)

2-(3-Trifluoromethylphenyl) isothioazol-3(2H)-one	$C_{10}H_6F_3NOS$ 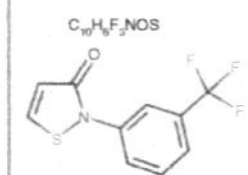	Mutant IDH1	Suppresses 2-HG production in IDH1 mutant cells. It is a IDH1-R132H inhibitor	IDH1 mutant glioma cells	NA	Kim et al. (2015)
Enasidenib (AG-221)	$C_{19}H_{17}F_6N_7O$ 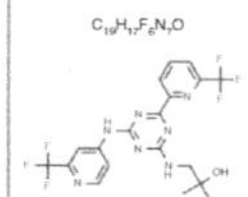	Mutant IDH2	Suppresses 2-HG production in IDH1 mutant cells. AG-221 induces demethylation of DNA and reverses histone hypermethylation profile in IDH2- mutant leukaemia cell lines	Acute myeloid leukaemia, solid tumour, glioma, angioimmunoblastic T-cell lymphoma, intrahepatic cholangiocarcinoma, chondrosarcoma	Phase I/III	Celgene (2018, 2020), Groupe Francophone des Myelodysplasi es (2020)
Vorasidenib (AG-881)	$C_{14}H_{13}ClF_6N_6$ 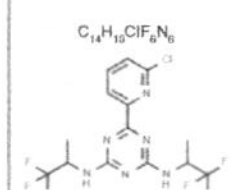	Mutant IDH2	Suppresses 2-HG production in IDH1 mutant cells	Acute myeloid leukaemia, myelodysplastic syndrome, haematologic malignancies, glioma	Phase I	Agios Pharmaceutica ls, Inc. (2020a, c)
AGI-6780	$C_{21}H_{18}F_3N_3O_3S_2$ 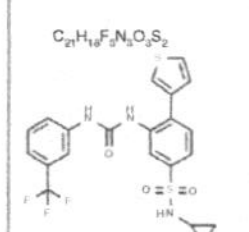	Mutant IDH2	Suppresses 2-HG production in IDH1 mutant cells. AGI-6780 induces demethylation of DNA and histone marks in IDH2- mutant leukaemia cell lines	Acute myeloid leukaemia	NA	Wang et al. (2013), Kernytsky et al. (2015), Chen et al. (2017)
DZNep (3-deazaneplanocin A)		SAH hydrolase	Increases the SAH to SAM ratio. Inhibits trimethylation of H3K27me3 and	Multiple cancer cell lines	NA	Miranda et al. (2009)

(continued)

Table 16.1 (continued)

Drug	Chemical structures	Target	Method of action	Used in the treatment of	Stage of trial	References
	$C_{12}H_{16}N_4O_3$ 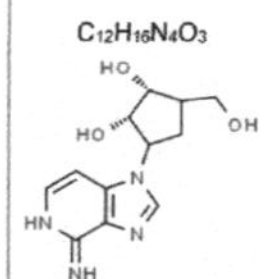		H4K20me3 and reactivates silenced genes			
Adenosine dialdehyde (AdOx)	$C_{10}H_{11}N_5O_4$ 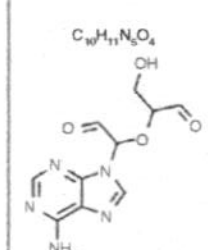	SAH hydrolase	Increases the SAH to SAM ratio. It is an indirect methyltransferase inhibitor and inhibited DNA and histone methylation	Multiple cancer cell lines	NA	Chen et al. (2004)
N-methylnicotinamide	$C_7H_8N_2O$ 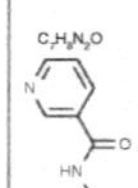	Nicotinamide N-methyltransferase (NNMT)	N-methylnicotinamide increased SAM level. It impairs the methylation potential of cancer cells	NNMT-overexpressing cancer cells	NA	Bayer (2019, 2020)

Cancer metabolism pathways can dysregulate epigenetic processes. Metabolic reprogramming in tumour cells promote accumulation of particular metabolites which interfere with epigenetic effectors. This table identifies metabolic substrates that target enzymes which in turn dysregulate epigenetic processes and gene expression. Drugs targeting metabolic enzymes leading to epigenetic dysregulation

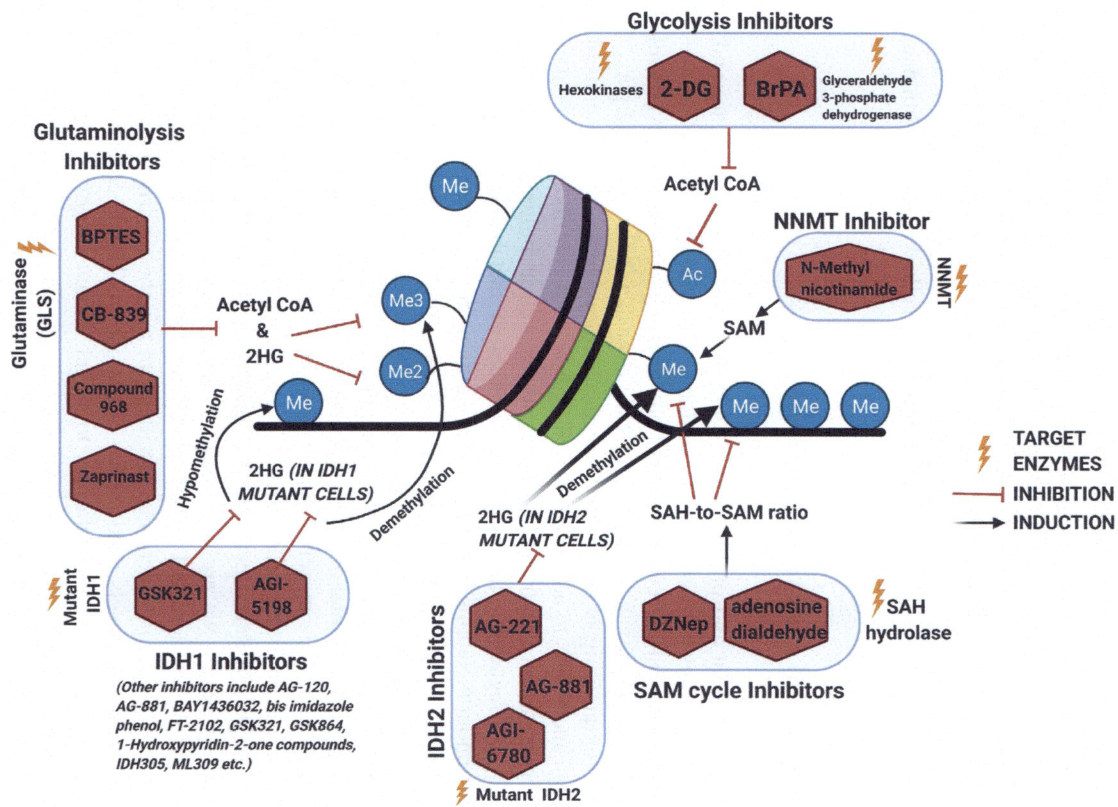

Fig. 16.2 Drugs targeting tumour metabolome that impact epigenetic landscape. The cancer metabolic pathways modulate the epigenetic processes like histone acetylation, DNA methylation DNA/histone demethylation, N-linked glycosylation, etc., in the context of tumour progression. The altered epigenetic scenario in turn deregulates gene expression. (linked to Table 16.1)

Therefore, we have summarized only those drugs that have been shown to target epigenetic modifiers involved in metabolic deregulation and consequently have metabolic targets (Table 16.2 and Fig. 16.3). The full range of effects of these drugs is still under study, and further research may well elucidate other metabolic targets. In addition, other small-molecule inhibitors that are under development, but have not been understood for their role in regulating metabolic targets, will likely also be understood as such. As with all drugs, one major concern with the use of the small-molecule inhibitors targeting metabolic function will be the potential off-target effects as well as toxicity to healthy cells. Hence, the deeper and more nuanced our understanding of the differences between healthy cells and malignant ones, as well as our understanding of the full range of effects and targets that a drug has, the more effectively we can design drugs to target only those cells we wish to eliminate.

Conclusion and Future Perspectives

Cancer development and progression involve mutation of specific genes and dysfunction of important biological processes that lead to transformation of normal cells to malignant ones. In this context, "metabolic rewiring" has taken a pivotal role for providing selective advantage to tumour cells over their normal counterparts. Metabolic reprogramming is an active driver of carcinogenesis and is responsible for favouring cancer cells to adapt in the modified niche. It contributes to regulation of tumour environment and overall homeostasis, as well as influences the epigenetic machinery by regulating the steady-state levels of metabolites. Epigenetics can promote tumour metabolic rewiring by facilitating reprogramming of genes involved in metabolic functions. Thus, the dynamic and reciprocal crosstalks between epigenetics and metabolism are functionally critical to tumour progression.

We have summarized in the present chapter the current understanding of the links between epigenetic regulation and cancer metabolism and reviewed all the small-molecule drugs that are currently under development or in use to target either the deregulated metabolic enzymes which affect epigenetic mechanisms or deregulated epigenetic regulators that affect metabolic enzymes and processes. However, despite appreciable progress, we are still challenged with several unanswered questions which need to be addressed before developing potential therapeutic strategies. Firstly, the mechanisms underlying specificity of metabolite-driven epigenetic regulation of particular targets need further investigation. This is critical when determining how to use these metabolites to target certain enzymes and modifiers. Secondly, the conditions which are required to cue the same metabolite to target different chromatin regions remain unclear. Characterizing these cues is important for drug development. Thirdly, identification of transcription factors activated in metabolic challenges will help understand how the crosstalk between epigenomics and metabolism is integrated. Finally, it will be necessary to gain a deeper understanding of how effective these drugs are in regulating other oncogenic phenotypes,

Table 16.2 Drugs targeting epigenetic regulators of metabolic enzymes

Drug	Target	Metabolic targets/effects	Used in the treatment of	Phase of trial	References
Azacitidine (Vidaza®)	DNMT	Restoration of PTEN expression Restoration of VHL expression Restoration of FBP1 expression	MDS, AML	(As monotherapy) FDA approved (As polytherapy) phase I/II/III	Christman (2002), Derissen et al. (2013), Song et al. (2014), Strand et al. (2020)
Decitabine (5-Aza- 2-′-deoxycytidine/ Dacogen®)	DNMT	Restoration of PTEN expression Increase in VHL (in combination with HDAC inhibitors) Increased FBP1 expression Restoration of LKB1 expression	MDS, AML, HER2-negative BC, TCL	(As monotherapy) FDA approved (As polytherapy) phase I/II/III	Qanungo et al. (2003), Chen et al. (2011), Kalac et al. (2011)
CUDC-101	HDAC	Inhibits Akt signalling	HNSCC, LC, BC, NSCLC, GC	(As monotherapy/ polytherapy) phase I	Lai et al. (2010), Galloway et al. (2015)
Belinostat (Beleodaq/ PXD101)	HDAC	Inhibits PI3K/Akt/mTOR signalling Inhibits HIF1α signalling	CS, MDS, AML, TCL, BC, OC, lymphoma	(As monotherapy) FDA approved (As polytherapy) phase I/II	(Chien et al. (2014), Lee et al. (2015), Sawas et al. (2015)
Panobinostat (LBH-589	HDAC	Inhibition of hexokinase II Restoration of FBP1 expression Downregulation of HIF1α, GLUT1	DIPG, MM	FDA approved (As monotherapy/ polytherapy) phase I/II	Hainsworth et al. (2011), Richardson et al. (2015), Raedler (2016)
Valproic acid (VPA)	HDAC	Downregulates glycolysis, GPI and PGK1 expression through E2F1 inhibition Activation of AMPK	PC, MDS, AML, melanoma, glioma, HNSCC	(As polytherapy) phase II	Sharma et al. (2008), Avery and Bumpus (2014), Caponigro et al. (2016), Fang et al. (2019)

(continued)

Table 16.2 (continued)

Drug	Target	Metabolic targets/effects	Used in the treatment of	Phase of trial	References
		Inhibition of HDAC1/2-mediated activation of myc Inhibition of GLUT1 expression Inhibition of HK1 activity			
Vorinostat (SAHA)	HDAC	Inhibition of HDAC1/2-mediated activation of myc Inhibition of GLUT1 expression Inhibition of HK1 activity	CTCL, AML, NSCLC, BC, CRC,	FDA approved (As polytherapy) phase I/II/III	Haematology-Oncology (2013), H. Lee Moffitt Cancer Center and Research Institute (2014)
Romidepsin (depsipeptide/ FK22 8)	HDAC	Inhibition of HDAC1/2-mediated activation of c- myc	CTCL, PTCL, relapsed/ refractory TCL	FDA approved (As monotherapy) phase II (As polytherapy) phase I/II/III	Nguyen et al.; Celgene (2019)
OTX-015 (birabresib/MK-8628)	BET	Downregulation of c-myc	AML, ALL, BCL, MM, NMC, PC, NSCLC, PDAC, triple-negative BC	Phase I	Boi et al. (2015), Coudé et al. (2015), Amorim et al. (2016), Berthon et al. (2016), Stathis et al. (2016), Berenguer-Daizé et al. (2016), Vázquez et al. (2016), Oncoethix GmbH (2018, 2019)
CPI-0610	BET	Downregulation of myc	AML, MM, myelofibrosis	(As polytherapy) phase I/II/III	Moros et al. (2014), Albrecht et al. (2016), Siu et al. (2017), Constellation Pharmaceuticals (2020)
I-BET762 (molibresib/ GSK525762)	BET	Downregulation of myc	NMC	Phase I/II	Wyce et al. (2013), Leal et al. (2017), Xie et al. (2018)
CPI203	BET	Downregulation of myc	MCL	Preclinical	Siegel et al. (2014)

GS-5829	BET	Downregulation of p-Akt and myc	ER-positive/HER2-negative BC, solid tumours, lymphomas	Phase I/II	Gilead Sciences (2019), Kim et al. (2020)
MRG106 (anti-miR-155/cobomarsen)	miRNA	Re-expression of miR143 and inhibition of HKII	CTCL, CLL, MF, BCL, TCL	Phase I	Querfeld et al. (2016), Foss et al. (2018), Seto et al. (2018), miRagen Therapeutics, Inc. (2020)

Epigenetic modifiers of metabolic enzymes for important therapeutic target and many small-molecule drugs are under clinical trials for the treatment of various forms of cancer. While many metabolic enzymes have been shown to be deregulated by aberrant epigenetic marks and could, as a result, be restored in principle by small-molecule drugs targeting these epigenetic modifiers, not all metabolic targets of these drugs have been adequately studied or identified. Here, we identify only those direct links between epigenetic drugs and metabolism, particularly glucose metabolism. It is entirely possible that other epigenetic drugs, not mentioned here, also target some of these metabolic targets, as well as others, but these links have not been directly identified and therefore have been omitted. (Abbreviations: *ALL* acute lymphoid leukaemia, *AML* acute myeloid leukaemia, *BC* breast cancer, *BCL* B-cell lymphoma, *CLL* chronic lymphocytic leukaemia, *CRC* colorectal cancer, *CS* chondrosarcoma, *CTCL* cutaneous T-cell lymphoma, *DIPG* diffuse intrinsic pontine glioma, *ER* oestrogen receptor, *GC* gastric cancer, *HER2* human epidermal growth factor receptor 2, *HNSCC* head and neck squamous cell carcinoma, *LC* liver cancer, *MDS* myelodysplastic syndrome, *MF* mycosis fungoides, *MM* multiple myeloma, *NMC* nuclear protein in testis midline carcinoma, *NSCLC* non-small cell lung cancer, *OC* ovarian cancer, *PC* prostate cancer, *PDAC* pancreatic ductal adenocarcinoma, *PTCL* peripheral T-cell lymphoma, *TCL* T-cell lymphoma)

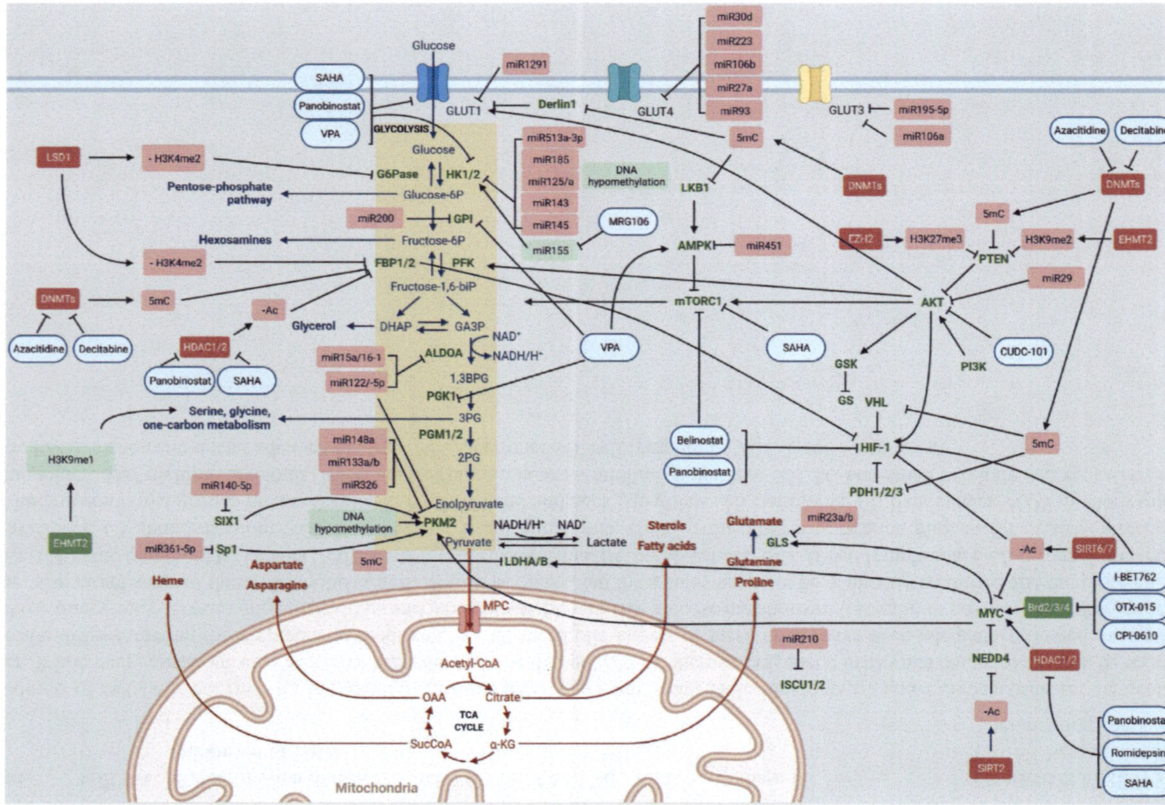

Fig. 16.3 Drugs targeting epigenetic regulators of metabolic enzymes. Given the importance of epigenetic modifiers in regulating or deregulating the expression of various components of metabolic pathways, these modifiers are proving to be increasingly valuable targets for cancer therapy. Here, we summarize some of the main metabolic pathways, particularly those linked to glucose metabolism, which are known to be deregulated through epigenetic mechanisms. Also shown are some drugs that have been shown to affect these modifiers and, consequently, their metabolic targets (linked to Table 16.2)

such as aggressiveness and cancer stem cell properties, or how effectively they can be used in combination with other treatments directed against these phenotypes.

Advancements in epigenome-mapping projects and genome sequencing techniques in different cancers will contribute to development of epi-drug therapies. Structural studies on histone-modifying enzymes and investigating their interacting partners are also essential to elucidate drug targets. Furthermore, microRNAs represent an exciting new means of targeting deregulated processes. Ways to deliver them safely and effectively to tumour tissues hold tremendous potential. Personalized medicine based on genomic and epigenomic changes will probably unfold in the next decade.

Acknowledgements Work in the RT lab is supported by the National Medical Research Council (NMRC) Grant NMRC/OFIRG/0073/201 and the Ministry of Education (MOE) MOE2019-T2-1-024. All images were created with BioRender.com.

References

Agios Pharmaceuticals, Inc. (2018a) A phase 1, open-label, single-dose study to evaluate the pharmacokinetics, safety, and tolerability of AG-120 (Ivosidenib) in subjects with mild or moderate hepatic impairment or normal hepatic function. clinicaltrials.gov

Agios Pharmaceuticals, Inc. (2018b) Expanded access program for Ivosidenib (AG-120) monotherapy in patients with relapsed or refractory acute myeloid leukemia with an IDH1 mutation. clinicaltrials.gov

Agios Pharmaceuticals, Inc. (2019) A phase I, multicenter, open-label, dose-escalation and expansion, safety, pharmacokinetic, pharmacodynamic, and clinical activity study of orally administered AG-881 in patients with advanced hematologic malignancies with an IDH1 and/or IDH2 mutation. clinicaltrials.gov

Agios Pharmaceuticals, Inc. (2020a) A phase 1, multicenter, randomized, controlled, open-label, perioperative study of AG-120 and AG-881 in subjects with recurrent, non-enhancing, IDH1 mutant, low grade glioma. clinicaltrials.gov

Agios Pharmaceuticals, Inc. (2020b) A phase 1, multicenter, open-label, dose-escalation and expansion, safety, pharmacokinetic, pharmacodynamic, and clinical activity study of orally administered AG-881 in patients with advanced solid tumors, including gliomas, with an IDH1 and/or IDH2 mutation. clinicaltrials.gov

Agios Pharmaceuticals, Inc. (2020c) A phase 3, multicenter, randomized, double-blind, placebo-controlled study of AG-881 in subjects with residual or recurrent grade 2 glioma with an IDH1 or IDH2 mutation. clinicaltrials.gov

Ahmad A, Aboukameel A, Kong D et al (2011) Phosphoglucose isomerase/autocrine motility factor mediates epithelial-mesenchymal transition regulated by miR-200 in breast cancer cells. Cancer Res 71:3400–3409. https://doi.org/10.1158/0008-5472.CAN-10-0965

Albrecht BK, Gehling VS, Hewitt MC et al (2016) Identification of a Benzoisoxazoloazepine inhibitor (CPI-0610) of the Bromodomain and extra-terminal (BET) family as a candidate for human clinical trials. J Med Chem 59:1330–1339. https://doi.org/10.1021/acs.jmedchem.5b01882

Amorim S, Stathis A, Gleeson M et al (2016) Bromodomain inhibitor OTX015 in patients with lymphoma or multiple myeloma: a dose-escalation, open-label, pharmacokinetic, phase 1 study. Lancet Haematol 3:e196–e204. https://doi.org/10.1016/S2352-3026(16)00021-1

Avery LB, Bumpus NN (2014) Valproic acid is a novel activator of AMP-activated protein kinase and decreases liver mass, hepatic fat accumulation, and serum glucose in obese mice. Mol Pharmacol 85:1–10. https://doi.org/10.1124/mol.113.089755

Bajor D (2020) Phase I/II study of CB-839 and capecitabine in patients with advanced solid tumors and fluoropyrimidine resistant PIK3CA mutant colorectal cancer. clinicaltrials.gov

Bayer (2019) An open-label, non-randomized, multicenter phase I study to determine the maximum tolerated and/or recommended phase II dose of oral mutant IDH1 (mIDH1) inhibitor BAY1436032 and to characterize its safety, tolerability, pharmacokinetics, pharmacodynamics, and preliminary clinical efficacy in patients with mIDH1-R132X advanced Acute Myeloid Leukemia (AML). clinicaltrials.gov

Bayer (2020) An open-label, non-randomized, multicenter phase i study to determine the maximum tolerated or recommended phase II dose of oral mutant IDH1 inhibitor BAY1436032 and to characterize its safety, tolerability, pharmacokinetics and preliminary pharmacodynamic and anti-tumor activity in patients with IDH1-R132X-mutant advanced solid tumors. clinicaltrials.gov

Baylin SB, Jones PA (2011) A decade of exploring the cancer epigenome - biological and translational implications. Nat Rev Cancer 11:726–734. https://doi.org/10.1038/nrc3130

Baylin SB, Jones PA (2016) Epigenetic determinants of cancer. Cold Spring Harb Perspect Biol 8: a019505. https://doi.org/10.1101/cshperspect.a019505

Berenguer-Daizé C, Astorgues-Xerri L, Odore E et al (2016) OTX015 (MK-8628), a novel BET inhibitor, displays in vitro and in vivo antitumor effects alone and in combination with conventional therapies in glioblastoma models. Int J Cancer 139:2047–2055. https://doi.org/10.1002/ijc.30256

Berthon C, Raffoux E, Thomas X et al (2016) Bromodomain inhibitor OTX015 in patients with acute leukaemia: a dose-escalation, phase 1 study. Lancet Haematol 3:e186–e195. https://doi.org/10.1016/S2352-3026(15)00247-1

Best JD, Carey N (2010) Epigenetic opportunities and challenges in cancer. Drug Discov Today 15: 65–70. https://doi.org/10.1016/j.drudis.2009.10.010

Boi M, Gaudio E, Bonetti P et al (2015) The BET Bromodomain inhibitor OTX015 affects Pathogenetic pathways in preclinical B-cell tumor models and synergizes with targeted drugs. Clin Cancer Res 21:1628–1638. https://doi.org/10.1158/1078-0432.CCR-14-1561

Calin GA, Cimmino A, Fabbri M et al (2008) MiR-15a and miR-16-1 cluster functions in human leukemia. Proc Natl Acad Sci U S A 105:5166–5171. https://doi.org/10.1073/pnas.0800121105

Calithera Biosciences, Inc (2017) A phase 1 study of the safety, pharmacokinetics, and pharmacodynamics of escalating oral doses of the glutaminase inhibitor CB-839 in patients with relapsed and/or treatment-refractory leukemia. clinicaltrials.gov

Calithera Biosciences, Inc (2018) A phase 1 study of the safety, pharmacokinetics, and pharmacodynamics of escalating oral doses of the glutaminase inhibitor CB-839 in patients with advanced and/or treatment-refractory hematological malignancies. clinicaltrials.gov

Calvert AE, Chalastanis A, Wu Y et al (2017) Cancer-associated IDH1 promotes growth and resistance to targeted therapies in the absence of mutation. Cell Rep 19:1858–1873. https://doi.org/10.1016/j.celrep.2017.05.014

Cantor JR, Sabatini DM (2012) Cancer cell metabolism: one hallmark, many faces. Cancer Discov 2:881–898. https://doi.org/10.1158/2159-8290.CD-12-0345

Caponigro F, Di Gennaro E, Ionna F et al (2016) Phase II clinical study of valproic acid plus cisplatin and cetuximab in recurrent and/or metastatic squamous cell carcinoma of head and neck-V- CHANCE trial. BMC Cancer 16:918. https://doi.org/10.1186/s12885-016-2957-y

Celgene (2018) A phase 1/2, multicenter, open-label, dose-escalation study of AG-221 in subjects with advanced solid tumors, including glioma, and with angioimmunoblastic T-cell lymphoma, that harbor an IDH2 mutation. clinicaltrials.gov

Celgene (2019) A single agent phase II study of depsipeptide (FK228) in the treatment of cutaneous T-cell lymphoma. clinicaltrials.gov

Celgene (2020) A phase 1/2, multicenter, open-label, dose-escalation and expansion, safety, pharmacokinetic, pharmacodynamic, and clinical activity study of orally administered AG-221 in subjects with advanced hematologic malignancies with an IDH2 mutation. clinicaltrials.gov

Chen W, Guéron M (1992) The inhibition of bovine heart hexokinase by 2-deoxy-d-glucose-6-phosphate: characterization by 31P NMR and metabolic implications. Biochimie 74:867–873. https://doi.org/10.1016/0300-9084(92)90070-U

Chen D-H, Wu K-T, Hung C-J et al (2004) Effects of adenosine dialdehyde treatment on in vitro and in vivo stable protein methylation in HeLa cells. J Biochem 136:371–376. https://doi.org/10.1093/jb/mvh131

Chen M, Zhang J, Li N et al (2011) Promoter hypermethylation mediated downregulation of FBP1 in human hepatocellular carcinoma and colon cancer. PLoS One 6:e25564. https://doi.org/10.1371/journal.pone.0025564

Chen Q, Chen Y, Bian C et al (2013a) TET2 promotes histone O-GlcNAcylation during gene transcription. Nature 493:561–564. https://doi.org/10.1038/nature11742

Chen Y-H, Heneidi S, Lee J-M et al (2013b) miRNA-93 inhibits GLUT4 and is overexpressed in adipose tissue of polycystic ovary syndrome patients and women with insulin resistance. Diabetes 62:2278–2286. https://doi.org/10.2337/db12-0963

Chen J, Yang J, Sun X et al (2017) Allosteric inhibitor remotely modulates the conformation of the orthestric pockets in mutant IDH2/R140Q. Sci Rep 7:16458. https://doi.org/10.1038/s41598-017-16427-w

Chien W, Lee DH, Zheng Y et al (2014) Growth inhibition of pancreatic cancer cells by histone deacetylase inhibitor belinostat through suppression of multiple pathways including HIF, NFkB, and mTOR signaling in vitro and in vivo. Mol Carcinog 53:722–735. https://doi.org/10.1002/mc.22024

Christman JK (2002) 5-Azacytidine and 5-aza-2′-deoxycytidine as inhibitors of DNA methylation: mechanistic studies and their implications for cancer therapy. Oncogene 21:5483–5495. https://doi.org/10.1038/sj.onc.1205699

Chu C-S, Lo P-W, Yeh Y-H et al (2014) O-GlcNAcylation regulates EZH2 protein stability and function. Proc Natl Acad Sci U S A 111:1355–1360. https://doi.org/10.1073/pnas.1323226111

Chuang T-Y, Wu H-L, Chen C-C et al (2015) MicroRNA-223 expression is upregulated in insulin resistant human adipose tissue. J Diabetes Res 2015:943659. https://doi.org/10.1155/2015/943659

Cluntun AA, Huang H, Dai L et al (2015) The rate of glycolysis quantitatively mediates specific histone acetylation sites. Cancer Metab 3:10. https://doi.org/10.1186/s40170-015-0135-3

Constellation Pharmaceuticals (2020) A phase 1/2 study of CPI-0610, a small molecule inhibitor of BET proteins: phase 1 (in patients with hematological malignancies) and phase 2 (dose expansion of CPI-0610 with and without ruxolitinib in patients with myelofibrosis). clinicaltrials.gov

Coudé M-M, Braun T, Berrou J et al (2015) BET inhibitor OTX015 targets BRD2 and BRD4 and decreases c-MYC in acute leukemia cells. Oncotarget 6:17698–17712. https://doi.org/10.18632/oncotarget.4131

Currie E, Schulze A, Zechner R et al (2013) Cellular fatty acid metabolism and cancer. Cell Metab 18:153–161. https://doi.org/10.1016/j.cmet.2013.05.017

Dai D-W, Lu Q, Wang L-X et al (2013) Decreased miR-106a inhibits glioma cell glucose uptake and proliferation by targeting SLC2A3 in GBM. BMC Cancer 13:478. https://doi.org/10.1186/1471-2407-13-478

Dang CV (2010) Glutaminolysis: supplying carbon or nitrogen or both for cancer cells? Cell Cycle 9:3884–3886. https://doi.org/10.4161/cc.9.19.13302

Dang L, White DW, Gross S et al (2009) Cancer-associated IDH1 mutations produce 2-hydroxyglutarate. Nature 462:739–744. https://doi.org/10.1038/nature08617

Davis MI, Gross S, Shen M et al (2014) Biochemical, cellular, and biophysical characterization of a potent inhibitor of mutant isocitrate dehydrogenase IDH1. J Biol Chem 289:13717–13725. https://doi.org/10.1074/jbc.M113.511030

Dawson MA, Kouzarides T (2012) Cancer epigenetics: from mechanism to therapy. Cell 150:12–27. https://doi.org/10.1016/j.cell.2012.06.013
Delmore JE, Issa GC, Lemieux ME et al (2011) BET Bromodomain inhibition as a therapeutic strategy to target c-Myc. Cell 146:904–917. https://doi.org/10.1016/j.cell.2011.08.017
Deng G, Shen J, Yin M et al (2015) Selective inhibition of mutant isocitrate dehydrogenase 1 (IDH1) via disruption of a metal binding network by an allosteric small molecule. J Biol Chem 290:762–774. https://doi.org/10.1074/jbc.M114.608497
Derissen EJB, Beijnen JH, Schellens JHM (2013) Concise drug review: azacitidine and decitabine. Oncologist 18:619–624. https://doi.org/10.1634/theoncologist.2012-0465
Desai S, Ding M, Wang B et al (2013) Tissue-specific isoform switch and DNA hypomethylation of the pyruvate kinase PKM gene in human cancers. Oncotarget 5:8202–8210
Ding J, Li T, Wang X et al (2013) The histone H3 methyltransferase G9A epigenetically activates the serine-glycine synthesis pathway to sustain cancer cell survival and proliferation. Cell Metab 18:896–907. https://doi.org/10.1016/j.cmet.2013.11.004
Dong Y, Huaying S, Danying W et al (2018) Significance of methylation of FBP1 gene in non-small cell lung cancer. Biomed Res Int 2018:3726091. https://doi.org/10.1155/2018/3726091
Elhammali A, Ippolito JE, Collins L et al (2014) A high-throughput fluorimetric assay for 2-hydroxyglutarate identifies Zaprinast as a glutaminase inhibitor. Cancer Discov 4:828–839. https://doi.org/10.1158/2159-8290.CD-13-0572
Fang R, Xiao T, Fang Z et al (2012) MicroRNA-143 (miR-143) regulates cancer glycolysis via targeting hexokinase 2 gene. J Biol Chem 287:23227–23235. https://doi.org/10.1074/jbc.M112.373084
Fang E, Wang J, Hong M et al (2019) Valproic acid suppresses Warburg effect and tumor progression in neuroblastoma. Biochem Biophys Res Commun 508:9–16. https://doi.org/10.1016/j.bbrc.2018.11.103
Fei X, Qi M, Wu B et al (2012) MicroRNA-195-5p suppresses glucose uptake and proliferation of human bladder cancer T24 cells by regulating GLUT3 expression. FEBS Lett 586:392–397. https://doi.org/10.1016/j.febslet.2012.01.006
Fong JJ, Nguyen BL, Bridger R et al (2012) β-N-Acetylglucosamine (O-GlcNAc) is a novel regulator of mitosis-specific Phosphorylations on histone H3. J Biol Chem 287:12195–12203. https://doi.org/10.1074/jbc.M111.315804
Forma Therapeutics, Inc. (2019) A phase 1/2, multicenter, open-label study of FT-2102 as a single agent and in combination with azacitidine or cytarabine in patients with acute myeloid leukemia or myelodysplastic syndrome with an IDH1 mutation. clinicaltrials.gov
Forma Therapeutics, Inc. (2020) A phase 1b/2 study of FT 2102 in participants with advanced solid tumors and gliomas with an IDH1 mutation. clinicaltrials.gov
Foss FM, Querfeld C, Kim YH et al (2018) Ph 1 study of MRG-106, an inhibitor of miR-155, in CTCL. JCO 36:2511–2511. https://doi.org/10.1200/JCO.2018.36.15_suppl.2511
Fujiki R, Hashiba W, Sekine H et al (2011) GlcNAcylation of histone H2B facilitates its monoubiquitination. Nature 480:557–560. https://doi.org/10.1038/nature10656
Galloway TJ, Wirth LJ, Colevas AD et al (2015) A phase I study of CUDC-101, a multitarget inhibitor of HDACs, EGFR, and HER2, in combination with Chemoradiation in patients with head and neck squamous cell carcinoma. Clin Cancer Res 21:1566–1573. https://doi.org/10.1158/1078-0432.CCR-14-2820
Gambichler T, Sand M, Skrygan M (2013) Loss of 5-hydroxymethylcytosine and ten-eleven translocation 2 protein expression in malignant melanoma. Melanoma Res 23:218–220. https://doi.org/10.1097/CMR.0b013e32835f9bd4
Gan L, Yang Y, Li Q et al (2018) Epigenetic regulation of cancer progression by EZH2: from biological insights to therapeutic potential. Biomarker Research 6:10. https://doi.org/10.1186/s40364-018-0122-2

Gao P, Tchernyshyov I, Chang T-C et al (2009) C-Myc suppression of miR-23a/b enhances mitochondrial glutaminase expression and glutamine metabolism. Nature 458:762–765. https://doi.org/10.1038/nature07823

García JM, Silva J, Peña C et al (2004) Promoter methylation of the PTEN gene is a common molecular change in breast cancer. Genes Chromosom Cancer 41:117–124. https://doi.org/10.1002/gcc.20062

Gilead Sciences (2019) A phase 1b/2 study of GS-5829 in combination with fulvestrant or exemestane in subjects with advanced estrogen receptor positive, HER2 negative-breast cancer. clinicaltrials.gov

Glazer RI, Hartman KD, Knode MC et al (1986) 3-deazaneplanocin: a new and potent inhibitor of S- adenosylhomocysteine hydrolase and its effects on human promyelocytic leukemia cell line HL-60. Biochem Biophys Res Commun 135:688–694. https://doi.org/10.1016/0006-291x(86)90048-3

Godlewski J, Bronisz A, Nowicki MO et al (2010) microRNA-451: a conditional switch controlling glioma cell proliferation and migration. Cell Cycle 9:2742–2748

Goel A, Mathupala SP, Pedersen PL (2003) Glucose metabolism in cancer. Evidence that demethylation events play a role in activating type II hexokinase gene expression. J Biol Chem 278: 15333–15340. https://doi.org/10.1074/jbc.M300608200

Golub D, Iyengar N, Dogra S et al (2019) Mutant Isocitrate dehydrogenase inhibitors as targeted cancer Therapeutics. Front Oncol 9:417. https://doi.org/10.3389/fonc.2019.00417

Greer EL, Shi Y (2012) Histone methylation: a dynamic mark in health, disease and inheritance. Nat Rev Genet 13:343–357. https://doi.org/10.1038/nrg3173

Gregersen LH, Jacobsen A, Frankel LB et al (2012) MicroRNA-143 down-regulates hexokinase 2 in colon cancer cells. BMC Cancer 12:232. https://doi.org/10.1186/1471-2407-12-232

Gross MI, Demo SD, Dennison JB et al (2014) Antitumor activity of the glutaminase inhibitor CB-839 in triple-negative breast cancer. Mol Cancer Ther 13:890–901. https://doi.org/10.1158/1535-7163.MCT-13-0870

Groupe Francophone des Myelodysplasies (2020) A single-arm phase II multicenter study of IDH2 (AG-221) inhibitor in patients with IDH2 mutated myelodysplastic syndrome. clinicaltrials.gov

Guo JU, Su Y, Zhong C et al (2011) Hydroxylation of 5-methylcytosine by TET1 promotes active DNA demethylation in the adult brain. Cell 145:423–434. https://doi.org/10.1016/j.cell.2011.03.022

Gupta V, Gopinath P, Iqbal MA et al (2014) Interplay between epigenetics & cancer metabolism. Curr Pharm Des 20:1706–1714. https://doi.org/10.2174/13816128113199990536

H. Lee Moffitt Cancer Center and Research Institute (2014) Phase II trial of Suberoylanilide Hydroxamic acid (SAHA, Vorinostat) in combination with tamoxifen for patients with advanced breast cancer who have failed prior anti-hormonal therapy. clinicaltrials.gov

Haematology-Oncology (2013) Phase I/II clinical trial of vorinostat in patients with recurrent and/or metastatic breast cancer. clinicaltrials.gov

Hainsworth JD, Infante JR, Spigel DR et al (2011) A phase II trial of Panobinostat, a histone deacetylase inhibitor, in the treatment of patients with refractory metastatic renal cell carcinoma. Cancer Investig 29:451–455. https://doi.org/10.3109/07357907.2011.590568

Hanahan D, Weinberg RA (2011) Hallmarks of cancer: the next generation. Cell 144:646–674. https://doi.org/10.1016/j.cell.2011.02.013

Herman JG, Latif F, Weng Y et al (1994) Silencing of the VHL tumor-suppressor gene by DNA methylation in renal carcinoma. Proc Natl Acad Sci U S A 91:9700–9704. https://doi.org/10.1073/pnas.91.21.9700

Issa J-PJ, Kantarjian HM, Kirkpatrick P (2005) Azacitidine. Nat Rev Drug Discov 4:275–276. https://doi.org/10.1038/nrd1698

Ito S, D'Alessio AC, Taranova OV et al (2010) Role of Tet proteins in 5mC to 5hmC conversion, ES- cell self-renewal and inner cell mass specification. Nature 466:1129–1133. https://doi.org/10.1038/nature09303

Ito R, Katsura S, Shimada H et al (2014) TET3-OGT interaction increases the stability and the presence of OGT in chromatin. Genes Cells 19:52–65. https://doi.org/10.1111/gtc.12107

Jarome TJ, Perez GA, Hauser RM et al (2018) EZH2 methyltransferase activity controls Pten expression and mTOR signaling during fear memory reconsolidation. J Neurosci 38:7635–7648. https://doi.org/10.1523/JNEUROSCI.0538-18.2018

Jiang S, Zhang L-F, Zhang H-W et al (2012) A novel miR-155/miR-143 cascade controls glycolysis by regulating hexokinase 2 in breast cancer cells. EMBO J 31:1985–1998. https://doi.org/10.1038/emboj.2012.45

Jin L, Alesi GN, Kang S (2016) Glutaminolysis as a target for cancer therapy. Oncogene 35:3619–3625. https://doi.org/10.1038/onc.2015.447

Jin F, Wang Y, Zhu Y et al (2017) The miR-125a/HK2 axis regulates cancer cell energy metabolism reprogramming in hepatocellular carcinoma. Sci Rep 7:3089. https://doi.org/10.1038/s41598-017-03407-3

Johannessen T-CA, Mukherjee J, Viswanath P et al (2016) Rapid conversion of mutant IDH1 from driver to passenger in a model of human Gliomagenesis. Mol Cancer Res 14:976–983. https://doi.org/10.1158/1541-7786.MCR-16-0141

Johnson C, Warmoes MO, Shen X, Locasale JW (2015) Epigenetics and cancer metabolism. Cancer Lett 356:309–314. https://doi.org/10.1016/j.canlet.2013.09.043

Kalac M, Scotto L, Marchi E et al (2011) HDAC inhibitors and decitabine are highly synergistic and associated with unique gene-expression and epigenetic profiles in models of DLBCL. Blood 118:5506–5516. https://doi.org/10.1182/blood-2011-02-336891

Kernytsky A, Wang F, Hansen E et al (2015) IDH2 mutation-induced histone and DNA hypermethylation is progressively reversed by small-molecule inhibition. Blood 125:296–303. https://doi.org/10.1182/blood-2013-10-533604

Kim J, Dang CV (2006) Cancer's molecular sweet tooth and the Warburg effect. Cancer Res 66: 8927–8930. https://doi.org/10.1158/0008-5472.CAN-06-1501

Kim H-J, Choi BY, Keum Y-S (2015) Identification of a new selective chemical inhibitor of mutant isocitrate dehydrogenase-1. J Cancer Prev 20:78–83. https://doi.org/10.15430/JCP.2015.20.1.78

Kim E, ten Hacken E, Sivina M et al (2020) The BET inhibitor GS-5829 targets chronic lymphocytic leukemia cells and their supportive microenvironment. Leukemia 34:1588–1598. https://doi.org/10.1038/s41375-019-0682-7

Knoepfler PS, Zhang X, Cheng PF et al (2006) Myc influences global chromatin structure. EMBO J 25:2723–2734. https://doi.org/10.1038/sj.emboj.7601152

Kooistra SM, Helin K (2012) Molecular mechanisms and potential functions of histone demethylases. Nat Rev Mol Cell Biol 13:297–311. https://doi.org/10.1038/nrm3327

Kudo Y, Tateishi K, Yamamoto K et al (2012) Loss of 5-hydroxymethylcytosine is accompanied with malignant cellular transformation. Cancer Sci 103:670–676. https://doi.org/10.1111/j.1349-7006.2012.02213.x

Lai C-J, Bao R, Tao X et al (2010) CUDC-101, a multitargeted inhibitor of histone deacetylase, epidermal growth factor receptor, and human epidermal growth factor receptor 2, exerts potent anticancer activity. Cancer Res 70:3647–3656. https://doi.org/10.1158/0008-5472.CAN-09-3360

Leal AS, Williams CR, Royce DB et al (2017) Bromodomain inhibitors, JQ1 and I-BET 762, as potential therapies for pancreatic cancer. Cancer Lett 394:76–87. https://doi.org/10.1016/j.canlet.2017.02.021

Lee H-Z, Kwitkowski VE, Valle PLD et al (2015) FDA approval: Belinostat for the treatment of patients with relapsed or refractory peripheral T-cell lymphoma. Clin Cancer Res 21:2666–2670. https://doi.org/10.1158/1078-0432.CCR-14-3119

Li X, Kazgan N (2011) Mammalian sirtuins and energy metabolism. Int J Biol Sci 7:575–587. https://doi.org/10.7150/ijbs.7.575

Li B, Qiu B, Lee DSM et al (2014) Fructose-1,6-bisphosphatase opposes renal carcinoma progression. Nature 513:251–255. https://doi.org/10.1038/nature13557

Li H, Zhang X, Jin Z et al (2019) MiR-122 promotes the development of colon cancer by targeting ALDOA in vitro. Technol Cancer Res Treat 18:1533033819871300. https://doi.org/10.1177/1533033819871300
Li C, Yu Z, Ye J (2020a) MicroRNA-513a-3p regulates colorectal cancer cell metabolism via targeting hexokinase 2. Exp Ther Med 20:572–580. https://doi.org/10.3892/etm.2020.8727
Li L, Yang L, Fan Z et al (2020b) Hypoxia-induced GBE1 expression promotes tumor progression through metabolic reprogramming in lung adenocarcinoma. Signal Transduct Target Ther 5:54. https://doi.org/10.1038/s41392-020-0152-8
Ling Z, Liu D, Zhang G et al (2017) miR-361-5p modulates metabolism and autophagy via the Sp1-mediated regulation of PKM2 in prostate cancer. Oncol Rep 38:1621–1628. https://doi.org/10.3892/or.2017.5852
Liu PY, Xu N, Malyukova A et al (2013) The histone deacetylase SIRT2 stabilizes Myc oncoproteins. Cell Death Differ 20:503–514. https://doi.org/10.1038/cdd.2012.147
Liu X-S, Little JB, Yuan Z-M (2015) Glycolytic metabolism influences global chromatin structure. Oncotarget 6:4214–4225. https://doi.org/10.18632/oncotarget.2929
Liu C, Cai L, Li H (2019) miR-185 regulates the growth of osteosarcoma cells via targeting hexokinase 2. Mol Med Rep 20:2774–2782. https://doi.org/10.3892/mmr.2019.10534
Lopez-Serra P, Marcilla M, Villanueva A et al (2014) A DERL3-associated defect in the degradation of SLC2A1 mediates the Warburg effect. Nat Commun 5. https://doi.org/10.1038/ncomms4608
Lu H, Buchan RJ, Cook SA (2010) MicroRNA-223 regulates Glut4 expression and cardiomyocyte glucose metabolism. Cardiovasc Res 86:410–420. https://doi.org/10.1093/cvr/cvq010
Lu C, Ward PS, Kapoor GS et al (2012) IDH mutation impairs histone demethylation and results in a block to cell differentiation. Nature 483:474–478. https://doi.org/10.1038/nature10860
Luo L, Xiao L, Lian G et al (2020) miR-125a-5p inhibits glycolysis by targeting hexokinase-II to improve pulmonary arterial hypertension. Aging (Albany NY) 12:9014–9030. https://doi.org/10.18632/aging.103163
Mentch SJ, Mehrmohamadi M, Huang L et al (2015) Histone methylation dynamics and gene regulation occur through the sensing of one-carbon metabolism. Cell Metab 22:861–873. https://doi.org/10.1016/j.cmet.2015.08.024
Mertz JA, Conery AR, Bryant BM et al (2011) Targeting MYC dependence in cancer by inhibiting BET bromodomains. PNAS 108:16669–16674. https://doi.org/10.1073/pnas.1108190108
miRagen Therapeutics, Inc. (2020) A phase 1 dose-ranging study to investigate the safety, tolerability, and pharmacokinetics of MRG-106 following local intratumoral, subcutaneous, and intravenous administration in subjects with various lymphomas and leukemias. clinicaltrials.gov
Miranda TB, Cortez CC, Yoo CB et al (2009) DZNep is a global histone methylation inhibitor that reactivates developmental genes not silenced by DNA methylation. Mol Cancer Ther 8:1579–1588. https://doi.org/10.1158/1535-7163.MCT-09-0013
Moros A, Rodríguez V, Saborit-Villarroya I et al (2014) Synergistic antitumor activity of lenalidomide with the BET bromodomain inhibitor CPI203 in bortezomib-resistant mantle cell lymphoma. Leukemia 28:2049–2059. https://doi.org/10.1038/leu.2014.106
Moussaieff A, Rouleau M, Kitsberg D et al (2015) Glycolysis-mediated changes in acetyl-CoA and histone acetylation control the early differentiation of embryonic stem cells. Cell Metab 21:392–402. https://doi.org/10.1016/j.cmet.2015.02.002
Myers SA, Panning B, Burlingame AL (2011) Polycomb repressive complex 2 is necessary for the normal site-specific O-GlcNAc distribution in mouse embryonic stem cells. Proc Natl Acad Sci U S A 108:9490–9495. https://doi.org/10.1073/pnas.1019289108
Nakajima K, Kawashima I, Koshiisi M et al (2019) Glycolytic enzyme hexokinase II is a putative therapeutic target in B-cell malignant lymphoma. Exp Hematol 78:46–55.e3. https://doi.org/10.1016/j.exphem.2019.09.023
National Cancer Institute (NCI) (2021) A phase 1b trial of CB-839 in combination with radiation therapy and temozolomide in patients with IDH-mutated diffuse astrocytoma and anaplastic astrocytoma. clinicaltrials.gov

Nguyen TTT, Zhang Y, Shang E et al (2020a) HDAC inhibitors elicit metabolic reprogramming by targeting super-enhancers in glioblastoma models. J Clin Investig 130:3699–3716. https://doi.org/10.1172/JCI129049

Nguyen TTT, Zhang Y, Shang E et al (2020b) HDAC inhibitors elicit metabolic reprogramming by targeting super-enhancers in glioblastoma models. J Clin Invest 130:3699–3716. https://doi.org/10.1172/JCI129049

Nie Z-Y, Liu X-J, Zhan Y et al (2019) miR-140-5p induces cell apoptosis and decreases Warburg effect in chronic myeloid leukemia by targeting SIX1. Biosci Rep 39:BSR20190150. https://doi.org/10.1042/BSR20190150

Novartis Pharmaceuticals (2020) A phase I study of IDH305 in patients with advanced malignancies that harbor IDH1R132 mutations. clinicaltrials.gov

Oncoethix GmbH (2018) A phase I, dose-finding study of the Bromodomain (Brd) Inhibitor OTX015/MK-8628 in Haematological Malignancies. clinicaltrials.gov

Oncoethix GmbH (2019) A phase IB trial with OTX015/MK-8628, a small molecule inhibitor of the Bromodomain and Extra-Terminal (BET) proteins, in patients with selected advanced solid tumors. clinicaltrials.gov

Pan D, Mao C, Wang Y-X (2013) Suppression of gluconeogenic gene expression by LSD1-mediated histone demethylation. PLoS One 8:e66294. https://doi.org/10.1371/journal.pone.0066294

Pavlova NN, Thompson CB (2016) The emerging hallmarks of cancer metabolism. Cell Metab 23:27–47. https://doi.org/10.1016/j.cmet.2015.12.006

Peschiaroli A, Giacobbe A, Formosa A et al (2013) miR-143 regulates hexokinase 2 expression in cancer cells. Oncogene 32:797–802. https://doi.org/10.1038/onc.2012.100

Pinho SS, Reis CA (2015) Glycosylation in cancer: mechanisms and clinical implications. Nat Rev Cancer 15:540–555. https://doi.org/10.1038/nrc3982

Pollard PJ, Wortham NC, Tomlinson IPM (2003) The TCA cycle and tumorigenesis: the examples of fumarate hydratase and succinate dehydrogenase. Ann Med 35:632–639. https://doi.org/10.1080/07853890310018458

Pollyea DA, Stevens BM, Jones CL et al (2018) Venetoclax with azacitidine disrupts energy metabolism and targets leukemia stem cells in patients with acute myeloid leukemia. Nat Med 24:1859–1866. https://doi.org/10.1038/s41591-018-0233-1

Popovici-Muller J, Lemieux RM, Artin E et al (2018) Discovery of AG-120 (Ivosidenib): a first-in-class mutant IDH1 inhibitor for the treatment of IDH1 mutant cancers. ACS Med Chem Lett 9:300–305. https://doi.org/10.1021/acsmedchemlett.7b00421

Qanungo S, Haldar S, Basu A (2003) Restoration of silenced Peutz-Jeghers syndrome gene, LKB1, induces apoptosis in pancreatic carcinoma cells. Neoplasia 5:367–374. https://doi.org/10.1016/S1476-5586(03)80030-1

Querfeld C, Pacheco T, Foss FM et al (2016) Preliminary results of a phase 1 trial evaluating MRG-106, a synthetic microRNA antagonist (LNA antimiR) of microRNA-155, in patients with CTCL. Blood 128:1829–1829. https://doi.org/10.1182/blood.V128.22.1829.1829

Raedler LA (2016) Farydak (Panobinostat): first HDAC inhibitor approved for patients with relapsed multiple myeloma. Am Health Drug Benefits 9:84–87

Rathore MG, Saumet A, Rossi J-F et al (2012) The NF-κB member p65 controls glutamine metabolism through miR-23a. Int J Biochem Cell Biol 44:1448–1456. https://doi.org/10.1016/j.biocel.2012.05.011

Rawluszko AA, Bujnicka KE, Horbacka K et al (2013) Expression and DNA methylation levels of prolyl hydroxylases PHD1, PHD2, PHD3 and asparaginyl hydroxylase FIH in colorectal cancer. BMC Cancer 13:526. https://doi.org/10.1186/1471-2407-13-526

Richardson PG, Laubach JP, Lonial S et al (2015) Panobinostat: a novel pan-deacetylase inhibitor for the treatment of relapsed or relapsed and refractory multiple myeloma. Expert Rev Anticancer Ther 15:737–748. https://doi.org/10.1586/14737140.2015.1047770

Robertson KD, Uzvolgyi E, Liang G et al (1999) The human DNA methyltransferases (DNMTs) 1, 3a and 3b: coordinate mRNA expression in normal tissues and overexpression in tumors. Nucleic Acids Res 27:2291–2298. https://doi.org/10.1093/nar/27.11.2291

Robinson MM, McBryant SJ, Tsukamoto T et al (2007) Novel mechanism of inhibition of rat kidney- type glutaminase by bis-2-(5-phenylacetamido-1,2,4-thiadiazol-2-yl)ethyl sulfide (BPTES). Biochem J 406:407–414. https://doi.org/10.1042/BJ20070039

Röhrig F, Schulze A (2016) The multifaceted roles of fatty acid synthesis in cancer. Nat Rev Cancer 16:732–749. https://doi.org/10.1038/nrc.2016.89

Rothbart SB, Strahl BD (2014) Interpreting the language of histone and DNA modifications. Biochimica et Biophysica Acta (BBA) - Gene Regulatory Mechanisms 1839:627–643. https://doi.org/10.1016/j.bbagrm.2014.03.001

Sakabe K, Hart GW (2010) O-GlcNAc transferase regulates mitotic chromatin dynamics. J Biol Chem 285:34460–34468. https://doi.org/10.1074/jbc.M110.158170

Santos CR, Schulze A (2012) Lipid metabolism in cancer. FEBS J 279:2610–2623. https://doi.org/10.1111/j.1742-4658.2012.08644.x

Sawas A, Radeski D, O'Connor OA (2015) Belinostat in patients with refractory or relapsed peripheral T-cell lymphoma: a perspective review. Ther Adv Hematol 6:202–208. https://doi.org/10.1177/2040620715592567

Sebastián C, Zwaans BMM, Silberman DM et al (2012) The histone deacetylase SIRT6 is a tumor suppressor that controls cancer metabolism. Cell 151:1185–1199. https://doi.org/10.1016/j.cell.2012.10.047

Seto AG, Beatty X, Lynch JM et al (2018) Cobomarsen, an oligonucleotide inhibitor of miR-155, co- ordinately regulates multiple survival pathways to reduce cellular proliferation and survival in cutaneous T-cell lymphoma. Br J Haematol 183:428–444. https://doi.org/10.1111/bjh.15547

Sharma S, Symanowski J, Wong B et al (2008) A phase II clinical trial of Oral Valproic acid in patients with castration-resistant prostate cancers using an intensive biomarker sampling strategy. Transl Oncol 1:141–147

Shin J, He M, Liu Y et al (2013) SIRT7 represses Myc activity to suppress ER stress and prevent fatty liver disease. Cell Rep 5:654–665. https://doi.org/10.1016/j.celrep.2013.10.007

Siegel MB, Davare MA, Liu SQ et al (2014) The Bromodomain inhibitor CPI203 demonstrates preclinical synergistic activity with Bortezomib in drug resistant myeloma. Blood 124:4702–4702. https://doi.org/10.1182/blood.V124.21.4702.4702

Siu KT, Ramachandran J, Yee AJ et al (2017) Preclinical activity of CPI-0610, a novel small-molecule bromodomain and extra-terminal protein inhibitor in the therapy of multiple myeloma. Leukemia 31:1760–1769. https://doi.org/10.1038/leu.2016.355

Song D, Ni J, Xie H et al (2014) DNA demethylation in the PTEN gene promoter induced by 5-azacytidine activates PTEN expression in the MG-63 human osteosarcoma cell line. Exp Ther Med 7:1071–1076. https://doi.org/10.3892/etm.2014.1571

Stathis A, Zucca E, Bekradda M et al (2016) Clinical response of carcinomas harboring the BRD4-NUT Oncoprotein to the targeted Bromodomain inhibitor OTX015/MK-8628. Cancer Discov 6: 492–500. https://doi.org/10.1158/2159-8290.CD-15-1335

Strand KA, Sizhao L, Mutryn MF et al (2020) High throughput screen identifies the DNMT1 (DNA Methyltransferase-1) inhibitor, 5-Azacytidine, as a potent inducer of PTEN (phosphatase and Tensin homolog). Arterioscler Thromb Vasc Biol 40:1854–1869. https://doi.org/10.1161/ATVBAHA.120.314458

Sun X, Zhang L (2017) MicroRNA-143 suppresses oral squamous cell carcinoma cell growth, invasion and glucose metabolism through targeting hexokinase 2. Biosci Rep 37: BSR20160404. https://doi.org/10.1042/BSR20160404

Sun W, Zhao L, Song X et al (2017a) MicroRNA-210 modulates the cellular energy metabolism shift during H2O2-induced oxidative stress by repressing ISCU in H9c2 cardiomyocytes. CPB 43:383–394. https://doi.org/10.1159/000480417

Sun Z, Zhang W, Li Q (2017b) miR-125a suppresses viability and glycolysis and induces apoptosis by targeting hexokinase 2 in laryngeal squamous cell carcinoma. Cell Biosci 7:51. https://doi.org/10.1186/s13578-017-0178-y

Tahiliani M, Koh KP, Shen Y et al (2009) Conversion of 5-methylcytosine to 5-hydroxymethylcytosine in mammalian DNA by MLL partner TET1. Science 324:930–935. https://doi.org/10.1126/science.1170116

Tang S-W, Yang T-C, Lin W-C et al (2011) Nicotinamide N-methyltransferase induces cellular invasion through activating matrix metalloproteinase-2 expression in clear cell renal cell carcinoma cells. Carcinogenesis 32:138–145. https://doi.org/10.1093/carcin/bgq225

Teng Y, Zhang Y, Qu K et al (2015) MicroRNA-29B (mir-29b) regulates the Warburg effect in ovarian cancer by targeting AKT2 and AKT3. Oncotarget 6:40799–40814. https://doi.org/10.18632/oncotarget.5695

Thomas MG, Saldanha M, Mistry RJ et al (2013) Nicotinamide N-methyltransferase expression in SH-SY5Y neuroblastoma and N27 mesencephalic neurones induces changes in cell morphology via ephrin-B2 and Akt signalling. Cell Death Dis 4:e669. https://doi.org/10.1038/cddis.2013.200

Turcan S, Rohle D, Goenka A et al (2012) IDH1 mutation is sufficient to establish the glioma hypermethylator phenotype. Nature 483:479–483. https://doi.org/10.1038/nature10866

Ulanovskaya OA, Zuhl AM, Cravatt BF (2013) NNMT promotes epigenetic remodeling in cancer by creating a metabolic methylation sink. Nat Chem Biol 9:300–306. https://doi.org/10.1038/nchembio.1204

Vander Heiden MG, Cantley LC, Thompson CB (2009) Understanding the Warburg effect: the metabolic requirements of cell proliferation. Science 324:1029–1033. https://doi.org/10.1126/science.1160809

Vázquez R, Riveiro ME, Astorgues-Xerri L et al (2016) The bromodomain inhibitor OTX015 (MK-8628) exerts anti-tumor activity in triple-negative breast cancer models as single agent and in combination with everolimus. Oncotarget 8:7598–7613. https://doi.org/10.18632/oncotarget.13814

Venneti S, Felicella MM, Coyne T et al (2013) Histone 3 lysine 9 Trimethylation is differentially associated with Isocitrate dehydrogenase mutations in Oligodendrogliomas and high-grade Astrocytomas. J Neuropathol Exp Neurol 72:298–306. https://doi.org/10.1097/NEN.0b013e3182898113

Wang J-B, Erickson JW, Fuji R et al (2010) Targeting mitochondrial glutaminase activity inhibits oncogenic transformation. Cancer Cell 18:207–219. https://doi.org/10.1016/j.ccr.2010.08.009

Wang F, Travins J, DeLaBarre B et al (2013) Targeted inhibition of mutant IDH2 in leukemia cells induces cellular differentiation. Science 340:622–626. https://doi.org/10.1126/science.1234769

Wang D, Meng G, Zheng M et al (2016) The Glutaminase-1 inhibitor 968 enhances Dihydroartemisinin-mediated antitumor efficacy in hepatocellular carcinoma cells. PLoS One 11:e0166423. https://doi.org/10.1371/journal.pone.0166423

Warburg O (1956) On the origin of cancer cells. Science 123:309–314. https://doi.org/10.1126/science.123.3191.309

Wardell SE, Ilkayeva OR, Wieman HL et al (2009) Glucose metabolism as a target of histone deacetylase inhibitors. Mol Endocrinol 23:388–401. https://doi.org/10.1210/me.2008-0179

Wise DR, DeBerardinis RJ, Mancuso A et al (2008) Myc regulates a transcriptional program that stimulates mitochondrial glutaminolysis and leads to glutamine addiction. Proc Natl Acad Sci U S A 105:18782–18787. https://doi.org/10.1073/pnas.0810199105

Wong CC, Qian Y, Yu J (2017) Interplay between epigenetics and metabolism in oncogenesis: mechanisms and therapeutic approaches. Oncogene 36:3359–3374. https://doi.org/10.1038/onc.2016.485

Wu Y, Siadaty MS, Berens ME et al (2008) Overlapping gene expression profiles of cell migration and tumor invasion in human bladder cancer identify metallothionein 1E and nicotinamide N-methyltransferase as novel regulators of cell migration. Oncogene 27:6679–6689. https://doi.org/10.1038/onc.2008.264

Wu H, Wang Y, Wu C et al (2016) Resveratrol induces cancer cell apoptosis through MiR- 326/ PKM2-mediated ER stress and mitochondrial fission. J Agric Food Chem 64:9356–9367. https://doi.org/10.1021/acs.jafc.6b04549

Wyce A, Degenhardt Y, Bai Y et al (2013) Inhibition of BET bromodomain proteins as a therapeutic approach in prostate cancer. Oncotarget 4:2419–2429. https://doi.org/10.18632/ oncotarget.1572

Xie F, Huang M, Lin X et al (2018) The BET inhibitor I-BET762 inhibits pancreatic ductal adenocarcinoma cell proliferation and enhances the therapeutic effect of gemcitabine. Sci Rep 8:8102. https://doi.org/10.1038/s41598-018-26496-0

Xu W, Yang H, Liu Y et al (2011) Oncometabolite 2-hydroxyglutarate is a competitive inhibitor of α- ketoglutarate-dependent dioxygenases. Cancer Cell 19:17–30. https://doi.org/10.1016/j.ccr. 2010.12.014

Yamasaki T, Seki N, Yoshino H et al (2013) Tumor-suppressive microRNA-1291 directly regulates glucose transporter 1 in renal cell carcinoma. Cancer Sci 104:1411–1419. https://doi.org/10. 1111/cas.12240

Yang H, Liu Y, Bai F et al (2013) Tumor development is associated with decrease of TET gene expression and 5-methylcytosine hydroxylation. Oncogene 32:663–669. https://doi.org/10. 1038/onc.2012.67

Yang J, Jin X, Yan Y et al (2017a) Inhibiting histone deacetylases suppresses glucose metabolism and hepatocellular carcinoma growth by restoring FBP1 expression. Sci Rep 7:43864. https:// doi.org/10.1038/srep43864

Yang J, Jin X, Yan Y et al (2017b) Inhibiting histone deacetylases suppresses glucose metabolism and hepatocellular carcinoma growth by restoring FBP1 expression. Sci Rep 7:43864. https:// doi.org/10.1038/srep43864

Yoshino H, Enokida H, Itesako T et al (2013) Tumor-suppressive microRNA-143/145 cluster targets hexokinase-2 in renal cell carcinoma. Cancer Sci 104:1567–1574. https://doi.org/10. 1111/cas.12280

Yu X, Li S (2017) Non-metabolic functions of glycolytic enzymes in tumorigenesis. Oncogene 36: 2629–2636. https://doi.org/10.1038/onc.2016.410

Yu G, Sun W, Shen Y et al (2018) PKM2 functions as a potential oncogene and is a crucial target of miR-148a and miR-326 in thyroid tumorigenesis. Am J Transl Res 10:1793–1801

Yuan L, Sheng X, Clark LH et al (2016) Glutaminase inhibitor compound 968 inhibits cell proliferation and sensitizes paclitaxel in ovarian cancer. Am J Transl Res 8:4265–4277

Zhang Y, Reinberg D (2001) Transcription regulation by histone methylation: interplay between different covalent modifications of the core histone tails. Genes Dev 15:2343–2360. https://doi. org/10.1101/gad.927301

Zhang J, Wang Y, Li G et al (2014) Down-regulation of nicotinamide N-methyltransferase induces apoptosis in human breast cancer cells via the mitochondria-mediated pathway. PLoS One 9: e89202. https://doi.org/10.1371/journal.pone.0089202

Zheng B, Yao Y, Liu Z et al (2013) Crystallographic investigation and selective inhibition of mutant Isocitrate dehydrogenase. ACS Med Chem Lett 4:542–546. https://doi.org/10.1021/ml400036z

Zhong L, D'Urso A, Toiber D et al (2010) The histone deacetylase SIRT6 regulates glucose homeostasis via Hif1α. Cell 140:280. https://doi.org/10.1016/j.cell.2009.12.041

Zhou T, Meng X, Che H et al (2016) Regulation of insulin resistance by multiple MiRNAs via targeting the GLUT4 Signalling pathway. Cell Physiol Biochem 38:2063–2078. https://doi.org/ 10.1159/000445565

Zwaans BMM, Lombard DB (2014) Interplay between sirtuins, MYC and hypoxia-inducible factor in cancer-associated metabolic reprogramming. Dis Model Mech 7:1023–1032. https://doi.org/ 10.1242/dmm.016287

Chapter 17
Modulation of DNA/RNA Methylation by Small-Molecule Modulators and Their Implications in Cancer

Ayushi Verma, Abhipsa Sinha, and Dipak Datta

Abstract Chromatin is an organized complex of DNA, histone proteins, and RNA. Chromatin modifications include DNA methylation, RNA methylation, and histone acetylation and methylation. The methylation of chromatin complexes predominantly alters the regulation of gene expression, and its deregulation is associated with several human diseases including cancer. Cancer is a disease characterized by dynamic changes in the genetic and epigenetic architecture of a cell. Altered DNA methylation by DNA methyltransferases (DNMTs) and m6A RNA methylation facilitate tumor initiation and progression and thus serve as critical targets for cancer therapy. Small-molecule modulators of these epigenetic targets are at the hotspots of current cancer drug discovery research. Indeed, recent studies have led to the discovery of several chemical modulators against these targets, some of which have already gained approval for cancer therapy while others are undergoing clinical trials. In this chapter, we will focus on the role of small-molecule modulators in regulating DNA/RNA methylation and their implications in cancer.

Keywords DNA methylation · DNMTs · RNA methylation · Epigenetics · Small-molecule modulator · Drug discovery

Introduction

Epigenetics is the study of somatically inherited changes that exert an alteration in the gene expression pattern without changing the DNA sequence. Epigenetic variations, unlike genetic variations, are dynamic and reversible. It majorly includes nuclear organization, DNA and RNA methylation, and histone modifications. Gene

Ayushi Verma and Abhipsa Sinha contributed equally to this article.

A. Verma · A. Sinha · D. Datta (✉)
Division of Cancer Biology, CSIR-Central Drug Research Institute (CDRI), Lucknow, India
e-mail: dipak.datta@cdri.res.in

T. K. Kundu, C. Das (eds.), *Metabolism and Epigenetic Regulation: Implications in Cancer*, Subcellular Biochemistry 100,
https://doi.org/10.1007/978-3-031-07634-3_17

regulation with the aid of chromatin modification complexes integrates environmental signals for modulating homeostatic responses. Epigenetic mechanisms play a pivotal role in the onset and progression of various diseases including cancer. The modification of the epigenome contributes to either activation of oncogenes or the repression of tumor suppressor genes in the course of tumorigenic processes. The global reprogramming of the epigenetic marks such as alterations in DNA and RNA methylation and histone modifications has now emerged as a leading player in carcinogenesis and can be effectively targeted for therapeutic purposes (Costello et al. 2000).

DNA methylation contributes significantly in controlling gene expression and maintaining the architecture and integrity cell nucleus. The process of methylation is catalyzed by methyltransferases ("writers") which donate the methyl group from S-adenosylmethionine to the cytosine residue of the DNA. The methyltransferases work in tandem with dedicated demethylases ("erasers") and methyl group recognizing effector proteins ("readers") (Michalak et al. 2019). In humans, DNA methylation occurs in the cytosine residues preceding guanines; these are referred to as dinucleotide CpG's. These CpG-rich regions, predominantly referred to as the "CpG islands," are localized at the 5′ regulatory end of many genes (Virani et al. 2012). Aberrant DNA methylation triggers the formation of malignant tumors majorly by DNA hyper- or hypomethylation. DNA hypermethylation is associated with transcriptional repression and reduced gene expression; on the other hand, DNA hypomethylation influences chromosome instability and enhances aneuploidy (Pan et al. 2018a). DNA methylation also includes chemical modifications of histone proteins.

RNA methylation is also a reversible and context-dependent epigenetic modulation similar to DNA and histone methylation in mammalian cells. More than 150 RNA modifications have been identified to date in eukaryotes (Boccaletto et al. 2018). Growing evidence advocates that RNA methylation is associated with numerous biological processes such as mRNA decay, including noncoding RNA (ncRNA) processing, nuclear export, translation modulation, RNA metabolism, pre-RNA splicing, 3′-end processing, etc. RNA methylation plays a critical role in various developmental processes; this includes embryonic development, sex determination, circadian rhythm, neurogenesis, and tumorigenesis. Recent advancement in the field of high throughput methods has empowered researchers to explore the consequences of RNA methylation in disease biology, particularly in cancer (Yue et al. 2015).

Histones, despite being DNA packaging proteins, also participate in the regulation of gene expression. They serve as a storehouse of epigenetic information through posttranslational modifications such as acetylation of lysine, methylation of lysine and arginine, and serine phosphorylation, thereby modulating gene transcription. The functional implications of methylation of histones depend on the target residue, i.e. lysine or arginine, and the specific site that methylation modifies (K4, K9, K27) (Bernstein et al. 2007; Mack 2006). H3K4 methylation is associated with transcriptional activation, whereas methylation of H3 at K9 or K27 and H4 at K20 is closely related to the suppression of transcription. Thus, these modifications also contribute to the physiological activities of cells and tissues (Esteller 2008; Karpf and Matsui 2005).

Here, our major point of discussion will be DNA and RNA methylation and how small molecule modulators against them are being used as cancer therapeutics.

DNA Methylation

DNA methylation is the covalent addition of methyl group by DNA methyltransferases (DNMTs) to the C-5 residue of the cytosine ring of the DNA (Jin et al. 2011). Methylation of CpG dinucleotide is commonly found in somatic cells but in non-CpG dinucleotide in embryonic stem cells (ESCs) (Lister et al. 2009). DNA methylation is broadly classified into maintenance methylation and de novo methylation. DNMT1 is primarily responsible for the maintenance of methylation by catalyzing the transfer of methyl groups to the hemimethylated DNA strand following replication (Bird 2002), whereas DNMT3a and DNMT3b direct de novo methylation by methylating the unmodified cytosine residues (Bird 2002; Jia et al. 2007). Methylation of CpG islands is followed by the recruitment of inhibitory proteins or hindrance in the interaction of transcription factors with DNA sequences (Singal et al. 2002). The classical example for this is the methyl-CpG binding protein which possesses a methyl cytosine binding domain (MBD) that preferentially binds to methylated CpGs and serves as transcriptional repressors (Fan and Hutnick 2005). Epigenetic changes involving DNA methylation play a crucial role in distinct biological processes such as genome imprinting, tissue-specific gene expression, repression of transposable elements, and X-chromosome inactivation, and disruption of normal DNA methylation gives rise to many diseases including cancer (Fedoriw et al. 2012; Michalak et al. 2019; Payer et al. 2011; Slotkin and Martienssen 2007). Abnormal DNA methylation in cells is correlated with spontaneous defects in DNA methyltransferases (DNMTs), leading to prominent nuclear abnormalities (Easwaran et al. 2010). The elevation in DNA hypomethylation is associated with the activation of oncogenes, instability of chromosomes and retrotransposable elements. For example, the hypomethylation of Alu and LINE-1 retrotransposons is shown to be associated with lung carcinoma (Daskalos et al. 2009).

DNA Methylation Profiles in Cancer

Abnormal DNA methylation is associated with the perturbation of signal transduction pathways, leading to the development of various diseases. Initial findings indicate that DNA methylation is one of the major causes for the initiation of cancer (Fig. 17.1). Deregulated DNA methylation leads to the growth of a variety cancer types such as breast, colon, ovarian, and cervical cancer (Pan et al. 2018a).

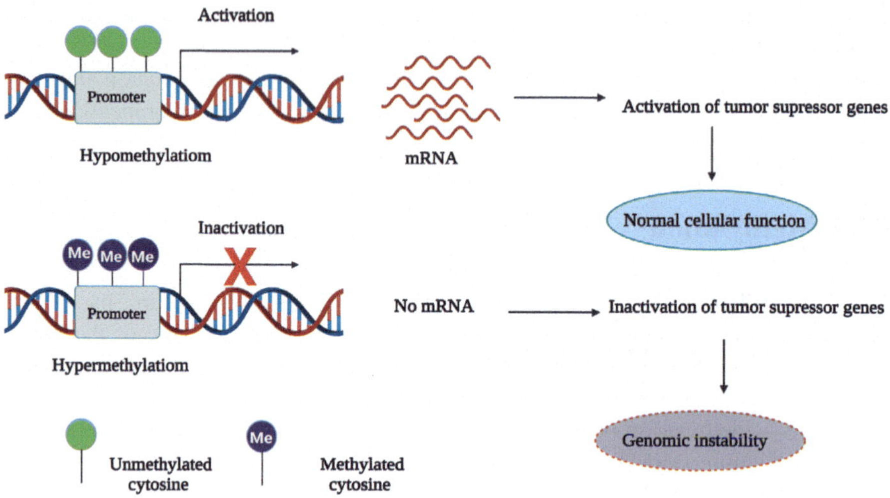

Fig. 17.1 Atypical DNA methylation status in cancer cells. (**a**) Loss of DNA methylation (hypomethylation) of CpG island at gene promoters is associated with active transcription of tumor suppressor genes and maintains normal cellular function. (**b**) CpG island-specific increased promoter methylation (hypermethylation) stalls gene expression and locks the tumor suppressors into an inactive state causing genomic instability. The green circles indicate unmethylated CpG sites, and pink circles show methylated CpG sites

Hypermethylation in the promoter regions of various tumor suppressor genes contributes significantly to the origin of many cancers (Esteller 2002). The phenomenon of promoter hypermethylation was initially reported in the retinoblastoma tumor suppressor gene (*Rb*) (Greger et al. 1989; Sakai et al. 1991). Further reports suggest that the hypermethylation of CpG island causes inactivation of other tumor suppressor genes such as *hMLH1* (a homologue of MutL *Escherichia coli*) and *BRCA1* (breast cancer susceptibility gene 1) (Esteller et al. 2000; Herman and Baylin 2003). Promoter hypermethylation is known to alter the functional activities of genes involved in cell cycle, apoptosis, DNA repair, cell-cell interaction, and angiogenesis, all of which are involved in the initiation and promotion of cancer (Esteller 2007). Transcriptional inactivation of DNA-repair genes such as *MLH1*, *MSH2*, *MSH3*, *BRCA1*, and *MGMT* (O6-methylguanine-DNA methyltransferase) impairs DNA repair, thereby paving the way to the development of neoplastic cells (Esteller 2008).

In tumorigenesis, DNA hypomethylation is a less common event compared to its hypermethylation. Some of the important mechanisms that contribute to cancer cell development are loss of imprinting, chromosomal stability, aberrant deletions, translocation of chromosome segments, and reactivation of transposable elements (Feinberg 1999). Hypomethylation leading to overexpression of certain imprinted genes such as *IGF11* and *H19* is known to be involved in carcinogenesis (Karpf and Matsui 2005). The loss of imprinting of *IGF2* (the insulin-like growth factor gene) increases the risk of cancer, and it is also found to be associated with Beckwith-Wiedemann syndrome (a disorder characterized by macrosomia, organomegaly, macroglossia, and developmental abnormalities) (Cui et al. 2003; Sakatani et al. 2005). The hypomethylation of gene regions contributes to oncogenesis by transforming classical oncogenes and activation of genes associated with tumor invasion and metastasis. Hypomethylation of tandem repeats and interspersed repeats such as L1 (long interspersed nuclear elements) and Alu repeats promotes tumor formation by rendering DNA rearrangements (Bestor 2005). Epigenetic changes leading to the loss of imprinting of *IGF2* and its overexpression contribute to a wide range of cancer, including colorectal cancer, particularly in Wilms' tumor (Feinberg 1999). Other prominent hypomethylation examples include the upregulation of *PAX2* which encodes a transcription factor that regulates cell proliferation, differentiation, and distinct cellular activities (Wu et al. 2005).

DNA Methylation: An Epigenetic Mark Storing Cellular Memory in DNA

DNA methylation is a stable epigenetic mark that can be transmitted and inherited through mitotic cell divisions. DNA methylation is a dynamic process during the course of development and cellular differentiation, but certain DNA methylation states can be retained in the form of epigenetic memory (Moore et al. 2013). The occurrence of these genome-wide and gene-specific alterations during cancer initiation and progression is a consequence of mutated or deregulated chromatin modulators (Sharma et al. 2010). The early aberrant modification in the DNA methylation patterns that occurs during transformation is conserved during tumor evolution. The variations in DNA methylation patterns among varied regions of a tumor manifest the history of cancer cells and their response to the tumor microenvironment.

The malignant transformation of a normal cell into cancerous cell shares intricate similarities with the epigenetic reprogramming of a somatic cell to a pluripotent cell (Suva et al. 2013). The chromatin reprogramming is a critical element that directs transcriptional response in cancer. The chromatin landscape, enhancer regions, and the various transcription factors (TFs) comprise of a complex network signaling and functional processes. Enhancer reprogramming plays a vital role in determining the transcriptional outcomes in a cell. Enhancers can broadly be classified as active, primed, poised, or silent. Active enhancers are characterized by nucleosome depletion and the presence of active histone modifications such as H3K4me1 and H3K27ac. Poised enhancers lack nucleosomes and encompass H3K4me1 and the repressive Polycomb H3K27me3 marks, whereas primed enhancers display the presence of H3K4me1 mark along with nucleosome depletion. Enhancer CpG density plays a major role in determining transcriptional fate of the targeted gene. Thus, improved technologies such as precise epigenome editing and single-cell genomic and transcriptomic assays should aim at defining the role of 5mC presence at distinct enhancer classes and its importance for interaction with the linked promoter and consequently their contribution in regulating gene expression (Angeloni and Bogdanovic 2019).

The Role of TET and TDG Pathway in Cancer

The ten-eleven translocation (TET) and thymine DNA glycosylase are important regulators of the DNA demethylation process. The TET proteins mediate the conversion of 5mC to 5-hydroxymethylcytosine (5hmC), 5-carboxylcytosine (5caC), and 5-formylcytosine (5fC), which can be further oxidized and repaired by thymine DNA glycosylase (TDG), resulting in demethylation. Evidences suggest that impairment in TET-mediated DNA demethylation contributes to carcinogenesis. The marked upregulation of TET1 protein has been reported in mixed lineage leukemia (MLL)-rearranged leukemia (Huang et al. 2013). Several mutations including nonsense/missense, frameshift, deletion, and somatic in TET2 are reported to be associated with myelodysplastic syndrome (MDS) and different types of leukemia (Delhommeau et al. 2009; Huang et al. 2013). The upregulation or downregulation of the TET gene and differential 5hmC levels has been detected in numerous solid tumors.

Targeting DNMTs for Cancer Therapy

DNMT1 methylates the newly synthesized DNA and plays a pivotal role in maintaining DNA methylation. Other DNMTs, especially DNMT3A, DNMT3B, and DNMT3L, are hyperactivated in breast (Girault et al. 2003), colon (el-Deiry et al. 1991), liver (Oh et al. 2007), leukemia (Melki et al. 1998), and prostate cancer (Patra et al. 2002) and are partially responsible for the inactivation of tumor suppressors in these cancer types. Several findings indicate the role DNMT3A and DNMT3B as crucial tumor suppressors, preventing lymphoid malignancies (Peters et al. 2014). Inhibition of DNMTs results in reduced tumor formation, partly due to the activation of tumor suppressor genes. Thus, DNMTs may serve as key therapeutic targets for the development of anticancer agents.

DNMT Inhibitors (DNMTi) in Cancer

DNMT inhibitors are prodrugs (Table 17.1) that form a covalent complex with the DNA and trigger proteasomal degradation of the enzymes (Patel et al. 2010). Genetic or pharmacological targeting of DNA methyltransferases contributes to epigenetic reprogramming of cancer cells (Xiao et al. 2016). Various strategies can be employed for designing DNA methyltransferase inhibitors such as identification of the SAM analogues. DNA binding agents can also function as DNMTi by

Table 17.1 List of DNMT inhibitors as anticancer agents

Types	Names	Targets	Clinical status	References
Nucleosides	AZA	DNMTs	Approved in 2004	Kaminskas et al. (2005)
	Decitabine	DNMTs	Approved in 2006	Gore et al. (2006)
	Zebularine	DNMTs	Preclinical	Cheng et al. (2004)
	5-F-CdR	DNMTs	Phase I	Guo et al. (2010)
Non-nucleosides	SGI-1027	DNMTs	Preclinical	Datta et al. (2009)
	EGCG	DNMTs	Phase II	Nandakumar et al. (2011)
	Procaine	DNMTs	Preclinical	Yang et al. (2010)
	Procainamide	DNMT1	Preclinical	Lee et al. (2005)
	Hydralazine	DNMT1	Phase III	Coronel et al. (2011)
	RG108	DNMTs	Preclinical	Stresemann et al. (2006)
Antisense oligonucleotides	miRNA-29b	DNMTs	Preclinical	Garzon et al. (2009)

competitively inhibiting DNA-enzyme binding. Another approach could be the "multi-substrate adduct" approach which involves the covalent interaction of two enzyme substrates to form a single entity, such as the combination of SAM and cytosine/deoxycytidine. On the other hand, targeting the interaction of DNMTs to its protein partners can be a suitable approach for inhibiting DNMT activity (Erdmann et al. 2015). DNMTi can be classified into nucleoside and non-nucleoside-derived inhibitors. DNMT inhibitors also include antisense oligonucleotides such as microRNA-29b (Garzon et al. 2009). Nucleoside-derived DNMT inhibitors include azacitidine (AZA), 5-F-CdR, decitabine, and zebularine. There are several non-nucleoside-derived DNMT inhibitors as well such as SGI-1027, RG108, epigallocatechin gallate (EGCG), etc.

Nucleoside Analogs as DNMT Inhibitors

The nucleoside analogs, azacitidine (5-aza-cytidine) and decitabine, have been tested in myelodysplastic syndrome (MDS), acute myeloid leukemia (AML), and chronic myelomonocytic leukemia (CMML) patients. Both these analogs act as hypomethylating agents and are irreversible DNMT inhibitors. Azacitidine treatment showed promising results in phase III clinical trials and improved overall survival of MDS patients (Fenaux et al. 2009). In the year 2004, azacitidine has rendered FDA approval for therapy in MDS patients. This was followed by the development and approval of decitabine in 2006. Although beneficial, these epi-drugs have certain disadvantages such as lack of chemical and metabolic stability due to degradation by activation-induced cytidine deaminase (AID), poor bioavailability, and inefficacy in solid tumors (Hauser et al. 2018). The dinucleotide analog SGI-110 (guadecitabine) is a prodrug of decitabine, has improved resistance to cleavage by AID, hence, and increased bioavailability. It is currently undergoing clinical trials for AML, MDS, and CMML. Another nucleoside analog, zebularine, originally considered as cytidine deaminase inhibitor (Cheng et al. 2003), is also a DNMTi found to be effective against bone marrow disorders (Khan et al. 2013). In comparison to other nucleoside analogs, zebularine has improved chemical stability and minimal cytotoxic effects. Zebularine is a known inducer of the p16 gene and prevents its methylation. In addition, zebularine treatment demethylates various hypermethylated regions by completely hindering DNA methyltransferase 1 (DNMT1) and partly inhibiting DNMT3A and DNMT3B activity (Cheng et al. 2004).

Non-nucleoside-Derived DNMT Inhibitors

Various non-nucleoside analogs have been reported to inhibit DNA methylation, including hydralazine, procainamide, EGCG, and others. In silico and high-throughput screening strategies have been used to discover non-nucleoside DNMT inhibitors which unlike nucleoside analogs could inhibit DNMTs without incorporation into the DNA (Hauser et al. 2018). The non-nucleoside DNMT inhibitors demonstrated adequate antitumor effects in a variety of prostate, leukemia, and breast cancer cell lines (Pan et al. 2018a). Despite promising results, none of the non-nucleoside-derived analogs have moved to the clinical trials, making it unclear that whether these reported agents are true DNMT inhibitors or it has off-target effects.

Antisense Oligonucleotides as DNMT Inhibitors

Aberrant DNA hypermethylation is associated with myeloid leukemogenesis by silencing genes involved in hematopoiesis. MicroRNAs (miRNAs) are noncoding RNAs that modulate transcription by targeting protein-coding mRNA. Recently, miRNAs have gained immense importance in regulating gene hypermethylation and reducing malignancy. For example, *miR-29b directs* reduced expression of DNA methyltransferases *DNMT1*, *DNMT2A*, *and DNMT3B*, respectively, by minimizing global DNA methylation and restoring p15^{INK4b} and ESR1 expression in acute myeloid leukemia cells (Garzon et al. 2009).

RNA Methylation and Its Types

RNA methylation is one of the key posttranscriptional RNA modifications which is reversible in nature and regulates multiple biological processes like RNA stability and mRNA translation (Jia et al. 2011; Wang et al. 2014) by multiple RNA methyltransferases, often via diverse catalytic strategies. Emerging studies have shown the importance of various types of RNA methylation in the context to tumorigenic processes.

N6-Methyl Adenosine (m6A)

The methylation that occurs at the sixth N position of adenosine in RNA is known as m6A RNA methylation. N6 methylation (m6A) was discovered in 1970 and the most frequently identified RNA modification in eukaryotic mRNA (Roundtree et al. 2017). The m6A methylation found in different types of RNA includes rRNA, tRNA, small nuclear RNA (snRNA), and various long noncoding RNA, such as Xist (Dominissini et al. 2012). m6A methylation plays a central role in numerous biological processes like stability, splicing, turnover, cap-independent translation, and nuclear export of mRNA and ncRNA (Meyer et al. 2015).

N1-Methyladenosine (m1A)

The incorporation of a methyl group at the N1 position of adenosine in RNA is known as m1A RNA methylation. A positive charge at Watson-Crick base interface alters the RNA secondary structures and the RNA-protein interaction by electrostatic effects (Roundtree et al. 2017). The m1A methylation is majorly identified in tRNA and rRNA and influences the structure and function of associated RNA. The m1A methylation also supports the mRNA processing and translation, as this mark is present at a unique translation start site and splice sites to facilitate the binding of noncanonical exon-exon junction complex (Dominissini et al. 2012; Romano et al. 2018; Roundtree et al. 2017).

2′-O-Methylation (2′-OMe/Nm)

The 2′-O-methylation is persisted in all forms of RNA including mRNA, rRNA, tRNA micro-RNA, and sn-RNA and required for the fundamental biogenesis and function of all types of RNA. This modification is usually found in the second and third nucleotides in mRNA, and the deletion of 2–3 position at A and P sites can intensely delay the pre-mRNA processing with impaired translation (Liang et al. 2009).

5-Methylcytosine (m5C)

The insertion of methyl group at the fifth carbon atom of cytosine in RNA is known as m5C methylation. The m5C methylation persists in poly (A) RNA, tRNA, rRNA, snRNA, and lncRNA (Amort et al. 2013; Lewis et al. 2017). The m5C methylation helps in the stabilization of the secondary structure of tRNA and also protects the mRNA transcript from degradation. Moreover, this modification prevents the cleavage and degradation of rRNA during translation (Esteller and Pandolfi 2017).

Writers, Erasers, and Readers for RNA Methylation

The proteins that can add or remove or recognize the m6A modified sites in RNA are commonly considered as writers, erasers, and readers, respectively (Fig. 17.2).

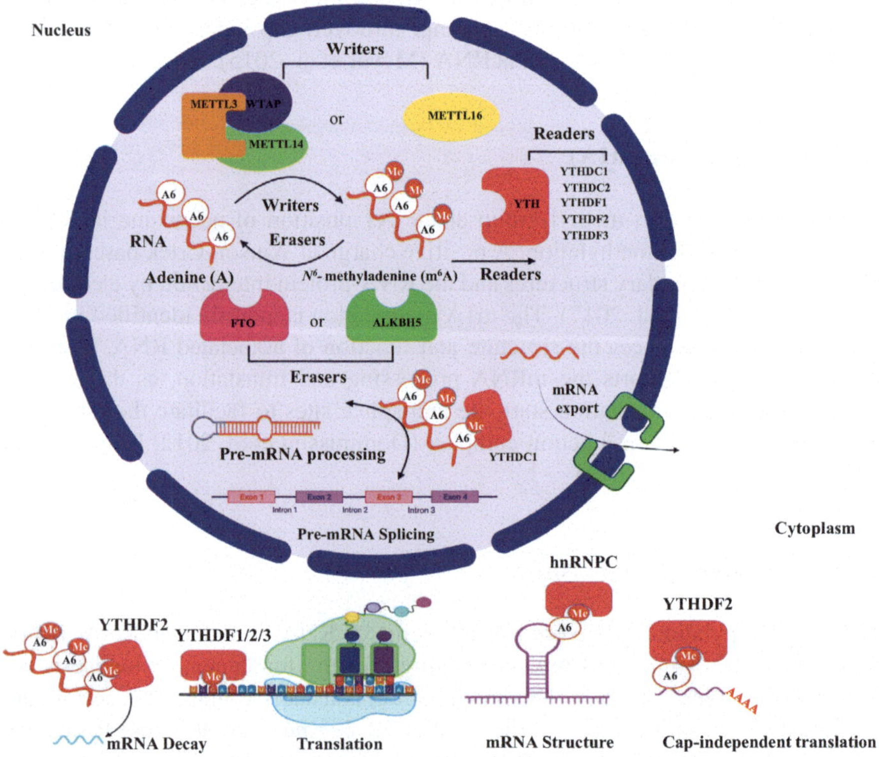

Fig. 17.2 Various biological functions of m6A methylation. m6A RNA methylation is regulated by different modulators such as writers, erasers, and readers. The writer complex which facilitates m6A methylation of RNA consists of METTL3, METTL14, or METTL16 and WTAP subunits. The erasers are m6A demethylase including FTO and ALKBH5. Readers are proteins which bind with the m6A-modified RNA and employ numerous biological functions; some of the examples of readers are YTHDF1, YTHDF2, YTHDF3, YTHDC1, YTHDC2, and hnRNPA2B. m6A RNA methylation is recognized to be involved in all stages in the life cycle of RNA including pre-mRNA splicing and pri-miRNA processing, through nuclear export, RNA translation modulation, and RNA degradation

Writers as Methyltransferase

Methyltransferase-like 3 (METTL3) is the first characterized constituent of the m6A methyltransferase complex that catalyzes m6A methylation by binding to S-adenosylmethionine (SAM) (Bokar et al. 1994). This component is vastly conserved in eukaryotes from yeast to humans (Bokar et al. 1997). The second component of the m6A methyltransferase complex is methyltransferase-like 14 (METTL14). METTL3 is a core catalytic subunit that facilitates the methylation of m6A, while METT14 assists as a structural care for RNA binding. Both METTL3 and METTL14 together form a heterodimeric complex and co-localized in a nuclear speckle (Liu et al. 2014; Wang et al. 2016). Another subunit of this multicomplex is WTAP (Wilms' tumor 1-associated protein); however, WTAP lacks the conserved methyltransferase domain and acts as an adaptor protein for the interaction between METTL3 and METTL14, thus helping in the maintenance of cellular m6A RNA load (Wang et al. 2016). An earlier report revealed methyltransferase-like 16 (METTL16) as a new m6A methyltransferase, homologous to METTL3. METTL16 has been found to control cellular SAM level and methylates U6 small nuclear RNA (Du et al. 2016; Jia et al. 2011).

Eraser as a Demethylase

The fat mass and obesity-associated protein (FTO) was discovered as the first m6A demethylase in 2011 which indicates that m6A methylation is reversible and potent (Jia et al. 2011). After the discovery of FTO, regulation of RNA methylation has emerged as a new paradigm in gene regulation in the last decade. Subsequently, the α-ketoglutarate-dependent dioxygenase homolog 5 (ALKBH5) was discovered as demethylase in 2013 (Zheng et al. 2013). Both FTO and ALKBH5 are associated with the α-ketoglutarate-dependent dioxygenase family and required Fe(II) and α-ketoglutarate for enzymatic activity.

Readers

YT521-B homology (YTH) domain-containing family members consisting of YTHDF1, YTHDF2, YTHDF3, YTHDC1, and YTHDC2 have m6A-binding conserved domain. The first characterized m6A reader was YTHDF2, which promotes the decay of m6A-modulated transcripts by recruitment of CCR4-NOT deadenylase complex in the m6A RNA (Wang et al. 2014). Also, the YTHDF1 has been identified to bind near stop codon in m6A sites and enhances the translation efficiency of target RNA in eukaryotes (Wang et al. 2015). YTHDF3 stimulates the methylated RNA translation via interaction with YTHDF1, while it promotes mRNA decay through interaction with YTHDF2 (Bokar et al. 1994; Li et al. 2017a). In conclusion, YTHDF3 functions as the refinement for the RNA accessibility of

YTHDF1 and YTHDF2. Moreover, YTHDC1, a nuclear m6A reader, enhances exon inclusion through restriction or recruitment of numerous splicing factors such as arginine and serine-rich splicing factors (SRSFs) on transcripts (Xiao et al. 2016). In addition, YTHDC1 also facilitates Xist (X-inactive specific transcript)-mediated transcriptional repression and X-chromosome inactivation by recognition of m6A methylation on Xist (Patil et al. 2016).

The Role of m6A Modification in Various Cancers

The development of advanced high-throughput techniques along with the identification and understanding of m6A as readers, writers, and erasers has helped us to find out the underlying mechanisms and biological regulations of m6A. The modification of m6A in RNA plays an important function in diverse physiological processes, including DNA damage response, heat shock responses, circadian rhythms, spermatogenesis, pluripotency, embryogenesis, and epigenetic reprogramming (Adhikari et al. 2016; Roundtree et al. 2017; Zhao et al. 2017). Moreover, emerging evidence has showed that methylation of m6A RNA also plays crucial role in tumor development. Also, the abnormal changes in m6A are associated with cancer (Pan et al. 2018b; Wang et al. 2018b). Herein, we briefly reviewed the recent studies of m6A methylation in different cancers (Table 17.2).

Small-Molecule Modulators for RNA Methylation

METTL14 Inhibition

METTL14 expression is high in the acute myeloid leukemia (AML) cells and normal hematopoietic stem/progenitor cells (HSPCs) carrying translocations in t(15;17), t (11;23), or t(8;21) (Weng et al. 2018). It was studied that METTL14 silencing inhibits the survival and proliferation of AML cells. The R298P hotspot mutation is persisting in the METTL14 in human endometrial cancer. Nearly 70% of the endometrial tumors have reduced m6A methyltransferase activity due to reduced METTL3 expression and mutation in METTL14. The in silico-based methods were used by Simona Selberg's team to figure out the small-molecule ligands that bind to the METTL3–14-WTAP complex. They utilized the molecular docking calculations to assure the interaction between METTL3 and SAM. Their compound 1 can bind in the SAM domain region of METTL3. The enzyme kinetics, ligand binding affinity, and their impact on catalytic function were experimentally determined. However, the computationally analyzed strong ligands that bind to the METTL3–14-WTAP complex in the enzymatic assay enhanced the activity of this complex despite the catalytic inhibition. To explain this mechanism, two important factors are deliberated (Gu et al. 2020). First, the probability of binding affinity of SAM might be

Table 17.2 M6A RNA methylation in various cancers; adapted from (Sun et al. 2019)

Molecule	Cancer	Role in cancer	Mode of action	References
(Writer) METTL3	GBM	Suppressor	Suppresses GSCs growth and self-renewal ability by enhancing m6A modification of ADAM19 and decreasing its expression	Cui et al. (2017)
	AML	Oncogene	Promotes tumorigenesis by promoting m6A modification of BCL2 and PTEN	Vu et al. (2017)
	BC	Oncogene	Enhances tumor growth by enhancing m6A modification and translation of HBXIP	Cai et al. (2018)
	HCC	Oncogene	Promotes GSCs growth by enhancing m6A modification of SOCS2 and its degradation	Chen et al. (2018)
	AML	Oncogene	Binds to the promoter region of mRNA transcript and enhances m6A modification and gene expression	Barbieri et al. (2017)
	GBM	Oncogene	Promotes GSC growth and self-renewal by upregulating POU3F2, SOX2, SALL2, and OLIG2 gene expression	Visvanathan et al. (2018)
	Endometrial cancer	Suppressor	METTL3 downregulation leads to increased cell proliferation and tumorigenicity through activation of AKT	Liu et al. (2018)
(Writers) METL14	AML	Oncogene	Promotes tumorigenesis by enhancing m6A modification and expression of MYC and MYB	Weng et al. (2018)
	HCC	Suppressor	Suppresses tumor metastasis by promoting mir126 processing in an m6A-dependent manner	Wang et al. (2020)
	GBM	Suppressor	Suppresses GSC growth and self-renewal by enhancing m6A modification of ADAM19 and decreasing its expression	Cui et al. (2017)
	Endometrial cancer	Suppressor	METTL14 mutation leads to increased cell proliferation and tumorigenicity through activation of AKT	Liu et al. (2018)
(Erasers) FTO	AML	Oncogene	Promotes tumorigenesis by decreasing m6A on ASB2 and RARA at UTRs, which destabilized ASB2 and RARA	Li et al. (2017b)
	AML	Oncogene	Promotes AML carcinogenesis by enhancing MYC and CEBPA mRNA stability	Su et al. (2018)
	GBM	Oncogene	FTO catalytic activity inhibition by MA2 suppressor glioblastoma progression	Cai et al. (2018)

(continued)

Table 17.2 (continued)

Molecule	Cancer	Role in cancer	Mode of action	References
	Cervical cancer	Oncogene	Enhance chemoradiotherapy resistance by decreasing m6A modification of β-catenin	Zhou et al. (2018)
(Erasers) ALKBH5	GBM	Oncogene	Promotes GSCs proliferation and tumor progression by positively regulating FOXM1 through m6A modification	Zhang et al. (2017)
	BC	Oncogene	ALKBH5 mediates enrichment of BCSCs by decreasing NANOG mRNA methylation and increasing its mRNA levels	Zhang et al. (2016a)
	BC	Oncogene	Promotes tumorigenesis by decreasing m6A methylation of KLF4 and NANOG and enhancing their mRNA stability	Zhang et al. (2016b)
	AML	Suppressor	ALKBH5 copy number loss is frequent in AML, which predicts poorer survival of AML patients	Kwok et al. (2017)
	Pancreatic cancer	Suppressor	Suppresses pancreatic cancer motility by decreasing lncRNA KCNK15-AS1 methylation	He et al. (2018)
(Readers) YTHDF2	Pancreatic cancer	Dual effect	Promotes tumor proliferation by activating AKT/GSK3β/cyclin D1 and suppresses metastasis by destabilizing YAP mRNA	Chen et al. (2017)
	Prostate cancer	Oncogene	Promotes prostate cancer growth and migration by enhancing m6A-containing mRNA degradation	Li et al. (2018)
	HCC	Oncogene	YTHDF2 was upregulated, and miR145 modulates m6A levels by targeting the 3′-UTR of YTHDF2 mRNA in HCC cells	Yang et al. (2017)

accelerated due to the contact between the carbonyl oxygen atoms present in the series of ligands, and the sulfur atom acts as a methylation center in the methionine group of it (Gu et al. 2020). Since the overexpression of METTL3 and METTL14 was observed to be related to improved longevity in glioma patients, there might be promising therapeutic application of small-molecule function as an m6A writer activators (Cui et al. 2017).

FTO Inhibition

FTO is a 2-oxoglutarate (2OG)-dependent N-methyl nucleic acid demethylase which requires 2OG for the demethylation process. An elegant study by Weishen Aik's group identified a series of 2OG analogues or relevant compounds and found both acyclic and cyclic 2OG analogues as FTO inhibitors (Rose et al. 2011). Notably, maximum 2OG-type inhibitors have been identified either as iron cheaters or chemical mimics of 2OG. The natural product Rhein was revealed as the first strong FTO inhibitor in 2012 by ssRNA substrate (Chen et al. 2012). However, Rhein has been showed some selectivity toward the AlkB subfamily. Therefore, to screen selective FTO inhibitors, Huang's lab from China used a high-throughput fluorescence polarization (FP) assay to measure the variation in the dislocation of M6A-ssDNA binding to FTO and ALKBH5. This group has identified meclofenamic acid (MA) as a specific potent inhibitor of FTO over ALKBH5 in the screening procedure (Huang et al. 2019). In addition, this group further developed a structure-guided design for the discovery of highly selective and potent FTO inhibitors, after the mechanistic analysis of the inhibition of M6A demethylation by its inhibitors. In 2019, the same group discovered FB23 as a highly selective FTO inhibitor that persists a half-maximal inhibitory concentration (IC50) of about 0.006 mM (Huang et al. 2019). To improve the bioavailability of FB23, the modified benzohydroxamic acid FB23e2 was synthesized, and it showed a better inhibitory effect against FTO demethylase in vitro and demonstrates robust antiproliferative activity against MONOMAC6 and NB4 cells with IC50 of 0.8 1.5 mM. Moreover, Joel D Toh's group from the National University of Singapore in 2014 also revealed a selective and bioactive FTO inhibitor by developing a dynamic combinatorial chemistry (DCC) method that can concurrently recognize subfamily-specific probes for both FTO andALKBH3 (Das et al. 2018; Toh et al. 2015). This group discovered N-(5-chloro-2,4-dihydroxyphenyl)-1-phenylcyclobutanecarboxamide (N-CDPCB) as a strong inhibitor of FTO consisting of a unique binding site in FTO (He et al. 2015). Subsequently, this group demonstrates the human FTO crystal structure with binding to N-CDPCB and confirms the structural base for their interaction. However, the N-CDPCB showed partial overlapping with the FTO-specific inhibitor MA, although N-CDPCB and MA have different chemical structures. Recently the same group have discovered 4-chloro-6-[60-chloro-70-hydroxy-20,40,40-trimethylchroman-20-yl]-benzene-1,3-diol (CHTB) as FTO inhibitor. To explore the novel binding site of N-CDPCB, further study has been taken place based on structural activity. This study identified 4-chloro-1,3-diol as a common structural motif shared by the FTO inhibitors such as N-CDPCB analogues and CHTB. Thus, the investigators accomplished virtual screening based on the structure of compounds that carried a 4-chloro-1,3-diol group based on the unique binding site recognized in the N-CDPCB-FTO complex. As an outcome, the natural compound radicicol was discovered as a strong FTO inhibitor (Wang et al. 2018a). In conclusion, FTO is the most considered m6A small-molecule target to date. In summary, the structure-guided design of FTO inhibitors has been verified to be an effective methodology for m6A methylation.

ALKBH5 and YTH Domain-Containing Proteins Inhibition

ALKBH5 is a ferrous iron-dependent nucleic acid oxygenase (NAOX) and 2OG that catalyzes m6A RNA demethylation. In contrast to FTO, to date, no ALKBH5-specific inhibitor has been discovered, because the 2OG-type inhibitors have pan-inhibitory activity. Form Zhengzhou University, the Dan-Dan-Shen group has established an HTS-amenable formaldehyde dehydrogenase coupled assay along with antibody-based assay by utilizing recombinant ALKBH5 to discover ALKBH5 inhibitors (Shen et al. 2019). Moreover, the crystal structure of the YTH domain of YTHDC1 with m6A RNA has been identified (Xu et al. 2014). This study demonstrated the YTHDC1 recognizes the GG (m6A) C sequences, and subsequently, the YTH domain binds with m6A RNA. The interaction and binding of the YTH domain of human YTHDF2 with m6A RNA have been confirmed by fluorescence polarization (FP) assays (Li et al. 2014). The RNA probe consisting of AUGG (m6A) CUCC has been found to bind with the YTH domain of YTHDF2, while the unmethylated RNA probe showed no binding with the YTH domain of YRHDF2. Researchers can use this probe as a powerful tool for the screening of small-molecule inhibitors against YTH domain-containing proteins.

Conclusion and Future Perspectives

Tumor development is accompanied by epigenetic alterations. The critical changes include DNA methylation and histone modifications in the chromatin, but the recently discovered mechanism of RNA methylation has also achieved equal importance. All of these events act in concert to silence the expression of genes that inhibit tumorigenesis. The aberrant DNA methylation profile in cells serves as an important biomarker for early diagnosis of cancer. Epigenetic changes are probable targets for therapeutic inventions. Currently, two DNMT inhibitors, azacytidine and decitabine, have shown to be effective as anticancer drugs. In the year 2004 and 2006, azacytidine and decitabine have been approved for use in treating acute myeloid leukemia and myelodysplastic syndrome, respectively. A newly identified DNMT inhibitor, zebularine, also possesses anticancer activities. It inhibits DNMTs and reverses gene silencing in cancer cells similar to the two FDA-approved drugs, azacytidine and decitabine. Zebularine has emerged as a promising epigenetic therapy compared to its other two counterparts due to better efficacy and bioavailability. Research has also been attributed to the discovery of non-nucleoside DNMT inhibitors, which are presently undergoing preclinical and clinical studies. These include SGI-1027, procaine, procainamide, RG108, EGCG, and hydralazine, respectively. Downregulation of DNMTs by antisense or miRNAs has been found to be efficient to restore the expression of hypermethylated genes, but due to its instability, cytotoxicity, and off-target effects, its administration as a therapy remains controversial.

The function and importance of m6A RNA modification in context to cellular biology remain unknown. However, the creation of m6A-seq or MeRIP-seq mapping in 2012 provided substantial information at the transcriptome level. After the discovery of FTO demethylase, the m6A got prime attention in the scientific community, and the researchers found that the level of m6A has been deregulated in various diseases including cancer. Several m6A modulators have been characterized as writers, erasers, and readers for m6A RNA methylation. Several structures-based basic research projects are in progress for the screening of inhibitors for m6A RNA modulators. The METTL3–14-WTAP complex is the fundamental core subunit of most m6A writers. Several studies revealed that the METTL3 or METTL14 is associated with the development of various tumors including CSCC and AML Besides, the overexpression of ALKBH5 has been found in breast and GSC tumor cells. Notably, FTO is presently the most explored and established m6A-related enzyme. Most of the discovered FTO inhibitors are highly potent and selective. The treatment of FTO inhibitors against AML has been found convincing and considered as a most promising drug target for AML. With the constant hi-tech advancements, screening of ALKBH5 inhibitors is also under development. Subsequently, the agonists screening for METTL3–14-WATP complex has also been started. Considering this, the molecular understanding of m6A needs great efforts of basic research for the identification of highly active and selective small-molecule modulators for disease treatment. The synergistic activity of epigenetic drugs with conventional anticancer therapeutics in combination also suggests that they have immense potential for chemotherapy in cancer.

References

Adhikari S, Xiao W, Zhao YL, Yang YG (2016) M(6)a: signaling for mRNA splicing. RNA Biol 13:756–759

Amort T, Souliere MF, Wille A, Jia XY, Fiegl H, Worle H, Micura R, Lusser A (2013) Long non-coding RNAs as targets for cytosine methylation. RNA Biol 10:1003–1008

Angeloni A, Bogdanovic O (2019) Enhancer DNA methylation: implications for gene regulation. Essays Biochem 63:707–715

Barbieri I, Tzelepis K, Pandolfini L, Shi J, Millan-Zambrano G, Robson SC, Aspris D, Migliori V, Bannister AJ, Han N et al (2017) Promoter-bound METTL3 maintains myeloid leukaemia by m (6)A-dependent translation control. Nature 552:126–131

Bernstein BE, Meissner A, Lander ES (2007) The mammalian epigenome. Cell 128:669–681

Bestor TH (2005) Transposons reanimated in mice. Cell 122:322–325

Bird A (2002) DNA methylation patterns and epigenetic memory. Genes Dev 16:6–21

Boccaletto P, Machnicka MA, Purta E, Piatkowski P, Baginski B, Wirecki TK, de Crecy-Lagard V, Ross R, Limbach PA, Kotter A et al (2018) MODOMICS: a database of RNA modification pathways. 2017 update. Nucleic Acids Res 46:D303–D307

Bokar JA, Rath-Shambaugh ME, Ludwiczak R, Narayan P, Rottman F (1994) Characterization and partial purification of mRNA N6-adenosine methyltransferase from HeLa cell nuclei. Internal mRNA methylation requires a multisubunit complex. J Biol Chem 269:17697–17704

Bokar JA, Shambaugh ME, Polayes D, Matera AG, Rottman FM (1997) Purification and cDNA cloning of the AdoMet-binding subunit of the human mRNA (N6-adenosine)-methyltransferase. RNA 3:1233–1247

Cai X, Wang X, Cao C, Gao Y, Zhang S, Yang Z, Liu Y, Zhang X, Zhang W, Ye L (2018) HBXIP-elevated methyltransferase METTL3 promotes the progression of breast cancer via inhibiting tumor suppressor let-7g. Cancer Lett 415:11–19

Chen B, Ye F, Yu L, Jia G, Huang X, Zhang X, Peng S, Chen K, Wang M, Gong S et al (2012) Development of cell-active N6-methyladenosine RNA demethylase FTO inhibitor. J Am Chem Soc 134:17963–17971

Chen J, Sun Y, Xu X, Wang D, He J, Zhou H, Lu Y, Zeng J, Du F, Gong A et al (2017) YTH domain family 2 orchestrates epithelial-mesenchymal transition/proliferation dichotomy in pancreatic cancer cells. Cell Cycle 16:2259–2271

Chen M, Wei L, Law CT, Tsang FH, Shen J, Cheng CL, Tsang LH, Ho DW, Chiu DK, Lee JM et al (2018) RNA N6-methyladenosine methyltransferase-like 3 promotes liver cancer progression through YTHDF2-dependent posttranscriptional silencing of SOCS2. Hepatology 67:2254–2270

Cheng JC, Matsen CB, Gonzales FA, Ye W, Greer S, Marquez VE, Jones PA, Selker EU (2003) Inhibition of DNA methylation and reactivation of silenced genes by zebularine. J Natl Cancer Inst 95:399–409

Cheng JC, Yoo CB, Weisenberger DJ, Chuang J, Wozniak C, Liang G, Marquez VE, Greer S, Orntoft TF, Thykjaer T et al (2004) Preferential response of cancer cells to zebularine. Cancer Cell 6:151–158

Coronel J, Cetina L, Pacheco I, Trejo-Becerril C, Gonzalez-Fierro A, de la Cruz-Hernandez E, Perez-Cardenas E, Taja-Chayeb L, Arias-Bofill D, Candelaria M et al (2011) A double-blind, placebo-controlled, randomized phase III trial of chemotherapy plus epigenetic therapy with hydralazine valproate for advanced cervical cancer. Preliminary results Med Oncol 28(Suppl 1): S540–S546

Costello JF, Fruhwald MC, Smiraglia DJ, Rush LJ, Robertson GP, Gao X, Wright FA, Feramisco JD, Peltomaki P, Lang JC et al (2000) Aberrant CpG-island methylation has non-random and tumour-type-specific patterns. Nat Genet 24:132–138

Cui H, Cruz-Correa M, Giardiello FM, Hutcheon DF, Kafonek DR, Brandenburg S, Wu Y, He X, Powe NR, Feinberg AP (2003) Loss of IGF2 imprinting: a potential marker of colorectal cancer risk. Science 299:1753–1755

Cui Q, Shi H, Ye P, Li L, Qu Q, Sun G, Sun G, Lu Z, Huang Y, Yang CG et al (2017) M(6)a RNA methylation regulates the self-renewal and tumorigenesis of glioblastoma stem cells. Cell Rep 18:2622–2634

Das M, Yang T, Dong J, Prasetya F, Xie Y, Wong KHQ, Cheong A, Woon ECY (2018) Multiprotein dynamic combinatorial chemistry: a strategy for the simultaneous discovery of subfamily-selective inhibitors for nucleic acid demethylases FTO and ALKBH3. Chem Asian J 13:2854–2867

Daskalos A, Nikolaidis G, Xinarianos G, Savvari P, Cassidy A, Zakopoulou R, Kotsinas A, Gorgoulis V, Field JK, Liloglou T (2009) Hypomethylation of retrotransposable elements correlates with genomic instability in non-small cell lung cancer. Int J Cancer 124:81–87

Datta J, Ghoshal K, Denny WA, Gamage SA, Brooke DG, Phiasivongsa P, Redkar S, Jacob ST (2009) A new class of quinoline-based DNA hypomethylating agents reactivates tumor suppressor genes by blocking DNA methyltransferase 1 activity and inducing its degradation. Cancer Res 69:4277–4285

Delhommeau F, Dupont S, Della Valle V, James C, Trannoy S, Masse A, Kosmider O, Le Couedic JP, Robert F, Alberdi A et al (2009) Mutation in TET2 in myeloid cancers. N Engl J Med 360: 2289–2301

Dominissini D, Moshitch-Moshkovitz S, Schwartz S, Salmon-Divon M, Ungar L, Osenberg S, Cesarkas K, Jacob-Hirsch J, Amariglio N, Kupiec M et al (2012) Topology of the human and mouse m6A RNA methylomes revealed by m6A-seq. Nature 485:201–206

Du H, Zhao Y, He J, Zhang Y, Xi H, Liu M, Ma J, Wu L (2016) YTHDF2 destabilizes m(6)A-containing RNA through direct recruitment of the CCR4-NOT deadenylase complex. Nat Commun 7:12626

Easwaran HP, Van Neste L, Cope L, Sen S, Mohammad HP, Pageau GJ, Lawrence JB, Herman JG, Schuebel KE, Baylin SB (2010) Aberrant silencing of cancer-related genes by CpG hypermethylation occurs independently of their spatial organization in the nucleus. Cancer Res 70:8015–8024

el-Deiry WS, Nelkin BD, Celano P, Yen RW, Falco JP, Hamilton SR, Baylin SB (1991) High expression of the DNA methyltransferase gene characterizes human neoplastic cells and progression stages of colon cancer. Proc Natl Acad Sci U S A 88:3470–3474

Erdmann A, Halby L, Fahy J, Arimondo PB (2015) Targeting DNA methylation with small molecules: what's next? J Med Chem 58:2569–2583

Esteller M (2002) CpG island hypermethylation and tumor suppressor genes: a booming present, a brighter future. Oncogene 21:5427–5440

Esteller M (2007) Cancer epigenomics: DNA methylomes and histone-modification maps. Nat Rev Genet 8:286–298

Esteller M (2008) Epigenetics in cancer. N Engl J Med 358:1148–1159

Esteller M, Pandolfi PP (2017) The Epitranscriptome of noncoding RNAs in cancer. Cancer Discov 7:359–368

Esteller M, Garcia-Foncillas J, Andion E, Goodman SN, Hidalgo OF, Vanaclocha V, Baylin SB, Herman JG (2000) Inactivation of the DNA-repair gene MGMT and the clinical response of gliomas to alkylating agents. N Engl J Med 343:1350–1354

Fan G, Hutnick L (2005) Methyl-CpG binding proteins in the nervous system. Cell Res 15:255–261

Fedoriw A, Mugford J, Magnuson T (2012) Genomic imprinting and epigenetic control of development. Cold Spring Harb Perspect Biol 4:a008136

Feinberg AP (1999) Imprinting of a genomic domain of 11p15 and loss of imprinting in cancer: an introduction. Cancer Res 59:1743s–1746s

Fenaux P, Mufti GJ, Hellstrom-Lindberg E, Santini V, Finelli C, Giagounidis A, Schoch R, Gattermann N, Sanz G, List A et al (2009) Efficacy of azacitidine compared with that of conventional care regimens in the treatment of higher-risk myelodysplastic syndromes: a randomised, open-label, phase III study. Lancet Oncol 10:223–232

Garzon R, Liu S, Fabbri M, Liu Z, Heaphy CE, Callegari E, Schwind S, Pang J, Yu J, Muthusamy N et al (2009) MicroRNA-29b induces global DNA hypomethylation and tumor suppressor gene reexpression in acute myeloid leukemia by targeting directly DNMT3A and 3B and indirectly DNMT1. Blood 113:6411–6418

Girault I, Tozlu S, Lidereau R, Bieche I (2003) Expression analysis of DNA methyltransferases 1, 3A, and 3B in sporadic breast carcinomas. Clin Cancer Res 9:4415–4422

Gore SD, Jones C, Kirkpatrick P (2006) Decitabine. Nat Rev Drug Discov 5:891–892

Greger V, Passarge E, Hopping W, Messmer E, Horsthemke B (1989) Epigenetic changes may contribute to the formation and spontaneous regression of retinoblastoma. Hum Genet 83:155–158

Gu J, Xu J, You Q, Guo X (2020) Recent developments of small molecules targeting RNA m(6)a modulators. Eur J Med Chem 196:112325

Guo D, Myrdal PB, Karlage KL, O'Connell SP, Wissinger TJ, Tabibi SE, Yalkowsky SH (2010) Stability of 5-fluoro-2′-deoxycytidine and tetrahydrouridine in combination. AAPS PharmSciTech 11:247–252

Hauser AT, Robaa D, Jung M (2018) Epigenetic small molecule modulators of histone and DNA methylation. Curr Opin Chem Biol 45:73–85

He W, Zhou B, Liu W, Zhang M, Shen Z, Han Z, Jiang Q, Yang Q, Song C, Wang R et al (2015) Identification of a novel small-molecule binding site of the fat mass and obesity associated protein (FTO). J Med Chem 58:7341–7348

He Y, Hu H, Wang Y, Yuan H, Lu Z, Wu P, Liu D, Tian L, Yin J, Jiang K et al (2018) ALKBH5 inhibits pancreatic cancer motility by decreasing long non-coding RNA KCNK15-AS1 methylation. Cell Physiol Biochem 48:838–846

Herman JG, Baylin SB (2003) Gene silencing in cancer in association with promoter hypermethylation. N Engl J Med 349:2042–2054

Huang H, Jiang X, Li Z, Li Y, Song CX, He C, Sun M, Chen P, Gurbuxani S, Wang J et al (2013) TET1 plays an essential oncogenic role in MLL-rearranged leukemia. Proc Natl Acad Sci U S A 110:11994–11999

Huang J, Dong X, Gong Z, Qin LY, Yang S, Zhu YL, Wang X, Zhang D, Zou T, Yin P et al (2019) Solution structure of the RNA recognition domain of METTL3-METTL14 N(6)-methyladenosine methyltransferase. Protein Cell 10:272–284

Jia D, Jurkowska RZ, Zhang X, Jeltsch A, Cheng X (2007) Structure of Dnmt3a bound to Dnmt3L suggests a model for de novo DNA methylation. Nature 449:248–251

Jia G, Fu Y, Zhao X, Dai Q, Zheng G, Yang Y, Yi C, Lindahl T, Pan T, Yang YG et al (2011) N6-methyladenosine in nuclear RNA is a major substrate of the obesity-associated FTO. Nat Chem Biol 7:885–887

Jin B, Li Y, Robertson KD (2011) DNA methylation: superior or subordinate in the epigenetic hierarchy? Genes Cancer 2:607–617

Kaminskas E, Farrell AT, Wang YC, Sridhara R, Pazdur R (2005) FDA drug approval summary: azacitidine (5-azacytidine, Vidaza) for injectable suspension. Oncologist 10:176–182

Karpf AR, Matsui S (2005) Genetic disruption of cytosine DNA methyltransferase enzymes induces chromosomal instability in human cancer cells. Cancer Res 65:8635–8639

Khan H, Vale C, Bhagat T, Verma A (2013) Role of DNA methylation in the pathogenesis and treatment of myelodysplastic syndromes. Semin Hematol 50:16–37

Kwok CT, Marshall AD, Rasko JE, Wong JJ (2017) Genetic alterations of m(6)a regulators predict poorer survival in acute myeloid leukemia. J Hematol Oncol 10:39

Lee BH, Yegnasubramanian S, Lin X, Nelson WG (2005) Procainamide is a specific inhibitor of DNA methyltransferase 1. J Biol Chem 280:40749–40756

Lewis CJ, Pan T, Kalsotra A (2017) RNA modifications and structures cooperate to guide RNA-protein interactions. Nat Rev Mol Cell Biol 18:202–210

Li F, Zhao D, Wu J, Shi Y (2014) Structure of the YTH domain of human YTHDF2 in complex with an m(6)a mononucleotide reveals an aromatic cage for m(6)a recognition. Cell Res 24: 1490–1492

Li A, Chen YS, Ping XL, Yang X, Xiao W, Yang Y, Sun HY, Zhu Q, Baidya P, Wang X et al (2017a) Cytoplasmic m(6)a reader YTHDF3 promotes mRNA translation. Cell Res 27:444–447

Li Z, Weng H, Su R, Weng X, Zuo Z, Li C, Huang H, Nachtergaele S, Dong L, Hu C et al (2017b) FTO plays an oncogenic role in acute myeloid leukemia as a N(6)-Methyladenosine RNA demethylase. Cancer Cell 31:127–141

Li J, Meng S, Xu M, Wang S, He L, Xu X, Wang X, Xie L (2018) Downregulation of N(6)-methyladenosine binding YTHDF2 protein mediated by miR-493-3p suppresses prostate cancer by elevating N(6)-methyladenosine levels. Oncotarget 9:3752–3764

Liang XH, Liu Q, Fournier MJ (2009) Loss of rRNA modifications in the decoding center of the ribosome impairs translation and strongly delays pre-rRNA processing. RNA 15:1716–1728

Lister R, Pelizzola M, Dowen RH, Hawkins RD, Hon G, Tonti-Filippini J, Nery JR, Lee L, Ye Z, Ngo QM et al (2009) Human DNA methylomes at base resolution show widespread epigenomic differences. Nature 462:315–322

Liu J, Yue Y, Han D, Wang X, Fu Y, Zhang L, Jia G, Yu M, Lu Z, Deng X et al (2014) A METTL3-METTL14 complex mediates mammalian nuclear RNA N6-adenosine methylation. Nat Chem Biol 10:93–95

Liu J, Eckert MA, Harada BT, Liu SM, Lu Z, Yu K, Tienda SM, Chryplewicz A, Zhu AC, Yang Y et al (2018) M(6)a mRNA methylation regulates AKT activity to promote the proliferation and tumorigenicity of endometrial cancer. Nat Cell Biol 20:1074–1083

Mack GS (2006) Epigenetic cancer therapy makes headway. J Natl Cancer Inst 98:1443–1444

Melki JR, Warnecke P, Vincent PC, Clark SJ (1998) Increased DNA methyltransferase expression in leukaemia. Leukemia 12:311–316

Meyer KD, Patil DP, Zhou J, Zinoviev A, Skabkin MA, Elemento O, Pestova TV, Qian SB, Jaffrey SR (2015) 5′ UTR m(6)a promotes cap-independent translation. Cell 163:999–1010

Michalak EM, Burr ML, Bannister AJ, Dawson MA (2019) The roles of DNA, RNA and histone methylation in ageing and cancer. Nat Rev Mol Cell Biol 20:573–589

Moore LD, Le T, Fan G (2013) DNA methylation and its basic function. Neuropsychopharmacology 38:23–38

Nandakumar V, Vaid M, Katiyar SK (2011) (−)-Epigallocatechin-3-gallate reactivates silenced tumor suppressor genes, Cip1/p21 and p16INK4a, by reducing DNA methylation and increasing histones acetylation in human skin cancer cells. Carcinogenesis 32:537–544

Oh BK, Kim H, Park HJ, Shim YH, Choi J, Park C, Park YN (2007) DNA methyltransferase expression and DNA methylation in human hepatocellular carcinoma and their clinicopathological correlation. Int J Mol Med 20:65–73

Pan Y, Liu G, Zhou F, Su B, Li Y (2018a) DNA methylation profiles in cancer diagnosis and therapeutics. Clin Exp Med 18:1–14

Pan Y, Ma P, Liu Y, Li W, Shu Y (2018b) Multiple functions of m(6)a RNA methylation in cancer. J Hematol Oncol 11:48

Patel K, Dickson J, Din S, Macleod K, Jodrell D, Ramsahoye B (2010) Targeting of 5-aza-2-′-deoxycytidine residues by chromatin-associated DNMT1 induces proteasomal degradation of the free enzyme. Nucleic Acids Res 38:4313–4324

Patil DP, Chen CK, Pickering BF, Chow A, Jackson C, Guttman M, Jaffrey SR (2016) M(6)a RNA methylation promotes XIST-mediated transcriptional repression. Nature 537:369–373

Patra SK, Patra A, Zhao H, Dahiya R (2002) DNA methyltransferase and demethylase in human prostate cancer. Mol Carcinog 33:163–171

Payer B, Lee JT, Namekawa SH (2011) X-inactivation and X-reactivation: epigenetic hallmarks of mammalian reproduction and pluripotent stem cells. Hum Genet 130:265–280

Peters SL, Hlady RA, Opavska J, Klinkebiel D, Pirruccello SJ, Talmon GA, Sharp JG, Wu L, Jaenisch R, Simpson MA et al (2014) Tumor suppressor functions of Dnmt3a and Dnmt3b in the prevention of malignant mouse lymphopoiesis. Leukemia 28:1138–1142

Romano G, Veneziano D, Nigita G, Nana-Sinkam SP (2018) RNA methylation in ncRNA: classes, detection, and molecular associations. Front Genet 9:243

Rose NR, McDonough MA, King ON, Kawamura A, Schofield CJ (2011) Inhibition of 2-oxoglutarate dependent oxygenases. Chem Soc Rev 40:4364–4397

Roundtree IA, Evans ME, Pan T, He C (2017) Dynamic RNA modifications in gene expression regulation. Cell 169:1187–1200

Sakai T, Toguchida J, Ohtani N, Yandell DW, Rapaport JM, Dryja TP (1991) Allele-specific hypermethylation of the retinoblastoma tumor-suppressor gene. Am J Hum Genet 48:880–888

Sakatani T, Kaneda A, Iacobuzio-Donahue CA, Carter MG, de Boom Witzel S, Okano H, Ko MS, Ohlsson R, Longo DL, Feinberg AP (2005) Loss of imprinting of Igf2 alters intestinal maturation and tumorigenesis in mice. Science 307:1976–1978

Sharma S, Kelly TK, Jones PA (2010) Epigenetics in cancer. Carcinogenesis 31:27–36

Shen DD, Suo FZ, Song QM, Chang J, Zhang T, Hong JJ, Zheng YC, Liu HM (2019) Development of formaldehyde dehydrogenase-coupled assay and antibody-based assays for ALKBH5 activity evaluation. J Pharm Biomed Anal 162:9–15

Singal R, Wang SZ, Sargent T, Zhu SZ, Ginder GD (2002) Methylation of promoter proximal-transcribed sequences of an embryonic globin gene inhibits transcription in primary erythroid cells and promotes formation of a cell type-specific methyl cytosine binding complex. J Biol Chem 277:1897–1905

Slotkin RK, Martienssen R (2007) Transposable elements and the epigenetic regulation of the genome. Nat Rev Genet 8:272–285

Stresemann C, Brueckner B, Musch T, Stopper H, Lyko F (2006) Functional diversity of DNA methyltransferase inhibitors in human cancer cell lines. Cancer Res 66:2794–2800

Su R, Dong L, Li C, Nachtergaele S, Wunderlich M, Qing Y, Deng X, Wang Y, Weng X, Hu C et al (2018) R-2HG exhibits anti-tumor activity by targeting FTO/m(6)a/MYC/CEBPA signaling. Cell 172(90–105):e123

Sun T, Wu R, Ming L (2019) The role of m6A RNA methylation in cancer. Biomed Pharmacother 112:108613

Suva ML, Riggi N, Bernstein BE (2013) Epigenetic reprogramming in cancer. Science 339:1567–1570

Toh JDW, Sun L, Lau LZM, Tan J, Low JJA, Tang CWQ, Cheong EJY, Tan MJH, Chen Y, Hong W et al (2015) A strategy based on nucleotide specificity leads to a subfamily-selective and cell-active inhibitor of N(6)-methyladenosine demethylase FTO. Chem Sci 6:112–122

Virani S, Colacino JA, Kim JH, Rozek LS (2012) Cancer epigenetics: a brief review. ILAR J 53: 359–369

Visvanathan A, Patil V, Arora A, Hegde AS, Arivazhagan A, Santosh V, Somasundaram K (2018) Essential role of METTL3-mediated m(6)a modification in glioma stem-like cells maintenance and radioresistance. Oncogene 37:522–533

Vu LP, Pickering BF, Cheng Y, Zaccara S, Nguyen D, Minuesa G, Chou T, Chow A, Saletore Y, MacKay M et al (2017) The N(6)-methyladenosine (m(6)a)-forming enzyme METTL3 controls myeloid differentiation of normal hematopoietic and leukemia cells. Nat Med 23:1369–1376

Wang X, Lu Z, Gomez A, Hon GC, Yue Y, Han D, Fu Y, Parisien M, Dai Q, Jia G et al (2014) N6-methyladenosine-dependent regulation of messenger RNA stability. Nature 505:117–120

Wang X, Zhao BS, Roundtree IA, Lu Z, Han D, Ma H, Weng X, Chen K, Shi H, He C (2015) N(6)-methyladenosine modulates messenger RNA translation efficiency. Cell 161:1388–1399

Wang P, Doxtader KA, Nam Y (2016) Structural basis for cooperative function of Mettl3 and Mettl14 methyltransferases. Mol Cell 63:306–317

Wang R, Han Z, Liu B, Zhou B, Wang N, Jiang Q, Qiao Y, Song C, Chai J, Chang J (2018a) Identification of natural compound Radicicol as a potent FTO inhibitor. Mol Pharm 15:4092–4098

Wang S, Chai P, Jia R, Jia R (2018b) Novel insights on m(6)a RNA methylation in tumorigenesis: a double-edged sword. Mol Cancer 17:101

Wang L, Hui H, Agrawal K, Kang Y, Li N, Tang R, Yuan J, Rana TM (2020) M(6) a RNA methyltransferases METTL3/14 regulate immune responses to anti-PD-1 therapy. EMBO J 39: e104514

Weng H, Huang H, Wu H, Qin X, Zhao BS, Dong L, Shi H, Skibbe J, Shen C, Hu C et al (2018) METTL14 inhibits hematopoietic stem/progenitor differentiation and promotes Leukemogenesis via mRNA m(6)a modification. Cell Stem Cell 22(191–205):e199

Wu H, Chen Y, Liang J, Shi B, Wu G, Zhang Y, Wang D, Li R, Yi X, Zhang H et al (2005) Hypomethylation-linked activation of PAX2 mediates tamoxifen-stimulated endometrial carcinogenesis. Nature 438:981–987

Xiao W, Adhikari S, Dahal U, Chen YS, Hao YJ, Sun BF, Sun HY, Li A, Ping XL, Lai WY et al (2016) Nuclear m(6)a reader YTHDC1 regulates mRNA splicing. Mol Cell 61:507–519

Xu C, Liu K, Tempel W, Demetriades M, Aik W, Schofield CJ, Min J (2014) Structures of human ALKBH5 demethylase reveal a unique binding mode for specific single-stranded N6-methyladenosine RNA demethylation. J Biol Chem 289:17299–17311

Yang X, Lay F, Han H, Jones PA (2010) Targeting DNA methylation for epigenetic therapy. Trends Pharmacol Sci 31:536–546

Yang Z, Li J, Feng G, Gao S, Wang Y, Zhang S, Liu Y, Ye L, Li Y, Zhang X (2017) MicroRNA-145 modulates N(6)-Methyladenosine levels by targeting the 3′-untranslated mRNA region of the N(6)-Methyladenosine binding YTH domain family 2 protein. J Biol Chem 292:3614–3623

Yue Y, Liu J, He C (2015) RNA N6-methyladenosine methylation in post-transcriptional gene expression regulation. Genes Dev 29:1343–1355

Zhang C, Samanta D, Lu H, Bullen JW, Zhang H, Chen I, He X, Semenza GL (2016a) Hypoxia induces the breast cancer stem cell phenotype by HIF-dependent and ALKBH5-mediated m(6) A-demethylation of NANOG mRNA. Proc Natl Acad Sci U S A 113:E2047–E2056

Zhang C, Zhi WI, Lu H, Samanta D, Chen I, Gabrielson E, Semenza GL (2016b) Hypoxia-inducible factors regulate pluripotency factor expression by ZNF217- and ALKBH5-mediated modulation of RNA methylation in breast cancer cells. Oncotarget 7:64527–64542

Zhang S, Zhao BS, Zhou A, Lin K, Zheng S, Lu Z, Chen Y, Sulman EP, Xie K, Bogler O et al (2017) M(6)a demethylase ALKBH5 maintains Tumorigenicity of glioblastoma stem-like cells by sustaining FOXM1 expression and cell proliferation program. Cancer Cell 31(591–606): e596

Zhao BS, Roundtree IA, He C (2017) Post-transcriptional gene regulation by mRNA modifications. Nat Rev Mol Cell Biol 18:31–42

Zheng G, Dahl JA, Niu Y, Fedorcsak P, Huang CM, Li CJ, Vagbo CB, Shi Y, Wang WL, Song SH et al (2013) ALKBH5 is a mammalian RNA demethylase that impacts RNA metabolism and mouse fertility. Mol Cell 49:18–29

Zhou S, Bai ZL, Xia D, Zhao ZJ, Zhao R, Wang YY, Zhe H (2018) FTO regulates the chemo-radiotherapy resistance of cervical squamous cell carcinoma (CSCC) by targeting beta-catenin through mRNA demethylation. Mol Carcinog 57:590–597

Chapter 18
Understanding the Crosstalk Between Epigenetics and Immunometabolism to Combat Cancer

Anuradha Seth and Susanta Kar

Abstract The interaction between metabolic and epigenetic events shapes metabolic adaptations of cancer cells and also helps rewire the proliferation and activity of surrounding immune cells in the tumor microenvironment (TME). Recent studies indicate that the TME imposes metabolic constraints on immune cells, inducing them to attain a tolerogenic state, incompetent of mounting effective tumor eradication. Owing to extensive mutations acquired over repeated cell divisions, tumor cells selectively accumulate metabolites that regulate the activity of key epigenetic enzymes to mediate activation/suppression of genes associated with T-cell function and macrophage polarization. Further, multiple modulators connecting epigenetic and metabolic pathways help dictate the preferential induction of cytokines and expression of lineage-specifying genes associated with immunosuppressive T-cell differentiation.

In this chapter, we attempt to discuss the mechanisms underpinning the metabolic and epigenetic interplay in immune cells of the TME and how modulating these events can boost the application of existing anticancer immunotherapy.

Keywords Epigenetics · Oncometabolites · Cancer immune cell · DNA methylation · Histone methylation · Histone acetylation

Introduction

The tumor microenvironment is populated by cancer cells, infiltrating immune cells, blood vessels, fibroblast, and signaling molecules prevailing in the extracellular matrix. To cope with the voracious energy needs of proliferation, cancer cells

---

A. Seth · S. Kar (✉)
Division of Molecular Microbiology and Immunology, CSIR-Central Drug Research Institute, Lucknow, India

Academy of Scientific and Innovative Research (AcSIR), AcSIR Headquarters CSIR-HRDC Campus, Ghaziabad, Uttar Pradesh, India
e-mail: susantakar@cdri.res.in

T. K. Kundu, C. Das (eds.), *Metabolism and Epigenetic Regulation: Implications in Cancer*, Subcellular Biochemistry 100,
https://doi.org/10.1007/978-3-031-07634-3_18

acquire extensive mutations and epigenetic modifications that enable increased consumption of nutrients such as glucose and amino acids, creating nutrient deprivation for surrounding immune cells in the TME. Heterogeneity in mutations adopted by different cancer cells contributes to their distinct metabolic adaptations (Franchina et al. 2017). Deregulation of cellular energy metabolism is recently recognized as a prerequisite to tumor immune escape (Villalba et al. 2012). Metabolism-induced malignancy manifests itself in several forms: lactate accumulation, glucose deprivation, and oncogene-driven metabolic deregulation as a result of mutations (e.g., loss of tumor suppressor gene, p53) (Vander Heiden and DeBerardinis 2017). The ability of tumor cells to modulate local concentration of vital nutrients coupled with the immunosuppressive effects of the waste products secreted by tumor cells helps determine the dynamics of metabolic deregulation imposed by the TME on infiltrating immune cells (Franchina et al. 2017). It has been widely recognized that metabolites and metabolic enzymes can have various nonmetabolic roles relevant in cancer (Kaelin and McKnight 2013). Studies have found that metabolites can contribute to tumorigenesis via epigenetic modifications. Epigenetic mechanisms include multifaceted changes in chromatin structure that influence gene expression without changing the DNA sequence (Mews et al. 2017; TeSlaa et al. 2016). These include DNA methylation, chromatin remodeling, and posttranslational modifications involving acetylation/deacetylation and methylation/demethylation of histones. Many chromatin modifications require common metabolic intermediates that orchestrate transcriptional and epigenetic events by serving as cofactors, substrates, or inhibitors of epigenetic modifiers. For instance, acetyl-CoA is critical for acetylation activities of histone acetyltransferase (HAT). Nicotinamide adenine dinucleotide (NAD^+) serves as cofactor for deacetylation reactions catalyzed by sirtuin family of histone deacetylases (HDACs). Likewise, α-ketoglutarate (α-KG) and S-adenosylmethionine (SAM) are indispensable for DNA/histone demethylase and DNA methyltransferase/histone methyltransferase group of enzymes, respectively (Kinnaird et al. 2016). Numerous energetic/metabolic sensors such as mammalian target of rapamycin complex1 (mTOR), as AMP-mediated protein kinase (AMPK), and hypoxia-induced factor 1α (HIF-1α), have been found to modulate critical signaling pathways leading to activation and differentiation of T cells and macrophage polarization (Wang and Green 2012; Sag et al. 2008). Several studies have linked abnormal metabolism with gliomas and breast cancer (Strickland and Stoll 2017). Mutations in metabolic enzymes cause aberrant accumulation of metabolites within tumor cells, which then perturb the normal balance of histone and DNA modifications, causing deregulated epigenetic control of gene expression. In the TME, persistent epigenetic dysregulation alters the phenotype of tumor-associated cells as well as surrounding immune cells, changing the TME from an immunopermissive environment to a protumor, immunosuppressive one. While immunological responses are often regulated by dynamic interactions between metabolic and epigenetic pathways, the various mechanisms underpinning the feedback modulation of these two primary biological pathways in immune cell functioning during tumor progression are attracting increasing attention. In this chapter, we attempt to discuss how the TME influences metabolic characteristics and energy consumption of various immune cell types and understand the key epigenetic mechanisms that facilitate such metabolic rewiring of cancer immune cells.

Macrophage Polarization in the TME

The versatility of macrophages allows them to exhibit a range of functional states dictated by environmental stimuli. Although the classification of macrophage phenotypes is difficult to demarcate, they can be broadly categorized into two major extremes: the classically activated M1 and alternately activated M2. However, this overly simplistic M1/M2 classification cannot be indiscriminately applied to in vivo situations, where tissue macrophages often display hybrid phenotypes with overlapping features of both M1 and M2 states (Sica and Mantovani 2012). In cancer too, while researchers have repeatedly assigned M2-like characteristics to tumor-associated macrophages (TAMs), the conventional M1/M2 paradigm fails to completely recapitulate the full breadth of their features, considering that TAMs must constantly adapt to the dynamic TME based on the availability of nutrients and oxygen (Netea-Maier et al. 2017).

Key Metabolic Features of M1 and M2 Macrophages

In response to proinflammatory signals (such as LPS/IFN-γ stimulation), macrophages acquire M1 phenotypic states characterized by increased rates of aerobic glycolysis (Freemerman et al. 2014). M1 macrophages depend on glycolysis and pentose phosphate pathway (PPP) to fuel their ATP requirements. They further exhibit a shunted Krebs cycle, which is broken at strategic points to facilitate accumulation of metabolic intermediates: succinate, itaconate, and citrate (Tannahill et al. 2013). Another metabolic characteristic of proinflammatory macrophages is that both oxidative phosphorylation (OXPHOS) and fatty acid oxidation processes (FAO) are found to be downregulated (Jha et al. 2015; Funk et al. 1993). On the other hand, M2 macrophages undergo a complete Krebs cycle and demonstrate increased rates of FAO and OXPHOS (Jha et al. 2015). Glycolysis is also essential for maintaining the M2-polarized macrophages (Suzuki et al. 2016); M2-polarized human macrophages seem to rely on glycolysis for the production of cytokines following LPS stimulation (Chiba et al. 2017). However, the exact role of glycolysis in M2-associated functions remains ambiguous. While studies have found that blockade of glycolysis via 2-deoxyglucose (2-DG) inhibits M2 polarization (Tan et al. 2015), recent studies have shown that glycolysis is indispensable for M2 differentiation so long as OXPHOS remains functional (Wang et al. 2018).

Pentose Phosphate Pathway (PPP) in M1/M2 Macrophages

In M1 macrophages, oxidative PPP is prevalent (Tannahill et al. 2013), where conversion of glycolytic intermediate, glucose-6-phosphate into ribulose-5-phosphate facilitates the production of NADPH. NADPH is then utilized by enzyme NADPH oxidase to generate reactive oxygen species (ROS) which is crucial for the phagocytic activity of macrophages (Yi et al. 2012). However, in M2 macrophages, the nonoxidative PPP is dominant (Haschemi et al. 2012), leading to increased ribose-5P production, which is necessary for synthesis of uridine diphosphate N-acetylglucosamine (UDP-GlcNAC) and subsequent N-glycosylation modification for several M2 surface maker proteins, such as CD206 (Jha et al. 2015).

Tricarboxylic Acid (TCA) cycle in M1/M2 Macrophages

Due to inhibition of multiple TCA cycle enzymes, M1 macrophages feature a stunted TCA cycle with multiple breakpoints characterized by increased accumulation of various TCA cycle intermediates such as citrate and succinate (Jha et al. 2015). Citrate accumulation that results from inhibition of isocitrate dehydrogenase (IDH) leads to the production of itaconic acid via enzyme, immunoresponsive gene 1 (Irg1) (Németh et al. 2015). Itaconic acid further inhibits enzyme, succinate dehydrogenase (SDH), leading to the accumulation of succinate (Freemerman et al. 2014). Succinate accumulation in LPS-stimulated (M1) macrophages has been associated with an increased glycolysis, ROS production, and proinflammatory cytokine production (Lampropoulou et al. 2016). Thus, while M2-polarized macrophages rely on TCA cycle for their energetic needs, M1-polarized macrophages depend on the TCA cycle mainly for ROS production and also for accumulation of metabolic intermediates that potential M1 macrophage functions.

FAO/FAS in M1/M2 Macrophages

As professional phagocytic cells, macrophages possess the capacity to uptake lipids from dying cells in the microenvironment (Tabas and Bornfeldt 2016). Ingested lipids are processed by lysosomes, resulting in free fatty acids (Huang et al. 2014) that are funneled into the mitochondria for their effective breakdown via the fatty acid oxidation (FAO) pathway.

IL-4-stimulated (M2) macrophages demonstrate increased uptake of fatty acids (Huang et al. 2014) and fatty acid oxidation, following activation by various M2-specific transcription factors such as STAT6, PPAR-γ, and PGC1β (Vats et al. 2006). Unlike M2 macrophages, where FAO is a crucial energy source, M1 macrophages rely on fatty acid synthesis (FAS) for energy production (Liu et al. 2021).

Inhibition of FAO via pharmacological approaches targeting carnitine palmitoyltransferase (CPT) enables fatty acid translocation into the mitochondria and has been found to prevent M2 polarization (Kerner and Hoppel 2000; Malandrino et al. 2015). Further, FAO inhibition in TAMS also promotes antitumorigenic differentiation, inhibiting tumor growth (Niu et al. 2017). However, while FAO has long been associated with protumoral functions, in a more recent study, it has been established that parallel induction of FAO and FAS favors antitumoral polarization of TAMs. In this study, the use of Toll-like receptor (TLR)-9 agonist, CpG-oligodeoxynucleotide, was found to engender a unique metabolic state in macrophages that varied from the canonical M1 (LPS)/M2 (IL-4) classification (Liu et al. 2019). This state was characterized by increased de novo lipogenesis and FAO, a feature characteristic to M2 macrophages yet without any discernible increase of exogenous fatty acid uptake.

Functionally, FAO fulfils the metabolic requirements of phagocytosis and also promotes increased breakdown of lipid loads following engulfment of target cells (Han and Ravichandran 2011), thereby potentiating antitumoral functions.

Role of Transcription Factors in TAM Polarization

The conventional paradigm assigns a protumoral M2 phenotype to TAMs. However, in a recent study, gene expression profiling of M2-TAMs infiltrating the tumors of patients of nonmetastatic renal cell carcinoma revealed a heterogeneous transcriptional profile for these macrophages that does not quite fit into the classical definitions assigned to M1-M2 (Nirschl et al. 2020). This increases the difficulty of attributing the same breadth of transcription factors that are already associated with M2 polarization to the differentiation of TAMs. The aura of mystery further deepens, considering the differences in human and animal models (Nirschl et al. 2020). Nonetheless, several M2-associated transcription factors such as PU.1 and c-Myc have been found to be involved in the progression of several tumors, including myeloma (Ueno et al. 2017), breast cancer (Lin et al. 2018), and hepatocellular carcinoma (Song et al. 2015). In vivo inhibition of c-Myc was found to attenuate protumoral functions including the expression of VEGF, HIF1α, and MMP9, which led to impaired angiogenesis and reduced tumor growth (Pello et al. 2012). A 2020 study by M Liu unraveled a novel role for another M2 transcription factor, c-Maf, in regulating TCA cycle and nonoxidative steps of the PPP for increasing biosynthesis of UDP-GlcNAc and production of N-glycosylated M2 surface markers (Liu et al. 2020). Notably, c-Maf was found to be widely expressed in TAMs of human patients of non-small cell lung carcinoma (NSCLC), and its deletion was associated with increased response to anti-PD-1 therapy and enhanced antitumoral T-cell responses in subcutaneous Lewis lung carcinoma (LLC) tumors. Expectantly, treatment with β-glucan, an inhibitor of c-Maf, seemed to reduce population of TAMs, leading to antitumor immunity in mice. These studies suggest that identifying transcription factors that are specifically overexpressed in

TAMs may forge new paths for the exploitation of well-known natural compounds or synthetic inhibitors that reduce or inhibit the expression of such transcription factors in such immunosuppressive macrophages, thereby boosting the efficacy of existing tumor immunotherapies.

Influence of the TME Signals on TAM Polarization

The altered metabolic profile of the TME is characterized by glycolysis end products, by-products of amino acid metabolism, release of tumor secretory exosomes, and physiological changes including hyperacidification and hypoxia. These act as extracellular messengers to confer upon functional alterations in TAMs associated with the acquisition of protumoral, immunosuppressive features. The rapid proliferation of tumor cells triggers hypoxia, a key distinguishing feature of the TME in many solid tumors. It arises from an imbalance between oxygen availability and oxygen consumption and fosters diverse alterations in metabolism and functioning of surrounding immune cells. Lung cancer cells, under hypoxic conditions, secrete a different complement of microRNAs as compared to normoxic cells. Preferential elevation of one of these microRNA, miR-103a, has been linked with the regulation of TAM polarization via uninterrupted activation of STAT3 and PI3K/Akt signaling pathways (Hsu et al. 2018). In oral squamous cell carcinoma, tumor-derived exosomes carrying miR-23a-3p have shown a similar effect in promoting M2-like phenotypes of macrophages by activating the SOCS1/STAT1 signaling (Ma et al. 2015). Hypoxia also enables crosstalk between tumor cells and TAMs, via activation of Jumonji domain-containing histone demethylases 1A (JMJD1A), which increases the abundance of these macrophages and stimulates pro-angiogenesis, leading to aggressive tumor growth (Osawa et al. 2013). Accumulation of ATP and its enzymatic by-product, adenosine, in the TME is another consequence of tumor-associated hypoxia. Adenosine regulates macrophage functions and phenotypes via adenosine receptors (A1, A2A, A2B, and A3). Downregulation of A2A receptor in myeloid cells facilitates a phenotypic switch from M2-like TAM to M1 polarization, leading to reduced melanoma tumor growth and metastasis (Cekic et al. 2014). Other tumor-derived signals in the TME such as lactate enable macrophages in the TME to adopt a tumor-promoting state marked by the upregulation of arginase-1 and vascular endothelial growth factor (VEGF) (Colegio et al. 2014). The upregulation of these genes then facilitates tumor growth by both providing nutrients like polyamines for tumor cell proliferation, as well as inducing vascularization (Chang et al. 2001; Qian and Pollard 2010). In murine model of breast cancer, tumor-derived lactate favors immunosuppressive TAM polarization by regulating the MEK/STAT3 axis (Mu et al. 2018). Expectantly, therefore, employing inhibitors of MEK/STAT3 such as selumetinib/static has shown promise in negating the effects of lactate on M2 polarization.

Epigenetic Regulation of Metabolism in Tumor-Associated Macrophages (TAMs)

Tumor-associated macrophages (TAMs) are a predominant population of immune cells residing in the tumor microenvironment. They are phenotypically diverse myeloid cells with a varied metabolic nature. TAMs have been associated with a predominant M2-like phenotype (Mantovani and Sica 2010) and promote tumor progression and aggressive metastasis (Ugel et al. 2015). Epigenetic regulation plays a key role in the differentiation of macrophages toward distinctive phenotypes, M1 or M2, also commonly known as classical or alternatively activated macrophages (Chen et al. 2019). Epigenetic control of macrophages is primarily mediated by DNA methylation and modification of histone proteins. DNA methylation is critical for macrophage differentiation, whereas methylation of cognate residues on histone proteins promotes timely activation of principal inflammatory factors during macrophage activation.

DNA Methylation and Macrophage Polarization in the TME

DNA methylation is an epigenetic modification that involves the covalent addition of methyl groups to cytosine residues. This process is regulated by DNA methyltransferase, or DNMT family, which includes enzymes, DNMT1, DNMT3A, and DNMT3B. DNA methylation occurs on CpG sites, which are linear CG sequences, present in the proximity of promoter regions. Hypermethylation of CpG sites by DNMT family members facilitates the transition of chromatin structure from open to condensed form, thereby restricting transcription factors from accessing and binding to gene promoters and leading to silencing of the corresponding gene. The reverse phenomenon, i.e., hypomethylation or DNA demethylation, is catalyzed by the ten-eleven translocation (TET) proteins (Rasmussen and Helin 2016). DNA methyltransferase enzymes also influence macrophage differentiation. DNMT3b knockdown drives M2 polarization (Yang et al. 2014b), while DNMT1 has been implicated in classical activation of macrophages via silencing of SOCS1 genes, leading to consequent increase in IL-6 and TNF-α production (Cheng et al. 2014a). In pancreatic ductal adenocarcinoma, tumor cells induce DNA methylation in metabolic genes, including aldehyde dehydrogenase family 1 member a3 (ALDH1a3) and NAD(P)H dehydrogenase (quinone-1) (NQO-1), leading to metabolic reprogramming and ultimately promoting M1-M2 transition. Inhibition of intrinsic EZH2/DNMT1 activities via a combination of small-molecule epigenetic inhibitors, EPZ011989 and decitabine, seemed to reversed inhibitory H3K27me3 in the enhancer regions and histone methylation marks on promoter regions of chemoattractant, CCL2 gene, leading to increased macrophage infiltration in human small cell lung cancer (SCLC) lesions (Zheng et al. 2020).

Upregulation of the TET family of proteins is found to assist in M2 polarization. Mutations pertaining to TET2 loss of function have been implicated in various myeloid malignancies. In particular, TET2 and TET3 were found to be overexpressed in intratumoral myeloid cells in both melanoma patients of murine models of the disease (Pan et al. 2017) and prevented the consistent production of inflammatory mediators, such as IL-6 (Zhang et al. 2015). Arginine methyltransferase enzyme, PRMT1, overexpression is associated with increased methylation of the histone mark, H4R3me2a, at the PPAR-γ (a key M2-transcription factor) promoter, leading to M2 polarization (Tikhanovich et al. 2017). Overall, TAM polarization is regulated by the dynamic coordination between metabolic intermediates, transcription factor, and epigenetic modifiers that act together in concert to drive expression of specific genes. Induction of classical macrophage activation via LPS and TLR4 ligands results in impaired mitochondrial respiration and a rapid shift toward glycolysis. The latter event is associated with a concomitant increase in lactate, which causes inhibition of class II histone deacetylases (Latham et al. 2012). Cancer cells employ an enzyme, ATP-citrate lyase (ACLY), which converts citrate into acetyl-CoA, the primary substrate for histone acetylation. Increase in acetyl-CoA levels and decrease in 2-oxoglutarate levels are important for favoring histone acetylation over histone demethylation, thereby compromising the activities of TET and Jmjc domain-containing histone demethylase (JHDM) proteins (Chowdhury et al. 2011). As such, the ratio of these two metabolites is critically responsible for facilitating the phenotypic switch to M1 or M2 differentiation. Immune-responsive gene (IRG1) promotes itaconate production, a metabolite highly expressed in classically activated macrophages (Michelucci et al. 2013). Itaconate modulates succinate levels by inhibiting succinate dehydrogenase-driven oxidation of succinate, leading to increased production of inflammatory cytokines. In alternately activated macrophages, the STAT6-PGC1-α-PPARγ axis is responsible for mitochondrial function and oxidative metabolism (Chawla 2010). In addition, an increased synthesis of polyamines is found to be associated with histone acetyltransferase enzymes, leading to subsequent hypomethylation of DNA and induction of M2 genes (Hobbs and Gilmour 2000; (Latour et al. 2019).

Histone Modification and Macrophage Polarization in the TME

Several studies have highlighted the importance of histone methylation and demethylation in regulating M1/M2 differentiation. These activities are under the control of different enzymes. While histone arginine methyltransferase, PRMT1, positively regulates M2 macrophages via induction of PPAR-γ (Tikhanovich et al. 2017) and negatively regulates M1 macrophages by repressing CIITA, SMYD3, another histone methyltransferase (HMT), solely acts as a positive regulator of M2 phenotype (Kittan et al. 2013). JMJD3, a demethylase, carries out H3K27 demethylation, inducing key markers of M2 polarization, Arg1, Irf4, and CD206 (Satoh et al. 2010). In a study that revealed metabolite regulation of histone methylation in macrophages, the authors found that pathways pertaining to glucose and amino acid metabolism synergize to enhance production of S-adenosylmethionine, the methyl group donor responsible for histone H3K36me3 on gene loci of IL-1β (Yu et al. 2019), thereby increasing its production on inflammatory macrophages. In a study by Tan et al., histone methylation was found to reprogram macrophage phenotype toward the antitumor M1 phenotype in breast cancer model (Tan et al. 2019). Classically activated (M1) macrophages featured subdued expression of lysine-specific demethylase (LSD1) in murine model of triple-negative breast cancer. Treatment with LSD1 inhibitor, phenelzine, seems to restore normalized expression of M1 markers, leading to acquisition of normal, M1-like signatures (Tan et al. 2019). This study indicates that modulating histone methylation may help manipulate macrophage programming in order to potentiate antitumoral M1-like macrophages for treating cancer. Among histone modifications, histone acetylation/deacetylation is also found in macrophages, though not as prevalent as histone methylation. Here, HDAC9 and HDAC11 negatively regulate M2 polarization (Villagra et al. 2008), while other HDACs, HDAC4 and SIRT2, positively regulate M2 phenotype by inducing expression of GATA3, Cd11c, and Arg1 in mouse BMDMs (Yang et al. 2014a). Recently, SIRT1/2 has been found to be preferentially upregulated during differentiation of macrophages. SIRT1/2 mediates methylation gains on inflammatory gene loci, resulting in the decrease of activating histone marks and consequent gene repression. Inhibition of SIRT1/2 upon LPS challenge abrogated those SIRT1/2-dependent methylation marks, leading to increased expression of inflammation-related genes and macrophage activation (Li et al. 2019). Taken together, these studies show that HDACs, specifically SIRT1/2, restrict aberrant activation of proinflammatory genes, and hence targeting them via inhibitor strategies can help in restoration of M1-like phenotype. Other HDACs are also found to link immunometabolism with epigenetics. In particular, HDAC7 tethers TLR-triggered aerobic glycolysis with polarized inflammatory responses in mice macrophages. Overexpression of HDAC7 accentuates lactate production, promoting a glycolysis-induced inflammatory phenotype. Glycolytic enzyme, pyruvate kinase M isoform 2 (PKM2), acts together with HDAC7 in a complex, where deacetylation by PKM2 is a prerequisite to proinflammatory reactions driven by HDAC7 (Das Gupta et al. 2020). Compromising the activity of this complex dampens M1-phenotype differentiation in murine macrophages as well as in vivo. Histone modifications have been widely studied in TAMs as well. Tumor cells enhance histone deacetylation at promoter regions ahead of genes encoding class II transactivator (CIITA), the primary regulator of MHC-II, resulting in decreased expression of MHC class-II and its dependent antigen presentation genes (Chang et al. 2008).

Another lesser known histone modification, histone lactylation, links immunometabolism with epigenetics by regulating homeostasis in M1 macrophages and preventing transition to M2 phenotype in tumors. Lactate-dependent lactylation of lysine resides on histones results in direct activation of gene transcription. In M1 macrophages challenged with bacteria, histone lactylation was found to increase in a temporal fashion and correlated with heightened expression of homeostatic markers of M1 macrophages associated with wound healing (Zhang et al. 2019). Therefore, targeting histone lactylation may be seen as a strategy for preventing M1-M2 transition in tumor macrophages as a way to boost antitumoral responses.

Overall, as summarized in Fig. 18.1, the roles of histone modifications in macrophages imply that targeting these enzymes can help reorient macrophage phenotype to one that ameliorates cancer progression.

Metabolic Programming of Dendritic Cells (DCs) in the TME

Dendritic cells (DCs) are specialized antigen-presenting cells (APCs) that are capable of mounting potent antitumoral T-cell responses. The metabolic state of DCs is a crucial regulator of its function. Upon activation, DCs depend on glycolysis for their increased energetic needs, while immature or tolerogenic DCs mainly carry out OXPHOS for their bioenergetic demands (Krawczyk et al. 2010). Increased lipid deposition in tumor-associated DCs favors FAO, leading to immune tolerance (Herber et al. 2010). How lipid alterations obliterate DC functionality was uncertain until recently when it was found that intracellular accumulation of lipids interferes with the transport of MHC class I peptide and suppresses antigen presentation functions of DCs (Ramakrishnan et al. 2014). Accumulating evidence has shown that several tumor-derived signals manipulate metabolic pathways of DCs in the TME, driving a phenotypic switch that induces tolerized states and creates an immunocompromised microenvironment. Aberrant metabolic changes in tumor cells evoke a plethora of changes in the TME involving the accumulation of lactate, lipids, tumor-derived exosomes, as well as acidification and hypoxia that suppress the functioning of DCs and activation of T-effector cell responses. For instance, tumor-secreted oxysterols inhibit the expression of C-C chemokine receptor type 7 (CCR7) on DCs, preventing their subsequent maturation and generation of antitumoral immunity (Villablanca et al. 2009).

The constant efflux of tumor-derived lactate in the TME can activate G protein-coupled receptor 81 (GPR81) on DCs, inhibiting MHCII expression and presentation of tumor-derived antigens for the effective curation of Teff cell responses in murine breast cancer model (Brown et al. 2020). Besides lactate-induced hyperacidification, another metabolic characteristic of the TME that drives DC tolerization is hypoxia. It has been found that nullifying the effects of hypoxia via pharmacological inhibition of HIF-1α can successfully enhance the efficacy of DC-based cancer vaccine in 4T1 breast cancer model (Kheshtchin et al. 2016). The high levels of ROS in the TME also trigger tolerized DC states, supposedly through the oxidation of high-mobility group protein B1 (HMGB1), a redox-sensitive protein which then adopts a neutralized state, incapable of inducing DC maturation (Apetoh et al. 2007).

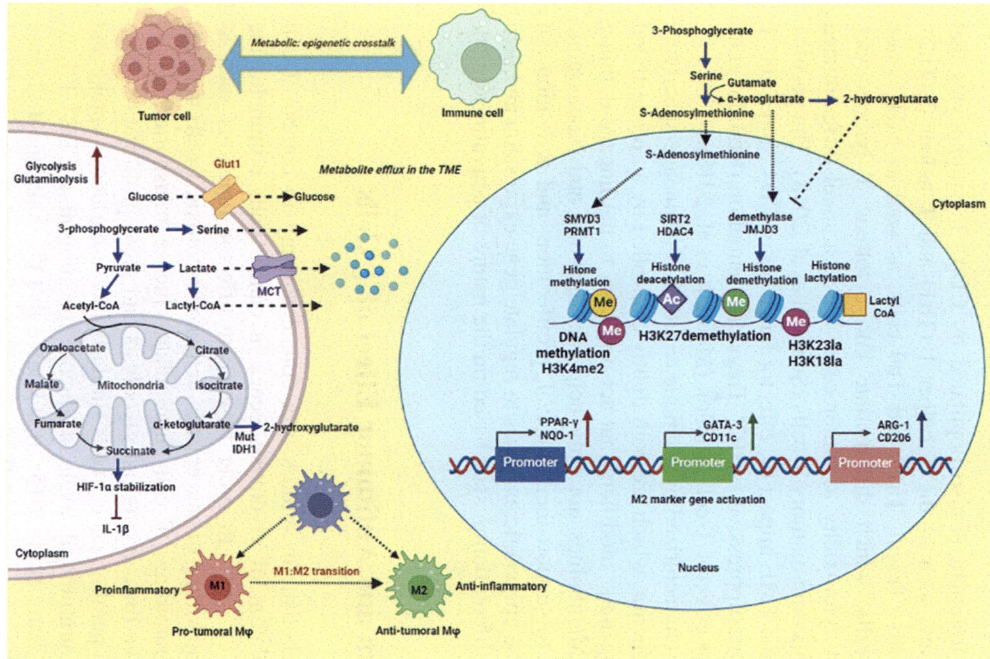

Fig. 18.1 Dysregulation in tumor metabolism regulates epigenetic reprogramming of M1/M2 macrophages: Rapidly proliferating cancer cells scavenge essential nutrients such as glucose, glutamine, and tryptophan to fuel glycolysis, limiting nutrient availability to immune cells. In the TME, oncometabolites like succinate, fumarate, and 2-hydroxyglutarat (2-HG) bear structural similarity to α-ketoglutarate (αKG) and thus act as competitive inhibitors of αKG-dependent dioxygenases, the KDMs and TET family of DNA hydroxylases, leading to genome-wide histone and DNA methylation on tumor suppressor genes and aiding in tumor progression. In protumoral M1 macrophages of the TME, M1:M2 transition is aided by epigenetic changes involving PRMT1, HDACs, SMYD3, and JMJD3 that upregulate key M2 marker genes, Arg1, CD206, GATA3, CD11c, and PPARγ. Increased DNA methylation at metabolic genes such as NQO-1 further promotes M2 polarization. In addition, efflux of metabolite, lactate in the TME, acts as an epigenetic regulator, inducing M2 polarization by upregulation of Arg-1

Metabolic and Epigenetic Regulation of T-Cell Subsets in TME

When it comes to launching antitumor immunity, T cells are the forefront players. Activated T cells mount targeted responses by producing cytotoxic components that directly kill cancer cells and also secrete signaling molecules that activate other naive immune cells.

The tumor microenvironment is also populated by $CD4^+$ T lymphocytes that include various categories, such as the T helper 1 (Th1) and T helper 2 (Th2), $FoxP3^+$ regulatory T cells (Tregs), Th17 and Th9 cells, as well as MHC class I-expressing $CD8^+$ T cells, which differentiate into cytotoxic T lymphocytes (CTLs) and exert direct cytotoxicity to tumor cells. Later, a small differentiated fraction of these CTLs become memory T cells (Schmidl et al. 2018). The peculiar metabolic signatures of the TME impact the effector functions of $CD4^+$ and $CD8^+$ T cells and promote immunosuppressive $FoxP3^+$ Tregs differentiation (Kouidhi et al. 2017) that further repress effector T-cell function (Maimela et al. 2018).

Each T-cell subset is characterized by a unique expression of lineage-specifying genes and cytokine production which are both tightly controlled by an orchestrated network of multiple epigenetic modifiers that act upstream to the lineage-defining transcriptional factors. Metabolites not only help fuel anabolic and bioenergetic requirements of T cells to support their survival, proliferation, and activation but also coordinate with these epigenetic enzymes to regulate gene expression and help them achieve a certain differentiation state based on the nature of activating signal.

Glucose Metabolism and Antitumor Effect of T Cells

Increased glucose metabolism in tumors is a crucial strategy to combat T-cell infiltration, and upregulation of glycolysis genes in cancer cells eliminates the antitumoral abilities of T cells. In order to execute effector functions such as secretion of cytokines and killing of cancer cells, activated cytotoxic T cells undergo rapid metabolic reprogramming. T cells upregulate glucose transporter, Glut1, increasing glucose uptake for carrying out glycolysis and rapid proliferation upon antigenic stimulation. In parallel, cancer cells similarly increase glucose uptake and glycolysis, decreasing intratumoral levels of glucose for infiltrating $CD8^+$ T lymphocytes in the TME (Chang et al. 2015). It is known that the consequent glucose deprivation abrogates the production of IFN-γ, a key mediator of antitumor $CD8^+$ T-cell responses. These effects are carried out by GAPDH, an enzyme involved in glycolysis that influences posttranscriptional modifications of IFN-γ mRNA. When T cells perform increased glycolysis, GAPDH functions are intrinsic to the glycolytic pathway; however, during low glycolysis, GAPDH serves to modify the translation of IFN-γ instead (Chang et al. 2013). Glucose deprivation further dampens NFAT signaling and T-cell receptor (TCR)-dependent blockage of sarco/

endoplasmic reticulum Ca2^{+-} ATPase, causing CD8^{+} T-cell hyporesponsiveness (Ho et al. 2015; Oh-hora and Rao 2008). Indeed, a prominent correlation has been found between increased intratumoral glucose metabolism and reduced infiltration of CD8^{+} T cells in various cancers such as squamous cell carcinoma (Ottensmeier et al. 2016) and renal cell carcinoma (Singer et al. 2010). Targeting regulators of aerobic glycolysis such as oncogenic epidermal growth factor receptor (EGFR) and HIF-1α may offer a means to selectively downregulate tumor metabolism, as observed in case of lung adenocarcinoma cells, without disrupting effector T-cell functioning (Makinoshima et al. 2014). Interestingly, emerging clinical data point toward tumor glucose metabolism as a mechanism of resistance to T-cell-mediated tumor rejection. As recently shown, overexpression of glycolysis-related genes in cancer cells impairs the antitumor activity of T cells, and, inversely, the inhibition of cancer glycolysis enhances T-cell-mediated tumor rejection (Cascone et al. 2018). From the therapeutic point of view, glucose supply in the tumor microenvironment helps rescue IFN-γ production by TILs. Further, PD-L1 blockade on tumor cells inhibits mTOR signaling, further reducing glycolytic flux of cancer cells to increase available glucose stores for optimal nutrient acquisition (Chang et al. 2015) and boost proliferation of tumor-infiltrating CD8^{+} T cells (Spranger et al. 2014). In adoptive cell transfer (ACT), inhibiting glucose metabolism via 2-deoxyglucose (2-DG), an analog of glucose, was found to impede glycolysis in CD8^{+} T cells, increasing production of memory cells (Sukumar et al. 2013). In parallel with the effects of low glucose in the TME, hypoxia-driven increase of transcription regulator, HIF-1α, too has several immunoregulatory effects, from suppression of TCR-mediated Ca^{2+} signaling and CD8^{+} T-cell activation (Damgaci et al. 2018) to promoting apoptosis of surrounding T cells in solid tumors (Mpekris et al. 2020). Moreover, it has been found that the hypoxia/HIF-1α axis can increase the accumulation of immune inhibitory ligand, PD-L1, on prostrate and human breast cancer cells (Barsoum et al. 2013), making them more resistant to CTL-mediated killing. Considering the monumental impact that hypoxic TME has in shaping T cell responses and T cell differentiation, increasing research is being geared toward understanding how to target hypoxia in order to leverage cancer immunotherapy.

Role of Lipid Metabolism in T Cells

Naive CD8^{+} T cells depend upon fatty acid oxidation (FAO) as the major energy source (van der Windt and Pearce 2012) and promptly switch to glycolysis upon antigen encounter for their effector functions (Chang et al. 2013). Inhibition of glycolysis via 2-DG has been found to drive CD8^{+} T cells toward development of memory CD8^{+} T cells for long-term antitumoral immunity (Sukumar et al. 2013). Thus, glucose uptake and glycolytic flux act as rheostats to regulate both fate and differentiation of CD8^{+} T cells in the TME. In response to the hypoglycemic conditions of the TME, CD8^{+} TILs switch to the catabolism of fatty acids as their main energy source. Promoting fatty acid catabolism therapeutically has been found to synergize with PD-1 blockade in mouse melanoma models (Zhang et al. 2017). Furthermore, modulating cholesterol metabolism via administration of avasimibe, an inhibitor of the key cholesterol esterification enzyme, acyl-CoA acyltransferase (ACAT1) potentiated function and proliferation of $CD8^+$ T cells specifically and not $CD4^+$ T cells. Mechanistically, inhibition of ACAT1 promoted biosynthesis of cholesterol and increased its accumulation at the plasma membrane, thus aiding in the clustering of T-cell receptor and resulting in more efficient formation of immunological synapses (Yang et al. 2016). In the TME of pancreatic ductal adenocarcinoma (PDA), $CD8^+$ T cells accumulate long-chain fatty acids over time, which eventually impairs mitochondrial functions and disrupts the normal catabolism of fatty acids (Manzo et al. 2020). The use of bezafibrate, an agonist of peroxisome proliferator-activated receptor γ (PPARγ) coactivator 1-α (PGC-1α), has been uncovered as a strategy to restore mitochondrial function in energetically compromised CTLs, restoring both OXPHOS and glycolysis for more effective utilization of nutrients and promoting their survival (Chowdhury et al. 2018). As such, strategies activating mitochondrial metabolism in CTLs that boost both FAO and glycolysis can increase and maintain the numbers of functional CTLs in the TME, thus offering a means to synergize with PD-1 blockade (Chowdhury et al. 2018).

Acetyl-CoA and Histone Acetylation in T-Cell Differentiation and Functioning

Glycolysis end product, acetyl-CoA, fuels histone acetylation reactions in tumor cells and T cells. Histone acetylation regulates $CD4^+$ T-cell differentiation in a synergistic manner with metabolic factors and cytokines produced in the TME. The high glycolytic rate of tumor cells leads to increased accumulation of lactate causing acidification in the TME. This lactic acid has an inhibitory effect on $CD8^+$ T-cell proliferation (Brand et al. 2016). A recent study demonstrated that glycolysis in $CD4^+$ Th1 cells is essential for maintaining adequate acetyl-CoA levels during differentiation. The enzyme, lactate dehydrogenase, LDHA, by converting pyruvate to lactate, helps divert further processing of glycolytic end products into the OXPHOS cycle, thereby promoting citrate accumulation and increasing acetyl-CoA levels within the cytosol. Mechanistically, acetyl-CoA production by LDHA activity enhances levels of H3K9Ac and H3K27Ac on IFN locus, leading to increased IFN-γ expression on activated Th1 cells (Peng et al. 2016). These findings underscore the importance of epigenetic mechanisms in governing aerobic glycolysis for the augmentation of Teff cell differentiation and functions. Pharmacological blockade of lactate transporters, MCT1 and MCT4, minimizes the immunosuppressive action of lactic acid in mouse tumor model (Renner et al. 2019). Additionally, LDHA knockdown in xenograft mouse model of hepatocellular carcinoma via interference RNA nanoparticles seemed to lower TME acidification, causing increased infiltration of $CD8^+$ T cells that were otherwise immobilized by lactate metabolites (Sheng et al. 2012).

Role of TCA Intermediates in Epigenetic Modulation of T-Cell Differentiation

Fluctuations in metabolites such as 2-hydroxyglutarate (2-HG), butyrate, and S-adenosylmethionine (SAM) have been found to alter epigenetic states that contribute to T-cell differentiation programs. By acting as donors, cofactors, substrates, and even antagonists to activities of critical epigenetic-modifying complexes, these metabolites direct the differentiation of $CD4^+$ T cells into more specialized categories, i.e., the effector or memory T-cell programs. A scrutiny of the links that connect epigenetics and metabolism will not only abrogate these crosstalks that promote tumorigenesis but also boost immune fitness. An outline of these mechanisms has been illustrated in Fig. 18.2.

In the following section, we discuss how nutrients process through key metabolic pathways such as glutaminolysis to mechanistically influence epigenetic events and alter programming of T cells in a dysregulating setting such as the tumor microenvironment.

Glutamine

The ability of T cells to take up glutamine relies on the expression of amino acid transporter type 2 (ASCT2) (Nakaya et al. 2014). Deficiency of ASCT2 hinders the development of Th17 cells, favoring the differentiation of Tregs instead (Klysz et al. 2015). However, during glutamine deprivation, α-ketoglutarate addition restores Th1 differentiation via induction of T-bet (Klysz et al. 2015). Overexpression of oncogenes like c-Myc enhances glutamine addiction in cancer cells. Thus glutamine deprivation can be a major contributing factor of tumor-induced immunosuppression (Wise et al. 2008). Antagonist of glutamine, 6-diazo-5-oxo-L-norleucine [DON], disrupts tumor cell metabolism, increasing nutrient availability and reducing hypoxia in the tumor microenvironment. Also, it has been found to provide leverage to antitumoral responses produced in response to anti-PD1 blockade (Sharma et al. 2019).

Connections Between α-Ketoglutarate and Histone Demethylases in T-Cell Functioning

Glutamine is the primary source of α-ketoglutarate, a metabolic intermediate, intrinsic to the TCA cycle. The latter moonlights as a cofactor for histone and demethylase (also known as DNA dioxygenase) complexes (Tahiliani et al. 2009). Histone demethylases can be broadly divided into two main classes: lysine-specific demethylase (LSD) and Jumonji C domain (JmjC). While flavin adenine

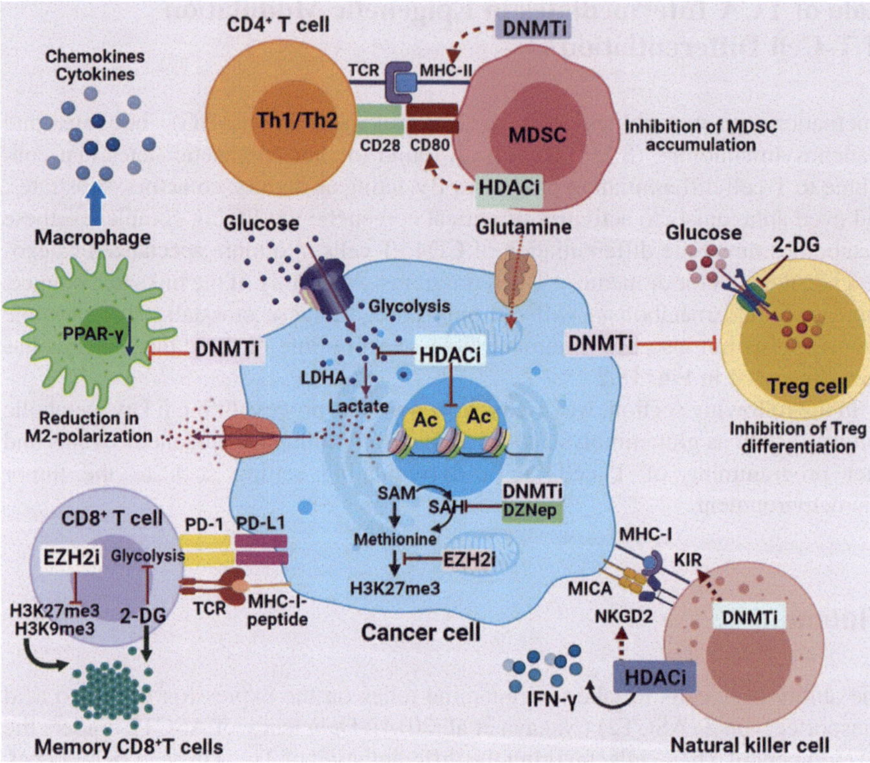

Fig. 18.2 Metabolic and epigenetic inhibitors targeting cancer immune cells. Epigenetic inhibitors, DNMTi and HDACi, increase the expression of surface molecules (MHC-II and CD80) on MDSCs in the TME, enabling antigen presentation to naive T cells for antitumoral Teff cell responses. Epigenetic and metabolic therapies like HDACi and DZNep that target tumor cells impede glucose uptake and glycolysis, reducing production of lactate, acetyl-CoA, and methionine that fuel aberrant epigenetic modifications. Epigenetic drugs, EZH2i and DNMTi, help reverse suppressive hypermethylation on tumor suppressor genes. In addition, these therapies have immunomodulatory potential in reduction of M2 macrophage polarization, inhibition of Treg differentiation and restore memory $CD8^+$ T-cell differentiation and cytotoxicity of NK cells by increasing surface expression of KIRs. In addition, HDACi also increases the surface expression of NKGD2 in NK cells, thereby inducing its cytotoxic effector responses by secreting IFN-γ

dinucleotide (FAD) serves as a cofactor for LSD group of enzymes, JmjC demethylases employ metabolites such as αKG for their enzymatic reactions. JMJD3, a member of the JmjC family utilizes αKG as cofactor to target H3K27me3 marks. It is also involved in the epigenetic regulation of STAT4 and T-bet-sensitive target genes in $CD4^+$ Th1 cells. JMJD3, in a T-bet-dependent manner, gets recruited to these genes, leading to a reversal of H3K27me3 modifications and a general increase in promoter accessibility. Therefore, αKG-dependent demethylase activities are a crucial aspect of JMJD3 functioning in $CD4^+$ Th1 cells (Li et al. 2014). In a study, Pan et al. (2016) showed that low glutamine in the core regions of solid tumors led to rampant histone hypermethylation as a result of lowered αKG insufficient levels of αKG, inducing dedifferentiation and drug resistance. Interestingly, the levels of H3K4me3 remain recalcitrant to αKG levels (Hwang et al. 2016). So far, it remains unclear whether αKG determines the specificity of chromatin modifications in demethylase reactions. Interestingly, succinate and fumarate, both TCA cycle intermediates, are antagonistic to αKG activities. Mutations in enzymes producing succinate and fumarate are found in several cancers (Sciacovelli et al. 2016; Xiao et al. 2012). While it was known that αKG serves as a cofactor in demethylase complexes, accumulation of αKG has been linked with overall reduction in global histone methylation specifically in relation to histone modifications associated with repression, e.g., H3K27me3 (Carey et al. 2014). TET family of proteins catalyze oxidation of 5-methyl-cytosine (5-mC) into 5-hydroxymethyl-cytosine (5-hmC), thereby reversing demethylation of abnormally methylating cytosine residues (Pastor et al. 2013). As TETs are inhibited by metabolite intermediates, succinate and fumarate, and require αKG as the cofactor, the ratio of αKG/succinate regulates chromatin status via TET1/TET2 and JMJD3-dependent demethylation of histone modifications, H3K9me3, H4K20me, and H3K27me3 (Xiao et al. 2012; Carey et al. 2014). TET family has roles in solid tumors, wherein loss of function leads to accumulation of aberrant DNA methylation. Further, absence of 5-hmC is a key epigenetic feature of melanoma (Lian et al. 2012).

2-Hydroxyglutarate (2-HG)

The metabolite, 2-HG, is a by-product of the TCA cycle. Being structurally similar to αKG, it acts as its antagonist in enzymatic reactions where αKG serves as cofactor to demethylase complexes (Xu et al. 2011). Notably, in cancers such as acute myeloid leukemia (AML) and gliomas, mutated forms of isocitrate dehydrogenase (IDH1/IDH2), the enzymes responsible for producing 2HG, cause overproduction of succinate, fumarate, and 2-HG (Medeiros et al. 2016; Dang et al. 2009). Current data suggests that fine-tuning the balance between 2HG and αKG is critical for regulating the functioning of demethylase complexes. Hence, 2-HG has been touted as an oncometabolite, partly because of its contribution in creating an aberrant epigenetic environment that triggers the cancer phenotype (Lu et al. 2012). IDH mutations cause cells to display dysregulated DNA and histone methylation patterns, as a result of 2HG antagonizing TET and JmjC activities that involve αKG (Figueroa et al. 2010; Huang and Rao 2014).

Metabolomic analysis by LC-MS has found that 2-hydroxyglutarate levels are significantly elevated in Th17 cells compared to Treg cells. As discussed before, 2-HG is antagonistic to TET1-/TET3-mediated demethylation of Foxp3 promoter. As such, 2-HG inhibits Foxp3 expression by increasing methylation levels of Foxp3 and other hypomethylated regions of Th17 and Treg cells, thereby preventing expression of the key marker, Foxp3, associated with these immunosuppressive

Treg cells (Xu et al. 2017). Another study has found 2-HG accumulations shortly after T-cell receptor stimulation in $CD8^+$ T cells. S-2-HG, an enantiomer of 2-HG, was found to regulate stability of HIF-1α and modulate histone and DNA methylation patterns across $CD8^+$ T cells (Tyrakis et al. 2016). Cancer cells bearing IDH1/2 mutations are characterized by significant accumulation of 2-HG. Whether it is involved in other aspects of T-cell differentiation such as immune tolerance or T-cell memory still requires further study. Also mechanisms behind its dramatic induction in differentiating Th17 cells, TCR-stimulated $CD8^+$ T cells, and their implications with cancer progression are unclear.

Methionine and Histone/DNA Methylation

Methylation of DNA and histones is catalyzed by methyltransferases that require SAM as the methyl group donor. As a result, metabolites and factors that are directly linked with SAM metabolism, such as methionine that have a profound effect on the methylation status of DNA and histones. Alterations in methionine modulate H3K4me3 levels by reducing SAM levels (Dai et al. 2018). Also, it impairs the functioning of H3K27 methyltransferase EZH2, which is responsible for bestowing repressive methylation marks on chemokines CXCL9 and CXCL10, thereby halting the trafficking of $CD8^+$ Teff cells into the TME (Peng et al. 2015). Dysregulated DNA methylation is a key characteristic of cancer cells. DNMT complexes influence the development of immune cell populations as well. DNMT1 has been found to be involved in T-cell development by aiding in the maintenance of DNA methylation states across cell divisions. Loss of de novo methyltransferase DNMT3a resulted in aberrant expression of Th1 cytokine, IFN-γ, in $CD4^+$ T-cell subtypes, Th2 and Th17 cells, implying that normal activity of DNMT3a is crucial in the silencing of IFN-γ gene in fates alternative to Th1 (Gamper et al. 2009). In human patients of ovarian cancer, expression of DNMT1 was negatively correlated with amount of tumor-infiltrating $CD8^+$ T cells (Li et al. 2017). Further, DNMT inhibitor, decitabine, was effective in carrying out DNA demethylation on epigenetic programs associated with T-cell exhaustion (Li et al. 2017). By inhibiting the expression of DNMT1 and DNMT3a, decitabine could restore Th1 polarization via upregulation of IFN–γactivated genes. Together, these findings build a strong rationale for the potential of DNA methylation modulation in controlling T-cell functioning for treatment of cancer.

Butyrate and Histone Deacetylases in Determination of Treg/Th17 Ratio in Cancer Progression

Butyrate, a type of short-chain fatty acid (SCFA), is produced by intestinal bacteria by fermenting dietary fiber in the colon. Butyrate has been found to inhibit class I and class II HDACs (Davie 2003). In particular, butyrate has been found to inhibit repressive histone deacetylation mark at the Foxp3 locus, promoting acetylation instead and permitting the expression of Foxp3, the signature surface molecule of Treg cells (Furusawa et al. 2013). In a study by Zhang et al., authors found that butyrate, by promoting Treg development, inhibited IL-17 generation by Th17 cells, reducing colitis and halting colon cancer development (Zhou et al. 2018). Several metabolic and epigenetic inhibitors have been found to regulate Treg cell differentiation. DNA methylation is an essential way of modulating expression of FoxP3, the surface marker on Tregs (Delacher et al. 2017). It was found that DNMTi, 5-azacytidine, significantly inhibited expression of FoxP3, thereby reducing number of Tregs (Costantini et al. 2012). Similarly, considering that glucose availability is another criterion for adequate Treg differentiation, glycolysis inhibitors hold promise in achieving their suppression (Delgoffe et al. 2009). Further, in mouse polyposis model, butyrate supplementation increases the levels of G protein-coupled receptor 109a (GPR109a) in colonic antigen-presenting cells, macrophages and dendritic cells, which then eventually induce the differentiation of Treg cells that counteract the proliferation of proinflammatory cells and inhibit widespread inflammation at early stages of carcinogenesis (Singh et al. 2014). Interestingly, levels of H3K9ac, an indicator of histone deacetylation, did not accompany the rise in GRP109a activation in responses to butyrate production, suggesting that butyrate-mediated immunomodulation is independent of its epigenetic effects (Singh et al. 2014).

Myeloid-Derived Suppressor Cells (MDSCs)

Myeloid-derived suppressor cells (MDSCs) are immunosuppressive bone marrow-derived cells that act as precursors to macrophages, dendritic cells, or granulocytes. Human MDSCs express a variety of markers, namely, CD14, CD33, alpha M-integrin CD11b, and/or CD15. MDSCs, like macrophages, turn "corrupt" upon accumulating in the TME and exert immunosuppression by secreting TGF-β and IL-10, which promote recruitment of Treg cells and tolerogenic TAM polarization (Beury et al. 2014; Schlecker et al. 2012). The presence of increased lactate and a hypoxic TME induces HIF-1α stabilization, resulting in continual expression of PD-L1, which then engages with PD-1 to downregulate Teff cell function (Daneshmandi et al. 2019). Furthermore, they indirectly inhibit immune cell proliferation by scavenging carbon and amino acid sources in the TME. The metabolic requirements of MDSCs are interestingly similar with those of cancer cells. MDSC maturation is accompanied with accelerated glycolysis, with lower than usual rates of pentose phosphate pathway and OXPHOS activity to ensure optimal levels of NAPDH for fueling their anabolic needs (Goffaux et al. 2017).

Epigenetic modifications found in MDSCs are of a diverse type: DNA methylation, histone acetylation, methylation, ubiquitination, and, the more recently identified, histone lactylation. Epigenetic enzyme inhibitors, HDACi, have been found to induce expression of surface markers, MHC-II, CD40, and CD86, on peripheral MDSCs of lung cancer patients, increasing antigen presentation of tumor antigens

(Adeegbe et al. 2017). Intraperitoneal treatment of leukemic mice with demethylating agent, decitabine, serves to specifically deplete MDSC populations via induction of apoptosis and restore the full complement of antitumoral T-cell responses (Zhou et al. 2017). As MDSCs are a mixed population of immature myeloid cells, they exhibit decent amount of plasticity, and further differentiation into DCs or macrophages is regulated by a host of epigenetic modifications, many of which are linked with metabolic pathways.

Epigenetic and Metabolic Crosstalk in MDSC: Roles of AMPK and HIF1α Adenosine 5′-Monophosphate (AMP)-Activated Protein Kinase (AMPK)

The energy sensor, AMPK, plays a critical role in mediating the tumor-promoting effect of MDSCs (De Veirman et al. 2018). For one, it regulates metabolism by histone acetylation. AMPK activation leads to increase in acetyl-CoA levels, thereby facilitating the action of lysine acetyltransferases (KATS) (De Veirman et al. 2018). Additionally, AMPK increases the activity of HDACs by regulating levels of its cofactor, NAD^+ (Vancura et al. 2018). Further, it helps promote histone acetylation and DNA methylation by phosphorylating key epigenetic factors like DNMT1 and HAT1 (Marin et al. 2017) Histone acetylation, an event regulated by the balance of HAT/HDAC enzymes, has important consequences on MDSC amplification. Inhibitors to HDAC11 have been found to negatively affect expansion and function of MDSCs (Sahakian et al. 2015). Recently, one study established that MDSC accumulation relies on the STAT3-DNMT epigenetic axis regulated by IL-6 in an autocrine manner to inhibit TNF-α expression and promote tumor development (Smith et al. 2020). DNMTs have been associated with hypermethylation across epigenome of MDSCs, and DNMT3a, in particular, was overexpressed in a PGE2-dependent fashion in ovarian cancer patient (Rodríguez-Ubreva et al. 2017). Consistent with these findings, downregulation of DNMTs via pharmacological inhibitor, decitabine (DAC), eliminates hypermethylation and accumulation of MDSCs, leading to activation of appropriate CTLs in mice model (Smith et al. 2020).

Hypoxia-Inducible Factor 1-Alpha (HIF-1α)

HIF-1α is the master regulator of the extensive glycolytic flux of MDSCs in the hypoxic TME. HIF-1α promotes increased expression of glycolysis-related genes and enhances uptake of glucose from the extracellular milieu. In addition, it regulates lactate dehydrogenase (LDHA), the enzyme responsible for converting pyruvate, the end product of glycolysis into lactate, thereby preventing OXPHOS. As such, HIF-1α is not only conducive for the proliferation and survival of MDSCs within the TME but also aids in their eventual differentiation into tumor-associated macrophages (Yu et al. 2018). HIF-1α-mediated regulation of glucose metabolism is modulated by other factors as well, such as SIRT1, a histone deacetylase that requires NAD^+ as the cofactor. SIRT1 targets HIF-1α, inhibiting glycolysis. Additionally, it can act alone as well to exert immune responses. HIF-1α plays a central role in metabolic rewiring of MDSCs via induction of aerobic glycolysis through the AKT-mTOR axis (Cheng et al. 2014b). Since MDSCs are immature myeloid cells, the tumor can display either a classically activated (M1) or an alternately activated (M2) phenotype. As mentioned before, SIRT1 targets HIF-1α to curb its glycolytic functions. As such, inhibition of SIRT1 activates the mTOR-HIF-1α glycolysis pathway, leading to differentiation of MDSCs across the M1 phenotype and paving the way for tumor protection (Liu et al. 2013). The phenotypic switching of MDSCs via their metabolic reprogramming through well-known epigenetic modifiers such as SIRT1 provides a rationale for cancer treatment.

Altered Epigenetic and Metabolic Features of NK Cells in the TME

Natural killer (NK) cells are potent lymphocytes with pronounced antitumoral capabilities. They kill malignant cells both directly and via secretion of cytokines and chemokines that potentiate antitumoral effector functions. Similar to T cells, increasing clinical evidence hint to the fact that the effector functions of NK cells are tied to their cellular energetics (O'Brien and Finlay 2019; Terrén et al. 2019). NK cells in the TME exhibit extensive metabolic rewiring. At resting state, NK cells display reduced rates of glycolysis and OXPHOS. Upon activation, they increase expression of glycolytic enzymes and upregulate glucose transporters, leading to increased OXPHOS in order to fuel rapid effector responses (Keppel et al. 2015). Interestingly, the pyruvate derived from glycolysis does not enter the TCA cycle of NK cells, rather it is diverted for NADH production for OXPHOS and ATP production. Because this OXPHOS production in NK cells is not reliant upon glutamine, and is fueled by glucose instead, therapeutic approaches involving glutaminase inhibition, such as CB-839, have been successfully implicated for boosting cytotoxic activity of NK cells in various solid tumors (ClinicalTrials.gov ID: NCT02071862 and NCT02861300). Similarly, glycogen synthase kinase-3 (GSK3) via inhibition of c-Myc degradation has been found to boost NK effector properties (Gardiner 2019; Terrén et al. 2019). It has been found that NK cells in the TME exhibit reduced glycolysis (Cong et al. 2018b). In murine model of lung cancer, tumors overexpress fructose-1,6-bisphosphatase (FBP1), the rate-limiting enzyme of gluconeogenesis, which then binds to HIF-1α and inhibits expression of glycolytic genes. In addition, inhibiting fructose-bisphosphatase 1 (FBP1) via MB05032 seemed to restore glycolysis flux in NK cells of the lung cancer TME, thereby restoring their cytotoxicity toward tumors (Cong et al. 2018a).

Aberrant epigenetic modifications on surface receptors of NK cells contribute to their dampened cytotoxic functions in the TME. For instance, overexpression of DNMT in certain cancers confers widespread hypermethylation on transcriptional regulatory sequence of killer cell immunoglobulin-like receptors (KIR), preventing its expression on NK cells and binding of its ligand, the MHC class I molecules (Küçük et al. 2016). DNMT inhibition via demethylating agents such as decitabine has been used to effectively increase KIR expression by removing these repressive methylation marks and consequently restore NK cell activity in killing tumor cells in treat AML (Kopp et al. 2013). Another surface receptor on NK cells, natural killer group 2 member D (NKG2D), is similarly regulated by epigenetic modifications. NKG2D upregulation by histone deacetylase inhibitors (HDACi), a prominent tenant of cancer immunotherapy, has been recently evaluated in primary NK cells of human and rhabdomyosarcoma cell lines (Idso et al. 2020). In one study that exemplified the involvement of regulatory metabolites, citrate efflux from mitochondria to cytosol seemed to promote expression of NKG2D ligand expression, MICA. Conversion of citrate into acetyl-CoA by ATP-citrate lyase facilitates histone acetylation on DNA regulatory sequences and enhances chromatin accessibility, thus dictating the basal level of MICA transcription (Møller et al. 2020). Defining the metabolic and epigenetic mechanisms behind increased expression of NKG2D and NKG2DLs may enable immune surveillance mediated by NK cells. The hypoxia-driven TME has been associated with the downregulation of several activating receptors, including NKG2D on NK cells (Balsamo et al. 2013). Further, in advanced stages, tumor cells can evade NKG2D-mediated cytotoxicity through shedding of cell-surface ligands, MICA and MICB, in response to hypoxia. Directing antibodies against the MICA α3 domain retained these immunostimulatory tags on human melanoma cells, restoring the full breadth of NK cell responses thereafter (Ferrari de Andrade et al. 2018).

An overview of clinically tested metabolic and epigenetic inhibitors has been presented in Table 18.1.

Conclusion

The intricate nodes connecting epigenetic and metabolic pathways can be exploited in a manner to selectively inhibit tumorigenesis and boost antitumoral responses. As such, these strategies may be touted as the future of cancer immunotherapy, especially when applied in synergy with immune checkpoint blockade. With increasing awareness of metabolomics, transcriptomics, and epigenomes, recent studies have repeatedly linked metabolites in alteration of epigenome, leading to dysregulated immune cell responses that favor tumor development. However, the lacunae lie in understanding how exactly these cellular metabolic events coordinate with epigenetic events to affect the functioning of specialized immune cells, such as macrophages and dendritic cells at the single-cell level in the dysregulated setting of tumor microenvironment. Furthermore, translation of experimental findings from in vitro

Table 18.1 Therapeutic opportunities that exploit metabolic/epigenetic crosstalk in the TME

Inhibitors	Target enzyme	Metabolic effect	Epigenetic implication	Ongoing clinical trials	References
Metabolic inhibitors for reversal of aberrant epigenetic modification in tumor cells					
2-Deoxyglucose (2-DG)	Hexokinases	Inhibition of glycolysis and reduction of acetyl-CoA levels	Reduced acetylation of histones H2A, H2B, H3 and H4, and H2B in cancer cell lines	Phase I/II	Liu et al. (2015)
3-Bromopyruvate (BrPA)	Glyceraldehyde 3-phosphate dehydrogenase	Inhibition of GAPDH and acetyl-CoA production	Suppression of histone acetylation	NA	Moussaieff et al. (2015)
Bis-2-(5-phenylacetamido-1,2,4-thiadiazol-2-yl) ethyl sulfide (BPTES)	Glutaminase (GLS)	Suppressed production of acetyl-CoA and 2-hydroxyglutarate	None	NA	Robinson et al. (2007)
CB-839, zaprinast, compound 968	Glutaminase (GLS)	Suppressed production of acetyl-CoA and 2-hydroxyglutarate	Compound 968: reversed histone H3K4me3 modifications in breast cancer cells Zaprinast: reduced H3K9me2/3 in IDH1-mutant cancer cells	CB-839, in phase I	Simpson et al. (2012), Elhammali et al. (2014)
AG-120, AG-881, IDH-305, AGI-5198	Mutant isocitrate dehydrogenase (IDH)-I	Suppressed production of 2-hydroxyglutarate	AGI-5198: reversed H3K9me3 and H3K27me3 modifications in chondrosarcoma cells	AG-120 (phase I/II), AG-881 (phase I), IDH-305 (phase I)	172-178 Golub et al. (2019), Okoye-Okafor et al. (2015)
AG-221	Mutant isocitrate dehydrogenase (IDH)-2	Suppressed production of 2-hydroxyglutarate	Demethylation of DNA and histone marks in leukemia cell lines bearing mutant IDH2	AG-221 (phase I/III)	Yen et al. (2017)
Adenosine dialdehyde, DZNep (3-deazaneplanoci A)	SAH hydrolase	Increase in SAH:SAM ratio	Inhibition of DNA/histone methylation in cancer cell lines	None	Glazer et al. (1986), Miranda et al. (2009)

(continued)

Table 18.1 (continued)

Inhibitors	Target enzyme	Metabolic effect	Epigenetic implication	Ongoing clinical trials	References
Epigenetic drugs that modulate cancer metabolism					
5-Aza-2-′-deoxycytidine, 5-azacytidine	DNA methyltransferases (DNMT1, DNMT3a, DNMT3b)	Reduced expression of genes involved in lipid and cholesterol metabolism (e.g., HMGCR, PCSK9, FASN)	Induced hypomethylation and reversed hypermethylation phenotype of glioma cells bearing IDH1 mutations	Approved for acute myeloid leukemia, in phase I–III for few malignancies	Stever Poirier 2014, JBC, Turcan et al. (2013)
Butyrate, vorinostat, romidepsin	Histone deacetylases (HDACs)	Reduced glucose uptake and glycolysis, reduced lactate accumulation	Restores the normal balance of histone acetylation/histone deacetylation and reversed aberrant regulation of gene expression by mutated HDACs	In use for cutaneous T-cell lymphoma	Wardell et al. (2008), Rodrigues et al. (2014)
Free fatty acids (FFA): linoleic acid, oleic acid, myristic acid,	Sirtuin 6 (SIRT6)	Inhibition of glycolysis via increase in tumor suppressor functions of SIRT6	Activation of SIRT6 represses antiapoptotic factor, survivin, by lowering histone H3K9 acetylation	NA	Feldman et al. (2013)

culture systems to in vivo may better reflect the complex intertwining of epigenetics and metabolics and make these findings clinically relevant. Likewise, preclinical studies can further help validate potential targets involved in such crosstalk for the development of novel therapeutics. Metabolic reprogramming of tumor cells and immune cells in the TME can be effectively achieved by epigenetic drugs like HDAC and DNMT inhibitors. Then again, targeting epigenetic machinery is envisioned to exert global changes in gene expression and more studies would be needed to attach more specificity to their effects in context of tumor metabolism. A prevailing challenge with such targeted therapies is to understand how relevant the target is in terms of driving carcinogenesis. One way to navigate this problem could be by identifying biomarkers, both epigenomic (DNA/modifications) and metabolomic (metabolite intermediates) in patient samples benefitting most from the targeted therapy. Clinical trials incorporating such extensive analysis can act as a great starting point to fully unravel the potential of multi-modular metabolic/epigenetic agents.

In summary, these studies shed light on the mechanistic understanding of metabolic control of immune cell fate by epigenetic machinery or vice versa and opens up new avenue for the prevention and treatment of immune cell-mediated cancer or other inflammatory disorders.

Acknowledgments Work done in our laboratory is supported by Council of Scientific and Industrial Research (MLP2031); Intramural grant from CSIR-CDRI (MLP0106) Department of Biotechnology, Govt. of India (BT/PR32490/MED/29/1457/2019); and Department of Science and Technology (DST), Govt. of India (CRG/2020/002932). AS is a senior research fellow of the University Grant Commission (UGC), India. This article has CSIR-Central Drug Research Institute communication number 10273.

References

Adeegbe DO, Liu Y, Lizotte PH, Kamihara Y, Aref AR, Almonte C, Dries R, Li Y, Liu S, Wang X, Warner-Hatten T, Castrillon J, Yuan GC, Poudel-Neupane N, Zhang H, Guerriero JL, Han S, Awad MM, Barbie DA, Ritz J, Jones SS, Hammerman PS, Bradner J, Quayle SN, Wong KK (2017) Synergistic immunostimulatory effects and therapeutic benefit of combined histone deacetylase and bromodomain inhibition in non-small cell lung cancer. Cancer Discov 7:852–867

Apetoh L, Ghiringhelli F, Tesniere A, Criollo A, Ortiz C, Lidereau R, Mariette C, Chaput N, Mira JP, Delaloge S, André F, Tursz T, Kroemer G, Zitvogel L (2007) The interaction between HMGB1 and TLR4 dictates the outcome of anticancer chemotherapy and radiotherapy. Immunol Rev 220:47–59

Balsamo M, Manzini C, Pietra G, Raggi F, Blengio F, Mingari MC, Varesio L, Moretta L, Bosco MC, Vitale M (2013) Hypoxia downregulates the expression of activating receptors involved in NK-cell-mediated target cell killing without affecting ADCC. Eur J Immunol 43:2756–2764

Barsoum IB, Smallwood CA, Siemens DR, Graham CH (2013) A mechanism of hypoxia-mediated escape from adaptive immunity in cancer cells. Cancer Res 74:665–674

Beury DW, Parker KH, Nyandjo M, Sinha P, Carter KA, Ostrand-Rosenberg S (2014) Cross-talk among myeloid-derived suppressor cells, macrophages, and tumor cells impacts the inflammatory milieu of solid tumors. J Leukoc Biol 96(6):1109–1118

Brand A, Singer K, Koehl GE, Kolitzus M, Schoenhammer G, Thiel A, Matos C, Bruss C, Klobuch S, Peter K, Kastenberger M, Bogdan C, Schleicher U, Mackensen A, Ullrich E, Fichtner-Feigl S, Kesselring R, Mack M, Ritter U, Schmid M, Blank C, Dettmer K, Oefner PJ, Hoffmann P, Walenta S, Geissler EK, Pouyssegur J, Villunger A, Steven A, Seliger B, Schreml S, Haferkamp S, Kohl E, Karrer S, Berneburg M, Herr W, Mueller-Klieser W, Renner K, Kreutz M (2016) LDHA-associated lactic acid production blunts tumor immunosurveillance by T and NK cells. Cell Metab 24:657–671

Brown TP, Bhattacharjee P, Ramachandran S, Sivaprakasam S, Ristic B, Sikder MOF, Ganapathy V (2020) The lactate receptor GPR81 promotes breast cancer growth via a paracrine mechanism involving antigen-presenting cells in the tumor microenvironment. Oncogene 39:3292–3304

Carey BW, Finley LW, Cross JR, Allis CD, Thompson CB (2014) Intracellular α-ketoglutarate maintains the pluripotency of embryonic stem cells. Nature 518:413–416

Cascone T, McKenzie JA, Mbofung RM, Punt S, Wang Z, Xu C, Williams LJ, Bristow CA, Carugo A, Peoples MD, Li L, Karpinets T, Huang L, Malu S, Creasy C, Leahey SE, Chen J, Chen Y, Pelicano H, Bernatchez C, Gopal YNV, Heffernan TP, Hu J, Wang J, Amaria RN, Garraway LA, Huang P, Yang P, Wistuba II, Woodman SE, Roszik J, Davis RE, Davies MA, Heymach JV, Hwu P, Peng W (2018) Increased tumor glycolysis characterizes immune resistance to adoptive T cell therapy. Cell Metab 27:977–987.e974

Cekic C, Day YJ, Sag D, Linden J (2014) Myeloid expression of adenosine A2A receptor suppresses T and NK cell responses in the solid tumor microenvironment. Cancer Res 74: 7250–7259

Chang CI, Liao JC, Kuo L (2001) Macrophage arginase promotes tumor cell growth and suppresses nitric oxide-mediated tumor cytotoxicity. Cancer Res 61:1100–1106

Chang YC, Chen TC, Lee CT, Yang CY, Wang HW, Wang CC, Hsieh SL (2008) Epigenetic control of MHC class II expression in tumor-associated macrophages by decoy receptor 3. Blood 111:5054–5063

Chang CH, Curtis JD, Maggi LB Jr, Faubert B, Villarino AV, O'Sullivan D, Huang SC, van der Windt GJ, Blagih J, Qiu J, Weber JD, Pearce EJ, Jones RG, Pearce EL (2013) Posttranscriptional control of T cell effector function by aerobic glycolysis. Cell 153:1239–1251

Chang CH, Qiu J, O'Sullivan D, Buck MD, Noguchi T, Curtis JD, Chen Q, Gindin M, Gubin MM, van der Windt GJ, Tonc E, Schreiber RD, Pearce EJ, Pearce EL (2015) Metabolic competition in the tumor microenvironment is a driver of cancer progression. Cell 162:1229–1241

Chawla A (2010) Control of macrophage activation and function by PPARs. Circ Res 106:1559–1569

Chen S, Yang J, Wei Y, Wei X (2019) Epigenetic regulation of macrophages: from homeostasis maintenance to host defense. Cell Mol Immunol 17:36–49

Cheng C, Huang C, Ma TT, Bian EB, He Y, Zhang L, Li J (2014a) SOCS1 hypermethylation mediated by DNMT1 is associated with lipopolysaccharide-induced inflammatory cytokines in macrophages. Toxicol Lett 225:488–497

Cheng SC, Quintin J, Cramer RA, Shepardson KM, Saeed S, Kumar V, Giamarellos-Bourboulis EJ, Martens JH, Rao NA, Aghajanirefah A, Manjeri GR, Li Y, Ifrim DC, Arts RJ, van der Veer BM, Deen PM, Logie C, O'Neill LA, Willems P, van de Veerdonk FL, van der Meer JW, Ng A, Joosten LA, Wijmenga C, Stunnenberg HG, Xavier RJ, Netea MG (2014b) mTOR- and HIF-1α-mediated aerobic glycolysis as metabolic basis for trained immunity. Science 345: 1250684

Chiba S, Hisamatsu T, Suzuki H, Mori K, Kitazume MT, Shimamura K, Mizuno S, Nakamoto N, Matsuoka K, Naganuma M, Kanai T (2017) Glycolysis regulates LPS-induced cytokine production in M2 polarized human macrophages. Immunol Lett 183:17–23

Chowdhury R, Yeoh KK, Tian YM, Hillringhaus L, Bagg EA, Rose NR, Leung IK, Li XS, Woon EC, Yang M, McDonough MA, King ON, Clifton IJ, Klose RJ, Claridge TD, Ratcliffe PJ,

Schofield CJ, Kawamura A (2011) The oncometabolite 2-hydroxyglutarate inhibits histone lysine demethylases. EMBO Rep 12:463–469

Chowdhury PS, Chamoto K, Kumar A, Honjo T (2018) PPAR-induced fatty acid oxidation in T cells increases the number of tumor-reactive CD8(+) T cells and facilitates anti-PD-1 therapy. Cancer Immunol Res 6:1375–1387

Colegio OR, Chu NQ, Szabo AL, Chu T, Rhebergen AM, Jairam V, Cyrus N, Brokowski CE, Eisenbarth SC, Phillips GM, Cline GW, Phillips AJ, Medzhitov R (2014) Functional polarization of tumour-associated macrophages by tumour-derived lactic acid. Nature 513:559–563

Cong J, Wang X, Zheng X, Wang D, Fu B, Sun R, Tian Z, Wei H (2018a) Dysfunction of natural killer cells by FBP1-induced inhibition of glycolysis during lung cancer progression. Cell Metab 28(2):243–255.e5

Cong J, Wang X, Zheng X, Wang D, Fu B, Sun R, Tian Z, Wei H (2018b) Dysfunction of natural killer cells by FBP1-induced inhibition of glycolysis during lung cancer progression. Cell Metab 28:243–255.e245

Costantini B, Kordasti SY, Kulasekararaj AG, Jiang J, Seidl T, Abellan PP, Mohamedali A, Thomas NS, Farzaneh F, Mufti GJ (2012) The effects of 5-azacytidine on the function and number of regulatory T cells and T-effectors in myelodysplastic syndrome. Haematologica 98: 1196–1205

Dai Z, Mentch SJ, Gao X, Nichenametla SN, Locasale JW (2018) Methionine metabolism influences genomic architecture and gene expression through H3K4me3 peak width. Nat Commun 9: 1955

Damgaci S, Ibrahim-Hashim A, Enriquez-Navas PM, Pilon-Thomas S, Guvenis A, Gillies RJ (2018) Hypoxia and acidosis: immune suppressors and therapeutic targets. Immunology 154: 354–362

Daneshmandi S, Wegiel B, Seth P (2019) Blockade of Lactate Dehydrogenase-A (LDH-A) Improves Efficacy of Anti-Programmed Cell Death-1 (PD-1) Therapy in Melanoma, Cancers (Basel), p 11

Dang L, White DW, Gross S, Bennett BD, Bittinger MA, Driggers EM, Fantin VR, Jang HG, Jin S, Keenan MC, Marks KM, Prins RM, Ward PS, Yen KE, Liau LM, Rabinowitz JD, Cantley LC, Thompson CB, Vander Heiden MG, Su SM (2009) Cancer-associated IDH1 mutations produce 2-hydroxyglutarate. Nature 462:739–744

Das Gupta K, Shakespear MR, Curson JEB, Murthy AMV, Iyer A, Hodson MP, Ramnath D, Tillu VA, von Pein JB, Reid RC, Tunny K, Hohenhaus DM, Moradi SV, Kelly GM, Kobayashi T, Gunter JH, Stevenson AJ, Xu W, Luo L, Jones A, Johnston WA, Blumenthal A, Alexandrov K, Collins BM, Stow JL, Fairlie DP, Sweet MJ (2020) Class IIa histone deacetylases drive toll-like receptor-inducible glycolysis and macrophage inflammatory responses via pyruvate kinase M2. Cell Rep 30:2712–2728.e2718

Davie JR (2003) Inhibition of histone deacetylase activity by butyrate. J Nutr 133:2485s–2493s

De Veirman K, Menu E, Maes K, De Beule N, De Smedt E, Maes A, Vlummens P, Fostier K, Kassambara A, Moreaux J, Van Ginderachter JA, De Bruyne E, Vanderkerken K, Van Valckenborgh E (2018) Myeloid-derived suppressor cells induce multiple myeloma cell survival by activating the AMPK pathway. Cancer Lett 442:233–241

Delacher M, Imbusch CD, Weichenhan D, Breiling A, Hotz-Wagenblatt A, Träger U, Hofer AC, Kägebein D, Wang Q, Frauhammer F, Mallm JP, Bauer K, Herrmann C, Lang PA, Brors B, Plass C, Feuerer M (2017) Genome-wide DNA-methylation landscape defines specialization of regulatory T cells in tissues. Nat Immunol 18:1160–1172

Delgoffe GM, Kole TP, Zheng Y, Zarek PE, Matthews KL, Xiao B, Worley PF, Kozma SC, Powell JD (2009) The mTOR kinase differentially regulates effector and regulatory T cell lineage commitment. Immunity 30:832–844

Elhammali A, Ippolito JE, Collins L, Crowley J, Marasa J, Piwnica-Worms D (2014) A high-throughput fluorimetric assay for 2-hydroxyglutarate identifies Zaprinast as a glutaminase inhibitor. Cancer Discov 4:828–839

Fan Z, Li J, Li P, Ye Q, Xu H, Wu X, Xu Y (2017) Protein arginine methyltransferase 1 (PRMT1) represses MHC II transcription in macrophages by methylating CIITA. Sci Rep 7:40531. https://doi.org/10.1038/srep40531

Feldman JL, Baeza J, Denu JM (2013) Activation of the protein deacetylase SIRT6 by long-chain fatty acids and widespread deacylation by mammalian sirtuins. J Biol Chem 288:31350–31356

Ferrari de Andrade L, Tay RE, Pan D, Luoma AM, Ito Y, Badrinath S, Tsoucas D, Franz B, May KF Jr, Harvey CJ, Kobold S, Pyrdol JW, Yoon C, Yuan GC, Hodi FS, Dranoff G, Wucherpfennig KW (2018) Antibody-mediated inhibition of MICA and MICB shedding promotes NK cell-driven tumor immunity. Science 359:1537–1542

Figueroa ME, Abdel-Wahab O, Lu C, Ward PS, Patel J, Shih A, Li Y, Bhagwat N, Vasanthakumar A, Fernandez HF, Tallman MS, Sun Z, Wolniak K, Peeters JK, Liu W, Choe SE, Fantin VR, Paietta E, Löwenberg B, Licht JD, Godley LA, Delwel R, Valk PJ, Thompson CB, Levine RL, Melnick A (2010) Leukemic IDH1 and IDH2 mutations result in a hypermethylation phenotype, disrupt TET2 function, and impair hematopoietic differentiation. Cancer Cell 18:553–567

Franchina DG, He F, Brenner D (2017) Survival of the fittest: cancer challenges T cell metabolism. Cancer Lett 412:216–223

Freemerman AJ, Johnson AR, Sacks GN, Milner JJ, Kirk EL, Troester MA, Macintyre AN, Goraksha-Hicks P, Rathmell JC, Makowski L (2014) Metabolic reprogramming of macrophages: glucose transporter 1 (GLUT1)-mediated glucose metabolism drives a proinflammatory phenotype. J Biol Chem 289:7884–7896

Funk JL, Feingold KR, Moser AH, Grunfeld C (1993) Lipopolysaccharide stimulation of RAW 264.7 macrophages induces lipid accumulation and foam cell formation. Atherosclerosis 98:67–82

Furusawa Y, Obata Y, Fukuda S, Endo TA, Nakato G, Takahashi D, Nakanishi Y, Uetake C, Kato K, Kato T, Takahashi M, Fukuda NN, Murakami S, Miyauchi E, Hino S, Atarashi K, Onawa S, Fujimura Y, Lockett T, Clarke JM, Topping DL, Tomita M, Hori S, Ohara O, Morita T, Koseki H, Kikuchi J, Honda K, Hase K, Ohno H (2013) Commensal microbe-derived butyrate induces the differentiation of colonic regulatory T cells. Nature 504:446–450

Gamper CJ, Agoston AT, Nelson WG, Powell JD (2009) Identification of DNA methyltransferase 3a as a T cell receptor-induced regulator of Th1 and Th2 differentiation. J Immunol 183:2267–2276

Gardiner CM (2019) NK cell metabolism. J Leukoc Biol 105:1235–1242

Glazer RI, Hartman KD, Knode MC, Richard MM, Chiang PK, Tseng CK, Marquez VE (1986) 3-Deazaneplanocin: a new and potent inhibitor of S-adenosylhomocysteine hydrolase and its effects on human promyelocytic leukemia cell line HL-60. Biochem Biophys Res Commun 135: 688–694

Goffaux G, Hammami I, Jolicoeur M (2017) A dynamic metabolic flux analysis of myeloid-derived suppressor cells confirms immunosuppression-related metabolic plasticity. Sci Rep 7:9850

Golub D, Iyengar N, Dogra S, Wong T, Bready D, Tang K, Modrek AS, Placantonakis DG (2019) Mutant Isocitrate dehydrogenase inhibitors as targeted cancer therapeutics. Front Oncol 9:417

Han CZ, Ravichandran KS (2011) Metabolic connections during apoptotic cell engulfment. Cell 147:1442–1445

Haschemi A, Kosma P, Gille L, Evans CR, Burant CF, Starkl P, Knapp B, Haas R, Schmid JA, Jandl C, Amir S, Lubec G, Park J, Esterbauer H, Bilban M, Brizuela L, Pospisilik JA, Otterbein LE, Wagner O (2012) The sedoheptulose kinase CARKL directs macrophage polarization through control of glucose metabolism. Cell Metab 15:813–826

Herber DL, Cao W, Nefedova Y, Novitskiy SV, Nagaraj S, Tyurin VA, Corzo A, Cho HI, Celis E, Lennox B, Knight SC, Padhya T, McCaffrey TV, McCaffrey JC, Antonia S, Fishman M, Ferris RL, Kagan VE, Gabrilovich DI (2010) Lipid accumulation and dendritic cell dysfunction in cancer. Nat Med 16:880–886

Ho PC, Bihuniak JD, Macintyre AN, Staron M, Liu X, Amezquita R, Tsui YC, Cui G, Micevic G, Perales JC, Kleinstein SH, Abel ED, Insogna KL, Feske S, Locasale JW, Bosenberg MW,

Rathmell JC, Kaech SM (2015) Phosphoenolpyruvate is a metabolic checkpoint of anti-tumor T cell responses. Cell 162:1217–1228

Hobbs CA, Gilmour SK (2000) High levels of intracellular polyamines promote histone acetyltransferase activity resulting in chromatin hyperacetylation. J Cell Biochem 77:345–360

Hsu YL, Hung JY, Chang WA, Jian SF, Lin YS, Pan YC, Wu CY, Kuo PL (2018) Hypoxic lung-cancer-derived extracellular vesicle MicroRNA-103a increases the oncogenic effects of macrophages by targeting PTEN. Mol Ther 26:568–581

Huang Y, Rao A (2014) Connections between TET proteins and aberrant DNA modification in cancer. Trends Genet 30:464–474

Huang SC, Everts B, Ivanova Y, O'Sullivan D, Nascimento M, Smith AM, Beatty W, Love-Gregory L, Lam WY, O'Neill CM, Yan C, Du H, Abumrad NA, Urban JF Jr, Artyomov MN, Pearce EL, Pearce EJ (2014) Cell-intrinsic lysosomal lipolysis is essential for alternative activation of macrophages. Nat Immunol 15:846–855

Hwang IY, Kwak S, Lee S, Kim H, Lee SE, Kim JH, Kim YA, Jeon YK, Chung DH, Jin X, Park S, Jang H, Cho EJ, Youn HD (2016) Psat1-dependent fluctuations in α-ketoglutarate affect the timing of ESC differentiation. Cell Metab 24:494–501

Idso JM, Lao S, Schloemer NJ, Knipstein J, Burns R, Thakar MS, Malarkannan S (2020) Entinostat augments NK cell functions via epigenetic upregulation of IFIT1-STING-STAT4 pathway. Oncotarget 11:1799–1815

Jha AK, Huang SC, Sergushichev A, Lampropoulou V, Ivanova Y, Loginicheva E, Chmielewski K, Stewart KM, Ashall J, Everts B, Pearce EJ, Driggers EM, Artyomov MN (2015) Network integration of parallel metabolic and transcriptional data reveals metabolic modules that regulate macrophage polarization. Immunity 42:419–430

Kaelin WG Jr, McKnight SL (2013) Influence of metabolism on epigenetics and disease. Cell 153: 56–69

Keppel MP, Saucier N, Mah AY, Vogel TP, Cooper MA (2015) Activation-specific metabolic requirements for NK Cell IFN-γ production. J Immunol 194:1954–1962

Kerner J, Hoppel C (2000) Fatty acid import into mitochondria. Biochim Biophys Acta 1486:1–17

Kheshtchin N, Arab S, Ajami M, Mirzaei R, Ashourpour M, Mousavi N, Khosravianfar N, Jadidi-Niaragh F, Namdar A, Noorbakhsh F, Hadjati J (2016) Inhibition of HIF-1α enhances anti-tumor effects of dendritic cell-based vaccination in a mouse model of breast cancer. Cancer Immunol Immunother 65:1159–1167

Kim JH, Kim J, Fau-Kim SC, Kim Sc Fau-Yi Y-S, Yi Ys Fau-Yang WS, Yang Ws Fau-Yang Y, Fau-Kim HG YY, Kim Hg Fau-Lee JY, Lee Jy Fau-Kim K-H, Kim Kh Fau-Yoo BC, Yoo Bc Fau-Hong S, Fau-Cho JY HS, Cho JY (2013) Adenosine dialdehyde suppresses MMP-9-mediated invasion of cancer cells by blocking the Ras/Raf-1/ERK/AP-1 signaling pathway. Biochem Pharmacol 86(9):1285–1300. https://doi.org/10.1016/j.bcp.2013.08.022

Kinnaird A, Zhao S, Wellen KE, Michelakis ED (2016) Metabolic control of epigenetics in cancer. Nat Rev Cancer 16:694–707

Kittan NA, Allen RM, Dhaliwal A, Cavassani KA, Schaller M, Gallagher KA, Carson WF, Mukherjee S, Grembecka J, Cierpicki T, Jarai G, Westwick J, Kunkel SL, Hogaboam CM (2013) Cytokine induced phenotypic and epigenetic signatures are key to establishing specific macrophage phenotypes. PLoS One 8:e78045

Klysz D, Tai X, Robert PA, Craveiro M, Cretenet G, Oburoglu L, Mongellaz C, Floess S, Fritz V, Matias MI, Yong C, Surh N, Marie JC, Huehn J, Zimmermann V, Kinet S, Dardalhon V, Taylor N (2015) Glutamine-dependent α-ketoglutarate production regulates the balance between T helper 1 cell and regulatory T cell generation. Sci Signal 8:ra97

Kopp LM, Ray A, Denman CJ, Senyukov VS, Somanchi SS, Zhu S, Lee DA (2013) Decitabine has a biphasic effect on natural killer cell viability, phenotype, and function under proliferative conditions. Mol Immunol 54:296–301

Kouidhi S, Elgaaied AB, Chouaib S (2017) Impact of metabolism on T-cell differentiation and function and cross talk with tumor microenvironment. Front Immunol 8:270

Krawczyk CM, Holowka T, Sun J, Blagih J, Amiel E, DeBerardinis RJ, Cross JR, Jung E, Thompson CB, Jones RG, Pearce EJ (2010) Toll-like receptor-induced changes in glycolytic metabolism regulate dendritic cell activation. Blood 115:4742–4749

Küçük C, Hu X, Gong Q, Jiang B, Cornish A, Gaulard P, McKeithan T, Chan WC (2016) Diagnostic and biological significance of KIR expression profile determined by RNA-Seq in natural Killer/T-Cell lymphoma. Am J Pathol 186:1435–1441

Lampropoulou V, Sergushichev A, Bambouskova M, Nair S, Vincent EE, Loginicheva E, Cervantes-Barragan L, Ma X, Huang SC, Griss T, Weinheimer CJ, Khader S, Randolph GJ, Pearce EJ, Jones RG, Diwan A, Diamond MS, Artyomov MN (2016) Itaconate links inhibition of succinate dehydrogenase with macrophage metabolic remodeling and regulation of inflammation. Cell Metab 24:158–166

Latham T, Mackay L, Sproul D, Karim M, Culley J, Harrison DJ, Hayward L, Langridge-Smith P, Gilbert N, Ramsahoye BH (2012) Lactate, a product of glycolytic metabolism, inhibits histone deacetylase activity and promotes changes in gene expression. Nucleic Acids Res 40:4794–4803

Latour YL, Gobert AP, Wilson KT (2019) The role of polyamines in the regulation of macrophage polarization and function. Amino Acids 52:151–160

Li Q, Zou J, Wang M, Ding X, Chepelev I, Zhou X, Zhao W, Wei G, Cui J, Zhao K, Wang HY, Wang RF (2014) Critical role of histone demethylase Jmjd3 in the regulation of CD4+ T-cell differentiation. Nat Commun 5:5780

Li X, Zhang Y, Chen M, Mei Q, Liu Y, Feng K, Jia H, Dong L, Shi L, Liu L, Nie J, Han W (2017) Increased IFNγ(+) T cells are responsible for the clinical responses of low-dose DNA-demethylating agent decitabine antitumor therapy. Clin Cancer Res 23:6031–6043

Li T, Garcia-Gomez A, Morante-Palacios O, Ciudad L, Özkaramehmet S, Van Dijck E, Rodríguez-Ubreva J, Vaquero A, Ballestar E (2019) SIRT1/2 orchestrate acquisition of DNA methylation and loss of histone H3 activating marks to prevent premature activation of inflammatory genes in macrophages. Nucleic Acids Res 48:665–681

Lian CG, Xu Y, Ceol C, Wu F, Larson A, Dresser K, Xu W, Tan L, Hu Y, Zhan Q, Lee CW, Hu D, Lian BQ, Kleffel S, Yang Y, Neiswender J, Khorasani AJ, Fang R, Lezcano C, Duncan LM, Scolyer RA, Thompson JF, Kakavand H, Houvras Y, Zon LI, Mihm MC Jr, Kaiser UB, Schatton T, Woda BA, Murphy GF, Shi YG (2012) Loss of 5-hydroxymethylcytosine is an epigenetic hallmark of melanoma. Cell 150:1135–1146

Lin J, Liu W, Luan T, Yuan L, Jiang W, Cai H, Yuan W, Wang Y, Zhang Q, Wang L (2018) High expression of PU.1 is associated with Her-2 and shorter survival in patients with breast cancer. Oncol Lett 14:8220–8226

Liu G, Bi Y, Shen B, Yang H, Zhang Y, Wang X, Liu H, Lu Y, Liao J, Chen X, Chu Y (2013) SIRT1 limits the function and fate of myeloid-derived suppressor cells in tumors by orchestrating HIF-1α-dependent glycolysis. Cancer Res 74:727–737

Liu XS, Little JB, Yuan ZM (2015) Glycolytic metabolism influences global chromatin structure. Oncotarget 6:4214–4225

Liu M, O'Connor RS, Trefely S, Graham K, Snyder NW, Beatty GL (2019) Metabolic rewiring of macrophages by CpG potentiates clearance of cancer cells and overcomes tumor-expressed CD47-mediated 'don't-eat-me' signal. Nat Immunol 20:265–275

Liu M, Tong Z, Ding C, Luo F, Wu S, Wu C, Albeituni S, He L, Hu X, Tieri D, Rouchka EC, Hamada M, Takahashi S, Gibb AA, Kloecker G, Zhang HG, Bousamra M 2nd, Hill BG, Zhang X, Yan J (2020) Transcription factor c-Maf is a checkpoint that programs macrophages in lung cancer. J Clin Invest 130:2081–2096

Liu Y, Xu R, Gu H, Zhang E, Qu J, Cao W, Huang X, Yan H, He J, Cai Z (2021) Metabolic reprogramming in macrophage responses. Biomark Res 9:1

Lu C, Ward PS, Kapoor GS, Rohle D, Turcan S, Abdel-Wahab O, Edwards CR, Khanin R, Figueroa ME, Melnick A, Wellen KE, O'Rourke DM, Berger SL, Chan TA, Levine RL, Mellinghoff IK, Thompson CB (2012) IDH mutation impairs histone demethylation and results in a block to cell differentiation. Nature 483:474–478

Ma S, Liu M, Xu Z, Li Y, Guo H, Ge Y, Liu Y, Zheng D, Shi J (2015) A double feedback loop mediated by microRNA-23a/27a/24-2 regulates M1 versus M2 macrophage polarization and thus regulates cancer progression. Oncotarget 7:13502–13519

Maimela NR, Liu S, Zhang Y (2018) Fates of CD8+ T cells in tumor microenvironment. Comput Struct Biotechnol J 17:1–13

Makinoshima H, Takita M, Matsumoto S, Yagishita A, Owada S, Esumi H, Tsuchihara K (2014) Epidermal growth factor receptor (EGFR) signaling regulates global metabolic pathways in EGFR-mutated lung adenocarcinoma. J Biol Chem 289:20813–20823

Malandrino MI, Fucho R, Weber M, Calderon-Dominguez M, Mir JF, Valcarcel L, Escoté X, Gómez-Serrano M, Peral B, Salvadó L, Fernández-Veledo S, Casals N, Vázquez-Carrera M, Villarroya F, Vendrell JJ, Serra D, Herrero L (2015) Enhanced fatty acid oxidation in adipocytes and macrophages reduces lipid-induced triglyceride accumulation and inflammation. Am J Physiol Endocrinol Metab 308:E756–E769

Mantovani A, Sica A (2010) Macrophages, innate immunity and cancer: balance, tolerance, and diversity. Curr Opin Immunol 22:231–237

Manzo T, Prentice BM, Anderson KG, Raman A, Schalck A, Codreanu GS, Nava Lauson CB, Tiberti S, Raimondi A, Jones MA, Reyzer M, Bates BM, Spraggins JM, Patterson NH, McLean JA, Rai K, Tacchetti C, Tucci S, Wargo JA, Rodighiero S, Clise-Dwyer K, Sherrod SD, Kim M, Navin NE, Caprioli RM, Greenberg PD, Draetta G, Nezi L (2020) Accumulation of long-chain fatty acids in the tumor microenvironment drives dysfunction in intrapancreatic CD8+ T cells. J Exp Med 217:e20191920

Marin TL, Gongol B, Zhang F, Martin M, Johnson DA, Xiao H, Wang Y, Subramaniam S, Chien S, Shyy JY (2017) AMPK promotes mitochondrial biogenesis and function by phosphorylating the epigenetic factors DNMT1, RBBP7, and HAT1. Sci Signal 10(464):eaaf7478

Medeiros BC, Fathi AT, DiNardo CD, Pollyea DA, Chan SM, Swords R (2016) Isocitrate dehydrogenase mutations in myeloid malignancies. Leukemia 31:272–281

Mews P, Donahue G, Drake AM, Luczak V, Abel T, Berger SL (2017) Acetyl-CoA synthetase regulates histone acetylation and hippocampal memory. Nature 546:381–386

Michelucci A, Cordes T, Ghelfi J, Pailot A, Reiling N, Goldmann O, Binz T, Wegner A, Tallam A, Rausell A, Buttini M, Linster CL, Medina E, Balling R, Hiller K (2013) Immune-responsive gene 1 protein links metabolism to immunity by catalyzing itaconic acid production. Proc Natl Acad Sci U S A 110:7820–7825

Miranda TB, Cortez CC, Yoo CB, Liang G, Abe M, Kelly TK, Marquez VE, Jones PA (2009) DZNep is a global histone methylation inhibitor that reactivates developmental genes not silenced by DNA methylation. Mol Cancer Ther 8:1579–1588

Møller SH, Mellergaard M, Madsen M, Bermejo AV, Jepsen SD, Hansen MH, Høgh RI, Aldana BI, Desler C, Rasmussen LJ, Sustarsic EG, Gerhart-Hines Z, Daskalaki E, Wheelock CE, Hiron TK, Lin D, O'Callaghan CA, Wandall HH, Andresen L, Skov S (2020) Cytoplasmic citrate flux modulates the immune stimulatory NKG2D ligand MICA in cancer cells. Front Immunol 11: 1968

Moussaieff A, Rouleau M, Kitsberg D, Cohen M, Levy G, Barasch D, Nemirovski A, Shen-Orr S, Laevsky I, Amit M, Bomze D, Elena-Herrmann B, Scherf T, Nissim-Rafinia M, Kempa S, Itskovitz-Eldor J, Meshorer E, Aberdam D, Nahmias Y (2015) Glycolysis-mediated changes in acetyl-CoA and histone acetylation control the early differentiation of embryonic stem cells. Cell Metab 21:392–402

Mpekris F, Voutouri C, Baish JW, Duda DG, Munn LL, Stylianopoulos T, Jain RK (2020) Combining microenvironment normalization strategies to improve cancer immunotherapy. Proc Natl Acad Sci U S A 117:3728–3737

Mu X, Shi W, Xu Y, Xu C, Zhao T, Geng B, Yang J, Pan J, Hu S, Zhang C, Zhang J, Wang C, Shen J, Che Y, Liu Z, Lv Y, Wen H, You Q (2018) Tumor-derived lactate induces M2 macrophage polarization via the activation of the ERK/STAT3 signaling pathway in breast cancer. Cell Cycle 17:428–438

Nakaya M, Xiao Y, Zhou X, Chang JH, Chang M, Cheng X, Blonska M, Lin X, Sun SC (2014) Inflammatory T cell responses rely on amino acid transporter ASCT2 facilitation of glutamine uptake and mTORC1 kinase activation. Immunity 40:692–705

Németh B, Doczi J, Csete D, Kacso G, Ravasz D, Adams D, Kiss G, Nagy AM, Horvath G, Tretter L, Mócsai A, Csépányi-Kömi R, Iordanov I, Adam-Vizi V, Chinopoulos C (2015) Abolition of mitochondrial substrate-level phosphorylation by itaconic acid produced by LPS-induced Irg1 expression in cells of murine macrophage lineage. Faseb j 30:286–300

Netea-Maier RT, Smit JWA, Netea MG (2017) Metabolic changes in tumor cells and tumor-associated macrophages: A mutual relationship. Cancer Lett 413:102–109

Nirschl TR, El Asmar M, Ludwig WW, Ganguly S, Gorin MA, Johnson MH, Pierorazio PM, Drake CG, Allaf ME, Zarif JC (2020) Transcriptional profiling of tumor associated macrophages in human renal cell carcinoma reveals significant heterogeneity and opportunity for immunomodulation. Am J Clin Exp Urol 8:48–58

Niu Z, Shi Q, Zhang W, Shu Y, Yang N, Chen B, Wang Q, Zhao X, Chen J, Cheng N, Feng X, Hua Z, Ji J, Shen P (2017) Caspase-1 cleaves PPARγ for potentiating the pro-tumor action of TAMs. Nat Commun 8:766

O'Brien KL, Finlay DK (2019) Immunometabolism and natural killer cell responses. Nat Rev Immunol 19:282–290

Oh-hora M, Rao A (2008) Calcium signaling in lymphocytes. Curr Opin Immunol 20:250–258

Okoye-Okafor UC, Bartholdy B, Cartier J, Gao EN, Pietrak B, Rendina AR, Rominger C, Quinn C, Smallwood A, Wiggall KJ, Reif AJ, Schmidt SJ, Qi H, Zhao H, Joberty G, Faelth-Savitski M, Bantscheff M, Drewes G, Duraiswami C, Brady P, Groy A, Narayanagari SR, Antony-Debre I, Mitchell K, Wang HR, Kao YR, Christopeit M, Carvajal L, Barreyro L, Paietta E, Makishima H, Will B, Concha N, Adams ND, Schwartz B, McCabe MT, Maciejewski J, Verma A, Steidl U (2015) New IDH1 mutant inhibitors for treatment of acute myeloid leukemia. Nat Chem Biol 11:878–886

Osawa T, Tsuchida R, Muramatsu M, Shimamura T, Wang F, Suehiro J, Kanki Y, Wada Y, Yuasa Y, Aburatani H, Miyano S, Minami T, Kodama T, Shibuya M (2013) Inhibition of histone demethylase JMJD1A improves anti-angiogenic therapy and reduces tumor-associated macrophages. Cancer Res 73:3019–3028

Ottensmeier CH, Perry KL, Harden EL, Stasakova J, Jenei V, Fleming J, Wood O, Woo J, Woelk CH, Thomas GJ, Thirdborough SM (2016) Upregulated glucose metabolism correlates inversely with CD8+ T-cell infiltration and survival in squamous cell carcinoma. Cancer Res 76:4136–4148

Pan M, Reid MA, Lowman XH, Kulkarni RP, Tran TQ, Liu X, Yang Y, Hernandez-Davies JE, Rosales KK, Li H, Hugo W, Song C, Xu X, Schones DE, Ann DK, Gradinaru V, Lo RS, Locasale JW, Kong M (2016) Regional glutamine deficiency in tumours promotes dedifferentiation through inhibition of histone demethylation. Nat Cell Biol 18:1090–1101

Pan W, Zhu S, Qu K, Meeth K, Cheng J, He K, Ma H, Liao Y, Wen X, Roden C, Tobiasova Z, Wei Z, Zhao J, Liu J, Zheng J, Guo B, Khan SA, Bosenberg M, Flavell RA, Lu J (2017) The DNA Methylcytosine dioxygenase Tet2 sustains immunosuppressive function of tumor-infiltrating myeloid cells to promote melanoma progression. Immunity 47:284–297.e285

Pastor WA, Aravind L, Rao A (2013) TETonic shift: biological roles of TET proteins in DNA demethylation and transcription. Nat Rev Mol Cell Biol 14:341–356

Pello OM, Chèvre R, Laoui D, De Juan A, Lolo F, Andrés-Manzano MJ, Serrano M, Van Ginderachter JA, Andrés V (2012) In vivo inhibition of c-MYC in myeloid cells impairs tumor-associated macrophage maturation and pro-tumoral activities. PLoS One 7:e45399

Peng D, Kryczek I, Nagarsheth N, Zhao L, Wei S, Wang W, Sun Y, Zhao E, Vatan L, Szeliga W, Kotarski J, Tarkowski R, Dou Y, Cho K, Hensley-Alford S, Munkarah A, Liu R, Zou W (2015) Epigenetic silencing of TH1-type chemokines shapes tumour immunity and immunotherapy. Nature 527:249–253

Peng M, Yin N, Chhangawala S, Xu K, Leslie CS, Li MO (2016) Aerobic glycolysis promotes T helper 1 cell differentiation through an epigenetic mechanism. Science 354(6311):481–484

Qian BZ, Pollard JW (2010) Macrophage diversity enhances tumor progression and metastasis. Cell 141:39–51
Ramakrishnan R, Tyurin VA, Veglia F, Condamine T, Amoscato A, Mohammadyani D, Johnson JJ, Zhang LM, Klein-Seetharaman J, Celis E, Kagan VE, Gabrilovich DI (2014) Oxidized lipids block antigen cross-presentation by dendritic cells in cancer. J Immunol 192:2920–2931
Rasmussen KD, Helin K (2016) Role of TET enzymes in DNA methylation, development, and cancer. Genes Dev 30:733–750
Renner K, Bruss C, Schnell A, Koehl G, Becker HM, Fante M, Menevse AN, Kauer N, Blazquez R, Hacker L, Decking SM, Bohn T, Faerber S, Evert K, Aigle L, Amslinger S, Landa M, Krijgsman O, Rozeman EA, Brummer C, Siska PJ, Singer K, Pektor S, Miederer M, Peter K, Gottfried E, Herr W, Marchiq I, Pouyssegur J, Roush WR, Ong S, Warren S, Pukrop T, Beckhove P, Lang SA, Bopp T, Blank CU, Cleveland JL, Oefner PJ, Dettmer K, Selby M, Kreutz M (2019) Restricting glycolysis preserves T cell effector functions and augments checkpoint therapy. Cell Rep 29:135–150.e139
Robinson MM, McBryant SJ, Tsukamoto T, Rojas C, Ferraris DV, Hamilton SK, Hansen JC, Curthoys NP (2007) Novel mechanism of inhibition of rat kidney-type glutaminase by bis-2-(5-phenylacetamido-1,2,4-thiadiazol-2-yl)ethyl sulfide (BPTES). Biochem J 406:407–414
Rodrigues MF, Carvalho É, Pezzuto P, Rumjanek FD, Amoêdo ND (2014) Reciprocal modulation of histone deacetylase inhibitors sodium butyrate and trichostatin A on the energy metabolism of breast cancer cells. J Cell Biochem 116:797–808
Rodríguez-Ubreva J, Català-Moll F, Obermajer N, Álvarez-Errico D, Ramirez RN, Company C, Vento-Tormo R, Moreno-Bueno G, Edwards RP, Mortazavi A, Kalinski P, Ballestar E (2017) Prostaglandin E2 leads to the acquisition of DNMT3A-dependent tolerogenic functions in human myeloid-derived suppressor cells. Cell Rep 21(1):154–167
Sag D, Carling D, Stout RD, Suttles J (2008) Adenosine 5′-monophosphate-activated protein kinase promotes macrophage polarization to an anti-inflammatory functional phenotype. J Immunol 181:8633–8641
Sahakian E, Powers JJ, Chen J, Deng SL, Cheng F, Distler A, Woods DM, Rock-Klotz J, Sodre AL, Youn JI, Woan KV, Villagra A, Gabrilovich D, Sotomayor EM, Pinilla-Ibarz J (2015) Histone deacetylase 11: a novel epigenetic regulator of myeloid derived suppressor cell expansion and function. Mol Immunol 63:579–585
Satoh T, Takeuchi O, Vandenbon A, Yasuda K, Tanaka Y, Kumagai Y, Miyake T, Matsushita K, Okazaki T, Saitoh T, Honma K, Matsuyama T, Yui K, Tsujimura T, Standley DM, Nakanishi K, Nakai K, Akira S (2010) The Jmjd3-Irf4 axis regulates M2 macrophage polarization and host responses against helminth infection. Nat Immunol 11:936–944
Schlecker E, Stojanovic A, Fau-Eisen C, Eisen C, Fau-Quack C, Quack C, Fau-Falk CS, Falk Cs Fau-Umansky V, Umansky V, Fau-Cerwenka A, Cerwenka A (2012) Tumor-infiltrating monocytic myeloid-derived suppressor cells mediate CCR5-dependent recruitment of regulatory T cells favoring tumor growth. J Immunol 189(12):5602–5611. https://doi.org/10.4049/jimmunol.1201018
Schmidl C, Delacher M, Huehn J, Feuerer M (2018) Epigenetic mechanisms regulating T-cell responses. J Allergy Clin Immunol 142:728–743
Sciacovelli M, Gonçalves E, Johnson TI, Zecchini VR, da Costa AS, Gaude E, Drubbel AV, Theobald SJ, Abbo SR, Tran MG, Rajeeve V, Cardaci S, Foster S, Yun H, Cutillas P, Warren A, Gnanapragasam V, Gottlieb E, Franze K, Huntly B, Maher ER, Maxwell PH, Saez-Rodriguez J, Frezza C (2016) Fumarate is an epigenetic modifier that elicits epithelial-to-mesenchymal transition. Nature 537:544–547
Sharma NS, Gupta VK, Garrido VT, Hadad R, Durden BC, Kesh K, Giri B, Ferrantella A, Dudeja V, Saluja A, Banerjee S (2019) Targeting tumor-intrinsic hexosamine biosynthesis sensitizes pancreatic cancer to anti-PD1 therapy. J Clin Invest 130:451–465
Sheng SL, Liu JJ, Dai YH, Sun XG, Xiong XP, Huang G (2012) Knockdown of lactate dehydrogenase a suppresses tumor growth and metastasis of human hepatocellular carcinoma. Febs j 279:3898–3910

Sica A, Mantovani A (2012) Macrophage plasticity and polarization: in vivo veritas. J Clin Invest 122:787–795

Simpson NE, Tryndyak VP, Pogribna M, Beland FA, Pogribny IP (2012) Modifying metabolically sensitive histone marks by inhibiting glutamine metabolism affects gene expression and alters cancer cell phenotype. Epigenetics 7:1413–1420

Singer K, Kastenberger M, Gottfried E, Hammerschmied CG, Büttner M, Aigner M, Seliger B, Walter B, Schlösser H, Hartmann A, Andreesen R, Mackensen A, Kreutz M (2010) Warburg phenotype in renal cell carcinoma: high expression of glucose-transporter 1 (GLUT-1) correlates with low CD8(+) T-cell infiltration in the tumor. Int J Cancer 128:2085–2095

Singh N, Gurav A, Sivaprakasam S, Brady E, Padia R, Shi H, Thangaraju M, Prasad PD, Manicassamy S, Munn DH, Lee JR, Offermanns S, Ganapathy V (2014) Activation of Gpr109a, receptor for niacin and the commensal metabolite butyrate, suppresses colonic inflammation and carcinogenesis. Immunity 40:128–139

Smith AD, Lu C, Payne D, Paschall AV, Klement JD, Redd PS, Ibrahim ML, Yang D, Han Q, Liu Z, Shi H, Hartney TJ, Nayak-Kapoor A, Liu K (2020) Autocrine IL6-mediated activation of the STAT3-DNMT axis silences the TNFα-RIP1 necroptosis pathway to sustain survival and accumulation of myeloid-derived suppressor cells. Cancer Res 80:3145–3156

Song LJ, Zhang WJ, Chang ZW, Pan YF, Zong H, Fan QX, Wang LX (2015) PU.1 is identified as a novel metastasis suppressor in hepatocellular carcinoma regulating the miR-615-5p/IGF2 axis. Asian Pac J Cancer Prev 16:3667–3671

Spranger S, Koblish HK, Horton B, Scherle PA, Newton R, Gajewski TF (2014) Mechanism of tumor rejection with doublets of CTLA-4, PD-1/PD-L1, or IDO blockade involves restored IL-2 production and proliferation of CD8(+) T cells directly within the tumor microenvironment. J Immunother Cancer 2:3

Strickland M, Stoll EA (2017) Metabolic reprogramming in glioma. Front Cell Dev Biol 5:43

Sukumar M, Liu J, Ji Y, Subramanian M, Crompton JG, Yu Z, Roychoudhuri R, Palmer DC, Muranski P, Karoly ED, Mohney RP, Klebanoff CA, Lal A, Finkel T, Restifo NP, Gattinoni L (2013) Inhibiting glycolytic metabolism enhances CD8+ T cell memory and antitumor function. J Clin Invest 123:4479–4488

Suzuki H, Hisamatsu T, Chiba S, Mori K, Kitazume MT, Shimamura K, Nakamoto N, Matsuoka K, Ebinuma H, Naganuma M, Kanai T (2016) Glycolytic pathway affects differentiation of human monocytes to regulatory macrophages. Immunol Lett 176:18–27

Tabas I, Bornfeldt KE (2016) Macrophage phenotype and function in different stages of atherosclerosis. Circ Res 118:653–667

Tahiliani M, Koh KP, Shen Y, Pastor WA, Bandukwala H, Brudno Y, Agarwal S, Iyer LM, Liu DR, Aravind L, Rao A (2009) Conversion of 5-methylcytosine to 5-hydroxymethylcytosine in mammalian DNA by MLL partner TET1. Science 324:930–935

Tan Z, Xie N, Cui H, Moellering DR, Abraham E, Thannickal VJ, Liu G (2015) Pyruvate dehydrogenase kinase 1 participates in macrophage polarization via regulating glucose metabolism. J Immunol 194:6082–6089

Tan AHY, Tu W, McCuaig R, Hardy K, Donovan T, Tsimbalyuk S, Forwood JK, Rao S (2019) Lysine-specific histone demethylase 1A regulates macrophage polarization and checkpoint molecules in the tumor microenvironment of triple-negative breast cancer. Front Immunol 10: 1351

Tannahill GM, Curtis AM, Adamik J, Palsson-McDermott EM, McGettrick AF, Goel G, Frezza C, Bernard NJ, Kelly B, Foley NH, Zheng L, Gardet A, Tong Z, Jany SS, Corr SC, Haneklaus M, Caffrey BE, Pierce K, Walmsley S, Beasley FC, Cummins E, Nizet V, Whyte M, Taylor CT, Lin H, Masters SL, Gottlieb E, Kelly VP, Clish C, Auron PE, Xavier RJ, O'Neill LA (2013) Succinate is an inflammatory signal that induces IL-1β through HIF-1α. Nature 496:238–242

Terrén I, Orrantia A, Vitallé J, Zenarruzabeitia O, Borrego F (2019) NK cell metabolism and tumor microenvironment. Front Immunol 10:2278

TeSlaa T, Chaikovsky AC, Lipchina I, Escobar SL, Hochedlinger K, Huang J, Graeber TG, Braas D, Teitell MA (2016) α-Ketoglutarate accelerates the initial differentiation of primed human pluripotent stem cells. Cell Metab 24:485–493

Tikhanovich I, Zhao J, Olson J, Adams A, Taylor R, Bridges B, Marshall L, Roberts B, Weinman SA (2017) Protein arginine methyltransferase 1 modulates innate immune responses through regulation of peroxisome proliferator-activated receptor γ-dependent macrophage differentiation. J Biol Chem 292:6882–6894

Turcan S, Fabius AW, Borodovsky A, Pedraza A, Brennan C, Huse J, Viale A, Riggins GJ, Chan TA (2013) Efficient induction of differentiation and growth inhibition in IDH1 mutant glioma cells by the DNMT Inhibitor Decitabine. Oncotarget 4:1729–1736

Tyrakis PA, Palazon A, Macias D, Lee KL, Phan AT, Veliça P, You J, Chia GS, Sim J, Doedens A, Abelanet A, Evans CE, Griffiths JR, Poellinger L, Goldrath AW, Johnson RS (2016) S-2-hydroxyglutarate regulates CD8(+) T-lymphocyte fate. Nature 540:236–241

Ueno N, Nishimura N, Ueno S, Endo S, Tatetsu H, Hirata S, Hata H, Matsuoka M, Mitsuya H, Okuno Y (2017) PU.1 acts as tumor suppressor for myeloma cells through direct transcriptional repression of IRF4. Oncogene 36:4481–4497

Ugel S, De Sanctis F, Mandruzzato S, Bronte V (2015) Tumor-induced myeloid deviation: when myeloid-derived suppressor cells meet tumor-associated macrophages. J Clin Invest 125:3365–3376

van der Windt GJ, Pearce EL (2012) Metabolic switching and fuel choice during T-cell differentiation and memory development. Immunol Rev 249:27–42

Vancura A, Nagar S, Kaur P, Bu P, Bhagwat M, Vancurova I (2018) Reciprocal regulation of AMPK/SNF1 and protein acetylation. Int J Mol Sci 19(11):3314

Vander Heiden MG, DeBerardinis RJ (2017) Understanding the intersections between metabolism and cancer biology. Cell 168:657–669

Vats D, Mukundan L, Odegaard JI, Zhang L, Smith KL, Morel CR, Wagner RA, Greaves DR, Murray PJ, Chawla A (2006) Oxidative metabolism and PGC-1beta attenuate macrophage-mediated inflammation. Cell Metab 4:13–24

Villablanca EJ, Raccosta L, Zhou D, Fontana R, Maggioni D, Negro A, Sanvito F, Ponzoni M, Valentinis B, Bregni M, Prinetti A, Steffensen KR, Sonnino S, Gustafsson JA, Doglioni C, Bordignon C, Traversari C, Russo V (2009) Tumor-mediated liver X receptor-alpha activation inhibits CC chemokine receptor-7 expression on dendritic cells and dampens antitumor responses. Nat Med 16:98–105

Villagra A, Cheng F, Wang HW, Suarez I, Glozak M, Maurin M, Nguyen D, Wright KL, Atadja PW, Bhalla K, Pinilla-Ibarz J, Seto E, Sotomayor EM (2008) The histone deacetylase HDAC11 regulates the expression of interleukin 10 and immune tolerance. Nat Immunol 10:92–100

Villalba M, Rathore MG, Lopez-Royuela N, Krzywinska E, Garaude J, Allende-Vega N (2012) From tumor cell metabolism to tumor immune escape. Int J Biochem Cell Biol 45:106–113

Wang R, Green DR (2012) Metabolic checkpoints in activated T cells. Nat Immunol 13:907–915

Wang F, Zhang S, Vuckovic I, Jeon R, Lerman A, Folmes CD, Dzeja PP, Herrmann J (2018) Glycolytic stimulation is not a requirement for M2 macrophage differentiation. Cell Metab 28: 463–475.e464

Wardell SE, Ilkayeva OR, Wieman HL, Frigo DE, Rathmell JC, Newgard CB, McDonnell DP (2008) Glucose metabolism as a target of histone deacetylase inhibitors. Mol Endocrinol 23: 388–401

Wise DR, DeBerardinis RJ, Mancuso A, Sayed N, Zhang XY, Pfeiffer HK, Nissim I, Daikhin E, Yudkoff M, McMahon SB, Thompson CB (2008) Myc regulates a transcriptional program that stimulates mitochondrial glutaminolysis and leads to glutamine addiction. Proc Natl Acad Sci U S A 105:18782–18787

Xiao M, Yang H, Xu W, Ma S, Lin H, Zhu H, Liu L, Liu Y, Yang C, Xu Y, Zhao S, Ye D, Xiong Y, Guan KL (2012) Inhibition of α-KG-dependent histone and DNA demethylases by fumarate and succinate that are accumulated in mutations of FH and SDH tumor suppressors. Genes Dev 26: 1326–1338

Xu W, Yang H, Liu Y, Yang Y, Wang P, Kim SH, Ito S, Yang C, Xiao MT, Liu LX, Jiang WQ, Liu J, Zhang JY, Wang B, Frye S, Zhang Y, Xu YH, Lei QY, Guan KL, Zhao SM, Xiong Y (2011) Oncometabolite 2-hydroxyglutarate is a competitive inhibitor of α-ketoglutarate-dependent dioxygenases. Cancer Cell 19:17–30

Xu T, Stewart KM, Wang X, Liu K, Xie M, Ryu JK, Li K, Ma T, Wang H, Ni L, Zhu S, Cao N, Zhu D, Zhang Y, Akassoglou K, Dong C, Driggers EM, Ding S (2017) Metabolic control of T (H)17 and induced T(reg) cell balance by an epigenetic mechanism. Nature 548:228–233

Yang Q, Wei J, Zhong L, Shi M, Zhou P, Zuo S, Wu K, Zhu M, Huang X, Yu Y, Zhang H, Yin H, Zhou J (2014a) Cross talk between histone deacetylase 4 and STAT6 in the transcriptional regulation of arginase 1 during mouse dendritic cell differentiation. Mol Cell Biol 35:63–75

Yang X, Wang X, Liu D, Yu L, Xue B, Shi H (2014b) Epigenetic regulation of macrophage polarization by DNA methyltransferase 3b. Mol Endocrinol 28:565–574

Yang W, Bai Y, Xiong Y, Zhang J, Chen S, Zheng X, Meng X, Li L, Wang J, Xu C, Yan C, Wang L, Chang CC, Chang TY, Zhang T, Zhou P, Song BL, Liu W, Sun SC, Liu X, Li BL (2016) Potentiating the antitumour response of CD8(+) T cells by modulating cholesterol metabolism. Nature 531:651–655

Yen K, Travins J, Wang F, David MD, Artin E, Straley K, Padyana A, Gross S, DeLaBarre B, Tobin E, Chen Y, Nagaraja R, Choe S, Jin L, Konteatis Z, Cianchetta G, Saunders JO, Salituro FG, Quivoron C, Opolon P, Bawa O, Saada V, Paci A, Broutin S, Bernard OA, de Botton S, Marteyn BS, Pilichowska M, Xu Y, Fang C, Jiang F, Wei W, Jin S, Silverman L, Liu W, Yang H, Dang L, Dorsch M, Penard-Lacronique V, Biller SA, Su SM (2017) AG-221, a first-in-class therapy targeting acute myeloid leukemia harboring oncogenic IDH2 mutations. Cancer Discov 7:478–493

Yi L, Liu Q, Orandle MS, Sadiq-Ali S, Koontz SM, Choi U, Torres-Velez FJ, Jackson SH (2012) p47(phox) directs murine macrophage cell fate decisions. Am J Pathol 180:1049–1058

Yu Q, Dong L, Li Y, Liu G (2018) SIRT1 and HIF1α signaling in metabolism and immune responses. Cancer Lett 418:20–26

Yu W, Wang Z, Zhang K, Chi Z, Xu T, Jiang D, Chen S, Li W, Yang X, Zhang X, Wu Y, Wang D (2019) One-carbon metabolism supports S-adenosylmethionine and histone methylation to drive inflammatory macrophages. Mol Cell 75:1147–1160.e1145

Zhang Q, Zhao K, Shen Q, Han Y, Gu Y, Li X, Zhao D, Liu Y, Wang C, Zhang X, Su X, Liu J, Ge W, Levine RL, Li N, Cao X (2015) Tet2 is required to resolve inflammation by recruiting Hdac2 to specifically repress IL-6. Nature 525:389–393

Zhang Y, Kurupati R, Liu L, Zhou XY, Zhang G, Hudaihed A, Filisio F, Giles-Davis W, Xu X, Karakousis GC, Schuchter LM, Xu W, Amaravadi R, Xiao M, Sadek N, Krepler C, Herlyn M, Freeman GJ, Rabinowitz JD, Ertl HCJ (2017) Enhancing CD8(+) T cell fatty acid catabolism within a metabolically challenging tumor microenvironment increases the efficacy of melanoma immunotherapy. Cancer Cell 32:377–391.e379

Zhang D, Tang Z, Huang H, Zhou G, Cui C, Weng Y, Liu W, Kim S, Lee S, Perez-Neut M, Ding J, Czyz D, Hu R, Ye Z, He M, Zheng YG, Shuman HA, Dai L, Ren B, Roeder RG, Becker L, Zhao Y (2019) Metabolic regulation of gene expression by histone lactylation. Nature 574:575–580

Zheng Y, Wang Z, Wei S, Liu Z, Chen G (2020) Epigenetic silencing of chemokine CCL2 represses macrophage infiltration to potentiate tumor development in small cell lung cancer. Cancer Lett 499:148–163

Zhou J, Yao Y, Shen Q, Li G, Hu L, Zhang X (2017) Demethylating agent decitabine disrupts tumor-induced immune tolerance by depleting myeloid-derived suppressor cells. J Cancer Res Clin Oncol 143:1371–1380

Zhou L, Zhang M, Wang Y, Dorfman RG, Liu H, Yu T, Chen X, Tang D, Xu L, Yin Y, Pan Y, Zhou Q, Zhou Y, Yu C (2018) Faecalibacterium prausnitzii produces butyrate to maintain Th17/Treg balance and to ameliorate colorectal colitis by inhibiting histone deacetylase 1. Inflamm Bowel Dis 24:1926–1940

Index

T. K. Kundu, C. Das (eds.), *Metabolism and Epigenetic Regulation: Implications in Cancer*, Subcellular Biochemistry 100,
https://doi.org/10.1007/978-3-031-07634-3

The manufacturer's authorised representative in the EU is Springer Nature Customer Service Centre GmbH, Europaplatz 3, 69115 Heidelberg, Germany. If you have any concerns regarding our products, please contact ProductSafety@springernature.com

Printed and bound by CPI Group (UK) Ltd, Croydon, CR0 4YY
23/09/2026
02221538-0004